# PARTIAL DIFFERENTIAL EQUATIONS

# PARTIAL DIFFERENTIAL EQUATIONS

## Sources and Solutions

Arthur David Snider

University of South Florida

DOVER PUBLICATIONS, INC.
Mineola, New York

*Bibliographical Note*

This Dover edition, first published in 2006, is an unabridged corrected republication of the work originally published by Prentice-Hall, Inc., Upper Saddle River, New Jersey, 1999.

An errata list for this book may be downloaded from the author's web site: http://www.eng.usf.edu/EE/people/snider2.html

*Library of Congress Cataloging-in-Publication Data*

Snider, Arthur David, 1940–
    Partial differential equations : sources and solutions / Arthur David Snider.
        p. cm.
    Originally published: Upper Saddle River, N.J. : Prentice Hall, c1999.
    Includes bibliographical references and index.
    ISBN 0-486-45340-5 (pbk.)
    1. Differential equations, Partial. I. Title.

QA377.S62 2006
518'.64—dc22

                                                                2006047432

Manufactured in the United States of America
Dover Publications, Inc., 31 East 2nd Street, Mineola, N.Y. 11501

# Contents

# Preface

In writing a book on mathematical methods one must decide on the proper balance between rigor and ease of comprehension. Most engineering readers have no stomach for the bulky "n-times-continuously-differentiable" type of hypothesis modifiers which ensure complete mathematical generality, particularly when the delimiting counterexamples are unlikely to occur in a lifetime of engineering practice. But there are practical physical situations where such oddities as limited differentiability inhibit accuracy - as, for example, flows with reentrant corners. And more often than not when these pragmatic instances of noncontinuity arise, their analyses prove to be physically significant.

The compromises struck in this textbook reflect the author's experiences as a student/teacher/practitioner of mathematics, physics, and engineering. Every technique presented - without exception - is motivated by a heuristic discussion demonstrating the plausibility or, perhaps, inevitability of the procedure. Over 400 figures appear, in support of these arguments. Specialists may recognize that these heuristics are often "neutered" versions of the classical, rigorous analyses - as, for example, in our treatment of the existence of the Sturm-Liouville eigenfunctions. As a result we have been able to endow the separation-of-variable methodology with a systematic, unifying structure which should dispel the impression, implicitly conveyed by most authors, that it is an ad hoc procedure.

Additionally we present derivations for virtually all of the classical Sturm-Liouville expansions, even the singular ones which usually appear only as cookbook formulas in tabulations or as hideously formidable examples in ponderous tomes on the subject. A classic case in point is the problem of heat flow in a cylindrical wedge. Most texts only treat the situation where the cylindrical side is heated externally - it leads to an orderly Fourier series expansion. This author is aware of no other textbook in the English language which analyzes the configuration where the flat sides are heated (entailing the obscure Lebedev expansions). What is an engineering reader to think when an "applied mathematics" author sidesteps issues because they are not mathematically tidy?

Despite this commitment to thoroughness, however, we recognize that complete coverage of all aspects of eigenfunction applications is not feasible in the classroom. Therefore the book is peppered with optional reading sections from which the instructor can select, and with exhaustive tabulations which the user can exploit without tracing the associated derivations.

A chapter-by-chapter description will best convey our philosophy. Chapter One reviews the elementary differential equation procedures, but it emphasizes those methods which are important in engineering applications and suppresses the others. The delta function and Green's functions are introduced, with a brief optional discourse on the theory of distributions.

The power series method for solving analytic differential equations is described in Chapter Two. The reader is also introduced to Bessel functions, and a user's guide to the literature of special functions is presented. A very brief introduction to numerical methods fits in here as well; it is quite incomplete, serving only to provide perspective for the analytic methods.

In Chapter Three the classic "method of undetermined coefficients" is acknowledged as the foundation of frequency-space methodology and modal analysis. The description of the properties of the Fourier Integral is facilitated by the prior introduction of generalized functions. The presence and significance of resonances and transients are treated, and the pros/cons of the alternative Laplace transform formalism are explored. Fourier series and its attendant notions of orthogonality are introduced.

The first three chapters, then, are preliminary to the study of partial differential equations. Many of the students in my engineering analysis classes are practicing engineers, whose skills in ordinary differential equations have gotten a little rusty, and they appreciate a review from a different perspective. Well-prepared readers may wish to start with Chapter Four. The physical underpinnings of the wave, heat, Laplace, and Schrodinger equations are traced in this chapter. These are intended not as optional background reading, but as motivation for why the various types of partial differential equations each come equipped with their own peculiar variety of well-posed initial and boundary conditions. The mathematical classifications - elliptic, parabolic, and hyperbolic - are much easier to comprehend when the underlying physical situations are recognized. Also this chapter includes a discourse on the calculus of variations as a source of another important differential equation, that of Euler and Lagrange. By opening with this topic we are able to address the Lagrange formalism in a subsequent section on classical mechanics. Some special elementary solution techniques are surveyed in this chapter.

The structure of the separation of variables technique is introduced with some transparent examples in Chapter Five, which serves to unify the subsequent three chapters. The actual separation of the classical partial differential equations is extensively demonstrated in the final section.

In Chapter Six the methodology for the attendant Sturm-Liouville subproblems (regular and singular) is expounded, with an exhaustive list of examples worked out and tabulated. Most of the expansions are motivated herein through the two-dimensional Laplace equation. The chapter is arranged so that, once the basic ideas are understood, the reader can jump to Chapter Seven, culling particular expansions from Table 6.6 as needed.

Chapter Seven shows how the basic procedures are extended to handle higher dimensions and time dependence. As a result of this unified approach, we are able to develop a compact table of solutions to the Laplace, heat, and wave equations in various coordinate systems. This saves a lot of work in practice. Indeed, Dr. S. Kadamani and the author have developed an expert system, $USFKAD$, which implements these solution expansions on the computer; it is available over the internet (see below).

Chapter Eight addresses the use of Green's functions for nonhomogeneous partial differential equations. Attention is focused on eigenfunction expansion methods, but a compilation of most of the known closed forms is also included. Time dependent nonhomogeneities and/or boundary conditions usually require some preliminary Fourier transformation, and the discussion is coupled with an extensive analysis of the proper formulation of radiation problems. Section 8.4 shows how facility with Green's functions allows one to mollify troublesome discontinuities which would otherwise inhibit rapid convergence.

Chapter Nine discusses perturbation methods and their use in the various disciplines of applied mathematics. This leads to an exposition of small-wave analysis, demonstrating the concepts of signal-front, group, and phase velocities.

For the reader's convenience Appendix A includes a sketch of the basics of numerical differentiation, integration, and solution of equations. Appendix B merely provides elaboration on the evaluation of the Bromwich integral for some enticing radiation problems discussed in Chapter Eight.

This book is obviously too long to be covered in a semester; it is intended to remain on the user's desk as a reference long after formal coursework is complete. I have on occasion constructed a minimal course, for well-prepared students, by covering sections 3-6 of Chapter Four, all of Chapter Five, sections 1-4 of Chapter Six, sections 1-3 and 5 of Chapter Seven, and sections 1 and 5 of Chapter Eight.

This text is quite table-intensive, and there will inevitably be errors. The author maintains an errata sheet on his Web page at
http://ee.eng.usf.edu/people/snider2.html.
Selected answers can also be found there. We invite readers to submit their suggestions by email to
snider@eng.usf.edu.
Details on obtaining the expert system are also available at the web site.

In this author's opinion, the expositor of engineering mathematics whose obsession with rigor inhibits his students from toying with "illegal" mathematics for fear of getting burnt does a disservice to the academic community. On the other hand I share my colleagues' distaste and mistrust of cookbook expositions. It is hoped that the present work contains the appropriate mixture of rigor and heuristic for the audience whom it is intended to serve.

*Dave Snider*

# Chapter 1

# Basics of Differential Equations

We assume that the student has had a basic course in elementary differential equations prior to reading this book. Nonetheless the author has found that a brief review of some topics is very helpful in establishing their relative priorities, for practical applications. Also a reevaluation, from a different author's point of view, of material previously mastered often provides a fresh perspective.

Without a doubt the most commonly used mathematical model for engineering systems is the second order linear ordinary differential equation. In Section 1.1 we survey, mainly by example, the basic theoretical structure for such equations and the techniques for finding explicit solutions in the constant-coefficient and equidimensional cases. Readers seeking a more thoroughgoing review will discover that the problem set contains, in addition to the usual drill-type questions, a set of theoretically oriented exercises intended to guide them through the derivations and generalizations of the procedures.

Section 1.2 conducts a brief review of the mechanics for changing variables in a differential equation. This is such an important tool in the practice of applied mathematics that a little extra attention is warranted.

The most widely used procedure for handling nonhomogeneities in engineering analysis is the Green's function, and its heuristic exposition through the delta function makes its visualization very immediate. These notions are explored for ordinary differential equations in Sections 1.3 and 1.4. Since the concepts may be unfamiliar to some readers, the development proceeds at a more leisurely pace.

Section 1.5 is included for the edification of those students whose sense of logical order is uncomfortable with the heuristics of Section 1.3. It surveys the logical underpinnings of the delta–function calculus (the theory of distributions). Also it derives some of the subtler properties of generalized functions which resurface in the theory of the Fourier transform in Chapter 3.

## 1.1    Review: Theory of Linear Ordinary Differential Equations

**Summary**    A linear homogeneous ordinary differential equation has the form

$$a_n(x)\frac{d^n y}{dx^n} + a_{n-1}(x)\frac{d^{n-1}y}{dx^{n-1}} + \cdots + a_1(x)\frac{dy}{dx} + a_0(x)y = 0.$$

Its general solution looks like

$$Y(x; c_1, c_2, \ldots, c_n) = c_1 y_1(x) + c_2 y_2(x) + \cdots + c_n y_n(x)$$

where the $y_i(x)$ are an independent set of particular solutions. The constants $c_i$ can be adjusted so that the solution satisfies specified initial values

$$
\begin{aligned}
y(x_0) &= b_0, \\
y'(x_0) &= b_1, \\
y''(x_0) &= b_2, \\
&\vdots \\
y^{n-1}(x_0) &= b_{n-1}
\end{aligned}
$$

at any point $x_0$ where the equation coefficients $a_j(x)$ are continuous and $a_n(x)$ is nonzero. When the equation coefficients are constants the independent solutions can be deduced by inserting the ansatz $y = e^{px}$. For equidimensional equations, where $a_j(x) = (constant) \times x^j$, the appropriate ansatz is $y = x^r$.

The differential equation that occurs most commonly in engineering analysis is the *second order linear ordinary differential equation*; its general form is

$$a_2(x)y'' + a_1(x)y' + a_0(x)y = f(x). \tag{1}$$

The coefficients $a_i(x)$ and the right-hand side $f(x)$ are given functions of $x$, and the function $y(x)$ is unknown. Since the reader is presumed to have completed a first course in differential equations, most of the present section will comprise a series of examples intended to stimulate his (her) recollection of the basic theory and techniques.

First let us review the precise meaning of the italicized words above.

**a.** The *order* of a differential equation is the order of the highest derivative of the unknown function which appears in the equation. Since $y''$ appears in (1), but not $y'''$ or $y''''$, the equation is of second order.

**b.** The terminology "linear" could be criticized. The differential *operator*

$$L = a_2(x)\frac{d^2}{dx^2} + a_1(x)\frac{d}{dx} + a_0(x) \tag{2}$$

appearing on the left-hand side of eq. (1) is a bona fide *linear operator* - that is, for any two (twice differentiable) functions $y_1(x)$ and $y_2(x)$ and any two constants $c_1$ and $c_2$, it satisfies

$$L\{\,c_1 y_1(x) + c_2 y_2(x)\,\} = c_1 L y_1(x) + c_2 L y_2(x). \tag{3}$$

However it does *not* follow that the *linear combination* $\{c_1 y_1(x) + c_2 y_2(x)\}$ is a solution to the equation (1), even if $y_1$ and $y_2$ are. After all, from $Ly_1 = f(x)$ and $Ly_2 = f(x)$ we deduce

$$L\{c_1 y_1(x) + c_2 y_2(x)\} = [c_1 + c_2]f(x) \tag{4}$$

which, as a rule, does not coincide with $f(x)$. Only if $f(x)$ is identically zero can we guarantee that $\{c_1 y_1(x) + c_2 y_2(x)\}$ satisfies (1), for all possible $c_i$. Nonetheless, it is traditional to call (1) a linear differential equation, even if $f(x) \neq 0$. We call the equation *homogeneous* when $f$ is zero, and *nonhomogeneous* otherwise.

An *nth* order differential equation is said to be linear if it can be written in the form

$$Ly = f(x), \tag{5}$$

where $L$ is a linear operator (i.e., satisfies (3)), $y$ is the unknown function, and $f(x)$ is known. The equation is homogeneous if $f(x) = 0$; otherwise it is nonhomogeneous, and $f(x)$ is called the nonhomogeneity. The general form of the *nth* order linear ordinary differential equation is

$$a_n(x)\frac{d^n y}{dx^n} + a_{n-1}(x)\frac{d^{n-1}y}{dx^{n-1}} + \cdots + a_1(x)\frac{dy}{dx} + a_0(x)y = f(x). \tag{6}$$

We stress that only linear *homogeneous* equations possess the property that any linear combination of solutions is, again, a solution. Note, however, that the sum of a solution to a nonhomogeneous differential equation and a solution to the "associated" homogeneous equation obtained by replacing $f(x)$ with 0 is, again, a solution to the nonhomogeneous differential equation:

$$Ly_1 = f \quad \text{and} \quad Ly_2 = 0 \quad \text{implies} \quad L\{y_1 + y_2\} = f.$$

Conversely, any two solutions to the nonhomogeneous equation differ by a solution to the associated homogeneous equation:

$$Ly_1 = f \quad \text{and} \quad Ly_2 = f \quad \text{implies} \quad L\{y_1 - y_2\} = 0.$$

The solution structure for a linear differential equation is typified by the trivial example

$$y'' = \sin x, \tag{7}$$

which can be solved for $y$ by integrating twice:

$$y' = \int \sin x \, dx + c_1 = -\cos x + c_1$$

where $c_1$ is any constant, and

$$y = \int [-\cos x + c_1] \, dx + c_2 = -\sin x + c_1 x + c_2. \tag{8}$$

As (8) illustrates, one expects the *general solution* to an $n$th order differential equation to contain $n$ arbitrary parameters (here $c_1$ and $c_2$), and to specify them one must append $n$ *auxiliary conditions* to the equation.

Commonly we specify, as auxiliary conditions, the values of the first $n-1$ derivatives (including the zeroth) at some point. For example if we require that the solution to (7) also satisfy

$$y(0) = 2, \quad y'(0) = 3, \tag{9}$$

then the *unique* solution is

$$y(x) = -\sin x + 4x + 2, \tag{10}$$

as we ascertain by solving the equations

$$\begin{aligned}
y(0) &= -\sin 0 + c_1 0 + c_2 &= 2, \\
y'(0) &= -\cos 0 + c_1 &= 3
\end{aligned}$$

for $c_1$ and $c_2$. This illustrates the initial value problem.

The *initial value problem* for the $n$th order linear differential equation (6) calls for finding a solution to (6) satisfying also the "initial conditions"

$$\begin{aligned}
y(x_0) &= b_0, \\
y'(x_0) &= b_1, \\
y''(x_0) &= b_2, \\
&\ \vdots \\
y^{(n-1)}(x_0) &= b_{n-1};
\end{aligned} \tag{11}$$

here $x_0$ is some specific value of $x$, and the $b_i$ are given numbers.
*The mathematical theory shows that if all the coefficients $a_i(x)$
are continuous at $x = x_0$, and if in addition the leading coefficient $a_n(x)$ is not zero at $x_0$, then there will be precisely one solution to the initial value problem.* This "existence-uniqueness"
theorem is proved in the textbooks by Birkhoff and Rota or
Coddington and Levinson.

In other words the enforcement of the $n$ initial conditions (11) exhausts the
$n$ "degrees of freedom" provided by the $n$ arbitrary constants in the general
solution.

**Example 1.** Consider the initial value problem

$$\begin{aligned} x^2 y'' - 2y &= 0; \\ y(1) &= 2, \\ y'(1) &= 1. \end{aligned} \tag{12}$$

It is easy to check that a general solution of the differential equation is

$$y = c_1 x^2 + c_2/x \tag{13}$$

(we'll review the solution procedure for this type of equation shortly). Enforcement of (12) results in

$$\begin{aligned} y(1) &= c_1 \times 1^2 + c_2/1 &= 2, \\ y'(1) &= 2c_1 \times 1 - c_2/1^2 &= 1, \end{aligned}$$

whose solution is $c_1 = c_2 = 1$, or

$$y(x) = x^2 + 1/x. \qquad\blacksquare$$

**Example 2.** If these same initial conditions are posed at $x = 0$ rather than
at $x = 1$,

$$\begin{aligned} x^2 y'' - 2y &= 0; \\ y(0) &= 2, \\ y'(0) &= 1, \end{aligned} \tag{14}$$

they cannot be met with the general solution form (13):

$$\begin{aligned} y(0) &= c_1 \times 0^2 + c_2/0 &= 2, \ (meaningless!) \\ y'(0) &= 2c_1 \times 0 - c_2/0^2 &= 1. \end{aligned}$$

This illustrates the "escape clause" in the existence-uniqueness theorem;
the leading coefficient, $a_2(x) \equiv x^2$ in this case, is zero at $x = 0$, so the
initial value problem is not well posed there. (In fact $x = 0$ is a "regular
singular point" of the differential equation, and we shall study such points
in some detail in Section 2.4.) $\qquad\blacksquare$

If we had been alert, we could have anticipated trouble with the system (14). At the point $x = 0$ the differential equation states

$$0^2 y''(0) - 2y(0) = 0,$$

but the initial conditions require

$$0^2 y''(0) - 2y(0) = 0 - 2 \times 2 = -4 \quad (not\, 0!)$$

Notice that even though we call (11) "initial conditions," they are not necessarily imposed at $x = x_0 = 0$; the point $x_0$ can be any point where the coefficients are continuous and $a_n(x)$ is not zero. In most applications where the independent variable $x$ is *time*, the appropriate auxiliary conditions for the differential equation are the initial position and velocity - hence, "initial" data. When $x$ represents a spacial coordinate, however, it is more common to impose auxiliary conditions at the ends of the solution interval. These are known as "boundary conditions", and they give rise to *boundary value problems*.

**Example 3.** To solve the boundary value problem

$$
\begin{aligned}
x^2 y'' - 2y &= 0, \\
y(1) &= 2, \\
y(2) &= 1,
\end{aligned}
\tag{15}
$$

we again employ the general solution (13). Enforcement of the boundary conditions in (15) leads to

$$
\begin{aligned}
y(1) &= c_1 \times 1^2 + c_2/1 = 2, \\
y(2) &= c_1 \times 2^2 + c_2/2 = 1,
\end{aligned}
$$

and the result - $c_1 = 0$, $c_2 = 2$ - is the solution $y(x) = 2/x$. ∎

Although the above example is straightforward, boundary value problems are significantly more complicated than initial value problems. The existence-uniqueness condition, for instance, is much more involved. Simple-looking boundary value problems may have no solutions, or an infinity of solutions.

**Example 4.** Consider the differential equation

$$y'' + y = 0 \tag{16}$$

together with the three sets of boundary conditions:

$$
\begin{cases} (a) & y(0) = 1 \\ & y(\frac{\pi}{2}) = 2 \end{cases}
\quad
\begin{cases} (b) & y(0) = 1 \\ & y(\pi) = 2 \end{cases}
\quad
\begin{cases} (c) & y(0) = 1 \\ & y(\pi) = -1. \end{cases}
\tag{17}
$$

The general solution to (16) is $c_1 \cos x + c_2 \sin x$ (again, we'll review this construction later). In all three cases, the boundary condition at $x = 0$,

$$y(0) = c_1 \cos 0 + c_2 \sin 0 = 1,$$

dictates that $c_1 = 1$.

For case (a), the second boundary condition causes no difficulty:

$$y(\frac{\pi}{2}) = \cos\frac{\pi}{2} + c_2 \sin\frac{\pi}{2} = 2$$

implies $c_2 = 2$, and the unique solution is $\cos x + 2\sin x$.

For case (b), *no* value of $c_2$ will meet the condition at $x = \pi$:

$$y(\pi) = \cos\pi + c_2 \sin\pi = -1 \ (not\,2!)$$

Thus case (b) *has no solution!*

On the other hand the second boundary condition for case (c) is met for *any* value of $c_2$:

$$y(\pi) = \cos\pi + c_2 \sin\pi = -1 \ (regardless\,of\,c_2).$$

Therefore case (c) has an *infinite* number of solutions:

$$\cos x + \sin x, \cos x + 2\sin x, \cos x + 3\sin x, \ldots, \cos x + c_2 \sin x. \qquad \blacksquare$$

We'll study this curious phenomenon for boundary value problems further in Chapter 6, where we'll find that it can be *exploited* in analyzing engineering systems. In the remainder of this section we confine our attention to initial value problems. Also for the moment we presume that the given differential equation is homogeneous; the nonhomogeneous case will be addressed later.

Given the linear homogeneous differential equation

$$a_n(x)\frac{d^n y}{dx^n} + a_{n-1}(x)\frac{d^{n-1}}{dx^{n-1}} + \cdots + a_1(x)\frac{dy}{dx} + a_0(x)y = 0, \qquad (18)$$

how do we characterize a format for its *general solution*? That is, we seek a formula containing $n$ constants $c_1, c_2, \ldots, c_n$, which

(*i*) satisfies the differential equation for all values of the $c_i$'s, and

(*ii*) has the flexibility to satisfy, by proper specification of the $c_i$'s, any set of initial conditions

$$y(x_0) = b_0,$$
$$y'(x_0) = b_1,$$
$$y''(x_0) = b_2,$$
$$\vdots$$
$$y^{n-1}(x_0) = b_{n-1}. \qquad (19)$$

The linearity properties suggest one possible answer;[1] one takes $n$ different solutions to the equation $y_1(x)$, $y_2(x)$, ..., $y_n(x)$ and forms the linear combination

$$Y(x; c_1, c_2, \ldots, c_n) = c_1 y_1(x) + c_2 y_2(x) + \cdots + c_n y_n(x) \qquad (20)$$

Clearly $Y$ satisfies the differential equation (since it's linear and homogeneous), and the constants can be selected to match the initial data if the conditions

$$
\begin{aligned}
c_1 y_1(x_0) &+ c_2 y_2(x_0) &+ \cdots + & c_n y_n(x_0) &= b_0 \\
c_1 y_1'(x_0) &+ c_2 y_2'(x_0) &+ \cdots + & c_n y_n'(x_0) &= b_1 \\
c_1 y_1''(x_0) &+ c_2 y_2''(x_0) &+ \cdots + & c_n y_n''(x_0) &= b_2 \\
&\quad\vdots \\
c_1 y_1^{(n-1)}(x_0) &+ c_2 y_2^{(n-1)}(x_0) &+ \cdots + & c_n y_n^{(n-1)}(x_0) &= b_{n-1}
\end{aligned}
\qquad (21)
$$

can be met. Conditions (21) form a linear system of algebraic equations for the $c_i$'s, and the general condition for its solvability is the nonvanishing of the determinant of the coefficient matrix, which is known as the *Wronskian*:

$$
Wr(x_0; y_1, y_2, \ldots, y_n) =
\begin{vmatrix}
y_1(x_0) & y_2(x_0) & \cdots & y_n(x_0) \\
y_1'(x_0) & y_2'(x_0) & \cdots & y_n'(x_0) \\
y_1''(x_0) & y_2''(x_0) & \cdots & y_n''(x_0) \\
\vdots \\
y_1^{(n-1)}(x_0) & y_2^{(n-1)}(x_0) & \cdots & y_n^{(n-1)}(x_0)
\end{vmatrix}
\neq 0
$$

$$(22)$$

To summarize: the linear combination of solutions (20) provides a *general* solution to the homogeneous differential equation (18) - in the sense that the constants can be chosen to satisfy any initial conditions at a point $x = x_0$ - if the Wronskian of the solutions is nonzero at $x_0$.

In exercise 4 you will be invited to show that the Wronskian of $n$ solutions either vanishes *identically*, or it *never* vanishes. In the latter case the solutions enjoy a property called *linear independence* (exercise 5), and as we have seen they can be used to construct a general solution. Other properties are developed in exercise 5.

How does one *obtain* a set of $n$ linearly independent solutions to a given homogeneous differential equation? Using elementary methods we can answer this question for two special cases: equations with constant coefficients and equidimensional equations. More general methods are derived in Chapter 2.

If the coefficients $a_i(x)$ in the linear homogeneous differential equation (18) are constants -

$$a_n \frac{d^n y}{dx^n} + a_{n-1} \frac{d^{n-1} y}{dx^{n-1}} + \cdots + a_1 \frac{dy}{dx} + a_0 y = 0 \qquad (23)$$

---

[1] There are others; see exercise 3.

- then the substitution of the *ansatz* ("trial solution")

$$y = e^{px}, \ y' = pe^{px}, \ y'' = p^2 e^{px}, \ldots, \ y^{(n)} = p^n e^{px} \tag{24}$$

will reduce (23) to an *algebraic* equation, since the only $x$-dependence will be a common factor of $e^{px}$ in every term:

$$a_n p^n e^{px} + a_{n-1} p^{n-1} e^{px} + \cdots + a_1 p e^{px} + a_0 e^{px} = 0. \tag{25}$$

Thus we get a solution to (23) for every root of the *characteristic equation*

$$a_n p^n + a_{n-1} p^{n-1} + \cdots + a_1 p + a_0 = 0; \tag{26}$$

and since (26) is a polynomial equation of degree $n$, as a rule we can expect this procedure to give us our desired $n$ different solutions to the differential equation. (Their independence is verified in exercise 6.) Presuming that the reader has some familiarity with this elementary method, we proceed with examples which demonstrate its various aspects. (Exercise 7 elaborates on the theory.)

**Example 5.** For the differential equation

$$y'' + 7y' + 12y = 0$$

the substitution $y = e^{px}$ leads to the characteristic equation

$$p^2 + 7p + 12 = 0,$$

whose roots are $p = -3$ and $p = -4$. The general solution is thus

$$Y(x) = c_1 e^{-3x} + c_2 e^{-4x} = c_1 y_1(x) + c_2 y_2(x).$$

The Wronskian of these solutions is

$$Wr(x_0; e^{-3x}, e^{-4x}) = \begin{vmatrix} y_1(x_0) & y_2(x_0) \\ y_1'(x_0) & y_2'(x_0) \end{vmatrix} = \begin{vmatrix} e^{-3x_0} & e^{-4x_0} \\ -3e^{-3x_0} & -4e^{-4x_0} \end{vmatrix} = -e^{-7x_0}$$

and is never zero; the solutions $y_1$ and $y_2$ are linearly independent.    ∎

**Example 6.** The differential equation

$$y'' + 6y' + 9y = 0$$

has

$$p^2 + 6p + 9 = (p+3)^2 = 0$$

as its characteristic equation. Thus

$$y_1(x) = e^{-3x}$$

is one solution. The theory says (exercise 7) that in the case of a double root $p$ to the characteristic equation, the function $xe^{px}$ is also a solution.[2] Therefore the general solution is

$$Y(x) = c_1 e^{-3x} + c_2 x e^{-3x} = c_1 y_1(x) + c_2 y_2(x).$$

Its Wronskian is

$$Wr(x_0; e^{-3x}, xe^{-3x}) = \begin{vmatrix} y_1(x_0) & y_2(x_0) \\ y_1'(x_0) & y_2'(x_0) \end{vmatrix} = \begin{vmatrix} e^{-3x_0} & x_0 e^{-3x_0} \\ -3e^{-3x_0} & e^{-3x_0} - 3x_0 e^{-3x_0} \end{vmatrix}$$

$$= e_0^{-6x}$$

and the solutions, again, are linearly independent.                          ■

**Example 7.** The characteristic equation for

$$y'' + 2y' + 5y = 0$$

is

$$p^2 + 2p + 5 = 0$$

and its roots are complex conjugates:

$$p_1 = -1 + 2i , \ p_2 = -1 - 2i.$$

This gives rise to the solution

$$y_1(x) = e^{-x} e^{i2x} = e^{-x} \{\cos 2x + i \sin 2x\} \tag{27}$$

and its conjugate. It is easy to see (exercise 8) that both the real and imaginary parts of complex solutions such as (27) are, separately, solutions (when the coefficients in the differential equation are real); thus the general solution can be expressed as

$$Y(x) = c_1 e^{-x} \cos 2x + c_2 e^{-x} \sin 2x.$$

The reader can verify that the Wronskian of these solutions is

$$Wr(x_0; e^{-x} \cos 2x, e^{-x} \sin 2x) = 2e^{-2x_0} \neq 0.$$                          ■

**Example 8. (VERY IMPORTANT!)** Throughout this book we shall see that the differential equation

$$y'' = \lambda y, \tag{28}$$

---

[2]For a triple root, $x^2 e^{px}$ is also a solution; and so on.

where $\lambda$ is an unspecified (constant) parameter, arises repeatedly in engineering applications. Thus it is worthwhile to devote a little extra attention to this example. We shall demonstrate in Section 3.2 that if $x$ is interpreted as time and $y$ as displacement, then eq. (28) describes the motion of an undamped, unforced mass on a spring.

Equation (28) has constant coefficients, and its characteristic equation is easily seen to be $p^2 = \lambda$, with roots $p = \pm\sqrt{\lambda}$. The analysis then depends on the nature of $\lambda$.

($i$) If $\lambda$ is positive, a general solution is

$$Y(x) = c_1 e^{\sqrt{\lambda}x} + c_2 e^{-\sqrt{\lambda}x}.$$

However, it is usually more convenient to use a different form. Note that if we set $c_1 = c_2 = 1/2$, the particular solution

$$y_1(x) = \frac{e^{\sqrt{\lambda}x} + e^{-\sqrt{\lambda}x}}{2} = \cosh\sqrt{\lambda}x$$

results, while the choice $c_1 = 1/2, c_2 = -1/2$ yields

$$y_2(x) = \frac{e^{\sqrt{\lambda}x} - e^{-\sqrt{\lambda}x}}{2} = \sinh\sqrt{\lambda}x.$$

Thus

$$Y(x) = d_1 \cosh\sqrt{\lambda}x + d_2 \sinh\sqrt{\lambda}x \tag{29}$$

is an equally acceptable general solution. The advantage of (29) is that the enforcement of initial conditions at $x = 0$ is quite transparent with this format; the solution satisfying

$$y(0) = b_0 \, , \, y'(0) = b_1 \tag{30}$$

is readily seen to be

$$y(x) = b_0 \cosh\sqrt{\lambda}x + \frac{b_1}{\sqrt{\lambda}} \sinh\sqrt{\lambda}x.$$

More generally, the choice $c_1 = e^{-\sqrt{\lambda}x_0}/2, c_2 = e^{\sqrt{\lambda}x_0}/2$ yields the particular solution

$$y_1(x) = \frac{e^{\sqrt{\lambda}(x-x_0)} + e^{-\sqrt{\lambda}(x-x_0)}}{2} = \cosh\sqrt{\lambda}(x - x_0),$$

while the solution $y_2(x) = \sinh\sqrt{\lambda}(x - x_0)$ results from taking the opposite sign for $c_2$; then the combination

$$y(x) = b_0 \cosh\sqrt{\lambda}(x - x_0) + \frac{b_1}{\sqrt{\lambda}} \sinh\sqrt{\lambda}(x - x_0)$$

satisfies initial conditions at a generic point $x_0$:

$$y(x_0) = b_0, \; y'(x_0) = b_1. \tag{31}$$

(*ii*) If $\lambda$ is negative, the solutions $e^{\pm\sqrt{\lambda}x} = e^{\pm i\sqrt{-\lambda}x}$ are immediately replaced by the real functions $\cos\sqrt{-\lambda}x$ and $\sin\sqrt{-\lambda}x$, as in the previous example. (Keep in mind that since $\lambda < 0$, $\sqrt{-\lambda} = \sqrt{|\lambda|}$.) The solution satisfying the initial conditions (30) is given by

$$y(x) = b_0 \cos\sqrt{-\lambda}x + \frac{b_1}{\sqrt{-\lambda}}\sin\sqrt{-\lambda}x,$$

while the more general form

$$y(x) = b_0 \cos\sqrt{-\lambda}(x - x_0) + \frac{b_1}{\sqrt{-\lambda}}\sin\sqrt{-\lambda}(x - x_0)$$

satisfies (31). When (28) is used to model a mass on a spring, the parameter $\lambda$ is negative and the solutions are the oscillatory functions $\cos\sqrt{-\lambda}x$ and $\sin\sqrt{-\lambda}x$. Thus it is traditional to refer to eq. (28) as the *Harmonic Oscillator Equation*.[3]

(*iii*) If $\lambda = 0$ in (28), the general solution is simply $Y(x) = c_1 + c_2 x$. ∎

To summarize:

The general solution to the harmonic oscillator equation

$$y'' = \lambda y$$

can be expressed as

$$Y(x) = \begin{cases} c_1 \cosh\sqrt{\lambda}(x - x_0) + c_2 \sinh\sqrt{\lambda}(x - x_0) & \text{if } \lambda > 0, \\ c_1 + c_2 x & \text{if } \lambda = 0, \\ c_1 \cos\sqrt{-\lambda}(x - x_0) + c_2 \sin\sqrt{-\lambda}(x - x_0) & \text{if } \lambda < 0. \end{cases}$$

For the equidimensional or *Cauchy-Euler* equation each appearance of the differential operator $\frac{d}{dx}$ is "compensated" by a factor $x$; thus the coefficient $a_k(x)$ can be written as $a_k x^k$, and the generic form is

$$a_n x^n \frac{d^n y}{dx^n} + a_{n-1} x^{n-1} \frac{d^{n-1}y}{dx^{n-1}} + \cdots + a_1 x \frac{dy}{dx} + a_0 y = 0 \tag{32}$$

Now the substitution of the ansatz

$$y = x^r, y' = rx^{r-1}, y'' = r(r-1)x^{r-2}, \ldots, \tag{33}$$

---

[3]Herein we use this nomenclature regardless of the sign of $\lambda$.

reduces the differential equation to an algebraic equation, since each term will contain a factor $x^r$:

$$a_n x^n r(r-1)(r-2)\cdots(r-[n-1])x^{r-n}$$
$$+a_{n-1}x^{n-1}r(r-1)\cdots(r-[n-2])x^{r-[n-1]}$$
$$+\cdots+a_2 x^2 r(r-1)x^{r-2}+a_1 x r x^{r-1}+a_0 x^r=0.$$

Thus each of the $n$ roots of the equation

$$a_n r(r-1)(r-2)\cdots(r-[n-1])+a_{n-1}r(r-1)\cdots(r-[n-2])$$
$$+\cdots+a_2 r(r-1)+a_1 r+a_0=0$$

gives rise to a solution to the equidimensional equation.

> The possibility that $r$ might not be an integer adds a troublesome wrinkle to the equidimensional equation - namely, $x^r$ may not be defined for negative $x$. Thus we restrict our solutions to the domain $x > 0$ only. If solutions for negative $x$ are imperative, one can simply change the independent variable to $\xi = -x$ by the methods of Sec. 1.2.

Again relegating the discussion of fine points to the exercises (see exercise 7), we illustrate with examples.

**Example 9.** The differential equation considered in Example 1,

$$x^2 y'' - 2y = 0,$$

generates, through the substitution (33), the algebraic equation

$$r(r-1)-2=r^2-r-2=(r-2)(r+1)=0.$$

Clearly its roots are 2 and -1 and the general solution is

$$Y(x)=c_1 x^2+c_2/x,$$

corresponding to a Wronskian

$$Wr(x_0;x^2,x^{-1})=-3 \qquad\qquad\blacksquare$$

**Example 10.** The differential equation

$$x^2 y'' + xy' = 0$$

produces an algebraic equation with a double root:

$$r(r-1)+r=r^2=0$$

Certainly the function $y_1(x)=x^0=1$ (*constant*) is one solution. For the equidimensional equation the theory (exercise 7) says that a multiple root $r$ gives rise to solutions

$$x^r,\ x^r \ln x,\ x^r(\ln x)^2,\ x^r(\ln x)^3,\ \ldots$$

(up to the order of the multiplicity). Thus the general solution of this equation is

$$Y(x) = c_1 + c_2 \ln x,$$

with Wronskian

$$Wr(x_0; 1, \ln x) = \begin{vmatrix} 1 & \ln x_0 \\ 0 & \frac{1}{x_0} \end{vmatrix} = 1/x_0. \qquad \blacksquare$$

**Example 11.** The equidimensional equation

$$x^2 y'' + 5xy' + 5y = 0$$

yields

$$r(r-1) + 5r + 5 = r^2 + 4r + 5 = 0,$$

which has the complex conjugate roots $r = -2 \pm i$. Thus one solution is

$$y_1(x) = x^{-2} x^i.$$

To interpret $x^i$ we use the identity $x = e^{\ln x}$ to obtain

$$x^i = e^{i \ln x} = \cos [\ln x] + i \sin [\ln x].$$

Once again, both the real and imaginary parts of $y_1(x)$ are, separately, solutions (exercise 8) and we have a general solution

$$Y(x) = c_1 x^{-2} \cos [\ln x] + c_2 x^{-2} \sin [\ln x].$$

The Wronskian is

$$Wr(x_0; x^{-2} \cos [\ln x], x^{-2} \sin [\ln x]) = x_0^{-5}. \qquad \blacksquare$$

In this section we have surveyed the theoretical underpinnings of linear ordinary differential equations, and we have seen how to find general solutions for constant-coefficient and equidimensional homogeneous equations. This leaves two issues to be resolved: homogeneous equations with arbitrary coefficients, and nonhomogeneous equations (in general).

When the coefficients are neither constant nor equidimensional, but are *analytic* (smooth enough to possess Taylor series expansions), the power series methods of Chapter 2 apply. More generally, approximate numerical methods - discussed briefly in Section 2.7 - can be utilized. Some examples where the coefficients are *piecewise* analytic arise occasionally in applications; the appropriate analyses are illustrated in this book for the quantum-mechanical square well potential (Section 7.4) and the fiber optic cable (Section 8.5).

For nonhomogeneous equations the method of Green's functions is universally applicable. It is described, for ordinary differential equations, in Section 1.4; its extension to partial differential equations comprises Chapter 8. This procedure is so much more powerful than the customary "Variation of Parameters" formula that the latter is relegated to an exercise in this book. For constant coefficient equations the "method of undetermined coefficients" finds its generalization in the Fourier methods of Chapter 3.

Some aspects of *nonlinear* equations are described in exercise 9.

# EXERCISES 1.1

1. Find the general solution:

   (a) $y'' + y' - 2y = 0$

   (b) $4y'' + 4y' + y = 0$

   (c) $10y'' + 6y' + y = 0$

   (d) $y'''' + y = 0$

   (e) $x^2 y'' - 2xy' + 2y = 0$

   (f) $2x^2 y'' + 5xy' + y = 0$

   (g) $x^4 y'''' + 6x^3 y''' + 15x^2 y'' + 9xy' - 9y = 0$

2. Show that the general solution to the differential equation

$$x^2 y'' + xy' = \lambda y$$

can be written as

$$Y(x) = \begin{cases} c_1 x^{\sqrt{\lambda}} + c_2 x^{-\sqrt{\lambda}} & \text{if } \lambda > 0, \\ c_1 + c_2 \ln x & \text{if } \lambda = 0, \\ c_1 \cos\left(\sqrt{-\lambda}\ln x\right) + c_2 \sin\left(\sqrt{-\lambda}\ln x\right) & \text{if } \lambda < 0, \end{cases}$$

3. A *general solution* to an $n$th order linear homogeneous differential equation like (18) is one containing $n$ constants which can be used to satisfy the initial conditions (19). In this section we have emphasized the format (20), where the constants $c_i$ appear as *coefficients* in a linear combination of independent solutions. Other formats are possible, however. Consider the equation

$$y'' + y = 0.$$

   (a) Show that the format $y = A\cos(x + \phi)$, for constant $A$ and $\phi$, provides a general solution; i.e., it solves the differential equation and its constants can be chosen to satisfy initial conditions of the form

$$y(a) = b_0, \quad y'(a) = b_1.$$

(b) What are the formulas for $A$ and $\phi$ in terms of $a$, $b_0$, and $b_1$?

(c) If this $y$ is rewritten as $c_1 \cos x + c_2 \sin x$, what are the expressions for the $c_i$ in terms of $A$, $\phi$, and $a$?

4. (*Wronskian Theory*)

(a) Show that the Wronskian $Wr(x; y_1, y_2)$ of two solutions to the second order homogeneous equation

$$a_2(x)y'' + a_1(x)y' + a_0(x)y = 0$$

satisfies the equation

$$\frac{d}{dx}Wr = -\frac{a_1(x)}{a_2(x)}Wr. \qquad (34)$$

(Hint: differentiate $Wr = y_1 y_2' - y_1' y_2$ and use the differential equation for $y_i$.)

(b) From (34) derive *Abel's identity*

$$Wr(x; y_1, y_2) = (constant)\, e^{-\int [a_1(x)/a_2(x)]\,dx}. \qquad (35)$$

(c) From (35) argue that the Wronskian of two solutions is always positive, always negative, or identically zero.

(d) (*A challenge!*) Show that the $n$th order Wronskian (22) of $n$ solutions to the homogeneous equation (18), as a function of $x$, satisfies

$$\frac{d}{dx}Wr = -\frac{a_{n-1}(x)}{a_n(x)}Wr. \qquad (36)$$

(Hint: first you should argue that the derivative of a determinant is the sum of the determinants wherein each row is differentiated separately. For the Wronskian, only one of these is nonzero. Then substitute the differential equation and simplify.) Thus the preceding statement (c) holds for higher order equations as well.

(e) Show by example that statement (c) is not necessarily true *unless the functions $y_1$ and $y_2$ are solutions to the same differential equation.*

5. In mathematical parlance one says that an *arbitrary* set of functions

$$y_1(x), y_2(x), \ldots, y_n(x)$$

is *linearly independent* if, and only if, the only linear combination of them

$$c_1 y_1(x) + c_2 y_2(x) + \cdots + c_n y_n(x)$$

which vanishes for all $x$ is the trivial one (each $c_i = 0$). Show that *if the $y_i$ are solutions to an nth order linear homogeneous differential equation,* this condition is equivalent to the vanishing of the Wronskian at some (and hence every) point. (Hint: argue that if the Wronskian is zero, there are nontrivial solutions to (19) with each $b_i = 0$ - giving rise to nontrivial combinations of the $y_i$ *which vanish everywhere, because of the uniqueness theorem.* Conversely, a nontrivial linear combination vanishing everywhere would provide a nontrivial solution to equations (19) with each $b_i = 0$ - which, in turn, would require the vanishing of the Wronskian.)

6. Show that the Wronskian of the functions $e^{p_1 x}, e^{p_2 x}, \ldots, e^{p_n x}$ at $x = 0$ is the *Vandermonde determinant*

$$
\begin{vmatrix}
1 & 1 & \cdots & 1 \\
p_1 & p_2 & \cdots & p_n \\
p_1^2 & p_2^2 & \cdots & p_n^2 \\
& & \vdots & \\
p_1^{n-1} & p_2^{n-1} & \cdots & p_n^{n-1}
\end{vmatrix}
$$

which is known to be $\prod_{i \neq j}(p_i - p_j)$ and thus is nonzero if the $p_i$ are distinct.

7. Suppose that $y_1(x)$ is a solution of the differential equation

$$ a_2(x)y'' + a_1(x)y' + a_0(x)y = 0. $$

(a) Show that the ansatz $y_2(x) = u(x)y_1(x)$ will satisfy the differential equation if $u$ satisfies

$$ a_2(x)y_1(x)u'' + [2a_2(x)y_1'(x) + a_1(x)y_1(x)]u' = 0. \qquad (37) $$

(b) Regarding (37) as a *first* order equation for $u'$, show that its solution can be expressed

$$ u'(x) = K\, e^{-\int [\frac{a_1}{a_2} + 2\frac{y_1'}{y_1}]dx} = K\, \frac{e^{-\int [\frac{a_1}{a_2}]dx}}{y_1^2(x)} $$

(where the integrals are indefinite).

(c) Conclude that a second solution to the original differential equation is given by the formula

$$ y_2(x) = y_1(x) \int \frac{e^{-\int [a_1(x)/a_2(x)]\,dx}}{y_1^2(x)}\, dx \qquad (38) $$

(d) For the constant coefficient equation

$$ a_2\, y'' + a_1 y' + a_0\, y = 0 $$

having $y_1 = e^{p_1 x}$ as a solution, demonstrate how formula (38) displays a second solution as (*a constant times*) $e^{p_2 x}$ if the roots of the characteristic equation are distinct, and as $xe^{p_1 x}$ if they are equal. (Hint: recall - from the quadratic formula, e.g. - that the sum of the roots equals $-a_1/a_2$.)

(e) For the equidimensional case

$$a_2 x^2 y'' + a_1 x y' + a_0 y = 0$$

having $y_1 = x^{r_1}$ as a solution, demonstrate how formula (38) displays a second solution as (*a constant times*) $x^{r_2}$ if the roots of

$$a_2 r^2 + (a_1 - a_2)r + a_0 = 0$$

are distinct, and as $x^{r_1} \ln x$ if they are equal. (See the previous hint.)[4]

**8.** Show that if a *complex* function

$$y(x) = u(x) + iv(x)$$

satisfies a linear homogeneous differential equation with real coefficients then the real and imaginary parts, $u(x)$ and $v(x)$ respectively, are solutions also. (Hint: $Ly = Lu + iLv = 0$; now what are the real and imaginary parts of zero?)

**9.** *(Nonlinear Differential Equations)*

(a) Show that the function

$$y(x) = \frac{1}{1-x} \tag{39}$$

solves the *nonlinear* initial value problem

$$y' = y^2, \ y(0) = 1. \tag{40}$$

(b) Show that *both* functions

$$y_1(x) = (x/2)^2, \ y_2(x) \equiv 0 \tag{41}$$

solve the nonlinear initial value problem

$$y' = \sqrt{y}, \ y(0) = 0. \tag{42}$$

---

[4]The generalization of the results of exercise 7 to *nth* order equations is discussed in the textbooks by Birkhoff and Rota or Coddington and Levinson.

The analysis of nonlinear differential equations is more subtle than the linear case. Note that there is no hint, in the formulation of initial value problem (40), which would indicate that its solution (39) blows up at $x = 1$. And clearly problem (42) is not even subject to the uniqueness theorem! Small wonder that most mathematical works on nonlinear equations focus on the existence, uniqueness, and qualitative properties; except for some very special cases, no solution techniques are even known.

(c) A first order differential equation in the form

$$\frac{dy}{dx} = g(x)h(y)$$

is said to be *separable* because it can be rewritten in the form

$$\frac{dy}{h(y)} = g(x)\, dx$$

and solved by (indefinite) integration of both sides. Derive the solution (39) and $y_1$ in (41) by this technique.

10. Note that the *sequence* $\{x_n\} = \{1, 2, 4, 8, 16, \ldots\}$ satisfies the equation $x_n = 2x_{n-1}$. In general an equation of the form

$$x_n = h_1 x_{n-1} + h_2 x_{n-2} + \cdots + h_r x_{n-r} \tag{43}$$

is known as a *difference equation* of order $r$ (primarily because such equations arise in the finite-difference approximations to ordinary differential equations: see Section 2.7).[5] Clearly if one is given a set of *initial values* $x_0, x_1, x_2, \ldots, x_{r-1}$, one can use (43) to compute $x_r, x_{r+1}, \ldots$ . Investigate how the insertion of the *ansatz* $x_j = \lambda^j$ leads to a general solution of (43) of the form

$$x_j = c_1 \lambda_1^j + c_2 \lambda_2^j + \cdots + c_r \lambda_r^j \tag{44}$$

if the roots $\{\lambda_s\}$ of the polynomial

$$\lambda^r - h_1 \lambda^{r-1} - h_2 \lambda^{r-2} - \cdots - h_r$$

are distinct, while if $\lambda_2 = \lambda_1$ one substitutes $j\lambda_1^j$ for $\lambda_2^j$ in (44), etc.

---

[5]To be precise, (43) is a linear homogeneous constant-coefficient difference equation of order $r$.

## 1.2   Change of Variable

**Summary**   A change of independent variable $x = f(t)$ in a function $y(x) = y(f(t)) = Y(t)$ dictates the following relations among the derivatives. With

$$\frac{dy}{dx} = y', \ \frac{d^2y}{dx^2} = y'', \ \frac{dY}{dt} = Y', \text{ and } \frac{d^2Y}{dt^2} = Y'',$$

we have

$$y' = Y' / \left(\frac{dx}{dt}\right), \ y'' = \frac{1}{[dx/dt]^2}Y'' - \frac{d^2x/dt^2}{[dx/dt]^3}Y'.$$

Almost all continuum physical and engineering systems are modeled by differential equations, and a mastery of the various techniques for analyzing "DEs" is a valuable asset for any analyst.

However it is important to keep these efforts in perspective. The mathematicians of the nineteenth and early twentieth centuries directed a lot of energy to studying properties and tabulating solutions of the common ordinary differential equations arising in applications, and we owe it to them - and to ourselves - to take advantage of their labors. Therefore besides studying the more advanced methods which are used to solve DEs, we are going to survey the contributions of the old masters; as professionals, we don't want to "reinvent the wheel" every time we are confronted with an unfamiliar DE.

Of course a catalog of every individual differential equation that has been studied is out of the question. But many times the same DE occurs *in different forms* in engineering. The Bessel equation, for instance, arises in many distinct situations, as we shall see; but its recurrence only becomes apparent after we change variables in some way. Thus we devote this preliminary section to a study of the various forms that a DE can take when it is expressed in new variables. The transformation procedure is straightforward, but it convenient to have the identities spelled out explicitly.

The simplest approach is to start with a *solution* to a DE, change variables, and then observe what new DE the altered solution satisfies. In this way we can observe and characterize some of the more obscure forms a given DE can take.

Thus suppose we have a solution $y(x)$ to a differential equation of the form

$$a_2(x)y'' + a_1(x)y' + a_0(x)y = 0 \tag{1}$$

We specify a change of the independent variable from $x$ to (say) $t$, which is related to $x$ by a *transformation* equation

$$x = f(t) \tag{2}$$

The *inverse* of the transformation, obtained by solving (2) for $t$ in terms of $x$, will be denoted

$$t = g(x). \tag{3}$$

Examples of such a transformation could be

$$
\begin{aligned}
x &= 2t + 4, \ \ t = \frac{x}{2} - 2 \ \ (\text{all } x, \ t) \\
x &= t^2, \ \ t = \sqrt{x} \ \ (x \geq 0, \ t \geq 0) \\
x &= 1/t, \ \ t = 1/x \ \ (x > 0, \ t > 0 \ \text{ or } \ x < 0, \ t < 0) \\
x &= \sin t, \ \ t = \arcsin x \ \ (-1 \leq x \leq 1, \ -\pi/2 \leq t \leq \pi/2)
\end{aligned}
$$

(The notation $t = f^{-1}(x)$ is very nice but cumbersome in the present context.)

Now suppose we substitute $f(t)$ for $x$ in the expression $y(x)$ to obtain the new expression $y(f(t))$; for example, if we substitute $x = t^2$ into $\cos x$ we get $\cos t^2$. We obtain a new mathematical function, although "physically" it corresponds to the same variable ("$y$"); so let's employ the suggestive notation

$$Y(t) = y(f(t)). \tag{4}$$

Note that it would be quite hazardous to denote $y(f(t))$ by $y(t)$, although it is common practice to do so. If $y(x)$ is $\cos x$, then $y(f(t))$ is $\cos t^2$, while $y(t)$ is $\cos t$.

If we are given that $y(x)$ satisfies the differential equation (1), we ask what differential equation does $Y(t)$ satisfy? The key to answering this question is the relationship between the derivatives $dy/dx$ and $dY/dt$, and it is revealed through the *Inverse Function Rule* and the *Chain Rule* of calculus.

*Inverse Function Rule.* The derivative of (2) with respect to $t$,

$$\frac{dx}{dt} = f'(t)$$

is related to the derivative of (3) with respect to $x$

$$\frac{dt}{dx} = g'(x)$$

by

$$\frac{dx}{dt} = 1 / \left( \frac{dt}{dx} \right), \ \text{ or } \ f'(t) = 1/g'(x) \tag{5}$$

as long as neither derivative is zero (an occurrence that could hardly be missed if we tried to apply (5)!).

*Chain Rule.* The derivative of a composite function obeys

$$\frac{dY}{dt} = \frac{d}{dt}y(f(t)) = y'(f(t))\, f'(t). \tag{6}$$

Here $y'(f(t))$ denotes the result of differentiating the expression $y(x)$ with respect to $x$, and then inserting the substitution $x = f(t)$. Equation (6) can thus be abbreviated

$$\frac{dY}{dt} = \frac{dy}{dx}\frac{dx}{dt},$$

and since $Y$ and $y$ are "physically" the same quantity we write symbolically

$$\frac{d}{dt} = \frac{dx}{dt}\frac{d}{dx} \quad \left(\text{and } \frac{d}{dx} = \frac{dt}{dx}\frac{d}{dt}\right). \tag{7}$$

Thus the relationship between $y' = \frac{dy}{dx}$ and $Y' = \frac{dY}{dt}$ is

$$y' = \left(\frac{dt}{dx}\right) \cdot Y' = Y' / \left(\frac{dx}{dt}\right) \tag{8}$$

To relate $y'' = d^2y/dx^2$ and $Y'' = d^2Y/dt^2$ we invoke (7, 8) again:

$$y'' = \frac{d}{dx}y' = \left(\frac{dt}{dx}\right)\frac{d}{dt}\left\{Y'/\left(\frac{dx}{dt}\right)\right\} = \frac{dt}{dx}\left\{Y''/\left(\frac{dx}{dt}\right) - Y'\frac{d^2x}{dt^2}/\left(\frac{dx}{dt}\right)^2\right\}$$

or

$$y'' = \frac{1}{[dx/dt]^2}Y'' - \frac{d^2x/dt^2}{[dx/dt]^3}Y' \tag{9}$$

Substitution of (4), (8), and (9) into the original differential equation (1), and replacement of $x$ throughout by $f(t)$, will then result in a differential equation for $Y(t)$.

**Example 1.** The function $y(x) = \cos x$ satisfies $y'' + y = 0$. If $x = 2t$, what equation does $Y(t) = \cos 2t$ satisfy?

We have $x = 2t$, so $dx/dt = 2$ and $d^2x/dt^2 = 0$. Thus by (9)

$$y'' = Y''/4 - 0$$

and substitution into $y'' + y = 0$ leads to

$$Y''/4 + Y = 0 \quad \text{or} \quad Y'' + 4Y = 0.$$

Obviously $Y = y(2t) = \cos 2t$ is a solution. ∎

**Example 2.** Again let $y = \cos x$, satisfying $y'' + y = 0$. If $x = t^2$, what DE does $\cos t^2$ satisfy?

We have $dx/dt = 2t$, $d^2x/dt^2 = 2$. Thus by (9)

$$y'' = Y''/(2t)^2 - [2/(2t)^3]Y'$$

and substitution into $y'' + y = 0$ yields

$$Y''/4t^2 - 2Y'/8t^3 + Y = 0, \quad \text{or} \quad tY'' - Y' + 4t^3Y = 0$$ ∎

**Example 3.** In Sec. 2.5 we shall see that the *Bessel function of the first kind of order* $\nu$, denoted by $J_\nu(x)$, satisfies the equation

$$x^2y'' + xy' + (x^2 - \nu^2)y = 0 \quad (y = J_\nu(x)) \tag{10}$$

If $x = bt$, $b$ a constant, what equation does $Y(t) = J_\nu(bt)$ satisfy?

We have $dx/dt = b$, $d^2x/dt^2 = 0$, so by (8) and (9)

$$y' = Y'/b, \quad y'' = Y''/b^2,$$

and thus $J_\nu(bt)$ satisfies

$$(bt)^2Y''/b^2 + (bt)Y'/b + [(bt)^2 - \nu^2]Y = 0,$$

or

$$t^2Y'' + tY' + (b^2t^2 - \nu^2)Y = 0 \tag{11}$$

∎

A simple transformation of the form $x = bt$ is called a *change of scale*. By comparing (11) with (10) we see that a change of scale in the Bessel equation has the net effect of introducing a factor $b^2$ in front of the "$x^2$" multiplying the (undifferentiated) term $y$.

A *shift* is a transformation of the form $x = a + t$ (since the zero-point of the new variable has shifted to $x = a$). The combination of these two,

$$x = a + bt, \quad t = (x - a)/b,$$

is known as a *linear transformation* (because of its resemblance to the equation of a straight line).

If $y(x)$ satisfies an equation of the form

$$y'' + A(c + dx)^r y' + B(c + dx)^s y = 0 \tag{12}$$

then obviously it is desirable to introduce the linear transformation $t = c + dx$, or $x = (t - c)/d$. The resulting equation for $Y(t) = y(x(t)) = y(\frac{t-c}{d})$ is easily seen to be (exercise 1)

$$Y'' + [At^r/d]Y' + [Bt^s/d^2]Y = 0. \tag{13}$$

So the linear transformation applied to eq. (12) has the obvious consequence of replacing $(c + dx)$ by $t$ in the coefficients - and it also introduces a factor $\frac{1}{d}$ into the first-derivative term and $\frac{1}{d^2}$ into the zeroth-derivative term.

**Example 4.** The transformation $t = -x$ (or $t = c + dx$ with $c = 0$ and $d = -1$), applied to the constant coefficient equation

$$y'' + Ay' + By = 0, \tag{14}$$

changes the sign of the $y'$ term: taking $r = s = 0$ in (12) we have

$$Y'' - AY' + BY = 0. \tag{15}$$

Physically speaking, we can say that (15) is the "time-reversed" form of (14).  ∎

**Example 5.** The time-reversed form of *Airy's* equation

$$y'' - xy = 0 \tag{16}$$

is easily seen to be

$$Y'' + tY = 0.$$

Two independent solutions to $y'' - xy = 0$ will be identified in Section 2.5 as the Airy functions $Ai(x)$ and $Bi(x)$. We immediately conclude that independent solutions to $y'' + xy = 0$ are given by $Ai(-x)$ and $Bi(-x)$.  ∎

So far we have only considered the effect on the differential equation of a transformation of the *independent* variable. The effect of changing the *dependent* variable is easy to analyze, as the following example shows.

**Example 6.** We return to the solution $y = \cos x$ of $y'' + y = 0$. What DE does the function

$$z(x) = x \cos x \tag{17}$$

satisfy?

We express the original dependent variable $y$ in terms of the new one:

$$z(x) = x\, y(x), \quad \text{so } y(x) = \frac{z(x)}{x}.$$

Then the relations between the derivatives are simply

$$y' = \frac{z'}{x} - \frac{z}{x^2}\, , \; y'' = \frac{z''}{x} - 2\frac{z'}{x^2} + 2\frac{z}{x^3}, \tag{18}$$

and inserting into $y'' + y = 0$ produces

$$\frac{z''}{x} - \frac{2z'}{x^2} + \frac{2z}{x^3} + \frac{z}{x} = 0, \text{ or}$$
$$x^2 z'' - 2xz' + (2 + x^2)z = 0. \qquad ∎$$

**Example 7.** As we mentioned, the Bessel equation (10) arises in many different forms. By combining a change of independent *and* dependent variables using the theory developed in this section, one can show that if $y = J_\nu(x)$ denotes a solution of the Bessel equation of order $\nu$

$$x^2 y'' + x y' + (x^2 - \nu^2) y = 0 \tag{19}$$

then

$$z(x) = x^a J_\nu(bx^c) \tag{20}$$

satisfies

$$z'' - \left(\frac{2a-1}{x}\right) z' + \left(b^2 c^2 x^{2c-2} + \frac{a^2 - \nu^2 c^2}{x^2}\right) z = 0 \tag{21}$$

Equation (21) is thus a very general form of Bessel's equation. The verification of (21), and a further generalization, are left to the exercises. ∎

**Example 8.** We can express the solutions to Airy's equation

$$y'' - x y = 0 \quad \text{(Eq. (16) repeated)} \tag{22}$$

in terms of Bessel functions! In order to match up eq. (21) with eq. (16), first we get rid of the $z'$ term in the former by taking $a = 1/2$. To force the coefficient of $z$ to be $(-x)$ we need

$$
\begin{aligned}
b^2 c^2 &= -1, \\
2c - 2 &= 1, \\
\text{and } a^2 - \nu^2 c^2 &= 0.
\end{aligned}
$$

Thus $c = 3/2$, $b = \pm\frac{2}{3}i$, and $\nu = \pm\frac{1}{3}$. From eq. (20), then, we now know that the Airy functions are combinations of the (complex) Bessel functions $x^{1/2} J_{\pm 1/3}(\pm 2ix^{3/2}/3)$. ∎

On a personal note, this author has always been fascinated with the power of differential equation theory. In the preceding example, we deduced the existence of a relationship between the Airy functions and the Bessel functions, before having defined either one! And perhaps you have seen some differential equation texts which derive all the trigonometric identities from differential equations - without drawing a single triangle! But the supreme example has to be a calculation that I tested in 1984, and I beg the reader's indulgence as I share it with you. Without the slightest idea of how to cook and without explicitly solving any partial differential equations, I worked out the correct scaling law for the roasting time for a turkey!

Because the proper interpretation of all the effects of rescaling can be subtle, the computation is presented in meticulous detail.

**Example 9.** The process of cooking a roast involves taking a piece of meat at an initial temperature $T_{cold}$ and placing it in an oven at a constant temperature $T_{oven}$ until a meat thermometer indicates that the temperature at a specific location, say $(x, y, z) = (0, 0, 0)$, has reached the value $T_{done}$. This requires a cooking time $t_{done}$.

We will use a change-of-variable argument to derive how the cooking time scales with the size of the roast. First we must formulate the problem mathematically. Let $T(x, y, z, t)$ denote the temperature of the roast. As described in Section 4.5, if the *thermal diffusivity* of the meat is denoted $k$ then its temperature evolves according to the heat equation

$$\frac{\partial T}{\partial t} = k\nabla^2 T = k\left\{\frac{\partial^2 T}{\partial x^2} + \frac{\partial^2 T}{\partial y^2} + \frac{\partial^2 T}{\partial z^2}\right\} \tag{23}$$

The initial temperature is $T_{cold}$:

$$T(x, y, z, 0) = T_{cold}. \tag{24}$$

The temperature on the skin or "boundary" $B$ of the roast is maintained by the oven at $T_{oven}$:

$$T(x, y, z, t) = T_{oven}, \quad (x, y, z) \text{ on boundary } B, \text{ all } t > 0. \tag{25}$$

And the cooking time is specified by the condition

$$T(0, 0, 0, t_{done}) = T_{done}. \tag{26}$$

If a larger roast is placed in the oven, what equations does its temperature $\Theta$ satisfy? Because we hope to use a change of scale in the previous solution $T(x, y, z, t)$ let us employ new symbols for the independent variables in $\Theta$:

$$\Theta = \Theta(\xi, \eta, \zeta, \tau). \tag{27}$$

Assuming the same type of meat is used, $k$ remains unchanged in the heat equation:

$$\frac{\partial \Theta}{\partial \tau} = k\left\{\frac{\partial^2 \Theta}{\partial \xi^2} + \frac{\partial^2 \Theta}{\partial \eta^2} + \frac{\partial^2 \Theta}{\partial \zeta^2}\right\} \tag{28}$$

The initial temperature will also be the same:

$$\Theta(\xi, \eta, \zeta, 0) = T_{cold}. \tag{29}$$

But the oven temperature condition (25) will be maintained on the new skin-boundary $B'$, which is a rescaled version of the former boundary:

$$\Theta(\xi, \eta, \zeta, \tau) = T_{oven}, \quad (\xi, \eta, \zeta) \text{ on } B', \text{ all } \tau > 0. \tag{30}$$

Let us be very explicit about this. Suppose that the larger roast is bigger by a linear factor of $\lambda$. Then if $(x, y, z)$ locates a point on the former boundary $B$, $(\lambda x, \lambda y, \lambda z)$ lies on the boundary $B'$. Thus (30) can be rewritten

$$\Theta(\lambda x, \lambda y, \lambda z, t) = T_{oven}, \quad (x, y, z) \text{ on } B, \text{ all } t > 0. \tag{31}$$

Now we hope to find a function $\Theta(\xi, \eta, \zeta, \tau)$ satisfying (28, 29, 31) by simply rescaling the variables in the old solution $T(x, y, z, t)$. Thus consider the simple transformations

$$x = \alpha\xi, \ y = \alpha\eta, \ z = \alpha\zeta, \ t = \beta\tau,$$
$$\Theta(\xi, \eta, \zeta, \tau) = T(x, y, z, t) = T(\alpha\xi, \alpha\eta, \alpha\zeta, \beta\tau) \tag{32}$$

where $\alpha$ and $\beta$ are constants to be determined later. According to the rules (8, 9) the derivatives of $T$ in (23) become replaced by

$$\frac{\partial T}{\partial t} = \frac{\partial \Theta}{\partial \tau} \Big/ \frac{dt}{d\tau} = \frac{1}{\beta}\frac{\partial \Theta}{\partial \tau}, \quad \frac{\partial^2 T}{\partial x^2} = \frac{\partial^2 \Theta}{\partial \zeta^2} \Big/ \left(\frac{dx}{d\zeta}\right)^2 = \frac{1}{\alpha^2}\frac{\partial^2 \Theta}{\partial \zeta^2}, \tag{33}$$

and similarly for $y$ and $z$. Thus $\Theta(\xi, \eta, \zeta, \tau)$ satisfies

$$\frac{1}{\beta}\frac{\partial \Theta}{\partial \tau} = \frac{k}{\alpha^2}\left\{\frac{\partial^2 \Theta}{\partial \xi^2} + \frac{\partial^2 \Theta}{\partial \eta^2} + \frac{\partial^2 \Theta}{\partial \zeta^2}\right\}, \tag{34}$$

which agrees with the heat equation (28) if we take

$$\beta = \alpha^2. \tag{35}$$

Moreover, the initial condition (29) is satisfied automatically because

$$\Theta(\xi, \eta, \zeta, 0) = T(\alpha\xi, \alpha\eta, \alpha\zeta, 0) = T_{cold}$$

by (24). Similarly the boundary condition (31) will follow from (25) if we set

$$\alpha = 1/\lambda \ \ (\text{and hence } \ \beta = 1/\lambda^2, \text{ by (35)}) \tag{36}$$

because

$$\Theta(\lambda x, \lambda y, \lambda z, \tau) = T(\alpha\lambda x, \alpha\lambda y, \alpha\lambda z, \beta t) = T(x, y, z, t/\lambda^2)$$
$$= T_{oven} \text{ for } (x, y, z) \text{ on } B.$$

To summarize: *the temperature of the larger roast evolves according to*

$$\Theta(\xi, \eta, \zeta, \tau) = T(\xi/\lambda, \eta/\lambda, \zeta/\lambda, \tau/\lambda^2). \tag{37}$$

The larger roast is done at time $\tau_{done}$ when the relation

$$\Theta(0, 0, 0, \tau_{done}) = T_{done} \tag{38}$$

is met. Since

$$\Theta(0,0,0,\tau_{done}) = T(0,0,0,\tau_{done}/\lambda^2)$$

and, by eq. (26), $T(0,0,0,t) = T_{done}$ when $t = t_{done}$, eq. (38) is satisfied for

$$\tau_{done}/\lambda^2 = t_{done}, \quad \text{or} \quad \tau_{done}/t_{done} = \lambda^2. \tag{39}$$

According to *Betty Crocker's Cookbook* (General Mills, Inc., Minneapolis, Minnesota, 1969), in a $450°F$ oven a 10 pound turkey cooks in about $2\frac{3}{4}$ hours, while a 20 pounder takes $4\frac{1}{2}$ hours. That gives a cooking time ratio of

$$\tau_{done}/t_{done} = 4.5/2.75 \approx 1.64.$$

A turkey twice as heavy has twice the volume, and hence its linear dimensions are bigger by a factor $\lambda = 2^{1/3}$. Thus the *theory* predicts a cooking time ratio of

$$\tau_{done}/t_{done} = 2^{2/3} \approx 1.59.$$

That's excellent agreement! (Evidently the mathematician's turkey turns out slightly undercooked; *mathematicians do make errors - but they are rare!*)                                           ∎

# Exercises 1.2

1. (a) Verify that the linear transformation changes eq. (12) to (13).

   (b) Verify that a constant-coefficient differential equation is left invariant by a shift of the independent variable.

   (c) Verify that an equidimensional differential equation is left invariant by a change of scale of the independent variable.

2. (a) What form does the equation

   $$y' + p(x)y = q(x)$$

   take when the substitution

   $$y(x) = e^{-\int^x p(\xi)d\xi} z(x)$$

   is made?

   (b) From this result, derive the solution formula

   $$y(x) = e^{-\int^x p(\xi)d\xi} \left[ \int^x \left\{ e^{\int^\eta p(\xi)d\xi} q(\eta) \right\} d\eta + c \right],$$

   where all integrals are indefinite.

(c) Verify this formula by using it to solve

$$y' + y/x = x^2,$$

and check your answer by direct substitution.

**3.** Show that the *equidimensional* equation

$$a_2 x^2 y'' + a_1 x y' + a_0 y = 0$$

transforms to an equation with constant coefficients if we set $x = e^t$.

**4.** Show that the nonlinear *Bernoulli* equation

$$y' = p(x)y + q(x)y^n$$

transforms to a linear equation through the substitution $z = y^{1-n}$.

**5.** Show that if $y(x)$ satisfies the *Legendre* differential equation

$$(1 - x^2)y'' - 2xy' - \lambda y = 0 \quad (\lambda = \text{ constant})$$

then the function $\Phi(\phi) = y(\cos\phi)$ satisfies

$$(\sin^2\phi)\Phi'' + (\sin\phi\cos\phi)\Phi' - (\lambda\sin^2\phi)\Phi = 0$$

**6.** Show that if $y(x)$ satisfies *Weber's* equation

$$y'' - x^2 y = (\mu - 1)y$$

then $z(x) = e^{x^2/2}y$ satisfies *Hermite's* equation

$$z'' - 2xz' - \mu z = 0$$

**7.** Later we shall see how the following transformation is important in atomic physics. Given that $y(x)$ satisfies the equation

$$x^2 y'' + 2xy' + \{qx - \ell(\ell+1) - \lambda x^2\}y = 0$$

(which governs the radial dependence of solutions to the quantum mechanical hydrogen atom), show that the substitutions

$$
\begin{aligned}
x &= \frac{t}{2\sqrt{\lambda}}, \\
y(x) &= t^\ell e^{-t/2} z(t), \\
k &= 2\ell + 1, \quad \mu = \ell + 1 - \frac{q}{2\sqrt{\lambda}}
\end{aligned}
$$

result in the *Associated Laguerre* equation

$$tz'' + (k + 1 - t)z' - \mu z = 0$$

8. Verify eq. (21).

9. Show that if $J_{1/2}(x)$ satisfies the Bessel equation of order one-half then $y = \sqrt{x} J_{1/2}(x)$ satisfies $y'' + y = 0$.

10. Use eq. (21) to express the solutions to the following equations in terms of solutions to Bessel's equation (19).

   (a) $y'' + x^m y = 0$

   (b) $xy'' + 2y' + 4xy = 0$

   (c) $y'' + y = 0$

   (d) $(3x + 2)y'' + 3y' - y = 0$  (*Hint:* first let $t = 3x + 2$.)

   (e) $y'' + 4e^{bx} y = 0$ (*Hint:* let $t = e^{bx/2}$.)

   (f) $y'' + ky' + (A + Be^{bx})y = 0$ (*Hint:* let $t = e^{bx}$.)

11. (a) Show that if $J_\nu(x)$ is a solution of Bessel's equation of order $\nu$ and

$$\nu^2 = \left\{ \left( \frac{1-A}{2} \right)^2 - C \right\} / a^2$$

   then

$$y(x) = x^{(1-A)/2} e^{-Bx^b/b} J_\nu\left( \frac{\sqrt{D}}{a} x^a \right)$$

   satisfies[6]

$$x^2 y'' \quad + \quad (A + 2Bx^b)xy'$$
$$+ \quad [C + Dx^{2a} - B(1 - A - b)x^b + B^2 x^{2b}]y = 0.$$

   (b) Express the solution to

$$x^2 y'' + (2x^2 + x)y' + (x^2 + 3x - 1)y = 0$$

   in terms of solutions to Bessel's equations.

12. (*Classification of Partial Differential Equations*) When contemplating a change of coordinates in a *two*-dimensional system one must acknowledge the increased complexity of the chain rule.

---

[6]For further generalizations of Bessel's equation see the reference by Hildebrand.

If $(x, y)$ coordinates are changed to $(\xi, \eta)$ coordinates through the transformation equations

$$x = x(\xi, \eta), \quad y = y(\xi, \eta), \tag{40}$$

inducing changes in the dependent variable according to

$$\phi(x, y) = \phi(x(\xi, \eta), y(\xi, \eta)) = \psi(\xi, \eta),$$

then the partial derivatives are related by the rule

$$\frac{\partial \phi}{\partial x} = \frac{\partial \xi}{\partial x}\frac{\partial \psi}{\partial \xi} + \frac{\partial \eta}{\partial x}\frac{\partial \psi}{\partial \eta}, \quad \frac{\partial \phi}{\partial y} = \frac{\partial \xi}{\partial y}\frac{\partial \psi}{\partial \xi} + \frac{\partial \eta}{\partial y}\frac{\partial \psi}{\partial \eta}. \tag{41}$$

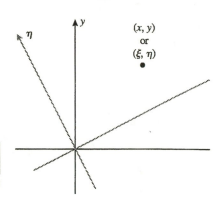

Suppose that the transformation (40) is a simple rotation of the axes of the coordinate system, so that in matrix notation (see Fig. 1)

$$\begin{bmatrix} x \\ y \end{bmatrix} = \begin{bmatrix} \cos\theta & -\sin\theta \\ \sin\theta & \cos\theta \end{bmatrix} \begin{bmatrix} \xi \\ \eta \end{bmatrix}, \quad \begin{bmatrix} \xi \\ \eta \end{bmatrix} = \begin{bmatrix} \cos\theta & \sin\theta \\ -\sin\theta & \cos\theta \end{bmatrix} \begin{bmatrix} x \\ y \end{bmatrix} \tag{42}$$

(a) Show that the results of the chain rule (41) in this case can be written as follows with $\alpha = \cos\theta$, $\beta = \sin\theta$:

$$\begin{bmatrix} \phi_x \\ \phi_y \end{bmatrix} = \begin{bmatrix} \alpha & -\beta \\ \beta & \alpha \end{bmatrix} \begin{bmatrix} \psi_\xi \\ \psi_\eta \end{bmatrix}$$

**Figure 1**  *Axis rotation*

$$\begin{aligned}
\phi_{xx} &= \alpha^2 \psi_{\xi\xi} + \beta^2 \psi_{\eta\eta} - 2\alpha\beta\psi_{\xi\eta}, \\
\phi_{yy} &= \beta^2 \psi_{\xi\xi} + \alpha^2 \psi_{\eta\eta} + 2\alpha\beta\psi_{\xi\eta}, \\
\phi_{xy} &= \alpha\beta\psi_{\xi\xi} - \alpha\beta\psi_{\eta\eta} + (\alpha^2 - \beta^2)\psi_{\xi\eta}.
\end{aligned}$$

(b) Consider a linear second order partial differential operator of the form

$$L[\phi] = a\phi_{xx} + b\phi_{xy} + c\phi_{yy}. \tag{43}$$

Show that after the transformation (42) is performed the expression (43) takes the form

$$L[\phi] = a\phi_{xx} + b\phi_{xy} + c\phi_{yy} = A\psi_{\xi\xi} + B\psi_{\eta\xi} + C\psi_{\eta\eta} = \mathcal{L}[\psi] \tag{44}$$

where the new coefficients can be expressed as

$$\begin{bmatrix} A & B/2 \\ B/2 & C \end{bmatrix} = \begin{bmatrix} \alpha & \beta \\ -\beta & \alpha \end{bmatrix} \begin{bmatrix} a & b/2 \\ b/2 & c \end{bmatrix} \begin{bmatrix} \alpha & -\beta \\ \beta & \alpha \end{bmatrix} \tag{45}$$

(c) Use determinant theory to show that $AC - B^2/4 = ac - b^2/4$. (One says "The *discriminant* of the coefficients is not changed by the rotation (42).").

(d) It is desirable to select the angle $\theta$ in the rotation so that the resulting differential operator $\mathcal{L}[\psi]$ is simplified; by examining the expression for $B$, show that the choice

$$\theta = \frac{1}{2}\arctan\frac{b}{a-c}$$

renders $\mathcal{L}[\psi]$ in the form

$$\mathcal{L}[\psi] = A\psi_{\xi\xi} + C\psi_{\eta\eta}. \qquad (46)$$

Recall from analytic geometry that an equation of the form $ax^2 + bxy + cy^2 + dx + ey = K$ describes a *conic section* - and that one way to see whether its graph is an ellipse, a hyperbola, or a parabola is to rotate to the *principal axis* system wherein the cross term ($\xi\eta$) does not appear. Then $A\xi^2 + C\eta^2 + D\xi + E\eta = K$ describes an ellipse if $AC$ is positive, a hyperbola if $AC$ is negative, and a parabola if $AC$ is zero (and thus either $A$ or $C$ is missing).

Note further that the *sign* of $AC$ - and hence the classification as ellipse, hyperbola, or parabola - can be deduced directly from the original coefficients $(a, b, c)$ without performing the rotation, since the discriminant is invariant:

$$ac - b^2/4 = AC - B^2/4 = AC.$$

Accordingly, in advanced works one says that the differential equation $L[\phi] = f$ is

**elliptic** if $ac - b^2/2 > 0$,

**hyperbolic** if $ac - b^2/4 < 0$, and

**parabolic** if $ac = b^2/4$ .

(Note that the classification of an equation can vary from point to point, if the coefficients are nonconstant.)

(e) Classify Laplace's equation

$$\frac{\partial^2\phi}{\partial x^2} + \frac{\partial^2\phi}{\partial y^2} = 0,$$

the wave equation

$$\frac{\partial^2\phi}{\partial x^2} = \frac{\partial^2\phi}{\partial t^2},$$

and the heat equation

$$\frac{\partial^2 \phi}{\partial x^2} = \frac{\partial \phi}{\partial t}.$$

(In the last two, we have changed $y$ to $t$.)

## 1.3   Delta Functions

**Summary**   The (Dirac) delta function $\delta(x)$ is a hypothetical function which is interpreted to be zero everywhere except for $x = 0$, where it is "so infinite" that it subtends an area of unity; as a result, for any continuous function

$$\int_{-\infty}^{\infty} \delta(x - x_0) f(x)\, dx = f(x_0).$$

Also known as the unit impulse function, $\delta(x)$ is the formal derivative of the (Heaviside) unit step function, $\delta(x) = H'(x)$, and its derivative, in turn, is the doublet function $\delta'(x)$ satisfying

$$\int_{-\infty}^{\infty} \delta'(x - x_0) f(x)\, dx = -f'(x_0),$$

in accordance with integration by parts.  The delta function quantifies the nature of the singularity of $1/x^n$, in the sense that the generalized solution to the equation $x^n f(x) = g(x)$ is given by

$$f(x) = \frac{g(x)}{x^n} + c_1 \delta(x) + c_2 \delta'(x) + \cdots + c_{n-1} \delta^{(n-1)}(x).$$

In applied mathematics it is often useful to call upon a function $\delta(x)$ which, when integrated "against" an arbitrary smooth function $f(x)$ -

$$\int_{-\infty}^{\infty} \delta(x) f(x)\, dx$$

- reproduces the value of $f$ at $x = 0$:

$$\int_{-\infty}^{\infty} \delta(x) f(x)\, dx = f(0) \ \text{ (all } f). \tag{1}$$

Such a function "$\delta(x)$" does not exist in the usual sense, since it must possess the following contradictory properties:

(a) Because the integral (1) is immune to $f's$ behavior away from the origin, $\delta(x)$ has to be zero for $x \neq 0$

$$\delta(x) = 0, \ \ x \neq 0; \tag{2}$$

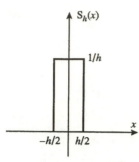

**Figure 1** *Step function approximant*

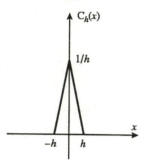

**Figure 2** *Triangle function approximant*

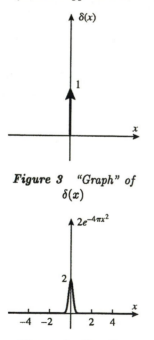

**Figure 3** *"Graph" of $\delta(x)$*

**Figure 4** *Gaussian density*

(b) The graph of $\delta(x)$ has to enclose unit area, since if we insert $f(x) = 1$ into (1) we find

$$\int_{-\infty}^{\infty} \delta(x)dx = 1. \tag{3}$$

So if we try to picture $\delta(x)$, we come up with graphs like Fig. 1, which depicts a rectangular pulse function $S_h(x)$ with base width $h$ and height $1/h$, or perhaps a "triangle" function $C_h(x)$ as in Fig. 2, with base width $2h$ and height $1/h$. Of course, it is absurd to claim that

$$\delta(x) = \lim_{h\downarrow 0} S_h(x) \;\; \text{or} \;\; \lim_{h\downarrow 0} C_h(x) \tag{4}$$

since these limits don't exist - although they *do* suggest the diagram in Fig. 3 as an artist's conception of the "graph" of $\delta(x)$. However a purist could cautiously observe that

$$f(0) = \lim_{h\downarrow 0} \int_{-\infty}^{\infty} S_h(x)f(x)\,dx = \lim_{h\downarrow 0} \int_{-\infty}^{\infty} C_h(x)f(x)\,dx \tag{5}$$

(at least, for continuous $f$; see exercise 1), and regard (5) as the rigorous interpretation of (1, 4). (In other words in practice one does not take the limit in (4) until after the integral in (1) is performed.)

Figure 3 motivates the terminology "impulse function" and "point singularity function" for $\delta(x)$. Also if Figs. 1 and 2 are interpreted as linear *mass densities*, Fig. 3 would denote the "density" for a *point mass*.

Other functions can be said to approach $\delta(x)$ in the sense of (5); essentially all that is required is that the functions be non-negative, subtend unit area, and converge to zero for all $x \neq 0$. Some examples are

the Gaussian density $\dfrac{e^{-x^2/\sigma^2}}{\sigma\sqrt{\pi}}$ as $\sigma \downarrow 0$ (Fig. 4);

the Cauchy density $\dfrac{n}{\pi(1 + n^2 x^2)}$ as $n \to \infty$ (Fig. 5);

the function $\dfrac{\sin nx}{\pi x}$ as $n \to \infty$ (Fig. 6).

Although the latter function (known as the *Dirichlet kernel* in some contexts) does not literally approach zero for $x \neq 0$ (nor is it nonnegative), it oscillates so rapidly that the contributions to the integral cancel, for smooth functions $f$.

Historically the delta function was invented by the electrical engineer Oliver Heaviside, who manipulated it as if it were a bona fide mathematical function (to the horror, evidently, of his mathematical contemporaries!). Later L. Schwartz earned a place in mathematical history by devising a theory which justified most of Heaviside's calculations. He called the new class of generalized functions "distributions," and in fact his *theory of distributions*, itself, has been generalized in several directions; see the references by Kammler, Lighthill, and Zemanian. (If there is a moral to be found in the history of distribution theory, it would have to be "don't abandon a mathematical construction simply because it violates prevailing standards of rigor.") A sketch of Schwartz's theory is given in Section 1.5.

Herein we shall proceed heuristically and manipulate $\delta(x)$ as a legitimate function, as did Heaviside. We shall encounter some rather startling identities, but hopefully the overall consistency and plausibility of the distribution calculus will become evident.

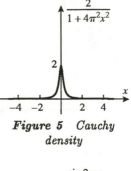

*Figure 5    Cauchy density*

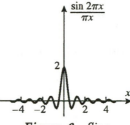

*Figure 6    Sine function*

It is instructive to consider the integral of the delta function, from $-\infty$ up to a *variable* upper limit. From Fig. 3 it is clear that if we stop integrating at some negative value, the integral will be zero; but if we integrate beyond $x = 0$, we pick up the area under the spike and the integral is one. Thus with a minor change of notation we have

$$\int_{-\infty}^{x} \delta(\xi)\,d\xi \;=\; 0 \text{ if } x < 0, \; 1 \text{ if } x > 0$$
$$= \; H(x) \tag{6}$$

where $H(x)$ (for Heaviside) denotes the unit step function depicted in Fig. 7. We do not assign a value to $H(0)$. Note that eq. (6) says $\delta(x)$ can be interpreted as the derivative of the Heaviside function (recall the Fundamental Theorem of calculus):

$$\delta(x) = \frac{d}{dx} H(x) = H'(x) \tag{7}$$

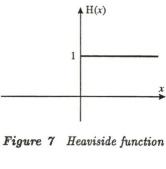

*Figure 7    Heaviside function*

In passing we also note that the Heaviside function is, in turn, the derivative of the ramp function depicted in Fig. 8.

With the revelation (7) that $\delta(x)$ gives an interpretation to the derivative of a discontinous function (see exercises 2, 3), it is enlightening to contemplate the derivative of the delta function. First note that if we evaluate the slopes of the triangle function $C_h(x)$ in Fig. 2, we get the graph in Figure 9.

Figure 9 suggests the physical interpretation of $\delta'(x)$, or $\frac{d}{dx}\delta(x)$, as the *dipole* of electromagnetism or the *pure torque* in mechanics; it is called a "doublet." To see how it acts under an integral, we recall the centered

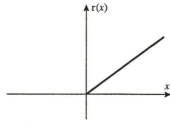

*Figure 8    Ramp function*

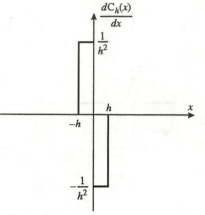

**Figure 9**  *Derivative of triangle function*

difference approximation (Appendix A.3) to the derivative:

$$g'(x) = \lim_{h \to 0} \frac{g(x+h) - g(x-h)}{2h} \tag{8}$$

for smooth functions $g(x)$. Formally applying (8) to the delta function we have

$$\int_{-\infty}^{\infty} \delta'(x) f(x)\, dx = \int_{-\infty}^{\infty} \lim_{h \to 0} \frac{\delta(x+h) - \delta(x-h)}{2h} f(x)\, dx$$

Interpreting this in the spirit of (4, 5) - that is, reversing the order of the limit and integration operations - we derive

$$\int_{-\infty}^{\infty} \delta'(x) f(x)\, dx = \lim_{h \to 0} \int_{-\infty}^{\infty} \frac{\delta(x+h) - \delta(x-h)}{2h} f(x)\, dx.$$

(The rigorous basis for reversing these operations will be discussed in Section 1.5; for now we continue on faith.) Now since $\delta(x+h)$ represents an impulse applied at $x = -h$, under integration it reproduces the value of $f$ at $-h$:

$$\int_{-\infty}^{\infty} \delta(x+h) f(x)\, dx = f(-h). \tag{9}$$

Thus we find

$$\int_{-\infty}^{\infty} \delta'(x) f(x)\, dx = \lim_{h \to 0} \frac{f(-h) - f(h)}{2h} = -f'(0). \tag{10}$$

Exercise 5 demonstrates how integration by parts can be used to reconfirm (10) and develop the following extensions (note the change in notation to shift the site of the impulse from "0" to "$x$"):

$$\int_{-\infty}^{\infty} \delta(\xi - x) f(\xi)\, d\xi \;=\; f(x) \tag{11}$$

$$\int_{-\infty}^{\infty} \delta'(\xi - x) f(\xi)\, d\xi \;=\; -f'(x) \tag{12}$$

$$\int_{-\infty}^{\infty} \delta''(\xi - x) f(\xi)\, d\xi \;=\; f''(x) \tag{13}$$

$$\vdots$$

$$\int_{-\infty}^{\infty} \delta^{(n)}(\xi - x) f(\xi)\, d\xi \;=\; (-1)^n f^{(n)}(x). \tag{14}$$

**Example 1.** A change of variable in the basic identity (1) turns up an unexpected wrinkle. Consider $\int_{-\infty}^{\infty} \delta(2x) f(x)\, dx$. By letting $t = 2x$ we derive

$$
\begin{aligned}
\int_{x=-\infty}^{x=\infty} \delta(2x)\, f(x)\, dx
&= \int_{t=-\infty}^{t=\infty} \delta(t)\, f\!\left(\frac{t}{2}\right) d\!\left(\frac{t}{2}\right) \\
&= \left(\frac{1}{2}\right) \int_{-\infty}^{\infty} \delta(t)\, f(t/2)\, dt \\
&= \frac{f(0)}{2}.
\end{aligned}
\tag{15}
$$

On the other hand in $\int_{-\infty}^{\infty} \delta(-2x)\, f(x)\, dx$ we could set $t = -2x$ and find

$$
\begin{aligned}
\int_{x=-\infty}^{x=\infty} \delta(-2x) f(x)\, dx
&= \int_{t=+\infty}^{t=-\infty} \delta(t)\, f(-t/2)\, d(-t/2) \\
&= -\left(\frac{1}{2}\right) \int_{+\infty}^{-\infty} \delta(t)\, f(-t/2)\, dt \\
&= +\left(\frac{1}{2}\right) \int_{-\infty}^{+\infty} \delta(t)\, f(-t/2)\, dt \\
&= \frac{f(0)}{2}.
\end{aligned}
\tag{16}
$$

also. The rather complicated generalization

$$
\int_{-\infty}^{\infty} \delta[g(x)] f(x)\, dx = f(x_0)/|g'(x_0)| \quad \text{where} \quad g(x_0) = 0
\tag{17}
$$

(note the absolute value) is derived, and quantified, in exercise 6.   ∎

It is very enlightening to look at the generalized functions created when powers of $x$ are multiplied by $\delta(x)$ and its derivatives. For instance one would expect that $x\delta(x)$, $x^2\delta(x)$, $x\delta'(x)$, $x^2\delta'(x)$, etc., would be identically zero, since one factor or the other vanishes at each point. However it turns out that $x^n \delta^{(m)}(x)$ is zero only if the order $n$ of the zero of $x^n$ exceeds the order $m$ of the derivative of the delta function.

**Example 2.** Simplify $x\delta(x)$, $x^2\delta(x)$, $x\delta'(x)$, and $x^2\delta'(x)$.

According to the basic rule (1) for the delta function,

$$
\int_{-\infty}^{\infty} x\delta(x) f(x)\, dx = x f(x)\big|_{x=0} = 0
$$

for any continuous function $f(x)$, so (as we expect)

$$
x\delta(x) = 0.
$$

Similarly

$$
\int_{-\infty}^{\infty} x^2 \delta(x) f(x)\, dx = x^2 f(x)\big|_{x=0} = 0
$$

and $x^2\delta(x) = 0$. For $x\delta'(x)$, however, the rule (10) applies and we find

$$
\begin{aligned}
\int_{-\infty}^{\infty} x\delta'(x)f(x)\,dx &= -[xf(x)]'_{x=0} \\
&= -[f(0)] \\
&= -\int_{-\infty}^{\infty} \delta(x)f(x)\,dx
\end{aligned}
$$

for arbitrary (smooth) $f$; therefore we obtain the surprising identity

$$x\delta'(x) = -\delta(x). \tag{18}$$

By the same token

$$\int_{-\infty}^{\infty} x^2\delta'(x)f(x)\,dx = -[x^2 f(x)]'_{x=0} = 0$$

so $x^2\delta'(x) = 0$.

The generalization,

$$
x^n \delta^{(m)}(x) = \begin{cases} (-1)^n \dfrac{m!}{(m-n)!}\delta^{(m-n)}(x) & \text{for } m \geq n, \\ 0 & \text{for } m < n, \end{cases} \tag{19}
$$

is proffered as exercise 9.                                                                ∎

**Example 3.** With the admission of delta functions, the equation

$$xf(x) = 0, \tag{20}$$

as an equation for an unknown function $f$, has new significance. If $x \neq 0$ then clearly $f(x) = 0$. But $f(0)$ is undetermined. *Thus apparently we can add to the solution any generalized function which, like the delta function, is zero for nonzero $x$:*

$$f(x) = c_1\delta(x) + c_2\delta'(x) + c_3\delta''(x) + \cdots$$

*However* the previous example shows that $x\delta'(x)$ is *not* zero (it is $-\delta(x)$), and neither is $x\delta''(x)$ nor $x\delta'''(x)$, etc. Thus the correct generalized solution to $xf(x) = 0$ is

$$f(x) = c_1\delta(x) \tag{21}$$

for arbitrary $c_1$. As an extension of this we have the following:

*The (generalized) solutions of the equation*

$$x^n f(x) = g(x) \tag{22}$$

are

$$f(x) = \frac{g(x)}{x^n} + c_0\delta(x) + c_1\delta'(x) + \cdots + c_{n-1}\delta^{(n-1)}(x) \qquad (23)$$

*for any constants* $c_0, c_1, \ldots, c_{n-1}$ . (Here $g$ is assumed to be an ordinary function. See exercise 10 for a broader generalization). ■

As we shall see in the following section and in Chapter 3, the generalized functions introduced by Heaviside provide considerable economy in the mathematical analysis of differential equations. Granted, there are subtleties that must be acknowledged - eqs. (17), (19), and (22, 23), in particular. But acquiring the facility to deal with delta functions is well worth the effort for any practicing analyst.

# Exercises 1.3

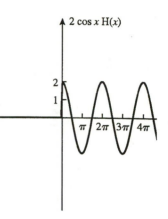

**Figure 10**   $2\cos x H(x)$

**1.** Prove eqs. (5) for continuous $f(x)$. What is the limit if $f(x) = H(x)$?

**2.** See Fig. 10. Let

$$f(x) = 2\cos x H(x) = \begin{cases} 0 & \text{for } x < 0, \\ 2\cos x & \text{for } x > 0. \end{cases} \qquad (24)$$

Derive the formula $f'(x) = 2\delta(x) - 2\sin x\, H(x)$,

(a) by analyzing the graph;

(b) by applying the product rule $(uv)' = u'v + uv'$.

**3.** Generalize exercise 2 to a rule for differentiating functions with jump discontinuities:

Let

$$f(x) = \begin{cases} f_1(x) & \text{for } x < \xi; \\ f_2(x) & \text{for } x > \xi. \end{cases}$$

Then

$$f'(x) = [f_2(\xi) - f_1(\xi)]\,\delta(x - \xi) + \begin{cases} f_1'(x) & \text{for } x < \xi; \\ f_2'(x) & \text{for } x > \xi. \end{cases}$$

if $f_1$ and $f_2$ are differentiable.

(a) Demonstrate this rule graphically.

(b) Show that this rule is consistent with application of the product rule to the expression $f(x) = f_1(x)H(\xi - x) + f_2(x)H(x - \xi)$.

**4.** What is the distribution corresponding to

(a) $\frac{d^2}{dx^2}|x|$ ;

(b) $\frac{d^2}{dx^2}|2x|$ ;

(c) $\frac{d^2}{dx^2}|\sinh x|$ ?

**5.** (a) "Derive" formula (10) using integration by parts.

   (b) Generalize to derive formulas (12, 13, 14).

**6.** In this exercise we are going to analyze formula (17).

   (a) Suppose $g(x) = 3x + 6$. Justify each step in the following:

   $$\int_{x=-\infty}^{\infty} \delta(3x+6)f(x)\,dx = \int_{\xi=-\infty}^{\infty} \delta(\xi)f\left(\frac{\xi-6}{3}\right)\frac{d\xi}{3} = \frac{f(-2)}{3}.$$

   (b) Now generalize. Suppose $\xi = g(x)$, $x = h(\xi)$ is an invertible transformation, i.e. a change of variables as described in Section 1.2, defined for $-\infty < x < \infty$, $-\infty < \xi < \infty$. Suppose also that the transformation preserves order, in the sense that $x_1 < x_2$ implies $g(x_1) < g(x_2)$; thus $g(-\infty) = -\infty$ and $g(+\infty) = +\infty$. Finally let $x_0 = h(0)$. By retracing the analogous steps in part (a), justify the following:

   $$\begin{aligned}\int_{x=-\infty}^{\infty} \delta[g(x)]f(x)\,dx &= \int_{\xi=-\infty}^{\infty} \delta(\xi)f[h(\xi)]h'(\xi)\,d\xi \\ &= f[h(0)]\,h'(0) \\ &= f(x_0)/g'(x_0).\end{aligned}$$

   (c) Now suppose the transformation $\xi = g(x)$, $x = h(\xi)$ *reverses* the order: $x_1 < x_2$ implies $g(x_1) > g(x_2)$ and $g(-\infty) = +\infty$, $g(+\infty) = -\infty$. Justify

   $$\begin{aligned}\int_{x=-\infty}^{\infty} \delta[g(x)]f(x)\,dx &= \int_{\xi=+\infty}^{-\infty} \delta(\xi)f[h(\xi)]\,h'(\xi)\,d\xi \\ &= -\int_{\xi=-\infty}^{+\infty} \delta(\xi)f[h(\xi)]\,h'(\xi)\,d\xi \\ &= -f[h(0)]\,h'(0) = -f(x_0)/g'(x_0) \\ &= f(x_0)/|g'(x_0)|\end{aligned}$$

   (d) Suppose that the transformation $\xi = g(x)$, $x = h(\xi)$ is an invertible change of variables over some interval $a \le x \le b$, $c \le \xi \le d$. Suppose further that 0 is *not* contained in the interval $c \le \xi \le d$. (For example, consider $\xi = g(x) = x^2 + 1$, $x = h(\xi) = \sqrt{\xi - 1}$; $0 \le x \le 2$, $1 \le \xi \le 5$.) Argue that

   $$\int_a^b \delta[g(x)]f(x)\,dx = 0.$$

   (e) (*Full generalization*) Suppose that as $x$ runs from $a$ to $b$, $g(x)$ takes the value 0 at the points $x_0, x_1, \dots, x_m$, but $g'$ is nonzero there. Argue that

$$\int_a^b \delta[g(x)] f(x)\, dx = \sum_{j=0}^{m} f(x_j)/|g'(x_j)|.$$

**7.** What is $\int_{-\infty}^{\infty} \delta(x^2 - 1) e^{-(x-1)^2} dx$?

**8.** Show that

$$
\begin{aligned}
(\sin x)\, \delta(x - n\pi) &= 0 \quad \text{and} \\
(\sin x)\, \delta'(x - n\pi) &= (-1)^{n+1} \delta(x - n\pi)
\end{aligned}
$$

by multiplying both sides by an arbitrary (differentiable) function and integrating.

**9.** (a) Verify eq. (19) for $m = 2$, $n = 1$; for $m = 2$, $n = 2$; for $m = 3$, $n = 2$.

(b) Prove eq. (19) in general.

**10.** (a) Verify that formula (23) solves (22). (In the references it is proved that (23) is the *only* solution.)

(b) Show that the equation

$$(x - x_1)(x - x_2) \cdots (x - x_n) f(x) = g(x)$$

has the distribution solutions

$$
\begin{aligned}
f(x) &= c_1 \delta(x - x_1) + c_2 \delta(x - x_2) + \cdots + c_n \delta(x - x_n) \\
&\quad + \frac{g(x)}{(x - x_1)(x - x_2) \cdots (x - x_n)}
\end{aligned}
$$

for any $c_1, c_2, \ldots, c_n$ if the numbers $x_1, x_2, \ldots, x_n$ are distinct.

(c) Show that the equation

$$(x - x_1)(x - x_2)^2 (x - x_3)^3 f(x) = g(x)$$

has the solutions

$$
\begin{aligned}
c_1 \delta(x - x_1) \ &+ \ c_2 \delta(x - x_2) + c_3 \delta'(x - x_2) + c_4 \delta(x - x_3) \\
&+ \ c_5 \delta'(x - x_3) + c_6 \delta''(x - x_3) \\
&+ \ g(x)/(x - x_1)(x - x_2)^2 (x - x_3)^3
\end{aligned}
$$

for any $c_i$, if $\{x_1, x_2, x_3\}$ are distinct. State the generalization.

**11.** What is the distribution solution of the equation

$$x^2(x^2 - 1) f(x) = \cos x \ ?$$

# 1.4  Green's Functions

**Summary**  The general solution to a linear nonhomogeneous differential equation can be decomposed into a particular solution of the nonhomogeneous equation plus a general solution of the associated homogeneous equation. Green's functions provide a method for finding particular nonhomogeneous solutions. If the equation in operator form is $\mathcal{L}y = f(x)$, then a Green's function $G(x,\xi)$ is a solution to the equation $\mathcal{L}G = \delta(x-\xi)$, for fixed $\xi$; it follows from linearity that

$$y(x) = \int_a^b G(x,\xi)\, f(\xi)\, d\xi$$

is a particular solution to $\mathcal{L}y = f$. One method for computing Green's functions is to piece together a pair of solutions to the associated homogeneous equation in such a way that the mismatch in their derivatives generates, through differentiation, the delta function.

Now we are going to turn to the problem of finding the general solution to a *non*homogeneous linear differential equation. We shall see that the delta function provides us with a neat mechanism for accomplishing this. First, however, it is worthwhile to discuss some theoretical considerations, in order to simplify the task as much as possible.

As we saw in Section 1.2, if $y_1(x)$ is any solution to the nonhomogeneous equation

$$\mathcal{L}y_1 = a_2(x)y_1'' + a_1(x)y_1' + a_0(x)y_1 = f(x), \tag{1}$$

and $y_2(x)$ is any solution to the associated homogeneous equation

$$\mathcal{L}y_2 = a_2(x)y_2'' + a_1(x)y_2' + a_0(x)y_2 = 0, \tag{2}$$

then $[y_1 + y_2]$ is, again, a solution to the nonhomogeneous equation:

$$\mathcal{L}[y_1 + y_2] = f(x) + 0 = f(x). \tag{3}$$

Now we are looking for a *general solution* to (1), i.e. a solution formula containing two arbitrary constants which can be used to satisfy initial conditions. Suppose we only know *one*, particular, solution $y_1$ to (1) - containing no free constants - but we also know a *general* solution $y_2$ to (2), with the constants. Then the combination $[y_1 + y_2]$ will be exactly what we need - it solves the nonhomogeneous equation (1), and it contains two arbitrary constants! In short,

$$y_{particular,nonhomog} + y_{general,homog} = y_{general,nonhomog} \tag{4}$$

In the beginning of Section 1.1 we found the general solution to the equation

$$y'' = \sin x$$

to be $y = -\sin x + c_1 x + c_2$. We can identify $(-\sin x)$ as a particular solution and $c_1 x + c_2$ as the general solution to the associated homogeneous equation $y'' = 0$.

For this reason the methodology for solving a nonhomogeneous equation naturally breaks up into two phases:

1. finding a *general* solution to the associated homogeneous equation, and

2. finding a *particular* solution to the nonhomogeneous equation.

Phase 1 can be solved by the methods of Section 1.1 for equations with constant or equidimensional coefficients, or by the methods of Chapter 2 for most other cases. So to characterize nonhomogeneous equations we don't need to find the general solution - only some particular solution. Let's see how delta functions facilitate this.

Equation (11) from the previous section,

$$f(x) = \int_{\xi=-\infty}^{\infty} \delta(\xi - x) f(\xi) \, d\xi \qquad (5)$$

has a very provocative interpretation. If we keep in mind that an integral is, in reality, a limiting form of a sum, then it suggests

$$f(x) \;\approx\; \sum_i \delta(\xi_i - x) f(\xi_i) \, \Delta\xi_i = \sum_i f(\xi_i) \, \delta(x - \xi_i) \, \Delta\xi_i \qquad (6)$$

(since $\delta(-x) = \delta(x)$). A typical term in (6), plotted as a function of $x$, looks like a delta function supported at $\xi_i$ and scaled by $f(\xi_i) \, \Delta\xi_i$; see Fig. 1. The superposition (6) then depicts the graph of $f(x)$ being decomposed into a series of impulses (Fig. 2). Thus *eq. (5) identifies every function with a superposition of a continuous series of impulses.*

This is enormously helpful in analyzing linear nonhomogeneous differential equations. Suppose we seek a solution of

$$a_2(x)y'' + a_1(x)y' + a_0(x)y = f(x) \qquad (7)$$

in, say, the interval $a \leq x \leq b$. We regard $f(x)$ as a superposition of impulse functions:

$$a_2(x)y'' + a_1(x)y' + a_0(x)y = f(x) \;=\; \int_a^b \delta(x - \xi) f(\xi) \, d\xi$$
$$\approx \sum_i \delta(x - \xi_i) f(\xi_i) \, \Delta\xi_i. \qquad (8)$$

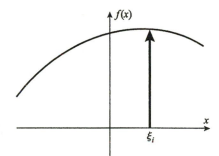

**Figure 1**   *Delta function*

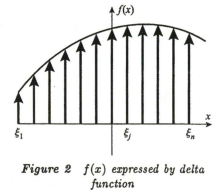

**Figure 2**   *f(x) expressed by delta function*

Now suppose we could, efficiently, find individual solutions to the differential equation corresponding to (8) *when the nonhomogeneity is replaced by a single impulse* $\delta(x - \xi_i)$, *applied at the point* $x = \xi_i$. These "impulse responses," which are functions of $x$ but also depend on the point $\xi_i$ where the impulse is applied, will be denoted $G(x; \xi_i)$: and the equation (8) simplifies to

$$a_2(x)\frac{\partial^2}{\partial x^2}G(x; \xi_i) + a_1(x)\frac{\partial}{\partial x}G(x; \xi_i) + a_0(x)G(x; \xi_i) = \delta(x - \xi_i). \quad (9)$$

(Note that we use partial derivatives to emphasize that $\xi_i$ is a parameter, not a variable, at this point. Note also that $G(x; \xi)$ is not unique, since we have not affixed initial or boundary conditions to (9)). Then, because of the linearity of the left-hand side of (9), we would have

$$a_2(x)\frac{\partial^2}{\partial x^2}\sum_i G(x; \xi_i)f(\xi_i)\,\Delta\xi_i + a_1(x)\frac{\partial}{\partial x}\sum_i G(x; \xi_i)f(\xi_i)\,\Delta\xi_i$$

$$+a_0(x)\sum_i G(x; \xi_i)f(\xi_i)\,\Delta\xi_i = \sum_i \delta(x - \xi_i)f(\xi_i)\,\Delta\xi_i,$$

which in the limit $(\Delta\xi_i \to 0)$ becomes

$$a_2(x)\frac{\partial^2}{\partial x^2}\int_a^b G(x; \xi)f(\xi)\,d\xi \quad + \quad a_1(x)\frac{\partial}{\partial x}\int_a^b G(x; \xi)f(\xi)\,d\xi$$

$$+a_0(x)\int_a^b G(x; \xi)f(\xi)\,d\xi \quad = \quad \int_a^b \delta(x - \xi)f(\xi)\,d\xi$$

$$= \quad f(x). \quad (10)$$

In other words,

$$\int_a^b G(x; \xi)f(\xi)\,d\xi \quad\quad\quad (11)$$

is a formula for a particular solution $y(x)$ of the nonhomogeneous differential equation!

The impulse response $G(x; \xi)$ which solves (9) is called a *Green's function* for the problem (7). It turns out that Green's functions are fairly easy to compute. In the remainder of this section we construct Green's functions for second order ordinary differential equations based on the graphical interpretation of the delta function. In Chapter 9 a more general procedure, based on eigenfunctions, will be derived.

Suppose $y_1(x)$ and $y_2(x)$ are independent solutions to the homogeneous equation associated with (7). Then, on either side of the point $x = \xi$ the Green's function $G(x; \xi)$ must be a linear combination of $y_1$ and $y_2$, *because* *eq. (9) itself is a homogeneous equation in the regions $x < \xi$ and $x > \xi$.* Therefore

$$\begin{aligned} G(x;\xi) &= Ay_1(x) + By_2(x), \quad x < \xi; \\ &= Cy_1(x) + Dy_2(x), \quad x > \xi. \end{aligned} \tag{12}$$

$A$, $B$, $C$, and $D$ are independent of $x$, of course; they *will* depend on $\xi$. To establish the relation between $\{C,D\}$ and $\{A,B\}$, keep in mind that $G$ gets differentiated *twice* in (9), and that we get a delta function by a *single* differentiation of a function with a unit jump (the Heaviside function). Thus we satisfy eq. (9) by prescribing a jump of size $1/a_2(\xi)$ in the first derivative of $G(x;\xi)$, at the point $x = \xi$. ($G$ itself remains continuous.) See Fig. 3.

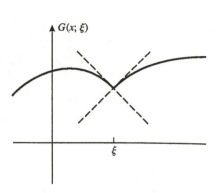

**Figure 3**  *Green's functions*

The coefficients $A$, $B$, $C$, and $D$ will implement this derivative-jump condition if we enforce

$$\frac{\partial}{\partial x} G(x = \xi + 0;\ \xi) = \frac{\partial}{\partial x} G(x = \xi - 0;\ \xi) + \frac{1}{a_2(\xi)}$$

or

$$Cy_1'(\xi) + Dy_2'(\xi) = Ay_1'(\xi) + By_2'(\xi) + \frac{1}{a_2(\xi)}. \tag{13}$$

The continuity condition, of course, states

$$Cy_1(\xi) + Dy_2(\xi) = Ay_1(\xi) + By_2(\xi). \tag{14}$$

By solving (13, 14) for $C, D$ in terms of $A, B$, we derive the equations expressing the coefficients to the right of the jump in terms of those to the left (exercise 6):

$$C = A - \frac{y_2(\xi)/a_2(\xi)}{Wr(\xi; y_1, y_2)}, \quad D = B + \frac{y_1(\xi)/a_2(\xi)}{Wr(\xi; y_1, y_2)}. \tag{15}$$

To summarize: for any $A$ and $B$ independent of $x$, the insertion of the following formula -

$$G(x;\xi) = \begin{cases} Ay_1(x) + By_2(x) & \text{for } x < \xi, \\ Ay_1(x) + By_2(x) + \frac{-y_1(x)y_2(\xi) + y_2(x)y_1(\xi)}{a_2(\xi)Wr(\xi; y_1, y_2)} & \text{for } x > \xi \end{cases} \tag{16}$$

- into the integral

$$y(x) = \int_a^b G(x;\xi) f(\xi)\, d\xi$$

will produce a particular solution to the nonhomogeneous differential equation (7):

$$\begin{aligned} y(x) &= \int_a^x \left\{ Ay_1(x) + By_2(x) + \frac{-y_1(x)y_2(\xi) + y_2(x)y_1(\xi)}{a_2(\xi)Wr(\xi; y_1, y_2)} \right\} f(\xi)\, d\xi \\ &\quad + \int_x^b \{ Ay_1(x) + By_2(x) \} f(\xi)\, d\xi. \end{aligned} \tag{17}$$

(Note how the order of the expressions becomes reversed since $x$ is greater than $\xi$ in the first integral, and less in the second.)

The choice $A = B = 0$ yields the simple Green's function

$$G(x;\xi) = \begin{cases} 0 & \text{for } x < \xi, \\ \frac{-y_1(x)y_2(\xi)+y_2(x)y_1(\xi)}{a_2(\xi)Wr(\xi;y_1,y_2)} & \text{for } x > \xi, \end{cases} \tag{18}$$

and the corresponding particular solution

$$y(x) = \int_a^b G(x;\xi)f(\xi)\,d\xi = \int_a^x \left\{ \frac{-y_1(x)y_2(\xi)+y_2(x)y_1(\xi)}{a_2(\xi)Wr(\xi;y_1,y_2)} \right\} f(\xi)\,d\xi \tag{19}$$

**Example 1.** Consider the task of finding a particular solution to

$$y'' + y = x. \tag{20}$$

Note that the functions $y_1 = \cos x$, $y_2 = \sin x$ solve the associated homogeneous equation $y'' + y = 0$. Their Wronskian equals 1. Thus we can concoct a Green's function from (18):

$$G(x;\xi) = \begin{cases} 0 & \text{for } x < \xi, \\ \frac{-(\cos x)(\sin \xi)+(\sin x)(\cos \xi)}{(1)} & \text{for } x > \xi. \end{cases}$$

Formula (19) then yields the particular solution

$$\begin{aligned} y(x) &= -\cos x \int_0^x (\sin \xi)(\xi)\,d\xi + \sin x \int_0^x (\cos \xi)(\xi)\,d\xi \\ &= -\cos x\,[\sin x - x\cos x] + \sin x\,[\cos x + x\sin x - 1] \\ &= x - \sin x. \end{aligned}$$

Formula (19) is derived in some textbooks by a procedure called "variation of parameters" (exercise 8). However, other choices for $A, B$ may be more convenient in special situations (exercise 11.)

# Exercises 1.4

**1.** Find Green's functions for the following differential equations:

(a) $y'' = f(x)$ (Answer: $G(x;\xi) = x - \xi$ *for* $x > \xi$, 0 *otherwise*)
(b) $y'' - y = f(x)$
(c) $xy'' + y' = f(x)$
(d) $x^2 y'' + xy' - y = f(x)$

2. (a) Find $y(x)$ in (a) through (d) of exercise 1 if $f(x) = 1$. Verify your solutions.

   (b) Find $y(x)$ in (a) through (d) of exercise 1 if $f(x) = x^3$. Verify your solutions.

3. Show that $G(x; \xi) = -\frac{1}{2k} e^{-k|x-\xi|}$ is a Green's function for the equation

$$y'' - k^2 y = f(x).$$

   (Hint: verify that $G$ is a solution of the associated homogeneous equation on each side of $x = \xi$, and that the derivative of $G$ jumps by the proper amount.)

4. Show that $G(x; \xi) = -\left\{ \text{ Real part of } \frac{i}{2k} e^{ik|x-\xi|} \right\}$ is a Green's function for the equation

$$y'' + k^2 y = f(x).$$

   (See hint for exercise 3.)

5. Construct a Green's function for the first order equation

$$y' + p(x)y = f(x).$$

6. Construct a Green's function for the third order equation

$$y''' = f(x).$$

7. Derive eq. (15).

8. This exercise derives a procedure known as *Variation of Parameters*, or more whimsically, *Variation of Constants*, for constructing an integral formula for a particular solution to a nonhomogeneous differential equation. It is illustrated here for the second order equation

$$a_2(x)y'' + a_1(x)y' + a_0(x)y = f(x). \qquad (21)$$

   (Generalizations to higher orders are possible.) We assume we have a general solution

$$Y(x) = c_1 y_1(x) + c_2 y_2(x) \qquad (22)$$

   to the associated homogeneous equation

$$a_2(x)y'' + a_1(x)y' + a_0(x)y = 0 \qquad (23)$$

   and we seek a solution to (21) in the form of (22), *but with the constants replaced by functions* (hence the name!):

$$y(x) = c_1(x)y_1(x) + c_2(x)y_2(x). \qquad (24)$$

   Since we have a *pair* of functions to exploit in (24), we can manipulate $y(x)$ to meet two conditions. The first is the obvious one - that $y(x)$ should solve (21).

(a) Show that imposing the additional condition

$$c_1'(x)\,y_1(x) + c_2'(x)\,y_2(x) = 0 \qquad (25)$$

reduces the condition (21) to

$$a_2[c_1'y_1' + c_2'y_2'] = f(x). \qquad (26)$$

(b) Now solve (25, 26) for $c_1'(x)$ and $c_2'(x)$ to derive

$$c_1'(x) = \frac{-f(x)y_2(x)}{a_2(x)Wr(x;y_1,y_2)}$$

$$c_2'(x) = \frac{f(x)y_1(x)}{a_2(x)Wr(x;y_1,y_2)}$$

(c) Integrate and substitute into (24) to derive (19).

9. The simplest form of *Leibniz's Rule* in calculus is

$$\frac{d}{dx}\int_a^x f(\xi)\,d\xi = f(x). \qquad (27)$$

After all, the Fundamental Theorem of Calculus says that the integral in (27) can be evaluated using an antiderivative $\mathcal{F}$ for $f$ -

$$\frac{d}{d\xi}\mathcal{F}(\xi) = f(\xi) \qquad (28)$$

- and substituting:

$$\int_a^x f(\xi)\,d\xi = \mathcal{F}(x) - \mathcal{F}(a); \qquad (29)$$

now (27) follows immediately from (28) and (29). Similarly we get

$$\frac{d}{dx}\int_x^b f(\xi)\,d\xi = -f(x). \qquad (30)$$

In this exercise we will consider generalizations of this rule.

(a) By applying the chain rule to (27) show that

$$\frac{d}{dx}\int_a^{g(x)} f(\xi)\,d\xi = f(g(x)) \times g'(x). \text{ (Hint: let } t = g(x).)$$

(b) Generalize this result to

$$\frac{d}{dx}\int_{g_1(x)}^{g_2(x)} f(\xi)\,d\xi = f(g_2(x)) \times g_2'(x) - f(g_1(x)) \times g_1'(x).$$

In case the variable $x$ appears in the integrand, but is not the variable of integration, $\int f(x,\xi)\,d\xi$ is a "partial integral". Since an integral is, at bottom, a limit of sums, and since the derivative of a sum is the sum of the derivatives, the differentiation is taken inside the integral sign:

$$\frac{d}{dx}\int_a^b f(x,\xi)\,d\xi = \frac{d}{dx}\lim\sum_i f(x,\xi)\,\Delta\xi_i$$

$$= \lim\sum_i \frac{\partial}{\partial x}f(x,\xi_i)\,\Delta\xi_i = \int_a^b \frac{\partial f}{\partial x}(x,\xi)\,d\xi.$$

The generalized Leibniz's Rule encompasses all these cases. It states

$$\frac{d}{dx}\int_{g_1(x)}^{g_2(x)} f(x,\xi)\,d\xi = \int_{g_1(x)}^{g_2(x)} \frac{\partial f}{\partial x}(x,\xi)\,d\xi$$
$$+ f(x,g_2(x))g_2'(x) - f(x,g_1(x))g_1'(x) \qquad (31)$$

It is valid when $f$, $\partial f/\partial x$, $g_1'$, and $g_2'$ are continuous over the various intervals involved.[7]

10. Use Leibniz's rule to differentiate $\int_{x^2}^{x^3}\sin x\xi\,d\xi$. Verify by carrying out the integration directly and then differentiating.

11. (a) Using Leibniz's rule, show that the particular solution defined by (19) satisfies the initial conditions $y(a)=0$, $y'(a)=0$.

(b) Show that by selecting $y_1(x)$ as a solution vanishing at $x=a$, and $y_2(x)$ vanishing at $x=b$, the choice

$$A = \frac{y_2(\xi)}{a_2(\xi)Wr(\xi;y_1,y_2)}, \quad B=0$$

in (17) yields a particular solution

$$y(x) = y_1(x)\int_x^b \frac{y_2(\xi)f(\xi)}{a_2(\xi)Wr(\xi;y_1,y_2)}\,d\xi$$
$$+ y_2(x)\int_a^x \frac{y_1(\xi)f(\xi)}{a_2(\xi)Wr(\xi;y_1,y_2)}\,d\xi$$

satisfying the boundary conditions $y(a)=y(b)=0$.

---

[7]If either of the limits is infinity, (31) will still hold if $|\partial f/\partial x|$ is bounded, independently of $x$, by an integrable function $\mathcal{M}(\xi)$, $\int_{g_1(x)}^{g_2(x)}\mathcal{M}(\xi)\,d\xi < \infty$.

# 1.5   Generalized Functions and Distributions

**Summary**   A continuous function $f(x)$ generates a linear functional on the space of infinitely differentiable finitely supported test functions, through the formula

$$\mathcal{L}_f[\phi] = \int_{-\infty}^{\infty} f(x)\,\phi(x)\,dx.$$

Schwartz's theory of distributions equips any such function with a generalized "derivative" through the linear functional obtained formally by integrating by parts:

$$\mathcal{L}_{f'}[\phi] = -\int_{-\infty}^{\infty} f(x)\,\phi'(x)\,dx.$$

In such a manner the step, impulse, and doublet "functions" are identified as the generalized derivatives of the ramp function $f(x) = \max(x, 0)$, and a rigorous and mathematically consistent calculus of generalized functions results.

As we mentioned in Section 1.3, Heaviside's incorporation of the delta function and its derivatives into his "operational calculus" was both a blessing and a curse. Certainly the interpretation of $\delta(x)$ as an impulse (and $\delta'(x)$ as a doublet) is physically very appealing; we have seen how it lends enormous insight into the solution of linear nonhomogeneous differential equations, and Section 3.7 will demonstrate how the concept of the Fourier transform would be aesthetically incomplete without it. On the other hand, the taking of such outrageous liberties with mathematical logic and rigor cannot go unpunished; Heaviside himself encountered inconsistencies and once, out of frustration, declared mathematics *an experimental science*!

The salvaging of this invaluable analysis tool was accomplished by Laurent Schwartz, who discovered a mathematically sound interpretation of Heaviside's ideas and a resolution of his paradoxes. The key to Schwartz's theory is a scheme for associating, with each *function*, a linear mapping from the *space of functions* to the *space of scalars* - in mathematical terms, each function is identified with a *linear functional*.

A prime example of a linear functional is the definite integral: it assigns to a function $\phi(x)$ the scalar value

$$\mathcal{L}[\phi] = \int_{-\infty}^{\infty} \phi(x)\,dx \tag{1}$$

$\mathcal{L}$ is clearly linear in its dependence on $\phi$, and it is defined on the space of all functions integrable over $(-\infty, \infty)$. For our present purposes, however, we restrict its domain to "$C_0^{\infty}$" functions;

the superscript indicates that these functions are infinitely often differentiable, and the subscript denotes that they are nonzero only on a bounded set (such functions are said to have "finite support"). (See Exercise 3.) Certainly the definite integral (1) is well defined for all $\mathcal{C}_0^\infty$ functions.

As a broader example of a linear functional consider the integral of $\phi(x)$ *times* some fixed function $f(x)$:

$$\mathcal{L}[\phi] = \int_{-\infty}^{\infty} f(x)\phi(x)\,dx$$

Now the value of the scalar $\mathcal{L}[\phi]$ depends on the choice of the auxiliary function $f$, and we incorporate this fact into the notation:

$$\mathcal{L}_f[\phi] = \int_{-\infty}^{\infty} f(x)\phi(x)\,dx \qquad (2)$$

Notice that we need not require that $f(x)$ itself be $\mathcal{C}_0^\infty$, nor even approach zero at infinity; because of the restriction on $\phi(x)$, (2) will be well defined as long as $f$ is merely continuous. In fact, with $f(x) = 1$ we recover the example (1).

Equation (2) provides a very natural way of associating a linear functional with each continuous function $f$, because it corresponds to the way we "measure" a function physically. If $f(x)$ represents, say, temperature, there is no way to determine physically the value of $f$ at a single point (particularly when we recall the molecular theory of heat!). So in practice we use a test probe (thermometer) which averages the temperature over a small sensing surface or volume, and we apply it at different locations (Fig. 1).

Thus all of our information about $f$ is really derived from samples of integrals of the form (2), with a variety of $\mathcal{C}_0^\infty$ "test functions" like those in Fig. 2. Physically speaking, one could make a case that the linear functional associated with $f$ through (2) has more significance than the pointwise prescription of $f(x)$.[8]

As we indicated, $f(x)$ need not be $\mathcal{C}_0^\infty$ in formula (2); the linear functional will be well defined as long as the test function $\phi$ itself is in this class. If $f(x)$ *is* differentiable, however, then there is a linear functional associated with its derivative:

$$\mathcal{L}_{f'}[\phi] = \int_{-\infty}^{\infty} f'(x)\phi(x)\,dx \qquad (3)$$

For the class of test functions $\phi$ that we are considering ($\mathcal{C}_0^\infty$), the linear

**Figure 1**
*Temperature probe*

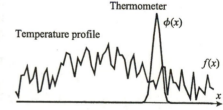

**Figure 2**   *Model of temperature probe*

---

[8]Students of advanced mathematics will recognize this statement as an argument for "equating" two functions which differ only on a "set of measure zero."

functional associated with $f'$ is strongly related to the one associated with $f$:

$$\mathcal{L}_{f'}[\phi] = \int_{-\infty}^{\infty} f'(x)\phi(x)\,dx$$

$$= f(x)\phi(x)\big|_{-\infty}^{\infty} - \int_{-\infty}^{\infty} f(x)\phi'(x)\,dx \quad (integration\ by\ parts)$$

$$= (0) - \mathcal{L}_f[\phi'] \quad (since\ \phi(\pm\infty) = 0)$$

$$\mathcal{L}_{f'}[\phi] = -\mathcal{L}_f[\phi'] \tag{4}$$

(Note that $\phi'$ is also in $C_0^{\infty}$.)

Schwartz observed that if we embrace the notion that a function is effectively defined by its action on $C_0^{\infty}$ test functions under integration, identity (4) could be employed to extend the concept of differentiability! For even if $f$ itself is not differentiable, one can nonetheless use (4) to specify the values of the linear functional $\mathcal{L}_{f'}$ *which would correspond to $f$'s derivative if it had one.* That is, we take

$$\mathcal{L}_{f'}[\phi] = -\mathcal{L}_f[\phi'] = -\int_{-\infty}^{\infty} f(x)\phi'(x)\,dx \tag{5}$$

as the definition, effectively, of $f'$. Thus Schwartz assigned a "generalized derivative" to every continuous function:

$$\text{``}\int_{-\infty}^{\infty} f'(x)\phi(x)\,dx\text{''} \quad \text{is defined to equal} \quad -\int_{-\infty}^{\infty} f(x)\phi'(x)\,dx$$

for all $C_0^{\infty}$ test functions $\phi$.

By the same token if $f$ is *twice* differentiable then two integrations by parts reveal that

$$\mathcal{L}_{f''}[\phi] = (+)\mathcal{L}_f[\phi''] \tag{6}$$

for $C_0^{\infty}$ test functions (exercise 1). Equation (6) therefore provides a formula for the "generalized second derivative" of any continuous function. Indeed, under the linear functional interpretation *every continuous function is infinitely often "differentiable"*!

**Example 1.** In Sec. 1.3 we argued that the step function $H(x)$ is the derivative of the ramp. Since, classically speaking, this makes no sense at $x = 0$, let us verify this claim, in the context of Schwartz's formalism.

If $r(x)$ is the ramp function, then the linear functional associated with it is

$$\mathcal{L}_r[\phi] = \int_{-\infty}^{\infty} r(x)\phi(x)\,dx = \int_0^{\infty} x\phi(x)\,dx.$$

The linear functional associated with $r'(x)$ is, according to (5),

$$
\begin{aligned}
\mathcal{L}_{r'}[\phi] &= -\int_{-\infty}^{\infty} r(x)\phi'(x)\,dx \\
&= -\int_{0}^{\infty} x\phi'(x)\,dx \\
&= \{x\phi(x)\}_{0}^{\infty} + \int_{0}^{\infty} (1)\,\phi(x)\,dx \; (integration\; by\; parts) \\
&= \int_{-\infty}^{\infty} H(x)\,\phi(x)\,dx \;(\text{since } \phi \text{ is } C_0^{\infty}) \\
&= \mathcal{L}_H(\phi).
\end{aligned}
$$

Since $r'(x)$ and $H(x)$ generate the same linear functional on all test functions, Schwartz equates them: $r'(x) = H(x)$. ∎

Where does the delta function fit into this formalism? In Sec. 1.3 we argued that $\delta(x)$ is the derivative of $H(x)$. To authenticate this we must show that $\delta(x)$ and $H'(x)$ give rise to the same linear functional on all $C_0^{\infty}$ test functions. Now the functional associated with $\delta$ is

$$
\text{``}\int_{-\infty}^{\infty} \delta(x)\phi(x)\,dx\text{''} = \mathcal{L}_\delta[\phi] = \phi(0). \tag{7}
$$

For $H'$ we have

$$
\begin{aligned}
\text{``}\int_{-\infty}^{\infty} H'(x)\,\phi(x)\,dx\text{''} = \mathcal{L}_{H'}[\phi] &= -\mathcal{L}_H[\phi'] \\
&= -\int_{-\infty}^{\infty} H(x)\,\phi'(x)\,dx \\
&= -\int_{0}^{\infty} \phi'(x)\,dx \\
&= -\phi(x)|_{0}^{\infty} \\
&= \phi(0). \tag{8}
\end{aligned}
$$

Since (7) and (8) agree, we have $\delta(x) = H'(x)$. Similarly the doublet $\delta'$ is the generalized derivative of the impulse $\delta$, and so on (exercise 2). In this way Schwartz placed all of Heaviside's generalized functions on a firm mathematical basis.

Schwartz defined a "distribution" to be any linear functional which is *continuous* over the space $C_0^{\infty}$ of infinitely often differentiable functions of finite support, and he showed that every distribution can be identified with a generalized ($n$th order) derivative of some continuous function. We direct the interested reader to the references for the fine points regarding "continuity" in this function space (it is a subtle matter). Indeed different notions of "distribution" arise as one alters the interpretation of "continuity". (In fact subsequent mathematicians have varied the space of test functions, as well. A particularly remarkable example is discussed in exercise 8.)

Historically, the practitioners of Heaviside's operational calculus, while extremely successful in obtaining solutions to some very difficult problems, were plagued by inconsistencies in the system. One troublesome point was the validity of the operation of multiplying two generalized functions. Another was the apparently erratic behavior of the function $1/x$ and its powers, acting under the integral sign. In the course of exploring the consequences of Schwartz's scheme, we shall see how it provides an unambiguous, though sometimes counterintuitive, resolution of these matters.

If $f(x)$ is a generalized function and $g(x)$ is infinitely often differentiable (not necessarily $C_0^\infty$), then the product $f(x)g(x)$ can be associated with a well-defined linear functional on $C_0^\infty$ simply by absorbing the factor $g(x)$ into the test function $\phi(x)$:

$$\text{``} \int_{-\infty}^{\infty} [f(x)\,g(x)]\,\phi(x)\,dx \text{''} = \int_{-\infty}^{\infty} f(x)\,[g(x)\,\phi(x)]\,dx$$

or

$$\mathcal{L}_{fg}[\phi] = \mathcal{L}_f[g\phi]. \tag{9}$$

Note that $\mathcal{L}_{fg}$ is well defined because $g(x)\phi(x)$ belongs to the class $C_0^\infty$. Therefore the expressions $x\delta(x)$, $x^2 H(x)$, and $(\cos x)\delta'(x)$ are all valid generalized functions.

However the logic of (9) fails if $g$ is not infinitely often differentiable (for then $g(x)\phi(x)$ will not belong to $C_0^\infty$). What value, for instance, can we sensibly assign to

$$\mathcal{L}_{\delta H}[\phi] = \int_{-\infty}^{\infty} \delta(x)H(x)\phi(x)\,dx \stackrel{(?)}{=} H(0)\phi(0)?$$

(After all, $H(0)$ is undefined!) Similarly,

$$\mathcal{L}_{r\delta'}[\phi] = \int_{-\infty}^{\infty} r(x)\delta'(x)\phi(x)\,dx \stackrel{(?)}{=} [r(0)\phi(0)]'$$

has no meaning because $r'(0)$ is undefined. So the product of two generalized functions is not defined, in general.

We are now going to investigate the distribution interpretation of the function $1/x$, when it occurs under the integral sign. As you read the subsequent discussion, keep in mind that each of the functions $1/x, 1/x^2, 1/x^3, \ldots,$ has infinite area under its graph for $x = 0$:

$$\int_\epsilon^1 \frac{dx}{x^n} \text{ diverges as } \epsilon \downarrow 0, \text{ for } n \geq 1.$$

Therefore in classical terms the integrals

$$\int_{-\infty}^\infty \frac{\phi(x)}{x^n} \, dx, \, n \geq 1,$$

are, as a rule, divergent.

According to L'Hopital's rule, the function

$$f(x) = x \ln|x| - x \tag{10}$$

is continuous for all $x$ if we set

$$f(0) = \lim_{x \to 0} \{x \ln|x| - x\} = 0 \tag{11}$$

(exercise 5). Therefore it is associated with a linear functional on $C_0^\infty$:

$$\mathcal{L}_{x \ln|x| - x}[\phi] = \int_{-\infty}^\infty [x \ln x - x] \, \phi(x) \, dx. \tag{12}$$

Accordingly, Schwartz's theory equips $f(x)$ with a sequence of generalized derivatives $f', f'', f''', \ldots,$ associated with the functionals

$$\mathcal{L}_{[x \ln|x| - x]'}[\phi] = -\int_{-\infty}^\infty [x \ln|x| - x] \, \phi'(x) \, dx$$

$$\mathcal{L}_{[x \ln|x| - x]''}[\phi] = \int_{-\infty}^\infty [x \ln|x| - x] \, \phi''(x) \, dx$$

$$\mathcal{L}_{[x \ln|x| - x]'''}[\phi] = -\int_{-\infty}^\infty [x \ln|x| - x] \, \phi'''(x) \, dx \text{ etc.}$$

But now note that the classical derivatives of these functions are given by

$$
\begin{aligned}
(x \ln|x| - x)' &= \ln|x| + x/x - 1 = \ln|x|, \\
(x \ln|x| - x)'' &= 1/x, \\
(x \ln|x| - x)''' &= -1/x^2, \text{ etc.}
\end{aligned}
$$

Therefore in the context of distribution theory one adopts the conventions

$$\int_{-\infty}^\infty [\ln|x|] \, \phi(x) \, dx \stackrel{def}{=} \mathcal{L}_{[x \ln|x| - x]'}[\phi]$$

$$= -\int_{-\infty}^\infty [x \ln x - x] \, \phi'(x) \, dx, \tag{13}$$

$$\int_{-\infty}^{\infty} \left[\frac{1}{x}\right] \phi(x)\, dx \overset{def}{=} \mathcal{L}_{[x\ln|x|-x]''}[\phi]$$

$$= \int_{-\infty}^{\infty} [x\ln x - x]\, \phi''(x)\, dx, \qquad (14)$$

$$\int_{-\infty}^{\infty} \left[\frac{1}{x^2}\right] \phi(x)\, dx \overset{def}{=} \mathcal{L}_{[x\ln|x|-x]'''}[\phi]$$

$$= -\int_{-\infty}^{\infty} [x\ln x - x]\, \phi'''(x)\, dx, \qquad (15)$$

and so on.

**Example 2.** Some of these integrals can be interpreted classically. The function $\ln|x|$, for example, is integrable around $x = 0$:

$$\int_a^b \ln|x|\, dx = b\ln|b| - b - a\ln|a| + a \;\to\; b\ln|b| - b$$

as $a \downarrow 0$. Therefore the identity (13) can be taken literally, since the integration-by-parts step in its derivation (eq. (4)) is valid.

If we manipulate the integral on the right of eq. (14) by taking the limit

$$\int_{-\infty}^{\infty} [x\ln x - x]\, \phi''(x)\, dx = \lim_{\epsilon_1,\epsilon_2 \downarrow 0} \left\{ \int_{-\infty}^{-\epsilon_1} + \int_{\epsilon_2}^{\infty} \right\} [x\ln|x| - x]\phi''(x)\, dx$$

and integrating each piece by parts, we get

$$\lim_{\epsilon_1,\epsilon_2 \downarrow 0} \left\{ \int_{-\infty}^{-\epsilon_1} + \int_{\epsilon_2}^{\infty} \right\} [-\ln|x|]\phi'(x)\, dx$$
$$+ \lim_{\epsilon_1,\epsilon_2 \downarrow 0} [x\ln|x| - x]\phi'(x)\Big|_{\epsilon_2}^{-\epsilon_1}.$$

The latter term goes to zero, thanks to (11). The former term is integrated by parts once again to produce

$$\lim_{\epsilon_1,\epsilon_2 \downarrow 0} \left\{ \int_{-\infty}^{-\epsilon_1} + \int_{\epsilon_2}^{\infty} \right\} \frac{\phi(x)}{x}\, dx - \lim_{\epsilon_1,\epsilon_2 \downarrow 0} [\ln|x|\,\phi(x)]_{\epsilon_2}^{-\epsilon_1}.$$

Now *if we require* $\epsilon_1 = \epsilon_2$, then the integrated term also goes to zero:

$$[\ln|x|\,\phi(x)]_{\epsilon_1}^{-\epsilon_1} = \{\phi(-\epsilon_1) - \phi(\epsilon_1)\}\ln\epsilon_1$$
$$\approx \{-(2\epsilon_1)\phi'(0)\}\ln\epsilon_1$$
$$\to \; 0 \quad \text{as} \quad \epsilon_1 \downarrow 0.$$

In other words, the distribution interpretation of

$$\int_{-\infty}^{\infty} \frac{\phi(x)}{x}\, dx \qquad (16)$$

coincides with the classical interpretation of

$$\lim_{\epsilon \downarrow 0} \left\{ \int_{-\infty}^{-\epsilon} + \int_{\epsilon}^{\infty} \right\} \frac{\phi(x)}{x} \, dx, \tag{17}$$

where we omit a *symmetric* interval around the singularity at $x = 0$. The limiting procedure specified in eq. (17) is known as the *Cauchy Principal Value of the integral.*

With a little more work it can be shown that (15) has a classical interpretation also; see exercise 6. ∎

We close out our discussion of distribution theory with some remarks on the significance of *limits* of generalized functions. The reader may recall that in Section 1.3 we loosely defined the doublet $\delta'(x)$ by a difference quotient of the form

$$\delta'(x) = \lim_{h \to 0} \frac{\delta(x+h) - \delta(h)}{2h}, \tag{18}$$

but this was rendered more meaningful by the characterization of the action of $\delta'$ under the integral:

$$
\begin{aligned}
\int_{-\infty}^{\infty} \delta'(x)\phi(x)\, dx &= \int_{-\infty}^{\infty} \lim_{h \to 0} \frac{\delta(x+h) - \delta(x-h)}{2h} \phi(x)\, dx \\
&\overset{(?)}{=} \lim_{h \to 0} \int_{-\infty}^{\infty} \frac{\delta(x+h) - \delta(x-h)}{2h} \phi(x)\, dx \\
&= \lim_{h \to 0} \frac{\phi(-h) - \phi(h)}{2h} \\
&= -\phi'(0). \tag{19}
\end{aligned}
$$

Is the exchange of order of the integration operation and the limit operation justified? To decide this issue one must have a clear understanding of both operations; but the limit operation (18) is anything but clear! In no way does the right-hand side get "arbitrarily close" to the left; no $\epsilon, \delta$ criteria can be formulated to describe what we mean by (18).

The fact of the matter is that, since distributions are effectively *defined* by their action on $C_0^\infty$ functions, their limits must be defined that way also.

If $f_1(x), f_2(x), f_3(x), \ldots$ is a sequence of generalized functions with the property that for every test function $\phi(x)$ in $C_0^\infty$, the values of the linear functionals

$$\mathcal{L}_{f_1}[\phi], \mathcal{L}_{f_2}[\phi], \mathcal{L}_{f_3}[\phi], \ldots$$

approach a limit $\alpha[\phi]$, then we define the limit of $f_1, f_2, f_3, \ldots$ to be the distribution $f$ corresponding to the linear functional $\lim_{n \to \infty} \mathcal{L}_{f_n}[\phi]$ :

$$\mathcal{L}_f[\phi] \overset{def}{=} \alpha[\phi] = \lim_{n \to \infty} \mathcal{L}_{f_n}[\phi].$$

In other words,

$$\int_{-\infty}^{\infty} f(x)\phi(x)\,dx \;\equiv\; \int_{-\infty}^{\infty} \lim_{n\to\infty} f_n(x)\phi(x)\,dx$$

$$\stackrel{def}{=}\; \lim_{n\to\infty} \int_{-\infty}^{\infty} f_n(x)\phi(x)\,dx$$

by definition.

Similarly if $f(x;h)$ is a family of linear functionals characterized by the variable $h$, we say "$f(x) = \lim_{h\to 0} f(x;h)$" and

$$\int_{-\infty}^{\infty} f(x)\phi(x)\,dx \;\equiv\; \int_{-\infty}^{\infty} \lim_{h\to 0} f(x;h)\phi(x)\,dx$$

$$\stackrel{def}{=}\; \lim_{h\to 0} \int_{-\infty}^{\infty} f(x;h)\phi(x)\,dx,$$

provided the latter exists for all $\phi$ in $\mathcal{C}_0^\infty$.

*The issue of exchange of order of integration and limit is thus finessed, by its incorporation into this definition.*

**Example 3.** Now we can reclaim our original interpretation of the delta function as the limit of the triangle function $C_h(x)$ in Fig. 3 (recall Section 1.3, Fig. 2).

Since the test functions $\phi(x)$ are continuous, for $-h < x < h$ we can write $|\phi(x) - \phi(0)| < \epsilon$, where $\epsilon \to 0$ as $h \to 0$. Then the statement

$$\delta(x) = \lim_{h\to 0} C_h(x)$$

is authenticated by the calculation

$$\int_{-\infty}^{\infty} \lim_{h\downarrow 0} C_h(x)\phi(x)\,dx \;\stackrel{def}{=}\; \lim_{h\downarrow 0} \int_{-\infty}^{\infty} C_h(x)\phi(x)\,dx$$

$$= \lim_{h\downarrow 0} \int_{-h}^{h} C_h(x)\{\phi(0) + [\phi(x) - \phi(0)]\}\,dx$$

$$= \phi(0) + \lim_{h\downarrow 0} \eta(h)$$

with $|\eta(h)| < \epsilon \int_{-h}^{h} C_h(x)\,dx = \epsilon \to 0$ in the limit. *(qed)*  ∎

**Example 4.** The "comb function" depicted in Fig. 4 is the distribution specified by

$$f(x) = \sum_{n=-\infty}^{\infty} \delta(x-n) = \lim_{N\to\infty} \sum_{n=-N}^{N} \delta(x-n)$$

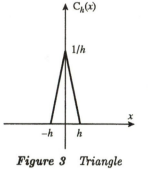

*Figure 3*   *Triangle function*

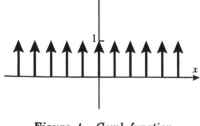

*Figure 4*   *Comb function*

It is well defined because the limit

$$\lim_{N \to \infty} \int_{-\infty}^{\infty} \sum_{n=-N}^{N} \delta(x - n)\, \phi(x)\, dx = \lim_{N \to \infty} \sum_{n=-N}^{N} \phi(n)$$

always exists for any $\mathcal{C}_0^\infty$ test function (only a finite number of the terms in the final sum are nonzero!). The operation of "sampling" a continuous function $g(x)$ can be characterized mathematically by multiplying it with the comb function (Fig. 5).

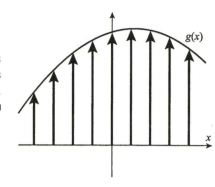

**Figure 5**  $g(x)$ *sampled*

Some, but not all, of the linear functionals associated with distributions can be extended to test functions $\phi(x)$ not in $\mathcal{C}_0^\infty$. For example $\mathcal{L}_\delta[\phi] = \phi(0)$ makes sense for any *continuous* function $\phi$, and $\mathcal{L}_{\delta'}[\phi] = -\phi'(0)$ is well defined for any continuously differentiable function $\phi$. Note however that the comb function distribution $\sum_{n=-\infty}^{\infty} \phi(n)$ is undefined if $\phi$ is, say, a polynomial. (The sum diverges.)

Suffice it to say that a distribution $\mathcal{L}[\phi]$ can be extended to any function $\phi$, not in $\mathcal{C}_0^\infty$, if $\phi(x)$ is a limit of functions $\{\phi_n\}$ in $\mathcal{C}_0^\infty$ and $\lim_{n \to \infty} \mathcal{L}[\phi_n]$ is well defined.

Because the topology of $\mathcal{C}_0^\infty$ is somewhat complicated, we direct the interested reader to the texts by Lighthill, Zemanian, and Kammler for details.

# Exercises 1.5

1. Verify eq. (6).

2. Use the Schwartz formalism to show that the doublet $\delta'(x)$, defined through the linear functional

$$\mathcal{L}_{\delta'}[\phi] = -\phi'(0),$$

   is the generalized derivative of the delta function.

3. Show that the function

$$\phi(x) = \begin{cases} \exp\left\{\frac{-1}{(x-1)^2(x+1)^2}\right\} & , \quad |x| < 1 \\ 0 & , \quad |x| \geq 1 \end{cases}$$

   belongs to the class $\mathcal{C}_0^\infty$. This function is displayed as the "probe" in Fig. 2.

4. Verify the limit in eq. (11). (Recall L'Hopital's rule.)

5. (a) Show that (15) has the classical interpretation

$$\text{``}\int_{-\infty}^{\infty} \frac{\phi(x)}{x^2}\, dx\text{''} = \int_{0}^{\infty} \frac{\phi(x) - 2\phi(0) + \phi(-x)}{x^2}\, dx.$$

(b) Show that (17) is equivalent to $\lim_{\epsilon \downarrow 0} \int_\epsilon^\infty \frac{\phi(x) - \phi(-x)}{x} \, dx$.

6. Prove that if $f$ is a generalized function and $g$ is infinitely often differentiable, then $(fg)' = f'g + fg'$.

7. Prove that $\sum_{n=-\infty}^\infty a_n \delta(x - x_n)$ is a well-defined distribution, regardless of the numbers $a_n$, if the set of points $\{x_n\}$ accumulates only at infinity.

8. (*This discussion presupposes analytic function theory.*) In advanced works the theory of distributions is extended to complex numbers and functions. Thus

$$\mathcal{L}_{\delta(z-z_0)}[\phi] = \int_{-\infty}^\infty \delta(z - z_0)\phi(z) \, dz = \phi(z_0) \tag{20}$$

is a valid continuous linear functional for *entire* functions $\phi$. Show that the "Taylor series" expansion of the delta function

$$\delta(z + h) = \sum_{n=0}^\infty \delta^{(n)}(z) \, h^n/n! \tag{21}$$

is valid when each side is interpreted as a linear functional over the space of entire test functions.

Equation (21) presents an enigma. From our picture of the delta function as described in Section 1.3 it would seem that

$$\int_{-\infty}^1 \delta(z - 2)\phi(z) \, dz = 0 \tag{22}$$

(because of the upper limit). However if $\phi$ is entire then (21) can be used to derive a contradiction:

$$
\begin{aligned}
\int_{-\infty}^1 \delta(z - 2)\phi(z) \, dz &= \int_{-\infty}^1 \sum_{n=0}^\infty \delta^{(n)}(z) \, (-2)^n \phi(z)/n! \, dz \\
&= \sum_{n=0}^\infty \phi^{(n)}(0) \, 2^n/n! \\
&= \phi(2) \ (\textit{not zero!}).
\end{aligned}
$$

The puzzle is resolved as follows. Strictly speaking, integrals of generalized functions over *subintervals* of $(-\infty, \infty)$ — as indicated in eq. (22) — are not directly addressed by the theory of distributions. They must be recast as integrals over the whole line before such rules as (20) can be applied. Thus

$$\int_{-\infty}^1 \delta(z - 2)\phi(x) \, dz$$

must be interpreted as

$$\int_{-\infty}^{\infty} \delta(z-2)\phi(z)\chi(z)\,dz,$$

where $\chi(z)$ is one for $-\infty < z < 1$ and zero for $1 < z < \infty$. But $\phi(z)\chi(z)$ is not entire (even though $\phi$ is); hence (20) does not pertain.

# Chapter 2

# Series Solutions for Ordinary Differential Equations

The solution of the linear second order ordinary differential equation can be effected by the methods of Chapter 1 if its coefficients are constant or equidimensional. Although such conditions prevail in many applications in engineering, other situations do arise and more robust methods have to be employed.

Fortunately it seems to be the case that all of the important differential equations of physics and engineering have coefficients which can be represented - at least locally - by convergent Taylor series expansions (or minor variants of such). This in turn implies that their *solutions* can also be so expressed. Moreover, these solution expansions can be obtained in a relatively straightforward manner. The disclosure of this methodology is the main subject of Chapter 2. We begin with preliminary sections reviewing, first, the factorial function and, second, the logic underlying the construction of a Taylor series.

Before the advent of computers much of the effort in applied mathematics was directed toward obtaining, analyzing, and tabulating these *series solutions* for every conceivable physical situation. Therefore no exposition of this topic would be complete without an acknowledgement of, and homage to, the mathematicians of the past and the "special functions" designed by them for this purpose. We examine the Bessel functions in some detail and provide a directory steering the reader to further information on the others.

Although these Taylor series calculations can easily be implemented on modern computers possessing symbol manipulation capability, for numerical results other methods are more efficient. Section 2.7 provides a very brief glimpse at this topic.

## 2.1   The Gamma Function

**Summary**   The gamma function $\Gamma(x) = \int_0^\infty t^{x-1} e^{-t}\, dt$ provides a useful extension of the factorial function $(n-1)! = \Gamma(n)$ to *all* real and complex numbers; it preserves the functional identity $\Gamma(x+1) = x\Gamma(x)$. As a result, however, $\Gamma(x)$ diverges when $x$ is zero or a negative integer.

The factorial function is defined for positive integers $n$ by

$$n! = n(n-1)(n-2)...3 \times 2 \times 1 \tag{1}$$

It arises frequently in combinatorical analysis, where it is found convenient to extend this definition so that

$$0! = 1. \tag{2}$$

[For instance, the number of ways of selecting $r$ objects from a collection of $n$ objects is known to be $n!/r!(n-r)!$, and this formula gives the correct answer (one) when $r = n$ as long as the convention (2) is adopted.]

In this chapter we shall see that it is also convenient to have a formula for $n!$ which extends the definition (1) to *fractional, negative,* and even *complex* values of $n$. Historically this led to the formulation of the *Gamma function* $\Gamma(x)$.

DEFINITION The gamma function is defined for $x > 0$ by

$$\Gamma(x) = \int_0^\infty t^{x-1} e^{-t}\, dt. \tag{3}$$

$\Gamma(x)$ is depicted in Fig. 1 and tabulated in Table 2.1.

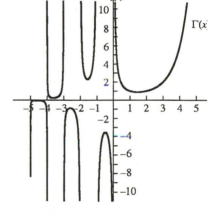

**Figure 1**   *Gamma function*

*TABLE 2.1* $\Gamma(x)$

| $x$ | $\Gamma(x)$ |
|-----|-------------|
| 1.0 | 1.00000 |
| 1.1 | 0.95135 |
| 1.2 | 0.91817 |
| 1.3 | 0.89747 |
| 1.4 | 0.88726 |
| 1.5 | 0.88623 |
| 1.6 | 0.89352 |
| 1.7 | 0.90864 |
| 1.8 | 0.93138 |
| 1.9 | 0.96177 |

The feature that relates the gamma function to the factorial is the *recursive relation*:

$$\Gamma(x+1) = x\Gamma(x) \ \ for\ x > 0. \tag{4}$$

PROOF. This functional equation is proved by applying the formula for integration by parts:

$$\int_a^b uv'dt = uv\big|_a^b - \int_a^b vu'dt. \tag{5}$$

Setting $u = t^x$ and $v' = e^{-t}$, we observe $u' = xt^{x-1}$ and $v = -e^{-t}$, so

$$\Gamma(x+1) = \int_0^\infty t^x e^{-t}dt = t^x(-e^{-t})\big|_0^\infty + x\int_0^\infty t^{x-1}e^{-t}dt$$

$$= 0 + x\Gamma(x). \tag{6}$$

If $n$ is a positive integer the functional equation (4) can be iterated to obtain

$$\Gamma(n) = (n-1)\Gamma(n-1) = (n-1)(n-2)\Gamma(n-2) = \cdots = (n-1)!\Gamma(1), \tag{7}$$

and this would be just what we want if $\Gamma(1)$ happened to equal 1 - which it does:

$$\Gamma(1) = \int_0^\infty t^0 e^{-t}dt = 1.$$

Thus we have established that the Gamma function is a bona fide extension of the factorial:

$$\Gamma(n) = (n-1)! \tag{8}$$

It is perhaps regrettable that $\Gamma(x)$ was defined, historically, to give $(n-1)!$ instead of $n!$ when $x = n$. At any rate we can now extend eq. (8) to give meaning to such expressions as $\frac{1}{3}!$ by *defining*

$$x! \equiv \Gamma(x+1) \tag{9}$$

for *all* $x$. The notion of $x!$ for noninteger $x$ will prove very helpful in subsequent sections of this chapter; it enables us to replace cumbersome expressions like

$$(3\tfrac{1}{3})(4\tfrac{1}{3})(5\tfrac{1}{3})(6\tfrac{1}{3})(7\tfrac{1}{3})$$

by the more compact $(7\tfrac{1}{3})!/(2\tfrac{1}{3})!$. See exercises 2 through 4.

It is important to note that the recursion equation (4) can be used to reduce the evaluation of $\Gamma(x)$, for *any* positive $x$, to values between 1 and 2 (or between 0 and 1, but this is inconvenient for reasons to be seen). An example will illustrate the procedure.

**Example 1.**

a. $\Gamma(3.2) = 2.2 \times \Gamma(2.2)$ $= 2.2 \times 1.2 \times \Gamma(1.2)$ (by eq.(4))

$\approx 2.2 \times 1.2 \times .91817$ (Table 2.1)

$\approx 2.424$

b. $\Gamma(.5)$ $= \Gamma(1.5)/.5$ (by eq.(4))

$\approx .88623/.5 \approx 1.772$

c. $|\Gamma(0)| = \infty$, or more precisely $\lim\limits_{h \to 0} |\Gamma(h)| = \infty$, because

$$\lim_{h \to 0} h\Gamma(h) = \lim_{h \to 0} \Gamma(1 + h) = \Gamma(1) = 1.$$

For this reason, the values of $\Gamma(x)$ are ungainly for $0 < x < 1$; thus the interval $[1, 2)$ is used for tabulation. ∎

By the same reasoning as in the example, eq.(4) can be used to *define* values of $\Gamma(x)$ for negative $x$. (Note that for $x < 0$ the integral in (3) diverges, so it cannot be used.)

**Example 2.**

a. $\Gamma(-.25) = \Gamma(.75)/(-.25)$ $= \Gamma(1.75)/(.75)(-.25)$ (eq.(4))

$\approx .92001/(.75)(-.25)$ (Table 2.1)

$= -4.902$

b. $|\Gamma(-1)| = |\Gamma(0)/(-1)| = \infty$, $|\Gamma(-2)| = |\Gamma(-1)/(-2)| = \infty$ , etc. (These relations can be shown more carefully by tracing the limit arguments as in the previous example, of course.) ∎

Thus the Gamma function diverges at the negative integers and zero.

Using advanced methods one can show that the following approximation is valid for large $x$:

$$x! = \Gamma(x + 1) \approx (2\pi x)^{1/2}(x/e)^x \qquad (10)$$

This approximation, which is quite good for $x > 9$, is called *Stirling's formula.*

# Exercises 2.1

**1.** Using a calculator, test Stirling's approximation for $x = 5, 10, 15, 30,$ and 60.

**2.** Show that $(1 + p)(2 + p)(3 + p) \cdots (j + p) = \Gamma(j + p + 1)/\Gamma(1 + p)$.

**3.** Show that $(2+p)(4+p)(6+p) \cdots (2j+p) = 2^j \, \Gamma(j+1+p/2)/\Gamma(1+p/2)$.

4. Show that $(m+p)(m+1+p)(m+2+p)\cdots(m+j+p) = \Gamma(m+j+p+1)/\Gamma(m+p)$.

5. Show that $\Gamma(1/2) = \sqrt{\pi}$. (Hint: first show that $\Gamma(1/2) = 2\int_0^\infty e^{-u^2}du$ by letting $x = u^2$ in (3). Then redefine dummy variables to show that

$$\left[\int_0^\infty e^{-u^2}du\right]^2 = \int_0^\infty e^{-x^2}dx\int_0^\infty e^{-y^2}dy = \int_0^\infty\int_0^\infty e^{-(x^2+y^2)}dx\,dy.$$

Finally, recast the latter form into polar coordinates to evaluate it.)

6. The Gamma function is defined for *complex* arguments $z = x+iy$ also; $\Gamma(x+iy)$ is tabulated in the handbook of Abramowitz and Stegun for $1 \le x \le 2$ and $y \ge 0$.

   (a) Argue that if $x > 0$ the following integral, derived from (3), converges:

   $$\Gamma(z) = \Gamma(x+iy) = \int_0^\infty t^{z-1}e^{-t}dt = \int_0^\infty t^{x-1}e^{iy\ln t}e^{-t}dt.$$
   $$(11)$$

   (b) Show that if $x = Re\,z > 0$ then the proof of relation (4) can be adapted to demonstrate that

   $$\Gamma(z+1) = z\Gamma(z) \qquad (12)$$

   Thus $\Gamma(z)$ can be extended to all complex values of $z$ (other than zero and the negative integers) by the methods illustrated in the text.

   (c) Show that $\Gamma(\bar{z}) = \overline{\Gamma(z)}$ (complex conjugate).

   (d) Given $\Gamma(1+i) = .4980 - .1550i$, find $\Gamma(2+1)$, $\Gamma(3-i)$, and $\Gamma(-2+i)$.

## 2.2  Taylor Series and Polynomials

**Summary**  A function $f(x)$ can be approximated near the point $x_0$ by its Taylor polynomial

$$p_n(x) = \sum_{j=0}^n f^{(j)}(x_0)(x-x_0)^j/j!,$$

which matches $f$ and its derivatives up to order $n$ at $x_0$. The Taylor *series* $\sum_{j=0}^\infty f^{(j)}(x_0)(x-x_0)^j/j!$ converges to $f(x)$ under some conditions. The error in the polynomial approximation can usually be estimated by the size of the next term in the series:

$$error \approx |f^{(n+1)}(x_0)(x-x_0)^{n+1}/(n+1)!| \;.$$

Probably the most important tool for numerically approximating functions is the *Taylor polynomial*. Recall the construction from calculus: given a function $f(x)$ possessing at least $n$ derivatives at $x = x_0$, the *Taylor polynomial of degree $n$*, $p_n(x)$, is the polynomial having the property that $p_n(x)$ and $f(x)$ match at $x_0$, as do all their derivatives up to order $n$:

$$
\begin{aligned}
p_n(x_0) &= f(x_0) \,, \\
p'_n(x_0) &= f'(x_0) \,, \\
p''_n(x_0) &= f''(x_0) \,,
\end{aligned}
$$

$$
\vdots
$$

$$
p_n^{(n)}(x_0) = f^{(n)}(x_0) \,. \tag{1}
$$

(Of course, $p_n^{(n+1)}(x)$ is identically zero.) Consequently, the formula for $p_n(x)$ is

$$
\begin{aligned}
p_n(x) &= f(x_0) + f'(x_0)(x - x_0) + f''(x_0)(x - x_0)^2/2! \\
&+ f'''(x_0)(x - x_0)^3/3! + \cdots + f^{(n)}(x_0)(x - x_0)^n/n! \\
&= \sum_{j=0}^{n} f^{(j)}(x_0)(x - x_0)^j/j! \tag{2}
\end{aligned}
$$

**Example 1.** The fourth-degree Taylor polynomials matching the functions $e^x$, $\cos x$, and $\sin x$ at $x_0$ are given by

$$
e^x \approx e^{x_0} + e^{x_0}(x - x_0) + e^{x_0}\frac{(x - x_0)^2}{2!} + e^{x_0}\frac{(x - x_0)^3}{3!} + e^{x_0}\frac{(x - x_0)^4}{4!} \,,
$$

$$
\begin{aligned}
\cos x \approx{} & \cos x_0 - \sin x_0 \, (x - x_0) - \cos x_0 \, \frac{(x - x_0)^2}{2!} + \sin x_0 \, \frac{(x - x_0)^3}{3!} \\
& + \cos x_0 \, \frac{(x - x_0)^4}{4!} \,,
\end{aligned}
$$

$$
\begin{aligned}
\sin x \approx{} & \sin x_0 + \cos x_0 \, (x - x_0) - \sin x_0 \, \frac{(x - x_0)^2}{2!} - \cos x_0 \, \frac{(x - x_0)^3}{3!} \\
& + \sin x_0 \, \frac{(x - x_0)^4}{4!} \,, \tag{3}
\end{aligned}
$$

■

The notation that we employed in eqs. (3) suggests that one can use $p_n(x)$ to approximate $f(x)$ for $x$ near $x_0$. Indeed, if we let $\varepsilon_n(x)$ measure the accuracy of the approximation -

$$
\varepsilon_n(x) = f(x) - p_n(x) \tag{4}
$$

- then calculus provides us with several formulas for estimating $\varepsilon_n$. The most transparent is due to Lagrange:

$$\varepsilon_n(x) = f^{(n+1)}(\xi)\frac{(x-x_0)^{n+1}}{(n+1)!} \tag{5}$$

where $\xi$, although unspecified, is guaranteed to lie between $x_0$ and $x$. (Equation (5) is proved by invoking the Mean Value Theorem; see exercise 8. Of course, it presupposes that the $(n+1)$st derivative of $f$ *exists* (and is continuous, as well) in the interval $[x_0, x]$.)

The form of eq. (5) suggests two possible strategies for reducing the error $\varepsilon_n(x)$:

(1) increasing the degree $n$ of the polynomial (i.e., taking more terms), thereby increasing the factor $(n+1)!$ in the denominator; or

(2) taking $x$ closer to $x_0$, thereby diminishing the factor $(x-x_0)$ in the numerator. Calculus students are usually more familiar with strategy #1, so let us review that procedure first.

The sequence of polynomials $p_n(x)$, for increasing values of $n$, constitutes the *Taylor series for $f(x)$, expanded around $x_0$.* (Of course $f$ must be infinitely often differentiable at $x_0$ in order for one to construct the Taylor series.) The series *converges* to $f(x)$ if the error $\varepsilon_n(x)$ approaches zero as $n \to \infty$.

$$f(x) = \sum_{j=0}^{n} f^j(x_0)(x-x_0)^j/j! + \varepsilon_n(x) \tag{6}$$

$$= \sum_{j=0}^{\infty} f^j(x_0)(x-x_0)^j/j! \quad \text{if the series converges.} \tag{7}$$

Now in Lagrange's formula (5) the ratio $(x-x_0)^{n+1}/(n+1)!$ always goes to zero for large $n$ (when $n$ increases the numerator accumulates factors of fixed size, but the denominator acquires larger and larger factors). Therefore $\varepsilon_n(x)$, too, will approach zero unless the higher derivatives of $f$ increase too rapidly with $n$.

For the exponential function, $f(x) = e^x$, the derivatives do not grow with $n$ at all $(f^{(n)}(x) = f(x))$; therefore the corresponding Taylor series converges for all $x$. Similarly the series for cosine and sine functions converge everywhere, since all the higher derivatives lie in [-1, 1]. On the other hand the derivatives of $1/(1-x)$, evaluated at $x_0 = 0$, grow rapidly:

$$1/(1-x_0) = 1,$$
$$1/(1-x_0)^2 = 1,$$
$$2/(1-x_0)^3 = 2,$$
$$6/(1-x_0)^4 = 6,$$

$$\vdots$$

$$n!/(1-x_0)^{n+1} = n!$$

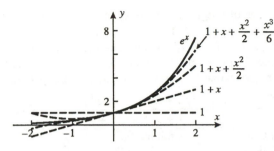

**Figure 1**   *Convergence of Taylor series*

As a result, the convergence of the Taylor series of $1/(1-x)$ expanded around $x_0 = 0$,

$$1 + 1x/1! + 2x^2/2! + 6x^3/3! + \cdots + (n!)x^n/n! + \cdots,$$

turns out to be limited to the interval $-1 < x < 1$. (This can be verified using the *ratio test*. Details of the convergence properties of Taylor series are best analyzed in the context of *complex variables*; see Saff and Snider.)

When the expansion point $x_0$ is taken to be zero, the Taylor series is called the *Maclaurin series*. The following Maclaurin series arise frequently and should be committed to memory:

(i) $e^x = 1 + x + x^2/2! + x^3/3! + \cdots$      (converges for all $x$)

(ii) $\sin x = x - x^3/3! + x^5/5! - \cdots$      (converges for all $x$)

(iii) $\cos x = 1 - x^2/2! + x^4/4! - \cdots$      (converges for all $x$)

(iv) $\dfrac{1}{1-x} = 1 + x + x^2 + x^3 + \cdots$      (converges for $|x| < 1$)

Figure 1 depicts the Taylor polynomials of degree 0 through 3 which approximate $f(x) = e^x$ around $x_0 = 0$ (see (i)). The convergence of the polynomials, as $n$ is increased, is evident; the error eventually goes to zero at every point.

Figure 1 also illustrates the *second* strategy for error control. For any *fixed* Taylor polynomial, taking $x$ closer to $x_0$ (zero, in the figure) reduces the error. But as the diagram demonstrates, when $x \to x_0$, $\varepsilon_n(x)$ approaches zero much faster along the curves for the *higher*-degree polynomials. Indeed, Lagrange's estimate (5) predicts that $\varepsilon_n$ goes to zero like the $(n+1)$st power of the presumedly small parameter $(x - x_0)$, in the sense that the ratio $\varepsilon_n(x)/(x - x_0)^{n+1}$ remains bounded as $x \to x_0$. This type of behavior is called **asymptotic convergence** and we write

$$\varepsilon_n(x) = \mathcal{O}([x - x_0]^{n+1}) \tag{8}$$

or "$\varepsilon_n(x)$ has order of magnitude $(x - x_0)^{n+1}$." (This notation is defined more precisely in Appendix A.1.) For example, from (i) above we have $e^x = 1 + x + x^2/2 + x^3/6 + \mathcal{O}(x^4)$; i.e. the difference between $e^x$ and its 3*rd* degree Maclaurin polynomial approaches zero at least as fast as the 4*th* power of $x$, as $x$ goes to zero. Table 2.2 confirms this prediction.

*TABLE 2.2*: $e^x = 1 + x + x^2/2 + x^3/6 + \mathcal{O}(x^4)$

| $x$ | $e^x$ | $1 + x + x^2/2 + x^3/6$ | $\frac{(difference)}{x^4}$ |
|---|---|---|---|
| 1 | 2.718281828 | 2.666666667 | .05162 |
| 0.5 | 1.648721271 | 1.645833337 | .04621 |
| 0.1 | 1.105170918 | 1.105166667 | .04251 |
| 0.05 | 1.051271096 | 1.051270833 | .04208 |
| | | | $\downarrow$ |
| | | | 1/24 (why?) |

The notion of asymptotic convergence is exploited in numerical analysis, as we shall see in Sec. 2.7 (and Appendix A.1).

Although the Taylor *polynomials* (6) are identical with the finite sections of the Taylor *series* (7), the underlying interpretation is somewhat different. In this book, we shall use the generic term *Taylor development* to describe both concepts. The context will convey whether we are speaking about a series or a polynomial and - in the latter case - the degree of the polynomial. The mathematical details are summarized in the following table:

*Comparison of Properties of Taylor Series and Taylor Polynomials*

| Series | Polynomials |
|---|---|
| 1. As $n$ approaches $\infty$, the Taylor *series* converges to $f(x)$ for all $x$ in some symmetric interval about $x_0$. The interval may be infinite, and it may be empty. (See exercise 7.) | Since there are only a finite number of terms in a polynomial there is no question of convergence as $n \to \infty$. |
| 2. Within the interval of convergence, the accuracy of approximation to $f(x)$ by finite truncations of the *series* is usually improved by taking more terms. | The accuracy of approximation of the *polynomial* to $f(x)$ is improved by taking $x$ nearer to $x_0$. |
| 3. The Taylor *series* presupposes the existence of all derivatives of all orders for $f(x)$ at $x = x_0$. | The Taylor *polynomial* of degree $n$ presumes only the existence of derivatives up to $f^{(n)}(x_0)$. |

The convergence of the *series* can be stated

$$\lim_{n \to \infty} \{f(x) - \sum_{j=0}^{n} f^{(j)}(x_0)(x - x_0)^j / j!\} = 0, \text{ at points of convergence,}$$

and for the polynomial,

$$\lim_{x \to x_0} \frac{f(x) - \sum_{j=0}^{n} f^{(j)}(x_0)(x - x_0)^j / j!}{(x - x_0)^n} = 0.$$

From the above, two very practical remarks can be made about the utility of the series and the polynomials:

(i) Although a Taylor *series* may converge as the number of terms $n \to \infty$, in practice one is usually inclined to compute a *fixed* number of terms, seldom more than 5 or 6 (or, say, 40 on a computer). Thus the polynomial interpretation would seem to be more relevant.

(ii) If the Taylor series doesn't converge, the expansion can be useful nonetheless, because the polynomials still exhibit the $\mathcal{O}([x - x_0]^n)$ asymptotic convergence.

The third property in the list above, concerning the possibility that $f(x)$ may have some, but not all, of its derivatives at $x = x_0$, may strike the reader as academic. Are we likely to encounter *in practice* (as opposed to mathematics textbooks) functions which have, say, 2 derivatives but not 3? The following example shows that the answer is emphatically *yes*.

**Example 2.** Compare the successive derivatives, at $x = 0$, of $f(x) = x^2$ and $f(x) = x^{8/3}$ .

| | | *at x = 0* | | | *at x = 0* |
|---|---|---|---|---|---|
| $f(x)$ | $= x^2$ | $= 0$ | $f(x)$ | $= x^{8/3}$ | $= 0$ |
| $f'(x)$ | $= 2x$ | $= 0$ | $f'(x)$ | $= (8/3)\, x^{5/3}$ | $= 0$ |
| $f''(x)$ | $= 2$ | $= 2$ | $f''(x)$ | $= (40/9)\, x^{2/3}$ | $= 0$ |
| $f^{(3)}(x)$ | $= 0$ | $= 0$ | $f^{(3)}(x)$ | $= (80/27)\, x^{-1/3}$ | $(\pm\infty)$ |
| $f^{(4)}(x)$ | $= 0$ | $= 0$ | $f^{(4)}(x)$ | $= -(80/81)\, x^{-4/3}$ | $(\pm\infty)$ |

Fractional exponents always cause loss of derivatives and consequently they limit the accuracy of approximation of the Taylor polynomials. From the above we see that no Taylor polynomial of degree 3 can be constructed for $x^{8/3}$ around $x_0 = 0$, and in fact the trivial polynomial "0" already has accuracy $\mathcal{O}(x^2)$ there. ∎

# Exercises 2.2

**1.** Compute the Taylor polynomials of degree 5, centered around $x_0 = 0$, for the functions

    (a) $\sin x$         (e) $\sin x^2$

    (b) $\cos x$        (f) $\cos x^2$

    (c) $e^x$           (g) $e^{-x^2}$

    (d) $x^4$          (h) $x^{14/3}$

2. Compute the Taylor polynomials of degree 2, centered at $x_0 = 1$, for the functions
   (a) $e^x$
   (b) $\ln x$
   (c) $x^4$
   (d) $x^2$

3. (a) Fill out a table like Table 2.2 for the approximations in exercises 1 and 2.

   (b) Explain the entry "1/24" in Table 2.2.

4. Compute the Taylor polynomials of degree 2, centered at $x_0 = \pi/4$, for $\sin x$ and $\cos x$.

5. *l'Hopital's Rule* states that when the ratio $\frac{f(x)}{g(x)}$ takes the limiting form $\frac{0}{0}$ as $x \to x_0$, then $\lim_{x \to x_0} \frac{f(x)}{g(x)} = \frac{f'(x_0)}{g'(x_0)}$ if the latter ratio is well defined. Show how this follows from Lagrange's assessment of the accuracy of the approximation of $p_1(x)$, the first-degree Taylor polynomial approximating $f(x)$ around $x_0$.

6. The first $n$ derivatives of the Taylor polynomial of degree $n$ agree with those of $f(x)$, at the expansion point $x = x_0$. Thus the graph of the linear polynomial is a straight line tangent to the graph of $f$, and that of the second degree polynomial is a parabola tangent to the graph of $f$ and also possessing the same second derivative as $f$ at $x = x_0$. The *osculating circle* at $x = x_0$ is the *circle* whose tangent and second derivative match those of $f$ at $x = x_0$. Write the equation of the osculating circle in terms of the data $\{x_0,\ f(x_0),\ f'(x_0),\ f''(x_0)\}$.[1]

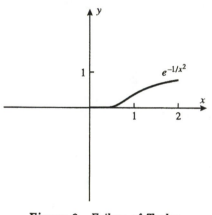

**Figure 2**  *Failure of Taylor representation*

7. (*A mathematical anomaly*) The function $f(x) = e^{-1/x^2}$ is graphed in Fig. 2. Obviously for the sake of continuity we define $f(0)$ to be zero. As a matter of fact, $f$ approaches zero so rapidly that one can prove *all of its derivatives at $x = 0$ are zero!*

   (a) Using the basic definitions

   $$f'(0) = \lim_{x \to 0} \frac{f(x) - f(0)}{x}, \quad f''(0) = \lim_{x \to 0} \frac{f'(x) - f'(0)}{x}, \ \ldots,$$

   (9)

   try to demonstrate this for the first and second derivatives; you will see from the pattern why the subsequent derivatives are zero also.

---

[1]The word *tangent* comes from the Latin *tangere*, to touch. *Osculate* comes from *osculare*, to kiss. (In the interests of good taste the mathematics community has not attempted to extend this picturesque terminology to curves with higher degrees of contact.)

(b) If the computations in part (a) are not to your taste, use a calculator to estimate $f'(0)$ and $f''(0)$ by evaluating eq. (9) with decreasing values of $x$.

(c) From these considerations, we see that the Taylor series for $f(x)$ expanded around $x = 0$ is given by

$$0 + 0 \cdot x + 0 \cdot x^2/2! + 0 \cdot x^3/3! + 0 \cdot x^4/4! + \cdots$$

It certainly converges for all $x$, but not to $f(x)$! Nonetheless the Taylor polynomial of degree $n$ - namely, zero - does approximate $f$ to an order of accuracy $\mathcal{O}(x^n)$:

$$e^{-1/x^2} = 0 + \mathcal{O}(x^n) \qquad \text{(all } n\text{)}$$

Demonstrate numerically that $e^{-1/x^2}$ does indeed approach zero faster than any power of $x$ by completing the following table.

| $x$ | $e^{-1/x^2}$ | $e^{-1/x^2}/x$ | $e^{-1/x^2}/x^2$ | $e^{-1/x^2}x^3$ | $e^{-1/x^2}/x^4$ | $e^{-1/x^2}/x^5$ |
|---|---|---|---|---|---|---|
| 1.0 | .368 | .368 | .368 | .368 | .368 | .368 |
| 0.7 | .123 | .185 | .265 | .379 | .541 | .773 |
| 0.4 | $1.93 \times 10^{-3}$ | ---- | ---- | ---- | ---- | ---- |
| 0.1 | $3.72 \times 10^{-44}$ | ---- | ---- | ---- | ---- | ---- |
| 0.05 | ---- | ---- | ---- | ---- | ---- | ---- |
| $\downarrow$ | $\downarrow$ | $\downarrow$ | $\downarrow$ | $\downarrow$ | $\downarrow$ | $\downarrow$ |
| | 0 | ? | ? | ? | ? | ? |

8. Follow these steps to derive Lagrange's form (5) for the accuracy of a Taylor polynomial approximant:

(a) Imagine $x$, as well as $x_0$, fixed. Then show that the function $F(t) = f(t) - p_n(t) - (t - x_0)^{n+1}[f(x) - p_n(x)]/(x - x_0)^{n+1}$ satisfies $F(x_0) = 0$, $F(x) = 0$.

(b) Conclude from the Mean Value Theorem of calculus that there must be a number $t_1$ between $x_0$ and $x$ where $F'(t_1) = 0$.

(c) Show that $F'(x_0) = 0$.

(d) Conclude, as before, that there must be a number $t_2$ between $x_0$ and $t_1$ where $F''(t_2) = 0$.

(e) Show that $F''(x_0) = 0$.

(f) Conclude that there must be number $t_3$ where $F'''(t_3) = 0$.

(g) Continue this argument until you demonstrate the existence of a number $\xi$, lying between $x_0$ and $x$, where $F^{(n+1)}(\xi) = 0$.

(h) Show that $F^{(n+1)}(\xi) = f^{(n+1)}(\xi) - (n+1)! \frac{f(x) - p(x)}{(x - x_0)^{n+1}}$.

(i) Combine (g) and (h) to derive Lagrange's formula (5).

## 2.3  Series Solutions at Ordinary Points

**Summary**   If an ordinary differential equation can be solved for its highest derivative in terms of lower order derivatives and other known functions,

$$y^{(n)}(x) = f(x, y, y', y'', \ldots, y^{(n-1)}),$$

then by successive differentiation all the coefficients for the Taylor series for $y$ at $x_0$ can be calculated from the initial data $y(x_0), y'(x_0), \ldots, y^{(n-1)}(x_0)$. This "power series" procedure is sometimes facilitated by formally inserting the ansatz $y(x) = \sum_{n=0}^{\infty} c_n(x - x_0)^n$ and finding recursion relations for the coefficients $c_n$.

In this section we will see how an ordinary differential equation can be interpreted as a prescription for constructing the Taylor development of its solutions. Besides providing a very general method for computing solutions to the equation, this notion also gives insight into the role of the constants in the "general solution". There are two mathematically equivalent ways to implement the procedure, and here we shall call them Method One and Method Two.

As a preliminary example illustrating Method One consider the nonlinear (!) first order equation

$$y' = y^2. \tag{1}$$

Since (1) is presumed to hold for a range of values of $x$, we can differentiate it to learn

$$y'' = \frac{dy^2}{dx^2} = 2y\frac{dy}{dx} = 2yy', \tag{2}$$

which expresses the second derivative in terms of $y$ and $y'$. Continuing in this manner we can express *every* derivative of $y$ in terms of lower derivatives:

$$
\begin{aligned}
y''' &= 2y'^2 + 2yy'' \\
y'''' &= 4y'y'' + 2y'y'' + 2yy''' = 6y'y'' + 2yy''' \\
&\ \vdots
\end{aligned}
\tag{3}
$$

Now let's seek to use this scheme to compute the Taylor development of the solution $y(x)$ expanded around the point $x = 0$. Equations (1) through (3) do not specify $y(0)$, but they *do* specify all of the higher derivatives once $y(0)$ is known. Thus if we set

$$y(0) = c,$$

then from (1, 2, 3) we find in succession

$$
\begin{aligned}
y'(0) &= y(0)^2 = c^2\,, \\
y''(0) &= 2y(0)y'(0) = 2cc^2 = 2c^3\,, \\
y'''(0) &= 2y'(0)^2 + 2y(0)y''(0) = 6c^4 = 3 \cdot 2c^4\,, \\
y''''(0) &= 6y'(0)y''(0) + 2y(0)y'''(0) = 24c^5 = 4 \cdot 3 \cdot 2c^5\,, \qquad (4)
\end{aligned}
$$

and the pattern is clearly

$$
y^{(n)}(0) = n!\,c^{n+1} \qquad (5)
$$

Therefore we can write down the Taylor development for the solution in terms of an arbitrary constant $(c)$ *which can be interpreted as the initial value $y(0)$ of the solution*:

$$
\begin{aligned}
y(x) &= c + (c^2)x + (2!\,c^3)x^2/2! + (3!\,c^4)x^3/3! + (4!\,c^5)x^4/4! + \cdots \\
&= c\,[1 + cx + (cx)^2 + (cx)^3 + (cx)^4 + \cdots]\,. \qquad (6)
\end{aligned}
$$

In this simple example we recognize (6) as the Maclaurin series for

$$
y(x) = c/(1 - cx) \qquad (7)
$$

(Sec. 2.2), converging for $|cx| < 1$. As a check note that

$$
y' = c^2/(1 - cx)^2 = y^2
$$

in agreement with the original differential equation (1). Note also that for $y(0) = c = 1$ we have reproduced the answer derived in exercise 9, Sec. 1.1 (by separating variables) for this same equation.

It is easy to generalize this example and formulate the overall solution strategy. An $n$th order differential equation

$$
y^{(n)} = f(x, y, y', y'', \ldots, y^{(n-1)})
$$

will, by successive differentiation, yield formulas for all the higher derivatives of the solution $y(x)$ at any point in terms of the first $n-1$ derivatives, and $y$ itself, at that point (unless some mathematical abnormality arises which foils the scheme. This ominous *caveat* is the subject of the next section.).

The second order equation

$$
y'' - xy' - (1 - x)y = 0 \quad \text{or} \quad y'' = xy' + (1 - x)y \qquad (8)
$$

provides second and higher order derivatives in terms of the lower ones, so the first step for expanding solutions around $x = 0$ is to set $y(0)$ *and* $y'(0)$ equal to undetermined constants:

$$
y(0) = c_0\,, \quad y'(0) = c_1 \qquad (9)
$$

Then the computations go as follows:

$$
\begin{aligned}
y'' &= xy' + y - xy = c_0 \quad (\text{at } x = 0) \\
y''' &= xy'' + 2y' - xy' - y = 2c_1 - c_0 \\
y'''' &= xy''' + 3y'' - 2y' - xy'' = 3c_0 - 2c_1 \\
y''''' &= xy'''' + 4y''' - 3y'' - xy''' = 4(2c_1 - c_0) - 3c_0 = 8c_1 - 7c_0
\end{aligned}
$$

and the solution is given by

$$
\begin{aligned}
y(x) &= c_0 + c_1 x + c_0 x^2/2! + (2c_1 - c_0)x^3/3! + (3c_0 - 2c_1)x^4/4! \\
&\quad + (8c_1 - 7c_0)x^5/5! + \cdots \\
&= c_0\{1 + x^2/2! - x^3/3! + 3x^4/4! - 7x^5/5! + \cdots\} \\
&\quad + c_1\{x + 2x^3/3! - 2x^4/4! + 8x^5/5! + \cdots\}
\end{aligned} \tag{10}
$$

Before we elaborate on the significance of (10), we want to illustrate an alternative method for computing the Taylor development of the solution. Later we will compare the merits of the two procedures.

Method Two begins with the presumption that the solution *has* a convergent Taylor series expansion; thus we write

$$
y(x) = \sum_{n=0}^{\infty} c_n x^n \tag{11}
$$

Notice that the customary $1/n!$ factor *is absorbed into the coefficient* $c_n$ with this formulation. (However the roles of $c_0$ and $c_1$ in (11) remain consistent with eq. (9); $c_0$ is the value, and $c_1$ the slope, of $y(x)$ at $x = 0$.) The derivatives of $y$ are then expressed

$$
y'(x) = \sum_{n=0}^{\infty} nc_n x^{n-1}, \quad y'' = \sum_{n=0}^{\infty} n(n-1)c_n x^{n-2} \tag{12}
$$

Now we insert (11) and (12) into (8):

$$
\sum_{n=0}^{\infty} n(n-1)c_n x^{n-2} - x \sum_{n=0}^{\infty} nc_n x^{n-1} - (1-x) \sum_{n=0}^{\infty} c_n x^n = 0
$$

or

$$
\sum_{n=0}^{\infty} n(n-1)c_n x^{n-2} - \sum_{n=0}^{\infty} nc_n x^n - \sum_{n=0}^{\infty} c_n x^n + \sum_{n=0}^{\infty} c_n x^{n+1} = 0 \tag{13}
$$

To combine the sums we must reindex so that common powers of $x$ are displayed. The easiest way to do this is to rewrite every term as a multiple of $x^m$. Thus in the first term $m$ equals $n - 2$ (and thus $n = m + 2$), in the second and third terms $m = n$, and in the final term $m = n+1$ $(n = m-1)$. Make sure you understand the significance of every symbol in the following:

$$\sum_{m=-2}^{\infty} (m+2)(m+1)c_{m+2}x^m - \sum_{m=0}^{\infty} mc_m x^m - \sum_{m=0}^{\infty} c_m x^m + \sum_{m=1}^{\infty} c_{m-1}x^m = 0 \tag{14}$$

Note that the terms with $m = -2$ and $-1$ are zero. Thus (14) can be reorganized as

$$\underbrace{(2)(1)c_2 - 0c_0 - c_0}_{(m=0 \text{ terms})} + \sum_{m=1}^{\infty}\{(m+2)(m+1)c_{m+2} - (m+1)c_m + c_{m-1}\}x^m = 0. \tag{15}$$

The display (15) can be interpreted as a Maclaurin expansion of the function $f(x) = 0$; consequently, *each term* must be zero. Therefore

$$c_2 = c_0/2 \quad \text{and}$$
$$c_{m+2} = \frac{(m+1)c_m - c_{m-1}}{(m+2)(m+1)} \quad \text{for each } m \geq 1 \tag{16}$$

As we anticipated, eqs. (16) do not specify $c_0$ or $c_1$, but express subsequent $c_m$ in terms of earlier ones; such formulas are called *recursion relations*. We work out a few terms:

$$(m=1) \quad c_3 = \frac{2c_1 - c_0}{3 \cdot 2} = \frac{2c_1 - c_0}{3!},$$
$$(m=2) \quad c_4 = \frac{3c_2 - c_1}{4 \cdot 3} = \frac{3c_0 - 2c_1}{4!},$$
$$(m=3) \quad c_5 = \frac{4c_3 - c_2}{5 \cdot 4} = \frac{8c_1 - 7c_0}{5!}$$

and so on. The solution is given by

$$\begin{aligned}
y(x) &= c_0 + c_1 x + c_2 x^2 + c_3 x^3 + c_4 x^4 + c_5 x^5 + \cdots \\
&= c_0 + c_1 x + c_0 x^2/2! + (2c_1 - c_0)x^3/3! + (3c_0 - 2c_1)x^4/4! + \\
&\qquad (8c_1 - 7c_0)x^5/5! + \cdots, \tag{17}
\end{aligned}$$

which is identical with (10).

From the results of either method we see that the general solution is expressed as the sum of constant multiples of two functions:

$$\begin{aligned}
y(x) &= c_0 y_1(x) + c_1 y_2(x) \\
&= c_0\{1 + x^2/2! - x^3/3! + 3x^4/4! - 7x^5/5! + \cdots\} \\
&\quad + c_1\{x + 2x^3/3! - 2x^4/4! + 8x^5/5! + \cdots\} \tag{18}
\end{aligned}$$

Clearly the functions $y_1$ and $y_2$ are, themselves, particular solutions to the differential equation satisfying the initial conditions

$$y_1(0) = 1 , \qquad y_1'(0) = 0 ;$$
$$y_2(0) = 0 , \qquad y_2'(0) = 1 . \tag{19}$$

An evaluation of their Wronskian at $x = 0$ demonstrates their independence:

$$W(x = 0; y_1, y_2) = \det \begin{vmatrix} y_1(0) & y_2(0) \\ y_1'(0) & y_2'(0) \end{vmatrix} = \det \begin{vmatrix} 1 & 0 \\ 0 & 1 \end{vmatrix} = 1 .$$

Neither series in (18) is recognizable as a readily identifiable function. This is usually the case; *differential equations with variable coefficients seldom have elementary function solutions.* However, it may be comforting to the reader to note that if $c_0 = c_1 = 1$, the solution $(10, 17)$ reduces to

$$y(x) = 1 + x + x^2/2! + x^3/3! + x^4/4! + x^5/5! + \cdots ,$$

which we identify as the Maclaurin series for $e^x$ (Sec. 2.2). And sure enough, $y = e^x$ does satisfy the differential equation (8)).

A few remarks at this point will be helpful for the subsequent examples.

1. Method One is usually more convenient when the equation is non-linear or when the coefficients in the linear equation have "long" Taylor series.

2. Method Two is occasionally superior for recognizing patterns in the recursion relations for the unknown coefficients.

3. For the linear second order equation

$$a_2(x)y'' + a_1(x)y' + a_0(x)y = 0 , \tag{20}$$

Method One cannot be implemented until we divide by $a_2(x)$, producing the form

$$y'' + p(x)y' + q(x)y = 0 \tag{21}$$

Thus in order for us to compute all the successive derivatives $y^{(j)}(0)$ $(j \geq 2)$, it is necessary that $p(x)$ and $q(x)$ be infinitely differentiable. *When $p(x)$ and $q(x)$ in (21) have convergent Taylor series at $x = 0$, then we say that $x = 0$ is an **ordinary point** for the equation.* (In the next section we shall see how to proceed when this condition is not met.)

4. In applying Method Two to the linear equation (20), it is not necessary to divide by $a_2(x)$ (although the requirement that $p(x)$ and $q(x)$ be infinitely differentiable still holds; see exercise 7). In fact, the calculations go much easier if the equation can be cleared of denominators at the outset.

5. It can be shown in general[2] that for the linear equation (21) the Taylor series for $y(x)$ will *converge* at any value of $x$ for which the Taylor series of both the coefficient functions $p(x)$ and $q(x)$ converge. For nonlinear equations there is no such guarantee of convergence. For example, we have seen that the solution to $y' = y^2$ approaches $\infty$ as $x$ approaches $1/c = 1/y(0)$ (recall eq. (7)); there is no clue in the original equation as to why this should happen. However the Taylor *polynomials* for the solutions can nonetheless be obtained through Method One.

6. In using Method Two to expand the solution around a *nonzero* point $x = a \neq 0$, the notation $y(x) = \sum_{n=0}^{\infty} c_n(x - a)^n$ becomes cumbersome. Experience has shown that it is always wise to begin by shifting the expansion point to zero, via a linear change of variable $t = x - a$ (recall Sec. 1.2).

7. The evaluation of the Taylor coefficients can be done automatically on computers possessing symbol manipulation capability. The "human touch," however, may be necessary to detect and exploit any systematic patterns in the recursion relations.

As a model for your computations we illustrate Method Two with an example demonstrating how a *parameter* in the equation coefficients can be carried through the calculations.

**Example 1.** Later in this book we shall see how *Legendre's differential equation* arises in the analysis of many physical systems possessing spherical symmetry. The equation has the form

$$(1 - x^2)y'' - 2xy' - \lambda y = 0 , \qquad (22)$$

with an unspecified parameter $\lambda$. (The role of $\lambda$ will be explained in Sec. 6.4. Some authors write the Legendre equation with $-\lambda$ instead of $\lambda$.)

Directly inserting the forms (11), (12) into (22) produces

$$\sum_{n=0}^{\infty} n(n-1)c_n x^{n-2} - \sum_{n=0}^{\infty} n(n-1)c_n x^n - 2\sum_{n=0}^{\infty} nc_n x^n - \lambda \sum_{n=0}^{\infty} c_n x^n = 0$$

(Notice that we did *not* divide eq. (22) by $(1 - x^2)$ before starting; the fractions would make the computations more complicated.) We reindex by letting $m = n - 2$ in the first term and $m = n$ in the others:

$$\sum_{m=-2}^{\infty} (m+2)(m+1)c_{m+2} x^m - \sum_{m=0}^{\infty} m(m-1)c_m x^m - 2\sum_{m=0}^{\infty} mc_m x^m$$

$$-\lambda \sum_{m=0}^{\infty} c_m x^m = 0$$

---

[2]See the references by Coddington and Levinson, Birkhoff and Rota, or Ince.

The first two terms ($m = -2, -1$) are trivial, and we have for $m \geq 0$

$$(m+2)(m+1)c_{m+2} - m(m-1)c_m - 2mc_m - \lambda c_m = 0, \quad \text{or}$$

$$c_{m+2} = \frac{m(m+1) + \lambda}{(m+2)(m+1)}c_m \tag{23}$$

To see the pattern, we write out a few coefficients:

$$c_2 = \tfrac{\lambda}{2}c_0 \qquad\qquad\qquad c_3 = \tfrac{1\cdot2+\lambda}{3\cdot2}c_1$$

$$c_4 = \tfrac{2\cdot3+\lambda}{4\cdot3}c_2 = \tfrac{(2\cdot3+\lambda)(\lambda)}{4\cdot3\cdot2}c_0 \quad c_5 = \tfrac{(3\cdot4+\lambda)}{5\cdot4}c_3 = \tfrac{(3\cdot4+\lambda)(1\cdot2+\lambda)}{5\cdot4\cdot3\cdot2}c_1$$

and observe

$$c_{2j} = \frac{\lambda(2\cdot3+\lambda)(4\cdot5+\lambda)\ldots([2j-2][2j-1]+\lambda)}{(2j)!}c_0$$

$$c_{2j+1} = \frac{(1\cdot2+\lambda)(3\cdot4+\lambda)\ldots([2j-1][2j]+\lambda)}{(2j+1)!}c_1$$

The general solution is displayed as

$$y(x) = \sum_{j=0}^{\infty} c_{2j}x^{2j} + \sum_{j=0}^{\infty} c_{2j+1}x^{2j+1}$$

Notice from the recursion relation (23) that *if the parameter $\lambda$ happens to take a value which is the negative of the product of two consecutive integers*

$$\lambda = -p(p+1), \quad p = \text{integer},$$

then one of the series terminates:

$$c_{p+2} = \frac{p(p+1) - p(p+1)}{(p+2)(p+1)}c_p = 0, \; c_{p+4} = 0, \; c_{p+6} = 0, \text{ etc.}$$

For example, if $\lambda = 0$ ($p = 0$) one solution to the Legendre equation is given by

$$P_0(x) = 1 \quad (\text{take } c_0 = 1, \, c_1 = 0).$$

If $\lambda = -2$ ($p = 1$) a solution is

$$P_1(x) = x \quad (\text{take } c_0 = 0, \, c_1 = 1)$$

If $\lambda = -6$ ($p = 2$) a solution is

$$P_2(x) = \frac{3}{2}x^2 - \frac{1}{2} \quad (\text{take } c_0 = -\frac{1}{2}, \, c_1 = 0)$$

If $\lambda = -12$ ($p = 3$) a solution is

$$P_3(x) = \frac{5}{2}x^3 - \frac{3}{2}x \quad (\text{take } c_0 = 0\,,\ c_1 = -\frac{3}{2}) \qquad (24)$$

These solutions are known as the *Legendre polynomials*. See exercise 3. ∎

**Example 2.** Although we only need to use the Taylor series methodology for *homogeneous* linear equations (since Green's functions can be used to take care of the nonhomogeneity), the procedure can handle the inhomogeneous term also. We close this section by demonstrating with the equation

$$y'' + x^2 y' + e^x y = 1/(1+x) \qquad (25)$$

The use of Method One on this problem proceeds as follows:

$$\begin{aligned}
y'' &= -x^2 y' - e^x y + (1+x)^{-1}\,; \\
y''(0) &= -0 - e^0 y(0) + 1 = -c_0 + 1. \\
y''' &= -2xy' - x^2 y'' - e^x y - e^x y' - (1+x)^{-2}\,; \\
y'''(0) &= 0 + 0 - y(0) - y'(0) - 1 = -c_0 - c_1 - 1. \\
y'''' &= -2y' - 2xy'' - 2xy'' - x^2 y''' - e^x y - e^x y' - e^x y' - e^x y'' \\
&\quad + 2(1+x)^{-3}\,; \\
y''''(0) &= -2c_1 - 0 - 0 - 0 - c_0 - c_1 - c_1 - [-c_0 + 1] + 2 \\
&= -4c_1 + 1.
\end{aligned}$$

Thus to order $\mathcal{O}(x^5)$ we have

$$\begin{aligned}
y(x) &= c_0 + c_1 x + \frac{-c_0 + 1}{2}x^2 - \frac{c_0 + c_1 + 1}{3!}x^3 + \frac{-4c_1 + 1}{4!}x^4 + \mathcal{O}(x^5) \\
&= c_0[1 - \frac{x^2}{2} - \frac{x^3}{6} + \mathcal{O}(x^5)] + c_1[x - \frac{x^3}{6} - \frac{x^4}{6} + \mathcal{O}(x^5)] \\
&\quad + \frac{x^2}{2} - \frac{x^3}{6} + \frac{x^4}{24} + \mathcal{O}(x^5)
\end{aligned}$$

(The curious reader is invited to try Method Two on eq. (25) to see why the author selected Method One!) ∎

# Exercises 2.3

1. Probably most people would agree that the simplest nontrivial second order differential equation with constant coefficients is $y'' = y$. Use Method One to find Taylor series expansions for two independent solutions, expanded around $x = 0$. Identify the series in terms of elementary functions. (This is trivial, of course - you already know how to obtain closed-form solutions to constant-coefficient equations,

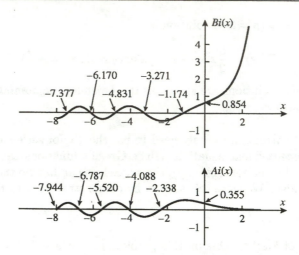

*Figure 1   Airy functions*

and in this case the solutions are $\exp(x)$ and $\exp(-x)$. However, these are *not* the solutions $y_1(x)$ and $y_2(x)$ that the method gives you, because they do not satisfy the initial conditions (19).) *Answer:* $c_0 \cosh x + c_1 \sinh x$

2. The simplest second order equation with variable coefficients would have to be *Airy's equation* $y'' = xy$. Obtain the Taylor series for the solutions to this equation, expanded around $x = 0$. (*Answer:*

$$c_0\{1 + x^3/2 \cdot 3 + x^6/2 \cdot 3 \cdot 5 \cdot 6 + x^9/2 \cdot 3 \cdot 5 \cdot 6 \cdot 8 \cdot 9 + \cdots\} +$$
$$c_1\{x + x^4/3 \cdot 4 + x^7/3 \cdot 4 \cdot 6 \cdot 7 + x^{10}/3 \cdot 4 \cdot 6 \cdot 7 \cdot 9 \cdot 10 + \cdots\})$$

>    *Note*: traditionally one employs the *Airy functions* $Ai(x)$ and $Bi(x)$ as the two independent solutions of this equation. $Ai(x)$ results if $c_0$ is taken to be $[\Gamma(2/3)3^{2/3}]^{-1} \approx .3550$ and $c_1$ to be $-[\Gamma(1/3)3^{1/3}]^{-1} \approx -.2588$. $Bi(x)$ results from $c_0 = [\Gamma(2/3)3^{1/6}]^{-1} \approx .6149$, $c_1 = 3^{1/6}/\Gamma(1/3) \approx .4483$. The Airy functions are plotted in Fig. 1.

3. Derive the formulas for the first four Legendre polynomials (eqs. (24)) by inserting the appropriate values for $\lambda, c_0$, and $c_1$ and working out the recursion relations. Then verify that the polynomials do, in fact, solve eq. (22).

4. Obtain the Taylor series, expanded around $x = 0$, for the following:

    (a) $(1 - x^2)y'' - xy' + \lambda y = 0$ (*Chebyshev equation*; show that if $\lambda = n^2$, $n$ an integer, there are polynomial solutions.)

    (b) $y'' + xy' - y = 0$

(c) $y'' - 3y' + 2y = 0$

(d) $(1 - x^2/2)y'' + xy' - y = 0$

5. Obtain the Taylor series expansions around $x = 0$ for solutions to the following *nonlinear* equations. As a check, solve the equations by separating variables (exercise 9, Sec. 1.1); then work out the Taylor series directly for these (closed-form) solutions, and compare.

(a) $y' = \sqrt{y}$

(b) $y' = 1/y$

(c) $y' = 1/y^3$

6. Find series expansions for two independent solutions to the Airy equation $y'' = -xy$ *around the point* $x = 1$.

7. Attempt to carry out the solution to $x^2 y'' + xy' + y = 0$ using the solution form $y = \sum_{n=0}^{\infty} a_n x^n$. What goes wrong?

## 2.4  Frobenius's Method

**Summary**  The power series method fails when applied to a linear homogeneous second order equation of the form $y'' = -p(x)y' - q(x)y$ if the coefficients $p$ and/or $q$ are not defined (infinite) at the expansion point. However the equation has a regular singular point at (say) $x = 0$ when the coefficients have expansions of the form

$$p(x) = \frac{p_{-1}}{x} + \sum_{n=0}^{\infty} p_n x^n, \quad q(x) = \frac{q_{-2}}{x^2} + \frac{q_{-1}}{x} + \sum_{n=0}^{\infty} q_n x^n.$$

A solution to the equation can then be found in the Frobenius format

$$y_1(x) = x^r \sum_{n=0}^{\infty} c_n x^n$$

where $r$ is the larger root to the indicial equation

$$r^2 + (p_{-1} - 1)r + q_{-2} = 0.$$

Unless the two roots of the indicial equation differ by an integer, an independent solution can be found in the same form, with $r$ equal to the smaller root. However in the exceptional case the independent solution takes the form

$$y_2(x) = x^r \sum_{n=0}^{\infty} d_n x^n + A \ln |x| \, y_1(x)$$

with $r$ equal to the smaller root. Although $A$ may or may not be zero in general, if the roots of the indicial equation are equal then $A$ will definitely be nonzero.

The kind of ordinary differential equation that arises most frequently in continuum engineering is the linear second order equation

$$a_2(x)y'' + a_1(x)y' + a_0(x)y = 0. \tag{1}$$

In the previous section we studied two methods for obtaining the Taylor developments of the solutions. Method One begins by dividing (1) by $a_2(x)$ to obtain the "standard form"

$$y'' = -p(x)y' - q(x)y, \tag{2}$$

and differentiates (2) successively to derive formulas expressing the higher derivatives of $y(x)$ at the expansion point $x_0$ in terms of lower ones (ultimately, in terms of the first two). Method Two inserts the expression

$$y(x) = \sum_{n=0}^{\infty} c_n(x - x_0)^n \tag{3}$$

into either (1) or (2), whichever is convenient, and generates recursion relations for the coefficients $c_n$.

We also pointed out that the coefficients $p(x)$ and $q(x)$ must be infinitely often differentiable at $x = x_0$ for the implementation of either method. Now it turns out that in many engineering applications this condition is not met. Thus we must extend the Taylor development methodology to cover a broader class of problems. In most cases that arise in practice the coefficient $p(x)$ diverges no faster than $1/(x - x_0)$, and $q(x)$ no faster than $1/(x - x_0)^2$, so the following definition encompasses all the situations that we shall see in this book. Throughout this section we presume that the independent variable has been shifted so that the expansion point is $x_0 = 0$.

**Definition.** If the functions $xp(x)$ and $x^2q(x)$ each have Taylor expansions which converge in some (nonempty) interval around $x = 0$

$$xp(x) = \sum_{n=0}^{\infty} \rho_n x^n \tag{4}$$

$$x^2q(x) = \sum_{n=0}^{\infty} \gamma_n x^n, \tag{5}$$

then we say the differential equation (2) has *a regular singular point* at $x = 0$.

Notice that (4) and (5) are equivalent to the representations

$$p(x) \;=\; \frac{p_{-1}}{x} + \sum_{n=0}^{\infty} p_n x^n \qquad (6)$$

$$q(x) \;=\; \frac{q_{-2}}{x^2} + \frac{q_{-1}}{x} + \sum_{n=0}^{\infty} q_n x^n \qquad (7)$$

(with $p_i = \rho_{i+1}$, $q_i = \gamma_{i+2}$). The terms singled out in (6) and (7) are called the "singular parts" of $p(x)$ and $q(x)$ respectively. If the singular parts are both zero, then $x = 0$ is called an *ordinary point*.

The terminology *irregular singular point* is used when $x = 0$ is neither ordinary nor regular singular. Some aspects of irregular singular points are explored in exercise 4.

The beauty of regular singular points is that convergent expansions for the solutions can be obtained around them by a minor modification of Method Two. To motivate this modification, notice that the following *equidimensional* equation (which was solved in Sec. 1.1) has a regular singular point at $x = 0$;

$$x^2 y'' + 5xy' + 6y = 0 \quad \text{or} \quad y'' + \frac{5}{x} y' + \frac{6}{x^2} y = 0. \qquad (8)$$

The solution was obtained in that section using the ansatz $y = x^r$ for $x > 0$. The proper generalization of this idea for the present situation of equation (2) is, again, to restrict attention to positive $x$ and let

$$
\begin{aligned}
y(x) \;&=\; x^r \sum_{n=0}^{\infty} c_n x^n \\
&=\; \sum_{n=0}^{\infty} c_n x^{n+r} \qquad (9)
\end{aligned}
$$

The flexibility afforded by attaching an extra exponent $r$ to the earlier format (3) results in a procedure known as the *method of Frobenius*; it will *always* generate expansions of solutions around a regular singular point, as we demonstrate with examples. Note that from (9) the first and second derivatives of $y$ are expressed as

$$y' = \sum_{n=0}^{\infty} (n+r) c_n x^{n+r-1}, \quad y'' = \sum_{n=0}^{\infty} (n+r)(n+r-1) c_n x^{n+r-2}. \quad (10)$$

**Example 1.** The differential equation

$$x^2 y'' + xy' + \left(x^2 - \frac{1}{9}\right) y = 0 \qquad (11)$$

is known as *Bessel's equation of order 1/3.* (We shall elaborate on this equation and its solutions in the next section; for now just note that 1/3 is the square root of 1/9.)

Observe that if we put (11) into standard form by dividing by $x^2$, $x = 0$ is revealed as a regular singular point. However we retain the form (11) in our work in order to avoid polynomial fractions. Inserting (9) and (10) into (11) we find

$$\sum_{n=0}^{\infty}(n+r)(n+r-1)c_n x^{n+r} + \sum_{n=0}^{\infty}(n+r)c_n x^{n+r} + \sum_{n=0}^{\infty}c_n x^{n+r+2}$$

$$-\sum_{n=0}^{\infty}\frac{1}{9}c_n x^{n+r} = 0 \qquad (12)$$

Reindexing with $m = n + 2$ in the third term and $m = n$ in the others, we get

$$\sum_{m=0}^{\infty}(m+r)(m+r-1)c_m x^{m+r} + \sum_{m=0}^{\infty}(m+r)c_m x^{m+r}$$

$$+ \sum_{m=2}^{\infty}c_{m-2} x^{m+r} - \sum_{m=0}^{\infty}\frac{1}{9}c_m x^{m+r} = 0 \qquad (13)$$

For $m = 0$ this says

$$r(r-1)c_0 + rc_0 - \frac{1}{9}c_0 = 0, \text{ or } \left(r^2 - \frac{1}{9}\right)c_0 = 0 \qquad (14)$$

Thus either $r^2 = 1/9$ or $c_0 = 0$. *Now, the latter alternative makes no sense; by* definition, *the series in (9) begins with $x^r$, so we never have $c_0$ equal to zero.* Consequently the condition (14) dictates the acceptable values of $r$:

$$r^2 - \frac{1}{9} = 0, \text{ or } r = \pm\frac{1}{3}. \qquad (15)$$

In general, when (9) is inserted into a differential equation with a regular singular point at $x = 0$ and the summations reindexed, the lowest degree term will always have the form

$$\text{(quadratic in } r) \cdot c_0 = 0. \qquad (16)$$

Since $c_0 = 0$ is never allowed, the values of the exponent $r$ are determined by the quadratic equation

$$\text{(quadratic in } r) = 0 \qquad (17)$$

which is known as the *Indicial Equation.*

Exercise 2 demonstrates this fact in general and shows that in terms of the singular parts of the coefficients displayed in eqs. (6) and (7) the indicial equation is

$$r^2 + (p_{-1} - 1)r + q_{-2} = 0. \qquad (18)$$

Note that (18) reduces to (15) for the Bessel equation. Note also that because of the quadratic nature of the indicial equation we can normally expect two solutions for $r$.

Let us set $r = 1/3$ for definiteness and proceed with our example. For $m = 1$,

$$(1 + \frac{1}{3})\frac{1}{3}c_1 + (1 + \frac{1}{3})c_1 - \frac{1}{9}c_1 = \frac{5}{3}c_1 = 0, \text{ or } c_1 = 0 \qquad (19)$$

(This is something new; $c_1$ was an arbitrary constant in Sec. 2.3! We'll return to this point.) For $m \geq 2$,

$$(m + \frac{1}{3})(m + \frac{1}{3} - 1)c_m + (m + \frac{1}{3})c_m + c_{m-2} - \frac{1}{9}c_m = 0, \text{ or}$$

$$(m^2 + \frac{2}{3}m)c_m = m(m + \frac{2}{3})c_m = -c_{m-2}, \text{ or}$$

$$c_m = -c_{m-2}/m(m + \frac{2}{3}) \qquad (20)$$

Since $c_1 = 0$, eq. (20) implies that all $c_m$ with odd subscripts are zero. Moreover, $c_0$ is arbitrary and the remaining $c_m$'s are proportional to $c_0$; the pattern is

$$c_2 = \frac{-c_0}{2(2 + 2/3)}$$

$$c_4 = \frac{-c_2}{4(4 + 2/3)} = \frac{c_0}{4 \cdot 2(4 + 2/3)(2 + 2/3)}$$

$$c_6 = \frac{-c_4}{6(6 + 2/3)} = \frac{-c_0}{6 \cdot 4 \cdot 2(6 + 2/3)(4 + 2/3)(2 + 2/3)}$$

$$\vdots$$

$$c_{2j} = \frac{(-1)^j c_0}{2^{2j} j!(1 + 1/3)\dots(2 + 1/3)(j + 1/3)}$$

$$= \frac{(-1)^j , (4/3)}{2^{2j} j!, (j + 4/3)}c_0 \quad , j > 0 \qquad (21)$$

(exercise 5). With the choice $c_0 = [2^{1/3}, (4/3)]^{-1}$ the series

$$\sum_{j=0}^{\infty} c_{2j} x^{2j+1/3}$$

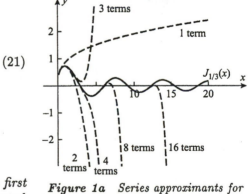

*Figure 1a* *Series approximants for Bessel functions*

represents a function known historically as *the Bessel function of the first kind of order 1/3*: $J_{1/3}(x)$. The convergence of the series is demonstrated in Fig. 1(a).

Although $c_0$ is arbitrary we have seen that $c_1$ is fixed (zero), so we have found only one solution of the differential equation (up to a constant multiple). Recall that in the previous section - when $x = 0$ was an *ordinary* point - both $c_0$ and $c_1$ were arbitrary, resulting in two independent solutions. Now, to find a second solution we must turn to the other root of the indicial equation, $r = -1/3$.

Thus we return to eq. (13); but to emphasize that we are computing a different solution, we change the unknown coefficients from $c$'s to $d$'s (and stipulate that $d_0 \neq 0$).

For $m = 1$ we get

$$\frac{2}{3}(-\frac{1}{3})d_1 + \frac{2}{3}d_1 - \frac{1}{9}d_1 = \frac{1}{3}d_1 = 0$$

and observe that $d_1$ suffers the same fate as $c_1$; it is zero. For $m \geq 2$,

$$(m^2 - \frac{2}{3}m)d_m = m(m - \frac{2}{3})d_m = -d_{m-2}\,, \text{ or } d_m = -d_{m-2}/m(m - 2/3) \tag{22}$$

and the pattern is easily seen to be (exercise 5)

$$d_{2j} = \frac{(-1)^j d_0}{2^{2j}j!(1 - 1/3)(2 - 1/3)\ldots(j - 1/3)} = \frac{(-1)^j \Gamma(2/3)}{2^{2j}j!\,\Gamma(j + 2/3)}d_0\,, j > 0 \tag{23}$$

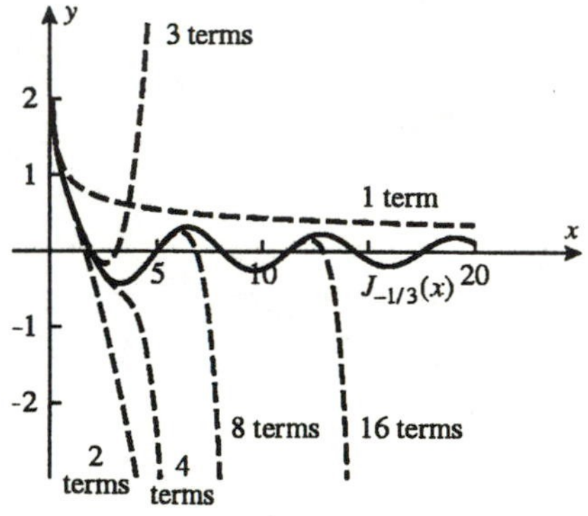

which, with the choice $d_0 = 2^{1/3}/\Gamma(2/3)$, yields the series for *the Bessel function of the first kind of order -1/3*: $J_{-1/3}(x)$. See Fig. 1(b).

To summarize the findings of this example: the general solution of (11) is expressed as

$$y(x) = c_0 y_1(x) + d_0 y_2(x) \tag{24}$$

where

$$y_1(x) = x^{1/3}\sum_{j=0}^{\infty}\frac{(-1)^j \Gamma(4/3)}{2^{2j}j!\,\Gamma(j + 4/3)}x^{2j}\,, \tag{25}$$

$$y_2(x) = x^{-1/3}\sum_{j=0}^{\infty}\frac{(-1)^j \Gamma(2/3)}{2^{2j}j!\,\Gamma(j + 2/3)}x^{2j}\,. \tag{26}$$

Notice that the exponents, $1/3$ in (25) and $-1/3$ in (26), were revealed by the indicial equation. Thus *even before we solved the recurrence relations* we knew that one solution would be infinite at the origin, while the other would be zero (but with infinite slope). ■

**Example 2.** The equation

$$4xy'' + 2y' + y = 0 \tag{27}$$

*Figure 1b*

has a regular singular point at $x = 0$. Inserting (9) and (10) we have

$$\sum_{n=0}^{\infty}(n+r)(n+r-1)4c_n x^{n+r-1} + \sum_{n=0}^{\infty}(n+r)2c_n x^{n+r-1} + \sum_{n=0}^{\infty}c_n x^{n+r} = 0. \tag{28}$$

Reindexing produces

$$\sum_{m=0}^{\infty}(m+r)(m+r-1)4c_m x^{m+r-1} + \sum_{m=0}^{\infty}(m+r)2c_m x^{m+r-1}$$
$$+ \sum_{m=1}^{\infty}c_{m-1} x^{m+r-1} = 0. \tag{29}$$

For $m = 0$ we get the *indicial equation*

$$r(r-1)4c_0 + r2c_0 = (4r^2 - 2r)c_0 = 0, \text{ or } r = 0,\ 1/2 \tag{30}$$

(and thus both solutions will be finite at $x = 0$). The recursion relation for $r = 0$ is

$$m(m-1)4c_m + 2mc_m + c_{m-1} = 0, \text{ or}$$
$$c_m = -\frac{c_{m-1}}{4m^2 - 2m} = -\frac{c_{m-1}}{2m(2m-1)} \tag{31}$$

and the pattern is

$$c_1 = -c_0/2 \cdot 1,\ c_2 = -c_1/4 \cdot 3 = c_0/4!,\ c_3 = -c_2/6 \cdot 5 = -c_0/6!,$$
$$\ldots, c_n = (-1)^n c_0/(2n)!. \tag{32}$$

So the solution is recognizable (recall Sec. 2.2, formula (*iii*)):

$$y_1(x) = c_0\{1 - \frac{x}{2!} + \frac{x^2}{4!} - \frac{x^3}{6!} + \cdots\} = c_0 \cos\sqrt{x}. \tag{33}$$

For $r = 1/2$ the relation is

$$(m+1/2)(m-1/2)4d_m + (m+1/2)2d_m + d_{m-1} = 0,$$
$$\text{or } d_m = -\frac{d_{m-1}}{2m(2m+1)} \tag{34}$$

and the pattern is

$$d_1 = -d_0/2 \cdot 3,\ d_2 = -d_1/4 \cdot 5 = d_0/5!,\ d_3 = -d_2/6 \cdot 7 = -d_0/7!,$$
$$\ldots, d_n = \frac{(-1)^n d_0}{(2n+1)!} \tag{35}$$

producing

$$y_2(x) = d_0 x^{1/2}\{1 - \frac{x}{3!} + \frac{x^2}{5!} - \frac{x^3}{7!} + \cdots\}$$

$$= d_0\{\sqrt{x} - \frac{\sqrt{x^3}}{3!} + \frac{\sqrt{x^5}}{5!} - \frac{\sqrt{x^7}}{7!} + \cdots\} = d_0 \sin\sqrt{x}. \quad (36)$$

Since $y_1$ and $y_2$ have familiar closed form expressions, their status as solutions can be verified directly (exercise 6). ∎

The Frobenius procedure as depicted by these examples is a fairly straightforward extension of Method Two of the previous section. However there is an exceptional circumstance which requires special treatment, and it arises in applications more often than not. The next three examples illustrate the "symptom" and its "cure."

**Example 3.** The *Bessel equation of order 1* is given by

$$x^2 y'' + xy' + (x^2 - 1)y = 0. \quad (37)$$

When the form (9), (10) is inserted, the resulting equation is

$$\sum_{m=0}^{\infty} (m+r)(m+r-1)c_m x^{m+r} + \sum_{m=0}^{\infty} (m+r)c_m x^{m+r}$$

$$+ \sum_{m=2}^{\infty} c_{m-2} x^{m+r} - \sum_{m=0}^{\infty} c_m x^{m+r} = 0. \quad (38)$$

The indicial equation is

$$r^2 - 1 = 0, \text{ or } r = \pm 1. \quad (39)$$

Before we begin computing, let's look ahead and anticipate a problem engendered by this root pattern. Presumably (39) will lead to two solutions of the form

$$y_1(x) = \sum_{n=0}^{\infty} c_n x^{n+1} = c_0 x^1 + c_1 x^2 + c_2 x^3 + c_3 x^4 + \cdots, \quad (40)$$

$$y_2(x) = \sum_{n=0}^{\infty} d_n x^{n-1} = d_0 x^{-1} + d_1 + d_2 x^1 + d_3 x^2 + d_4 x^3$$

$$+ d_5 x^4 + \cdots, \quad (41)$$

but when we reassemble these into the general solution we see that *the series overlap*, starting from the $x^1$ term:

$$y(x) = d_0 x^{-1} + d_1 + \{d_2 + c_0\}x + \{d_3 + c_1\}x^2 + \{d_4 + c_2\}x^3 + \cdots, \quad (42)$$

What does this mean? It means that when we go through the recursion relations for the lower root ($r = -1$) the value of the coefficient $d_2$ *cannot be determined* from $d_1$ and $d_0$ - there is an indeterminant amount of $c_0$ inseparably "mixed in" with $d_2$. In other words a knowledge of $d_0$ alone will not be sufficient to determine the higher $d_j$ because there is no way of preventing a multiple of the other ($r = +1$) solution from "leaking" into the series (41).

Clearly this overlapping will occur whenever the roots of the indicial equation differ by an integer, or are equal.

Whenever the roots $r_1$ and $r_2$ of the indicial equation differ by a positive integer or zero

$$r_1 = r_2 + s, \quad s = 0, 1, 2, \ldots, \tag{43}$$

then the coefficient $d_s$ in the series for the solution corresponding to the lower root $r_2$ cannot be determined by the recursion relations.

Let us proceed with the example at hand and the phenomenon will become clearer. Since there should be no problem with the larger root, we start by taking $r = +1$. The recursion relations yield

$$3c_1 = 0, \quad \text{or} \quad c_1 = 0, \tag{44}$$

and for $m \geq 2$

$$(m^2 + 2m)c_m = m(m+2)c_m = -c_{m-2}, \quad \text{or} \quad c_m = -c_{m-2}/m(m+2). \tag{45}$$

A little work reveals that this solution is given by

$$y_1(x) = \sum_{j=0}^{\infty} c_{2j+1} x^{2j+1} = c_0 \sum_{j=0}^{\infty} \frac{(-1)^j}{2^{2j} j!(j+1)!} x^{2j+1} \tag{46}$$

(see exercise 7). With the choice $c_0 = 1/2$ eq. (46) defines the *Bessel function of the first kind of order one*: $J_1(x)$. Graphs will be displayed in the next section.

The second root, $r = -1$, foils us. The recursion relations are

$$-1d_1 = 0 \quad \text{or} \quad d_1 = 0, \tag{47}$$

and for $m \geq 2$

$$(m^2 - 2m)d_m = m(m-2)d_m = -d_{m-2} \tag{48}$$

Thus the equation for $d_2$, which we anticipated from (42) would be troublesome, is

$$2(2-2)d_2 = 0 \cdot d_2 = -d_0. \tag{49}$$

Since we have agreed that $d_0$ is not zero, there is no value of $d_2$ that will enable us to satisfy (49). We need a little more slack, and it is provided by the following prescription:

**Frobenius's Rule.** Whenever the roots of the indicial equation (18) differ by a positive integer or zero ($r_1 = r_2 + s$, $s = 0, 1, 2, \ldots$), then one solution of the differential equation can be found in the form

$$y_1(x) = \sum_{n=0}^{\infty} c_n x^{n+r_1} \tag{50}$$

and an independent solution can be found in the form

$$y_2(x) = \sum_{n=0}^{\infty} d_n x^{n+r_2} + A \ln|x|\, y_1(x) \tag{51}$$

$$= \sum_{n=0}^{\infty} d_n x^{n+r_2} + A \ln|x| \sum_{n=0}^{\infty} c_n x^{n+r_1}$$

(see exercise 10).

The log term and the extra constant $A$ (which may turn out to be zero) give us the flexibility to overcome the difficulty produced by the overlapping series. To use (50) to find a second solution of our eq. (37?) we compute (taking $x > 0$ for simplicity; see exercise 9)

$$y_2'(x) = \sum_{n=0}^{\infty} (n+r_2) d_n x^{n+r_2-1} + A y_1(x)/x + A \ln x\, y_1'(x)$$

$$y''(x) = \sum_{n=0}^{\infty} (n+r_2)(n+r_2-1) d_n x^{n+r_2-2} - A y_1/x^2 + 2A y_1'/x$$

$$+ A \ln x\, y_1'' \tag{52}$$

Inserting (51) and (52) into (37) and setting $r_2 = -1$ leads to

$$\left\{ \sum_{m=0}^{\infty} (m-1)(m-2) d_m x^{m-1} + \sum_{m=0}^{\infty} (m-1) d_m x^{m-1} + \sum_{m=2}^{\infty} d_{m-2} x^{m-1} \right.$$

$$\left. - \sum_{m=0}^{\infty} d_m x^{m-1} \right\} + \left\{ A \ln x\, \underbrace{[x^2 y_1'' + x y_1' + (x^2-1) y_1]}_{= 0 \text{ (original eq. (37)!)}} \right\}$$

$$+ \{2A x y_1'\} = 0 \tag{53}$$

The first bracketed term in eq. (53) has the same form as (38), but with $r = r_2 = -1$. The second term is identically zero, since it contains a factor obtained by inserting $y_1$ into the original differential equation (37). The final term is $2Ax(\sum_{n=0}^{\infty} c_n x^{n+1})' = 2A\sum_{n=0}^{\infty}(n+1)c_n x^{n+1}$, which we re-index to match the others by setting $n+1 = m-1$   $(m = n+2, \; n = m-2)$:

$$2Axy_1' = \sum_{m=2}^{\infty} 2A(m-1)c_{m-2}x^{m-1}\,. \tag{54}$$

Now as we retrace our steps through the recursion relations for eq. (53) and (54) we will see how the extra flexibility afforded by "$A$" allows us to bypass the difficulty with $d_2$. For $m = 0$ and $m = 1$, of course, the "$A$" terms haven't yet come into the picture; so there is no change. Thus $d_0$ is arbitrary and (recall (47))

$$d_1 = 0\,. \tag{55}$$

For $m \geq 2$,

$$m(m-2)d_m + d_{m-2} + 2A(m-1)c_{m-2} = 0\,. \tag{56}$$

The troublesome equation for $m = 2$ takes the form

$$0 \cdot d_2 + d_0 + 2Ac_0 = 0 \tag{57}$$

(compare with (49).) We still can't solve this equation for $d_2$, but we can *satisfy* it by setting

$$A = -d_0/2c_0\,. \tag{58}$$

Then for $m \geq 3$ the recursion computations proceed without incident:

$$d_m = -\frac{d_{m-2} + 2A(m-1)c_{m-2}}{m(m-2)} = -\frac{d_{m-2}}{m(m-2)} + \frac{(m-1)}{m(m-2)}\frac{c_{m-2}}{c_0}d_0 \tag{59}$$

Recall that $d_1$ is zero, as are all the $c_m$'s with odd $m$ (eqs. (44) and (45)). Thus (59) dictates that $d_m$ also must be zero for odd $m$. A few of the even terms are given by

$$d_4 = -\frac{1}{8}d_2 - \frac{3}{64}d_0\,, \quad d_6 = \frac{1}{192}d_2 + \frac{7}{2304}d_0\,,\ldots \tag{60}$$

(where we have employed (45) to eliminate $c_{m-2}/c_0$).

Since the equation (57) for $d_2$ has been satisfied without specifying any particular value for $d_2$, we may set it equal to zero for convenience and display the *general solution* to the Bessel equation of order one as

$$\begin{aligned}
y(x) &= c_0 \sum_{j=0}^{\infty} \frac{(-1)^j}{2^{2j}j!(j+1)!}x^{2j+1} + d_0\Big\{-\frac{\ln x}{2}\sum_{j=0}^{\infty}\frac{(-1)^j}{2^{2j}j!(j+1)!}x^{2j+1} \\
&\qquad + x^{-1} - \frac{3}{64}x^3 + \frac{7}{2304}x^5 - \cdots\Big\}\,. \tag{61} \\
&= c_0 y_1(x) + d_0 y_2(x)\,.
\end{aligned}$$

As we mentioned, the Bessel function of the first kind of order one, $J_1(x)$, is obtained with the choice $c_0 = 1/2$, $d_0 = 0$. The *Bessel function of the second kind of order one*, $Y_1(x)$, results if $c_0 = (-2 \cdot \ln 2 - 1 + 2\gamma)/2\pi$, $d_0 = -2/\pi$, and $\gamma$ is *Euler's constant* $\approx .57721\ldots$ . ∎

Although these computations are laborious, be assured that they cannot fail to generate a second solution.[3] Often in engineering practice one can summarily dismiss any solution with a logarithmic singularity and thus save some effort (this will occur frequently in later chapters). Note that $d_2$ is never determined - there always remains the possibility of mixing the $y_1$ solution in with the $y_2$ - but all higher $d$'s can be computed in terms of $d_0$ and $d_2$ via eq. (59). Another instance of this phenomenon, with quite a different outcome, is illustrated in the next example.

**Example 4.** The *Bessel equation of order 1/2* is given by

$$x^2 y'' + xy' + (x^2 - \frac{1}{4})y = 0 \tag{62}$$

When the form (9), (10) is inserted, the resulting equation is

$$\sum_{m=0}^{\infty} (m+r)(m+r-1)c_m x^{m+r} + \sum_{m=0}^{\infty} (m+r)c_m x^{m+r}$$

$$+ \sum_{m=2}^{\infty} c_{m-2} x^{m+r} - \sum_{m=0}^{\infty} \frac{1}{4} c_m x^{m+r} = 0. \tag{63}$$

The indicial equation is

$$r^2 - \frac{1}{4} = 0, \quad \text{or} \quad r = \pm \frac{1}{2} \tag{64}$$

and since the roots differ by an integer the Frobenius series will again overlap:

$$y_1(x) = \sum_{n=0}^{\infty} c_n x^{n+1/2} = c_0 x^{1/2} + c_1 x^{3/2} + c_2 x^{5/2}$$

$$+ c_3 x^{7/2} + \cdots, \tag{65}$$

$$y_2(x) = \sum_{n=0}^{\infty} d_n x^{n-1/2} = d_0 x^{-1/2} + d_1 x^{1/2} + d_2 x^{3/2}$$

$$+ d_3 x^{5/2} + \cdots, \tag{66}$$

$$y_1 + y_2 = d_0 x^{-1/2} + \{c_0 + d_1\}x^{1/2} + \{c_1 + d_2\}x^{3/2}$$

$$+ \{c_2 + d_3\}x^{5/2} + \cdots. \tag{67}$$

In particular we see that the equation for $d_1$ will be troublesome.

---

[3]See the references by Coddington and Levinson, Birkhoff and Rota, or Ince.

If we take $r = 1/2$ then the recursion relations are

$$2c_1 = 0 \quad \text{or} \quad c_1 = 0,$$

$$(m^2 + m)c_m = m(m+1)c_m = -c_{m-2} \quad (m \geq 2), \tag{68}$$

and a little work reveals that this solution is given by

$$y_1(x) = \sum_{j=0}^{\infty} c_{2j} x^{2j+1/2} = c_0 \sum_{j=0}^{\infty} \frac{(-1)^j}{(2j+1)!} x^{2j+1/2} = c_0 \, x^{-1/2} \sin x. \tag{69}$$

(Recall Sec. 2.2, formula (*ii*).) In fact the *Bessel function of the first kind of order* 1/2, $J_{1/2}(x)$, results when $c_0$ is taken to be $\sqrt{2/\pi}$.

We seek the second solution for $x > 0$ in the form

$$y_2(x) = \sum_{n=0}^{\infty} d_n x^{n-1/2} + A \ln x \sum_{j=0}^{\infty} c_{2j} x^{2j+1/2} \tag{70}$$

The result of substituting (70) into (62) and reindexing is

$$\sum_{m=0}^{\infty} (m - 1/2)(m - 3/2) d_m x^{m-1/2} + \sum_{m=0}^{\infty} (m - 1/2) d_m x^{m-1/2}$$

$$+ \sum_{m=2}^{\infty} d_{m-2} x^{m-1/2} - (1/4) \sum_{m=0}^{\infty} d_m x^{m-1/2}$$

$$+ 2A \sum_{m=1}^{\infty} c_{m-1}(m - 1/2) x^{m-1/2} = 0. \tag{71}$$

The $m = 0$ term is zero because of the indicial equation. For $m = 1$ we find

$$(1/2)(-1/2)d_1 + (1/2)d_1 - (1/4)d_1 + 2Ac_0(1/2) = 0d_1 + Ac_0 = 0. \tag{72}$$

We cannot *solve* for $d_1$, but we can *satisfy* the equation with the choice $A = 0$. *Thus no log term is needed in this case.*

For $m \geq 2$ the recursion relation is

$$d_m = -\frac{1}{m(m-1)} d_{m-2} \tag{73}$$

and if we take $d_1 = 0$ the series looks like

$$y_2(x) = d_0 \sum_{j=0}^{\infty} \frac{(-1)^j}{(2j)!} x^{2j-1/2} = d_0 x^{-1/2} \cos x \tag{74}$$

(Sec. 2.2, formula (*iii*)). In fact, the *Bessel function of the first kind of order* $-1/2$, $J_{-1/2}(x)$, is $\sqrt{2/\pi x} \cos x$. ∎

**Example 5. .** In the previous section we studied the expansions of solutions to *Legendre's* differential equation

$$(1 - x^2)y'' - 2xy' - \lambda y = 0 \tag{75}$$

around the *ordinary* point $x = 0$. For future reference we now wish to expand around the *regular singular point* $x = -1$. It is always wise to shift the expansion center to the origin; thus we introduce

$$t = x + 1, \ x = t - 1; \quad y(x) = Y(t) = y(t - 1)$$

and by the identities (8, 9) of Sec. 1.2

$$\frac{dy}{dx} = \frac{dY}{dt}\frac{dt}{dx} = \frac{dY}{dt}, \quad \frac{d^2y}{dx^2} = \frac{d^2Y}{dt^2}$$

The transformed differential equation reads

$$(-t^2 + 2t)Y'' - 2(t - 1)Y' - \lambda Y = 0. \tag{76}$$

Omitting algebraic details we leave it for the reader to confirm that if the Frobenius form

$$Y = t^r \sum_{n=0}^{\infty} c_n t^n$$

is inserted into (76) and the result reindexed, one finds

$$-\sum_{n=0}^{\infty} c_n(n+r)(n+r-1)t^{n+r} + \sum_{n=1}^{\infty} 2c_{n+1}(n+1+r)(n+r)t^{n+r} -$$

$$\sum_{n=0}^{\infty} 2c_n(n+r)t^{n+r} + \sum_{n=-1}^{\infty} 2c_{n+1}(n+1+r)t^{n+r} - \sum_{n=0}^{\infty} \lambda c_n t^{n+r} = 0$$

The terms with $n = -1$ display the indicial equation:

$$2c_0 r(r - 1) + 2c_0 r = 2c_0 r^2 = 0,$$

so $r = 0$ is a double root. From this we immediately conclude that *the Legendre differential equation has a solution which is finite at $x = -1$:*

$$y_1(x) = \sum_{n=0}^{\infty} c_n(x+1)^n,$$

*and an independent solution diverging logarithmically at $x = -1$:*

$$y_2(x) = \sum_{n=0}^{\infty} d_n(x+1)^n + Ay_1(x)\ln|x+1|.$$

We go no further with this example except to note that the recurrence relation for the regular solution is given by

$$c_{n+1} = c_n \frac{n(n+1) + \lambda}{2(n+1)^2}.\tag{77}$$

∎

**Example 6.** The *modified Bessel equation of order $i\nu$* is given by

$$x^2 y'' + xy' - (x^2 - \nu^2)y = 0.$$

According to (6, 7, 18) its indicial equation is

$$r^2 + \nu^2 = 0,$$

which means that the series solutions begin with factors

$$x^{\pm i\nu} = e^{\pm i\nu \ln x} = \cos(\nu \ln x) \pm i \sin(\nu \ln x).$$

Recall that we encountered a similar situation in Example 11 of Sec. 1.1.

The introduction of complex numbers does not affect the procedure, however. It is a straightforward task to work out the calculations (exercise 13) and show that solutions are given by

$$y_1(x) = \{\cos(\nu \ln x) + i \sin(\nu \ln x)\} \sum_{n=0}^{\infty} c_n x^n$$

with

$$c_{n+2} = c_n / (n+2)(n+2+2i\gamma), \quad c_1 = 0,\tag{78}$$

and $y_2(x)$ equal to the complex conjugate of $y_1(x)$.

∎

# Exercises 2.4

**1.** Solve by Frobenius's method:

(a) $2xy'' + (1 - 2x)y' - y = 0$

(b) $3x^2 y'' - 2xy' - (2 + x^2)y = 0$

(c) $xy'' - (x+1)y' - y = 0$

(d) $4x^2 y'' + 4xy' - y = 0$

(e) $x^2 y'' - 3xy' + 4y = 0$

(f) $y'' - 2y' + (1 + \frac{1}{4x^2})y = 0$

(g) $xy'' + y' - xy = 0$

**2.** Derive the general form of the indicial equation (18).

**3.** What is the general nature of the solutions to the following equations around $x = 0$? (That is, for what $r$ is $y(x) = x^r + O(x^{r+1})$? Is there a log term?)

(a) $2x^2 y'' - xy' + (x - 5)y = 0$

(b) $x^2 y'' + (x^2 - 3x)y' + 3y = 0$

(c) $(x^2 - 1)y'' + xy' + xy = 0$

(d) $x^2 y'' + xy' - x^2 y = 0$

**4.** (a) Attempt to find two independent solutions using the Frobenius form (9) for the equation $x^2 y'' + y' = 0$, which has an irregular singular point at $x = 0$. What goes wrong?

(b) Verify directly that the equation

$$x^6 y'' - 4y' + 6x^2 y = 0 \,,$$

for which 0 is clearly an irregular singular point, has as one of its solutions $e^{-1/x^2}$. Recall (exercise 7, Sec. 2.2) that the Taylor series for this function converges, not to the function $e^{-1/x^2}$, but to zero. What happens when you try the Frobenius form for this equation?

**5.** Derive eqs. (21) and (23). (*Hint*: see exercises 2 through 4, Sec. 2.1.)

**6.** Verify directly that (33) and (36) solve eq. (27).

**7.** Derive eq. (46).

**8.** Show that Legendre's equation (75) has one solution which is continuous at $x = +1$, and another which diverges logarithmically there.

**9.** How are the computations in the examples of this section affected if $x$ (or $t$) is negative?

The next problem gives some theoretical background as to why the second solution takes the form (51) when the roots of the indicial equation differ by an integer. Recall from exercise 7 in Sec. 1.1 that when one solution $y_1(x)$ of the differential equation

$$y'' + p(x)y' + q(x)y = 0$$

is known, a second solution can be obtained from

$$y_2(x) = y_1(x) \int \frac{e^{-\int p(x)dx}}{y_1^2(x)} dx \,. \tag{79}$$

**10.** (a) Show that if $r_1$ and $r_2 = r_1 - s$ are the roots of the indicial equation (18), then

$$p_{-1} = 1 - r_1 - r_2 \qquad (80)$$

(b) From the display (6), conclude that

$$e^{-\int p(x)\,dx} = \frac{1}{x^{p_1}} g(x) \qquad (81)$$

where $g(x)$ is a function which is continuous and finite at $x = 0$.

(c) From the display

$$y_1(x) = x^{r_1} \sum_{n=0}^{\infty} c_n x^n \qquad (82)$$

conclude that

$$\frac{1}{y_1^2(x)} = \frac{1}{x^{2r_1}} h(x)$$

where $h(x)$ is a function which is also continuous and finite at $x = 0$.

(d) From (79) conclude that there is a series representation for $y_2(x)$ in the form

$$y_2(x) = y_1(x) \int \{ \frac{a_0 + a_1 x + a_2 x^2 + a_3 x^3 + \cdots}{x^{1+r_1-r_2}} \}\, dx \qquad (83)$$

with $a_0 \neq 0$.

(e) If $r_1 - r_2 = s$ is neither zero nor a positive integer show that term by term integration of (83), combined with (82), displays the second solution in the form

$$y_2(x) = x^{r_2} \sum_{n=0}^{\infty} d_n x^n \qquad (84)$$

(f) If $r_1 - r_2 = s$ *is* zero or a positive integer then show that term-by- term integration of (83), combined with (82), displays the second solution in the form

$$y_2(x) = a_{r_1-r_2}\, y_1(x) \ln x + x^{r_2} \sum_{n=0}^{\infty} d_n x^n \qquad (85)$$

in accordance with Frobenius's rule. Note in particular that if $r_1 = r_2$ then the coefficient of the log term, $a_0$, is guaranteed to be nonzero.

11. (*Point at Infinity*) In some cases it is desirable to have an expression for the solutions to a differential equation which is valid for *large* values of $x$. The procedure for accomplishing this is quite straightforward; first a change of variable, mapping $\infty$ to zero, is made in accordance with the theory of Sec. 1.2:

$$x = 1/t \quad (t = 1/x); \quad z(t) = y(1/t) \tag{86}$$

Then expansions are obtained, with the method of Frobenius if necessary, of the solutions of the transformed differential equation in powers of $t$.

(a) Show that the differential equation in the form (2) is transformed to the form

$$z'' + \{\frac{2}{t} - \frac{p(1/t)}{t^2}\}z' + \frac{q(1/t)}{t^4}z = 0. \tag{87}$$

(b) What are the conditions, analogous to (4, 5), on $p(x)$ and $q(x)$ in order that the point at $\infty$ be an ordinary point? (*Answer:*

$$\frac{p(1/t)}{t} = 2 + \sum_{n=1}^{\infty} \rho_n t^n, \quad \frac{q(1/t)}{t^4} = \sum_{n=0}^{\infty} \gamma\, t^n .) \tag{88}$$

(c) What are the conditions, analogous to (4, 5), on $p(x)$ and $q(x)$ in order that the point at $\infty$ be a regular singular point? (*Answer:*

$$\frac{p(1/t)}{t} = \sum_{n=0}^{\infty} \rho_n t^n, \quad \frac{q(1/t)}{t^2} = \sum_{n=0}^{\infty} \gamma_n t^n .) \tag{89}$$

(d) What is the nature of the point at infinity for Bessel's equation of order $\nu$? Legendre's equation?

(e) Work out a few terms in the expansion of solutions of Bessel's equation of order $\nu$ around the point at infinity.

12. Derive eqs. (77).

13. Derive eqs. (78).

## 2.5   Bessel Functions

**Summary**   The commonly tabulated solutions to Bessel's equation

$$x^2 y'' + xy' + (x^2 - \nu^2)y = 0$$

are

$J_\nu(x)$, the Bessel function of the first kind of order $\nu$, and

$Y_\nu(x)$, the Bessel function of the second kind of order $\nu$.
The commonly tabulated solutions of the modified Bessel equation

$$x^2 y'' + xy' - (x^2 + \nu^2)y = 0$$

are

$I_\nu(x)$, the modified Bessel function of the first kind of order $\nu$, and

$K_\nu(x)$, the modified Bessel function of the second kind of order $\nu$.

The modified equation results from changing variables from $x$ to $ix$ and $J_\nu(ix)$ is, up to a constant multiple, $I_\nu(x)$. A more elaborate change of variables renders Bessel's equation in the form

$$y'' - [(2a-1)/x]\,y' + \left[ b^2 c^2 x^{2c-2} + (a^2 - \nu^2 c^2)/x^2 \right] y = 0.$$

$J_\nu$ and $J_{-\nu}$ are independent solutions of Bessel's equation unless $\nu = n$, an integer; but $J_\nu$ and $Y_\nu$ are always independent. Similarly $I_\nu$ and $I_{-\nu}$ are independent solutions of the modified equation unless $\nu = n$, but $I_\nu$ and $K_\nu$ are always independent.

If $\nu$ is real, the only Bessel functions which are finite at $x = 0$ are $J_\nu(x)$ and $I_\nu(x)$, and these are only finite for $\nu \geq 0$. Furthermore if $\nu$ is real, $J_\nu$ and $Y_\nu$ are oscillatory and $I_\nu$ and $K_\nu$ are exponentially growing and decaying functions, respectively.

On several occasions we have referred to the Bessel differential equation and its solutions, the Bessel functions. In fact they will be recurring more frequently in the chapters to follow; they are very important in engineering applications. Of course they are less familiar than the *logs*, *sines*, and *cosh*'s that we feel comfortable with. In this section we shall cover only their most basic features so that the reader, while not expert, need not dread encountering them in practice.

Our starting point is the *Bessel differential equation of order $\nu$*:

$$x^2 y'' + xy' + (x^2 - \nu^2)y = 0$$

or

$$y'' + \frac{1}{x}y' + (1 - \frac{\nu^2}{x^2})y = 0. \tag{1}$$

The appearance of the constant parameter $\nu^2$ in the equation may seem a little awkward. Why, for instance, is it written as a *square* - why not simply "$\nu$", say? The reason is that in most engineering applications where Bessel's equation arises the variable $x$ stands for a radial (either cylindrical or spherical) coordinate, but the value of $\nu$ is directly determined by considerations involving the other coordinates. For problems with cylindrical symmetry $\nu$ is usually a nonnegative integer, and we frequently write "$n$" instead of $\nu$ in this case. For radial waves in three dimensions $\nu$ is usually a nonnegative half-integer, and the solutions are called "spherical Bessel functions." For "wedge-shaped" geometries, fractional values - and even imaginary values - of $\nu$ can occur. However only the square of $\nu$ affects the radial dependence.

For the moment we assume that $\nu^2$ is real and nonnegative, and that $\nu$ is its nonnegative square root ($\nu \geq 0$). (Situations involving negative values of $\nu^2$ will be studied in Sec. 6.3.)

Clearly $x = 0$ is a regular singular point of the Bessel equation. The indicial equation is

$$r^2 - \nu^2 = 0, \tag{2}$$

so the solutions take the Frobenius forms

$$y_1(x) = x^\nu \sum_{n=0}^{\infty} c_n x^n, \; y_2(x) = x^{-\nu} \sum_{n=0}^{\infty} d_n x^n \tag{3}$$

unless $\nu$ is an integer or a half-integer - in which case the roots of (2) differ by an integer (namely $2\nu$). Then, a log term *may* be needed in the second solution:

$$y_2(x) = x^{-\nu} \sum_{n=0}^{\infty} d_n x^n + A \ln|x| \, y_1(x). \tag{4}$$

(We shall presume that $x > 0$ in this section, so that the annoying absolute value symbol becomes unnecessary.) In the previous section we have seen that both solutions to the Bessel equation of order $1/3$ take the form (3) (as they must, since $2 \times 1/3$ is not an integer), and that the second solution to the equation of order 1 takes the form (4). For the Bessel equation of order $1/2$ the log term was unnecessary (despite the fact that $2 \times 1/2$ *is* an integer), and the form (3) sufficed for both solutions. We shall see shortly that the log term is only needed when $\nu$ is an integer, for the general Bessel equation:

*The Bessel equation of order $\nu$ has two independent solutions of the form (3) unless $\nu$ is a nonnegative integer, in which case the second solution must contain a log term of the form (4).*

The first solution to Bessel's equation is traditionally taken to be the *Bessel function of the first kind of order $\nu$,*

$$
\begin{aligned}
J_\nu(x) &= (x/2)^\nu \sum_{k=0}^\infty \frac{(-1)^k (x/2)^{2k}}{k!\Gamma(\nu+k+1)} \\
&= \frac{1}{2^\nu \Gamma(\nu+1)} x^\nu - \frac{1}{2^{\nu+2}\Gamma(\nu+2)} x^{\nu+2} + \frac{1}{2^{\nu+4}2!\Gamma(\nu+3)} x^{\nu+4} \\
&\quad + \cdots
\end{aligned}
\tag{5}
$$

If $\nu = n$, a nonnegative integer, these formulas become

$$
\begin{aligned}
J_n(x) &= (x/2)^n \sum_{k=0}^\infty \frac{(-1)^k (x/2)^{2k}}{k!(n+k)!} \\
&= \frac{1}{(2^n n!)} x^n - \frac{1}{2^{n+2}(n+1)!} x^{n+2} + \frac{1}{2^{n+4}2!(n+2)!} x^{n+4} \\
&\quad + \cdots
\end{aligned}
\tag{6}
$$

It is easy to show (exercise 7) that the coefficients defined above satisfy the recursion relations for the differential equation (1). In fact the computation reveals that the series for $J_\nu(x)$ formally satisfies (1) for *any* value of $\nu$, real or complex.

If $\nu$ *is* a negative integer, however, the infinite gamma functions alter the initial terms; thus the series for $J_{-n}(x)$ takes the form

$$
\begin{aligned}
J_{-n}(x) &= \frac{1}{2^{-n}\Gamma(-n+1)} x^{-n} - \frac{1}{2^{-n+2}\Gamma(-n+2)} x^{-n+2} + \\
&\quad \frac{1}{2^{-n+4}2!\Gamma(-n+3)} x^{-n+4} + \cdots \\
&= \frac{1}{2^{-n}\times\infty} x^{-n} - \frac{1}{2^{-n+2}\times\infty} x^{-n+2} + \frac{1}{2^{-n+4}\times\infty} x^{-n+4} \\
&\quad - \cdots
\end{aligned}
$$

which we interpret as

$$
J_{-n}(x) = 0 - 0 + 0 - \cdots + (x/2)^{-n} \sum_{k=n}^\infty \frac{(-1)^k (x/2)^{2k}}{k!\Gamma(-n+k+1)}.
$$

A little juggling reveals that $J_{-n}$ is just a disguised form of $J_n$ (exercise 5):

$$
J_{-n}(x) = (-1)^n J_n(x), \quad (n \geq 0).
\tag{7}
$$

Thus for $\nu = n$, $J_\nu(x)$ and $J_{-\nu}(x)$ are not independent solutions to (1), and the log form is dictated.

Retaining the dominant terms in the series (5) we see that except for the values $\nu = -1, -2, -3, \ldots$, the Bessel function of the first kind is approximated for small $x$ by

$$
J_\nu(x) = \frac{1}{2^\nu \Gamma(\nu+1)} x^\nu + \mathcal{O}(x^{\nu+2}) \approx \frac{1}{\nu!}(x/2)^\nu \quad (0 < x \ll 1).
\tag{8}
$$

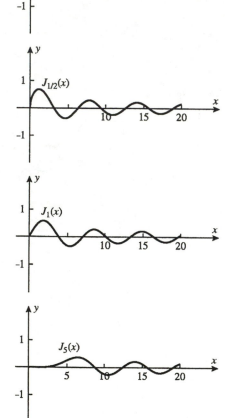

**Figure 1a**  *Bessel functions of the first kind*

Now observe:

1. $J_\nu(x)$ and $J_{-\nu}(x)$ both satisfy Bessel's equation;

2. for $\nu \neq 0, 1, 2, \ldots$, $J_\nu$ and $J_{-\nu}$ are obviously independent since (by (8)) one is finite at $x = 0$ and the other is infinite.[4]

We conclude

*If $\nu$ is not an integer, the general solution of Bessel's equation of order $\nu$ can be written as*

$$y(x) = c_1 J_\nu(x) + c_2 J_{-\nu}(x). \qquad (9)$$

When $\nu = n$ is an integer another solution, independent of $J_n$, must be used. Traditionally the second solution in this case is taken to be the *Bessel function of the second kind of order n* (sometimes known as *Weber's function*), whose series expansion is formidable (see exercise 2):

$$Y_n(x) = \frac{2}{\pi} \ln \frac{x}{2} J_n(x) - \frac{(x/2)^{-n}}{\pi} \sum_{k=0}^{n-1} \frac{(n-k-1)!(x/2)^{2k}}{k!}$$
$$- \frac{(x/2)^n}{\pi} \sum_{k=0}^{\infty} \{\Psi(k+1) + \Psi(n+k+1)\} \frac{(-1)^k (x/2)^{2k}}{k!(n+k)!} \quad (10)$$

where

$$\Psi(1) = -\gamma, \ \ \Psi(n) = -\gamma + \sum_{k=1}^{n-1} 1/k,$$

and $\gamma \approx .5772$ is known as "Euler's constant". Notice that the dominant term in $Y_n$ is $-(n-1)!(x/2)^{-n}/\pi$, for small $x$, unless $n = 0$ - whence the log term dominates:

$$Y_{n>0}(x) \approx -\frac{(n-1)!}{\pi}(x/2)^{-n} \quad (0 < x \ll 1),$$
$$Y_0(x) \approx \frac{2\ln x}{\pi} \quad (0 < x \ll 1). \qquad (11)$$

As Frobenius promised, the Bessel function of the second kind has a term $A \ln x J_n(x)$ in it. It can be shown that $Y_n(x)$ is the limit, as $\nu$ approaches $n$, of a certain combination of $J_\nu$ and $J_{-\nu}$:

$$Y_n(x) = \lim_{\nu \to n} \frac{J_\nu(x) \cos \nu\pi - J_{-\nu}(x)}{\sin \nu\pi} \quad (n = 0, 1, 2, \ldots). \qquad (12)$$

---

[4]The Wronskian can also be checked near zero, using (5).

If $\nu$ is *not* an integer the quotient on the right of (12) is still a solution of Bessel's equation, and it is independent of $J_\nu$; therefore, if we *define* $Y_\nu$ for $\nu \neq n$ to be this quotient,

$$Y_\nu(x) = \frac{J_\nu(x)\,\cos\,\nu\pi - J_{-\nu}(x)}{\sin\,\nu\pi} \quad (\nu \neq 0,1,2,\dots), \qquad (13)$$

we could use it as the second solution, instead of $J_{-\nu}$. This produces a convenient display of the *general* solution as

$$y(x) = c_1 J_\nu(x) + c_2 Y_\nu(x), \quad \text{any } \nu. \qquad (14)$$

*However $Y_\nu$ is commonly tabulated only for integer $\nu$.*

The forms (10), (12) for the second solution were chosen historically because of their limiting values for large $x$. It can be shown that the Bessel functions of the first kind behave like mildly damped sinusoids when $x$ is large:

$$J_\nu(x) = (2/\pi x)^{1/2} \cos(x - \nu\pi/2 - \pi/4), \quad x \gg 1. \qquad (15)$$

(See exercise 4). Thus it would be nice if the second solution approximated the "sin" form of (15) - and $Y_\nu(x)$ as defined by (10), (12) does just that:

$$Y_\nu(x) = (2/\pi x)^{1/2} \sin(x - \nu\pi/2 - \pi/4), \quad x \gg 1. \qquad (16)$$

Graphs of $J_\nu$ and $Y_\nu$ for some typical values of $\nu$ are depicted in Figures 1a, 1b, and 1c. Take note of the following features:

1. $J_\nu$ and $Y_\nu$ are both oscillatory; they have infinitely many zeros.

2. For small $x$, $J_\nu$ behaves like (constant)$\times x^\nu$ (except for $\nu = -1, -2, \dots$) and $Y_n$ behaves like (constant)$\times x^{-n}$ (except that $Y_0$ behaves like (constant)$\times \ln x$). For large $x$ both $J_\nu$ and $Y_\nu$ behave like slowly decaying sinusoids.

3. The only solutions to Bessel's equation which are finite at the origin are constant multiples of $J_\nu(x)$, for $\nu \geq 0$.

We shall see in Section 3.1 that it is often simpler to characterize oscillating systems by the complex functions $e^{ix}$ and $e^{-ix}$, rather than $\sin x$ and $\cos x$. Thus it should come as no surprise to learn that engineers dealing with radiating systems prefer to use the combinations $J_\nu(x) \pm iY_\nu(x)$. They are known historically as the *Hankel functions of the first and second kind*:

$$H_\nu^{(1)}(x) = J_\nu(x) + iY_\nu(x), \quad H_\nu^{(2)}(x) = J_\nu(x) - iY_\nu(x). \qquad (17)$$

Note that for large x

$$\begin{aligned}
H_\nu^{(1)}(x) &\approx (2/\pi x)^{1/2} e^{-i(\nu\pi/2 + \pi/4)} e^{ix}, \\
H_\nu^{(2)}(x) &\approx (2/\pi x)^{1/2} e^{i(\nu\pi/2 + \pi/4)} e^{-ix}.
\end{aligned}$$

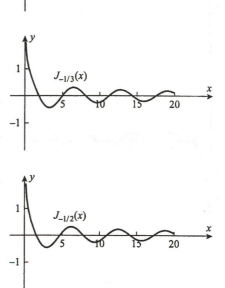

**Figure 1b**   *Bessel functions of the first kind*

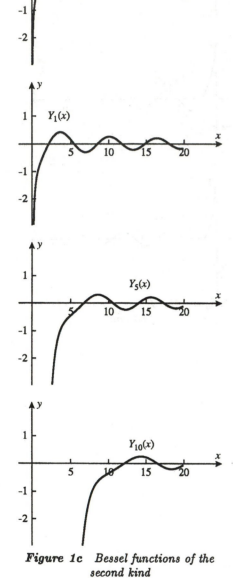

**Figure 1c**   *Bessel functions of the second kind*

The Hankel functions are, of course, complex.

We pointed out in Section 1.2 that if $\Xi_\nu(x)$ denotes *any* of the solutions to Bessel's equation, then the function

$$y(x) = x^a \Xi_\nu(bx^c) \tag{18}$$

satisfies the more general equation

$$y'' - [(2a-1)/x]\,y' + \left[b^2 c^2 x^{2c-2} + (a^2 - \nu^2 c^2)/x^2\right] y = 0. \tag{19}$$

An important special case of this is the *modified Bessel equation*

$$x^2 y'' + xy' - (x^2 + \nu^2)y = 0 \tag{20}$$

(let $a = 0$, $c = 1$, $b^2 = -1$, and multiply (19) by $x^2$). Taking (18) at face value, then, we deduce that $J_{\pm\nu}(ix)$ and $Y_\nu(ix)$ are solutions of (20), if they are well defined. As a matter of fact, it can be shown that the series (5), (10) converge for *all* values of $x$, real or complex, except for $x = 0$. (When $x = 0$ the convergence only works for $J_\nu$, $\nu \geq 0$, as we might expect from the graphs.)

Thus the *modified Bessel function of the first kind of order $\nu$*,

$$\begin{aligned}
I_\nu(x) &= (-i)^\nu J_\nu(ix) \\
&= (x/2)^\nu \sum_{k=0}^{\infty} \frac{(x/2)^{2k}}{k!\,\Gamma(\nu + k + 1)}, 
\end{aligned} \tag{21}$$

(which differs from $J_\nu(ix)$ only by a factor which renders it real) is traditionally taken to be the first solution of the modified Bessel equation. Unless $\nu$ is a positive integer or zero, the function $I_{-\nu}(x)$ is independent and, thus, serves as a second solution. When $\nu = n = 0, 1, 2, \ldots$, the second solution is traditionally taken to be the *modified Bessel function of the second kind of order $n$*:

$$K_\nu(x) = \frac{\pi}{2} i^{n+1} H_n^{(1)}(ix). \tag{22}$$

$K_n(x)$ is also real, and contains the typical log term.[5] Figure 2 suggests the approximations

$$\begin{aligned}
I_\nu(x) &\approx & (x/2)^\nu/\nu!, & \quad 0 < x \ll 1; \\
K_0(x) &\approx & -\ln(x), & \quad 0 < x \ll 1; \\
K_{n \neq 0}(x) &\approx & (n-1)!(x/2)^{-n}/2, & \quad 0 < x \ll 1; \\
I_\nu(x) &\approx & e^x/\sqrt{2\pi x}, & \quad x \gg 1 \\
K_n(x) &\approx & e^{-x}\sqrt{\pi/2x}, & \quad x \gg 1.
\end{aligned} \tag{23}$$

---

[5]Because of (21), it might seem to make more sense to take $K_n(x) = (-i)^n Y_n(ix)$, but this function turns out to be complex.

Note that $J_\nu(x)$ and $Y_n(x)$ are oscillatory, but $I_\nu(x)$ and $K_n(x)$ have no zeros. In this respect the former are like the cosine and sine, and the latter like the growing and decaying exponential.

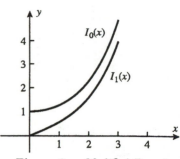

**Figure 2a**  *Modified Bessel functions of the first kind*

The Bessel functions satisfy many identities; they are similar to the trig functions in that respect. Table 2.3 at the end of this section lists most of the important properties, and the exercises guide you through proofs of some of them. (More extensive tables are easily available in the literature. The compilation by Abramowitz and Stegun is this author's favorite.) The letter "$z$" is used instead of "$x$" for the independent variable to emphasize that these functions are defined for complex arguments as well as real. (In the language of complex analysis, the Bessel functions are *analytic* in $x$. They are also analytic in the order parameter $\nu$.)

When Bessel functions are composed with other functions, the chain rule (Sec. 1.2) must be obeyed when applying these identities. This can be somewhat confusing, so the following detailed example is included as a model for your reference. It illustrates every subtlety that this author can imagine.

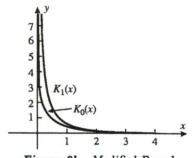

**Figure 2b**  *Modified Bessel functions of the second kind*

**Example 1.** Compute the value $y(.5)$ if $y$ satisfies

$$y'' + \frac{1}{x}y' - \left(\frac{4}{x^2} + \frac{1}{x^4}\right)y = 0, \quad y(1) = 1, \quad y'(1) = 0. \tag{24}$$

When we compare (24) to the generalized Bessel equation (19), we see that they agree if

$$a = 0, \quad c = -1, \quad b^2 = -1, \quad and \quad n^2 = 4.$$

Thus the equation is satisfied by Bessel functions of the form $x^0 J_{\pm 2}(\pm i/x)$. The "$i$" indicates that we should employ the modified Bessel function $I_{\pm 2}(\pm 1/x)$; since the Bessel functions are tabulated only for positive $x$, we choose $I_{\pm 2}(1/x)$; and since $I_2$ and $I_{-2}$ are not independent, we use $K_2$:

$$y(x) = c_1 I_2(1/x) + c_2 K_2(1/x).$$

The constants must be determined from the initial conditions. From Table 2.3,

$$
\begin{aligned}
1 &= y(1) = c_1 I_2(1/1) + c_2 K_2(1/1) \\
1 &\approx c_1 \times 1^{+2} \times .1357 + c_2 \times 1^{-2} \times 1.6248 = c_1 \times .1357 + \\
&\quad c_2 \times 1.6248
\end{aligned}
\tag{25}
$$

For the derivatives we must proceed carefully. By the chain rule

$$\frac{d}{dx}I_2(1/x) = I_2'(1/x) \times \frac{-1}{x^2}, \quad \frac{d}{dx}K_2(1/x) = K_2'(1/x) \times \frac{-1}{x^2}. \tag{26}$$

Also the table tells us

$$I_2'(x) = I_1(x) - \frac{2}{x} I_2(x);$$

thus

$$I_2'(1/x) = I_1(1/x) - \frac{2}{1/x} I_2(1/x) = I_1(1/x) - 2x I_2(1/x);$$

and at $x = 1$,

$$I_2'(1) = I_1(1) - 2 \times 1 \times I_2(1) \approx e^1 \times .2079 - 2 \times .1357 \approx .2937. \qquad (27)$$

For $K_2$ the recursion relation from Table 2.3 is (watch closely!)

$$(-1)^2 K_2'(x) = (-1)^1 K_1(x) - \frac{2}{x}(-1)^2 K_2(x);$$

thus

$$K_2'(1/x) = -K_1(1/x) - 2x K_2(1/x);$$

and at $x = 1$,

$$K_2'(1) \approx -e^{-1} \times 1.6362 - 2 \times 1.6248 \approx -3.8515. \qquad (28)$$

Combining (26, 27, 28) to meet the second initial condition we obtain

$$
\begin{aligned}
0 = y'(1) = \quad & c_1 \tfrac{d}{dx} I_2(1/x) + c_2 \tfrac{d}{dx} K_2(1/x) & \text{at } x = 1 \\
= \quad & c_1 I_2'(1/x)\tfrac{-1}{x^2} + c_2 K_2'(1/x)\tfrac{-1}{x^2} & \text{at } x = 1 \\
\approx \quad & c_1 \times .2937 \times (-1) + c_2 \times (-3.8515) \times (-1) \\
= \quad & -c_1 \times .2937 + c_2 \times 3.8515 & (29)
\end{aligned}
$$

The solution to (25, 29) is $c_1 \approx 3.8521$, $c_2 \approx .2936$. Thus our answer is

$$
\begin{aligned}
y(.5) &= c_1 I_2(1/.5) + c_2 K_2(1/.5) \\
&\approx 3.8521 \times 2^2 \times .1722 + .2936 \times 2^{-2} \times 1.0150 & (30) \\
&\approx 2.728. \qquad\blacksquare
\end{aligned}
$$

Conceptually this example is quite elementary; we merely fitted a general solution to a particular set of initial conditions. But we included it to alert the reader to the sometimes subtle workings of basic calculus.

*The most important facts for you to remember about Bessel functions are these:*

1. $J_\nu$ and $J_{-\nu}$ are independent solutions of Bessel's equation, unless $\nu = n$, an integer; $J_n$ and $Y_n$ are always independent. Similarly $I_\nu$ and $I_{-\nu}$ are independent solutions of the modified equation unless $\nu = n$; $I_n$ and $K_n$ are always independent.

2. If $\nu$ is real, the only Bessel functions which are finite at $x = 0$ are $J_\nu(x)$ and $I_\nu(x)$ - and *these* are only finite for $\nu \geq 0$.

3. If $\nu$ is real, $J_\nu$ and $Y_\nu$ are oscillatory; $I_\nu$ and $K_\nu$ are exponentially growing and decaying functions, respectively.

4. $J_\nu(ix)$ is, up to a constant multiple, $I_\nu(x)$.

# TABLE 2.3 Bessel Function Properties

Bessel function of the first kind of order $\nu$:

$$J_\nu(z) = (z/2)^\nu \sum_{k=0}^{\infty} \frac{(-1)^k (z/2)^{2k}}{k!\Gamma(\nu + k + 1)} \ ; \ 0 < |z| < \infty \ , \ \nu \neq -1, -2, -3, \ldots$$

Bessel function of the second kind of order $\nu$ ("Weber's function"):

$$Y_{\nu=n}(z) = \frac{2}{\pi} \ln \frac{z}{2} J_n(z) - \frac{(z/2)^{-n}}{\pi} \sum_{k=0}^{n-1} \frac{(n-k-1)!(z/2)^{2k}}{k!}$$

$$- \frac{(z/2)^n}{\pi} \sum_{k=0}^{\infty} \{\Psi(k+1) + \Psi(n+k+1)\} \frac{(-1)^k (z/2)^{2k}}{k!(n+k)!} \ ;$$

$$0 < |z| < \infty \ , \ n = 0, 1, 2, \ldots$$

$$\left[ \Psi(1) = -\gamma, \Psi(n) = -\gamma + \sum_{k=1}^{n-1} \frac{1}{k} \ , \ \gamma \approx .5772 \text{ (Euler's constant)} \right]$$

$$Y_\nu(z) = \frac{J_\nu(z)\cos\nu\pi - J_{-\nu}(z)}{\sin\nu\pi} \ , \ \nu \neq 0, 1, 2, \ldots \ ;$$

$$\left\{ Y_n(z) = \lim_{\nu \to n} Y_\nu(z) \right\}$$

Definitions of derived functions:

$$H_\nu^{(1)}(z) = J_\nu(z) + iY_\nu(z),$$
$$H_\nu^{(2)}(z) = J_\nu(z) - iY_\nu(z) \qquad \qquad \text{("Hankel functions")}$$

$$I_\nu(z) = (-i)^\nu J_\nu(iz),$$
$$K_\nu(z) = \frac{\pi}{2} i^{\nu+1} H_\nu^{(1)}(iz) \ , \ -\pi < arg(z) \leq \frac{\pi}{2}$$
$$\text{("Modified Bessel functions")}$$

$$Ai(z) = \frac{\sqrt{z}}{3} \left\{ I_{-1/3}(2z^{3/2}/3) - I_{1/3}(2z^{3/2}/3) \right\}$$
$$\text{("Airy function of the first kind")}$$

$$Bi(z) = \sqrt{z/3} \left\{ I_{-1/3}(2z^{3/2}/3) + I_{1/3}(2z^{3/2}/3) \right\}$$
$$\text{("Airy function of the second kind")}$$

$$j_\nu(z) = \sqrt{\pi/2z} J_{\nu+1/2}(z) \qquad \text{("Spherical Bessel function, first kind")}$$

$$y_\nu(z) = \sqrt{\pi/2z} Y_{\nu+1/2}(z) \qquad \text{("Spherical Bessel function, second kind")}$$

$$h_\nu^{(1,2)}(z) = \sqrt{\pi/2z} H_{\nu+1/2}^{(1,2)}(z) \qquad \text{("Spherical Bessel function, third kind")}$$

### Differential Equations

$$z^2 \Xi'' + z\Xi' + (z^2 - \nu^2)\Xi = 0 \; ; \; \Xi = J_\nu \text{ or } Y_\nu \text{ or } H_\nu^{(1)} \text{ or } H_\nu^{(2)}(z)$$

$$z^2 \Xi'' + z\Xi' - (z^2 + \nu^2)\Xi = 0 \; ; \; \Xi = I_\nu \text{ or } K_\nu(z)$$

$$y'' - [(2a-1)/x]\,y' + \left[b^2 c^2 x^{2c-2} + (a^2 - \nu^2 c^2)/x^2\right] y = 0 \; ;$$
$$y = x^a \, \Xi(bx^c) \; , \;\; \Xi = J \text{ or } Y \text{ or } H^{(1)} \text{ or } H^{(2)}$$

### Approximations

For small $z$:

$$J_\nu(z) \approx (z/2)^\nu/\nu! \; , \;\; Y_0(z) \approx (2\ln z)/\pi \; , \;\; Y_{\nu>0}(z) \approx -(\nu-1)!(z/2)^{-\nu}/\pi$$

$$I_\nu(z) \approx (z/2)^\nu/\nu! \; , \;\; K_0(z) \approx -\ln z \; , \;\;\;\;\;\;\;\; K_\nu(z) \approx (\nu-1)!(z/2)^{-\nu}/2$$

For large $z$:

$$J_\nu(z) = (2/\pi z)^{\frac{1}{2}} \cos(z - \nu\pi/2 - \pi/4) + \mathcal{O}(e^{Im(z)}/\,|z|)$$

$$Y_\nu(z) = (2/\pi z)^{\frac{1}{2}} \sin(z - \nu\pi/2 - \pi/4) + \mathcal{O}(e^{Im(z)}/\,|z|)$$

$$I_\nu(z) \approx e^z/(2\pi z)^{\frac{1}{2}} \; , \; K_\nu(z) \approx (\pi/2z)^{\frac{1}{2}} e^{-z} \; , \; |arg(z)| < \pi/2$$

### Symmetries

$$J_\nu(z) = z^\nu \times \text{Even}(z) \; , \;\; Y_\nu(z) = z^\nu \times \text{Even}(z) + (2/\pi)\ln(z/2)J_\nu(z)$$

$$I_\nu(z) = z^\nu \times \text{Even}(z) \; , \;\; K_\nu(z) = z^\nu \times \text{Even}(z) + (-1)^{\nu+1}\ln(z/2)I_\nu(z)$$

$$J_{-n}(z) = (-1)^n J_n(z) \; , \;\; Y_{-n}(z) = (-1)^n Y_n(z) \; ,$$

$$I_{-n}(z) = I_n(z) \; , \;\;\;\;\;\;\;\; K_{-n}(z) = K_n(z) \; ; \;\;\; n = 1, 2, \ldots$$

### Wronskians

$$J_\nu(z)J'_{-\nu}(z) - J'_\nu(z)J_{-\nu}(z) = -(2/\pi z)\sin\nu\pi$$

$$J_\nu(z)Y'_\nu(z) - J'_\nu(z)Y_\nu(z) = 2/\pi z$$

$$I_\nu(z)I'_{-\nu}(z) - I'_\nu(z)I_{-\nu}(z) = -(2/\pi z)\sin\nu\pi$$

$$I_\nu(z)K'_\nu(z) - I'_\nu(z)K_\nu(z) = -1/z$$

*Various Identities*

$$J_0' = -J_1 \ , \ Y_0' = -Y_1 \ , \ I_0' = I_1 \ , \ K_0' = -K_1$$

For $\Xi_\nu(z)$ = any linear combination of $J_\nu(z)$, $Y_\nu(z)$, $H_\nu^{(1)}(z)$, and $H_\nu^{(2)}(z)$ with coefficients independent of $z$ and $\nu$:

$$\Xi_{\nu-1}(z) + \Xi_{\nu+1}(z) = (2\nu/z)\,\Xi_\nu(z) \ , \ \ \Xi_{\nu-1}(z) - \Xi_{\nu+1}(z) = 2\,\Xi_\nu'(z)$$

$$\Xi_\nu'(z) = \Xi_{\nu-1}(z) - (\nu/z)\,\Xi_\nu(z) = -\Xi_{\nu+1}(z) + (\nu/z)\,\Xi_\nu(z)$$

$$\frac{d}{dz}\left[z^\nu\,\Xi_\nu(z)\right] = z^\nu\,\Xi_{\nu-1}(z) \ , \qquad \frac{d}{dz}\left[z^{-\nu}\,\Xi_\nu(z)\right] = -z^{-\nu}\,\Xi_{\nu+1}$$

For $\Xi_\nu(z)$ = any linear combination of $I_\nu(z)$ and $(-1)^\nu K_\nu(z)$ with coefficients independent of $z$ and $\nu$:

$$\Xi_{\nu-1}(z) + \Xi_{\nu+1}(z) = 2\,\Xi_\nu'(z) \ , \qquad \Xi_{\nu-1}(z) - \Xi_{\nu+1}(z) = (2\nu/z)\,\Xi_\nu(z)$$

$$\Xi_\nu'(z) = \Xi_{\nu-1}(z) - (\nu/z)\,\Xi_\nu(z) = \Xi_{\nu+1}(z) + (\nu/z)\,\Xi_\nu(z)$$

$$\frac{d}{dz}\left[z^\nu I_\nu(z)\right] = z^\nu I_{\nu-1}(z) \ , \qquad \frac{d}{dz}\left[z^{-\nu} I_\nu(z)\right] = z^{-\nu} I_{\nu+1}$$

$$\frac{d}{dz}\left[z^\nu K_\nu(z)\right] = -z^\nu K_{\nu-1}(z) \ , \qquad \frac{d}{dz}\left[z^{-\nu} K_\nu(z)\right] = -z^{-\nu} K_{\nu+1}$$

$$\frac{\partial}{\partial \nu} J_\nu(z) = J_\nu(z)\ln(z/2) - (z/2)^\nu \sum_{k=0}^{\infty} \frac{(-z^2/4)^k}{k!} \frac{\Gamma'(\nu+k+1)}{[\Gamma(\nu+k+1)]^2}$$

$$\frac{\partial}{\partial \nu} I_\nu(z) = I_\nu(z)\ln(z/2) - (z/2)^\nu \sum_{k=0}^{\infty} \frac{(z^2/4)^k}{k!} \frac{\Gamma'(\nu+k+1)}{[\Gamma(\nu+k+1)]^2}$$

For $\nu$ noninteger,

$$\frac{\partial}{\partial \nu} Y_\nu(z) = \cot(\nu\pi)\left\{\frac{\partial}{\partial \nu} J_\nu(z) - \pi Y_\nu(z)\right\} - \csc(\nu\pi)\left\{\frac{\partial}{\partial \nu} J_{-\nu}(z) - \pi J_\nu(z)\right\}$$

$$\frac{\partial}{\partial \nu} K_\nu(z) = -\pi\cot(\nu\pi)K_\nu(z) + \frac{\pi}{2}\csc(\nu\pi)\left\{\frac{\partial}{\partial \nu} I_{-\nu}(z) - \frac{\partial}{\partial \nu} I_\nu(z)\right\}$$

For $p$ = integer,

$$\frac{\partial}{\partial p} Y_p(z) = -\frac{\pi}{2} J_p(z) + \frac{p!(2/z)^p}{2} \sum_{k=0}^{p-1} \frac{(z/2)^k}{(p-k)k!} Y_k(z)$$

$$\frac{\partial}{\partial p} K_p(z) = \frac{p!(2/z)^p}{2} \sum_{k=0}^{p-1} \frac{(z/2)^k}{(p-k)k!} K_k(z)$$

### Integral Representations

$$J_0(z) = \int_0^\pi \cos(z \sin\theta) \, d\theta/\pi = \int_0^\pi \cos(z \cos\theta) \, d\theta/\pi$$

$$J_n(z) = \int_0^\pi \cos(z \sin\theta - n\theta) \, d\theta/\pi = \int_0^\pi e^{iz\cos\theta} \cos n\theta \, d\theta/\pi i^n$$

$$I_0(z) = \int_0^\pi \cosh(z \cos\theta) \, d\theta/\pi \;,\; I_n(z) = \int_0^\pi e^{z\cos\theta} \cos n\theta \, d\theta/\pi$$

### Normalization Integrals

For $\Xi_\nu(z), \Upsilon_\nu(z) = $ any linear combination of $J_\nu(z), Y_\nu(z), H_\nu^{(1)}(z)$, and $H_\nu^{(2)}(z)$ with coefficients independent of $z$ and $\nu$:

$$\int \Xi_\nu(\xi t) \, \Xi_\nu^*(\xi t) \, t \, dt = \tfrac{t^2}{4} \{ 2 \, \Xi_\nu(\xi t) \, \Xi_\nu^*(\xi t) - \Xi_{\nu-1}(\xi t) \, \Xi_{\nu+1}^*(\xi t)$$
$$- \Xi_{\nu+1}(\xi t) \, \Xi_{\nu-1}^*(\xi t) \}$$

$$\int \Xi_\nu^2(\xi t) \tfrac{dt}{t} = \tfrac{1}{2\nu} \Xi_\nu^2(\xi t) - \tfrac{\xi t}{2\nu} \left\{ \Xi_\nu(\xi t) \tfrac{\partial}{\partial\nu} \Xi_{\nu+1}(\xi t) - \Xi_{\nu+1}(\xi t) \tfrac{\partial}{\partial\nu} \Xi_\nu(\xi t) \right\}$$

$$\int \left\{ (a^2 - b^2)t - \tfrac{c^2-d^2}{t} \right\} \Xi_c(at) \Upsilon_d(bt) \, dt =$$
$$\{ a \, \Xi_{c+1}(at) \Upsilon_d(bt) - b \, \Xi_c(at) \Upsilon_{d+1}(bt) \} \, t - (c - d) \, \Xi_c(at) \Upsilon_d(bt)$$

$$\int_0^1 J_\nu^2(\mu t) \, t \, dt = J_{\nu+1}^2(\mu)/2 \qquad\qquad \text{if } J_\nu(\mu) = 0$$

$$\int_0^1 J_\nu^2(\mu t) \, t \, dt = J_\nu^2(\mu)(\mu^2 - \nu^2)/2\mu^2 \qquad \text{if } J_\nu'(\mu) = 0$$

$$\int_0^1 J_\nu^2(\mu t) \, t \, dt = J_\nu^2(\mu) \left\{ \tfrac{a^2}{b^2} + \mu^2 - \nu^2 \right\} /2\mu^2 \;\; \text{if } a J_\nu(\mu) + b\mu J_\nu'(\mu) = 0$$

### Generating Function

$$e^{(t-1/t)z/2} = \sum_{n=-\infty}^\infty J_n(z) t^n$$

### Addition Theorem

$$J_n(x + y) = \sum_{k=-\infty}^\infty J_{n-k}(x) J_k(y)$$

*Short Table of Values*

| $\backslash x$ | 0 | .5 | 1 | 1.5 | 2 | 3 | 4 |
|---|---|---|---|---|---|---|---|
| $J_0$ | 1 | .9385 | .7652 | .5118 | .2239 | -.2601 | -.3971 |
| $J_1$ | 0 | .2423 | .4401 | .5579 | .5767 | .3391 | -.0660 |
| $J_2$ | 0 | .0306 | .1149 | .2321 | .3528 | .4861 | .3641 |
| $e^{-x}I_0$ | 1 | .6450 | .4658 | .3674 | .3085 | .2430 | .2070 |
| $e^{-x}I_1$ | 0 | .1564 | .2079 | .2190 | .2153 | .1968 | .1788 |
| $x^{-2}I_2$ | $(\to .125)$ | .1276 | .1357 | .1501 | .1722 | .2495 | .4014 |
| $Y_0$ | $\infty$ | -.4445 | .0883 | .3824 | .5104 | .3769 | -.0169 |
| $Y_1$ | $\infty$ | -1.4715 | -.7812 | -.4123 | -.1070 | .3247 | .3979 |
| $Y_2$ | $\infty$ | -5.4414 | -1.6507 | -.9322 | -.6174 | -.1604 | .2159 |
| $e^{x}K_0$ | $\infty$ | 1.5241 | 1.1445 | .9582 | .8416 | .6978 | .6093 |
| $e^{x}K_1$ | $\infty$ | 2.7310 | 1.6362 | 1.2432 | 1.0335 | .8066 | .6816 |
| $x^{+2}K_2$ | $(\to 2)$ | 1.8875 | 1.6248 | 1.3132 | 1.0150 | .5536 | .2784 |

*Zeros of Bessel Functions*

Roots $j_{n,p}$ of $J_n(j_{n,p}) = 0$

| $\backslash p$ $n$ | 1 | 2 | 3 | 4 | 5 | 6 | 7 |
|---|---|---|---|---|---|---|---|
| 0 | 2.405 | 5.520 | 8.654 | 11.792 | 14.931 | 18.071 | 21.212 |
| 1 | 3.832 | 7.016 | 10.173 | 13.324 | 16.471 | 19.616 | 22.760 |
| 2 | 5.136 | 8.417 | 11.620 | 14.796 | 17.960 | 21.117 | 24.270 |
| 3 | 6.380 | 9.761 | 13.015 | 16.223 | 19.409 | 22.583 | 25.748 |
| 4 | 7.588 | 11.065 | 14.373 | 17.616 | 20.827 | 24.019 | 27.199 |

Roots $j'_{n,p}$ of $J'_n(j'_{n,p}) = 0$

| $\backslash p$ $n$ | 1 | 2 | 3 | 4 | 5 | 6 | 7 |
|---|---|---|---|---|---|---|---|
| 0 | 3.832 | 7.016 | 10.173 | 13.324 | 16.471 | 19.616 | 22.760 |
| 1 | 1.841 | 5.331 | 8.536 | 11.706 | 14.864 | 18.016 | 21.164 |
| 2 | 3.054 | 6.706 | 9.969 | 13.170 | 16.348 | 19.513 | 22.672 |
| 3 | 4.201 | 8.015 | 11.346 | 14.586 | 17.789 | 20.972 | 24.145 |
| 4 | 5.318 | 9.282 | 12.682 | 15.964 | 19.196 | 22.401 | 25.590 |

Roots of $J_n(v_{n,p}b)Y_n(v_{n,p}a) - J_n(v_{n,p}a)Y_n(v_{n,p}b)$
(The values of $[v_{n,p}a]$ are tabulated.)

For $b/a = 1.5$ :

| $\backslash p$ $n$ | 1 | 2 | 3 | 4 |
|---|---|---|---|---|
| 0 | 6.27 | 12.56 | 18.85 | 25.13 |
| 1 | 6.32 | 12.59 | 18.86 | 25.14 |
| 2 | 6.47 | 12.66 | 18.92 | 25.18 |

For $b/a = 2.0$:

| $\backslash p$ | 1 | 2 | 3 | 4 |
|---|---|---|---|---|
| $n$ | | | | |
| 0 | 3.12 | 6.27 | 9.42 | 12.56 |
| 1 | 3.20 | 6.31 | 9.44 | 12.58 |
| 2 | 3.40 | 6.43 | 9.52 | 12.64 |

For $b/a = 3.0$:

| $\backslash p$ | 1 | 2 | 3 | 4 |
|---|---|---|---|---|
| $n$ | | | | |
| 0 | 1.55 | 3.13 | 4.70 | 6.28 |
| 1 | 1.64 | 3.18 | 4.74 | 6.30 |

Roots of $J_n'(v_{n,p}b)Y_n'(v_{n,p}a) - J_n'(v_{n,p}a)Y_n'(v_{n,p}b)$
(The values of $[v_{n,p}a]$ are tabulated.)

For $b/a = 1.5$:

| $\backslash p$ | 1 | 2 | 3 |
|---|---|---|---|
| $n$ | | | |
| 0 | 0 | 6.32 | 12.59 |
| 1 | .81 | 6.38 | 12.61 |
| 2 | 1.61 | 6.54 | 12.69 |

For $b/a = 2.0$:

| $\backslash p$ | 1 | 2 | 3 |
|---|---|---|---|
| $n$ | | | |
| 0 | 0 | 3.20 | 6.31 |
| 1 | .68 | 3.28 | 6.35 |
| 2 | 1.34 | 3.50 | 6.47 |

For $b/a = 3.0$:

| $\backslash p$ | 1 | 2 | 3 |
|---|---|---|---|
| $n$ | | | |
| 0 | 0 | 1.64 | 3.18 |
| 1 | .51 | 1.76 | 3.24 |

(See Marcuvitz, Dwight, or Abramowitz and Stegun for more extensive tabulations.)

# Exercises 2.5

1. With the book closed sketch the graphs of $J_4(x)$, $Y_4(x)$, $I_4(x)$, $K_4(x)$, $J_{1/2}(x)$, and $J_{-1/2}(x)$ for $x > 0$.

2. From the discussion in the previous section, we would expect the second solution to the Bessel equation of integer order $n$ to have the following properties:

   (i) there should be a term (constant) $\times$ $(\ln x) \times J_n(x)$;

(ii) there should be a series starting with the terms $x^{-n}$, whose co-efficient pattern is interrupted at the $x^n$-term;

(iii) there may be an arbitrary multiple of $J_n(x)$ itself added in.

Show that these properties are realized by eq. (10).

3. (a) Show that the transformation $Z(x) = \sqrt{x}y(x)$ renders Bessel's equation in the form

$$Z'' + \left(1 + \frac{1 - 4\nu^2}{4x^2}\right) Z = 0 . \qquad (31)$$

(b) From part (a) show that all Bessel functions of order $\pm 1/2$ have the form

$$\{A \cos x + B \sin x\} / \sqrt{x} .$$

(c) Show from the series representation that

$$J_{1/2}(x) = \sqrt{2/\pi x} \sin x \text{ and } J_{-1/2}(x) = \sqrt{2/\pi x} \cos x . \qquad (32)$$

(d) Use the recursion relations in Table 2.3 to prove that all Bessel functions of half-integer order can be expressed in terms of elementary functions.

(e) Work out the formula for $J_{3/2}(x)$ from (32) and the recursion relations. [*Answer:* $J_{3/2}(x) = \sqrt{\frac{2}{\pi x}} \left(\frac{\sin x}{x} - \cos x\right)$ .]

4. For large $x$, equation (31) becomes approximately $Z'' + Z = 0$. How does this demonstrate the plausibility of the approximation formulas (15), (16)?

5. Derive eq. (7).

6. Derive the formulas

$$\frac{d}{dx} \left[x^\nu J_\nu(x)\right] = x^\nu J_{\nu-1}(x) , \quad \frac{d}{dx} \left[x^{-\nu} J_\nu(x)\right] = -x^{-\nu} J_{\nu+1}(x)$$

for $J_\nu$ by manipulating the power series expression (5). Use them to derive the recursion formulas

$$J_\nu'(x) = J_{\nu-1}(x) - (\nu/x)J_\nu(x) = -J_{\nu+1}(x) + (\nu/x)J_\nu(x)$$

$$J_{\nu-1}(x) + J_{\nu+1}(x) = (2\nu/x)J_\nu(x)$$

$$J_{\nu-1}(x) - J_{\nu+1}(x) = 2J_\nu'(z) .$$

7. Show that the coefficients in eq. (5) satisfy the recursion relations for the Bessel equation.

8. Express the general solution to Airy's equation $y'' - xy = 0$ in terms of Bessel functions. (Recall exercise 2, Sec. 2.3 and Example 8, Sec. 1.2.)

9. Find the value $y(2)$ if y satisfies the differential equation and initial conditions

$$y'' + \left(4 - \frac{3}{4x^2}\right) y = 0 \; ; \; y(1) = 1, \; y'(1) = 0 \; . \; (Answer : \; -.2983)$$

10. Find $y(2)$ if $xy'' + 3y' + xy = 0$, $y(1) = 0$, and $y'(1) = 1$.

11. Find $y(2)$ if $x^2y'' + 5xy' - \frac{16}{x^4}y = 0$ , $y(1) = 0$, $y'(1) = 1$.

## 2.6   Famous Differential Equations

**Summary**   The ordinary differential equations which arise most commonly in applications are, in descending order of importance,

- constant coefficient equations;
- equidimensional equations;
- Bessel's equation in various forms; and
- Legendre's equation.

Other equations can often be identified by considering their singular points.

Although you have now mastered the power-series method for solving differential equations, you may be relieved to know that the chances are very good that you will never have to use this technique in your professional career! The reason is that so much mathematical activity was focused on solving differential equations in the last century (before computers) that practically every equation that arises in engineering situations has already been studied. (Exhaustively!) In this section we give you a guide to the "famous" differential equations that have been analyzed. When you encounter an unfamiliar equation form, hopefully you can classify it as one of these classical types and, if so, go directly to the literature - and avoid "reinventing the wheel." Some material from previous sections will be repeated in this guide.

Unless we mention otherwise, the reference for further details about the equations below is the compendium by Abramowitz and Stegun.

For the sake of exposition we proceed roughly in order of complexity. Of course the simplest type of differential equation has *constant coefficients*.

$$y'' + \lambda y = 0 \tag{1}$$

describes the motion of a mass-spring system and is called the *harmonic oscillator* equation. Its solutions are typically $\cos\sqrt{\lambda}x$, $\sin\sqrt{\lambda}x$, $\exp(\pm\sqrt{-\lambda}x)$, $\cosh\sqrt{-\lambda}x$, and $\sinh\sqrt{-\lambda}x$ (but if $\lambda = 0$ the solution is $c_1 + c_2x$). The *damped* harmonic oscillator equation includes the effect of viscous friction forces:

$$y'' + \nu y' + \lambda y = 0. \tag{2}$$

Its solutions are typically exponentials or damped ($\nu > 0$) sinusoids.

Probably the simplest-looking second order equation with variable coefficients is the *Airy* equation:

$$y'' - xy = 0. \tag{3}$$

In fact (3) (as well $y'' + xy = 0$) is a special case of the generalized Bessel equation (below), so its solutions, the *Airy functions*, can be expressed in terms of Bessel functions:

$$A_i(x) = \frac{1}{3}\sqrt{x}\left[I_{-1/3}\left(\frac{2}{3}x^{3/2}\right) - I_{1/3}\left(\frac{2}{3}x^{3/2}\right)\right]$$

$$B_i(x) = \sqrt{\frac{x}{3}}\left[I_{-1/3}\left(\frac{2}{3}x^{3/2}\right) + I_{1/3}\left(\frac{2}{3}x^{3/2}\right)\right]. \tag{4}$$

The Airy equation arises in the description of caustics in optical systems, harmonic oscillators with aging springs, and quantum mechanical particles in a constant force field.

The easiest variable-coefficient equation to solve is known by three names - the *equidimensional, Euler-Cauchy,* or *homogeneous* equation:

$$y'' + \frac{a}{x}y' + \frac{b}{x^2}y = 0. \tag{5}$$

The first name derives from the observation that $y''$, $y'/x$, and $y/x^2$ all have the same physical dimensions. Typically the solutions are of the form $y = x^r$. The substitution $x = e^t$ reduces (5) to the form (2).

The *Bessel* equation is

$$x^2 y'' + xy' + (x^2 - \nu^2)y = 0. \tag{6}$$

It has a regular singular point at $x = 0$. Its solutions are *Bessel functions* $[J_\nu(x), Y_\nu(x)]$ *of the first and second kind of order $\nu$* and *Hankel functions* $[H_\nu^{(1)}(x), H_\nu^{(2)}(x)]$ *of the first and second kind of order $\nu$*. The substitution

$$z(x) = x^a y(bx^c)$$

produces a function that satisfies a generalized form of the Bessel equation

$$x^2 z'' - (2a - 1)xz' + [b^2 c^2 x^{2c} + a^2 - \nu^2 c^2]z = 0 \,; \qquad (7)$$

and more generally

$$u = x^{(1-a)/2} \exp(-bx^r/r) \, \Xi_p(\sqrt{d}x^s/s) \,,$$

where $\Xi_p$ is any of the Bessel functions of order $p$, produces functions satisfying

$$x^2 u'' + x(a + 2bx^r)u' + [c + dx^{2s} - b(1 - a - r)x^r + b^2 x^{2r}]u = 0 \,, \qquad (8)$$

when

$$p = \frac{1}{s}\sqrt{\left(\frac{1-a}{2}\right)^2 - c} \,.$$

In particular, the *modified Bessel functions of the first and second kind,*

$$I_\nu(x) = (-i)^\nu J_\nu(ix) \,, \quad K_\nu(x) = (\pi i/2)i^\nu H_\nu^{(1)}(ix) \,,$$

satisfy the *modified Bessel* equation:

$$x^2 y'' + xy' - (x^2 + \nu^2)y = 0 \,. \qquad (9)$$

The *associated Legendre* equation

$$(1 - x^2)y'' - 2xy' - \{\lambda + \frac{m^2}{1 - x^2}\}y = 0 \qquad (10)$$

sometimes appears in the alternative form

$$\sin^2 \phi \, \Phi'' + \sin \phi \, \cos \phi \, \Phi' - \{\lambda \sin^2 \phi + m^2\}\Phi = 0 \,; \qquad (11)$$

the two are equivalent if $x = \cos \phi$. Equation (10) has regular singular points at $\pm 1$, and its solutions are known as *associated Legendre functions,* usually denoted $P_\ell^m(x)$, $Q_\ell^m(x)$, where $\ell$ satisfies $\lambda = -\ell(\ell + 1)$. If $m = 0$ the *Legendre* equation

$$(1 - x^2)y'' - 2xy' - \lambda y = 0 \qquad (12)$$

has polynomial solutions $P_\ell(x)$ - the *Legendre polynomials* - when $\ell$ is a nonnegative integer. We shall see that the Legendre functions arise in the study of solutions to Laplace's equations in spherical coordinates (Sec. 6.4).

The *associated Laguerre* equation

$$xy'' + (k + 1 - x)y' - \mu y = 0 \qquad (13)$$

arises when we study the quantum mechanical hydrogen atom. It has a regular singular point at the origin, and when $k = 0$ (13) becomes the *Laguerre* equation. Its solutions, and the *Laguerre polynomials* in particular, are discussed in Section 7.4.

The *Weber* differential equation is

$$y'' - x^2 y - \lambda y = 0. \tag{14}$$

It occurs in the quantum mechanical description of the harmonic oscillator. The change of variables

$$y(x) = e^{-x^2/2} z(x), \quad \mu = \lambda + 1$$

transforms Weber's equation into *Hermite's* equation

$$z'' - 2xz' - \mu z = 0. \tag{15}$$

The most important solutions are the *Hermite* polynomials, and they are discussed in Section 7.4.

The *Chebyshev* differential equation

$$(1 - x^2)y'' - xy' + p^2 y = 0 \tag{16}$$

has few direct engineering applications, but it is noteworthy because its solutions include the *Chebyshev polynomials*, which are of the utmost importance in numerical approximation.

*Mathieu's* equation

$$y'' + (a + b \cos x)y = 0 \tag{17}$$

arises in some problems with elliptical geometries.

Several of the mathematicians of the last century attempted to concoct very general differential equations whose solutions would include, as special cases, many of the special functions occurring in applied mathematics. Gauss came up with the *hypergeometric* equation

$$x(1 - x)y'' + [c - (a + b + 1)x]y' - aby = 0. \tag{18}$$

It has regular singular points at 0, 1, and $\infty$ (recall exercise 11, Sec. 2.4). When $c$ is not an integer the general solution can be expressed

$$y(x) = c_1 F(a, b, c, x) + c_2 x^{1-c} F(a - c + 1, b - c + 1, 2 - c, x) \tag{19}$$

where $F$ is the *hypergeometric function*. The solutions of Gauss's equation include the log, Legendre, Chebyshev, Gegenbauer, incomplete beta, and complete elliptic functions. An interesting discussion of Gauss's equation appears in Simmons's text.

A change of variables in the hypergeometric equation renders it in the form

$$(1 - x^2)y'' + [b - a - (a + b + 2)x]y' + n(n + a + b + 1)y = 0, \qquad (20)$$

which is *Jacobi's* equation. Its regular singular points are at $\pm 1$ and $\infty$. If $a = b$ in Jacobi's equation, *Gegenbauer's ultraspherical* equation results; a useful form is

$$[(1 - x^2)^{a+1}y']' + n(n + 2a + 1)(1 - x^2)^a y = 0. \qquad (21)$$

It arises in the solutions to Laplace's equation in spherical coordinates *in a space of $2a + 3$ dimensions* (!). If $a = 0$ Legendre's equation results. If $a = -1/2$ the solutions are Chebyshev functions.

Solutions to the equation

$$[(1 - x^2)y']' + [A + B(1 - x^2) + C/(1 - x^2)]y = 0$$

are known as *spheroidal functions*. Special cases include the Legendre, Mathieu, and spherical Bessel functions; see the text of Jones.

The most general equation with regular singular points at 0, 1, and $\infty$ is *Riemann's* equation:

$$y'' + \left(\frac{1 - a_1 - a_2}{x} + \frac{1 - b_1 - b_2}{x - 1}\right)y' + \left[\frac{a_1 a_2}{x^2} + \frac{b_1 b_2}{(x - 1)^2}\right.$$
$$\left. + \frac{c_1 c_2 - a_1 a_2 - b_1 b_2}{x(x - 1)}\right]y = 0. \qquad (22)$$

The *confluent hypergeometric* equation is obtained mathematically from the hypergeometric equation by letting the regular singular point at 1 merge with that at $\infty$:

$$xy'' + (c - x)y' - ay = 0. \qquad (23)$$

It has a regular singular point at 0 and an irregular singular point at $\infty$. Its solutions, the *Kummer functions*, include (through simple transformations) the Bessel, Hermite, Laguerre, Fresnel, incomplete gamma, and error functions.

The simplest nontrivial first order *nonlinear* equation would have to be the *Ricatti* equation

$$y' = p(x) + q(x)y + r(x)y^2, \qquad (24)$$

which arises in circuit synthesis. Watson's book discusses explicit solutions for some special cases.

The *Bernoulli* equation

$$y' = p(x)y + q(x)y^n \tag{25}$$

is always mentioned in the elementary textbooks because it reduces to a linear equation with the substitution $z = y^{1-n}$.

*Rayleigh's* equation

$$y'' - \mu(1 - y'^2)y' + y = 0 \tag{26}$$

and *van der Pol's* equation

$$y'' - \mu(1 - y^2)y' + y = 0 \tag{27}$$

are studied in Birkhoff and Rota's text because they both exhibit a phenomenon peculiar to nonlinear equations, the "limit cycle." Van der Pol's equation, in fact, arises in the analysis of vacuum tubes.

*Duffing's* equation

$$y'' + y + \varepsilon y^3 = 0 \tag{28}$$

is a standard example in the study of nonlinear mechanics. A treatment is given in Greenberg's text.

The *Emden-Fowler* equation

$$(x^\rho y')' = x^\sigma y^P \tag{29}$$

is discussed in Bellman's book. Mehta and Aris's paper gives explicit solutions for some special cases arising in reaction-diffusion theory.

What should you do when you are confronted with a linear second order differential equation that you do not recognize? The following steps reflect your author's experience in such matters:

0. It goes without saying that the constant-coefficient case would be recognized at the outset.

1. Determine if multiplication by some power of $x$ produces an equidimensional form.

(If Steps 0 and 1 fail, there are probably no elementary-function solutions.)

2. Determine the (finite) singular points. If there is only one, check the equation against the generalized Bessel form (7)

or (8). If there are two, check against the associated Legendre equation[6].

3. If there are no finite singular points, check the equation against Airy's, Weber's, Hermite's, and Mathieu's equations.

4. Now determine the nature of the singular point at infinity (Sec. 2.4, exercise 11).

5. If there are, in all, three regular singular points, perform a change of variables if necessary so that they are mapped to 0, 1, and ∞. Check first against the hypergeometric equation, and then against Riemann's equation. If the result is positive, turn to Abramowitz and Stegun's handbook. The solutions may be expressible in terms of logs, Legendre functions, or Chebyshev functions.

6. If there is one regular singular point and one irregular singular point, map the former to 0 and the latter to ∞. Check the equation against the confluent hypergeometric equation. If it matches, turn to Abramowitz and Stegun; the solution may be expressible in terms of Bessel, Hermite, or Laguerre functions.

7. If all of the above fails, and the equation has significant application, you may become famous as its discoverer![7] Use the methods of this chapter to find out all you can about it.

8. Finally, if the equation is nonlinear, it should be checked against the forms (24) through (28) above.


Good luck!


## 2.7  Numerical Solutions

**Summary**   The Taylor approximations

$$y(x + \Delta x) \approx y(x) + \Delta x \cdot y'(x) + \frac{\Delta x^2}{2} \cdot y''(x)$$
$$y'(x + \Delta x) \approx y'(x) + \Delta x \cdot y''(x)$$

can be used to extend an estimate of a solution of a second order differential equation from the point $x$ to the point $x + \Delta x$. By iteration, then, the solution can be tabulated over any interval. As a rule, accuracy is enhanced by taking the step size $\Delta x$ smaller, but other considerations come into play.

In this section we will describe a way to obtain approximate numerical solutions to a second order ordinary differential equation. Our treatment

---

[6]Keep in mind that a slight numerical mismatch in the constants for a particular equation form may be resolvable by a simple change of scale; recall Sec. 1.2.

[7]Or you may want to name it for, say, some author who influenced you!

will be brief and cursory; we refer the reader to the specialized texts of Issacson and Keller, Gear, and Dahlquist, Bjork, and Anderson for more advanced methods and theoretical details; Nagel, Saff, and Snider contains an elementary exposition.

Suppose we have an initial value problem like

$$a_2(x)y'' + a_1(x)y' + a_0(x)y = f(x) , \quad y(x_0) = b_0 , \quad y'(x_0) = b_1. \quad (1)$$

Now recall that the Taylor development provides us with an approximation for $y(x + \Delta x)$ in terms of $y(x)$, $y'(x)$, and $y''(x)$:

$$y(x + \Delta x) \approx y(x) + \Delta x\, y'(x) + \frac{\Delta x^2}{2} y''(x) . \quad (2)$$

We can use (2) to estimate $y(x_0 + \Delta x)$; after all, $y(x_0)$ is known $(b_0)$, $y'(x_0)$ is known $(b_1)$, and $y''(x_0)$ can be found in terms of $y(x_0)$ and $y'(x_0)$ from the differential equation:

$$y''(x_0) = \frac{f(x_0) - a_1(x_0)b_1 - a_0(x_0)b_0}{a_2(x_0)} . \quad (3)$$

Thus we implement (2):

$$y(x_0 + \Delta x) \approx b_0 + \Delta x\, b_1 + \frac{\Delta x^2}{2} \frac{f(x_0) - a_1(x_0)b_1 - a_0(x_0)b_0}{a_2(x_0)} . \quad (4)$$

Of course the nature of the approximation (2) only promises us accurate solutions to the differential equation near the initial point $x = x_0$; the *step size* $\Delta x$ must be small. However, if we now restart the process from $x = x_0 + \Delta x$ we can estimate $y(x_0 + 2\Delta x)$, and then $y(x_0 + 3\Delta x)$, etc. Thus we can proceed step by step from $x = x_0$ to obtain the value of $y$ at any other point $x = x_1$ by continually updating the equation

$$y(x + \Delta x) \approx y(x) + \Delta x\, y'(x) + \frac{\Delta x^2}{2} \frac{f(x) - a_1(x)y'(x) - a_0(x)y(x)}{a_2(x)} . \quad (5)$$

We obtain higher accuracy by taking the step size, $\Delta x$, smaller. Note, however, that this will require more repetitions of (5) to reach a given terminal point $x = x_1$. Thus a trade-off between accuracy and computation time arises.

(The roundoff effect is also a factor in determining the accuracy of the result; if $\Delta x$ is too small, the accumulation of rounding errors due to the large number of computations may defeat the goal of increased accuracy. The references discuss this problem.)

We have overlooked one minor detail. In order to estimate $y(x + \Delta x)$ we need to know not only $y(x)$ but also $y'(x)$. Thus we must update $y'$, as well as $y$, as we step from point to point. This is easily accomplished by another Taylor development:

$$y'(x + \Delta x) \approx y'(x) + \Delta x\, y''(x)$$
$$\approx y'(x) + \Delta x\, \frac{f(x) - a_1(x)y'(x) - a_0(x)y(x)}{a_2(x)} \tag{6}$$

Combining (5) with (6) we can eventually arrive at estimates for $y(x)$ at any point.

**Example 1.** We shall try to estimate $y(.5)$ from the equations

$$x^2 y'' + xy' - (4 + \frac{1}{x^2})y = 0\,, \quad y(1) = 1\,, \quad y'(1) = 0 \tag{7}$$

(Recall Example 1, Sec. 2.5.) Starting from $x = 1$ we iterate

$$y'(x + \Delta x) \approx y'(x) + \Delta x\{(\frac{4}{x^2} + \frac{1}{x^4})y(x) - \frac{y'(x)}{x}\}\,,$$
$$y(x + \Delta x) \approx y(x) + \Delta x\, y'(x) + \frac{\Delta x^2}{2}\{(\frac{4}{x^2} + \frac{1}{x^4})y(x) - \frac{y'(x)}{x}\} \tag{8}$$

for (negative) $\Delta x$ until we reach $x = .5$ (i.e., $.5/\Delta x$ iterations). The results are

| $\Delta x$ | number of iterations | estimated $y(.5)$ |
|---|---|---|
| -.05 | 10 | 2.856 |
| -.01 | 50 | 2.766 |
| -.005 | 100 | 2.749 |
| -.001 | 500 | 2.732 |

∎

The numerical solution of *boundary* value problems is more difficult. The "shooting" method involves taking a guess at the correct initial derivative, running the above algorithm to evaluate the boundary value, and iteratively updating the guessed data. An example will make this clear.

**Example 2.** We shall try to estimate $y(.75)$ from the equations

$$x^2 y'' + xy' - (4 + \frac{1}{x^2})y = 0\,, \quad y(1) = 1\,, \quad y(.5) = 0\,. \tag{9}$$

We know from the previous example that if $y(1) = 1$ and we take the initial slope to be $y'(1) = 0$, then $y(.5)$ will be $\approx 2.732$. This value is too big; we want $y(.5) = 0$. So we run the iterations again, starting with the guess $y'(1) = 1$ and keeping $\Delta x = -.001$:

| $y'(1)$ | $y(.5)$ | $y(.75)$ |
|---|---|---|
| 0 | 2.732 (*too big*) | 1.226 |
| 1 | 1.662 (*better*) | 0.916 |

Continuing in this manner we vary $y'(1)$ until we home in on a value of $y'(1)$ that makes $y(.5) \approx 0$. The results are:

| $y'(1)$ | $y(.5)$ | | y(.75) |
|---------|---------|---|--------|
| 2 | 0.559 | | 0.605 |
| 2.6 | -0.103 | *(overshoot)* | 0.417 |
| 2.5 | 0.007 | | 0.450 |
| 2.51 | -0.004 | | 0.447 |

Therefore we conservatively estimate $y(.75) \approx 0.45$.                        ∎

We have glossed over many aspects of the numerical solution of differential equations, because that is not the focus of this book; we merely want to indicate the possibilities. The algorithm described here is due to Euler. More sophisticated formulations by Runge and Kutta lead to algorithms with much better performance. The interplay of accuracy, roundoff error, and other effects such as running time and stability is the focus of much ongoing research. Details may be found in the references quoted at the beginning of the section. Computer codes are readily available.

# Exercises 2.7

1. Experiment with the numerical solution of some of the equations in exercise 1, Sec. 1.1. Compare the accuracy with the exact solutions.

# Chapter 3

# Fourier Methods

Practically all physical systems, when driven by a sinusoidal oscillation, will eventually respond by oscillating at the same frequency as the driving force. The response may not always be in phase with the driver, and its amplitude may change as the driving frequency is increased; nonetheless, after an initial period of acclimation practically every system will vibrate "sympathetically" with a sinusoidal driver. In this chapter we are going to explain this widespread behavior and see how it can be exploited in engineering analysis.

The first section of this chapter surveys the various algebraic representations for sinusoidal functions; some are more convenient than others, depending on the application. In Section 2 we characterize the response of linear differential equations to nonhomogeneities which are pure sinusoids, and identify mathematically the precise attributes of these sympathetic vibrations. Readers may discern the kinship of this discussion with the "method of undetermined coefficients" in their introductory differential equations course.

If the nonhomogeneity is not a pure sinusoid, the analysis methods of Section 2 can still be used, thanks to *Fourier analysis*. Joseph Fourier argued that, practically speaking, every function can decomposed into a sum of sinusoids: the Fourier series. Sections 3 through 7 are each devoted to a specific aspect of Fourier methodology. The construction of the Fourier series, its convergence properties, and the related sine and cosine series are the subjects of Sections 3 through 5. Section 6 introduces the important generalization, the Fourier transform, and Section 7 extends the transform calculus to generalized functions. The identification of the Laplace transform as a particular adaptation of the Fourier transform to special situations comprises Section 8. Many readers will have had some introduction to the Laplace transform as a stand-alone tool for solving constant coefficient differential equations, but for advanced applications its true nature as a Fourier "cousin" is crucial.

Section 9 ties the various Fourier methodologies back to the topic of

Section 2 - the analysis of linear engineering systems. It discusses when to use which transform, the advantages and disadvantages of each, and their physical interpretations. The final section describes the numerical implementation of the Fourier transform.

## 3.1   Oscillations

**Summary**   The most general sinusoid of frequency $\omega$,

$$a_1 \sin \omega t + a_2 \cos \omega t + a_3 \sin(\omega t + \phi_1) + a_4 \cos(\omega t + \phi_2) + \cdots$$
$$+ a_{2n+2} \cos(\omega t + \phi_2 n)$$

can be expressed compactly as $Re\{Ae^{i\omega t}\}$. In this form differentiation reduces simply to multiplication by $i\omega$:

$$\frac{d}{dt} Re\left\{Ae^{i\omega t}\right\} = Re\left\{i\omega Ae^{i\omega t}\right\}. \qquad (1)$$

As mentioned in the introduction, practically all physical systems respond to a sinusoidal driving force by oscillating at the same frequency as the driver. (It is more accurate to say that the *steady state* responses to a *low-level* sinusoidal input are facsimiles of the input.) We shall see that these "sympathetic" vibrations arise as particular solutions to constant coefficient nonhomogeneous differential equations.

Before we begin, it is wise to settle a matter of notation. Since most applications of sinusoidal analysis are directed toward functions of *time*, we shall temporarily abandon the practice of denoting an unknown function as $y(x)$ in favor of the more suggestive $x(t)$. Using dots instead of primes to denote time differentiation, we thus are disposed to write a linear second order differential equation with constant coefficients as

$$a\ddot{x} + b\dot{x} + cx = f(t). \qquad (2)$$

The nonhomogeneity $f(t)$ will, for the moment, be a sinusoidal function with frequency $\omega$. There are many such sinusoids - $\sin \omega t$ and $\cos \omega t$, of course; another is $\sin(\omega t + \phi_1)$ for any constant "phase" angle $\phi_1$. In fact any linear combination

$$a_1 \sin \omega t + a_2 \cos \omega t + a_3 \sin(\omega t + \phi_1) + a_4 \cos(\omega t + \phi_2) + \cdots$$
$$+ a_{2n+2} \cos(\omega t + \phi_2 n) \qquad (3)$$

is, again, a sinusoid with frequency $\omega$; it can always be rewritten simply as

$$a \cos(\omega t + \phi). \qquad (4)$$

The easiest way to see this is to use Euler's formula

$$e^{i\theta} = \cos\theta + i\sin\theta. \tag{5}$$

Noting that $\sin\theta = \cos(\theta - \pi/2)$ we express (3) as the real part of complex exponentials:

$$a_1 Re\left[e^{i(\omega t - \pi/2)}\right] + a_2 Re\left[e^{i\omega t}\right] + a_3 Re\left[e^{(i\omega t + \phi_1 - \pi/2)}\right]$$

$$+ \cdots + a_{n+2} Re\left[e^{(i\omega t + \phi_n)}\right]$$

$$
\begin{aligned}
&= Re\left\{\left[-ia_1 + a_2 - ia_3 e^{i\phi_1} + \cdots + a_{2n+2} e^{i\phi_{2n}}\right] e^{i\omega t}\right\} \\
&= Re\left\{\left[ae^{i\phi}\right] e^{i\omega t}\right\} \\
&= a\cos(\omega t + \phi) \quad \text{(See exercise 1.)}
\end{aligned}
$$

In fact the above derivation shows that we can write the most general sinusoid of frequency $\omega$ as

$$Re\left\{Ae^{i\omega t}\right\} \tag{6}$$

if we let $A$ be complex (so that it encompasses both factors of $ae^{i\phi}$). This turns out to be the most convenient notation of all, because it simplifies differentiation: instead of

$$\frac{d}{dt}\cos\omega t = -\omega\sin\omega t, \quad \frac{d}{dt}\sin\omega t = \omega\cos\omega t, \tag{7}$$

we have simply

$$\frac{d}{dt}Re\left\{Ae^{i\omega t}\right\} = Re\left\{i\omega Ae^{i\omega t}\right\}. \tag{8}$$

(See exercise 2 if this is not clear.) Time differentiation of a sinusoid, in this notation, is equivalent to simple multiplication by $i\omega$.

There is another benefit of the notation (6) which works to our advantage. Rather than solve the differential equation

$$a\ddot{x} + b\dot{x} + cx = d\cos(\omega t + \phi) \tag{9}$$

we can, instead, solve

$$a\ddot{x} + b\dot{x} + cx = Ae^{i\omega t} \tag{10}$$

with $A = de^{i\phi}$; this, we shall see, is an easy task. Then *the solution to (9) equals the real part of the solution to (10).* (See exercise 3.)

EXAMPLE 1. To find a solution to

$$\ddot{x} + \dot{x} + x = \sin t \tag{11}$$

we observe

$$\sin t = Re\left\{-ie^{it}\right\}.$$

It is easily verified that a solution to

$$\ddot{x} + \dot{x} + x = -ie^{it}$$

is given by $-e^{it}$. Thus a solution to (11) is given by

$$x = Re\left\{-e^{it}\right\} = -\cos t. \qquad \blacksquare$$

Of course there are other ways of representing a sinusoid in terms of exponentials. One could, for example, use the *imaginary* part, but that's hardly worth mentioning. Another strategy is to use complex conjugation:

$$Re\left\{Ae^{i\omega t}\right\} = \frac{1}{2}\left\{Ae^{i\omega t} + \overline{A}e^{-i\omega t}\right\}. \tag{12}$$

Each of the forms (4, 6, 12) has its advantages and they will be used interchangeably in the sequel. *Observe that the frequencies $\omega$ in (4) and (6) are conventionally taken to be positive*, but both $\omega$ and the "unphysical" frequency $-\omega$ are required for the exponential description (12).

## Exercises 3.1

1. Express the following as $a\cos(\omega t + \phi)$:

   (a) $b\sin\omega t$

   (b) $b\cos\omega t + c\sin\omega t$

   (c) $b\cos\omega t + c\cos(\omega t + \theta)$

2. Let $A$ be a complex constant expressed in polar form as $|A|e^{i\phi}$. Work out the real part of the expression $Ae^{i\omega t}$, and take its derivative. Then work out the real part of $i\omega Ae^{i\omega t}$. Verify eq. (8).

3. Express the real part of each side of eq. (10) and verify that the real part of its solution satisfies eq. (9).

## 3.2 Transfer Functions

**Summary** The linear constant-coefficient second order ordinary differential equation

$$m\ddot{x} + \nu\dot{x} + kx = f(t)$$

can be identified with a forced damped mass-spring oscillator. A particular response to a sinusoidal forcing function

$$f(t) = Re\left\{Ae^{i\omega t}\right\}$$

is represented by multiplying $Ae^{i\omega t}$ by the transfer function

$$H(\omega) = \frac{1}{-m\omega^2 + i\nu\omega + k}$$

and taking the real part. If damping is present ($\nu > 0$) this solution, oscillating "sympathetically" with $f$ at the frequency $\omega$, is the steady state solution. If $\nu = 0$ the associated homogeneous solutions persist for all time. An alternative formula is available for undamped systems forced at resonance, but it has little physical validity.

In this section we shall consider in more detail our claim that most physical systems, externally driven by a sinusoidal excitation, eventually exhibit a sinusoidal response at the same frequency.

When we speak of a physical system in this sense, it is implicitly assumed that the system is divorced from the external agents which turn it on, adjust it, or utilize it. The word "autonomous" is commonly used. To us this means that the (presumably linear) differential equations describing its responses have coefficients which are time-independent. In analyzing a radio, for example, we assume that its capacitances are constant - we do not consider the person twiddling the knobs to be a part of the physical system.

We take as our model the damped harmonic oscillator depicted in Fig. 1. This simple model exhibits all the main features that we need to consider. More complex systems are studied in exercises 2 and 3.

The object is located at $x(t)$, measured relative to its equilibrium position, and has mass $m$. If the spring constant is $k$ and the coefficient of viscous damping is $\nu$, then Newton's second law equates its mass-times-acceleration with the sum of the spring force, the damping force, and the externally applied driving force:

$$m\ddot{x} = -kx - \nu\dot{x} + f(t) \text{ or } m\ddot{x} + \nu\dot{x} + kx = f(t). \tag{1}$$

As indicated in the previous section, we can denote an arbitrary applied sinusoidal force with frequency $\omega$ as

$$f(t) = Re\left\{Ae^{i\omega t}\right\} \tag{2}$$

and the resulting solution of (1) will be the real part of the solution to

$$m\ddot{x} + \nu\dot{x} + kx = Ae^{i\omega t}. \tag{3}$$

According to the discussion in Section 2.1 the general solution to (3) can be written as the general solution to the associated homogeneous equation

$$m\ddot{x} + \nu\dot{x} + kx = 0 \tag{4}$$

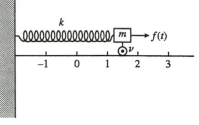

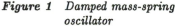

**Figure 1** *Damped mass-spring oscillator*

plus a particular solution to (3). Let us dispose of the homogeneous solution first. The ansatz $x = e^{pt}$ in (4) results in the characteristic equation

$$mp^2 e^{pt} + \nu p e^{pt} + k e^{pt} = 0 \qquad (5)$$

with roots

$$p = \frac{-\nu \pm \sqrt{\nu^2 - 4mk}}{2m}.$$

All real physical systems have positive damping coefficients $\nu$, in which case these homogeneous solutions are either damped sinusoids -

$$c_1 e^{-(\nu/2m)t} \cos \sqrt{k/m - \nu^2/4m^2}\, t + c_2 e^{-(\nu/2m)t} \sin \sqrt{k/m - \nu^2/4m^2}\, t \qquad (6)$$

- or decaying exponentials -

$$c_1 \exp\left\{\frac{-\nu + \sqrt{\nu^2 - 4mk}}{2m} t\right\} + c_2 \exp\left\{\frac{-\nu - \sqrt{\nu^2 - 4mk}}{2m} t\right\}. \qquad (7)$$

(If $\nu^2 = 4mk$ the solutions are $c_1 \exp(-\nu t/2m) + c_2 t \exp(-\nu t/2m)$, which still decay.)

Rather than use the variation of constants formalism or a Green's function to compute a particular *nonhomogeneous* solution, we note that insertion of the ansatz

$$x(t) = X e^{i\omega t} \qquad (8)$$

with an *undetermined coefficient* $X$ leads immediately to a solution of (3), since differentiation of the exponential function simply multiplies it by $i\omega$:

$$m(i\omega)^2 X e^{i\omega t} + \nu(i\omega) X e^{i\omega t} + k X e^{i\omega t} = A e^{i\omega t},$$

or

$$[-m\omega^2 + i\nu\omega + k] X = A,$$

$$X = \frac{A}{-m\omega^2 + i\nu\omega + k}. \qquad (9)$$

Thus the general solution to (3) is

$$x(t) = \frac{A}{-m\omega^2 + i\nu\omega + k} e^{i\omega t} + c_1 x_1(t) + c_2 x_2(t) \qquad (10)$$

where $x_1(t)$ and $x_2(t)$ are the homogeneous solutions (6 or 7).

The display (10) exemplifies the point we have been making about sympathetic vibrations. The last two terms in (10), for systems with damping, approach zero as time increases. They are called "transients." Thus the solution eventually decays down to the first term - a sinusoid oscillating at the same frequency as the driving term $Ae^{i\omega t}$.

This oscillation never dies out (as long as the driver $f(t)$ is applied); its amplitude (9) is fixed. It is called the "steady state sinusoidal response" to the force. We have shown that *all physical systems governed by a second order linear differential equation with positive, constant coefficients*

$$m\ddot{x} + \nu\dot{x} + kx = f(t) \tag{11}$$

*respond to a sinusoidal forcing term*

$$f(t) = Re\left\{Ae^{i\omega t}\right\} \tag{12}$$

*with a steady state sinusoid of the same frequency*

$$Re\left\{\frac{1}{-m\omega^2 + i\nu\omega + k}Ae^{i\omega t}\right\} \tag{13}$$

*plus a transient term which approaches zero as $t \to \infty$.* The steady state response is independent of the initial conditions (because the latter affect only the constants $c_1$, $c_2$ in the transient term (10)).

The ratio of the (complex) steady state response to the (complex) forcing term is called the *transfer function* of the system. For eq. (1) the transfer function is

$$H(\omega) = \frac{1}{-m\omega^2 + i\nu\omega + k} \tag{14}$$

It is a complex function of $\omega$. Clearly the steady state response can be obtained by multiplying the transfer function by the forcing sinusoid.[1]

How is all this modified if the system has no damping -

$$m\ddot{x} + (0)\dot{x} + kx = Ae^{i\omega t} \quad ? \tag{15}$$

Although we can regard *negative* damping as physically unrealistic, many systems are modeled as frictionless; so we have to address this idealization.

---

[1] It is obvious that, given the transfer function as a function of $\omega$, one can reconstruct the differential equation (1). The science of *system identification* and *modal analysis* is based on this idea. One experimentally measures the ratio of the force to the steady state response at different frequencies, and then chooses the parameters ($m$, $\nu$, $k$ - or more, for higher order systems) to fit these data. Thus a mass-spring oscillator model for the system is constructed mathematically.

There are two principle modifications that must be made to our previous analysis.

The first regards the nature of the solutions to the associated homogeneous equation -

$$m\ddot{x} + kx = 0 \quad ; \quad x = c_1 \cos\sqrt{k/m}\, t + c_2 \sin\sqrt{k/m}\, t \qquad (16)$$

No longer do these "transients" approach zero. Instead they persist, just like the particular solution - which now takes the form

$$Xe^{i\omega t} = \frac{1}{-m\omega^2 + k}Ae^{i\omega t} = \frac{-1/m}{\left[\omega - \sqrt{k/m}\right]\left[\omega + \sqrt{k/m}\right]}Ae^{i\omega t}. \qquad (17)$$

Despite the fact that the general solution does not decay down to this "sympathetic" one, we shall continue to designate (17) as the "steady state" solution and the coefficient therein as the transfer function. The general solution to (15) is, still, the sympathetic steady state sinusoid (17) plus the homogeneous solution (16).

The second modification becomes necessary as the forcing frequency $\omega$ approaches the values $\pm\sqrt{k/m}$. The amplitude of the steady state oscillations (17) increases without bound. And when $\omega = \pm\sqrt{k/m}$ the denominator in (17) goes to zero and the formula becomes useless. Now this frequency, $\sqrt{k/m}$ (recall that the unphysical negative frequency $-\sqrt{k/m}$ is an artifact of the exponential notation), is the frequency appearing in the homogeneous solution (16). Thus we are *driving* the system with a sinusoidal force oscillating at the system's natural frequency - the frequency at which the system would vibrate anyway, *if no force were present!*

It should come as no surprise, then, that the system offers no resistance to such a force and its response is singular. This phenomenon is known as "resonance", and there have been many spectacular demonstrations of its occurrence (actually, its "near-occurrence" in low-friction systems). The reference by Braun attributes the collapse of the Tacoma Narrows bridge in 1940 to wind patterns forcing it near its resonant frequency; also described is the collapse of the Broughton suspension bridge in 1831, when a troop of soldiers marched across it at a cadence close to resonance. Less tragic is the old television commercial proclaiming the performance of a brand of audio tape, which reproduces the artist's tones so faithfully that a wineglass is shattered by resonant vibrations produced either by the artist or by the recording.

The zero in the denominator of (17) has revealed an inadequacy in our transfer-function model when an undamped spring is forced at resonance. Can we remedy this? It *is* possible to concoct a mathematical solution to the undamped, resonant-forced equation

$$m\ddot{x} + kx = Ae^{i\sqrt{k/m}\, t} \qquad (18)$$

by using the methods of Section 1.1. As one can easily check (exercise 4), the function

$$x(t) = \frac{t}{i2\sqrt{km}} A e^{i\sqrt{k/m}\,t} \tag{19}$$

is a particular solution to (18). And the general solution

$$\frac{t}{i2\sqrt{km}} A e^{i\sqrt{k/m}\,t} + c_1 \cos\sqrt{k/m}\,t + c_2 \sin\sqrt{k/m}\,t \tag{20}$$

can be adjusted to match any initial conditions.

But we would be disinclined to call (19) a "steady state" solution since it is not a sinusoid. Its oscillations increase with time, and it does not settle down. *The fact of the matter is that an undamped system forced at its resonant frequency simply does not evolve into a steady state (mathematically or physically). Once such a force is applied, the system oscillates ever more violently until it breaks apart.*

Realistically speaking, one should acknowledge that any physical system has some friction, so eq. (15) and its solution (17) are only approximations. They work for most values of $\omega$, but when $\omega \approx \pm\sqrt{k/m}$ the damping coefficient $\nu$ must be acknowledged. Thus at this "resonant" value of $\omega$ the steady state response is the large (but finite) sinusoid derived from (9)[2]

$$\frac{1}{i\nu\omega} A e^{i\omega t}. \tag{21}$$

If the system can't support oscillations this big without breaking up then the differential equation model (1) has limited validity anyway.

It goes without saying that one of the tasks of engineering design is to adjust masses, stiffnesses, and damping so that environmental vibrations do not excite resonant responses.

## Exercises 3.2

1. Construct the transfer functions for the differential equations appearing in exercise 1a, 1b, and 1c of Section 1.1.

2. The transfer function concept can be applied to a constant coefficient differential equation of any order. Construct transfer functions for

   (a) $\dot{x} + 2x = f(t)$ ;

---

[2] As a matter of fact for $\nu \neq 0$ the peak response does not occur exactly at $\omega = \sqrt{k/m}$ ; see exercise 5.

(b) $\dddot{x} + 2\ddot{x} + 3\dot{x} + 4x = f(t)$ .

3. Transfer functions can also be constructed for *coupled systems of differential equations* with constant coefficients. Try the substitutions $x = Xe^{i\omega t}$, $y = Ye^{i\omega t}$ into the system

$$
\begin{aligned}
3\dot{x} + x + 4\dot{y} + 2y &= Ae^{i\omega t} \\
2\dot{x} + 3x + \dot{y} + y &= Be^{i\omega t}
\end{aligned}
$$

and obtain expressions for $X$ and $Y$.

4. (a) Verify that (19) is a solution to eq. (18).

   (b) Using the methods of Section 1.1, derive expression (19).

5. For which frequency $\omega$ does the magnitude of the transfer function $H(\omega)$ in (14) achieve its maximum?

6. Although the theory in this section pertains only to second order equations with constant coefficients, we can use the damped oscillator model to make some qualitative predictions about more general equations. Recall that $m\ddot{x} + \nu\dot{x} + kx = 0$ describes an oscillator with mass $m$, damping coefficient $\nu$ and spring constant $k$.

   (a) *Duffing's* equation

   $$\ddot{x} + x + \epsilon x^3 = 0$$

   can be interpreted as an undamped oscillator with a spring "constant" $(1+\epsilon x^2)$ that gets stronger as the excursion $x(t)$ increases. Thus we would expect to see sinusoidal-like oscillations which "sharpen" somewhat at the maxima and minima due to the additional stiffness of the spring.

   (b) *Airy's* equation

   $$\ddot{x} - tx = 0$$

   describes an undamped oscillator with a spring "constant" $(-t)$ which becomes increasingly negative for $t > 0$. Such a "spring" accelerates the mass away from the equilibrium position ($x = 0$). Therefore masses with outwardly directed initial velocities are driven away, and inbound masses are slowed down. If we track the motion through negative time, we see a positive spring constant which stiffens as $|t|$ increases. Thus the motion looks sinusoidal with increasing frequency and diminishing amplitude as $t$ goes negative.

   (c) The *van der Pol* equation

   $$\ddot{x} - \mu(1 - x^2)\dot{x} + x = 0$$

exhibits negative damping for $|x| < 1$. This means that the energy of the oscillations increases, rather than being damped out. For $|x| > 1$ the motion is damped, of course. We would expect, then, to see small motions grow and large motions die down, with some intermediate stable motions sandwiched in between; and the latter should have $|x| = 1$, on the average. As a matter of fact, it can be shown that the van der Pol equation has a periodic solution, called a "limit cycle," and all other solutions tend toward this limit cycle (except for $x = 0$). Similar considerations apply for the Raleigh equation

$$\ddot{x} - \mu(1 - \dot{x}^2)\dot{x} + x = 0,$$

which has negative damping for low velocities and positive damping for high velocities.

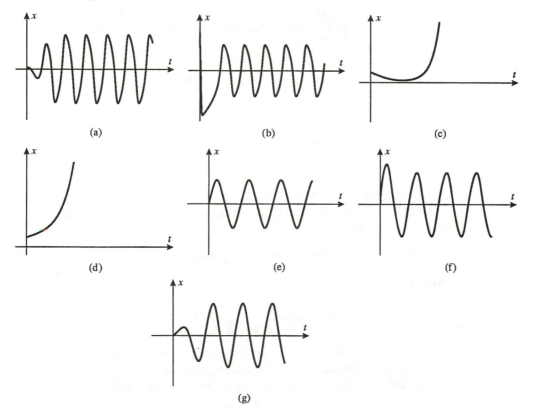

(a)  (b)  (c)

(d)  (e)  (f)

(g)

**Figure 2**  *Match the trajectories with the equations*

(d) Figure 2 depicts some solutions to these equations. Try to match the graphs with the equations that they solve. (You probably won't be able to distinguish between the Rayleigh and van der Pol solutions.)

(e) Interpret the Bessel equation in terms of a mass-spring oscillator for positive times. What features of the graphs of $J$, $Y$, $I$, and $K$ does this "explain"?

## 3.3   Fourier Series

**Summary**   Virtually any function $f(t)$ of period $T$ can be expressed as a Fourier series in either of the forms

$$f(t) = \sum_{n=1}^{\infty} \{c_n \cos \frac{2\pi n}{T}t + d_n \sin \frac{2\pi n}{T}t\} + c_0 = \sum_{n=-\infty}^{\infty} A_n e^{i2\pi nt/T}.$$

Because of orthogonality the coefficients can be obtained by simple integration:

$$c_0 = \frac{1}{T}\int_0^T f(t)\,dt, \; c_{n>0} = \frac{2}{T}\int_0^T f(t)\cos\frac{2\pi n}{T}t\,dt,$$

$$d_n = \frac{2}{T}\int_0^T f(t)\sin\frac{2\pi n}{T}t\,dt,$$

$$A_n = \frac{1}{T}\int_0^T f(t)e^{-i2\pi nt/T}\,dt.$$

In the previous section we saw how to characterize the steady state solution to a linear differential equation with constant coefficients and a sinusoidal forcing term,

$$m\ddot{x} + \nu\dot{x} + kx = f(t) = Ae^{i\omega t} : \tag{1}$$

namely,

$$x(t) = \frac{1}{-m\omega^2 + i\nu\omega + k}Ae^{i\omega t} = H(\omega)Ae^{i\omega t}, \tag{2}$$

where $H(\omega)$ denotes the transfer function. (We ignore resonance complications.) A very nice feature of the *linearity* of the differential equation (1) is that if the forcing term happens to be a sum of sinusoids

$$m\ddot{x} + \nu\dot{x} + kx = f(t) = A_1 e^{i\omega_1 t} + A_2 e^{i\omega_2 t} + \cdots + A_n e^{i\omega_n t}, \tag{3}$$

then the steady state solution is the sum of the solutions for each sinusoid separately:

$$x(t) = \frac{1}{-m\omega_1^2 + i\nu\omega_1 + k}A_1 e^{i\omega_1 t} + \frac{1}{-m\omega_2^2 + i\nu\omega_2 + k}A_2 e^{i\omega_2 t} + \cdots$$

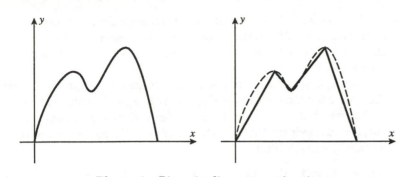

*Figure 1    Piecewise linear approximation*

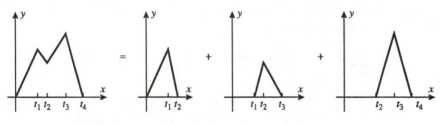

*Figure 2    Decomposition into triangle functions*

$$+\frac{1}{-m\omega_n^2 + i\nu\omega_n + k}A_n e^{i\omega_n t}$$

$$=\sum_{j=1}^{n} H(\omega_j) A_j e^{i\omega_j t}. \tag{4}$$

This is of the utmost importance, because in 1807 Fourier realized that virtually any function $f(t)$ could be expanded into a sum - actually an infinite sum, or series - of sinusoids: $f(t) = \sum_{\omega_n} A_n e^{i\omega_n t}$. Moreover, the expansion is easy to obtain; present day computers can approximate it in a fraction of a second.[3]

In the next few sections we are going to explore various mathematical aspects of this expansion. We shall return to its applications in differential equation theory in Section 3.9.

The proof of Fourier's result is rather difficult and tedious (see Kammler's book), but perhaps the following will be enlightening. Suppose $f(t)$ is a *continuous* function defined on $[0, T]$. Then because of the continuity it is fairly clear that $f$ can be approximated with arbitrary precision by piecewise linear functions: see Fig. 1.

Now any piecewise linear function can be written as a sum of triangle functions, as depicted in Fig. 2.

---

[3]Fourier was unable to convince his colleagues; the validity of the Fourier expansion was not accepted by many mathematicians until Dirichlet proved it in 1829.

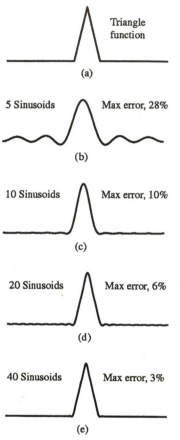

Triangle function

(a)

5 Sinusoids — Max error, 28%

(b)

10 Sinusoids — Max error, 10%

(c)

20 Sinusoids — Max error, 6%

(d)

40 Sinusoids — Max error, 3%

(e)

**Figure 3**   *Approximation of triangle function*

So if we are able to show that a triangle function can be approximated with arbitrary precision by a sum of sinusoids, then Fourier's claim follows by superposition (for continuous functions).

As graphical evidence of the efficacy of sinusoidal approximation, then, we offer the displays in Fig. 3. Although the convergence is sluggish at the corners, where the derivatives are discontinuous, it seems evident that an arbitrarily close fit can be achieved by taking enough terms.

Before we delve further into the nature of the convergence of Fourier's sinusoidal sums, let us investigate the claim that they are easy to compute. We take for granted that, in some limiting sense, the sums $\sum A_n e^{i\omega_n t}$ converge to $f$ over an interval $0 \leq t \leq T$:

$$f(t) = \sum_{\omega_n} A_n e^{i\omega_n t} = \sum_{\omega_n} c_n \cos \omega_n t + \sum_{\omega_n} d_n \sin \omega_n t, \ 0 \leq t \leq T \qquad (5)$$

(It is convenient for the moment to employ the cosine/sine representation.) Now Fourier realized that not only was the expansion (5) possible, but that the convergence could be achieved using only frequencies $\omega_n$ *which were multiples of a fundamental frequency* $\omega_1$, that corresponds to one cycle ($2\pi$ radians) over the basic interval $T$ (if $t$ is time, measured in seconds, then $\omega_1 = 2\pi/T$ radians per second). In other words (5) takes the specific form

$$f(t) = \sum_{n=0}^{\infty} c_n \cos n\frac{2\pi}{T}t + \sum_{n=0}^{\infty} d_n \sin n\frac{2\pi}{T}t, \ 0 \leq t \leq T \qquad (6)$$

or, still more specifically (since $\cos 0 = 1$ and $\sin 0 = 0$)

$$f(t) = \sum_{n=1}^{\infty} \left[ c_n \cos n\frac{2\pi}{T}t + d_n \sin n\frac{2\pi}{T}t \right] + c_0, \ 0 \leq t \leq T \qquad (7)$$

When frequencies are related by $\omega_n = n\omega_1$ we say that $\omega_n$ is the $n$th *harmonic* of the fundamental frequency $\omega_1$. Now because the frequencies appearing in (7) are all harmonics of $\omega_1$, there is a neat trick which enables us to find the coefficients in the expansion. It is based on the easily verified observation that any two distinct functions $\phi_1(t)$ and $\phi_2(t)$ in the set

$$\{1, \cos \frac{2\pi}{T}t, \cos 2\frac{2\pi}{T}t, \cos 3\frac{2\pi}{T}t, ...;$$
$$\sin \frac{2\pi}{T}t, \sin 2\frac{2\pi}{T}t, \sin 3\frac{2\pi}{T}t, ...\} \qquad (8)$$

have the property

$$\int_0^T \phi_1(t)\phi_2(t)\, dt = 0 \qquad (9)$$

(see exercise 1). Thus to expose, say, $c_5$ in the expansion (7), we multiply both sides by $\cos 5 \frac{2\pi}{T} t$,

$$f(t) \cos 5 \frac{2\pi}{T} t = \sum_{n=1}^{\infty} [c_n \cos n \frac{2\pi}{T} t \, \cos 5 \frac{2\pi}{T} t$$

$$+ d_n \sin n \frac{2\pi}{T} t \, \cos 5 \frac{2\pi}{T} t] + c_0 \cos 5 \frac{2\pi}{T} t \, ,$$

and integrate over the interval [0,T]:

$$\int_0^T f(t) \cos 5 \frac{2\pi}{T} t \, dt = \sum_{n=1}^{\infty} [c_n \int_0^T \cos n \frac{2\pi}{T} t \, \cos 5 \frac{2\pi}{T} t \, dt$$

$$+ d_n \int_0^T \sin n \frac{2\pi}{T} t \, \sin 5 \frac{2\pi}{T} t \, dt] + c_0 \int_0^T \cos 5 \frac{2\pi}{T} t \, dt \qquad (10)$$

(assuming that (7) converges fast enough to justify term-by-term integration). According to (9) the only term that survives on the right hand side of (10) is

$$c_5 \int_0^T \cos 5 \frac{2\pi}{T} t \, \cos 5 \frac{2\pi}{T} t \, dt$$

and thus we obtain $c_5$:[4]

$$c_5 = \frac{\int_0^T f(t) \cos 5 \frac{2\pi}{T} t \, dt}{\int_0^T \cos^2 5 \frac{2\pi}{T} t \, dt} = \frac{2}{T} \int_0^T f(t) \cos 5 \frac{2\pi}{T} t \, dt.$$

The other coefficients are isolated by the same trick and we find

$$c_0 = \frac{1}{T} \int_0^T f(t) \, dt, \quad c_{n>0} = \frac{2}{T} \int_0^T f(t) \cos n \frac{2\pi}{T} t \, dt,$$

$$d_n = \frac{2}{T} \int_0^T f(t) \sin n \frac{2\pi}{T} t \, dt. \qquad (11)$$

Thus the coefficients are determined by integration (numerically, if necessary; see Appendix A.6).

---

[4]The integral in the denominator is easy if you remember that the average value of $\cos^2 nt$, *over a period*, is the same as the average of $\sin^2 nt$; and both averages sum to 1 ($= \cos^2 nt + \sin^2 nt$). Thus their average value is 1/2 each, and their integral is 1/2 times the length of the interval. But beware of the exception: the average value of $\cos^2 0t$ is 1.

There is a widely used terminology for these operations, based on the following idea. Suppose we were to compute the integrals in (9) by the crude rectangular approximation (discussed in the beginning of Appendix A.6). Specifically, take $T = 2\pi$, and consider the computation of

$$\int_0^{2\pi} \cos t \sin t \, dt.$$

We would start with a table of values of $\cos t$ and $\sin t$:

| t | cos t | sin t |
|---|---|---|
| 0 | 1.000 | .000 |
| $\pi/6$ | .866 | .500 |
| $2\pi/6$ | .500 | .866 |
| $3\pi/6$ | .000 | 1.000 |
| $4\pi/6$ | -.500 | .866 |
| $5\pi/6$ | -.866 | .500 |
| $\pi$ | -1.000 | .000 |
| $7\pi/6$ | -.866 | -.500 |
| $8\pi/6$ | -.500 | -.866 |
| $9\pi/6$ | .000 | -1.000 |
| $10\pi/6$ | .500 | -.866 |
| $11\pi/6$ | .866 | -.500 |

Then we would multiply corresponding entries from each column, add up the products, and multiply by $\Delta t = \pi/6$.

*Except for the last step - multiplication by $\Delta t$ - we would be performing the same operation as taking the scalar product of the two vectors denoted by the columns in the table.*

By analogy then, the product-integral

$$\int_a^b \phi_1(t)\, \phi_2(t)\, dt$$

is commonly called the *scalar product* of the two functions $\phi_1$ and $\phi_2$, and a "bra-ket" symbol, introduced by P. A. M. Dirac, is frequently used:

$$\int_a^b \phi_1(t)\phi_2(t)\, dt = \langle \phi_1, \phi_2 \rangle \qquad (12)$$

Extending this analogy, we say that the functions $\phi_1$ and $\phi_2$ are *orthogonal* when this scalar product is zero. Exercise 1 shows that the trig functions

$$1, \cos t, \cos 2t, \cos 3t, \dots \, ; \, \sin t, \sin 2t, \sin 3t, \dots$$

constitute an *orthogonal set* over the interval $[0, 2\pi]$ because of the harmonic relation between the frequencies. It is the orthogonality property which makes the evaluation of the constants in the expansion (7) so easy.

With this picturesque terminology the representation

$$f(t) = \sum_n a_n \phi_n(t) \tag{13}$$

can be interpreted as an expansion of a given "function/vector" in terms of a set of orthogonal function/vectors. The coefficients are obtained by using the scalar product to "project" the given function/vector along the directions of the expansion function/vectors:

$$a_n = \frac{\langle f, \phi_n \rangle}{\langle \phi_n, \phi_n \rangle} = \frac{\langle f, \phi_n \rangle}{\|\phi_n\|^2}, \tag{14}$$

where the "length" or "norm" $\|\phi\|$ of a function/vector is inherited in the usual way from the scalar product:

$$\|\phi\| = \langle \phi, \phi \rangle^{1/2} = \{ \int_a^b \phi(t)^2 dt \}^{1/2}. \tag{15}$$

(Compare (14) with (11).)[5]

The series on the right hand side of (7), with coefficients given by (11), is known as the *Fourier series* for $f(t)$ over the interval $[0, T]$. It is easily rescaled to the interval $[a, b]$:

$$f(t) = \sum_{n=1}^{\infty} \{ c_n \cos \frac{2\pi n}{b-a}(t-a) + d_n \sin \frac{2\pi n}{b-a}(t-a) \} + c_0, \tag{16}$$

$$c_0 = \frac{1}{b-a} \int_a^b f(t) \, dt, \quad c_{n>0} = \frac{2}{b-a} \int_a^b f(t) \cos \frac{2\pi n}{b-a}(t-a) \, dt,$$

$$d_n = \frac{2}{b-a} \int_a^b f(t) \sin \frac{2\pi n}{b-a}(t-a) \, dt \tag{17}$$

Notice that the *fundamental frequency* for the interval $[a, b]$ is $\omega_1 = 2\pi/(b-a)$ radians per second or 1 cycle per $(b-a)$ seconds. All the frequencies occurring in the Fourier series are harmonics of $\omega_1$.

---

[5]When functions are regarded as vectors with norms defined through scalar products the resulting structure is known mathematically as a *Hilbert Space*.

**Example 1.** The coefficients in the Fourier series for the function $e^t$, for the interval $[-\pi, \pi]$, are

$$c_0 = \frac{1}{2\pi} \int_{-\pi}^{\pi} e^t \, dt = \frac{1}{2\pi}(e^\pi - e^{-\pi}) = \frac{\sinh \pi}{\pi};$$

$$c_n = \frac{1}{\pi} \int_{-\pi}^{\pi} e^t \cos n(t + \pi) \, dt = \frac{2}{\pi} \frac{\sinh \pi}{n^2 + 1};$$

$$d_n = \frac{1}{\pi} \int_{-\pi}^{\pi} e^t \sin n(t + \pi) \, dt = -\frac{2n}{\pi} \frac{\sinh \pi}{n^2 + 1} \tag{18}$$

(exercise 2). The Fourier series for $e^t$ in this interval is thus

$$\frac{\sinh \pi}{\pi} \left\{ 1 + \cos(t + \pi) + \frac{2}{5} \cos 2(t + \pi) + \frac{1}{5} \cos 3(t + \pi) + \cdots \right.$$

$$\left. - \sin(t + \pi) - \frac{4}{5} \sin 2(t + \pi) - \frac{3}{5} \sin 3(t + \pi) + \cdots \right\}.$$

For the exponential form of the Fourier series, the display (16, 17) looks like (exercise 3)

$$f(t) = \sum_{n=-\infty}^{\infty} A_n e^{i2\pi n \frac{t-a}{b-a}}, \ a \le t \le b; \tag{19}$$

$$A_n = \frac{1}{b-a} \int_a^b f(t) e^{-i2\pi n \frac{t-a}{b-a}} dt \ \ (\text{all } n). \tag{20}$$

Note the presence of the descending, as well as ascending, series in (19). Note also that both forms (16, 17) and (19, 20) are valid for *complex*, as well as real, functions $f$. If $f$ happens to be real then $A_{-n}$ is the complex conjugate of $A_n$ (exercise 4).                               ∎

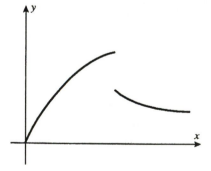

*Figure 4    Jump discontinuity*

We close with a brief remark about convergence of Fourier series. As Figs. 1 through 3 demonstrate, we expect the series to approach $f(t)$ where the latter is continuous (although discontinuities in the derivatives slow down the convergence). If $f$ has a jump discontinuity (Fig. 4) then convergence on both sides of the jump is not possible. After all the sums in (7) - if truncated at any *finite* level - are *continuous* functions; they can't be two places at once. In such a case the Fourier series converges democratically to the average of the left- and right-hand limits. This is illustrated in the first graph of Fig. 5, where several useful expansions are displayed. In the next section we will return to the question of convergence in more detail.

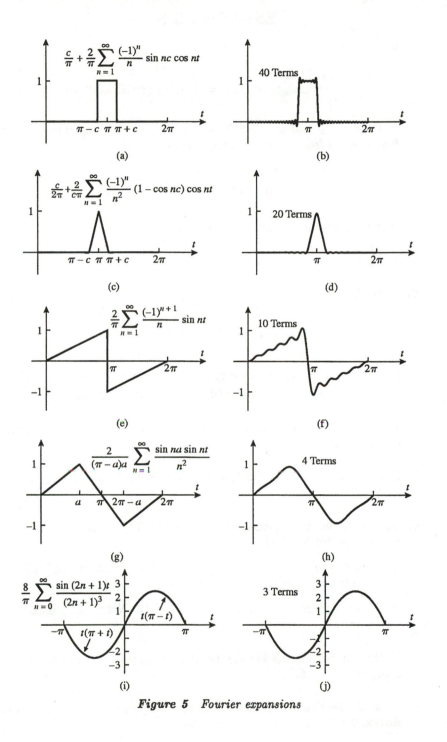

**Figure 5** *Fourier expansions*

# Exercises 3.3

**1.** Show that

$$\int_0^T \cos \omega_a t \, \cos \omega_b t \, dt = \int_0^T \sin \omega_a t \, \sin \omega_b t \, dt = 0$$

as long as $\omega_a$ and $\omega_b$ are *distinct* positive-integer multiples of the fundamental frequency $\omega = 2\pi/T$; also show that

$$\int_0^T \cos \omega_a t \, \sin \omega_b t \, dt = 0$$

even if $\omega_a$ and $\omega_b$ are not distinct.

**2.** Verify the integrals in (18).

**3.** Show that

$$\int_a^b e^{i2\pi n \frac{t-a}{b-a}} e^{-i2\pi n \frac{t-a}{b-a}} dt = 0$$

if $m \neq n$ (but each is an integer), and from this derive eq. (20). (*Hint:* multiply both sides of (19) by $e^{-i2\pi n \frac{t-a}{b-a}}$ and integrate.)

**4.** Use eq. (20) to show that if $f(t)$ is real then $A_{-n}$ is the complex conjugate of $A_n$.

**5.** Express $c_n$ and $d_n$ in eqs. (16, 17) in terms of $A_n$ and $A_{-n}$ of eq. (20).

**6.** (a) Write out the expansion equations (16, 17) for the case $a = -T/2, b = T/2$.

   (b) Suppose $f(t)$ is even - i.e. $f(-t) = f(t)$. What simplifications result in the expansion for part (a)?

   (c) Now suppose $f$ is odd: $f(-t) = -f(t)$. What simplifications result?

**7.** Verify the expressions for the Fourier coefficients for the functions depicted in Fig. 5.

**8.** What is the Fourier series for the function $\sin t, 0 \leq t \leq 2\pi$? for $\sin t/2, 0 \leq t \leq 2\pi$?

# 3.4 Convergence of Fourier Series

**Summary** The Fourier series for a square-integrable function $f(t)$ converges in the mean-squares sense to the function; the partial sums are best least-squares approximants. The pointwise convergence depends on the degree of smoothness of the function; roughly speaking, if $f$ is $\ell$ times continuously differentiable, the $N$th partial sum approximates $f$ to within an error proportional to $1/N^{\ell-1}$.

In this section we are going to survey the mathematical details concerning the nature of the convergence of the Fourier series. We have deferred discussion of such matters because we didn't want to interrupt our exposition of the formal computation of the series. Rigorous proofs of the results discussed here can be found in the references by Kammler, H. F. Davis, P. J. Davis, Edwards, and Ince.

For definiteness we work with expansions on the standard interval $[0, 2\pi]$; rescaling extends the results to other intervals. The Fourier series expansion takes the form

$$f(t) = \sum_{n=1}^{\infty} \{c_n \cos nt + d_n \sin nt\} + c_0, \ 0 \le t \le 2\pi; \qquad (1)$$

$$c_0 = \frac{1}{2\pi} \int_0^{2\pi} f(t)\,dt = \frac{\langle f, 1 \rangle}{\|1\|^2},$$

$$c_{n>0} = \frac{1}{\pi} \int_0^{2\pi} f(t) \cos nt\,dt = \frac{<f, \cos nt>}{\|\cos nt\|^2},$$

$$d_n = \frac{1}{\pi} \int_0^{2\pi} f(t) \sin nt\,dt = \frac{<f, \sin nt>}{\|\sin nt\|^2}. \qquad (2)$$

To get a feeling for the quality of the Fourier expansion, we begin by asking an important practical question. If, say, only 11 of the terms in the expansion (1) were to be used to approximate the "target" function $f(t)$, what values should be assigned to the coefficients $a_n, b_n$ in the sum

$$f(t) \approx \sum_{n=1}^{5} \{a_n \cos nt + b_n \sin nt\} + a_0 \qquad (3)$$

in order to get the best approximation in the *least squares* sense - that is, in the sense that the mean square integral of the residual

$$\int_a^b |f(t) - \sum_{n=1}^{5} \{a_n \cos nt + b_n \sin nt\} - a_0|^2\,dt$$

$$= \|f - \sum_{n=1}^{5} \{a_n \cos nt + b_n \sin nt\} - a_0\|^2 \qquad (4)$$

is minimal? Is it possible that some *other* formula than (2) might be better, if we knew we were only going to use 11 terms? The answer, perhaps surprisingly, is "No"; *the mean square error is minimized when the $a_n, b_n$'s are selected to coincide with the Fourier coefficients $c_n, d_n$ defined by (2).*

The proof of this statement is accomplished with a little bit of inspired mathematical juggling, beginning with the seemingly innocuous step of adding, and then subtracting, Fourier's terms

$$\sum_{n=1}^{5}\{c_n \cos nt + d_n \sin nt\} + c_0$$

to the generic error expression (4). To simplify the notation we abbreviate the Fourier terms as

$$P_{Fourier} \equiv \sum_{n=1}^{5}\{c_n \cos nt + d_n \sin nt\} + c_0$$

and the "generic" approximation as

$$P_{generic} \equiv \sum_{n=1}^{5}\{a_n \cos nt + b_n \sin nt\} + a_0.$$

Then we assess the error as follows:

$$mean\ square\ error = ||f - P_{generic}||^2$$
$$= ||(f - P_{Fourier}) + (P_{Fourier} - P_{generic})||^2$$

For the next step we invoke the Hilbert-space version of the Law of Cosines, proved in exercise 1:

$$||g + h||^2 = ||g||^2 + ||h||^2 + 2\langle g, h \rangle. \tag{5}$$

This results in

$$mean\ square\ error = ||f - P_{Fourier}||^2 + ||P_{Fourier} - P_{generic}||^2$$
$$+ 2\langle (f - P_{Fourier}), (P_{Fourier} - P_{generic}) \rangle. \tag{6}$$

But the inner product in (6) is zero, because $f - P_{fourier}$ is the sum of terms containing $\cos nt$ and $\sin nt$ with $n$ **greater than** 5 (recall eq. (1)), while $(P_{Fourier} - P_{generic})$ contains $\cos nt$ and $\sin nt$ with $n$ less than or equal to 5. So the two members of the above inner product are orthogonal.

As a result equation (6) says

$$mean\ square\ error = ||f - P_{Fourier}||^2 + ||P_{Fourier} - P_{generic}||^2$$

and clearly the error is minimal when the generic approximations agrees with Fourier's terms. The optimal choice for the coefficients is the Fourier inner product (2).

The precise statement of our deliberations goes as follows:

**Theorem 1.** The choice (2) for the coefficients $c_n, d_n$ minimizes the mean square error for finite-sum approximation

$$f(t) \approx \sum_{n=1}^{N} \{c_n \cos nt + d_n \sin nt\} + c_0$$

to any function $f(t)$ which is square-integrable over the interval $[0, 2\pi]$ (i.e.

$$\int_0^{2\pi} |f(t)|^2 dt < \infty). \tag{7}$$

This is most reassuring. In any realistic situation, of course, we can only compute with expansions involving a finite number of terms anyway. Our deliberations demonstrate that by following the Fourier procedure we are still taking the best course of action, even though we don't intend to use *all* the terms.

Note that *some* integrability condition, such as equation (7), is essential in order that the integrals defining the Fourier coefficients (2) be *finite*. It turns out that square-integrability is a natural "working hypothesis" for the Fourier theory. In fact one of the most general convergence results is the following:

**Theorem 2.** Let $f(t)$ be square-integrable on the interval $[0, 2\pi]$. Then the Fourier series (1,2) converges to $f(t)$ *in the mean square sense*. That is, the integral of the square of the *difference* between $f(t)$ and the finite sums of the expansion goes to zero, as the number of terms goes to infinity:[6]

$$\int_0^{2\pi} |f(t) - \sum_{n=1}^{N} \{c_n \cos nt + d_n \sin nt\} - c_0|^2 dt$$

$$= ||f(t) - \sum_{n=1}^{N} \{c_n \cos nt + d_n \sin nt\} - c_0||^2$$

$$\to 0 \quad as \; N \to \infty. \tag{8}$$

Roughly speaking this theorem implies that if we draw graphs of the finite sums of the Fourier series for higher and higher N, the area between these graphs and the graph of $f(t)$ goes to zero. It does *not* necessarily imply that at every point $t$ the numerical values of the sums approach those of $f(t)$, because the hypothesis imposed on $f$ is too broad. Consider, for example, the case when $f$ has a jump discontinuity; it's still square-integrable, but the (continuous) sums can't possibly match $f$ on both sides of the jump (recall Fig. 4 in the previous section).

---

[6]Mean square convergence is sometimes known as $L^2$ convergence in honor of the mathematician Lebesgue, who invented the modern theory of integration.

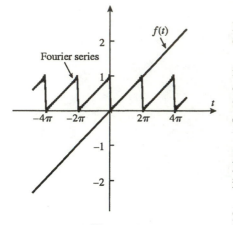

**Figure 1**

Before we consider the theorems governing convergence at individual points - "pointwise convergence" - we have to address an anomaly which occurs at the end points $0, 2\pi$ of the interval.

It is clear from (1) that the Fourier series defines a *periodic* function; it takes the same value at points spaced $2\pi$ apart. Consequently if we construct the expansion for a function $f$ which is not periodic, the graph of the *expansion* will reproduce itself in the intervals $[2\pi, 4\pi]$, $[4\pi, 6\pi]$, $[-2\pi, 0]$, etc. See Fig. 1.

Therefore *the Fourier series artificially extends $f$ as a periodic function*. This means that if the values of $f(t)$ do not match at 0 and $2\pi$ the Fourier series, in effect, sees a jump discontinuity at the end points. Thus it converges to $[f(0) + f(2\pi)]/2$ at the end points (i.e., to .5 in Fig. 1).

Similarly, if $f(0) = f(2\pi)$ but the *slopes* of $f$ do not match at 0 and $2\pi$, the Fourier series sees a discontinuous derivative and its convergence will be slowed there (recall the triangle function example in the previous section).

The facts governing pointwise convergence of the Fourier series for $f(t)$ on $[0,2\pi]$ are as follows:

(i) If $f(t)$, extended as a periodic function, is integrable, then the series converges to $[f(t^+) + f(t^-)]/2$ at every point $t$ where $f$ has both a right-hand derivative and a left-hand derivative (and thus to $f(t)$ if $f$ is continuous there).

(ii) If $f(t)$, extended as a periodic function, is continuous and has a piecewise continuous derivative, then the series converges *uniformly* to $f$ at every point.[7]

(iii) If $f(t)$, extended as a periodic function, is $\ell$ times continuously differentiable, then if $\ell > 1$

$$f(t) - \sum_{n=1}^{N} \{c_n \cos nt + d_n \sin nt\} - c_0 = \mathcal{O}(N^{-\ell+1}) \qquad (9)$$

uniformly as $N \to \infty$. (The order-of-magnitude notation is discussed in Appendix A.1; $\mathcal{O}(N^{-\ell+1})/N^{-\ell+1}$ remains bounded as $N \to \infty$.)

(iv) If the *generalized $\ell$th* derivative of $f(t)$, extended as a periodic function, equals a sum of a finite number of delta functions plus a bounded function, then

$$f(t) - \sum_{n=1}^{N} \{c_n \cos nt + d_n \sin nt\} - c_0 = \mathcal{O}(N^{-\ell+1}) \qquad (10)$$

uniformly as $N \to \infty$. (See exercise 3.)

---

[7]Uniform convergence means that, by taking a sufficiently large number of terms, we can make the difference between $f$ and the finite sum uniformly small along the entire interval $[0,2\pi]$. A study of Fig. 4 in the previous section shows that a discontinuous function can never be uniformly approximated by a continuous one.

In the course of manipulating Fourier expansions the questions of differentiation and integration arise. This issue is most conveniently addressed in the exponential format:

$$f(t) = \sum_{n=-\infty}^{\infty} A_n e^{int}, \ 0 \le t \le 2\pi. \tag{11}$$

When can we expect convergence of the term-by-term differentiated series

$$f'(t) \overset{(?)}{=} \sum_{n=-\infty}^{\infty} in A_n e^{int}, \ 0 \le t \le 2\pi, \tag{12}$$

and the integrated series

$$\int_0^t f(\tau)\,d\tau \overset{(?)}{=} A_0 t + \sum_{\substack{n=-\infty \\ n \ne 0}}^{\infty} \frac{A_n}{in}[e^{int} - 1] \ ? \tag{13}$$

One might expect the presence of the factor $n$ in the denominator of the integrated series (13) to *improve* its convergence, and this is true. The general theory shows that if $f$ is integrable the series (13) will converge pointwise to $\int f\,dt$ - even if the original series (11) did not converge to $f$!

The differentiated series (12) fares worse (as might be expected from the location of the factor $n$). Most of the convergence theorems guarantee only that if the derivative $f'(t)$ meets the conditions for its *own* Fourier series to converge, then that series will agree with the one obtained by term-by-term differentiation of the series for $f(t)$. The *rate* of convergence of (12) is governed by the theorems (i, ii, iii), applied to $f'$.

> To summarize: the Fourier series for a square-integrable function $f$ converges to $f$ in the mean square. Pointwise convergence of the series is governed by the smoothness of $f$: the smoother the function, the more rapidly convergent is the series. Term-by-term integration is justified for a convergent series and improves the rate of convergence. Term-by-term differentiation is justified only if the derived series convergences, and the rate of convergence is thereby slowed.

# Exercises 3.4

1. Prove identity (5). Why is it called the Law of Cosines? (*Hint*: interpret $g$ and $h$ as vectors.)

2. Obtain computer plots for the graphs of the (truncated) Fourier series of the functions depicted in Fig. 5 of the previous section; observe the convergence as more terms are included.

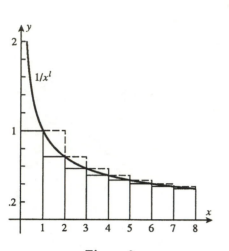

**Figure 2**

**3.** (a) By comparing the sum $\sum_{n=n_1}^{n_2} n^{-\ell}$ with the integral $\int_{n_1-1}^{n_2} x^{-\ell}\,dx$ (Fig. 2), establish the estimate

$$\sum_{n=N}^{\infty} n^{-\ell} = \mathcal{O}(N^{-\ell+1}), \quad \ell > 1. \tag{14}$$

(b) Suppose that $f(t)$ is a periodic function possessing $\ell$ continuous derivatives. Integrate by parts $\ell$ times to show that for $n > 0$

$$\int_0^T f(t)\,e^{i2\pi nt/T}dt = (-1)^{\ell}\{\frac{T}{2\pi in}\}^{\ell} \int_0^T f^{(\ell)}(t)\,e^{i2\pi nt/T}dt \tag{15}$$

and therefore that the coefficient $A_n$ in the series (11) satisfies

$$|A_n| \le \{\frac{T}{2\pi n}\}^{\ell}\, T\, max|f^{(\ell)}(t)| = \frac{(constant\,independent\,of\,n)}{n^{\ell}}, \tag{16}$$

where $T$ is the period of $f$.

(c) Use (14) and (16) to derive (9) by justifying the following:

$$
\begin{aligned}
|f(t) - \sum_{n=-N}^{N} A_n e^{i2\pi nt/T}| &= |\{\sum_{-\infty}^{-N-1} + \sum_{N+1}^{\infty}\} A_n e^{i2\pi nt/T}| \\
&\le \{\sum_{-\infty}^{-N-1} + \sum_{N+1}^{\infty}\}|A_n| \\
&= \mathcal{O}(N^{-\ell+1})
\end{aligned}
$$

(d) Let $f^{(\ell)}(t)$ be the $\ell$th derivative of $f$ in the distribution sense. Show from (15) that the final estimate in (16) is still valid if $f^{(\ell)}(t)$ equals a sum of delta functions plus a bounded function. (Explain why it fails if there is a $\delta'$ term in $f^{(\ell)}$.) Thus for the step function depicted in Fig. 5 of Section 3.3, the *first* derivative $f'$ contains delta functions; and, sure enough, the coefficients fall off like $\mathcal{O}(N^{-\ell})$ with $\ell=1$.

(e) Explain the decay of the coefficients in the other Fourier series depicted in Fig. 5 of Section 3.3.

## 3.5   Sine and Cosine Series

**Summary**   A square-integrable function $f(t)$ can be expressed either as a sine series or a cosine series:

$$f(t) = \sum_{n=1}^{\infty} b_n \sin \frac{\pi n}{T} t = a_0 + \sum_{n=1}^{\infty} a_n \cos \frac{\pi n}{T} t, \ 0 \le t \le T;$$

$$b_n = \frac{2}{T} \int_0^T f(t) \sin \frac{\pi n}{T} t \, dt;$$

$$a_0 = \frac{1}{T} \int_0^T f(t) \, dt, \ a_{n>0} = \frac{2}{T} \int_0^T f(t) \cos \frac{\pi n}{T} t \, dt.$$

The rate of convergence of each representation depends on the degree of smoothness of $f$ and on its end-point behavior; ideally $f(0) = f(T) = 0$ for the sine series, and $f'(0) = f'(T) = 0$ for the cosine series.

If we take a given function $f(t)$ and replace it by its Fourier series, computed over the interval $0 \le t \le T$ -

$$f(t) = \sum_{n=1}^{\infty} \{c_n \cos \frac{2\pi n}{T} t + d_n \sin \frac{2\pi n}{T} t\} + c_0; \tag{1}$$

$$c_0 = \frac{1}{T} \int_0^T f(t) \, dt, \ c_{n>0} = \frac{2}{T} \int_0^T f(t) \cos \frac{2\pi n}{T} t \, dt,$$

$$d_n = \frac{2}{T} \int_0^T f(t) \sin \frac{2\pi n}{T} t \, dt \tag{2}$$

- then, as discussed in Section 3.4, the result is the artificial extension of $f$ to a periodic function. See Fig. 1.

It is extremely beneficial to consider two other such (artificial) extensions. For the first, we initially extend $f(t)$ to the interval $-T \le t \le 0$ as an odd function, $f(-t) = -f(t)$ (Fig. 2). Then we construct the Fourier series for this new function, for $-T \le t \le T$:

$$f_{odd}(t) = \sum_{n=1}^{\infty} \{a_n \cos \frac{2\pi n}{2T} t + b_n \sin \frac{2\pi n}{2T} t\} + a_0; \tag{3}$$

$$a_0 = \frac{1}{2T} \int_{-T}^{T} f_{odd}(t) \, dt, \ a_{n>0} = \frac{2}{2T} \int_{-T}^{T} f_{odd}(t) \cos \frac{2\pi n}{2T} t \, dt,$$

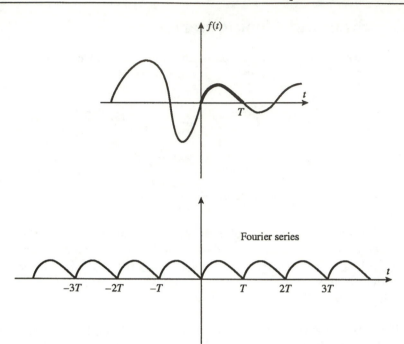

**Figure 1**   *Fourier series extension*

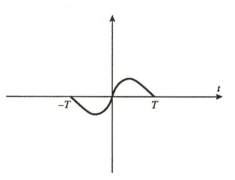

**Figure 2**   *Odd extension*

$$b_n = \frac{2}{2T} \int_{-T}^{T} f_{odd}(t) \sin \frac{2\pi n}{2T} t \, dt. \tag{4}$$

Notice that it gives an altogether different extension of $f$; see Fig. 3.

Now here's the payoff. Because the cosines are even functions, and $f_{odd}$ is odd, the coefficients $a_n$ in (4) are all zero! Also by symmetry the formula for the $b_n$ reduces to

$$b_n = \frac{2}{T} \int_{0}^{T} f(t) \sin \frac{\pi n}{T} t \, dt, \tag{5}$$

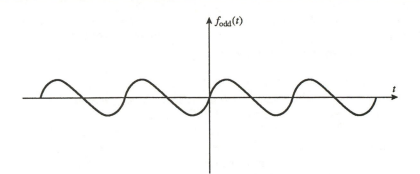

**Figure 3** *Periodic odd extension*

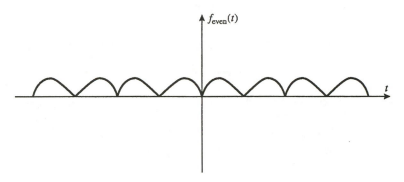

**Figure 4** *Periodic even extension*

and the new expansion, *valid only over the original interval* $[0, T]$, is

$$f(t) = \sum_{n=1}^{\infty} b_n \sin \frac{\pi n}{T} t, \; 0 \le t \le T. \tag{6}$$

Equations (5, 6) express the *Fourier sine series* expansion of $f(t)$. Its convergence is governed by theorems analogous to those for the full Fourier series discussed in the previous section - with one complication. Since $f$ is artificially extended as an *odd* function, $f(0)$ *and* $f(T)$ *must both be zero* for convergence at the end points.

If we repeat this construction with *even* extensions of $f$, instead of odd ones, we end up (exercise 1) with the *Fourier cosine series* over the original interval:

$$f(t) = \sum_{n=1}^{\infty} a_n \cos \frac{\pi n}{T} t + a_0, \; 0 \le t \le T, \tag{7}$$

$$a_0 = \frac{1}{T} \int_0^T f(t) \, dt, \quad a_{n>0} = \frac{2}{T} \int_0^T f(t) \cos \frac{\pi n}{T} t \, dt. \tag{8}$$

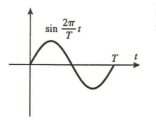

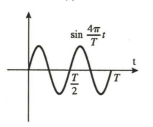

The extension of $f(t)$ induced by the cosine series is illustrated in Fig. 4. For fastest convergence of the cosine series $f'(0)$ and $f'(T)$ should both be zero (which is not true in the figure).

It may seem a little disparate that using, say, the sine series one can achieve the same degree of convergence as the full Fourier series over the interval $[0, T]$; after all, doesn't the sine series only have "half" as many terms? This "miscount" is revealed when we consider the individual sinusoids. The sine functions which appear in the full Fourier expansion, eq. (1), are plotted in Fig. 5. Those which arise in the sine series are illustrated in Fig. 6. Thus the sine series uses "twice" as many functions, and the balance is restored.

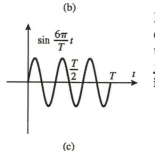

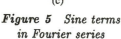

**Example 1.** We compute the sine series expansion for the function

$$f(t) = \cos t, \ 0 \le t \le 2\pi.$$

If one *mis*remembers that the sines are orthogonal to the cosines, this exercise might appear futile. However recall that in Section 3.3 we showed that $\sin \omega t$ and $\cos \omega t$ are orthogonal over $[0, 2\pi]$ *if $\omega$ is a harmonic of the fundamental frequency* - i.e. for the interval $[0, 2\pi]$, if $\omega$ is an integer. But in the sine series over $[0, 2\pi]$

$$\cos t = \sum_{n=1}^{\infty} b_n \sin \frac{n}{2} t, \tag{9}$$

**Figure 5** *Sine terms in Fourier series*

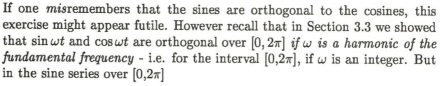

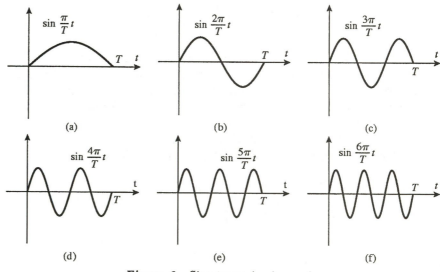

**Figure 6** *Sine terms in sine series*

the frequencies are integers and half-integers. Thus only the even-numbered sine terms are orthogonal to $\cos t$. In fact

$$b_{2m} = \frac{1}{\pi} \int_0^{2\pi} \cos t \sin \frac{2m}{2} t \, dt = 0,$$

$$b_{2m+1} = \frac{1}{\pi} \int_0^{2\pi} \cos t \sin(\frac{2m+1}{2}t) \, dt = \frac{4(2m+1)}{(2m-1)(2m+3)}.$$

The convergence is illustrated in Fig. 7.                                            ■

Note the discrepancy at the end points. The approximations *try* to match $\cos t$ along the entire interval, but the values of the sine series are constrained to be zero at the end points. The little overshoot that the series exhibits just prior to diving down to its end values is typical of Fourier approximations at jump discontinuities; it is called the *Gibbs phenomenon*.

The cosine series expansion for the function $f(t) = t$ is depicted in Fig. 8 (see exercise 4). Since it is the *slope* (and not the *value*) of the cosine series which is constrained at the end points, convergence is better.

# Exercises 3.5

**1.** Derive the equations (7, 8) for the Fourier cosine series.

**2.** Work out the coefficients for the sine series expansion of the function $f(t) = 1$, $0 \leq t \leq \pi$. Sketch the behavior of the series at the endpoints 0 and $\pi$.

**3.** Repeat exercise 2 for the function $e^t$.

**4.** Derive the expression

$$t = \frac{\pi}{2} - \frac{4}{\pi} \sum_{(odd\,n)} \frac{1}{n^2} \cos nt, \quad 0 < t < \pi.$$

**5.** Work out the coefficients for the cosine series expansion of the function $\sin t$, $0 \leq t \leq \pi$. Sketch the behavior of the series at the endpoints 0 and $\pi$.

**6.** Explain the rate of decay of the coefficients in the series (9), in terms of the ideas discussed in exercise 4 of Section 3.4.

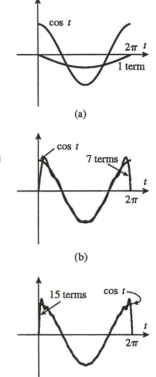

(a)

(b)

(c)

**Figure 7**  *Sine series for $\cos t$*

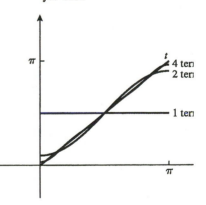

**Figure 8**  *Cosine series for $t$*

158                                                                    *Chapter 3. Fourier Methods*

## 3.6   The Fourier Transform

**Summary**   The Fourier expansion of a nonperiodic function, valid over the entire real line, requires the entire continuum of frequencies:

$$f(t) = \int_{-\infty}^{\infty} \mathcal{F}(\omega)e^{i\omega t}\, d\omega, \quad -\infty < t < \infty.$$

The coefficient of $e^{i\omega t}$ is the Fourier transform of $f$:

$$\mathcal{F}(\omega) = \frac{1}{2\pi} \int_{-\infty}^{\infty} f(t)e^{-i\omega t}\, dt.$$

If $f$ is square-integrable, the expansion converges in a mean-square sense. If $f$ is integrable, the pointwise convergence is governed by the degree of smoothness of $f$. Fourier sine and cosine transforms are available for the half-interval $[0,\infty)$.

A function and its transform share many properties which aid in computation and application. They involve reflection, differentiation, modulation, conjugation, time reversal, time dilation, and power scaling.

The exponential form of the Fourier series for a function $f(t)$ on the symmetric interval $[-T, T]$ is given by

$$f(t) \;=\; \sum_{n=-\infty}^{\infty} A_n e^{in\pi t/T}, \tag{1}$$

$$A_n \;=\; \frac{1}{2T} \int_{-T}^{T} f(t)e^{-in\pi t/T}\, dt. \tag{2}$$

As we have seen it converges (at least in the mean square) to $f(t)$ over the interval $-T \le t \le T$, and it reproduces $f$ periodically outside that interval.

The restriction of a time function to a *finite* interval - and its periodic extension beyond that interval - are undesirable for modeling many physical situations. Thus we now contemplate an expression for $f(t)$, in terms of sinusoids, which is valid for all time: $-\infty < t < \infty$. We shall achieve this by extending the interval $[-T, T]$ for the Fourier series expansion; as $T \to \infty$ we encompass *all* of the values of $f$ in the expansion, and the periodic copies are exiled to infinity. See Fig. 1.

To facilitate the extraction of the limit as $T \to \infty$ in (1, 2) we need to alter the notation. Let $\omega_n$ denote the frequencies occurring in (1, 2):

$$\omega_n = n\pi/T, \quad n = 0, 1, 2, \ldots.$$

The set of numbers $\{\omega_n\}$ is depicted in Fig. 2 for increasing values of $T$. Notice that as $T \to \infty$ the spacing

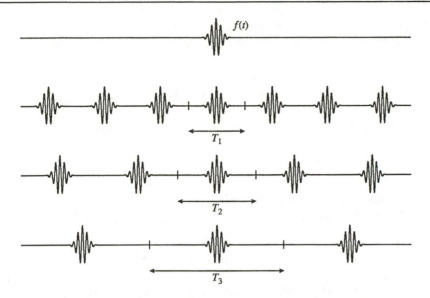

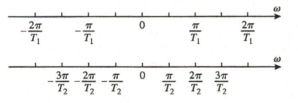

**Figure 1**   *Fourier expansions over increasing intervals*

**Figure 2**   *Refined spacing of frequencies*

$$\Delta\omega_n = \omega_{n+1} - \omega_n = \pi/T \tag{3}$$

gets finer and finer, and the set tends to fill up the positive real line. Thus we anticipate a *continuum* of frequencies arising in the expansion; the Fourier series will evolve into an *integral*.

To exhibit the metamorphosis of the Fourier series as $T$ increases, we rescale the coefficients $A_n$ in (2) by $\pi/T$, writing

$$A_n = \frac{\pi}{T}\, \mathcal{F}(\omega_n) = \Delta\omega_n\, \mathcal{F}(\omega_n) \tag{4}$$

where

$$\mathcal{F}(\omega) = \frac{1}{2\pi} \int_{-T}^{T} f(t) e^{-i\omega t}\, dt. \tag{5}$$

Then the Fourier expansion (1) becomes

$$f(t) = \sum_{n=-\infty}^{\infty} \mathcal{F}(\omega_n)\, e^{i\omega_n t} \Delta\omega_n, \quad -T \le t \le T. \tag{6}$$

Of course $\Delta\omega_n = \pi/T$ goes to zero as $T$ goes to infinity. So as we anticipated, (6) has the form of a sum approximating the (presumably convergent) integral

$$\int_{-\infty}^{\infty} \mathcal{F}(\omega)\, e^{i\omega t}\, d\omega.$$

Thus the limiting form of (6) is the expansion for $f(t)$ that we seek:

$$f(t) = \int_{-\infty}^{\infty} \mathcal{F}(\omega)\, e^{i\omega t}\, d\omega, \quad -\infty < t < \infty. \tag{7}$$

It expresses $f$ as a "sum" of sinusoids, over a continuum of frequencies. The "coefficients" in the expansion are displayed by the limiting form of (5):

$$\mathcal{F}(\omega) = \frac{1}{2\pi} \int_{-\infty}^{\infty} f(t)\, e^{-i\omega t}\, dt. \tag{8}$$

The function $\mathcal{F}(\omega)$ defined by (8) is known as the *Fourier transform* of $f(t)$. Equation (7) - the *Fourier Integral Theorem* - expresses $f$ in terms of its Fourier transform; it is the sinusoidal expansion that we have been seeking. It is also known as the "inverse Fourier transform" of $\mathcal{F}(\omega)$. Some authors define the transform so that the constant $1/2\pi$ appears in the Fourier integral (7), rather than the transform (8); others put $1/\sqrt{2\pi}$ in both equations and obtain a more symmetric form. We prefer (7) because it conspicuously expresses the format that we were seeking.

*Subject to conditions to be stated below, any function $f(t)$ can be expressed as a superposition of sinusoids for all $t$, $-\infty < t < \infty$.*

Our choice of language - superposition, rather than integral - emphasizes the interpretation we want to convey. The Fourier expansion is a generalization of the Fourier series. But in order that it converge to $f$ for *all* $t$ (so that there are no artificial periodic extensions induced on $f$), the entire continuum of frequencies is involved in the representation of $f$. It is very helpful to think of $\mathcal{F}(\omega)\, d\omega$ as the "(infinitesimal) coefficient" of $e^{i\omega t}$ in the "sum" (recall (6)).

**Example 1.** The Fourier transform of the function $e^{-|t|}$ is given by

$$\begin{aligned}
\frac{1}{2\pi} \int_{-\infty}^{\infty} e^{-|t|} e^{-i\omega t}\, dt &= \frac{1}{2\pi} 2 \int_{0}^{\infty} e^{-t} \cos\omega t\, dt \text{ (by symmetry)} \\
&= \frac{1}{\pi} \frac{e^{-t}}{1+\omega^2} \left\{ -\cos\omega t + \omega \sin\omega t \right\}_0^{\infty} \\
&= \frac{1}{\pi(1+\omega^2)}.
\end{aligned}$$

The integral theorem thus states

$$e^{-|t|} = \int_{-\infty}^{\infty} \frac{1}{\pi(1+\omega^2)} e^{i\omega t} d\omega.$$

This can be verified directly using complex function theory (e.g., see Saff and Snider). ∎

**Example 2.** The rectangular pulse function is given by

$$f(t) = \begin{cases} 1, & |t| < T \\ 0, & |t| > T \end{cases}$$

Its Fourier transform is

$$\mathcal{F}(\omega) = \frac{1}{2\pi} \int_{-T}^{T} (1) e^{-i\omega t} \, dt = \frac{\sin \omega T}{\pi \omega},$$

and the inverse transform equation becomes

$$\int_{-\infty}^{\infty} \frac{\sin \omega T}{\pi \omega} e^{i\omega t} \, d\omega = \begin{cases} 1, & |t| < T \\ 1/2, & |t| = T \\ 0, & |t| > T \end{cases}$$ ∎

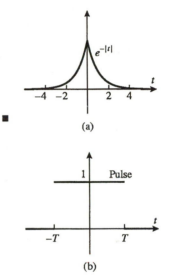

(a)

(b)

**Example 3.** The function

$$f(t) = \begin{cases} \sin t, & |t| < 6\pi \\ 0, & \text{otherwise} \end{cases}$$

is called a "finite wave train." Its transform is

$$\mathcal{F}(\omega) = \frac{1}{2\pi} \int_{-6\pi}^{6\pi} \sin t \, e^{-i\omega t} dt = \frac{i \sin 6\pi \omega}{\pi(1-\omega^2)},$$

The functions analyzed in these examples are depicted in Fig. 3. and we have, for the inverse transform,

$$\int_{-\infty}^{\infty} \frac{i \sin 6\pi\omega}{\pi(1-\omega^2)} e^{i\omega t} d\omega = \begin{cases} \sin t, & |t| < 6\pi \\ 0, & \text{otherwise} \end{cases} .$$ ∎

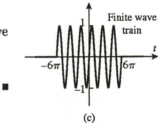

(c)

*Figure 3   Functions for Fourier transform*

The *Fourier sine* and *cosine transforms* are constructed by analogy to the corresponding series. Thus if we artificially redefine $f(t)$ for negative $t$ by $f(-t) = -f(t)$, $0 < t < \infty$, we can manipulate the transform equations (7, 8) to express $f$ for positive values by

$$f(t) = \int_0^{\infty} S(\omega) \sin \omega t \, d\omega, \quad 0 \le t < \infty, \tag{9}$$

where the *Fourier sine transform* is defined by

$$S(\omega) = \frac{2}{\pi} \int_0^\infty f(t) \sin \omega t \, dt \qquad (10)$$

(see exercise 1). Clearly $f(0)$ should be zero for (10) to hold pointwise.

Similarly the *Fourier cosine transform* equations

$$f(t) = \int_0^\infty C(\omega) \cos \omega t \, d\omega, \ 0 \le t < \infty, \qquad (11)$$

$$C(\omega) = \frac{2}{\pi} \int_0^\infty f(t) \cos \omega t \, dt \qquad (12)$$

model functions with $f'(0) = 0$.

The issue of *pointwise* convergence of the Fourier representation as more and more frequencies are admitted - that is, the validity of

$$\lim_{\Omega \to \infty} \int_{-\Omega}^{\Omega} \mathcal{F}(\omega) e^{i\omega t} \, d\omega = f(t) \qquad (13)$$

- is governed by the smoothness of $f$ in much the same manner as for the Fourier series in Section 3.4. A typical theorem states that if $f$ is integrable over $(-\infty, \infty)$ then its Fourier integral converges to $[f(t^+) + f(t^-)]/2$ at points where $f$ has both a right-hand and a left-hand derivative.

Notice, however, that the integrability condition

$$\int_{-\infty}^{\infty} |f(t)| \, dt < \infty \qquad (14)$$

is quite demanding. It is not sufficient for $f$ to be continuous, or bounded. *In particular, the basic function* $\sin t$ *is not covered by this theory!* This vexing situation will be repaired in the next section.

For *mean square* convergence of the Fourier integral to $f$, it is necessary that $f$ be square-integrable

$$\int_{-\infty}^{\infty} |f(t)|^2 \, dt < \infty. \qquad (15)$$

The precise statement of mean square convergence, however, is formidable:

$$\lim_{\Omega \to \infty} \int_{-\infty}^{\infty} |f(t) - \int_{-\Omega}^{\Omega} \mathcal{F}(\omega) e^{i\omega t} \, d\omega|^2 \, dt \to 0. \qquad (16)$$

It should be clear that assembling a table of Fourier transforms would be a valuable undertaking for engineering analysis. There are several rules which enhance the construction and usefulness of such a listing. Note first of all that if $f(t)$ and $\mathcal{F}(\omega)$ are a Fourier transform pair:

$$\mathcal{F}(\omega) = \frac{1}{2\pi} \int_{t=-\infty}^{\infty} f(t)e^{-i\omega t}dt, \; f(t) = \int_{\omega=-\infty}^{\infty} \mathcal{F}(\omega)e^{i\omega t}d\omega \qquad (17)$$

- and if we switch the names of the variables so that "$\omega$" becomes "$t$" and "$t$" becomes "$-\omega$":

$$\mathcal{F}(t) = \frac{1}{2\pi} \int_{-\omega=-\infty}^{\infty} f(-\omega)e^{i\omega t}d(-\omega), \; f(-\omega) = \int_{t=-\infty}^{\infty} \mathcal{F}(t)e^{-it\omega}dt \quad (18)$$

- the result can be rearranged to show that $f(-\omega)/2\pi$ is the Fourier transform of $\mathcal{F}(t)$:

$$\left[\frac{f(-\omega)}{2\pi}\right] = \frac{1}{2\pi} \int_{-\infty}^{\infty} \mathcal{F}(t)e^{-it\omega}dt, \; \mathcal{F}(t) = \int_{-\infty}^{\infty} \left[\frac{f(-\omega)}{2\pi}\right]e^{it\omega}d\omega. \quad (19)$$

This "reflection rule" allows us to double the number of entries in our transform table.

**Example 4.** It follows from the above that the calculations in Examples 1, 2, and 3 yield three more transforms:

the transform of $\frac{1}{\pi(1+t^2)}$ is $e^{-|\omega|}/2\pi$ ;

the transform of $\frac{\sin \Omega t}{\pi t}$ is $\begin{cases} 1/2\pi, & |\omega| < \Omega \\ 0, & |\omega| > \Omega \end{cases}$ ;

the transform of $\frac{i\sin 6\pi t}{\pi(1-t^2)}$ is $\begin{cases} -[\sin \omega]/2\pi, & |\omega| < 6\pi \\ 0, & \text{otherwise} \end{cases}$ .   ∎

The "derivative rule" holds when $f(t)$ and its derivative $f'(t)$ both have Fourier transforms; then differentiation under the integral sign can be justified and we find that

$$f(t) = \int_{-\infty}^{\infty} \mathcal{F}(\omega)e^{i\omega t}d\omega, \; f'(t) = \int_{-\infty}^{\infty} i\omega \mathcal{F}(\omega)e^{i\omega t}d\omega.$$

This demonstrates that *the Fourier transform of $f'(t)$ is $i\omega\mathcal{F}(\omega)$*. Similarly *the Fourier transform of $f^{(n)}(t)$ is $(i\omega)^n\mathcal{F}(\omega)$*.

The verification of the remaining rules will be left to the exercises:

("Modulation") *The Fourier transform of $f(t)\,e^{i\Omega t}$ is $\mathcal{F}(\omega - \Omega)$.*[8]

("Conjugation") *The Fourier transform of $\overline{f(t)}$ is $\overline{\mathcal{F}(-\omega)}$.*

("Time reversal") *The Fourier transform of $f(-t)$ is $\mathcal{F}(-\omega)$.*

("Time shift") *The Fourier transform of $f(t+T)$ is $e^{i\omega T}\mathcal{F}(\omega)$.*

("Time dilation") *The Fourier transform of $f(at)$ is $\mathcal{F}(\omega/a)/|a|$.*

---

[8]This rule finds application in electromagnetic wave transmission of information.

("Power scaling") *The Fourier transform of $t^n f(t)$ is $i^n F^{(n)}(\omega)$.*

Table 3.1 at the end of Sec. 3.7 summarizes these rules and contains most of the known formulas for transforms of common functions. The reference by Erdelyi et al. is more extensive.

However most inverse Fourier transforms that arise in practice are impossible to compute analytically. Residue theory, a complex variable technique, can be used in some cases. The next section discusses the Fourier analysis of generalized functions; naturally, simplifications result if the transforms contain delta functions.

The real workhorse of practical Fourier analysis, however, is its *numerical* implementation via the Fast Fourier Transform algorithm. Section 3.10 presents a survey of this technology from the vantage point of the theory we have studied.

## Exercises 3.6

1. Derive the forms (9-12) for the sine and cosine transforms.

2. Derive the six Fourier transform rules quoted in the text.

3. Derive the transform for the triangle function in Table 3.1.

4. Extend the computation in Example 1 to derive the Fourier transform of $e^{-a|t|}$ for $a > 0$.

5. Use exercise 4 and the rules to derive the transform of $\frac{a}{\pi(a^2+t^2)}$.

6. By differentiation with respect to $a$ in exercises 4 and 5, derive the transforms for $|t|e^{-a|t|}$ and $\frac{a^2-t^2}{\pi(a^2+t^2)^2}$.

7. If the real part of the complex number $a$ is positive, show that the Fourier transform of $t/(t^2+a^2)$ is given by $\frac{-i}{2}e^{-a|\omega|}\text{sgn}(\omega)$ (undefined for $\omega = 0$), by power scaling in the result in exercise 5. (*Most references quote $\frac{i\omega}{4a}e^{-a|\omega|}$, which is incorrect, for this Fourier transform!*)

## 3.7  Fourier Analysis of Generalized Functions

**Summary**  The Fourier transform can be extended to non-integrable functions through the machinery of the theory of distributions. This allows the important tools of Fourier analysis to be applied to sinusoids, exponentials, delta functions, etc., while preserving its basic properties.

The integrability conditions of Sec. 3.6 for the Fourier transform are too severe for our purposes. After all, things certainly seem out of kilter when the Fourier integral expansion, which was concocted to express functions in terms of sinusoids, is not valid for sinusoids themselves! Fortunately Heaviside's calculus and Schwartz's generalized function theory come to the rescue. A brief treatment of the *rigorous* description of Fourier transforms of distributions (along the lines of Sec. 1.5) is outlined in exercises 6 and 7. Herein we shall proceed formally, as we did in Sec. 1.4.

Observe that the identity

$$e^{i\omega_0 t} = \int_{-\infty}^{\infty} \delta(\omega - \omega_0) e^{i\omega t} d\omega \qquad (1)$$

has the form of a Fourier integral expansion, *if we identify the delta function as the Fourier transform of $e^{i\omega_0 t}$*. Thus we are led to postulate the transform equation

$$\delta(\omega - \omega_0) = \frac{1}{2\pi} \int_{-\infty}^{\infty} e^{i\omega_0 t} e^{-i\omega t} dt = \frac{1}{2\pi} \int_{-\infty}^{\infty} e^{i(\omega_0 - \omega)t} dt \qquad (2)$$

Is this sensible? That is, does the right-hand side of (2), which is obviously nonconvergent, possess the basic reproducing property of the delta function -

$$g(\omega_0) = \int_{-\infty}^{\infty} g(\omega)\, \delta(\omega - \omega_0)\, d\omega \ \text{ for all suitable } g? \qquad (3)$$

Yes it does, if we interpret the operations in the distribution sense. First we set up the integral (3) with (2) substituted for the delta function,

$$g(\omega_0) \stackrel{?}{=} \int_{\omega=-\infty}^{\infty} g(\omega) \left\{ \frac{1}{2\pi} \int_{t=-\infty}^{\infty} e^{i\omega_0 t} e^{-i\omega t} dt \right\} d\omega, \qquad (4)$$

*and reinterpret (4) with the t-integration deferred until the end:*

$$g(\omega_0) \stackrel{?}{=} \int_{t=-\infty}^{\infty} \left\{ \frac{1}{2\pi} \int_{\omega=-\infty}^{\infty} g(\omega) e^{-i\omega t} d\omega \right\} e^{i\omega_0 t} dt \qquad (5)$$

To facilitate recognition of (5) we rename the dummy variable "$\omega$" as "$t$" and vice versa, and we rename "$\omega_0$" as "$t_0$":

$$g(t_0) \stackrel{?}{=} \int_{\omega=-\infty}^{\infty} \left\{ \frac{1}{2\pi} \int_{t=-\infty}^{\infty} g(t) e^{-it\omega} dt \right\} e^{it_0 \omega} d\omega \qquad (6)$$

Now we observe that the quantity in the braces is the Fourier transform of $g$; so the outer integral expresses the Fourier integral for $g(t_0)$. The equation is valid!

Thus we have seen that $\delta(\omega - \omega_0)$ is the (generalized) Fourier transform of $e^{i\omega_0 t}$. Conversely, it is obvious that the Fourier transform of $\delta(t - t_0)$ is $e^{-i\omega t_0}/2\pi$, because

$$\frac{1}{2\pi} \int_{-\infty}^{\infty} \delta(t - t_0)e^{-i\omega t} dt = \frac{1}{2\pi}e^{-i\omega t_0}. \tag{7}$$

Using these generalized distribution transforms we can express a much larger class of functions with the Fourier integral.

**Example 1.** Any *periodic* function with a Fourier series

$$f(t) = \sum_{n=-\infty}^{\infty} A_n e^{i\omega_n t} = \sum_{n=-\infty}^{\infty} A_n e^{in\omega_1 t} \tag{8}$$

can clearly be rewritten as a Fourier integral with delta functions:

$$f(t) = \int_{-\infty}^{\infty} \left\{ \sum_{n=-\infty}^{\infty} A_n \delta(\omega - \omega_n) \right\} e^{i\omega t} d\omega. \tag{9}$$

The corresponding Fourier transform is

$$F(\omega) = \sum_{n=-\infty}^{\infty} A_n \delta(\omega - \omega_n).$$

In fact $f$ need not be periodic; whether the frequencies $\omega_n$ are harmonic or not, if the series (8) converges then (9) will hold. The mathematical theory of "almost periodic" functions is based on such representations.    ∎

**Example 2.** With the distribution calculus some functions which grow at $\infty$ can have Fourier transforms; the following formal integration by parts, for example, demonstrates that $te^{i\omega_0 t}$ has $i\delta'(\omega - \omega_0)$ as its Fourier transform:

$$\int_{-\infty}^{\infty} i\delta'(\omega - \omega_0)e^{i\omega t} d\omega = i\delta(\omega - \omega_0)e^{i\omega t}\big|_{-\infty}^{\infty}$$

$$- i \int_{-\infty}^{\infty} \delta(\omega - \omega_0)\frac{d}{d\omega}\left(e^{i\omega t}\right) d\omega$$

$$= -i \frac{d}{d\omega}\left(e^{i\omega t}\right)\bigg|_{\omega=\omega_0}$$

$$= te^{i\omega_0 t}.$$    ∎

**Example 3.** The modern theory of distributions extends the delta-function identity (3) to *complex* values of $\omega_0$, when the function $g$ is suitable (see exercise 9, Sec. 1.5, for a fuller description). The surprising equation

$$\int_{-\infty}^{\infty} \delta(\omega + i)e^{i\omega t} d\omega = e^{i\omega t}\big|_{\omega=-i} = e^t$$

thus assigns $\delta(\omega + i)$ as the Fourier transform of the exponential $e^t$.    ∎

It is evident from Example 3 that a rigorous treatment of all the aspects of generalized Fourier transforms is far beyond the scope of this book. However, the results of this theory are so useful in engineering analysis that it would be criminal to omit them. Therefore we will proceed with heuristic derivations for some of the rules and transforms and invite the reader to try some others in the exercises.

Remember that the main benefit of using sinusoidal functions in engineering analysis is the replacement of the differentiation operation by multiplication with $i\omega$ (Section 3.1). The distribution interpretation of convergence justifies the continuance of this practice; all Fourier integral expansions may be differentiated under the integral sign. Thus the identification in Sec. 3.6 of $i\omega F(\omega)$ as the Fourier transform of $f'(t)$ holds universally, without regard to convergence or integrability. The other transform rules of that section - reflection, time reversal, etc. - are also universally valid in the distribution sense.

**Example 4.** We have seen that the Fourier transform of $\delta(t)$ is $1/2\pi$ (set $t_0 = 0$ in eq. (7)). Also we know

$$\delta(t) = \frac{d}{dt}H(t),$$

where $H$ is the unit Heaviside (step) function (Section 1.3). Thus if $G(\omega)$ denotes the Fourier transform of $H(t)$, we have

$$\frac{1}{2\pi} = i\omega G(\omega). \tag{10}$$

The *generalized* solution of (10) (bear in mind Example 3 of Sec. 1.3) is given by

$$G(\omega) = \frac{1}{2\pi i\omega} + A\delta(\omega) \tag{11}$$

for some constant $A$. To evaluate $A$, recall that by the conjugation rule the transform of $H(-t)$ must be $G(-\omega)$ and observe that

$$1 = H(t) + H(-t). \tag{12}$$

Now since eq. (1) implies that the Fourier transform of 1 is $\delta(\omega)$, we take the transform of eq. (12) and get

$$\delta(\omega) = \frac{1}{2\pi i\omega} + A\delta(\omega) + \frac{1}{2\pi i(-\omega)} + A\delta(-\omega) = 2A\delta(\omega)$$

It follows that $A = 1/2$, and the Fourier transform of the Heaviside step function must be

$$G(\omega) = \frac{1}{2\pi i\omega} + \frac{1}{2}\delta(\omega) \tag{13}$$

∎

**Example 5.** We shall exploit (13) to derive the Fourier transform of $1/t$. Exchanging $t$ for $\omega$ in (13) we have

$$\frac{1}{t} = 2\pi i G(t) - \pi i \delta(t). \tag{14}$$

By the reflection rule the transform of $G(t)$ is $H(-t)/2\pi$, and the transform of $\delta(t)$ is $1/2\pi$. Therefore the transform of the function $1/t$ is given by

$$2\pi i H(-t)/2\pi - \pi i/2\pi = iH(-t) - i/2,$$

and the Fourier integral is expressed

$$\frac{1}{2\pi} \int_{-\infty}^{\infty} \frac{1}{t} e^{-i\omega t} dt = \begin{cases} -i/2, & \text{for } \omega > 0 \\ 0, & \text{for } \omega = 0 \\ i/2, & \text{for } \omega < 0 \end{cases} . \tag{15}$$

(In Sec. 1.5 we pointed out that this integral is to be interpreted as a Cauchy principal value).                                                                    ∎

**Example 6.** Let us construct the Fourier *series* for the delta function $\delta(t - t_0)$, with $t_0$ in the interval $(0, 2\pi)$. We compute

$$\delta(t - t_0) = \sum_{n=0}^{\infty} c_n \cos nt + \sum_{n=1}^{\infty} d_n \sin nt;$$

$$c_0 = \frac{1}{2\pi} \int_0^{2\pi} \delta(t - t_0)\, dt = \frac{1}{2\pi},$$

$$c_{n>0} = \frac{1}{\pi} \int_0^{2\pi} \delta(t - t_0) \cos nt\, dt = \frac{1}{\pi} \cos nt_0,$$

$$d_n = \frac{1}{\pi} \int_0^{2\pi} \delta(t - t_0) \sin nt\, dt = \frac{1}{\pi} \sin nt_0.$$

Therefore the expansion reads

$$\delta(t - t_0) = \frac{1}{\pi} \sum_{n=1}^{\infty} \{\cos nt_0 \cos nt + \sin nt_0 \sin nt\} + \frac{1}{2\pi}. \tag{16}$$

To be specific, if we set $t_0 = \pi$,

$$\delta(t - \pi) = \frac{1}{\pi} \sum_{n=1}^{\infty} (-1)^n \cos nt + \frac{1}{2\pi}. \tag{17}$$

∎

Certainly (17) does not converge in any pointwise manner - the individual terms don't even go to zero! This reflects the fact that the delta function does not satisfy any of the smoothness conditions that we hypothesized for convergence of a Fourier series in Sec. 3.4. Nonetheless, the series does have the reproducing property that we associate with the delta function, if we adopt the liberal rules of distribution calculus (as in Section 1.3). For, inserting (17) into the integral

$$\int_0^{2\pi} f(t)\delta(t-\pi)dt$$

and reversing the order of summation and integration, we derive

$$\int_0^{2\pi} f(t)\left\{\frac{1}{\pi}\sum_{n=1}^{\infty}(-1)^n\cos nt + \frac{1}{2\pi}\right\}dt$$
$$= \sum_{n=1}^{\infty}\frac{<f,\cos nt>}{\pi}(-1)^n + \frac{<f,1>}{2\pi}. \tag{18}$$

Now compare (18) with the (*de facto* convergent) Fourier series representation for $f(\pi)$: by eqs. (16, 17) of Sec. 3.3,

$$f(\pi) = \sum_{n=1}^{\infty}\left\{\frac{<f,\cos nt>}{\pi}\cos n\pi + \frac{<f,\sin nt>}{\pi}\sin n\pi\right\}$$
$$+ \frac{<f,1>}{2\pi} \tag{19}$$

(18) and (19) are exactly the same; the expansion (17) of the delta function is vindicated.

For curiosity's sake, we display the sum of the first 40 terms of (17) in Fig. 1.

**Example 7.** The Fourier series for the pulse function depicted in Fig. 2 is given by

$$f(t) = \frac{1}{2} - \frac{2}{\pi}\sum_{n=0}^{\infty}(-1)^n\frac{\cos(2n+1)t}{2n+1} \tag{20}$$

(exercise 5). Since it is discontinuous its derivative obeys none of the smoothness conditions of Sec. 3.4. And sure enough, the differentiated series

$$\frac{2}{\pi}\sum_{n=0}^{\infty}(-1)^n\sin(2n+1)t \tag{21}$$

converges nowhere but at 0, $\pi$, and $2\pi$.

However if we invoke Heaviside's calculus, we can say that $f(t)$ *is* differentiable; its derivative is a sum of delta functions:

$$f'(t) = \delta(t-\pi/2) - \delta(t-3\pi/2). \tag{22}$$

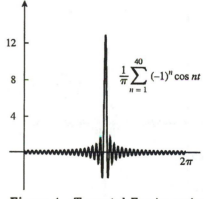

**Figure 1**   *Truncated Fourier series*
$$\delta(t-\pi)$$

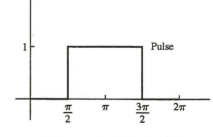

**Figure 2**   *Pulse function*

And if we compute the Fourier series for $\delta(t - \pi/2) - \delta(t - 3\pi/2)$ using (16), it is consistent with (21):

$$
\begin{aligned}
\delta(t - \pi/2) &= \frac{1}{\pi} \sum_{n=1}^{\pi} \left\{ \cos\frac{n\pi}{2}\cos t + \sin\frac{n\pi}{2}\sin t \right\} + \frac{1}{2\pi} \\
&= \frac{1}{\pi} \sum_{m=1}^{\infty} (-1)^m \cos 2mt + \frac{1}{\pi} \sum_{m=0}^{\infty} (-1)^m \sin(2m+1)t + \frac{1}{2\pi}; \\
\delta(t - 3\pi/2) &= \frac{1}{\pi} \sum_{n=1}^{\infty} \left\{ \cos\frac{3n\pi}{2}\cos t + \sin\frac{3n\pi}{2}\sin t \right\} + \frac{1}{2\pi} \\
&= \frac{1}{\pi} \sum_{m=1}^{\infty} (-1)^m \cos 2mt - \frac{1}{\pi} \sum_{m=0}^{\infty} (-1)^m \sin(2m+1)t + \frac{1}{2\pi};
\end{aligned}
$$

$\delta(t - \pi/2) - \delta(t - 3\pi/2) =$ same as (21).  ∎

One of the beautiful consequences of the theory of distributions is that it permits term-by-term differentiation of a Fourier series any number of times, if the result is interpreted in the distribution sense. That is, the result is not taken literally until it has been multiplied by a smooth function, integrated first, and then summed. Thus for example the wildly divergent series obtained by differentiating (17),

$$
-\frac{1}{\pi} \sum_{n=1}^{\infty} (-1)^n n \sin nt, \tag{23}
$$

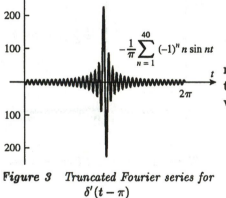

$$-\frac{1}{\pi} \sum_{n=1}^{40} (-1)^n n \sin nt$$

**Figure 3**  *Truncated Fourier series for* $\delta'(t - \pi)$

represents $\delta'(t-\pi)$; if it is multiplied by $f(t)$ and integrated before summing, the result is $-f'(\pi)$. The first 40 terms are plotted in Fig. 3. (Compare with the depiction of $\delta'$ in Fig. 9, Sec. 1.3.)

We summarize the convergence situation as follows. The Fourier series for a square-integrable function $f$ converges to $f$ in the mean square. Pointwise convergence of the series is governed by the smoothness of $f$: the smoother the function, the more rapidly convergent is the series. The Fourier series for a distribution like $\delta(t)$ typically diverges at every point; however it can be formally treated as convergent, with the understanding that the summations are taken literally only *after* it has been integrated termwise with a regular function.

Table 3.1 at the end of this section contains the Fourier transforms of most of the commonly used generalized functions.

# Exercises 3.7

**1.** Derive the Fourier transform of $t^n$ for a positive integer $n$.

**2.** Derive the Fourier transform of $sgn(t)$. From the result derive the distribution identity $\int_0^\infty \sin \omega t \, dt = 1/\omega$.

**3.** Derive the Fourier transform for $\delta^{(n)}(t)$.

**4.** Derive the Fourier transforms for $\cos \Omega t$ and $\sin \Omega t$.

**5.** Verify the expression for the coefficients in eq. (20).

**6.** Let $f(t)$, $g(t)$ be two square-integrable functions with Fourier transforms $F(\omega)$, $G(\omega)$ respectively. The basis for the rigorous theory of Fourier transforms of distributions is *Parseval's identity*:

$$\int_{-\infty}^\infty f(t)\overline{g(t)} \, dt = 2\pi \int_{-\infty}^\infty F(\omega)\overline{G(\omega)} \, d\omega. \qquad (24)$$

Derive this relation by inserting the Fourier integral for $g$ and reversing the order of integration.

Recall from Sec. 1.5 that distribution theory is premised on the identification of a function through its integrals with test functions (chosen from some family such as $C_0^\infty$), rather than through its pointwise values. Therefore if $g(t)$ is a test function, then the left side of (24) is well defined for any *generalized* function $f(t)$. Consequently if $G(\omega)$ is regarded as a test function also, (24) provides a "distribution" definition of $F(\omega)$, and as a result the generalized functions are endowed with Fourier transforms!

(In fact the theory does not proceed quite as smoothly as this, because a $C_0^\infty$ function does not have a $C_0^\infty$ Fourier transform in general - in other words, one cannot expect $g(t)$ and $G(\omega)$ both to be $C_0^\infty$ test functions in (24). Without going into details we simply state that with a suitable enlargement of the test function space, (24) can be used to identify a Fourier transform for the generalized function $f(t)$ in the manner we have suggested (see Kammler's book).)

Let us see how this formalism is used to prove some of the properties discussed in the text for the *generalized* transforms.

(i) The transform of $e^{i\omega_0 t}$ is $\delta(\omega - \omega_0)$:

For any (real) test function $g(t)$ with transform $G(\omega)$ we have

$$\int_{-\infty}^\infty e^{i\omega_0 t}\overline{g(t)} \, dt = 2\pi\overline{G(\omega_0)} \quad \text{(since } g(t) = \overline{g(t)}\text{).}$$

$$(25)$$

Now if we can express the right-hand side of (25) as $2\pi \int_{-\infty}^{\infty} F(\omega)\overline{G(\omega)}\, d\omega$ we can identify $F$ as the Fourier transform of $e^{i\omega_0 t}$ (thanks to the Parseval identity). But clearly we can employ the delta function to write the right-hand side as

$$2\pi \int_{-\infty}^{\infty} \delta(\omega - \omega_0)\overline{G(\omega)}\, d\omega$$

Thus we conclude that $\delta(\omega - \omega_0)$ is the Fourier transform of $e^{i\omega_0 t}$.

(ii) The derivative rule:

For any distribution $f$ and test function $g$ we have

$$\int_{-\infty}^{\infty} f'(t)\overline{g(t)}\, dt = -\int_{-\infty}^{\infty} f(t)\overline{g'(t)}\, dt.$$

Since we know that the transform of the *test* function $g'(t)$ is $i\omega G(\omega)$, by Parseval's rule this equals

$$= -2\pi \int_{-\infty}^{\infty} F(\omega)\overline{[i\omega G(\omega)]}\, d\omega$$

$$= (+)2\pi \int_{-\infty}^{\infty} i\omega F(\omega)\overline{G(\omega)}\, d\omega \qquad (26)$$

Considering the first and last members of (26), we conclude that the transform of $f'(t)$ is given by $i\omega F(\omega)$.

**7.** Demonstrate, in the above manner, the Fourier rules for

(a) reflection (b) modulation (c) conjugation (d) time shift
(d) time reversal (e) power scaling (f) time dilation

for generalized functions.

*Table 3.1 Fourier Transforms* **173**

# TABLE 3.1 FOURIER TRANSFORMS

$$f(t) = \int_{-\infty}^{\infty} F(\omega)e^{i\omega t}d\omega \qquad\qquad F(\omega) = \frac{1}{2\pi}\int_{-\infty}^{\infty} f(t)e^{-i\omega t}dt$$

---

*GENERAL RULES*

| | |
|---|---|
| $c_1 f_1(t) + c_2 f_2(t)$ | $c_1 F_1(\omega) + c_2 F_2(\omega)$   (Linearity) |
| $f(t)e^{i\omega_0 t}$ | $F(\omega - \omega_0)$   (Modulation) |
| $f^{(n)}(t)$ | $(i\omega)^n F(\omega)$   (Derivative) |
| $F(t)$ | $f(-\omega)/2\pi$   (Reflection)[9] |
| $\overline{f(t)}$ | $\overline{F(-\omega)}$   (Conjugation) |
| $f(at+b)$ | $e^{i\omega b/a}F(\omega/a)/|a|$   (Time shift/dilation) |
| $t^n f(t)$ | $i^n F^{(n)}(\omega)$   (Power scaling) |

*REGULAR TRANSFORMS*

| | |
|---|---|
| $(t^2 + a^2)^{-1}$ | $\dfrac{1}{2a}e^{-a|\omega|}$   $(\mathrm{Re}\{a\} > 0)$ |
| $t/(t^2 + a^2)$ | $\dfrac{-i}{2}e^{-a|\omega|}\,\mathrm{sgn}(\omega)$   $(\mathrm{Re}\{a\} > 0)$ <br> (Most references have $\frac{i\omega}{4a}e^{-a|\omega|}$, <br> which is incorrect.) |
| $pulse = \begin{cases} 1, & |t| < T \\ 0, & |t| > T \end{cases}$ | $\dfrac{\sin \omega T}{\pi\omega}$ |
| $triangle = \begin{cases} 0, & |t| > T \\ 1 - \dfrac{|t|}{T}, & |t| < T \end{cases}$ | $\dfrac{2\sin^2(\omega T/2)}{\pi T \omega^2}$ |
| $e^{-at}H(t)$ | $\dfrac{1}{2\pi(a + i\omega)}$   $(a > 0)$ |

---

[9]Because of this property one should search *both* columns of Table 3.1 when seeking a transform or an inverse transform.

| | |
|---|---|
| $wavetrain$ $= e^{i\omega_0 t} \times (pulse)$ | $\dfrac{\sin(\omega - \omega_0)T}{\pi(\omega - \omega_0)}$ |
| $e^{iat}e^{-bt}H[t\,\mathrm{sgn}(b)]$ | $\dfrac{\mathrm{sgn}(b)}{i2\pi(\omega - a - ib)}$ |
| $e^{-a|t|}$ | $\dfrac{a}{\pi(\omega^2 + a^2)} \quad (a > 0)$ |
| $|t|e^{-a|t|}$ | $\dfrac{a^2 - \omega^2}{\pi(a^2 + \omega^2)^2} \quad (a > 0)$ |
| $e^{-a^2 t^2}$ | $\dfrac{1}{2a\sqrt{\pi}}e^{-\omega^2/4a^2} \quad (a > 0)$ |
| $|t|^{-p}$ | $\dfrac{\Gamma(1 - p)\sin p\pi/2}{\pi}|\omega|^{p-1} \quad (0 < p < 1)$ |
| $\mathrm{sgn}(b)e^{iat-bt}H[t\,\mathrm{sgn}(b)]$ $- \mathrm{sgn}(t)/2$ | $\dfrac{a + ib}{i2\pi\omega(\omega - a - ib)}$ |
| $\mathrm{sgn}(b)e^{iat-bt}H[t\,\mathrm{sgn}(b)]$ $- \mathrm{sgn}(d)e^{ict-dt}H[t\,\mathrm{sgn}(d)]$ | $\dfrac{a - c + i(b - d)}{i2\pi(\omega - a - ib)(\omega - c - id)}$ |

## DISTRIBUTION TRANSFORMS

| | |
|---|---|
| $1$ | $\delta(\omega)$ |
| $t^n$ | $i^n\delta^{(n)}(\omega) \quad (\,n \text{ positive integer})$ |
| $t^n e^{i\omega_0 t}$ | $i^n\delta^{(n)}(\omega - \omega_0) \quad (n \text{ positive integer})$ |
| $1/t$ | $-(i/2)\,\mathrm{sgn}(\omega)$ |
| $1/t^n$ | $\dfrac{-i(-i\omega)^{n-1}}{2(n-1)!}\,\mathrm{sgn}(\omega) \quad (n \text{ positive integer})$ |
| $|t|$ | $-1/\pi\omega^2$ |

$$\text{sgn}(t) = \begin{cases} 1, & t > 0 \\ 0, & t = 0 \\ -1, & t < 0 \end{cases} \qquad 1/i\pi\omega$$

$$H(t) = \begin{cases} 1, & t > 0 \\ 1/2, & t = 0 \\ 0, & t < 0 \end{cases} \qquad \frac{\delta(\omega)}{2} + \frac{1}{i2\pi\omega}$$

$t^n \, \text{sgn}(t)$ $\qquad (i\omega)^{-n-1} n!/\pi \quad (n \text{ positive integer})$

$t^n H(t)$ $\qquad \dfrac{i^n}{2}\delta^{(n)}(\omega) - \dfrac{i^{n-1}n!}{2\pi}\omega^{-n-1}$

$\delta(t - t_0)$ $\qquad e^{-i\omega t_0}/2\pi$

$\delta^{(n)}(t - t_0)$ $\qquad e^{-i\omega t_0}(i\omega)^n/2\pi \quad (n \text{ positive integer})$

$e^{at}$ $\qquad \delta(\omega + ia) \quad (\text{any complex } a)$

$t^n e^{at}$ $\qquad i^n \delta^{(n)}(\omega + ia) \quad (\text{any complex } a)$

$\cos \Omega t$ $\qquad \dfrac{\delta(\omega - \Omega)}{2} + \dfrac{\delta(\omega + \Omega)}{2}$

$\sin \Omega t$ $\qquad \dfrac{\delta(\omega - \Omega)}{2i} - \dfrac{\delta(\omega + \Omega)}{2i}$

For more extensive listings see the references by Erdelyi et al., Sneddon, and Kammler.

## 3.8 The Laplace Transform

**Summary** The Laplace Transform is formally equivalent to $(2\pi)$ times the Fourier transform, applied to functions which are zero for negative values, and evaluated at imaginary frequencies:

$$\mathcal{L}(s; f) = 2\pi \, F(-is; f) = \int_0^\infty f(t)e^{-st}\, dt,$$

$$(s = i\omega; \; -\infty < \omega < \infty).$$

Like the Fourier transform it is useful for analyzing differential equations with constant coefficients, but the form of the differentiation rule

$$\mathcal{L}(s; f') = s\mathcal{L}(s; f) - f(0)$$

renders it more useful for solving initial-value problems than for obtaining steady-state solutions.

Many functions which have no (classical) Fourier transforms do have Laplace transforms, for the following reasons:

1. the Laplace transform can override exponential growth at $+\infty$ by assigning a positive real part to the variable $s$; and

2. the Laplace transform is immune to any kind of divergence at $-\infty$.

Accordingly, an inverse Laplace transform formula can be derived from the Fourier representation:

$$\frac{1}{2\pi i}\int_{s=\alpha-i\infty}^{\alpha+i\infty} \mathcal{L}(s;f)e^{st}\,ds = \begin{cases} f(t), & t>0 \\ 0, & t<0 \end{cases}$$

if $f(t)e^{-\alpha t}$ approaches zero at $\infty$.

The fact that the Fourier transformation changes the differentiation process into simple multiplication -

$$\begin{aligned} f(t) &\longleftrightarrow F(\omega) \\ f'(t) &\longleftrightarrow i\omega F(\omega) \\ &\vdots \qquad\quad \vdots \\ f^{(n)}(t) &\longleftrightarrow (i\omega)^n F(\omega) \end{aligned} \qquad (1)$$

- is the key to its success as an analysis tool; constant coefficient differential equations are changed into algebraic ones. Its main drawback is that it is intrinsically suited for modeling steady state responses *for all time*, from the distant past to the distant future:

$$f(t) = \int_{-\infty}^{\infty} F(\omega)e^{i\omega t}\,d\omega, \ -\infty < t < \infty; \ F(\omega) = \frac{1}{2\pi}\int_{-\infty}^{\infty} f(t)e^{-i\omega t}\,dt. \qquad (2)$$

The problem is that many engineering systems are better modeled as having been "turned on" at $t=0$ in a certain initial state, and evolving from that state into the future. Their past is irrelevant, possibly having been governed by a different set of physical laws. The *Laplace transform*, then, is a modification of the Fourier transform which preserves the advantages of the exchanges (1) without being burdened with the task of modeling the system's behavior for negative time.

Given a function $f(t)$ defined for $t \geq 0$, as indicated in Fig. 1, we create the function $f_{ext}(t)$ by extending $f$ artificially to be zero for negative $t$. Let

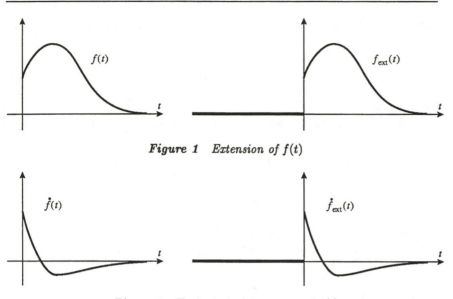

**Figure 1**   *Extension of $f(t)$*

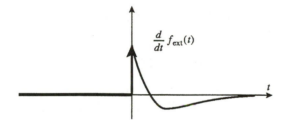

**Figure 2**   *Extension of derivative of $f(t)$*

**Figure 3**   *Derivative of $f_{ext}(t)$*

$F(\omega; f_{ext})$ be the Fourier transform of $f_{ext}$:

$$F(\omega; f_{ext}) = \frac{1}{2\pi} \int_{-\infty}^{\infty} f_{ext}(t)e^{-i\omega t}\, dt$$
$$= \frac{1}{2\pi} \int_{0}^{\infty} f(t)e^{-i\omega t}\, dt. \tag{3}$$

If we extend the derivative $\dot{f}$ of $f(t)$ as in Fig. 2, its Fourier transform reads

$$F(\omega; \dot{f}_{ext}) = \frac{1}{2\pi} \int_{0}^{\infty} \dot{f}(t)e^{-i\omega t}\, dt. \tag{4}$$

We must investigate the relationship of $F(\omega; \dot{f}_{ext})$ to $F(\omega; f_{ext})$. *It is not as simple as (1), because the function $\dot{f}_{ext}$ is not the derivative of $f_{ext}$.* Granted, $\dot{f}_{ext}(t) = \frac{d}{dt} f_{ext}(t)$ for $t > 0$ by construction; and $\dot{f}_{ext}(t) = \frac{d}{dt} f_{ext}(t)$ for $t < 0$ since both are zero; but since $f_{ext}$ jumps discontinuously by $f(0)$ at the origin, its derivative contains a delta function:

$$\frac{d}{dt} f_{ext}(t) = \dot{f}_{ext}(t) + f(0)\delta(t) \tag{5}$$

(see Fig. 3).

Thus, applying the differentiation rule (1) to eq. (5) we derive

$$
\begin{aligned}
i\omega F(\omega; f_{ext}) &= F(\omega; \dot{f}_{ext}) + \frac{1}{2\pi} \int_{-\infty}^{\infty} f(0)\delta(t)e^{-i\omega t}\, dt \\
&= F(\omega; \dot{f}_{ext}) + \frac{1}{2\pi} f(0).
\end{aligned} \tag{6}
$$

We tentatively define the "formal" Laplace transform $\mathcal{L}(s; f)$ of $f(t)$ to be $2\pi$ times the Fourier transform of $f_{ext}$, with the variable $\omega$ replaced by $-is$: that is,

$$\mathcal{L}(s; f) = 2\pi F(-is; f_{ext}) = \int_0^\infty f(t)e^{-st}\, dt \quad (s = i\omega;\ -\infty < \omega < \infty) \tag{7}$$

Eq. (6) then specifies how the Laplace transform exchanges differentiation for multiplication:

$$\mathcal{L}(s; \dot{f}) = s\mathcal{L}(s; f) - f(0). \tag{8}$$

It is more complicated than the Fourier transform exchange (1), because of the introduction of the initial value $f(0)$. Iteration of (8) results in (exercise 1)

$$\mathcal{L}(s; f^{(n)}) = s^n \mathcal{L}(s; f) - s^{n-1} f(0) - s^{n-2} \dot{f}(0) - \cdots - f^{(n-1)}(0). \tag{9}$$

Some modifications will be introduced into the definition (7) shortly, but for the moment we consider an example.

**Example 1.** The Laplace transform of $f(t) = e^{-t}$ is

$$\mathcal{L}(s; e^{-t}) = \int_0^\infty e^{-t} e^{-st}\, dt = \frac{-1}{s+1} e^{-t} e^{-(s=i\omega)t}\Big|_{t=0}^{t=\infty} = \frac{1}{s+1}.$$

Its derivative, $-e^{-t}$, of course has $\frac{-1}{s+1}$ as its transform. We can confirm eq. (8):

$$
\begin{aligned}
\frac{-1}{s+1} &\stackrel{?}{=} s\frac{1}{s+1} - e^{-0} \\
&= \frac{s}{s+1} - \frac{s+1}{s+1} \\
&= \frac{-1}{s+1}.
\end{aligned}
$$

$\blacksquare$

This simple example illustrates one of the advantages of the Laplace transform. The function $e^{-t}$ becomes indefinitely large as $t \to -\infty$, and its *Fourier* transform is a distribution (indeed, with a complex argument!):

$$F(\omega; e^{-t}) = \delta(\omega - i)$$

(see Table 3.1). But since the *Laplace* transform zeros out the negative-time record of $e^{-t}$ and since the latter decays in positive time, a smooth function is returned -

$$\mathcal{L}(s; e^{-t}) = \frac{1}{s+1}. \tag{10}$$

The next step in the development of this tool is to extend the domain of the independent variable $s = i\omega$ to include a real part -

$$s = \alpha + i\omega \tag{11}$$

We thereby introduce, mathematically, an extra convergence factor $e^{-\alpha t}$ into the transform:

$$
\begin{aligned}
\mathcal{L}(s; f) &= \mathcal{L}(\alpha + i\omega; f) \\
&= \int_0^\infty f(t) e^{-st}\, dt \\
&= \int_0^\infty f(t) e^{-\alpha t} e^{i\omega t}\, dt.
\end{aligned} \tag{12}
$$

As a result we can, by taking $\alpha = \text{Re}\{s\}$ large enough, accommodate some functions $f$ which *grow* at $+\infty$. For example, if $f(t) = e^{5t}$ then the integral $\int_0^\infty e^{5t} e^{-st}\, dt$ diverges for $s = i\omega$, but for $s = 6 + i\omega$ (say)

$$
\begin{aligned}
\int_0^\infty e^{5t} e^{-st}\, dt &= \int_0^\infty e^{5t} e^{-6t} e^{-i\omega t}\, dt \\
&= \int_0^\infty e^{(-1-i\omega)t}\, dt = \frac{1}{-1-i\omega} e^{-t} e^{-i\omega t}\Big|_{t=0}^{t=\infty} = \frac{1}{1+i\omega} \\
&= \frac{1}{s-5}.
\end{aligned}
$$

In fact the integral equals $1/(s-5)$ for *any* $s$ whose real part is greater than 5 (see exercise 2 if this is not clear).

With this in mind we can now extend our definition of the Laplace transform $\mathcal{L}(s; f)$ to include functions $f$ which grow exponentially at $+\infty$ and to include all complex values of $s = \alpha + i\omega$ for which the real part $\alpha$ compensates for the exponential growth rate of $f$:

*Given a function*[10] $f(t)$, $t \geq 0$, *which is bounded for large t by*

$$|f(t)| \leq M e^{pt} \tag{13}$$

---

[10]The function need not be continuous of course, but we assume that its discontinuities are tame enough for (14) to make sense.

*for some constants $M$ and $p$, the Laplace transform $\mathcal{L}(s; f)$ of $f$ is defined to be*

$$\mathcal{L}(s; f) = \int_0^\infty f(t)e^{-st}dt \tag{14}$$

*for all complex values $\alpha + i\omega$ of $s$ for which $\text{Re}\{s\} = \alpha > p$.*

**Example 2.** The Laplace transform for $\sin t$ is defined for $\text{Re}\{s\} > 0$, since $|\sin t| \leq e^{0t}$. Similarly $\mathcal{L}(s; t^n)$ is defined for $\text{Re}\{s\} > 0$ because

$$t^n < e^{pt} \quad \text{(for large } t\text{)}$$

for any $p > 0$ (exercise 3). A listing of common Laplace transforms and their basic properties and identities appears in Table 3.2 at the end of this section. Derivations are discussed in the exercises. ∎

The utilization of the Laplace transform in solving constant-coefficient differential equations is illustrated by the next example.

**Example 3.** To solve the initial value problem

$$\ddot{x} + 2\dot{x} + x = \sin t, \quad x(0) = 1, \quad \dot{x}(0) = 0,$$

we take the Laplace transform of both sides of the differential equation, incorporating the initial conditions according to (9).

$$s^2\mathcal{L}(s; x) - sx(0) - \dot{x}(0) + 2s\mathcal{L}(s; x) - 2x(0) + \mathcal{L}(s; x) = (s^2 + 1)^{-1}.$$

Solving for the unknown transform we find

$$\mathcal{L}(s; x) = \frac{s+2}{s^2 + 2s + 1} + \frac{1}{(s^2 + 2s + 1)(s^2 + 1)}. \tag{15}$$

To recover $f$ from its transform we first observe that (15) does not appear in Table 3.2. However if we reduce it by partial fractions (exercise 4)

$$\frac{s+2}{s^2 + 2s + 1} + \frac{1}{(s^2 + 2s + 1)(s^2 + 1)} = \frac{3/2}{s+1} + \frac{3/2}{(s+1)^2} - \frac{s/2}{s^2 + 1},$$

then we easily find

$$x(t) = \frac{3}{2}e^{-t} + \frac{3}{2}te^{-t} - \frac{1}{2}\cos t. \quad \blacksquare$$

The generalization to higher order equations is evident. For future reference we include the following example of an undamped system which is forced at resonance (exercise 5).

**Example 4.** The Laplace transform of the system

$$\ddot{x} + x = \sin t, \quad x(0) = 0, \quad \dot{x}(0) = 1$$

is found to be

$$s^2 \mathcal{L}(s;x) - s \cdot 0 - 1 + \mathcal{L}(s;x) = \frac{1}{s^2+1},$$

or

$$\mathcal{L}(s;x) = \frac{1}{s^2+1} + \frac{1}{(s^2+1)^2}.$$

From the table we read off the solution

$$x(t) = \sin t + \frac{1}{2}\sin t - \frac{1}{2}t\cos t = \frac{3}{2}\sin t - \frac{1}{2}t\cos t. \qquad \blacksquare$$

The reader should be somewhat familiar with Laplace transforms from a first course in differential equations, where $\mathcal{L}(s;f)$ is usually restricted to *real* values of $s$, sufficiently large to compensate for $f$'s growth. The inverse transform is then only available through the use of tables, which necessitates partial fraction expansions, etc. Herein we have derived the Laplace transform $\mathcal{L}$ as a special case of the Fourier transform $F$ - which has an inversion *formula*. So there must be an inversion formula for the Laplace transform also!

For *integrable* functions we have defined

$$\mathcal{L}(s;f) = 2\pi F(\omega = -is; f_{ext}).$$

But the inverse Fourier transform guarantees

$$f_{ext}(t) = \int_{-\infty}^{\infty} F(\omega; f_{ext})e^{i\omega t}\,d\omega$$

for integrable functions. Therefore making the formal substitution $\omega = -is$ we would expect

$$f_{ext}(t) = \frac{1}{2\pi i} \int_{s=-i\infty}^{i\infty} \mathcal{L}(s;f)e^{st}\,ds. \qquad (16)$$

And in fact this can be proved, for sufficiently smooth functions. (As one would expect, the integral converges to

$$\frac{f(t^+) + f(t^-)}{2}$$

at points $t > 0$ where $f$ has a jump discontinuity.)

But can we get an inverse transform if $f$ grows at infinity? Yes, because if $f$ has exponential growth bounded by $Me^{pt}$, then for $\alpha > p$ the function $[fe^{-\alpha t}]_{ext}$ *is* integrable, and hence in accordance with (16),

$$[fe^{-\alpha t}]_{ext} = \frac{1}{2\pi i} \int_{s=-i\infty}^{i\infty} \mathcal{L}(s;fe^{-\alpha t})e^{st}\,ds.$$

Now from Table 3.2 we have

$$\mathcal{L}(s; fe^{-\alpha t}) = \mathcal{L}(s + \alpha; f)$$

so

$$[fe^{-\alpha t}]_{ext} = \frac{1}{2\pi i} \int_{s=-i\infty}^{i\infty} \mathcal{L}(s + \alpha; f) e^{st}\, ds$$

or

$$f_{ext} = \frac{1}{2\pi i} \int_{s=-i\infty}^{i\infty} \mathcal{L}(s + \alpha; f) e^{(s+\alpha)t}\, ds,$$

and redefining "$s+\alpha$" to be "$s$" we arrive at *Bromwich's integral for the inverse Laplace transform*:

$$\frac{1}{2\pi i} \int_{s=\alpha-i\infty}^{\alpha+i\infty} \mathcal{L}(s; f) e^{st}\, ds = \begin{cases} f(t), & t > 0 \\ 0, & t < 0 \end{cases} \qquad (17)$$

for any real $\alpha$ such that $f(t)e^{-\alpha t} \to 0$ as $t \to \infty$. Again convergence is interpreted in the "Fourier" sense at jump discontinuities.

The Bromwich integral is difficult to evaluate without the tools of analytic function theory; we defer elaboration to the next section.

In Section 3.7 we saw how the Fourier transform formalism had to be supplemented with the distribution calculus to admit nonintegrable functions like $t$ and $e^{it}$ (whose Fourier transforms are $i\delta'(\omega)$ and $\delta(\omega - 1)$ respectively). But a glance at the transform column of Table 3.2 reveals no delta functions among the Laplace transforms for such functions! This can be explained in crude terms by looking at the following reparameterization of the Bromwich integral:

$$f(t) = \frac{1}{2\pi} \int_{\omega=-\infty}^{\infty} \mathcal{L}(s = \alpha + i\omega; f) e^{\alpha t} e^{i\omega t}\, d\omega, \quad t > 0. \qquad (18)$$

It demonstrates that the Laplace transform - unlike the Fourier transform - is not fettered by the constraint of modeling $f$ with mere sinusoids; the exponentially modulated sinusoids $e^{(\alpha+i\omega)t}$, which grow as time increases (for $\alpha > 0$), are available. Therefore the Laplace transform mechanism can accommodate many functions which grow at $+\infty$, without the necessity of invoking distributions. (Keep in mind, of course, that the Laplace transform is also relieved of the task of modeling $f$ at $-\infty$.)

We close the present discussion with two theorems which occasionally find application in engineering analysis: the initial-value and final-value theorems. Their derivation is based on the equation

$$\mathcal{L}(s; \dot{f}) = s\mathcal{L}(s; f) - f(0) \qquad (18 \text{ repeated})$$

and the identity

$$\mathcal{L}(s; \dot{f}) = \int_0^\infty \dot{f}(t) e^{-st} \, dt. \tag{19}$$

If we let $s$ go to $+\infty$ in (19) we see that $\mathcal{L}(s; \dot{f}) \to 0$ - unless $\dot{f}$ contains some delta functions supported at the origin. Subject to this limitation, then, we use this information in (8) to conclude the *initial value theorem*:

$$\lim_{s \to \infty} s\mathcal{L}(s; f) = f(0). \tag{20}$$

On the other hand recall that if $f$ does not grow at infinity but has a limit $f(\infty)$ there, then its Laplace transform is well-defined for any $s > 0$. In such a case (19) implies

$$\mathcal{L}(0; \dot{f}) = f(\infty) - f(0)$$

and this information in (8) results in the *final value theorem*:

$$\lim_{s \to 0} s\mathcal{L}(s; f) = f(\infty). \tag{21}$$

Illustrations of these identities are given in the exercises.

# Exercises 3.8

1. Derive eq. (9).

2. Evaluate the Laplace transform for the function $e^{5t}$, taking $s = \alpha + i\omega$ where $\alpha > 5$; confirm that the answer is $1/(s-5)$.

3. (a) Experiment with a calculator to show that for $t$ sufficiently large, $t^n < e^{pt}$ for any positive $n$ and $p$.

   (b) Prove $e^{pt}/t^n \to \infty$ as $t \to \infty$ for any positive $n$ and $p$. (Hint: look at the Taylor series for $e^{pt}$.)

4. Carry out the partial fraction expansion of the expression (15).

5. What is the natural frequency of oscillation for the equation $\ddot{x} + x = f(t)$? Show that the nonhomogeneity $f(t) = \sin t$ forces it at resonance.

6. Derive and interpret the "time-delay" property of the Laplace transform indicated in Table 3.2.

7. Derive the formula in Table 3.2 for the transform of $\int_0^t f(t) \, dt$.

8. Derive the formula in Table 3.2 for the transform of $t^n f(t)$.

9. Use Table 3.2 to find the Laplace transform of

(a) $f(t) = \begin{cases} \cos t, & 0 < t < 2\pi \\ 0, & 2\pi < t \end{cases}$

(b) $f(t) = \begin{cases} 0, & 0 < t \\ 1, & 1 < t < 2 \\ 0, & 2 < t \end{cases}$

(Hint: use the delay property.)

10. Find the inverse Laplace transforms of

   a. $\frac{1}{s^2 + 3s + 2}$   b. $\frac{1}{s^2 - 3s + 2}$   c. $\frac{1}{s^2(s^2+4)}$   d. $\frac{s+2}{s^2+16}$

   e. $\frac{e^{-2s}}{s+2}$   f. $\frac{3s+1}{s(s+1)(s+2)}$   g. $\frac{1}{(s^2-1)^3}$

11. Solve by Laplace transforms:

   (a) $\dot{x} + x = 3e^{2x}$,   $x(0) = 0$ .

   (b) $\ddot{x} + 2\dot{x} + 5x = 3e^{-x}\sin x$,   $x(0) = 0$,   $\dot{x}(0) = 3$.

   (c) $\ddot{x} + 2\dot{x} + 5x = \delta(t)$,   $x(0) = 0$,   $\dot{x}(0) = 1$ .

   (d) The answer to (c) is $x(t) = e^{-t}\sin 2t$; explain the discrepancy in the initial slope. (Hint: what is the effect of the delta function? Reread Section 1.4.)

12. Suppose $f(t)$ is periodic: $f(t+T) = f(t)$. Show that

$$\mathcal{L}(f(t)) = \int_0^T f(t)e^{-st}\, dt / [1 - e^{-sT}].$$

13. Show that

$$\mathcal{L}\left\{ e^t \frac{d^n}{dt^n}(t^n e^{-t}) \right\} = \frac{n!}{s}\left\{ \frac{s-1}{s} \right\}^n .$$

(The function $\{e^t \frac{d^n}{dt^n}(t^n e^{-t})\}$ is known as the Laguerre polynomial of degree $n$; see Section 7.4.)

14. Verify the initial value theorem for the functions

$$\sin t, \quad \cos t, \quad e^t, \quad e^{-t}\cos(2t + 2\pi/3).$$

15. (a) Verify the final value theorem for the functions

$$1, \quad e^{-t}, \quad \frac{\sin t}{t}$$

   (If you can't evaluate some of the limits analytically, do some calculator experimentation.)

   (b) Work out $\lim_{s \to 0} s\mathcal{L}(s; f)$ for $f(t) = e^t$ and for $f(t) = \sin t$. Note that the final value theorem does not apply, since $f(t)$ does not approach a limit at $\infty$.

Table 3.2   Laplace Transforms                                                185

# TABLE 3.2 LAPLACE TRANSFORMS

$f(t)$                $\mathcal{L}(s; f) = \int_0^\infty f(t)e^{-st}ds$

---

*GENERAL RULES*; for Re$\{s\}$ sufficiently large,

$c_1 f_1(t) + c_2 f_2(t)$    $c_1 \mathcal{L}(s; f_1) + c_2 \mathcal{L}(s; f_2)$

$f(t-T)H(t-T)$    $e^{-sT}\mathcal{L}(s; f)$ ("time delay")    $(T > 0)$

$f(ct)$            $\frac{1}{c}\mathcal{L}(\frac{s}{c}; f)$                      $(c > 0)$

$e^{-ct}f(t)$        $\mathcal{L}(s + c; f)$

$f'(t)$            $s\mathcal{L}(s; f) - f(0)$

$f^{(n)}(t)$        $s^n \mathcal{L}(s; f) - s^{n-1}f(0) - s^{n-2}f'(0) - \cdots - f^{(n-1)}(0)$

$\int_0^t f(t)dt$      $\mathcal{L}(s; f)/s$

$f(t)/t$          $\int_s^\infty \mathcal{L}(s; f)ds$ (if $f(t)/t$ has a transform)

$tf(t)$            $-\mathcal{L}'(s; f)$

$t^n f(t)$          $(-1)^n \mathcal{L}^{(n)}(s; f)$

*SPECIFIC TRANSFORMS*

$\delta(t)$          $1$

$1$              $\dfrac{1}{s}$                                  $(\text{Re}\{s\} > 0)$

$t$              $\dfrac{1}{s^2}$                                 $(\text{Re}\{s\} > 0)$

$t^n$            $n!/s^{n+1}$                            $(\text{Re}\{s\} > 0)$

$e^{-ct}$          $\dfrac{1}{s + c}$                             $(\text{Re}\{s\} > -c)$

$t^n e^{-ct}$        $n!/(s + c)^{n+1}$                      $(\text{Re}\{s\} > -c)$

| | | |
|---|---|---|
| $\sinh ct$ | $\dfrac{c}{s^2 - c^2}$ | $(\mathrm{Re}\{s\} > \|c\|)$ |
| $\cosh ct$ | $\dfrac{s}{s^2 - c^2}$ | $(\mathrm{Re}\{s\} > \|c\|)$ |
| $\sin \omega t$ | $\dfrac{\omega}{s^2 + \omega^2}$ | $(\mathrm{Re}\{s\} > 0)$ |
| $\cos \omega t$ | $\dfrac{s}{s^2 + \omega^2}$ | $(\mathrm{Re}\{s\} > 0)$ |
| $t \sin \omega t$ | $\dfrac{2\omega s}{(s^2 + \omega^2)^2}$ | $(\mathrm{Re}\{s\} > 0)$ |
| $t \cos \omega t$ | $\dfrac{s^2 - \omega^2}{(s^2 + \omega^2)^2}$ | $(\mathrm{Re}\{s\} > 0)$ |
| $\dfrac{\sin \omega t - \omega t \cos \omega t}{2}$ | $\dfrac{\omega^3}{(s^2 + \omega^2)^2}$ | $(\mathrm{Re}\{s\} > 0)$ |
| $\dfrac{\sin \omega t + \omega t \cos \omega t}{2}$ | $\dfrac{\omega s^2}{(s^2 + \omega^2)^2}$ | $(\mathrm{Re}\{s\} > 0)$ |
| $\cos \omega t - \dfrac{\omega t \sin \omega t}{2}$ | $\dfrac{s^3}{(s^2 + \omega^2)^2}$ | $(\mathrm{Re}\{s\} > 0)$ |
| $e^{-ct} \sin \omega t$ | $\dfrac{\omega}{(s + c)^2 + \omega^2}$ | $(\mathrm{Re}\{s\} > -c)$ |
| $e^{-ct} \cos \omega t$ | $\dfrac{s + c}{(s + c)^2 + \omega^2}$ | $(\mathrm{Re}\{s\} > -c)$ |
| $e^{(-c+i\omega)t}$ | $\dfrac{1}{s + c - i\omega} = \dfrac{s + c + i\omega}{(s + c)^2 + \omega^2}$ | $(\mathrm{Re}\{s\} > -c)$ |
| $e^{-ct} \cos(\omega t + \theta)$ | $\dfrac{(s + c) \cos \theta - \omega \sin \theta}{(s + c)^2 + \omega^2}$ | $(\mathrm{Re}\{s\} > -c)$ |
| $\dfrac{\sin t}{t}$ | $\mathrm{arccot}\, s$ | $(\mathrm{Re}\{s\} > 0)$ |
| $\dfrac{1 - e^{-t}}{t}$ | $\log \dfrac{s + 1}{s}$ | $(\mathrm{Re}\{s\} > 0)$ |
| $t^{-1/2} e^{-c/t}$ | $\sqrt{\pi/s}\; e^{-2\sqrt{cs}}$ | $(c, \mathrm{Re}\{s\} > 0)$ |
| $t^{-3/2} e^{-c/t}$ | $\sqrt{\pi/c}\; e^{-2\sqrt{cs}}$ | $(c, \mathrm{Re}\{s\} > 0)$ |

| | | |
|---|---|---|
| $e^{-c^2 t^2/4}$ | $\sqrt{\pi}\, e^{s^2/c^2}\, \text{Erfc}(s/b)/c$ | $(c>0)$[11] |
| $\text{Erf}(t/2c)$ | $e^{c^2 s^2}\, \text{Erfc}(cs)/s$ | $(c>0)$ |
| $\text{Erf}(c\sqrt{t})$ | $cs^{-1}(s+a^2)^{-1/2}$ | $(c>0)$ |
| $\text{Erfc}(c\sqrt{t})$ | $s^{-1}[1-c(s+c^2)^{-1/2}]$ | $(c>0)$ |
| $e^{c^2 t}\,\text{Erfc}(c\sqrt{t})$ | $s^{-1/2}(s^{1/2}+c)^{-1}$ | $(c>0)$ |
| $\text{Erfc}(c/\sqrt{t})$ | $s^{-1}e^{-2c\sqrt{s}}$ | $(c,\text{Re}\{s\}>0)$ |
| $t^{-1/2}\cos(c\sqrt{t})$ | $\sqrt{\pi/s}\; e^{-c^2/4s}$ | $(c,\text{Re}\{s\}>0)$ |
| $\sin c\sqrt{t}$ | $c\sqrt{\pi} s^{-3/2} e^{-c^2/4s}/2$ | $(c,\text{Re}\{s\}>0)$ |
| $\dfrac{\sin c\sqrt{t}}{t}$ | $\text{Erf}(cs^{-1/2}/2)$ | $(c,\text{Re}\{s\}>0)$ |
| $\int_0^t \dfrac{\sin \tau}{\tau}d\tau$ | $\dfrac{1}{s}\,\text{arccot}\,s$ | $(\text{Re}\{s\}>0)$ |
| $\int_t^\infty \dfrac{\cos \tau}{\tau}d\tau$ | $\dfrac{1}{s}\ln(s^2+1)^{1/2}$ | $(\text{Re}\{s\}>0)$ |
| $\int_t^\infty \dfrac{e^{-\tau}}{\tau}d\tau$ | $\dfrac{1}{s}\ln(s+1)$ | $(\text{Re}\{s\}>0)$ |

For more extensive listings see the references by Erdelyi et al. and Sneddon.

## 3.9 Frequency Domain and *s*-Plane Analysis

**Summary**  The classical Fourier series/transform methodology yields steady-state solutions for linear constant-coefficient differential equations when damping is present, while the Laplace transform procedure yields solutions conforming to initial conditions, and is unfettered by frictionless or resonant systems. The Fourier approach can be extended to produce general solutions for these cases as well, at the cost of invoking the distribution calculus. However, in physical systems damping is always present and, moreover, the type of equipment available for testing is more closely married to the sinusoids of the frequency do-

---

[11] $\text{Erf}(x) = \frac{2}{\sqrt{\pi}}\int_0^x e^{-\xi^2} d\xi$ ("error function"), $\text{Erfc}(x) = 1 - \text{Erf}(x)$

main than the exponentials of the $s$-plane. Another advantage of the frequency domain approach is the fact that large coupled systems, for which transient analysis would be impractical, can often be described compactly by transfer functions.

In Secs. 3.3 through 3.8 we have studied three powerful mathematical tools: the Fourier series (including the sine and cosine series), the Fourier transform (and the sine and cosine transform), and the Laplace transform. Now it is time to take inventory and sort out which tool should be used in specific situations. We address this issue from three viewpoints, reflecting the following goals:

1. the expansion of a given function,

2. the mathematical solution of a given differential equation with constant coefficients, and

3. the characterization of a given physical system (governed by a differential equation with constant coefficients).

## I. THE EXPANSION OF A GIVEN FUNCTION

The most obvious difference between the Fourier series, the Fourier transform, and the Laplace transform is the size of the interval over which they represent a given function. The Fourier series reproduces $f(t)$ over a finite interval $a < t < b$; the Fourier transform reproduces $f$ over the whole line $-\infty < t < \infty$; and the Laplace transform reproduces $f$ over the positive half-line $0 < t < \infty$.

**Example 1.** Let $f(t)$ be the rectangular pulse depicted in Fig. 1(a). The Fourier transform gives a representation of $f(t)$ (see exercise 1) -

$$f(t) = \int_{-\infty}^{\infty} \frac{\sin \omega \pi/2}{\omega \pi} e^{i\omega t} \, d\omega, \quad -\infty < t < \infty \tag{1}$$

- which is valid for all $t$; in particular it equals zero outside the interval $[-\pi/2, \pi/2]$ (Fig. 1(b)). The Fourier series which converges to $f(t)$ on the interval $-\pi < t < \pi$ is given by

$$f(t) = \frac{1}{2} + \frac{2}{\pi} \sum_{n=0}^{\infty} (-1)^n \frac{\cos(2n+1)t}{2n+1}, \quad -\pi < t < \pi, \tag{2}$$

but outside this interval it replicates these values periodically (Fig. 1(c)). The Laplace transform for $f$ - when inverted - yields $f(t)$ to the right of the origin, and zero to the left (Fig. 1(d)):

$$f(t) = \mathcal{L}^{-1}\{1 - e^{-\pi s/2}\}/s, \quad t > 0. \tag{3}$$

Thus the choice of tool - Fourier series, Fourier transform, or Laplace transform - for expanding a given function depends heavily on the size of the interval over which the expansion is required to be valid.                ∎

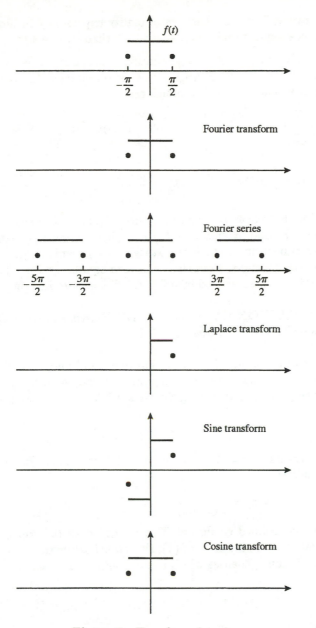

**Figure 1    Transform domains**

The Laplace transform, of course, is not the only transform devoted to representing $f$ on the half-infinite interval; the Fourier sine and cosine transforms also accomplish this. They differ in how they extend $f$ to the negative axis (see Figs. 1(e) and 1(f)), and this consideration may influence the choice of transform. Recall, too, that in Section 3.8 we pointed out that the inverse Laplace transform in reality expresses $f$ in terms of complex exponential functions (rather than simple sinusoids).

In a certain sense all of these expansions are special cases of the Fourier transform. After all, the Fourier series representation (2) can, itself, be expressed in terms of a Fourier transform:

$$\frac{1}{2} + \frac{2}{\pi} \sum_{n=0}^{\infty} (-1)^n \frac{\cos(2n+1)t}{2n+1} = \int_{-\infty}^{\infty} F(\omega) e^{i\omega t} \, d\omega, \quad (-\infty < t < \infty),$$

$$F(\omega) = \frac{1}{2}\delta(\omega) + \frac{2}{\pi} \sum_{n=0}^{\infty} (-1)^n \frac{\delta(\omega - 2n - 1) + \delta(\omega + 2n + 1)}{2(2n+1)} \qquad (4)$$

(exercise 2). Moreover the sine transform equals ($2i$ times) the Fourier transform of the odd extension of $f$; the cosine transform equals (2 times) the Fourier transform of the even extension of $f$ (which equals $f$ itself in the case of Fig. 1); and the Laplace transform equals ($2\pi$ times) the Fourier transform of the "zero extension" of $f$, when $s$ is pure imaginary ($s = i\omega$).

## II. THE SOLUTION OF A GIVEN DIFFERENTIAL EQUATION WITH CONSTANT COEFFICIENTS

We began this chapter by noting that one could find a solution to a linear constant-coefficient differential equation, forced with a sinusoidal nonhomogeneity $f(t)$, by considering sinusoids with the same frequency. Thus a solution to

$$\ddot{x} + 2\dot{x} + 5x = f(t) = e^{i\omega t} \qquad (5)$$

is found by trying $x(t) = Xe^{i\omega t}$ with an *undetermined coefficient X* in (1), and the result is

$$x(t) = \frac{1}{-\omega^2 + i2\omega + 5} e^{i\omega t} = H(\omega) e^{i\omega t}. \qquad (6)$$

Here the undetermined coefficient $X$ is identified as the *transfer function* $H(\omega)$. Because of the linearity of (1), this transfer-function formalism generalizes to nonhomogeneities which are expressible as *sums* of sinusoids: a solution to

$$\ddot{x} + 2\dot{x} + 5x = A_1 e^{i\omega_1 t} + A_2 e^{i\omega_2 t} + A_3 e^{i\omega_3 t} \qquad (7)$$

is given by

$$\begin{aligned}
x(t) &= \frac{A_1}{-\omega_1^2 + i2\omega_1 + 5} e^{i\omega_1 t} \\
&\quad + \frac{A_2}{-\omega_2^2 + i2\omega_2 + 5} e^{i\omega_2 t} + \frac{A_3}{-\omega_3^2 + i2\omega_3 + 5} e^{i\omega_3 t} \\
&= \sum_{n=1}^{3} H(\omega_n) A_n e^{i\omega_n t}
\end{aligned} \qquad (8)$$

The *general* solution to (7) is obtained from the transfer-function form by appending the general solution of the associated homogeneous equation:

$$x(t) = \sum_{n=1}^{3} H(\omega_n) A_n e^{i\omega_n t} + c_1 e^{(-1+2i)t} + c_2 e^{(-1-2i)t} \qquad (9)$$

(we prefer the complex form here).

Since we now know that virtually *any* function can be expressed in terms of sinusoids, the transfer function becomes a universal tool for analyzing constant-coefficient systems. Let us see how the three expansion tools relate to this task.

**Fourier series.** If $f(t)$ is expanded into a Fourier series over the *finite interval* $(a, b)$ -

$$f(t) = \sum_{n=-\infty}^{\infty} A_n e^{in\Omega t} \quad (\Omega = \frac{2\pi}{b-a}), \quad a < t < b \qquad (10)$$

- then a particular solution to

$$\ddot{x} + 2\dot{x} + 5x = f(t) \qquad (11)$$

*over this interval* is given by the obvious extension of the transfer-function form (8):

$$x(t) = \sum_{n=-\infty}^{\infty} \frac{1}{-n^2\Omega^2 + i2n\Omega + 5} A_n e^{in\Omega t}, \quad \text{or}$$

$$x(t) = \sum_{n=-\infty}^{\infty} H(n\Omega) A_n e^{in\Omega t}, \qquad (12)$$

and the general solution is again obtained by appending the associated homogeneous solutions:

$$
\begin{aligned}
x(t) &= \sum_{n=-\infty}^{\infty} \frac{1}{-n^2\Omega^2 + i2n\Omega + 5} A_n e^{in\Omega t} \\
&\quad + c_1 e^{(-1+2i)t} + c_2 e^{(-1-2i)t} \\
&= \sum_{n=-\infty}^{\infty} H(n\Omega) A_n e^{in\Omega t} \\
&\quad + c_1 e^{(-1+2i)t} + c_2 e^{(-1-2i)t}, \quad a < t < b.
\end{aligned} \qquad (13)
$$

**Fourier transform.** The Fourier transform for $f$ can be used if its values are known *for all time*:

$$f(t) = \int_{-\infty}^{\infty} F(\omega) e^{i\omega t} \, d\omega \quad (\text{all } t, -\infty < t < \infty), \qquad (14)$$

and the corresponding transfer-function form is the particular solution -
again, *valid for all time* -

$$x(t) = \int_{-\infty}^{\infty} \frac{1}{-\omega^2 + i2\omega + 5} F(\omega)e^{i\omega t}\, d\omega, \quad \text{or}$$

$$x(t) = \int_{-\infty}^{\infty} H(\omega)F(\omega)e^{i\omega t}\, d\omega. \tag{15}$$

The general solution is fashioned as before by appending the homogeneous
solutions by hand:

$$x(t) = \int_{-\infty}^{\infty} \frac{1}{-\omega^2 + i2\omega + 5} F(\omega)e^{i\omega t}\, d\omega + c_1 e^{(-1+2i)t} + c_2 e^{(-1-2i)t}$$

$$= \int_{-\infty}^{\infty} H(\omega)F(\omega)e^{i\omega t}\, d\omega + c_1 e^{(-1+2i)t} + c_2 e^{(-1-2i)t} \tag{16}$$

There's a better way of deriving the solution (16) using distribution calculus. Letting $F(\omega)$ and $X(\omega)$ denote the Fourier
transforms of the forcing function $f(t)$ and the unknown $x(t)$,
respectively, we *take the Fourier transforms of both sides of the
differential equation*

$$\ddot{x} + 2\dot{x} + 5x = f(t) \qquad (11\,repeated) \tag{17}$$

and obtain (recall the rule for transforming derivatives, Section
3.6)

$$-\omega^2 X(\omega) + 2i\omega X(\omega) + 5X(\omega) = F(\omega), \quad \text{or}$$

$$-(\omega - 2 - i)(\omega + 2 - i)X(\omega) = F(\omega). \tag{18}$$

*Equation (18) is called the* **frequency domain** *version of eq.
(11)*. Notice what happens if we solve (18) for $X(\omega)$, using the
distribution calculus (as is appropriate for Fourier transforms).
Since the coefficient of $X(\omega)$ vanishes when $\omega = i \pm 2$, the
solution contains delta functions:

$$X(\omega) = \frac{1}{-\omega^2 + i2\omega + 5} F(\omega) + c_1\delta(\omega - 2 - i) + c_2\delta(\omega + 2 - i) \tag{19}$$

for arbitrary $c_1$, $c_2$ (Sec. 1.3, eq. (23)). The inverse transform
of (19) is

$$x(t) = \int_{-\infty}^{\infty} \frac{1}{-\omega^2 + i2\omega + 5} F(\omega)e^{i\omega t}\, d\omega$$

$$+ \quad c_1 \int_{-\infty}^{\infty} \delta(\omega - 2 - i)e^{i\omega t}\, d\omega + c_2 \int_{-\infty}^{\infty} \delta(\omega + 2 - i)e^{i\omega t}\, d\omega$$

$$= \int_{-\infty}^{\infty} H(\omega)F(\omega)e^{i\omega t}\, d\omega$$

$$+ \quad c_1 e^{(-1+2i)t} + c_2 e^{(-1-2i)t} \tag{20}$$

(Example 3, Sec. 3.7). This agrees with (16), *but observe that the distribution calculus has generated the homogeneous solutions automatically*!

Because the Fourier transform procedure - coupled with the use of the distribution calculus to solve the frequency domain equation - produces the complete general solution automatically, it apparently has a definite advantage over the transfer-function procedure (eqs. (13) and (16)). However let us see how the Laplace transform scheme performs.

**Laplace transform.** Assuming that $f(t)$ is known *for $t \geq 0$*, we take the Laplace transform of both sides of (11) and find

$$s^2 \mathcal{L}(s;x) - sx(0) - \dot{x}(0) + 2s\mathcal{L}(s;x) - 2x(0) + 5\mathcal{L}(s;x) = \mathcal{L}(s;f). \quad (21)$$

*Equation (21) expresses the differential equation (11) in the s-plane domain.* Its solution is generated by solving for $\mathcal{L}(s;x)$ (without distributions) and inverse-transforming:

$$x(t) = \mathcal{L}^{-1} \left\{ \frac{\mathcal{L}(s;f) + (s+2)x(0) + \dot{x}(0)}{s^2 + 2s + 5} \right\}. \quad (22)$$

Thus, for instance, if $f(t) = e^{-t}$, (22) yields

$$x(t) = \frac{1}{4}e^{-t} + [x(0) - \frac{1}{4}]e^{-t}\cos 2t + \frac{x(0) + \dot{x}(0)}{2}e^{-t}\sin 2t \quad (23)$$

(exercise 3).

Not only does the Laplace transform procedure generate the homogeneous solutions automatically, it fits them to the initial conditions!

We summarize:

1. When the nonhomogeneity $f(t)$ for a linear constant coefficient differential equation takes the form $e^{i\omega t}$, there will be a particular solution of the form $Xe^{i\omega t}$ (unless $\omega$ is a natural resonant frequency; we return to this point later). The coefficient $X$, which depends on frequency, is called the *transfer function $H(\omega)$* for the equation.

2. If a general nonhomogeneity $f(t)$ is expressed as a Fourier series over an interval, insertion of the transfer function into the form (12) produces a particular solution over that interval. The general solution is obtained by appending the associated homogeneous solutions by hand (as in (13)).

3. If $f(t)$ is expressed in terms of its Fourier transform, insertion of the transfer function into the form (15) yields a particular solution valid for all time; again the general solution is attained by appending the homogeneous solutions.

4. If one Fourier-transforms the differential equation and solves it in the frequency domain using distribution calculus, the complete general solution is obtained.

5. If one Laplace-transforms the differential equation to the $s$-plane, distribution calculus is unnecessary and the solution generated automatically satisfies specified initial conditions.

It is interesting to compare these methods for the case of undamped resonant forcing.

**Example 2.** Consider the differential equation

$$\ddot{x} + x = f(t). \tag{24}$$

The general solution of the associated homogeneous equation is

$$c_1 e^{it} + c_2 e^{-it} \tag{25}$$

and its natural frequency is $\omega = 1$.                                                    ∎

**Fourier series.** If $f(t)$ is expanded in a Fourier series like (10) and if the natural frequency $\omega = 1$ occurs in the expansion -

$$f(t) = \sum_{\substack{n=-\infty \\ n \neq \pm N}}^{\infty} A_n e^{in\Omega t} + A_N e^{iN\omega t} + A_{-N} e^{-iN\omega t}$$

where $N\Omega = 1$ in this case - then the transfer function form (12) is invalid for those terms; they must be modified to acknowledge the resonance-solution form (eqs. (19, 20) of Section 3.2):

$$
\begin{aligned}
x(t) &= \sum_{\substack{n=-\infty \\ n \neq \pm N}}^{\infty} \frac{1}{-n^2\Omega^2 + 1} A_n e^{in\Omega t} + \frac{t}{2i} A_N e^{it} - \frac{t}{2i} A_{-N} e^{-it} \\
&\quad + c_1 e^{it} + c_2 e^{-it}
\end{aligned} \tag{26}
$$

Again, the associated homogeneous solution must be appended by hand.

For simplicity let us suppose $f(t) = e^{it}$ (a one-term Fourier series!) so that (24) reduces to

$$\ddot{x} + x = e^{it}. \tag{27}$$

The corresponding solution is then

$$x(t) = \frac{t}{2i} e^{it} + c_1 e^{it} + c_2 e^{-it}. \tag{28}$$

**Fourier transform**. How does the frequency-domain approach cope with this situation? We take Fourier transforms of both sides of eq. (27); from Table 3.1,

$$-\omega^2 X(\omega) + X(\omega) = -(\omega - 1)(\omega + 1)X(\omega) = \delta(\omega - 1). \qquad (29)$$

The distribution solution of (29) is (exercise 5)

$$X(\omega) = \frac{\delta'(\omega - 1)}{2} + c_1 \delta(\omega - 1) + c_2 \delta(\omega + 1) \qquad (30)$$

and its inverse transform is

$$
\begin{aligned}
x(t) &= \int_{-\infty}^{\infty} \{\frac{\delta'(\omega - 1)}{2} + c_1 \delta(\omega - 1) + c_2 \delta(\omega + 1)\} e^{i\omega t}\, d\omega \\
&= -\frac{it}{2} e^{it} + c_1 e^{it} + c_2 e^{-it}. \qquad (31)
\end{aligned}
$$

Both the resonant-growth term and the homogeneous solutions are generated automatically.

**Laplace transform**. The Laplace transform solution of (27) proceeds as follows: in the $s$-plane the equation becomes

$$s^2 \mathcal{L}(s; x) - sx(0) - \dot{x}(0) + \mathcal{L}(s; x) = \frac{1}{s - i}. \qquad (32)$$

Thus

$$\mathcal{L}(s; x) = \frac{1}{(s - i)(s^2 + 1)} + \frac{sx(0) + \dot{x}(0)}{s^2 + 1},$$

and from Table 3.2,

$$x(t) = -\frac{it}{2} e^{it} + x(0) \cos t + [\dot{x}(0) + \frac{i}{2}] \sin t \qquad (33)$$

(exercise 6). The resonant-growth term and the homogeneous solutions are generated automatically, the latter are fitted to the initial conditions, and the use of distribution calculus is avoided.

Which of the techniques - transfer function with Fourier series/integral, frequency domain with distribution calculus, or Laplace transform - should you use to solve differential equations? Surely the examples sway the decision toward the latter. After all, the transfer function procedure requires the *ad hoc* insertion of the associated homogeneous solutions as well as the resonance-growth terms; and while the frequency domain method begets both of these in an elegant manner, it does so at the cost of conjuring up the distribution calculus. But the

Laplace transform handles the resonance without invoking distributions, *and* it automatically delivers the appropriate combination of the homogeneous solutions which satisfies the initial data. For the solution of constant coefficient differential equations with specified initial data, then, the Laplace transform is the method of choice.

## III. THE ANALYSIS OF A GIVEN PHYSICAL SYSTEM GOVERNED BY A DIFFERENTIAL EQUATION WITH CONSTANT COEFFICIENTS

In the engineering analysis of a physical system there are two practical considerations which, while they may carry little weight with the theoretician, frequently dictate the *transfer function* approach as the method of choice - despite the preceding discussion. These are

1. the types of equipment available for laboratory testing; and

2. the presence of damping in all real systems.

Let's see how these circumstances affect the choice of analysis tool. Consider again the damped harmonic oscillator of Section 3.2. Its motion is governed by the differential equation

$$m\ddot{x} + \nu\dot{x} + kx = f(t). \tag{34}$$

Now a test engineer would seek to evaluate this system by applying a controlled force $f(t)$, observing the resulting motion, and comparing the latter to the mathematical solution. What kind of forces can he/she apply? The simplest answer is - sinusoids! Shakers and sine wave generators are standard laboratory instruments. *The most commonly cited characterization of a piece of physical equipment is its response to sinusoidal excitations.*

When damping is present (so that resonance can't occur), the response of the system to a sinusoidal force $f(t) = e^{i\omega t}$ is well handled by the transfer function formalism. The particular solution

$$\frac{1}{-m\omega^2 + i\nu\omega + k}e^{i\omega t} = H(\omega)e^{i\omega t} \tag{35}$$

vibrates at the same frequency as the force. And the associated homogeneous solutions

$$c_1 \exp\{[-\nu/2m + i\sqrt{k/m - \nu^2/4m^2}]t\}$$
$$+ c_2 \exp\{[-\nu/2m - i\sqrt{k/m - \nu^2/4m^2}]t\}$$

are *transient*; they decay exponentially in time (for $\nu > 0$). Therefore by driving the system with a sinusoidal excitation until the transients die out and the system vibrates sympathetically with the driver, the test engineer can observe and measure the transfer function $H(\omega)$.[12]

> Thus the values of the transfer function $H(\omega)$ are of immediate relevance to the test engineer; they are readily observable physical properties of the system. A theoretician whose concern is testability rather than the computation of startup transients may well opt to use the transfer function formalism instead of the Laplace transform.

In fact the transfer function approach - computing the sympathetic response $H(\omega)e^{i\omega t}$ to a single sinusoidal nonhomogeneity $e^{i\omega t}$ - is used in most of the engineering analysis literature, *even when the differential equation model contains no damping*. This immediately calls to mind the following objections:

- the test engineer cannot apply the theory because the transfer function response cannot be extracted from the real data (since the associated homogeneous-solution responses never die out); and

- the solution obtained is entirely invalid if the system is forced at resonance.

But of course the first is strictly academic since some damping is always present in the physical system (if not in the model), and thus the actual response will inevitably converge to the sympathetic solution $H(\omega)e^{i\omega t}$. And as far as the second objection is concerned, recall that in Section 3.2 we already conceded that near resonance a frictionless model has no credibility anyway.

At any rate, since the Laplace transform does not enjoy the same immediate physical significance as the transfer function, the latter is more widely used in engineering analysis.

**Example 3.** The rotation $x(t)$ produced by a voltage $f(t)$ applied to a armature-controlled dc motor (Fig. 2) satisfies the differential equation

$$\frac{LI}{K}\frac{d^3x}{dt^3} + K\frac{dx}{dt} = f(t), \tag{36}$$

where $L$ is the armature inductance, $I$ is the load inertia, and $K$ is the motor constant (see Dorf's text). Resistance and friction are ignored in eq. (36). For the purposes of illustration we take $L = I = 2$, $K = 4$:

$$\frac{d^3x}{dt^3} + 4\frac{dx}{dt} = f(t). \tag{37}$$

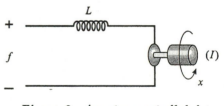

*Figure 2   Armature-controlled dc motor*

---

[12]The class of instruments known as "frequency analyzers" is designed to perform this task.

To obtain the transfer function for (37) we set $f(t) = e^{i\omega t}$ and seek a solution of the form $x = Xe^{i\omega t}$ :

$$\frac{d^3x}{dt^3} + 4\frac{dx}{dt} = (i\omega)^3 Xe^{i\omega t} + 4i\omega Xe^{i\omega t} = e^{i\omega t},$$

$$X = H(\omega) = \frac{1}{(i\omega)^3 + 4i\omega} = \frac{1}{-i\omega(\omega - 2)(\omega + 2)}. \qquad (38)$$

The natural frequencies of this system are 0 and $\pm 2$. If the frequency $\omega$ of the driving voltage $f(t) = e^{i\omega t}$ equals any of these values, resonance occurs and the model is invalid (the resistance and friction have to be incorporated). Otherwise the steady state response to this driver is

$$x(t) = \frac{1}{-i\omega(\omega - 2)(\omega + 2)}e^{i\omega t} \qquad (39)$$

(once the *de facto* resistance and friction have damped out the transients). More generally the steady state response to

$$f(t) = \sum_{n=-\infty}^{\infty} A_n e^{in\omega t} \quad \text{or} \quad \int_{-\infty}^{\infty} F(\omega)e^{i\omega t}\, d\omega, \qquad (40)$$

respectively, is

$$x(t) = \sum_{n=-\infty}^{\infty} \frac{1}{(in\omega)^3 + 4in\omega} A_n e^{in\omega t} \quad \text{or} \quad \int_{-\infty}^{\infty} \frac{F(\omega)}{(i\omega)^3 + 4i\omega}e^{i\omega t}\, d\omega \qquad (41)$$

if $f(t)$ contains no resonance components. ($A_0$, in particular, must be zero. See exercise 7 for further elaboration.) ∎

It may have occurred to the reader that if one has knowledge of the formula for the transfer function $H(\omega)$, then one can directly reconstruct the Fourier transform of the (non-resonant) *general* solution via

$$X(\omega) = H(\omega)F(\omega) + \sum c_\ell \delta(\omega - \omega_\ell), \qquad (42)$$

where $\{\omega_\ell\}$ are the zeroes of $1/H(\omega)$ (exercise 8(b)). A little more thought (exercise 8(c)) reveals that with the substitution of $(-is)$ for $\omega$, the Laplace transform of the solution *for zero initial conditions* can also be reconstructed:

$$\mathcal{L}(s; x) = H(-is)\mathcal{L}(s; f). \qquad (43)$$

**Example 4.** The equation for the armature (37) in the frequency domain is

$$(i\omega)^3 X + 4i\omega X = -i\omega(\omega - 2)(\omega + 2)X = \frac{1}{H(\omega)}X = F(\omega) \qquad (44)$$

Clearly the zeros of $1/H(\omega)$ generate the distribution solutions giving rise to the general solution of (37):

$$X(\omega) = H(\omega)F(\omega) + c_1\delta(\omega) + c_2\delta(\omega - 2) + c_3\delta(\omega + 2),$$

$$x(t) = \int_{-\infty}^{\infty} H(\omega)F(\omega)e^{i\omega t}\,d\omega + c_1 + c_2 e^{i2\omega t} + c_3 e^{-i2\omega t}, \qquad (45)$$

if no resonances are excited (see exercise 7).

Similarly the equation (37) becomes, in the $s$-plane,

$$s^3\mathcal{L}(s;x) + 4s\mathcal{L}(s;x) = (s^3 + 4s)\mathcal{L}(s;x) = \frac{1}{H(-is)}\mathcal{L}(s;x) = \mathcal{L}(s;f) \quad (46)$$

if the initial data $x(0)$, $\dot{x}(0)$, $\ddot{x}(0)$ are zero. The corresponding solution is given by

$$x(t) = \mathcal{L}^{-1}\left[\frac{\mathcal{L}(s;f)}{s^3 + 4s}\right] = \mathcal{L}^{-1}[H(-is)\mathcal{L}(s;f)]. \qquad \blacksquare$$

While eqs. (42, 43) are useful identities for the theoretician, they are less appealing to the test engineer. *The formulas (12, 15) for steady-state responses to arbitrary forcing functions can be implemented (numerically) from an* **experimental** *determination of the values of $H(\omega)$ - without knowledge of any of the details of the differential equation!* Since $\omega$ is real, however, one cannot determine the complex zeros experimentally - which disables (42) - nor can one implement the Bromwich integral, which generally requires knowledge of $\mathcal{L}(s;x)$ for complex $s$ (Section 3.8).[13]

There is one last point to be made in favor of the transfer function approach. If a system is complicated and has many components it may be impractical to model its overall transient behavior, but the description of its steady state sinusoidal responses can be relatively easy. Consider the following simple example.

**Example 5.** The coordinates of the mass-spring oscillator shown in Fig. 3 are measured from equilibrium locations $x_2 = 0$, $x_1 = 0$. The equation describing the motion of the mass when the left end of the spring is externally manipulated - $x_2 = x_2(t)$ - is given by

$$m_1\ddot{x}_1 = -\nu_1\dot{x}_1 - k_1(x_1 - x_2), \quad \text{or} \quad m_1\ddot{x}_1 + \nu_1\dot{x}_1 + k_1x_1 = k_1x_2(t). \quad (47)$$

The sympathic response of the mass $m_1$ to a sinusoidal excitation $x_2 = X_2 e^{i\omega t}$ is thus

$$x_1(t) = X_1 e^{i\omega t} = H(\omega)X_2 e^{i\omega t} = \frac{k_1}{-m_1\omega^2 + i\nu_1\omega + k_1}X_2 e^{i\omega t}. \qquad (48)$$

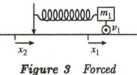
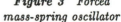

*Figure 3   Forced mass-spring oscillator*

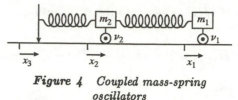

**Figure 4**  *Coupled mass-spring oscillators*

Now if a larger system is constructed by attaching the left end of this spring to another mass-spring oscillator ($m_2$, $\nu_2$, $k_2$) as in Fig. 4, the equation of motion of the mass $m_2$ is

$$m_2\ddot{x}_2 = -\nu_2\dot{x}_2 - k_2(x_2 - x_3(t)) + k_1(x_1 - x_2), \tag{49}$$

where $x_3(t)$ is the externally generated (given) motion of the left end of the second spring.

To analyze the motion of the mass $m_2$ we must find the simultaneous solutions of eqs. (47) and (49); and the direct solution of this coupled pair of equations is rather laborious (exercise 16). If we replace $x_3(t)$ with $Ae^{i\omega t}$ and restrict ourselves to steady state sinusoidal solutions, however, eq. (49) becomes

$$-\omega^2 m_2 X_2 e^{i\omega t} = -i\omega\nu_2 X_2 e^{i\omega t} - k_2(X_2 e^{i\omega t} - Ae^{i\omega t}) + k_1(X_1 e^{i\omega t} - X_2 e^{i\omega t}) \tag{50}$$

*and the incorporation of the equation (47) for the coupled motion of the mass $m_1$ is accomplished by the simple substitution of the transfer function (48) in eq. (50):*

$$
\begin{aligned}
-\omega^2 m_2 X_2 e^{i\omega t} &= -i\omega\nu_2 X_2 e^{i\omega t} - k_2(X_2 e^{i\omega t} - Ae^{i\omega t}) \\
&\quad + k_1\Big(\frac{k_1}{-m_1\omega^2 + i\nu_1\omega + k_1} X_2 e^{i\omega t} - X_2 e^{i\omega t}\Big)
\end{aligned}
$$

or

$$X_2 e^{i\omega t} = \Big[-\omega^2 m_2 + i\omega\nu_2 + k_2 - k_1\Big(\frac{k_1}{-m_1\omega^2 + i\nu_1\omega + k_1} - 1\Big)\Big]^{-1} k_2 Ae^{i\omega t} \tag{51}$$

In effect, the transfer function approach has allowed us to model the subsystem $\{m_1,\ \nu_1,\ k_1\}$ by the simple relation (48). In fact we could use (51) to construct a similar model for the *composite* system, and attach it to the end of yet another mass-spring oscillator. The savings in computational labor (compared to the direct solution of three coupled differential equations) would be considerable.                              ∎

---

[13]The science of *modal analysis and system identification* deals with the estimation of $H$ for complex values in terms of measured values of $H(\omega)$, but it presumes specific system models. See Ewen's book.

This approach is absolutely essential in the modeling of many physical systems. For example, consider the propagation of electromagnetic waves through a solid. If we had to account for the coupled transient responses of all $10^{23}$ molecules to the electric field the analysis would be impossible. However the steady state response of the solid to a *sinusoidal* electric field can be described simply by a frequency-dependent *index of refraction* (see Jackson's book; the "electric susceptibility" is, essentially, the transfer function for the polarization response to the electric field stimulus). With this device we are able to understand reflection and refraction phenomena.

The subject of Fourier analysis and the transforms is very rich in theory and applications. Frequency analysis is essential in signal processing and system identification. The Laplace transform, extended to large systems, is the basic tool of control theory. The reader is directed to the specialized literature for further investigations along these lines.

For now we suspend the study of Fourier methods. In Chapter 6 we shall see the expansions reemerge in the context of the Sturm-Liouville problem. The transforms will surface again in the context of partial differential equations in Chapter 9; the phenomenon of wave propagation, in particular, provides a very concrete illustration of the complementary nature of steady state and transient analysis.

Obviously there is much overlap between the transfer function, the frequency domain, and the $s$-plane. In the literature you may find these terms used interchangeably. Perhaps it may amuse you to speculate on the folly of attempting to trace the response of a system for all time - in particular, for the infinitely remote past; wouldn't any transients, introduced at $t = -\infty$, have settled down to zero by "finite time"? In this regard, the author begs your indulgence while he recounts a story he heard when he was a graduate student.

The story is told that when Einstein died, the first question that he asked God in heaven was whether or not time was infinite. God answered in the affirmative - "both forwards and backwards." Einstein was pleased to learn that the universe would continue forever into the future, but he said that he couldn't fathom the concept of time with no beginning. "How can one measure time from an infinite past?" he asked.

God replied that He left such matters to the archangel Thomas, and directed Einstein to him. When Einstein found the angel, Thomas was looking at a stopwatch and saying, "Minus four, minus three, minus two, minus one, zero, (sigh) one, two, three, ...."

# Exercises 3.9

**1.** Verify the expressions (1, 2, 3).

**2.** Verify eq. (4).

**3.** Derive eq. (23).

**4.** Verify eq. (26). (Note that $e^{\pm iN\Omega t} = e^{\pm it}$.)

**5.** Use the formula demonstrated in exercise 10(c), Section 1.3, to derive eq. (30). Then derive eq. (31).

**6.** Verify eq. (33).

**7.** The differential equation of an undamped harmonic oscillator is

$$m\ddot{x} + kx = f(t).$$

(a) Show that if $f(t)$ is a sinusoid at the frequency $\sqrt{k/m}$ then the solution contains unbounded oscillatory terms; specifically, the general solution to

$$m\ddot{x} + kx = e^{\pm i\sqrt{k/m}\,t}$$

is

$$x(t) = \frac{\mp it}{2\sqrt{k/m}}e^{\pm i\sqrt{k/m}\,t} + c_1 \cos\sqrt{k/m}\,t + c_2 \sin\sqrt{k/m}\,t.$$

(b) If the general forcing term $f(t)$ has a finite contribution from $\omega = \pm\sqrt{k/m}$ in its Fourier transform - that is, if $F(\omega)$ is of the form

$$F(\omega) = F_1(\omega) + A\,\delta(\omega - \sqrt{k/m}) + B\,\delta(\omega + \sqrt{k/m})$$

where $F_1(\omega)$ is continuous - show that the frequency domain equation for the oscillator takes the form

$$-m(\omega - \sqrt{k/m})(\omega + \sqrt{k/m})X(\omega)$$
$$= F_1(\omega) + A\delta(\omega - \sqrt{k/m}) + B\delta(\omega + \sqrt{k/m}).$$

(c) Using the formulas in exercise 10(c) of Section 1.3 verify that the distribution solutions of this equation are

$$X(\omega) = \frac{F_1(\omega)}{-m\omega^2 + k} + A\frac{\delta'(\omega - \sqrt{k/m})}{2\sqrt{km}} - B\frac{\delta'(\omega + \sqrt{k/m})}{2\sqrt{km}}$$
$$+ d_1\delta(\omega - \sqrt{k/m}) + d_2\delta(\omega + \sqrt{k/m})$$

where $d_1$ and $d_2$ are arbitrary.

(d) From the above obtain the solution[14]

$$
\begin{aligned}
x(t) \;=\;& \int_{-\infty}^{\infty} \frac{F_1(\omega)}{-m\omega^2 + k} e^{i\omega t}\, d\omega + A \int_{-\infty}^{\infty} \frac{\delta'(\omega - \sqrt{k/m})}{2\sqrt{km}} e^{i\omega t}\, d\omega \\
& - B \int_{-\infty}^{\infty} \frac{\delta'(\omega + \sqrt{k/m})}{2\sqrt{km}} e^{i\omega t}\, d\omega \\
& + \int_{-\infty}^{\infty} \{ d_1\delta(\omega - \sqrt{k/m}) + d_2\delta(\omega + \sqrt{k/m}) \} e^{i\omega t}\, d\omega \\
=\;& \int_{-\infty}^{\infty} \frac{F_1(\omega)}{-m\omega^2 + k} e^{i\omega t}\, d\omega - A \frac{it}{2\sqrt{km}} e^{i\sqrt{k/m}\,t} \\
& + B \frac{it}{2\sqrt{km}} e^{-i\sqrt{k/m}\,t} + d_1 e^{i\sqrt{k/m}\,t} + d_2 e^{-i\sqrt{k/m}\,t}.
\end{aligned}
$$

8. Consider a general $n$th order linear differential equation with constant coefficients

$$
a_n[\tfrac{d}{dt}]^n x + a_{n-1}[\tfrac{d}{dt}]^{n-1}x + \cdots + a_2 \frac{d^2}{dt^2}x + a_1 \frac{d}{dt}x + a_0 x = f(t).
$$

(a) The transfer function of this equation is the coefficient $H(\omega)$ in the particular solution $H(\omega)e^{i\omega t}$, when $f(t)$ equals $e^{i\omega t}$. Show that the transfer function equals

$$
H(\omega) = \{ a_n(i\omega)^n + a_{n-1}(i\omega)^{n-1} + \cdots + a_1(i\omega) + a_0 \}^{-1}.
$$

(b) Assuming that all of the zeros of $1/H(\omega)$ are simple (of order one) and non-real, show that the distribution solution of the frequency domain version of the differential equation is given by eq. (42).

(c) Show that the $s$-plane version of the differential equation is

$$
\{ a_n s^n + a_{n-1}s^{n-1} + \cdots + a_2 s^2 + a_1 s + a_0 \} \mathcal{L}(s; x) = \mathcal{L}(s; f)
$$

and that the Laplace transform of $x$ is given by

$$
\mathcal{L}(s; x) = H(-is)\mathcal{L}(s; f)
$$

when the initial conditions are all zero.

(d) Show that the Laplace transform of $x$ is given in general by the

---

[14]As was discussed in Section 1.5, divergent integrals such as the one below must be interpreted as distributions.

formula

$$
\begin{aligned}
\mathcal{L}(s;x) \;=\; & H(-is)\{\mathcal{L}(s;f) + x(0)[a_n s^{n-1} + \cdots + a_2 s + a_1] \\
& + \frac{d}{dt}x(0)[a_n s^{n-2} + a_{n-1}s^{n-3} + \cdots + a_3 s + a_2] \\
& + \frac{d^2}{dt^2}x(0)[a_n s^{n-3} + a_{n-1}s^{n-4} + \cdots + a_3] \\
& + \cdots \\
& + [\frac{d}{dt}]^{n-1}x(0)[a_n]\}
\end{aligned}
$$

**9.** (*Convolution Theorem*) The *convolution* $f_1 \circ f_2(t)$ of two functions $f_1(t)$ and $f_2(t)$ is defined by the operation

$$
f_1 \circ f_2(t) = \int_{-\infty}^{\infty} f_1(\tau) f_2(t - \tau)\, d\tau.
$$

Let $F_1(\omega)$ and $F_2(\omega)$ be the Fourier transforms of $f_1(t)$ and $f_2(t)$ respectively. By carrying out the steps indicated below, show that $2\pi F_1(\omega)F_2(\omega)$ is the Fourier transform of the convolution of $f_1(t)$ and $f_2(t)$.

(a) Insert the identities

$$
F_1(\omega) = \frac{1}{2\pi}\int_{-\infty}^{\infty} f_1(\tau)e^{-i\omega\tau}\, d\tau, \quad F_2(\omega) = \frac{1}{2\pi}\int_{-\infty}^{\infty} f_2(s)e^{-i\omega s}\, ds
$$

into the inverse Fourier transform of $2\pi F_1(\omega)F_2(\omega)$

$$
\int_{-\infty}^{\infty} 2\pi F_1(\omega)F_2(\omega)e^{i\omega t}\, d\omega
$$

and rearrange to obtain the expression

$$
\frac{1}{2\pi}\int_{-\infty}^{\infty} f_1(\tau)\, d\tau \int_{-\infty}^{\infty} f_2(s)\, ds \int_{-\infty}^{\infty} e^{i\omega(t-\tau-s)}\, d\omega.
$$

(b) Using Table 3.1 identify the final integral above as $2\pi\delta(t-\tau-s)$.

(c) Carry out the integration over $s$ and observe the convolution theorem.

**10.** (*Interpretation of the Transfer Function*) Let $H(\omega)$ be the transfer function for the general $n$th order linear differential equation with constant coefficients

$$
a_n[\frac{d}{dt}]^n x + a_{n-1}[\frac{d}{dt}]^{n-1} x + \cdots + a_2 \frac{d^2}{dt^2}x + a_1 \frac{d}{dt}x + a_0 x = f(t).
$$

(a) Let $h(t)$ be the inverse Fourier transform of $H(\omega)$:

$$h(t) = \int_{-\infty}^{\infty} H(\omega)e^{i\omega t}\, d\omega.$$

Show that $y(t) = h(t)/2\pi$ solves the equation

$$a_n\left[\frac{d}{dt}\right]^n y + a_{n-1}\left[\frac{d}{dt}\right]^{n-1} y + \cdots + a_2\frac{d^2}{dt^2}y + a_1\frac{d}{dt}y + a_0 y = \delta(t).$$

(Hint: what is the Fourier transform of $\delta(t)$?)

(b) Show that $y(t) = h(t - t_0)/2\pi$ solves the equation

$$a_n\left[\frac{d}{dt}\right]^n y + a_{n-1}\left[\frac{d}{dt}\right]^{n-1} y + \cdots + a_2\frac{d^2}{dt^2}y + a_1\frac{d}{dt}y + a_0 y = \delta(t - t_0).$$

(c) Recall that a solution of the equation in part (b) is known as a *Green's function* for the original equation (Section 1.4); thus

$$\frac{h(t - t_0)}{2\pi} = G(t; t_0)$$

in the terminology of that section. *The inverse Fourier transform of the transfer function equals $2\pi$ times the Green's function.* Apply the convolution theorem (exercise 9) to the transfer function display for the particular solution

$$x(t) = \int_{-\infty}^{\infty} H(\omega)F(\omega)e^{i\omega t}\, d\omega$$

and obtain the form

$$x(t) = \int_{-\infty}^{\infty} f(\tau)\frac{h(t - \tau)}{2\pi}\, d\tau = \int_{-\infty}^{\infty} f(\tau)G(t; \tau)\, d\tau.$$

Note that this reproduces formula (11) in Section 1.4!

**11.** Use the transfer function approach to find a Green's function for the equation

$$\ddot{x} + 2\dot{x} + 5x = f(t).$$

**12.** The equation

$$\ddot{x} - 4x = \delta(t)$$

describes the impulse response (or Green's function) of an undamped mass-spring system with a negative spring constant. Use distribution calculus in the frequency domain to obtain the general solution

$$x(t) = -e^{-2|t|}/4 + d_1\, e^{-2t} + d_2\, e^{2t}.$$

**13.** Note that the general solution for exercise 12 is nonzero for negative $t$, even though the impulse is applied at $t = 0$. It is a *noncausal* solution. The causal solution - for which the system is at rest until the force is applied - is implemented with the appropriate selection of $d_1, d_2$. Show that the causal solution is

$$x(t) = \begin{cases} \frac{\sinh 2t}{2}, & t \geq 0; \\ 0, & t \leq 0. \end{cases}$$

**14.** Solve exercise 13 with Laplace transforms, taking the initial conditions to be $x(0) = \dot{x}(0) = 0$.

**15.** Use distribution calculus in the frequency domain to find the general solution to

$$\ddot{x} + 4\dot{x} + 13x = H(t),$$

where $H(t)$ is the Heaviside function defined in Section 1.3.

**16.** Eliminate the variable $x_1$ from eqs. (47, 49) to obtain a single (higher-order) differential equation for $x_2(t)$. (You may need to consult your old elementary differential equations textbook if you've forgotten how to do this.) Compute the transfer function for $x_2(t)$ directly from this form and show that it agrees with eq. (51).

## 3.10   Numerical Fourier Analysis

**Summary**   The inverse Fourier transform $\int_{-\infty}^{\infty} F(\omega) e^{i\omega t} d\omega = f(t)$ can be approximated at the $N$ discrete points $t_j = \frac{j}{N}T$ in the interval $0 \leq t_j \leq \frac{N-1}{N}T$ by the formula

$$f(t_j) \approx \frac{2\pi}{T} \sum_{n=-N/2}^{N/2-1} F(n\omega_0) e^{inj2\pi/N}, \ \omega_0 = \frac{2\pi}{T}.$$

Accuracy is improved by taking $T$ and $N$ large. The Fast Fourier Transform algorithm takes advantage of trigonometric identities to simplify the computation of the right hand side.

The mathematics of Fourier analysis is a very fertile area. David Kammler's excellent book cites references which show how the investigation of the validity of the Fourier representation gave rise to the modern concept of a mathematical "function", the Riemann and Lebesgue integrals, bounded variation, Cantor's theory of sets, various notions of convergence, and the theory of distributions. Surveys of the different aspects of this subject are well documented in Kammler and in Papoulis.

In this section we are going to address only one issue: the numerical evaluation of the inverse Fourier transform of a given function $F(\omega)$. This task is of utmost practical concern to us for the following reasons:

(i) As we have seen (and as will become more apparent later in this book), some problems are simply too difficult to solve in the time domain, and we are forced to transform the time variable in order to get a tractible formulation. In such a case the theoretical analysis provides us with the Fourier transform of the answer, rather than the answer itself. Since the listings in Table 3.1 are relatively sparse, they can seldom be relied on to invert the transform and consummate the analysis.

(ii) Often we deal with systems which are too complicated for theoretical analysis, so their transfer functions $H(\omega)$ are tabulated experimentally. To use this information in predicting the response to a given input $f(t)$ with transform $F(\omega)$, we must take the inverse transform of a function $F(\omega)H(\omega)$ which is presented in tabular rather than analytic form.

(iii) In case (ii) the given input function $f(t)$ itself may not appear in Table 3.1, so we would have to compute the *forward* transform $G(\omega)$ numerically also; but note that any numerical algorithm for computing an inverse Fourier transform $\int_{-\infty}^{\infty} F(\omega) e^{i\omega t}\, d\omega$ can be easily adapted for computing the forward transform $\frac{1}{2\pi} \int_{-\infty}^{\infty} f(t) e^{-i\omega t}\, dt$.

Now the task of evaluating $f(t) = \int_{-\infty}^{\infty} F(\omega) e^{i\omega t}\, d\omega$ for a given value of $t$ may not seem all that complicated. Using one of the numerical integration algorithms, we would devise some strategy for truncating the infinite interval and then discretize the resulting (proper) integral, so that we end up with an approximation of the form

$$f(t) \approx \sum_{n=1}^{N} \mu_n F(\omega_n) e^{i\omega_n t} \tag{1}$$

where the $\{\mu_n\}$ are some set of weights. The problem is that although a particular choice of truncation strategy and discretization mesh may yield accurate estimates of $f(t_1)$, if we then require some other value $f(t_2)$, the truncation and discretization levels may need to be reset. In practice we need an approximation which is valid, not for a *single* value of $t$, but for all $t$ in some range of interest.

It is impossible for a finite sum like (1) to approximate $f(t)$ for all $t$ in $(-\infty, \infty)$, simply because the right-hand side of (1) does not approach 0 as $t \to \pm\infty$. This is easy to see if the frequencies are all multiples of some fundamental $\omega_0$, for then the right hand side is periodic (of period $T = 2\pi/\omega_0$). For non-commensurate $\{\omega_n\}$ the statement is a well-known consequence of the theory of aperiodic functions. Thus if $f(t) \to 0$ at $\infty$, (1) can't be accurate for large $|t|$.

Consequently when we seek a computational procedure for inverting the Fourier transform, we must be willing to compromise on two levels: the range of $t$ over which $f(t)$ is to be estimated, and the accuracy of the estimate in that range. We are going to describe a procedure which deals with these issues in a transparent way and exploits a very efficient computational algorithm known as the Fast Fourier Transform. Keep in mind that we are *given* the Fourier transform $F(\omega)$ (or tabulated values of $F(\omega)$), and our task is to find $f(t)$.

The first step that we will take is to replace the integral

$$f(t) = \int_{-\infty}^{\infty} F(\omega)\, e^{i\omega t}\, d\omega \tag{2}$$

by the discrete sum (series) suggested by the limiting process which *defines* (finite) integrals (Appendix A.6):

$$f_0(t) = \sum_{n=-\infty}^{\infty} F(n\,\Delta\omega)\, e^{in\,\Delta\omega\, t}\Delta\omega = \omega_0 \sum_{n=-\infty}^{\infty} F(n\omega_0)\, e^{in\omega_0 t}, \tag{3}$$

where we have set $\omega_0 = \Delta\omega$ in the final equality. Now what is the difference between $f(t)$ and $f_0(t)$?

Note first of all that the right-hand member of (3) is a *Fourier series*, and thus its sum $f_0(t)$ is a *periodic* function of period $2\pi/\omega_0$.

Next observe that if we dismantle the Fourier integral defining $F(n\omega_0)$ into pieces of length $T = 2\pi/\omega_0$ we can manipulate it to derive

$$
\begin{aligned}
F(n\omega_0) &= \frac{1}{2\pi} \int_{-\infty}^{\infty} f(\tau)\, e^{-in\omega_0\tau}\, d\tau \\
&= \frac{1}{2\pi} \sum_{m=-\infty}^{\infty} \int_0^{2\pi/\omega_0} f(\tau + m2\pi/\omega_0)\, e^{-in\omega_0(\tau + m2\pi/\omega_0)}\, d\tau \\
&= \frac{1}{2\pi} \int_0^{2\pi/\omega_0} \sum_{m=-\infty}^{\infty} f(\tau + m2\pi/\omega_0)\, e^{-in\omega_0\tau}\, (1)\, d\tau \\
&= \frac{1}{2\pi} \int_0^{T} \left[ \sum_{m=-\infty}^{\infty} f(\tau + mT) \right] e^{-in2\pi\tau/T}\, d\tau. 
\end{aligned} \tag{4}
$$

Thus

$$\frac{2\pi}{T} F(n\omega_0) = \frac{1}{T} \int_0^{T} \left[ \sum_{m=-\infty}^{\infty} f(\tau + mT) \right] e^{-in2\pi\tau/T}\, d\tau$$

*is the nth coefficient in the Fourier series expansion of the function* $\sum_{m=-\infty}^{\infty} f(t + mT)$. But from (3) $\frac{2\pi}{T} F(n\omega_0)$ is the $n$th coefficient in the Fourier series expansion of $f_0(t)$. Consequently we now know what $f_0(t)$ is:

$$f_0(t) = \sum_{m=-\infty}^{\infty} f(t + mT).$$

Therefore the function $f_0(t)$ is related to $f(t)$ by a process sometimes known as *time-domain aliasing*; we superpose all the copies of $f(t)$ obtained by shifting by $T = 2\pi/\omega_0$. Note that if $f(t)$ happens to equal zero for $t$ outside $[0, T]$ we obtain exact periodic replicas of $f(t)$ as in Fig. 1(a); otherwise the replicas are distorted as in Fig. 1(b). *In the interval $0 \le t \le T$ the difference between $f(t)$ and $f_0(t)$ equals*

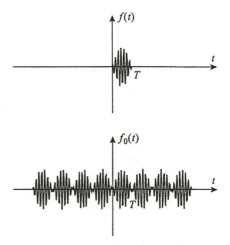

$$f(t) - f_0(t) = \left\{ \sum_{m=-\infty}^{-1} + \sum_{m=1}^{\infty} \right\} f(t + mT), \qquad (5)$$

and is small if $f(t)$ is insignificant outside this interval.

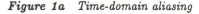

The right-hand side of (5) may be easy to estimate for certain $f(t)$. For instance if $f(t)$ decays exponentially

**Figure 1a**   *Time-domain aliasing*

$$|f(t)| \le C e^{-kt}, \quad t > T$$

(as in damped systems) and is zero for negative $t$, then we have

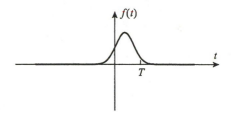

$$|f(t) - f_0(t)| \le \sum_{m=1}^{\infty} C e^{-kmT} = C e^{-kT}/[1 - e^{-kT}], \quad 0 \le t \le T; \qquad (6)$$

similarly if $f(t)$ falls off like a power of $t$,

$$|f(t)| \le C t^{-M}, \; t > T; \quad f(t) = 0, \; t < 0;$$

then

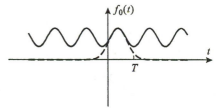

$$|f(t) - f_0(t)| \le \sum_{m=1}^{\infty} C (t + mT)^{-M} < CT^{-M+1}/(M-1) \quad (7)$$

**Figure 1b**   *Time-domain aliasing*

(exercise 1). In either of these cases $f_0(t)$ can be made arbitrarily close to $f(t)$ in $[0, T]$ by choosing $T$ sufficiently large.[15]

Now we turn to the formula (3) for $f_0(t)$. First we group the sum into pieces of $N$ terms each. $N$, which we presume to be even, will be specified later. The optimum grouping of the terms, for reasons to be seen below, is somewhat different from that employed in eq. (4):

$$f_0(t) = \omega_0 \sum_{m=-\infty}^{\infty} \sum_{n=-N/2}^{N/2-1} F[(mN + n)\omega_0] \, e^{i(mN+n)\omega_0 t}.$$

---

[15] For situations in which $f(t)$ *is* significant outside the interval $[0, T]$, "windowing" techniques can be used to estimate $f(t) - f_0(t)$; see Papoulis.

Next consider the values taken by $f_0(t)$ at the uniformly spaced points $t_j = j\Delta t$:

$$f_0(j\Delta t) = \omega_0 \sum_{m=-\infty}^{\infty} \sum_{n=-N/2}^{N/2-1} F[(mN+n)\omega_0] e^{i(mN+n)\omega_0 j\Delta t}. \qquad (8)$$

If we take $N$ such points in our fundamental interval $[0, T]$, so that $\Delta t = T/N = 2\pi/N\omega_0$ and $j = 0, 1, 2, \dots, N-1$, then we obtain a simplification in (8):

$$
\begin{aligned}
f_0(j\Delta t) &= \omega_0 \sum_{m=-\infty}^{\infty} \sum_{n=-N/2}^{N/2-1} F[(mN+n)\omega_0] e^{i(mN+n)\omega_0 j 2\pi/N\omega_0} \\
&= \omega_0 \sum_{m=-\infty}^{\infty} \sum_{n=-N/2}^{N/2-1} F[(mN+n)\omega_0] e^{imj2\pi} e^{inj2\pi/N} \\
&= \omega_0 \sum_{n=-N/2}^{N/2-1} \left\{ \sum_{m=-\infty}^{\infty} F[(mN+n)\omega_0] (1) \right\} e^{inj2\pi/N} \\
&= \omega_0 \sum_{n=-N/2}^{N/2-1} F_0(n\omega_0) e^{inj2\pi/N} \qquad\qquad (9)
\end{aligned}
$$

where $F_0(\omega)$ is the *frequency-aliased* version of $F(\omega)$ -

$$F_0(\omega) = \sum_{m=-\infty}^{\infty} F(\omega + mN\omega_0) \qquad (10)$$

- which superposes all the copies of $F(\omega)$ obtained by shifting by $\Omega_0 = N\omega_0$. Thus $F_0(\omega)$ bears the same relationship to $F(\omega)$ as $f_0(t)$ to $f(t)$, and the difference $F - F_0$ can be estimated as in eqs. (6, 7).

There is a slightly different feature of the aliased version of $F(\omega)$. Although it was perfectly reasonable to assume that $f(t)$ is significant only for $0 \le t \le T$, remember that $F(-\omega) = \overline{F(\omega)}$ when $f(t)$ is real, and the significant interval for the Fourier transform $F(\omega)$ is thus *symmetric* about the origin: $-\Omega_0/2 \le \omega \le \Omega_0/2$. See Fig. 2. This means that appropriate frequency-domain aliasing provides

$$F_0(\omega) \approx F(\omega) \quad \text{for} \quad -\Omega_0/2 \le \omega \le \Omega_0/2, \qquad (11)$$

which is why we grouped the sum in eq. (9) as we did. In other words, (11) is valid for the terms in (9).

Quantitative estimates for the error in (11) take the form $\mathcal{O}(e^{-\ell N\omega_0/2})$ if $F(\omega)$ decays like $e^{-\ell\omega}$, or $O[(N\omega_0)^{-P+1}]$ if $F$ decays like $\omega^{-P}$ (exercise 1). In particular, one controls the error by choosing $\Omega_0 = N\omega_0$ sufficiently large.

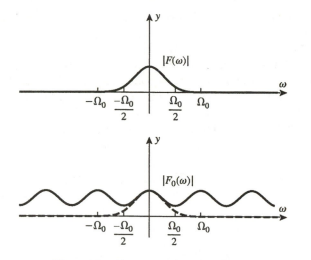

**Figure 2**    *Frequency-domain aliasing*

Now we are ready to formalize our procedure for inverting the Fourier transform $F(\omega)$ to obtain $f(t)$. For the $N$ discrete values $\{t_j = j\Delta t : j = 0, 1, 2, \ldots, N-1\}$ of $t$ lying in the range $0 \le t_j \le \frac{N-1}{N}T$ we piece together the following approximation:

$$f(t_j = j\Delta t) \approx f_0(j\Delta t) \quad = \quad \omega_0 \sum_{n=-N/2}^{N/2-1} F_0(n\omega_0)\, e^{inj2\pi/N}$$

$$\approx \quad \omega_0 \sum_{n=-N/2}^{N/2-1} F(n\omega_0)\, e^{inj2\pi/N}$$

or

$$f(t_j = j\Delta t) \approx \omega_0 \sum_{n=-N/2}^{N/2-1} F(n\omega_0)\, e^{inj2\pi/N}. \tag{12}$$

To estimate the accuracy of (12) we look at the pieces:

$$error = \quad [f(t_j) - f_0(t_j)] + [f_0(t_j) - \omega_0 \sum_{n=-N/2}^{N/2-1} F(n\omega_0)\, e^{inj2\pi/N}]$$

$$= \quad [f(t_j) - f_0(t_j)] + \omega_0 \sum_{n=-N/2}^{N/2-1} [F_0(n\omega_0) - F(n\omega_0)]\, e^{inj2\pi/N}$$

The first error term is controlled by choosing $T$ large enough so that the time-domain aliasing (Fig. 1) does not appreciably disturb the values of $f(t)$ for $0 \le t \le T$ (eqs. (5, 6, 7)). This sets the range $[0, T]$ of approximation for $f$ and the fundamental frequency $\omega_0 = 2\pi/T$. Then the remaining error is controlled by choosing $N$ large enough so that the frequency-domain aliasing (Fig. 2) (eq. (11)) does not appreciably disturb the values of $F(n\omega_0)$ for $-N/2 \le n \le N/2 - 1$; this sets the spacing $\Delta t = T/N$.

*Note the double payoffs: larger values of $T$ imply more accuracy and broader range of validity, and larger values of $N$ imply more accuracy and finer spacing of the tabulations $f(t_j)$.*

Of course larger values of $N$ also imply more computational effort. To estimate a single value of $f(t_j)$ by eq. (12) requires $N$ (complex) multiplications, and there are $N$ such values to compute - a total of $N^2$ multiplications. In applications it is often desirable to take $N$ to be several thousand, and the algorithm in this form is too computation-intensive. However, by judicious grouping of the terms in (12), the work can be reduced considerably.

Suppose, for instance, that $N = 16$ and that the values of $F(n\omega_0)$ are relabeled so that the computation of, say, $f(t_1)$ in (12) takes the symbolic form

$$f(t_1) \approx \sum_{n=-8}^{7} A_n \, e^{in(1)2\pi/16} = \sum_{n=-8}^{7} A_n \, e^{in\pi/8}.$$

The numerical values of $\{e^{in\pi/8} : n = -8, -7, -6, \cdots, 7\}$ are very redundant:

| $n$ | $e^{in\pi/8}$ |
|---|---|
| -8 | $-1.000 + 0.00i$ |
| -7 | $-.924 - .383i$ |
| -6 | $-.707 - .707i$ |
| -5 | $-.383 - .924i$ |
| -4 | $-1.000i$ |
| -3 | $.383 - .924i$ |
| -2 | $.707 - .707i$ |
| -1 | $.924 - .383i$ |
| 0 | $1.000 + 0.00i$ |
| 1 | $.924 + .383i$ |
| 2 | $.707 + .707i$ |
| 3 | $.383 + .924i$ |
| 4 | $1.000i$ |
| 5 | $-.383 + .924i$ |
| 6 | $-.707 + .707i$ |
| 7 | $-.924 + .383i$ |

So the computation of $f(\Delta t)$ can, in fact, be carried out with only *three* complex multiplications:

$$
\begin{aligned}
f(\Delta t) \;=\; & A_0 - A_{-8} + i(A_4 - A_{-4}) \\
& + .924[A_1 - A_7 - A_{-7} + A_{-1} + i(A_3 + A_5 - A_{-5} - A_{-3})] \\
& + .707[A_2 - A_6 - A_{-6} + A_{-2} + i(A_2 - A_6 - A_{-6} + A_{-2})] \\
& + .383[A_3 - A_5 - A_{-5} + A_{-3} + i(A_1 + A_7 - A_{-7} - A_{-1})].
\end{aligned}
$$

If we could achieve this same savings for all 16 values of $f(j\Delta t)$, the number of multiplications would be reduced from $16^2 = 256$ to $16 \times 3 = 48$.

The Fast Fourier Transform is an algorithm which systematically exploits these rearrangements of terms in the evaluation of (12). The emergence of the FFT in the late 1960s was a major milestone in modern system analysis and signal processing. For values of $N$ of the form $2^m$ the total number of multiplications required is reduced to $Nm/2 + N = (N/2)\log_2 N + N$ (or 48, when $n = 16$). Codes are readily available, and small computers can perform a 4096-point transform in seconds. We refer the interested reader to Kammler's book for details; a discussion of the basic strategy of the FFT is given in exercise 2.

For our purposes, the significance of the FFT is its utilization in computing inverse (or forward) Fourier transforms. To recapitulate: given an analytic formula for $F(\omega)$, we select $T$ (and consequently $\omega_0$) so that the range of interest of the time variable is covered and the accuracy of (5) is acceptable; then we choose $N$ so that the spacing $\Delta t$ and the accuracy of (11) is acceptable; and finally we implement (12) with an FFT code. If $F(\omega)$ is only available in tabulated form, then the choice of $T$ and $N$ is preempted, but the error estimates can be assessed.

## Exercises 3.10

1. Derive estimates (6) and (7) and the order-of-magnitude estimates below eq. (11). (Hint: for (6) you must sum a geometric series. For (7) you must estimate a sum by an integral; see exercise 4, Sec. 3.4).

2. (*Fast Fourier Transform*) Consider the evaluation of (12) with the data reindexed in the form

$$f(t_j) = f(j\Delta t) \approx \sum_{n=0}^{N-1} \alpha_n e^{ijn2\pi/N}. \tag{13}$$

As mentioned in the text, the computation of $N$ values of $f(j\Delta t)$ apparently entails $N^2$ complex multiplications.

   (a) Express the $\alpha_n$ in (13) in terms of the data $F(n\omega_0)$. (*Hint*: keep in mind that $e^{ijN2\pi/N} = 1$.)

   (b) Suppose that $N$ is even: $N = 2N_1$. Show that the formula for $f(j\Delta t)$ can be rewritten as

$$f(j\Delta t) \approx \sum_{n=0}^{N_1-1} \alpha_n e^{ijn2\pi/N} + \sum_{n=0}^{N_1-1} \alpha_{n+N_1} e^{ij[n+N_1]2\pi/N}$$

$$= \sum_{n=0}^{N_1-1} \{\alpha_n + (-1)^j \alpha_{n+N_1}\} e^{ijn2\pi/N} = \sum_{n=0}^{N_1-1} \beta_n e^{ijn2\pi/N}$$

(c) Now how many complex multiplications will it take to compute $N$ values of $f(j\Delta t)$? (*Answer:* $N$ coefficients times $N_1 = N/2$ multiplications per coefficient $= N^2/2$; the multiplications by (-1), of course, are not counted.)

> We seek to iterate this process, halving the number of multiplications again (assuming that $N_1$ is even). However the sum in part (b) does not have the same form as that in (13) - the "$N$" in the exponent does not match the "$N_1$" in the summation limits. So we have to back up.

(d) Show that if $j$ is even, $j = 2j_1$, then the sum formula in part (b) takes the form

$$f(j\Delta t) \approx \sum_{n=0}^{N_1-1} \beta_n e^{ijn2\pi/N} = \sum_{n=0}^{N_1-1} \beta_n e^{ij_1 n2\pi/N_1},$$

while if $j$ is odd, $j = 2j_1 + 1$, it can be written

$$f(j\Delta t) \approx \sum_{n=0}^{N_1-1} \{\beta_n e^{in2\pi/N}\} e^{ij_1 n2\pi/N_1}$$
$$= \sum_{n=0}^{N_1-1} \gamma_n e^{ij_1 n2\pi/N_1}.$$

(e) Noting that in part (d) the computation of the coefficients $\gamma_n$ requires an one-time "overhead" of $N_1 = N/2$ multiplications, how much work will it take to compute $N$ values $f(j\Delta t)$? (*Answer:* the $N/2$ overhead multiplications plus ($N$ coefficients) times $N/2$ multiplications per coefficient $= N/2 + N^2/2$.

(f) At this point the sums in part (d) have exactly the same form as the sum in (13) with $N$ replaced by $N_1 = N/2$. If each of the two new sums is manipulated as before, it will be replaced by a sum of $N_2 = N_1/2 = N/4$ terms, with an overhead of $N_2$ multiplications per sum to form the new coefficients. *Now* how much work will be required to compute $N$ values $f(j\Delta t)$? (*Answer:* the $N_1$ overhead multiplications to form the coefficients for (d), plus 2 times $N_2$ multiplications to perform the same overhead for each sum in (d), plus ($N$ coefficients) times $N_2$ multiplications per coefficient $= N/2 + 2(N/4) + N^2/4 = 2(N/2) + N^2/4$.)

(g) If the trick in (d) is implemented yet again for the sums therein, how much work will be required to compute $N$ values $f(j\Delta t)$? (*Answer:* $3(N/2) + N^2/8$.)

(h) If $N$ is a power of 2 and the trick in (d) is implemented to reduce the sums down to one term each, what is the net computational load to compute $N$ Fourier coefficients? (*Answer:* $(\log_2 N)\frac{N}{2} + N$ multiplications.)

# Chapter 4

# The Differential Equations
of Physics
and Engineering

In this chapter we are going to review the physical basis for the differential equations to be studied later in the book. An understanding of the source of the equations of engineering analysis is very helpful in devising solution strategies, and also in setting up the necessary conditions for formulating well-posed problems. Physical insight is often more revealing than mathematical demonstration, and we freely call upon the former in our exposition.

The natural language of physics is vector analysis, and we presume some familiarity with this topic on the part of the reader. In keeping with the text of Davis and Snider, we adopt the boldface type style for vectors: $\mathbf{r}$ denotes the position vector $x\mathbf{i} + y\mathbf{j} + z\mathbf{k}$ with $\mathbf{i}$, $\mathbf{j}$, and $\mathbf{k}$ as unit vectors along the coordinate axes, $\nabla$ is the "del" (*nabla*) operator

$$\nabla = \mathbf{i}\frac{\partial}{\partial x} + \mathbf{j}\frac{\partial}{\partial y} + \mathbf{k}\frac{\partial}{\partial z} \,,$$

the scalar product is written $\mathbf{u} \cdot \mathbf{v}$, and the vector product as $\mathbf{u} \times \mathbf{v}$. The reader may want to review the divergence theorem and the potential theorems to enhance his/her grasp of the material herein.

Subsequent chapters of this book will focus on eigenfunction methods for analyzing differential equations; this technique is very well adapted for engineering analysis, and is introduced in Chapter 5. However, for certain cases other solution forms are known and they, too, have occasional application in practical problems. Thus we shall also utilize the present chapter as a forum for surveying these special solution techniques.

# 4.1　The Calculus of Variations

**Summary**　The procedure for finding a number $x$ which minimizes a function $f(x)$ is generally known as variational calculus. The calculus of variations is an extension of this idea which addresses the task of finding a *function* $y(x)$, which minimizes a *functional*. When the functional $\mathcal{L}$ depends on $y(x)$, $y'(x)$, and $x$ (explicitly), the condition analogous to the vanishing of the first derivative is the Euler-Lagrange differential equation

$$\frac{d}{dx}\frac{\partial \mathcal{L}}{\partial y'} - \frac{\partial \mathcal{L}}{\partial y} = 0.$$

The condition can be generalized to cover higher order derivatives, higher dimensions, and auxiliary boundary conditions.

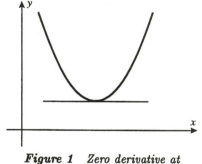

**Figure 1**　*Zero derivative at minimum*

A classic application of basic calculus is the *variational problem*: finding the minimum of a function $f(x)$. (Obviously the task of finding the maximum is mathematically equivalent to the minimization problem, so for the sake of efficiency we shall address only the latter.) For many situations the minimum is achieved at the point where the derivative $f'(x)$ is zero, as in Fig. 1. Of course there are several exceptional cases: a zero-slope point may be a maximum, or only a local minimum (Figs. 2a,b); the minimum may occur at an end point or a cusp or a point of discontinuity (Figs. 2c,d,e); or there may be *no* minimum (Fig. 2f). But in engineering practice one often knows *a priori* that the given function $f$ has the general shape[1] of Fig. 1, so the simple derivative criterion suffices.

The *calculus of variations* is the discipline that extends these notions to a higher class of "functions." As an example, consider the length of the graph of the curve $y(x)$, as $x$ runs from 0 to 1. The formula for arc length is $\int_0^1 \sqrt{dx^2 + dy^2} = \int_0^1 \sqrt{1 + y'(x)^2}\, dx$. If $y$ is a parabola $y(x) = x^2$, its arc length is $\int_0^1 \sqrt{1 + (2x)^2}\, dx$; for a sinusoid $y(x) = \sin x$, and the arc length is $\int_0^1 \sqrt{1 + \cos^2 x}\, dx$. Note that the input to the "arc length evaluator" is not a single number $x$; it is a whole function $y(x)$. *The arc length is a "function of a function", or a* **functional**.

Other examples of *functionals* - functions of functions - are

(i) the norm of $y(x)$: $\|y\| = \{\int_a^b |y(x)|^2\, dx\}^{1/2}$ (Sec. 3.3);

(ii) the average value of $y(x)$: $\int_a^b y(x)\, dx / (b - a)$;

(iii) the $m$th moment of a probability density function $y(x)$: $\int_a^b x^m y(x)\, dx$;

---

[1] In technical parlance, it is smooth and unimodal.

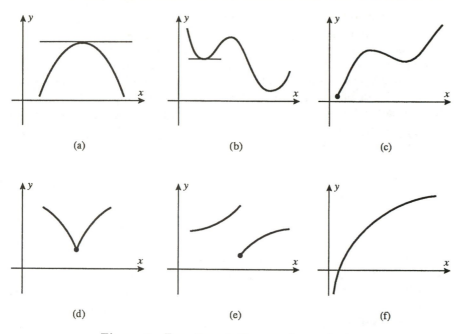

(a)                              (b)                              (c)

(d)                              (e)                              (f)

*Figure 2    Exceptions to the zero-slope criterion*

(iv)  the energy stored in an electric potential field $\psi(x, y, z)$:

$$\iiint |\nabla\psi(x, y, z)|^2 \, dx \, dy \, dz.$$

All of these functionals are expressible as integrals. The integrands can involve the function $y(x)$ itself (i, ii), its derivatives (arc length, (iv)), and the independent variable $x$ (iii). Thus we postulate the generic form of the functionals for our study to be

$$F[y(x)] = \int_a^b \mathcal{L}[y(x), y'(x), x] \, dx \tag{1}$$

or, in higher dimensions,

$$F[\psi(x, y, z)] = \int \int \int \mathcal{L}[\psi, \nabla\psi, x, y, z] \, dx \, dy \, dz. \tag{2}$$

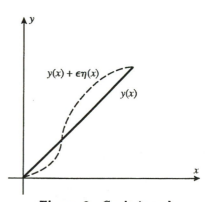

$y(x) + \epsilon\eta(x)$

$y(x)$

*Figure 3    Geodesic and perturbation*

The integrand $\mathcal{L}[\ldots]$ is traditionally called the *Lagrangian*.

*The basic task of the calculus of variations is to find, among a given class of functions $y(x)$ (or $\psi(x, y, z)$), the one which minimizes the functional $F[y(x)]$ (or $F[\psi(x, y, z)]$).*

**Example 1.** Let's start with a simple problem for which the answer is obvious. Find the shortest curve - i.e. the *geodesic* - $y(x)$ connecting the given points $[0, y_0]$ and $[1, y_1]$ (see Fig. 3).

The strategy will be to introduce a construction which reduces our task to that of minimizing a *function of a single variable*, and then applying the classical zero-slope rationale. Let $y(x)$ be the actual geodesic (ignoring the fact that we know it to be a straight line). Then consider the effect of *perturbing* $y(x)$ by a small function $\varepsilon\eta(x)$ to obtain a new curve $y(x)+\varepsilon\eta(x)$; here $\eta(x)$ can be any (differentiable) function, and $\varepsilon$ plays the role of a scaling factor (as will become clear shortly). Our functional is $F[y(x)] = \int_0^1 \mathcal{L}[y(x), y'(x), x]\, dx$ with Lagrangian $\mathcal{L}(y, y', x) = \sqrt{1 + y'(x)^2}$. The arc length of the perturbed curve is

$$
\begin{aligned}
F[y(x) + \varepsilon\eta(x)] &= \int_0^1 \mathcal{L}[y(x) + \varepsilon\eta(x), y'(x) + \varepsilon\eta'(x), x]\, dx \\
&= \int_0^1 \sqrt{1 + [y'(x) + \varepsilon\eta'(x)]^2}\, dx.
\end{aligned}
\tag{3}
$$

For a given but arbitrary $\eta(x)$, the formula (3) specifies a well-defined function of the scalar $\varepsilon$: $f(\varepsilon)$ equals the arc length of the curve $y(x)+\varepsilon\eta(x)$. The function $f(\varepsilon)$, however, has the important property that *it takes its minimum when $\varepsilon = 0$*, because the curve $y(x)$ itself has shortest length. So - if we can assume that none of the demons in Fig. 2 is present - the simple calculus rule tells us that

$$
\frac{d}{d\varepsilon} f(\varepsilon) = \frac{d}{d\varepsilon} F[y(x) + \varepsilon\eta(x)] = 0 \quad \text{at} \quad \varepsilon = 0.
\tag{4}
$$

With eq. (4) as our guiding principle, we analyze the functional (3). Its derivative is

$$
\frac{d}{d\varepsilon} f(\varepsilon) = \int_0^1 \frac{1}{2}\{1 + [y'(x) + \varepsilon\eta'(x)]^2\}^{-1/2}\, 2\, [y'(x) + \varepsilon\eta'(x)]\, \eta'(x)\, dx
$$

and at $\varepsilon = 0$,

$$
\frac{d}{d\varepsilon} f(0) = \int_0^1 [1 + y'(x)]^{-1/2} y'(x)\eta'(x)\, dx.
\tag{5}
$$

For reasons to be seen, we want to express (5) in terms of $\eta(x)$, undifferentiated. Thus we integrate by parts:

$$
\begin{aligned}
\frac{d}{d\varepsilon} f(0) &= \int_0^1 [1 + y'(x)^2]^{-1/2} y'(x) \frac{d}{dx}\eta(x)\, dx \\
&= -\int_0^1 \frac{d}{dx}\{[1 + y'(x)^2]^{-1/2} y'(x)\}\, \eta(x)\, dx \\
&\quad + \{[1 + y'(x)^2]^{-1/2} y'(x)\}\, \eta(x)\Big|_{x=0}^{x=1}.
\end{aligned}
\tag{6}
$$

Now since we are seeking the shortest curve between two *specified* terminals $[0, y(0)]$ and $[1, y(1)]$, we are only concerned with "competitors"

$y(x) + \varepsilon\eta(x)$ which pass through the same end points. Therefore we are in fact minimizing (3) over perturbations $\eta$ which vanish at the ends:

$$\eta(0) = 0, \ \eta(1) = 0.$$

Consequently the condition (4) is reduced via (6) to

$$\int_0^1 \frac{d}{dx}\{[1 + y'(x)^2]^{-1/2}y'(x)\}\eta(x)\,dx = 0. \tag{7}$$

Equation (7) states that the function $\dfrac{d}{dx}\{[1 + y'(x)^2]^{-1/2}y'(x)\}$, when multiplied by an *arbitrary* function $\eta(x)$, produces a zero integral. The only way this can happen is if $\dfrac{d}{dx}\{[1 + y'(x)^2]^{-1/2}y'(x)\}$, itself, is zero:

$$\frac{d}{dx}\{[1 + y'(x)^2]^{-1/2}y'(x)\} = 0. \tag{8}$$

The logic leading to (8) is sometimes identified as the Fundamental Lemma of the Calculus of Variations. A more compelling argument goes as follows: if there were some point where $\dfrac{d}{dx}\{[1 + y'(x)^2]^{-1/2}y'(x)\}$ were *not* zero, by continuity it would also be nonzero on a small interval around the point. But then if we chose $\eta(x)$ to be positive in that interval and zero elsewhere, (7) would be violated.

Expanding eq. 8, we find

$$
\begin{aligned}
0 &= \frac{d}{dx}\{[1 + y'(x)^2]^{-1/2}y'(x)\} \\
&= [1 + y'(x)^2]^{-1/2}y''(x) - \frac{1}{2}[1 + y'(x)^2]^{-3/2}2y'(x)^2y''(x) \\
&= [1 + y'(x)^2]^{-3/2}y''(x)
\end{aligned}
$$

or

$$y''(x) = 0. \tag{9}$$

The only functions satisfying (9) are of the form $y = ax + b$ - i.e., straight lines. We arrive at the gratifying conclusion that the shortest distance between the two points is a straight line. ∎

The derivation in this example generalizes readily to the case of an arbitrary functional of the form $F[y(x)] = \int_a^b \mathcal{L}[y(x), y'(x), x]\,dx$. Once again we let $y(x)$ denote the minimizing function, and consider the effect of perturbing $y$ by $\varepsilon\eta$ to get the function

$$f(\varepsilon) = F[y(x) + \varepsilon\eta(x)] = \int_a^b \mathcal{L}[y(x) + \varepsilon\eta(x), y'(x) + \varepsilon\eta'(x), x]\,dx.$$

Next we enforce the condition that $f'(\epsilon) = 0$ at $\varepsilon = 0$:

$$
\begin{aligned}
f'(0) &= \int_a^b \frac{d}{d\varepsilon} \mathcal{L}[y(x) + \varepsilon\eta(x), y'(x) + \varepsilon\eta'(x), x]\Big|_{\varepsilon=0} dx \\
&= \int_a^b \left\{ \frac{\partial}{\partial y}\mathcal{L}[y(x), y'(x), x]\, \eta(x) + \frac{\partial}{\partial y'}\mathcal{L}[y(x), y'(x), x]\, \eta'(x) \right\} dx.
\end{aligned}
$$

Integration by parts on the second term produces

$$
\begin{aligned}
f'(0) &= \int_a^b \left\{ \frac{\partial}{\partial y}\mathcal{L}[y(x), y'(x), x]\, \eta(x) - \frac{d}{dx}\frac{\partial}{\partial y'}\mathcal{L}[y(x), y'(x), x]\, \eta(x) \right\} dx \\
&\quad + \frac{\partial}{\partial y'}\mathcal{L}[y(x), y'(x), x]\, \eta(x) \Big|_{x=a}^{x=b}. \tag{10}
\end{aligned}
$$

Most applications of the calculus of variations are similar to the shortest-distance problem, in that one seeks the minimum of $F[y(x)]$ over all functions which take *specified* boundary values $y(a)$ and $y(b)$. For such applications the final term in (10) is zero (because $\eta(a) = \eta(b) = 0$). Then the condition $f'(0) = 0$ in (10) again leads to the conclusion that a certain function - namely,

$$
\frac{\partial}{\partial y}\mathcal{L}[y(x), y'(x), x] - \frac{d}{dx}\frac{\partial}{\partial y'}\mathcal{L}[y(x), y'(x), x], \tag{11}
$$

when multiplied by any *arbitrary* function $\eta(x)$, produces a zero integral over $[a, b]$. By the Fundamental Lemma we deduce that the function (11) itself is zero. This condition for a minimum of a functional is known as the *Euler-Lagrange equation*, and it is usually written

$$
\frac{d}{dx}\frac{\partial\mathcal{L}}{\partial y'} - \frac{\partial\mathcal{L}}{\partial y} = 0. \tag{12}
$$

Equation (12) is the generalization of eq. (8) for the class of functionals of the form $F[y(x)] = \int_a^b \mathcal{L}[y(x), y'(x), x]\, dx$. Note that as a rule the Euler-Lagrange equation, when expanded, is a second order ordinary differential equation for $y(x)$ (for example see eq. (9)). If the given values of $y(a)$ and $y(b)$ are conjoined with condition (12), we have a well-defined boundary value problem for the minimizing function $y(x)$.

The Euler-Lagrange condition for a functional is equivalent to the zero-slope condition for a function of one variable. As such, it is neither a necessary nor a sufficient condition for a minimum (recall Fig. 2). For the next three examples it is fairly clear from the form of the functional that a minimum does exist, and thus that the Euler-Lagrange equation will be valid there. In general, however, it is more precise to state that condition

(12) renders the functional *stationary* - analogous to the local minimum, local maximum, or point of inflection in ordinary calculus.

**Example 2.** The function $y(x)$ which minimizes the functional $F[y(x)] = \int_0^1 [y(x)^2 + y'(x)^2]dx$ and satisfies $y(0) = 0$ and $y(1) = 1$ must solve the differential equation

$$0 = \frac{d}{dx}\frac{\partial}{\partial y'}[y^2 + y'^2] - \frac{\partial}{\partial y}[y^2 + y'^2] = 2y'' - 2y ,$$

or $y'' = y$. Hence $y(x) = c\sinh x + d\cosh x$ and, for the boundary conditions given, $y(x) = \frac{\sinh x}{\sinh 1}$. ∎

**Example 3.** If the graph of a positive function $y(x)$, $-1 \le x \le 1$, is rotated around the $x$ axis, it generates a surface of revolution whose area is given by the functional

$$F[y(x)] = \text{area} = \int_{-1}^1 2\pi y(x)\sqrt{1 + y'(x)^2}dx. \tag{13}$$

Suppose that $y(-1) = y(1) = 2$. What curve $y(x)$ minimizes the surface area?

The Lagrangian is $\mathcal{L}[y, y', x] = 2\pi y(x)\sqrt{1 + y'(x)^2}$. The Euler-Lagrange equation then states that

$$0 = \frac{d}{dx}\frac{\partial \mathcal{L}}{\partial y'} - \frac{\partial \mathcal{L}}{\partial y} = \frac{d}{dx}\{2\pi y \frac{1}{2}[1 + y'^2]^{-1/2}2y'\} - 2\pi[1 + y'^2]^{1/2}. \tag{14}$$

In exercise 1 the reader will be invited to show that (14) can be reduced to the nonlinear differential equation

$$yy'' - y'^2 - 1 = 0, \tag{15}$$

whose general solution happens to be $y(x) = c\cosh(x/c + d)$. The particular solution satisfying the boundary conditions $y(\pm 1) = 2$ is easily found by iteration to be $y \approx 1.6967\cosh x/1.6967$, which generates the minimal surface area $F \approx 23.97$. As Fig. 4 demonstrates, this *catenary* economizes on surface area by pulling in its "waist." ∎

The fact that the nonlinear differential equation (15) arises as the Euler-Lagrange equation for a variational problem suggests an indirect method for approximating its solution $y(x)$. First we select a convenient functional form $p(x)$ for the approximation, containing some undetermined parameter(s). For example, since the conditions of Example 3 obviously dictate a *symmetric* curve, we take the crude approximation indicated in Fig. 5 - a piecewise linear "vee" passing through the given end points, with slope $\pm m$:

$$p(x) = (2 - m) + m|x|. \tag{16}$$

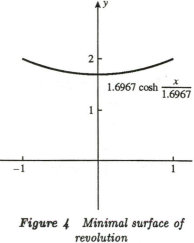

**Figure 4** *Minimal surface of revolution*

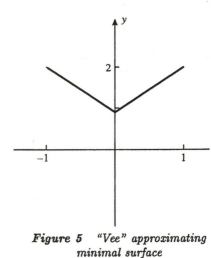

**Figure 5** *"Vee" approximating minimal surface*

Then we solve for the parameter(s) ($m$, in our case) by insisting that $p(x)$ minimize the same functional as $y(x)$. Thus $p(x)$ should minimize

$$
\begin{aligned}
F[p(x)] &= \int_{-1}^{1} 2\pi p(x) \sqrt{1 + p'(x)^2}\, dx \\
&= 2 \times 2\pi \int_{0}^{1} [(2-m) + m|x|] \sqrt{1+m^2}\, dx \\
&= \pi(8-2m)\sqrt{1+m^2}. \qquad (17)
\end{aligned}
$$

As a function of $m$, the graph of $F[p(x)]$ is depicted in Fig. 6. Its minimum occurs at the leftmost zero-slope point:

$$
\begin{aligned}
0 = \frac{\partial F}{\partial m} &= -2\pi\sqrt{1+m^2} + \pi(8-2m)\frac{1}{2}(1+m^2)^{-1/2} 2m \\
&= \pi(1+m^2)^{-1/2} - 4m^2 + 8m - 2
\end{aligned}
$$

or $m = 1 - \sqrt{2}/2$. The approximation $p(x) = \sqrt{2}/2 + (1 - \sqrt{2}/2)|x|$ and the exact solution $y(x) = 1.6967 \cosh x / 1.6967$ are compared in Fig. 7. Note that the surface area generated by the approximation is

$$
F[\sqrt{2}/2 + (1 - \sqrt{2}/2)\,|x|] \approx 24.27,
$$

which is quite close to the exact answer 23.97.

Clearly we could obtain better approximations to $y(x)$ by allowing more flexibility in our choice of $p(x)$. The *Finite Element Method* is premised on this observation (see Strang and Fix).

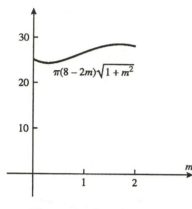

**Figure 6**    *Area functional*

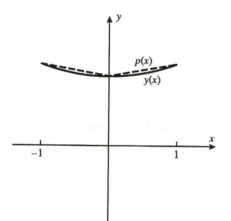

**Figure 7**    *Best "vee" approximant to minimal surface*

The derivation of the Euler-Lagrange condition for the minimization of higher dimensional functionals $F[\psi(x,y,z)] = \int\int\int \mathcal{L}[\psi, \nabla\psi, x, y, z]\, dx\, dy\, dz$ mimics the argument above, but it requires a little vector analysis for its implementation. In exercise 4 you will be guided through the analysis; the final result takes the form

$$
\frac{\partial}{\partial x}\frac{\partial \mathcal{L}}{\partial \psi_x} + \frac{\partial}{\partial y}\frac{\partial \mathcal{L}}{\partial \psi_y} + \frac{\partial}{\partial z}\frac{\partial \mathcal{L}}{\partial \psi_z} - \frac{\partial \mathcal{L}}{\partial \psi} = 0. \qquad (18)
$$

Here $\{\psi_x, \psi_y, \psi_z\}$ are the components of $\nabla\psi$.

**Example 4.** The *Dirichlet integral* is the functional defined by

$$
F[\psi] = \iiint_{D} |\nabla\psi|^2\, dx\, dy\, dz = \int\int\int_{D} \{\psi_x^2 + \psi_y^2 + \psi_z^2\}\, dx\, dy\, dz \qquad (19)
$$

where $D$ is some region of space. If we seek the function $\psi(x,y,z)$ which minimizes $F[\psi]$ over all functions taking specific values on the boundary of

$D, \psi$ will satisfy the equation obtained by applying (18) to the Lagrangian $\mathcal{L} = \psi_x^2 + \psi_y^2 + \psi_z^2$ in (19):

$$0 = \frac{\partial}{\partial x} 2\psi_x + \frac{\partial}{\partial y} 2\psi_y + \frac{\partial}{\partial z} 2\psi_z - (0) = 2\{\psi_{xx} + \psi_{yy} + \psi_{zz}\} = 2\nabla^2 \psi$$

or

$$\nabla^2 \psi = 0, \qquad (20)$$

which is known as *Laplace's equation*. As we shall see shortly, eq. (20) is of the utmost importance in applied mathematics, and the fact that it can be characterized by a variational principle has proved to be very valuable in obtaining approximations to its solutions (as in the discussion surrounding eq. (17). ∎

To go further into the study of the calculus of variations would take us far afield, so we direct the interested reader to the excellent texts by Smith and by Strang and Fix to explore how it is extended to handle more general functionals and boundary conditions. A few of these developments are surveyed in the exercises.

## Exercises 4.1

1. (a) Show that eq. (14) can be reduced to (15).

   (b) Verify that $y(x) = c\cosh(x/c + d)$ satifies (15).

   (c) Compute the values of $c$ and $d$ which enforce $y(\pm 1) = 2$, and verify the numbers reported in Example 3.

2. Work out the Euler-Lagrange equations for the following functionals:

   a. $\int_a^b \sqrt{1 + y'(x)^2}/y\, dx$ 
   
   b. $\int_0^1 (1 - x^2) y^2(x)\, dx$

   c. $\int_0^1 [y'^2 - y^2]\, dx$ 
   
   d. $\int_a^b [y^2 + y'^2 - yy']\, dx$

   e. $\iiint_D [\psi_x^2 + \psi_y^2 + \psi_z^2 + 2f(x,y,z)\psi]\, dx\, dy\, dz$   ($f$ a given function)

   f. $\iint_D \sqrt{1 + \psi_x^2 + \psi_y^2}\, dx\, dy$

3. Suppose that we have the task of minimizing the functional $F[y(x)] = \int_a^b \mathcal{L}[y(x), y'(x), x]\, dx$ over all functions $y(x)$ *whose value at a is predetermined but whose value at b is unconstrained.* The derivation of condition (10) proceeds as before, but now the only restriction imposed on $\eta(x)$ is that it be zero at $x = a$.

(a) By considering first only those functions $\eta(x)$ which are zero at *both* endpoints, reproduce the argument leading to the Euler-Lagrange condition on $y(x)$.

(b) Now by considering functions $\eta(x)$ which are zero throughout most of the interval $[a, b]$ but become positive near the endpoint $x = b$, argue that the minimizing function $y(x)$ must also satisfy the boundary condition

$$\frac{\partial}{\partial x}\mathcal{L}[y(x), y'(x), x] = 0 \quad \text{at} \quad x = b. \tag{21}$$

Equation (21) is very interesting. Usually when one applies the calculus of variations one has specified values for $y(a)$ and $y(b)$, and these provide boundary conditions supplementing the (second order) Euler-Lagrange differential equation. Now we see that when one of these boundary conditions is dropped, the variational principle itself provides the missing boundary condition, in the form of (21). The latter is called a *natural boundary condition* for the problem; see the references by Smith and by Strang and Fix.

(c) Show that the natural boundary condition at $x = 1$ for Example 2 is $y(1) = 0$.

(d) Work out the natural boundary conditions at the *left* end point for Examples 1 and 3.

4. Use the following steps as a guide to derive eq. (18).

(a) Consider the condition

$$\frac{d}{d\varepsilon}F[\psi(x, y, z) + \varepsilon\eta(x, y, z)] = 0 \quad \text{at} \quad \varepsilon = 0 \tag{22}$$

as the starting point for the derivation. Show that, if the triad $\left[\dfrac{\partial\mathcal{L}}{\partial\psi_x}, \dfrac{\partial\mathcal{L}}{\partial\psi_y}, \dfrac{\partial\mathcal{L}}{\partial\psi_z}\right]$ is regarded as a vector denoted by $\dfrac{\partial\mathcal{L}}{\partial\nabla\psi}$, then one can write

$$\frac{d}{d\varepsilon}F[\psi + \varepsilon\eta]\Big|_{\varepsilon=0} = \iiint_D \left\{\frac{\partial\mathcal{L}}{\partial\psi}\eta\right\} dx\, dy\, dz$$
$$+ \iiint_D \left\{\frac{\partial\mathcal{L}}{\partial\nabla\psi} \cdot \nabla\eta\right\} dx\, dy\, dz.$$

(b) Apply the divergence theorem to rewrite this as

$$\frac{d}{d\varepsilon}F[\psi + \varepsilon\eta]\Big|_{\varepsilon=0} = \iiint_D \left\{\frac{\partial\mathcal{L}}{\partial\psi} - \nabla \cdot \frac{\partial\mathcal{L}}{\partial\nabla\psi}\right\}\eta\, dx\, dy\, dz$$
$$+ \iint_S \mathbf{n} \cdot \frac{\partial\mathcal{L}}{\partial\nabla\psi}\eta\, dS \tag{23}$$

where $S$ is the surface bounding $D$ and $\mathbf{n}$ is its outward normal.

(c) Argue that if (22) must hold for every function $\eta$ which is zero on the boundary $S$, then the Euler-Lagrange condition (18) follows from (23).

(d) Argue that if (22) must hold for every function $\eta$ which is zero on *part of* the boundary $S$, then (18) holds inside $D$ and the *natural boundary condition*

$$\mathbf{n} \cdot \frac{\partial \mathcal{L}}{\partial \nabla \psi} = 0 \tag{24}$$

holds on the remaining portion of $S$.

(e) Show that the natural boundary condition for the functional of Example 4 is $\mathbf{n} \cdot \nabla \psi = 0$.

(f) Derive the natural boundary condition for the functionals of exercise 2(e) and 2(f).

5. Show that if the *Lagrangian* of the functional $F$ contains the second derivative of the unknown function - $\mathcal{L} = \mathcal{L}[y, y', y'', x]$ - then the Euler-Lagrange condition becomes

$$\frac{\partial \mathcal{L}}{\partial y} - \frac{d}{dx}\frac{\partial \mathcal{L}}{\partial y'} + \frac{d^2}{dx^2}\frac{\partial \mathcal{L}}{\partial y''} = 0.$$

This is ultimately a differential equation for $y(x)$; what is its order?

6. Show that the minimization of a functional of *two* functions

$$F[y(x), z(x)] = \int_a^b \mathcal{L}[y(x), y'(x), z(x), z'(x), x]\, dx$$

dictates the pair of Euler-Lagrange conditions

$$\frac{d}{dx}\frac{\partial \mathcal{L}}{\partial y'} - \frac{\partial \mathcal{L}}{\partial y} = 0, \qquad \frac{d}{dx}\frac{\partial \mathcal{L}}{\partial z'} - \frac{\partial \mathcal{L}}{\partial z} = 0.$$

7. Generalize exercises 5 and 6 for a functional of $m$ functions of $n$ independent variables, involving all partial derivatives up to order $s$.

## 4.2  Classical Mechanics

**Summary**  The Lagrangian of a mechanical system is the difference of its kinetic and potential energies. Newton's laws imply that the Lagrangian satisfies the Euler-Lagrange equation. The *sum* of the kinetic and potential energies is the Hamiltonion, and it satisfies a system of equations known as Hamilton's equations. These alternative formulations of classical mechanics are often convenient to use in curvilinear coordinate systems, and lead to deeper insights into the nature of the underlying physics.

Newton's second law - force equals mass times acceleration - is the basis for all of classical mechanics. Strictly speaking, it applies to isolated particles, or point masses; if $\mathbf{r} = \mathbf{r}(t)$ describes the trajectory of a particle and $\mathbf{F}$ is the force exerted upon it, then

$$\mathbf{F} = m\frac{d^2\mathbf{r}}{dt^2}. \tag{1}$$

For systems comprised of many particles - such as rigid bodies, fluids, plasmas, deformable structures, or composites of these - the equations of motion are derived by applying (1) to each point mass and manipulating the resulting equations. For example by summing (1) over all the points in the system one learns that the total force on the system equals the total mass times the acceleration of the center of mass (exercise 3). Similarly, if the forces between the particles obey the "equal and opposite reaction" postulate of Newton's third law, then the net *torque* equals the rate of change of *angular momentum* when both are measured about the center of mass. (See Davis and Snider for an exposition of these theorems; they are very significant for the characterization of rigid body motion.)

At bottom, then, the analyses of all classical mechanical systems come from applying (1) to their individual particles and making specific assumptions about the nature of the forces. Therefore in this section we shall focus on the consequences of (1).

An important instance of (1) occurs when the force on the particle is derivable from a *potential*; this means that $\mathbf{F}$ is a function of the particle's position $\mathbf{r}$ and that it can be calculated by taking the negative of the gradient of a scalar potential function $V(x, y, z)$:

$$\mathbf{F} = \mathbf{F}(\mathbf{r}) = \mathbf{F}(x, y, z) = -\nabla V(\mathbf{r}) = -\nabla V(x, y, z). \tag{2}$$

Such forces are called *conservative*, and vector analysis provides us with necessary and sufficient conditions for this condition to hold (line integrals independent of path, zero curl, etc. - see Davis and Snider). In such circumstances the sum, $E$, of the *kinetic energy*

$$K.E. = \frac{1}{2}m\left|\frac{d\mathbf{r}}{dt}\right|^2 \tag{3}$$

and the potential energy $V(\mathbf{r})$ is constant along the trajectory:

$$
\begin{aligned}
\frac{dE}{dt} = \frac{d}{dt}\{K.E. + V\} &= \frac{d}{dt}\{\frac{1}{2}m\left|\frac{d\mathbf{r}}{dt}\right|^2 + V(\mathbf{r})\} \\
&= \frac{1}{2}m \times 2 \times \frac{d\mathbf{r}}{dt} \cdot \frac{d^2\mathbf{r}}{dt^2} + \nabla V(\mathbf{r}) \cdot \frac{d\mathbf{r}}{dt} \\
&= \frac{d\mathbf{r}}{dt} \cdot \{\mathbf{F} - \mathbf{F}\} = 0.
\end{aligned}
$$

The net mechanical energy $E = K.E. + V$ is thus "conserved" in such motions.

The *Lagrangian formulation* of classical mechanics arises when one forms an "action" functional by integrating the *difference* of the kinetic and potential energies, with respect to time, along all possible trajectories $\mathbf{r} = \mathbf{r}(t)$ running from $\mathbf{r}(t_1)$ to $\mathbf{r}(t_2)$:

$$J[\mathbf{r}(t), \dot{\mathbf{r}}(t), t] = \int_{t_1}^{t_2} \{K.E. - V\} \, dt$$

or

$$J[x, y, z, \dot{x}, \dot{y}, \dot{z}, t] = \int_{t_1}^{t_2} \left\{ \frac{1}{2} m \left| \frac{d\mathbf{r}}{dt} \right|^2 - V(\mathbf{r}) \right\} dt$$

$$= \int_{t_1}^{t_2} \left\{ \frac{m}{2} [\dot{x}(t)^2 + \dot{y}(t)^2 + \dot{z}(t)^2] - V[x(t), y(t), z(t)] \right\} dt.$$

If one then asks which trajectories render the action integral stationary, the answer turns out to be - *the trajectories which obey Newton's Law*! For the Euler-Lagrange condition applied to the functional $J$ reads (recall exercise 7, Sec. 4.1)

$$\begin{aligned} 0 &= \frac{d}{dt} \frac{\partial}{\partial \dot{x}} (K.E. - V) - \frac{\partial}{\partial x} (K.E. - V) \\ &= \frac{d}{dt} m\dot{x} + \frac{\partial V}{\partial x} \end{aligned}$$

and similarly for the other components, so that

$$m \frac{d^2\mathbf{r}}{dt^2} + \nabla V = m \frac{d^2\mathbf{r}}{dt^2} - \mathbf{F} = 0$$

in agreement with eq. (1). Note that the *Lagrangian* $\mathcal{L}$ for the action integral equals the kinetic energy minus the potential energy.

The fact that the trajectories of classical mechanics can be characterized by what has become known as "The Principle of Least Action" (more precisely, the principle of *stationary* action) is extremely significant. The following example demonstrates one of the advantages.

**Example 1.** We wish to write the equations of motion for a particle constrained to move along a circular wire as in Fig. 1. This is a difficult task in the rectangular coordinate system. The gravitational force $-mg$ ($g$ is the gravitational acceleration) is straightforward, but we have no idea how much force the wire exerts on the particle. We only know that this force confines the particle's trajectory to the wire. Thus (1) is tricky to implement.

However it is easy to express the kinetic and potential energies in the polar coordinate system indicated in the figure:

$$K.E. = \frac{1}{2} ma^2 \dot{\theta}^2, \quad V = mgy = mga(1 - \cos\theta).$$

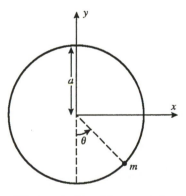

*Figure 1   Particle on a wire*

In this system the action integral becomes

$$J[\theta, \dot{\theta}, t] = \int_{t_1}^{t_2} \{\frac{1}{2}ma^2\dot{\theta}^2 - mga(1 - \cos\theta)\}\, dt,$$

for which the Euler-Lagrange equation reads

$$0 = \frac{d}{dt}\frac{\partial \mathcal{L}}{\partial \dot{\theta}} - \frac{\partial \mathcal{L}}{\partial \theta} = ma^2\frac{d^2\theta}{dt^2} + mga\sin\theta. \qquad (4)$$

Thus the variational formalism yields a single equation of motion which obviates the difficulties of the $F = ma$ approach. (Physics students may recognize (4) as equating the rate of change of angular momentum with the gravitational torque.)                                                                    ■

The *Hamitonian formulation* of classical mechanics is an alternative to the Lagrangian formulation. It is important because, as we shall see later in this chapter, it forms the basis for the Schrödinger equation when the system is modeled quantum-mechanically. The Hamiltonian $H$ of the system equals the *sum* of the kinetic energy and the potential energy,

$$H = K.E. + V,$$

with the kinetic energy expressed in terms of the *momentum*

$$\mathbf{p} = m\frac{d\mathbf{r}}{dt} \qquad (p_x = m\dot{x}, \text{etc.}) \qquad (5)$$

instead of the velocity $\frac{d\mathbf{r}}{dt}$:

$$K.E. = \frac{1}{2}m\left|\frac{d\mathbf{r}}{dt}\right|^2 = \frac{|\mathbf{p}|^2}{2m}.$$

Thus for a single particle,

$$H(\mathbf{p}, \mathbf{r}) = \frac{|\mathbf{p}|^2}{2m} + V(\mathbf{r}). \qquad (6)$$

The equations of motion in the Hamiltonian formulation are mathematically equivalent to the Lagrange equations. Here we shall merely state and verify them, directing the reader to Goldstein or Corben and Stehle for the derivations. The first set of Hamilton equations states that

$$\frac{d}{dt}\mathbf{p} = -\nabla H. \qquad (7)$$

Since $\dfrac{d}{dt}\mathbf{p} = m\dfrac{d^2\mathbf{r}}{dt^2}$ and $-\nabla H = -\nabla V = \mathbf{F}$, (7) is equivalent to Newton's second law (1). The second set of Hamilton equations,

$$\frac{dx}{dt} = \frac{\partial H}{\partial p_x}, \quad \frac{dy}{dt} = \frac{\partial H}{\partial p_y}, \quad \frac{dz}{dt} = \frac{\partial H}{\partial p_z},$$

merely confirm the formula (5) for the momentum; for example,

$$\frac{\partial H}{\partial p_x} = 2p_x/(2m) = p_x/m = \dot{x}, \quad \text{etc.}$$

**Example 2.** To implement the Hamiltonian formalism in exotic coordinate systems the "momentum" variable must be defined appropriately for the corresponding coordinate. Thus for the particle on a wire in the previous example, the momentum corresponding to the angle $\theta$ is the angular momentum

$$p_\theta = ma^2\dot{\theta}. \tag{8}$$

The Hamiltonian, expressed in terms of $p_\theta$ and $\theta$, is given by

$$H = p_\theta^2/2ma^2 + mga(1 - \cos\theta).$$

The first Hamilton equation is

$$\frac{d}{dt}p_\theta = -\frac{\partial H}{\partial \theta} \quad \text{or} \quad ma^2\frac{d^2\theta}{dt^2} = -mga\sin\theta,$$

in agreement with (4). The second equation is

$$\frac{d\theta}{dt} = \frac{\partial H}{\partial p_\theta} = p_\theta/(ma^2),$$

which confirms the definition (8). ∎

Since Newton's second law (1) is a second order differential equation, it requires the specification of initial position $\mathbf{r}(0)$ and initial velocity $\dfrac{d}{dt}\mathbf{r}(0)$ for its complete solution. The Lagrange equations such as (4) are also second order and call for initial position and velocity to be given. The Hamilton equations are first order, but there are equations for both the position and the momentum; thus they require specification of both $\mathbf{r}(0)$ and $\mathbf{p}(0)$, which are equivalent to initial position and velocity.

# Exercises 4.2

1. Consider a system of $n$ particles with masses $m_i$, located at position $\mathbf{r}_i$ $(i = 1, 2, \ldots, n)$. Suppose that there is a potential function $V(\mathbf{r}_1, \mathbf{r}_2, \ldots, \mathbf{r}_n)$ such that the force $\mathbf{F_i}$ on the $i$th particle is given

by the gradient of $V$ with respect to $\mathbf{r}_i$: $\mathbf{F}_i = -\nabla_i V$. Show that with the Lagrangian defined as the sum of the kinetic energies minus the potential energy, and the Hamiltonian defined as the sum of the kinetic energies plus the potential energy, Lagrange's equations and Hamilton's equations are each equivalent to Newton's second law.

In the subsequent problems, the *center of mass* of a system of $n$ particles as in exercise 1 is the point

$$\mathbf{R} = \{\sum_{i=1}^{n} m_i \mathbf{r}_i\}/M, \tag{9}$$

where $M$ is the total mass

$$M = \sum_{i=1}^{n} m_i. \tag{10}$$

The *motion of particle i in the center-of-mass system* is the trajectory of the vector $(\mathbf{r}_i - \mathbf{R})$.

2. Show that the total momentum of the system $\sum_{i=1}^{n} \mathbf{p}_i$ equals $M\dot{\mathbf{R}}$, the momentum of a fictitious particle of mass $M$ moving with the center of mass.

3. Show that the sum of all the forces on the system equals $M$ times the acceleration of the center of mass.

4. Show that the total kinetic energy of the system equals the sum of the kinetic energies of the particles in the center-of-mass system, plus the kinetic energy of a particle of mass $M$ moving with the center of mass.

## 4.3    The Wave Equation in One Dimension

**Summary**    The wave equation in one dimension

$$\frac{\partial^2 \psi}{\partial t^2} = v^2 \frac{\partial^2 \psi}{\partial x^2}$$

is most easily visualized as the law governing the vibrations of a taut string. For an infinite string, its general solution can be written as a superposition of functions of the form $f(x - vt) + g(x + vt)$, where $f$ and $g$ are arbitrary; such functions describe waves moving to the right or left, respectively, at the velocity $v$. Thus the value of $\psi(x, t)$ at each point is influenced only by previous values lying within a limited space-time region called the domain of dependence.

The simplest physical model of wave phenomena is the vibrating motion of a taut string. In Fig. 1, $\psi$ measures the vertical displacement of the string from equilibrium. Since the situation is dynamic, $\psi$ depends on time; and since the displacement varies along the string, $\psi$ also depends on the $x$ coordinate locating the point under consideration: $\psi = \psi(x,t)$.

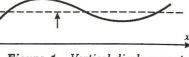

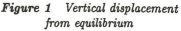

*Figure 1 Vertical displacement from equilibrium*

We shall isolate a tiny piece of the string, running from $x$ to $x+\triangle x$, and apply Newton's Second Law to it; see Fig. 2. The length of this piece is given by $\sqrt{\triangle x^2 + \triangle \psi^2}$, where $\triangle \psi = \psi(x+\triangle x, t) - \psi(x,t)$. If the density of the string is $\rho$ (mass per unit length), then the mass of the string "particle" is

$$\rho\sqrt{\triangle x^2 + \triangle \psi^2} = \rho\triangle x[1 + \mathcal{O}(\triangle\psi/\triangle x)^2] = \rho\triangle x[1 + \mathcal{O}(\partial\psi/\partial x)^2].$$

*We make the simplifying assumption that the displacement of the string is slight, so that we can neglect terms of order 2 or higher in $\psi$ and $\partial\psi/\partial x$.* The mass is then approximately $\rho\triangle x$.

The acceleration in the vertical direction is the second derivative of the displacement $\psi$ with respect to time, which becomes a partial derivative $\dfrac{\partial^2\psi}{\partial t^2}$ in this context, since we are focusing on a particular string particle - and hence a fixed value of $x$.

The dominant force on the particle is due to the tension $T$ in the string. We make the assumption that this tension is uniform; thus the contributions of $T$ from each end of this particle will cancel *unless there is curvature in the string*: see Fig. 2. In fact the tension *vector* **T** is directed along the the unit tangent to the string profile:

$$\mathbf{T}(x,t) = T\{\mathbf{i} + \frac{\partial\psi}{\partial x}\mathbf{j}\}/\sqrt{1 + (\frac{\partial\psi}{\partial x})^2},$$

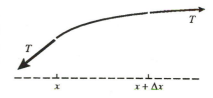

*Figure 2 Force diagram*

with vertical component given by

$$T^\uparrow(x,t) = T\frac{\partial\psi}{\partial x}/\sqrt{1 + (\frac{\partial\psi}{\partial x})^2} \approx T\frac{\partial\psi}{\partial x}.$$

The **net** vertical tensile force on the particle is then

$$T^\uparrow(x + \triangle x, t) - T_\uparrow(x,t) \approx T\frac{\partial^2\psi}{\partial x^2}\triangle x + \mathcal{O}(\triangle x^2). \tag{1}$$

Newton's second law now becomes (to first order in $\psi$ and $\partial\psi/\partial x$)

$$T\frac{\partial^2\psi}{\partial x^2}\triangle x + \mathcal{O}(\triangle x^2) = \rho\triangle x\frac{\partial^2\psi}{\partial t^2},$$

and if we divide by $\rho\triangle x$ and let $\triangle x \to 0$ we obtain

*the wave equation in one dimension:*

$$\frac{\partial^2\psi}{\partial t^2} = \frac{T}{\rho}\frac{\partial^2\psi}{\partial x^2} = v^2\frac{\partial^2\psi}{\partial x^2}, \tag{2}$$

where the constant $v = \sqrt{T/\rho}$ has dimensions of velocity.

We shall spend a good deal of effort in this book studying the solutions to eq. (2) under various circumstances. For the moment, we'd like to point out a special solution form which highlights some of the important properties.

Let $f(x)$ be any (twice-differentiable) function, and consider the function defined by $\psi(x,t) = f(x - vt)$. By the chain rule, the derivatives of $\psi$ are

$$\frac{\partial \psi}{\partial t} = \frac{\partial}{\partial t}f(x - vt) = f'(x - vt) \times (-v);$$

$$\frac{\partial^2 \psi}{\partial t^2} = \frac{\partial}{\partial t}f'(x - vt) \times (-v) = f''(x - vt) \times v^2;$$

$$\frac{\partial \psi}{\partial x} = \frac{\partial}{\partial x}f(x - vt) = f'(x - vt) \times (1);$$

$$\frac{\partial^2 \psi}{\partial x^2} = \frac{\partial}{\partial x}f'(x - vt) = f''(x - vt).$$

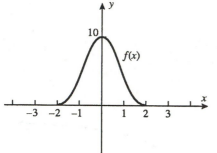

Thus $\dfrac{\partial^2 \psi}{\partial t^2} = v^2 \dfrac{\partial^2 \psi}{\partial x^2}$, i.e., $f(x - vt)$ satisfies the wave equation - *whatever the form of $f$.* If we sketch *any* graph $f(x)$ and form the family $f(x - vt)$, we get solutions of the wave equation! Figure 3 depicts a typical $f(x)$. To obtain the graph of $f(x - 1)$ as in Fig. 4, over any particular point $x$ we plot the value of $f$ that corresponded to $x - 1$ in Fig. 3; thus the peak value 10, which was plotted over $x = 0$ in Fig. 3, moves over to $x = 1$ in Fig. 4. *The graph of $f(x - 1)$ is exactly the graph of $f(x)$, translated one unit to the right.* Similarly the graph of $f(x - vt)$ is the graph of $f(x)$, translated $(vt)$ units to the right; see Fig. 5.

**Figure 3**  *Typical $f(x)$*

Now we know how to obtain solutions to the wave equation. We take any curve $f(x)$ and translate its graph to the right at velocity $v$. This, of course, is precisely what we mean intuitively by a "wave" (and hence the nomenclature for the equation). The identities above reveal that $f(x + vt)$ also solves the equation, representing a wave propagating to the left at speed $v$. And since eq. (2) is *linear*, we have uncovered a very general

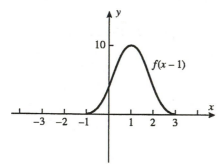

**Figure 4**  *$f(x)$ translated one unit to the right*

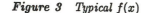

**Figure 5**  *Translates of $f(x)$*

solution form due to D'Alembert:

$$\psi(x,t) = f(x - vt) + g(x + vt), \quad f \text{ and } g \text{ arbitrary.} \tag{3}$$

Note the simplicity of formula (3). We fix two arbitrary functions ($f$ and $g$). At time $t$ we combine the value of $f$, evaluated $(vt)$ units to the left of $x$, with the value of $g$ evaluated at $(vt)$ units to the right. And this innocent prescription provides solutions to the wave equation at $(x, t)$.

How general is the form (3)? Can it be tailored to fit the auxiliary conditions which go with the differential equation (2)? Since (2) is, at bottom, a statement of Newton's second law, the principles of mechanics indicate that one needs to specify the initial position and velocity of each of the string particles in order to determine the motion completely. Thus, for each $x$, the values of $\psi(x, 0)$ and $\dfrac{\partial \psi}{\partial t}(x, 0)$ constitute the necessary auxiliary data.

*The initial conditions for the wave equation are the specifications of the functions*

$$\psi(x, 0) = F(x) \quad \text{and} \quad \frac{\partial \psi}{\partial t}(x, 0) = G(x). \tag{4}$$

**Example 1.** Let us use the form (3) to determine the solution to the wave equation (2) when the initial shape of the string is given as

$$\psi(x, 0) = e^{-x} \tag{5}$$

and the initial velocity profile is

$$\frac{\partial \psi}{\partial t}(x, 0) = \sin x. \tag{6}$$

If we try to enforce (5, 6) using the form (3) for $\psi$, we are left with the task of determining $f$ and $g$ so that

$$\psi(x, 0) = f(x - 0v) + g(x + 0v) = f(x) + g(x) = e^{-x}, \tag{7}$$

and

$$\frac{\partial \psi}{\partial t}(x, 0) = f'(x - 0v) \times (-v) + g'(x + 0v) \times v = -vf'(x) + vg'(x) = \sin x. \tag{8}$$

We eliminate $f$ by multiplying the derivative of eq. (7) by $v$ and adding to (8):

$$2vg'(x) = -ve^{-x} + \sin x.$$

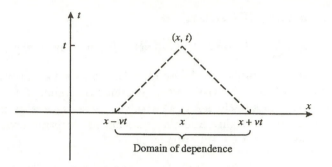

**Figure 6**  *Domain of dependence*

This is solved by an indefinite integral:

$$g(x) = \frac{e^{-x}}{2} - \frac{\cos x}{2v} + C.$$

Insertion into (7) then reveals

$$f(x) = e^{-x} - \frac{e^{-x}}{2} + \frac{\cos x}{2v} - C = \frac{e^{-x}}{2} + \frac{\cos x}{2v} - C,$$

and the solution form (3) becomes

$$\psi(x,t) = \frac{e^{-(x-vt)}}{2} + \frac{\cos(x-vt)}{2v} + \frac{e^{-(x+vt)}}{2} - \frac{\cos(x+vt)}{2v} \qquad (9)$$

(with the constant of integration $C$ disappearing).   ∎

The generalization for arbitrary initial conditions (4) is derived in exercise 1:

$$\psi(x,t) = \frac{F(x-vt) + F(x+vt)}{2} + \frac{1}{2v} \int_{x-vt}^{x+vt} G(\xi)\,d\xi. \qquad (10)$$

Equation (10) thus completely solves the initial value problem for the wave equation. Note how the effects of the initial conditions are propagated along the string:

> The solution at $(x,t)$ depends only on the initial data $F$ and $G$ at points within the distance $\pm vt$ from the point $x$. Values of $F$ and $G$ more remote than this have not had time to influence the solution at time $t$. See Fig. 6.

Unfortunately the initial value problem for the string seldom corresponds to physical reality. The string is not infinitely long; it has ends which are tied down or constrained in some way. Thus we typically have to solve the wave equation on a *finite* interval,

$$\frac{\partial^2 \psi}{\partial t^2} = v^2 \frac{\partial^2 \psi}{\partial x^2}, \qquad 0 < x < L, \qquad (11)$$

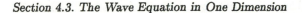

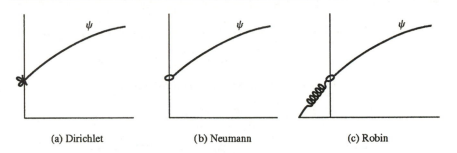

(a) Dirichlet      (b) Neumann      (c) Robin

**Figure 7**   *Boundary conditions*

with initial conditions given in this interval,

$$\psi(x,0) = F(x) \quad \text{and} \quad \frac{\partial \psi}{\partial t}(x,0) = G(x), \quad 0 < x < L, \qquad (12)$$

and *boundary conditions* specifying how the string is attached at the ends. If the left end of the string is tied down at some fixed height, we have a *Dirichlet boundary condition* (Fig. 7(a)):

$$\psi(0,t) = c. \qquad (13)$$

If the string is fastened so that its slope at $x = 0$ is maintained at a specific value, we have a *Neumann boundary condition*:

$$\frac{\partial \psi}{\partial x}(0,t) = d. \qquad (14)$$

(For example, if the string is tied to a massless ring which slides up and down a vertical pole without friction as in Fig. 7(b), it will always have zero slope at this end.) Somewhat more obscure is the *Robin boundary condition*, where a linear combination of the displacement and the slope at the endpoint is maintained constant:

$$\alpha \psi(0,t) + \beta \frac{\partial \psi}{\partial x}(0,t) = e. \qquad (15)$$

A condition like (15) results if the end of the string is tied to an elastic spring, as in Fig. 7(c).

The differential equation (11), initial conditions (12), and a boundary condition of the form (13, 14, 15) at $x = 0$ and $x = L$ constitute a *mixed initial-boundary value problem* for the wave equation. On physical grounds it is clear that the solution $\psi(x,t)$ is uniquely determined by these conditions[2]. The initial-boundary value problem is more typical of situations arising in engineering. Our form (3) for the solution to the (pure) initial value problem is difficult to adapt to the boundary conditions (13, 14,

---

[2]This can also be demonstrated mathematically, of course. See Courant and Hilbert, Garabedian, or Weinberger.

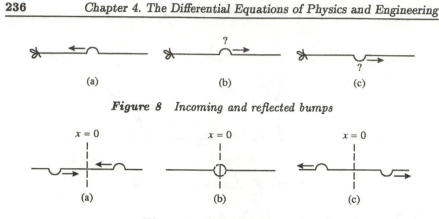

**Figure 8**  *Incoming and reflected bumps*

**Figure 9**  *Interacting bumps*

15), and in Chapters 5 through 8 we will develop an alternative solution procedure. However the following example is enlightening.

**Example 2.** For simplicity let us consider a "semi-infinite" string with its left endpoint tied down. That is, we add the Dirichlet condition

$$\psi(0, t) = 0 \tag{16}$$

to the equation

$$\frac{\partial^2 \psi}{\partial t^2} = v^2 \frac{\partial^2 \psi}{\partial x^2}, \quad \psi(x, 0) = F(x) \quad \frac{\partial \psi}{\partial t}(x, 0) = G(x), \quad 0 < x < \infty \tag{17}$$

Furthermore let us suppose that the initial conditions $F$ and $G$ have been chosen so as to give rise to a wave "bump" that propagates to the left, as depicted in Fig. 8(a). Our task is to figure how the bump reflects off the end; does it come back on the top (Fig. 8(b)), or the bottom (Fig. 8(c))?

We'll answer this question using mathematical, not physical, reasoning! Bear in mind that the conditions (16, 17) completely determine the behavior of the string; *there is only one possible motion which is consistent with (16, 17).*

Suppose hypothetically that the string were extended to $x = -\infty$, and the initial conditions were altered so that they generated the original bump moving to the left, plus a negative, mirror-reflected bump moving to the right, as in Fig. 9(a). Since (thanks to eq. (3)) we are experts on the motions of the *infinite string*, we know that the future states of this motion will be the continued propagation of the bumps. When they meet at $x = 0$ they exactly cancel by superposition, and afterward the lower bump propagates to the region $x > 0$ while the original bump continues to the left (Figs. 9(b,c)).

Now observe that *if we restrict attention to the region $x \geq 0$ for the hypothetical motion, we obtain a function $\psi(x, t)$ which meets all the conditions of the initial-boundary value problem (16, 17).* But, as we said, there is only one such motion! Thus the *hypothetical* motion coincides with the actual motion *to the right of* $x = 0$, and we have found the solution to the original problem; the bump is reflected back along the *bottom* of the string!

■

In the next section we will consider the extensions of these results for the wave equation in two and three dimensions.

# Exercises 4.3

1. Model the analysis of Example 1 to derive eq. (10).

2. Revise the argument of Example 2 to show that the Neumann condition $\dfrac{\partial \psi}{\partial x}(0,t) = 0$ gives rise to reflections along the top of the string.

3. The *inhomogeneous* wave equation

$$\frac{\partial^2 \psi}{\partial t^2} - v^2 \frac{\partial^2 \psi}{\partial x^2} = h(x,t)$$

arises if forces other than tension are considered in the string. Suppose that $h$ is known for all $x$ and $t$. Show that

$$\psi(x,t) = \int_{\tau=-\infty}^{t} \int_{\xi=x-v(t-\tau)}^{x+v(t-\tau)} \frac{h(\xi,\tau)}{2v} \, d\xi \, d\tau$$

is a solution. Note the propagation effect: at time $t$ the values of $h(\xi,\tau)$ for $\xi$ more remote than $\pm v(t-\tau)$ do not influence the solution at $x$.

4. What conditions on $F$ and $G$ in (4) will ensure that waves traveling only to the left are excited by the initial conditions?

5. A *standing wave* is a solution of (2) which can be expressed in factored form $\psi(x,t) = X(x)T(t)$. It does not propagate, but retains the same shape for all time (although it may shrink and swell).

    (a) Show that for any $k$ the function $\cos kx \sin kvt$ is a standing wave solution to (2). (Your computations should suggest three other standing wave solutions to (2).)

    (b) Find the corresponding functions $f$ and $g$ which render this solution in the form of two propagating waves (3).

    (c) Find the initial conditions ($F, G$ in (4)) giving rise to the standing wave in (a).

## 4.4     The Wave Equation in Higher Dimensions

**Summary**   In higher dimensions the wave equation takes the form

$$\frac{\partial^2 \psi}{\partial t^2} = v^2 \nabla^2 \psi.$$

It describes vibrating membranes and sonic and electromagnetic waves. The solution forms are more complicated than for one dimension, but they still exhibit the finite velocity-of-propagation effects. Unique solutions are obtained when the wave equation is supplemented with initial data $\psi(\mathbf{r}, 0)$ and $\frac{\partial \psi}{\partial t}(\mathbf{r}, 0)$ and boundary conditions of the form

$$(Dirichlet): \ \psi(\mathbf{r}, t) \ specified;$$

$$(Neumann): \ \frac{\partial \psi(\mathbf{r}, t)}{\partial n} = \mathbf{n} \cdot \nabla \psi(\mathbf{r}, t) \ specified;$$

$$(Robin): \ \alpha \psi(\mathbf{r}, t) + \beta \frac{\partial \psi(\mathbf{r}, t)}{\partial n} \ specified.$$

One of these three conditions must be specified at each point on the boundary; here $\mathbf{n}$ denotes the unit outward normal direction. Dirichlet conditions correspond to rigid rims for membranes or perfect conductors for tangential electrical field components. Neumann conditions apply at rigid walls for sound waves or perfect conductors for normal electric field components. Robin conditions indicate elastic or resistive walls.

The description of wave phenomena becomes more complicated in two and three dimensions. In this section we shall derive physical models for vibrating membranes, sound waves, and electromagnetics. In addition we present a brief survey of some general mathematical properties.

**Example 1.** The vibrating drum provides a good model for visualizing the two-dimensional wave equation. As depicted in Fig. 1, the drumhead is a membrane stretched taut over the rim of the drum. If we consider the forces on a small piece of the membrane lying above the $\Delta x$-by-$\Delta y$ rectangle, the tensile forces can be attributed to two "strings," one in each of the $x$ and $y$ directions. Thus if $\psi = \psi(x, y, t)$ denotes the vertical displacement as in the figure, the net tension is the sum of two second-derivative terms as analyzed in the preceding section (see eq. (1) therein) -

$$\text{force} = T_x \frac{\partial^2 \psi}{\partial x^2} \Delta x \ + T_y \frac{\partial^2 \psi}{\partial y^2} \Delta y \tag{1}$$

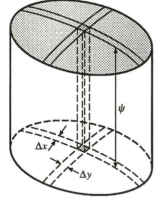

*Figure 1    Vibrating drum*

where $T_x$ is the tensile force on the constant-$x$ edges, and similarly for $T_y$. Note that the constant-$x$ edge is $\Delta y$ units long, and that obviously the net force $T$ acting on this edge increases with the length of the edge. For our model we postulate that there is a constant, isotropic, tension *per unit length* $T$, giving rise to $T_x$ and $T_y$ through the equations

$$T_x = T\Delta y + \mathcal{O}(\Delta y^2),\, T_y = T\Delta x + \mathcal{O}(\Delta x^2).\qquad(2)$$

Now if $\rho$ denotes the membrane mass per unit area and if we neglect higher powers of $\psi$ and $\nabla\psi$ as in the previous section, the mass of the piece of drumhead we are considering is given by $\rho\,\Delta x\,\Delta y$. Hence Newton's second law states

$$T\{\frac{\partial^2\psi}{\partial x^2} + \frac{\partial^2\psi}{\partial y^2}\}\{\Delta x\Delta y + \mathcal{O}(\Delta x^2 + \Delta y^2)\} = \rho\,\Delta x\,\Delta y\frac{\partial^2\psi}{\partial t^2},$$

and dividing by $\Delta x\,\Delta y$ and taking the limit we derive *the wave equation in two dimensions*:

$$\frac{\partial^2\psi}{\partial t^2} = \frac{T}{\rho}\{\frac{\partial^2\psi}{\partial x^2} + \frac{\partial^2\psi}{\partial y^2}\} = v^2\{\frac{\partial^2\psi}{\partial x^2} + \frac{\partial^2\psi}{\partial y^2}\},\qquad(3)$$

where the constant $v$ is the velocity $\sqrt{\frac{T}{\rho}}$.

Clearly the initial conditions which go with (3) are the position and velocity:

$$\psi(x,y,0) = F(x,y),\, \frac{\partial\psi}{\partial t}(x,y,0) = G(x,y).\qquad(4)$$

The usual boundary condition for the drumhead would be the specification of $\psi(x,y,t)$ for points $x$, $y$ on the rim:

$$\psi(x,y,t) = H(x,y),\, (x,y) \text{ on the boundary.}\qquad(5)$$

Normally we would expect to have $H(x,y) =$ constant; any other $H$ indicates that the rim of the drum is warped. As in the previous section, the direct specification of the values of $\psi$ on the boundary is called a *Dirichlet* boundary condition.

If a part of the rim is missing, leaving a portion of the edge of the membrane free to "flap," a detailed stress analysis (which we omit) shows that the directional derivative of $\psi$ in the direction **n** *normal* to the free edge will be zero:

$$\mathbf{n}\cdot\nabla\psi(x,y,t) = \frac{\partial\psi(x,y,t)}{\partial n} = 0,\, (x,y) \text{ on the boundary.}\qquad(6)$$

The specification of $\frac{\partial\psi}{\partial n}$ on the boundary constitutes a generalization of the *slope* condition on $\frac{\partial\psi}{\partial x}$ in the string problem (Sec. 4.3, eq. (14)).

Accordingly, a relation of the form

$$\mathbf{n} \cdot \nabla \psi(x, y, t) = \frac{\partial \psi(x, y, t)}{\partial n} = J(x, y), \quad (x, y) \text{ on the boundary,} \qquad (7)$$

is known as a *Neumann* boundary condition in higher dimensions.

The *Robin* boundary condition specifies some linear combination of $\psi$ and its normal derivative on the boundary:

$$\alpha \psi(x, y, t) + \beta \frac{\partial \psi(x, y, t)}{\partial n} = K(x, y), \quad (x, y) \text{ on the boundary.} \qquad (8)$$

It can be used to describe a flexible rim for the drumhead membrane. ∎

**Example 2.** The three-dimensional wave equation can be visualized as describing pressure (sound) waves in a fluid. If we focus on a small "element" of fluid contained in the rectangular solid shown in Fig. 2, we see that the net force in the $x$ direction due to the pressure $p(x, y, z)$ is the difference of the pressures on the two faces indicated:

$$F_x = p(x, y, z) \, \Delta y \, \Delta z - p(x + \Delta x, y, z) \, \Delta y \, \Delta z \approx -\frac{\partial p}{\partial x} \Delta x \, \Delta y \, \Delta z. \qquad (9)$$

If $\rho(x, y, z)$ denotes the fluid density and $\mathbf{V}(x, y, z)$ its velocity, then $\rho \, \Delta x \, \Delta y \, \Delta z$ is the mass of the element and Newton's second law implies

$$\rho \, \Delta x \, \Delta y \, \Delta z \frac{d}{dt} V_x = -\frac{\partial p}{\partial x} \Delta x \, \Delta y \, \Delta z,$$

or with the $y$ and $z$ pressure contributions included,

$$\rho \frac{d\mathbf{V}}{dt} = -\nabla p. \qquad (10)$$

Another important equation expresses the conservation of mass. Since the divergence of a vector field represents the net outflux per unit volume and an outflux of mass causes a decrease in density, the *continuity equation* states that (see Davis and Snider)

$$\frac{\partial \rho}{\partial t} = -\nabla \cdot \rho \mathbf{V}. \qquad (11)$$

Now the usual physical situation for sound waves involves very small velocities $\mathbf{V}$ and pressure and density variations which are small perturbations of uniform background levels:

$$p(x, y, z) = p_0 + \eta(x, y, z), \quad \rho(x, y, z) = \rho_0 + \psi(x, y, z). \qquad (12)$$

If we neglect products of the small quantities, then eqs. (10, 11) take the approximate form (perturbation computations are discussed in more detail

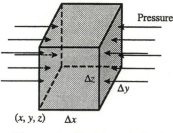

**Figure 2**  *Pressure on fluid element*

in Chapter 9)

$$\rho_0 \frac{d\mathbf{V}}{dt} = -\nabla\eta, \tag{13}$$

$$\frac{\partial\psi}{\partial t} = -\rho_0\nabla\cdot\mathbf{V}. \tag{14}$$

(See exercise 1.) Furthermore because we assume that the velocities are small, we can say that the derivative $\frac{d\mathbf{V}}{dt}$ - which is to be taken moving with fluid element along its trajectory (the so-called *convective derivative*) - is approximately equal (to the same order of accuracy) to the partial derivative $\frac{\partial\mathbf{V}}{\partial t}$, which is taken at a fixed location.[3] Consequently (13) can be approximated as

$$\rho_0 \frac{\partial\mathbf{V}}{\partial t} = -\nabla\eta\ . \tag{15}$$

Finally we assume that the thermal effects which occur during the motion are so minor that the pressure can be regarded as a function solely of the density: $p = p(\rho)$, or $\eta = \eta(\psi)$. Then (15) can be rewritten using the chain rule

$$\rho_0 \frac{\partial\mathbf{V}}{\partial t} = -\frac{d\eta}{d\psi}\nabla\psi\ . \tag{16}$$

The three-dimensional wave equation results when we take the divergence of (16) and combine it with the time derivative of (14) to obtain

$$\frac{\partial^2\psi}{\partial t^2} = \frac{d\eta}{d\psi}\nabla^2\psi = v^2\nabla^2\psi = v^2\{\frac{\partial^2\psi}{\partial x^2} + \frac{\partial^2\psi}{\partial y^2} + \frac{\partial^2\psi}{\partial z^2}\}. \tag{17}$$

*where the velocity $v$ is the square root of the derivative $d\eta/d\psi = dp/d\rho$, evaluated at $\rho = \rho_0$.* (As in Section 4.3 we drop second-order terms in $\psi$ and $\nabla\psi$.)

Since our analysis is based on Newton's second law for the fluid "particles," one would anticipate their initial positions and velocities to come into play in some way. But from our experience with the one- and two-dimensional wave equations, mathematically speaking we expect that *the initial conditions appropriate for (17) would be the specification of the density function $\psi$ and its derivative $\partial\psi/\partial t$ at $t = 0$.* So naturally these two descriptions are related: the initial positions would determine the density $\psi(x, y, z, 0)$, which could then be inserted into (14) with the initial velocity profile to determine $\partial\psi/\partial t$ at $t = 0$.

The most common boundary condition for sound waves occurs at rigid walls. There the velocity can have no normal component, since the fluid

---

[3]The convective derivative and the partial derivative are related by $\frac{d}{dt} = \frac{\partial}{\partial t} + \mathbf{V}\cdot\nabla$. See Jones for details on this derivation.

does not penetrate the wall. Equation (16) then tells us that the normal component of $\nabla\psi$ is zero. Thus we write the *Neumann* boundary condition as

$$\mathbf{n} \cdot \nabla\psi(x,y,z,t) = \frac{\partial\psi(x,y,z,t)}{\partial n} = 0 \text{ on rigid walls.} \qquad (18)$$

∎

**Example 3.** Another, extremely important, example of a phenomenon governed by the wave equation is the electromagnetic field. Maxwell's equations for the electric field $\mathbf{E}$ and the magnetic flux density $\mathbf{B}$ in a sourceless medium with uniform permittivity $\epsilon$ and permeability $\mu$ are

$$\nabla \cdot \mathbf{E} = 0 \qquad (19)$$

$$\nabla \times \mathbf{E} = -\partial\mathbf{B}/\partial t \qquad (20)$$

$$\nabla \cdot \mathbf{B} = 0 \qquad (21)$$

$$\nabla \times \mathbf{B} = \mu\epsilon\partial\mathbf{E}/\partial t . \qquad (22)$$

It is a matter of vector algebra (exercise 2) to show that each rectangular vector component of $\mathbf{E}$ satisfies the wave equation

$$\frac{\partial^2\psi}{\partial t^2} = \frac{1}{\mu\epsilon}\nabla^2\psi = v^2\nabla^2\psi \qquad (23)$$

with velocity $v = 1/\sqrt{\mu\epsilon}$. (Some other quantities of interest, such as the components of $\mathbf{B}$ or various potentials associated with the fields, also satisfy (23), but in this book we will only address examples where $\psi$ represents a component of $\mathbf{E}$. See Jackson, Panofsky and Phillips, Ramo, Whinnery, and van Duzer, or Davis and Snider.) According to the physical setup, eq. (23) can be one, two, or three dimensional. Let us see how the initial and boundary conditions appropriate to (23) arise.

The initial conditions for the wave equation call for $\psi$ and $\partial\psi/\partial t$ at $t = 0$. Now the initial values for $\mathbf{E}$ and $\mathbf{B}$ would be given as part of the specifications for an electromagnetic initial value problem. Thus if $\psi$ represents an electric field component in (23), $\psi(t = 0)$ is given directly. Then to find $\partial\psi/\partial t$ at $t = 0$, one simply takes the curl of the initial magnetic flux density $\mathbf{B}$ and employs (22).

Boundary conditions in electromagnetic problems are usually imposed along conducting walls. If the conductivity is modeled as infinite, then the references cited show that any component of the electric field within the conductor itself must be zero. Consistency with the curl equation (20) then demands that, next to the wall, the component of $\mathbf{E}$ tangent to the wall boundary must also be zero. Thus when $\psi$ represents an electric field component *parallel* to a conducting wall, it obeys the *Dirichlet* boundary condition

$$\psi(\mathbf{r},t) = 0 \text{ along the wall.} \qquad (24)$$

The divergence condition (19), applied next to a conducting wall, equates the normal derivative of the normal component of **E** with the tangential derivatives of the tangential components. Since the latter are zero along the wall, we have the *Neumann* boundary condition

$$\frac{\partial \psi}{\partial n}(\mathbf{r}, t) \;=\; 0 \text{ along the wall} \tag{25}$$

whenever $\psi$ represents an electric field component *normal* to a conducting wall.

Inside a cavity, then, $\psi$ will typically obey Dirichlet conditions along parts of the boundary and Neumann conditions along other parts. (The Robin boundary condition arises in situations where the conductivity is finite.) ∎

As in the case of the one-dimensional wave equation, there is a special solution for the three-dimensional equation which sheds some light on its basic properties. Recall that for waves on a string we found that a solution $\psi(x, t)$ could be obtained by adding the values of an arbitrary function $f(x')$, evaluated $vt$ units to the left ($x' = x - vt$), to the values of a function $g(x')$ evaluated $vt$ units to the right:

$$\psi(x, t) = f(x - vt) + g(x + vt)$$

(Sec. 4.3, eq. (3)). The analogous three-dimensional construction is to form $\psi(\mathbf{r}, t) = \psi(x, y, z, t)$ by averaging the values of a function $f(x', y', z') = f(\mathbf{r}')$ at all points at the distance $vt$ from $\mathbf{r} = (x, y, z)$. If we write $d\omega_{\mathbf{u}}$ for the differential of solid angle in the direction of the unit vector $\mathbf{u}$, this average is the integral over all directions, divided by $4\pi$:

$$M_t[f] = \frac{1}{4\pi} \int \int f(\mathbf{r} + vt\mathbf{u}) \, d\omega_{\mathbf{u}} \;. \tag{26}$$

*A solution of the wave equation (17) is then given by the formula*

$$\psi(\mathbf{r}, t) \;=\; \psi(x, y, z, t) = t M_t[f] \tag{27}$$

for arbitrary $f$. We omit the proof; see Courant and Hilbert (vol. II). However in exercise 4 you will be invited to show that the combination

$$\psi(r, t) \;=\; t M_t[f] + \frac{\partial}{\partial t}\{t M_t[g]\} \tag{28}$$

satisfies (17) and the initial conditions

$$\psi(\mathbf{r}, 0) \;=\; g(\mathbf{r}), \quad \frac{\partial \psi}{\partial t}(\mathbf{r}, 0) = f(\mathbf{r}). \tag{29}$$

Formula (28), then, provides the general solution to the (pure) initial value problem for the three-dimensional wave equation.

As in the one-dimensional case, eqs. (26, 28) demonstrate how the effects of the initial conditions are propagated throughout the medium. The solution at $\mathbf{r}, t$ depends only on the initial data $f$ and $g$ at points a distance $vt$ from $\mathbf{r}$; values of $f$ and $g$ more remote than this *domain of dependence* have not had time to influence the solution at time $t$.

The corresponding solution for the two-dimensional wave equation (3) is surprisingly more complicated. Given any function $f(x, y)$, Courant and Hilbert show that the integral

$$
\begin{aligned}
\psi(x, y, t) &= N_t[f] \\
&= \frac{1}{2\pi} \int \int_{[\sqrt{\xi^2 + \eta^2} \leq vt]} \frac{f(x+\xi, y+\eta)}{\sqrt{v^2 t^2 - \xi^2 - \eta^2}} \, d\xi \, d\eta,
\end{aligned} \tag{30}
$$

taken over the *interior* of the circle of radius $vt$ centered at $(x, y)$, provides a solution to eq. (3) satisfying the initial conditions

$$
\psi(x, y, 0) = 0, \quad \frac{\partial \psi}{\partial t}(x, y, 0) = f(x, y). \tag{31}
$$

Thus the value of $\psi(x, y, t)$ depends on the initial data at points a distance $vt$ *or closer* to $(x, y)$. More remote points, however, still have not had time to influence $\psi(x, y, t)$. Evidently the propagation of waves in two dimensions is more complicated than in one or three dimensions. See Courant and Hilbert for a detailed discussion of this point.

In the last two sections we have seen how the wave equation governs vibrations of a string, a drumhead, a fluid, and electromagnetic fields. We have also seen what types of initial-boundary conditions must be imposed to make its solution well-defined. Now we close the discussion of this equation with a brief summary.

The wave equation for $\psi(\mathbf{r}, t)$ is given by

$$
\frac{\partial^2 \psi}{\partial t^2} = v^2 \nabla^2 \psi, \tag{32}
$$

where the Laplacian $\nabla^2$ may be one, two, or three dimensional. By introducing a change of scale and redefining "$t$" to be "$vt$", we obtain a *nondimensional* form

$$
\frac{\partial^2 \psi}{\partial t^2} = \nabla^2 \psi, \tag{33}
$$

*which will be used in all subsequent chapters.*

To determine a solution to (32) uniquely, one must specify the initial conditions $\psi(\mathbf{r}, 0)$ and $\partial \psi(\mathbf{r}, 0)/\partial t$ throughout the domain of interest. If this domain has finite boundaries, then on each portion of the boundary one of three types of boundary conditions must also be prescribed:

$$\text{Dirichlet conditions: } \psi(\mathbf{r}, t) \text{ is specified;} \qquad (34)$$

$$\text{Neumann conditions: } \frac{\partial \psi}{\partial n}(\mathbf{r}, t) \text{ is specified;} \qquad (35)$$

$$\text{Robin conditions: } \alpha\psi(\mathbf{r}, t) + \beta\frac{\partial \psi}{\partial n}(\mathbf{r}, t) \text{ is specified.} \qquad (36)$$

Finally, the effects of the initial conditions propagate at a speed $v$, in the sense that the initial values of $\psi$ and $\partial\psi(\mathbf{r}, 0)/\partial t$ at a point $\mathbf{r}$ cannot influence the solution at point $\mathbf{r}'$ prior to the time $t = |\mathbf{r} - \mathbf{r}'|/v$.

# Exercises 4.4

1. Fill in the details of the derivation of the wave equation for sound waves.

2. (a) Show that each component of **E** obeys the wave equation (23). (Hint: take the curl of eq. (20) and combine with (19, 22).)

   (b) Show that each component of **B** also obeys the wave equation (23).

3. For electromagnetic fields propagating inside a uniform sourceless medium with finite conductivity $\sigma$, the last of Maxwell's equations (22) is replaced by

$$\nabla \times \mathbf{B} = \mu\epsilon\frac{\partial \mathbf{E}}{\partial t} + \mu\sigma\mathbf{E}.$$

   Show that the rectangular components of both the **E** and **B** fields obey the *telegrapher's equation* (also known as the *damped wave equation*)

$$\frac{\partial^2 \psi}{\partial t^2} + \frac{\sigma}{\mu}\frac{\partial \psi}{\partial t} = \frac{1}{\mu\epsilon}\nabla^2\psi.$$

4. (a) Show that the function $\eta(\mathbf{r}, t) = tM_t[f]$ satisfies the conditions

$$\eta(\mathbf{r}, 0) = 0, \quad \frac{\partial \eta}{\partial t}(\mathbf{r}, 0) = f(\mathbf{r}).$$

   (b) Given that $\eta(\mathbf{r}, t)$ satisfies the wave equation (17), show that $\partial\eta/\partial t$ also satisfies the wave equation.

   (c) Show that $\nabla^2\eta(\mathbf{r}, 0) = 0$, and thus $\frac{\partial^2\eta}{\partial t^2}(\mathbf{r}, 0) = 0$.

   (d) Show that the combination (28) satisfies the initial conditions (29).

## 4.5   The Heat Equation

**Summary**   Fick's law states that in diffusive processes fluxes are driven by concentration gradients. As a result, the concentrations are governed by the diffusion equation

$$\frac{\partial \psi}{\partial t} = \gamma \nabla^2 \psi,$$

also known as the heat equation when $\psi$ is identified with temperature. For a unique solution one must also specify the initial values $\psi(\mathbf{r}, 0)$ and boundary conditions similar to those for the wave equation.  Dirichlet conditions correspond to boundary surfaces held at given temperatures, Neumann conditions correspond to given heat flow rates (e.g., insulation) on the boundary, and Robin conditions correspond to imperfect insulation. Unlike the wave equation, the heat equation predicts (unrealistically) that diffusive effects propagate with infinite speed.

In all natural thermodynamic processess, heat flows from points at high temperatures $\psi$ to points at lower temperatures. The rate of heat flow is driven, then, by the temperature gradient $\nabla\psi(\mathbf{r}, t)$. We quantify the heat flow rate by a vector $\mathbf{q}(\mathbf{r}, t)$ known as the *heat flux*; at any location $\mathbf{r}$ it points in the direction of heat flow, and its magnitude equals the rate at which heat is conducted across a unit area normal to the direction of the flow. *Fourier's law*, an experimental observation (which can be understood through statistical mechanics), postulates that $\mathbf{q}$ is proportional to $\nabla\psi$:

$$\mathbf{q} = -k\nabla\psi \tag{1}$$

where $k$ is the *thermal conductivity* (and the minus sign arises because heat flows from hot to cold).

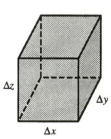

***Figure 1*** *Heat conductor volume element*

Consider, then, the rate at which heat flows into the rectangular box depicted in Fig. 1. For the face at the left, the cross section area is $\Delta y \Delta z$ and heat enters at a rate

$$-k\frac{\partial\psi(x, y, z)}{\partial x}\Delta y\,\Delta z.$$

A similar formula gives the heat flow *out* of the face on the right, but with $x$ replaced by $x + \Delta x$. The net heat influx through these two faces is then

$$k\frac{\partial\psi(x+\Delta x, y, z)}{\partial x}\Delta y\,\Delta z - k\frac{\partial\psi(x, y, z)}{\partial x}\Delta y\,\Delta z \approx k\frac{\partial^2\psi}{\partial x^2}\Delta x\,\Delta y\,\Delta z.$$

Adding in the contributions of the other four faces, we find the heat flow rate into the box to be

$$k\{\frac{\partial^2\psi}{\partial x^2} + \frac{\partial^2\psi}{\partial y^2} + \frac{\partial^2\psi}{\partial z^2}\}\Delta x\,\Delta y\,\Delta z = k\nabla^2\psi\Delta x\,\Delta y\,\Delta z. \tag{2}$$

Now the effect of this heat input is a rise in the temperature of the box. The *specific heat c* of a body is the amount of heat required to raise the temperature of a unit mass by one degree. Therefore if $\rho$ denotes the density, the heat flow rate of eq. (2) generates a rise in temperature at the rate

$$k\left\{\frac{\partial^2\psi}{\partial x^2} + \frac{\partial^2\psi}{\partial y^2} + \frac{\partial^2\psi}{\partial z^2}\right\}\Delta x\,\Delta y\,\Delta z = c\rho\Delta x\,\Delta y\,\Delta z\,\frac{\partial\psi}{\partial t}$$

or

$$\frac{\partial\psi}{\partial t} = \gamma\nabla^2\psi \tag{3}$$

where the *thermal diffusivity* $\gamma = k/\rho c$. Equation (3) is known as the *heat equation*.

It is of interest to note that (3) also governs many other diffusive processes. If $\psi$ denotes the concentration of some impurity species suspended in a liquid or solid, the species naturally tends to diffuse from regions of high concentration to sparser regions. This generalization of Fourier's law is known as *Fick's law*:

$$(\text{flux of species}) \propto (-\nabla\psi) \tag{4}$$

and by similar reasoning we reckon that $\psi(\mathbf{r}, t)$ is governed by the *diffusion equation*

$$\frac{\partial\psi}{\partial t} = \gamma\nabla^2\psi$$

where $\gamma$ denotes the *diffusivity* of the concentrate.

What kind of auxiliary conditions are needed to determine, uniquely, a solution to the heat equation? Physically it is easy to see that a knowledge of temperature growth rates is only helpful if we know the initial temperature. Thus (3) must be augmented with *initial conditions*:

$$\psi(\mathbf{r}, 0) = F(\mathbf{r}), \tag{5}$$

for some (given) function $F$.

Also we seldom address heat flows in infinite bodies, so some *boundary conditions* are needed to describe the thermal interaction of the body with its environment. The *Dirichlet* boundary condition is the specification of the temperature $\psi$, itself on the surface of the body:

$$\psi(\mathbf{r}, t) = G(\mathbf{r}) \tag{6}$$

where $G$ is a surface-temperature profile maintained by heat sources. (It is possible that $G$ is time dependent, but we shall ignore this complication for now.)

If we happen to know the rate at which heat flows across the surface, then according to Fourier's law the normal component of $k\nabla\psi$ is determined there. This is the *Neumann* boundary condition

$$\mathbf{n}\cdot\nabla\psi(\mathbf{r},t) = \frac{\partial\psi(\mathbf{r},t)}{\partial n} = H(\mathbf{r}) \ , \ \mathbf{r} \text{ on the boundary} \qquad (7)$$

For instance, if a portion of the surface is thermally insulated, then $H = 0$ there.

Finally, if the surface is shielded from a uniform environmental temperature ($\psi_0$), but the insulating material "leaks," the surface heat flow rate will be roughly proportional to the difference between the body surface temperature and the "outside" temperature (*Newton's law of cooling*):

$$-k\frac{\partial\psi}{\partial n} = C(\psi - \psi_0) \ .$$

Then we have an example of a *Robin* boundary condition

$$\alpha\psi(\mathbf{r},t) + \beta\frac{\partial\psi(\mathbf{r},t)}{\partial n} = J(\mathbf{r}) \qquad (8)$$

where, in our example, $\alpha = C$, $\beta = k$, and the given function $J(\mathbf{r}) = -C\psi_0$.

Thus on physical grounds we reason that the solution of the heat equation (3) is completely determined by the prescription of the initial temperature profile in the interior of the body and the specification of either Dirichlet, Neumann, or Robin boundary conditions at all points on its boundary. (This can also be proved mathematically; see Courant and Hilbert.)

We can deduce some of the mathematical properties of the heat equation from the formula for the Laplacian:

$$\nabla^2 f = \frac{\partial^2 f}{\partial x^2} + \frac{\partial^2 f}{\partial y^2} + \frac{\partial^2 f}{\partial z^2}. \qquad (9)$$

Certainly $\nabla^2 f$ is a generalization of the second derivative of a function of one variable. Recall that the second derivative determines the convexity of a function; if $f''$ is positive, the graph of $f$ "holds water," and if it is negative it spills water (Fig. 2). In other words *in a region where $f'' < 0$, the value of $f$ at $x$ lies above the secant line connecting the values of $f$ at $x + \Delta x$ and $x - \Delta x$:*

$$f(x) > \frac{f(x + \Delta x) + f(x - \Delta x)}{2}, \qquad (10)$$

and when $f'' > 0$ the opposite occurs. The extent to which $f$ exceeds the average of its neighbors, as quantified by (10), appears in Fig. 2 to be proportional to the negative of the second derivative. (In fact the expression

$$\left\{f(x) - \frac{f(x + \Delta x) + f(x - \Delta x)}{2}\right\}\frac{2}{\Delta x^2}$$

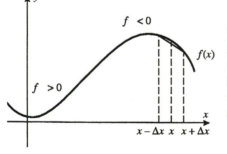

**Figure 2**   *Second derivative*

is an excellent difference approximation for $-f''(x)$; see Appendix A Sec. 3.)

Thus we might reasonably expect that some variant of this property would hold for the "higher-dimensional second derivative," $\nabla^2 f$; it should measure (in a negative sense) the extent to which $f(x, y, z)$ exceeds the averaged neighboring values $\bar{f}$:

$$\nabla^2 f \approx -M[f(x, y, z) - \bar{f}]. \tag{11}$$

In exercise 1 you will be invited to show that an approximation like (11) holds when f is averaged over a small cube centered at $(x, y, z)$, with the constant $M$ equal to 24 divided by the square of the length of one side.

If we insert the approximation (11) into the heat equation,

$$\frac{\partial \psi}{\partial t} = \gamma \nabla^2 \psi \approx -\gamma M[\psi(\mathbf{r}) - \bar{\psi}]$$

the physical interpretation is gratifying: the temperature of a point $\mathbf{r}$ falls at a rate proportional to the extent by which $\psi(r)$ exceeds the average temperature of its surroundings. Note in particular that if the maximum temperature at time $t$ occurs at an interior point $\mathbf{r}$, then $\partial \psi / \partial t \leq 0$ there. This observation is the basis for the *Maximum Principle for the heat equation*, and in advanced work (Protter and Weinberger) it is developed and exploited to analyze the solutions to (3) extensively.

As a second observation about the heat equation, in exercise 3 you will be invited to verify that the formula

$$\psi(x, t) = \frac{1}{2\sqrt{\pi \gamma t}} \int_{-\infty}^{\infty} F(\xi) e^{-(x-\xi)^2 / 4\gamma t} \, d\xi \tag{12}$$

defines an explicit solution to the one-dimensional version of (3) on the entire x-axis, $-\infty < x < \infty$. One can also show (but we refer the reader to Courant and Hilbert) that the initial value of $\psi$ is the function $F$:

$$\psi(x, 0) = F(x). \tag{13}$$

These formulas demonstrate that the solution $\psi$ at $x$, $t$ depends on the whole range of values of $F(\xi)$ from $-\infty$ to $+\infty$. This differs markedly from the wave equation, where the initial values cannot influence the solution at points more remote than $t/v$ (where $v$ is the velocity parameter; recall eq. (32), Sec. 4.4). Thermal effects governed by eq. (3) propagate with infinite speed!

Of course, this is physical nonsense. It is symptomatic of the oversimplification in modeling the diffusive process by Fourier's law. However, eq. (3) describes thermal phenomena quite accurately when it is applied to bodies of finite extent.

We close our discussion of the heat equation by recapping the auxiliary conditions which are necessary for a complete specification of a thermal problem. The initial value of the temperature $\psi(\mathbf{r}, 0)$ must be specified throughout the interior of the body, and at each point on its surface one must prescribe (for all time) either

the temperature $\psi$ itself - a *Dirichlet* condition, or
the normal heat flux $\mathbf{n} \cdot \nabla \psi$ - a *Neumann* condition, or
a linear combination $\alpha \psi + \beta \mathbf{n} \cdot \nabla \psi$ - a *Robin* condition.

If we redefine "$t$" to be "$\gamma t$" in the heat equation (3) we obtain a *nondimensional* form

$$\frac{\partial \psi}{\partial t} = \nabla^2 \psi \tag{14}$$

*which will be used in all subsequent chapters.*

# Exercises 4.5

1. Using the second order Taylor polynomial (exercise 5, Appendix A.4), estimate the difference between the value of $\psi$ at $\mathbf{r}$ and the average of its values throughout the interior of a cube of side $a$ centered at $\mathbf{r}$. Confirm the property mentioned below eq. (11).

2. In the text it was argued heuristically that the Laplacian of a function $\psi$ must be nonpositive at an interior maximum point $\mathbf{r_0}$ of $\psi$. For a more careful analysis, expand $\psi$ in a Taylor series to second order around $\mathbf{r} = \mathbf{r_0}$, with $y$, $z$, and $t$ held fixed. Noting that $\frac{\partial \psi}{\partial x} = 0$ at an interior maximum, argue that $\frac{\partial^2 \psi}{\partial x^2}$ cannot be positive, for otherwise your display demonstrates that $\psi(x, y_0, z_0, t)$ would exceed $\psi(x_0, y_0, z_0, t)$, the maximal value. Extend the argument to the other second partials of $\psi$, and conclude that $\nabla^2 \psi(\mathbf{r}, t) \le 0$ at any interior maximum of $\psi$.

3. Verify by formal differentiation that formula (12) solves eq. (3).

# 4.6   Laplace's Equation

**Summary**   Laplace's equation

$$\nabla^2 \psi = 0$$

governs the time-independent behavior of solutions to the wave and heat equations, and the electrostatic potential. Its solutions, known as harmonic functions, are unique when they are constrained by Dirichlet, Neumann, or Robin boundary conditions, with one exception - the pure Neumann problem only determines $\psi$ up to a constant. Solutions to the two-dimensional Laplace equation are the real and imaginary parts of analytic functions: in particular, $\theta$ and $\ln r$ are useful solutions. A non-constant harmonic function cannot have a maximum or minimum at an interior point.

In our physical world we observe that all thermal systems evolve in time toward a "steady state" or equilibrium configuration in which the temperature $\psi$ is independent of time. That is, as $t \to \infty$ the solutions of the heat equation

$$\frac{\partial \psi}{\partial t} = \gamma \nabla^2 \psi \tag{1}$$

tend to limiting forms,

$$\psi(\mathbf{r}, t) \to \psi_\infty(\mathbf{r}),$$

that are *stationary*. These *equilibrium solutions* then obviously satisfy the time-independent form of (1), which is known as

*Laplace's equation*

$$\nabla^2 \psi(\mathbf{r}) = 0 . \tag{2}$$

(We drop the subscript "$\infty$" in (2) to remain notationally consistent with the rest of this chapter.)

The time-independent solutions of the wave equation also clearly satisfy Laplace's equation.[4] Thus we have at our disposal a number of physical analogies to call upon in studying (2): steady state temperature profiles, drumheads at equilibrium, and electrostatic and magnetostatic fields.

There is another example which we shall use: the electrostatic potential. Recall from Maxwell's equations (Sec. 4.4) that in a sourceless uniform medium the *time-independent* electric field $\mathbf{E}$ satisfies $\nabla \times \mathbf{E} = 0$ and $\nabla \cdot \mathbf{E} = 0$. From vector analysis (see Davis and Snider) we know that the curl condition guarantees that $\mathbf{E}$ can be expressed as the gradient of a scalar *electric potential* function $\psi(\mathbf{r})$:

$$\mathbf{E}(\mathbf{r}) = -\nabla \psi(\mathbf{r}) . \tag{3}$$

(We introduce the minus sign to be consistent with eq. (2), Sec. 4.2. We also assume that the domain of interest is "simply connected" - see references.) The remaining divergence condition then states that $\psi$ is a solution of Laplace's equation $\nabla^2 \psi(\mathbf{r}) = 0$.

---

[4] In all physical situations waves tend to die out due to damping and lead to solutions of (2). Mathematically speaking, however, there is no damping in the wave equation; thus its dynamic solutions do not evolve towards solutions of (2). (See exercise 3, Sec. 4.4).

Solutions of Laplace's equation (2) are known as *harmonic functions.* Since they are independent of time, there are no initial conditions to go with (2). The boundary conditions which are needed to determine a solution uniquely are the Dirichlet, Neumann, and/or Robin conditions, which we have described in the preceding sections. (Usually Dirichlet conditions are specified for the electric potential; it is constant on conducting walls, since the tangential component of its gradient **E** is zero there (recall eq. (24), Sec. 4.4).)

The *pure Neumann* problem, wherein the value of $\partial\psi/\partial n$ is specified everywhere on the surface of the solution region, is exceptional. Note that if $\psi_1(r)$ is a solution of

$$\nabla^2\psi_1(\mathbf{r}) = 0 \text{ inside } D, \ \partial\psi_1/\partial n = G(\mathbf{r}) \text{ on the boundary of } D,$$
$$(4)$$

then $\psi = \psi_1 + (constant)$ also solves (4) - for any value of the constant! So solutions to the pure Neumann problem for Laplace's equation are only determined up to a constant. Physically, this is not surprising. For example, if (4) described a heat flow situation, the only data given explicitly (i.e. the values of the function $G$) refer to heat flow *rates*; there are no direct references to the temperature itself. Thus there is no way of telling what temperature is to be taken as the zero level (e.g., the solution could be stated in degrees Kelvin or degrees Celsius!). Similarly, if (4) constituted a problem for the electric potential, there is no reference to "electrical ground" (the zero level for $\psi$) in the data. Indeed, $\psi$ truly is only defined up to a constant for such situations! (See exercise 6 for further insight on this problem.)

We can immediately write down some solutions to the two-dimensional Laplace equation by exploiting its similarity to the two-dimensional wave equation:

$$\text{Laplace: } \frac{\partial^2\psi}{\partial x^2} + \frac{\partial^2\psi}{\partial y^2} = 0 \ ; \quad \text{Wave: } \frac{\partial^2\psi}{\partial x^2} - \frac{1}{v^2}\frac{\partial^2\psi}{\partial t^2} = 0 \ .$$

They become formally equivalent if we set "$t$" equal to "$y$" and *choose the velocity $v$ to be the imaginary unit* $i = \sqrt{-1}$. Now we have seen (Sec. 4.3) that any (twice differentiable) function of the form $f(x+vt)$ (or $f(x-vt)$) solves the wave equation. It follows that *any function of the form*

$$f(x + iy) = f(z)$$

*solves Laplace's equation, if $f(z)$ and the derivatives $df(z)/dz$ and $d^2f(z)/dz^2$ are well-defined functions of the complex variable $z = x + iy$.*

The classification of differentiable functions of a complex variable is the subject of *analytic function theory*. Conditions for differentiability of complex functions are well known - as, for example, the Cauchy-Riemann equations. For the following example we presume some familiarity with the basic analytic functions; readers unfamiliar with these issues are referred to Saff and Snider for an exposition.

**Example 1.** The functions $z^2$, $z^3$, $e^z$, and $\ln z$ are analytic (with the exception of certain points for the latter); their derivatives are $2z$, $3z^2$, $e^z$, and $1/z$, respectively. As functions of $x$ and $y$, then, they satisfy Laplace's equation. This is obviously true for $z^2$:

$$\nabla^2 z^2 = \{\frac{\partial^2}{\partial x^2} + \frac{\partial^2}{\partial y^2}\}[x^2 - y^2 + i2xy] = 0.$$

In exercise 1 you will be invited to verify directly that

$$\begin{aligned} z^3 &= (x^3 - 3xy^2) + i(3x^2 y - 3y^3)\ , \\ e^z &= (e^x \cos y) + i(e^x \sin y)\ , \\ \ln z &= (\ln \sqrt{x^2 + y^2}) + i(\arg z) = \ln r + i\theta \end{aligned}$$

are also harmonic functions ($r$ and $\theta$ are the polar coordinates of the point $(x, y)$).

It should be clear from this example that the real and imaginary parts of $f(z)$, separately, each solve Laplace's equation. ■

**Example 2.** Consider the Dirichlet problem for Laplace's equation indicated in Fig. 1; we need to find a function which is constant on each circle and harmonic in between.

Now the polar coordinate $r = \sqrt{x^2 + y^2}$ is constant on circles centered at the origin, and thus so is any function of $r$. In Example 1 we learned that $\ln r$ is harmonic, and from the nature of Laplace's equation it is clear that any combination of the form $A \ln r + B$ is also harmonic. Consequently we fit the constants $A$ and $B$ to the boundary conditions:

$$\begin{aligned} A \ln 1/2 + B &= 1 \\ A \ln 2 + B &= 3\ . \end{aligned} \tag{5}$$

Equations (5) imply that $A = (\ln 2)^{-1}$, $B = 2$. Thus the solution is given by

$$\psi(x, y) = \frac{\ln r}{\ln 2} + 2 = \frac{\ln \sqrt{x^2 + y^2}}{\ln 2} + 2\ . \qquad ■$$

**Example 3.** To find a harmonic function satisfying the Dirichlet conditions on the wedge depicted in Fig. 2, we'd like to employ some version of the polar angle $\theta$, shifted to an origin at $z_0 = 2 + i$.

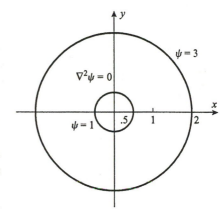

**Figure 1** *Equilibrium temperature in an annulus*

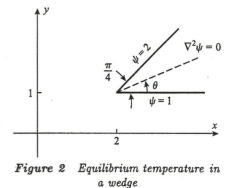

**Figure 2** *Equilibrium temperature in a wedge*

The indicated angle $\tilde{\theta}$ measured at $z_0$ is harmonic, because it arises as the imaginary part of the analytic function $\ln(z - z_0)$. Thus we try to fit $A\tilde{\theta} + B$ to the boundary data, and readily find the solution to be

$$\psi(x, y) = \frac{4}{\pi}\tilde{\theta} + 1 = \frac{4}{\pi}\arctan\frac{y-1}{x-2} + 1. \qquad \blacksquare$$

The fact that the remote border of the wedge in Fig. 2 lies "at infinity" and is not governed by boundary conditions suggests that there may be other solutions to Example 3. Exercise 3 reveals that this is indeed the case. Physically we can think of Fig. 2 as attempting to model the equilibrium temperature profile in a very long (but finite) wedge, and it is clear that this temperature will not be uniquely determined until we establish control of the thermal processes at the remote edge. Infinite regions are discussed further in Sec. 6.4.

The reference by Saff and Snider contains many other applications of analytic function theory in solving Laplace's equation in two dimensions. One of the most profound is *Poisson's formula*, which gives the values of the harmonic function $\psi$ inside the circle of radius $R$ centered at the origin, in terms of its values $F$ on the rim:

$$\psi(z) = \psi(re^{i\theta}) = \frac{R^2 - r^2}{2\pi}\int_0^{2\pi}\frac{F(Re^{i\phi})}{R^2 + r^2 - 2rR\cos(\phi - \theta)}\,d\phi\ . \qquad (6)$$

Equation (6) is thus a universal solution formula for the Dirichlet problem in the disk. It can be generalized to the three-dimensional sphere of radius $R$ as follows:

$$\psi(\mathbf{r}) = \frac{R^2 - |\mathbf{r}|^2}{4\pi R}\iint\frac{F(\mathbf{r'})}{|\mathbf{r} - \mathbf{r'}|^3}\,dA \qquad (7)$$

where the variable $\mathbf{r'}$ is integrated over the area $dA$ of the sphere. See Hellwig's text.

If we evaluate eqs. (6, 7) at $r = |\mathbf{r}| = 0$ we find

$$\psi(0) = \frac{1}{2\pi}\int_0^{2\pi}F(Re^{i\phi})\,d\phi\ \text{ or }\ \frac{1}{4\pi R^2}\iint F(\mathbf{r'})\,dA \qquad (8)$$

Equations (8) say that the value of any harmonic function at the center of a sphere equals the average of its values on the sphere itself. This *Mean Value Property* reinforces our interpretation of the Laplacian as a measure of the difference between the value of $\psi$ and its averaged neighboring values (eq. (11), Sec. 4.5); if $\nabla^2\psi = 0$, there is no difference.

But suppose that $\mathbf{r}$ is an interior maximum point for the harmonic function $\psi$; how can the *highest* value be the average of the neighboring values? The only possibility is that $\psi$ *always* takes its maximum value - i.e., $\psi$ is constant! This is the *Maximum Principle for Harmonic Functions*:

> *If a harmonic function achieves its maximum (or minimum)*
> *value at an interior point of a domain, it is constant throughout*
> *the domain. By default, a harmonic function always takes its*
> *maximum and minimum values on the boundary of the domain.*

It is evident that there can be no "hottest point" in the interior of a body at thermal equilibrium, because heat would flow away from this point. The drumhead interpretation is also clear: if a membrane is stretched over a warped rim, the highest and lowest points of the membrane will be on the rim (at equilibrium).

More insight into the nature of the relationship of a function $\psi$ with its Laplacian $\nabla^2\psi$ is provided by the vector identities of Green. We refer the reader to Davis and Snider for further details. The methodology and analogies we have covered herein will suffice for our subsequent discussions.

# Exercises 4.6

1. Verify directly that the functions listed in Example 1 are harmonic.

2. Verify the solution to Example 2 directly.

3. (a) Verify the solution to Example 3 directly.

   (b) Show that if $\psi_1(x,y)$ is a solution to Example 3, so is $\psi_1(x,y) + Im\{(z-2-i)^4\}$, where $Im$ denotes the imaginary part.

   (c) Show that if $\psi_1(x,y)$ is a solution to Example 2, so is $\psi_1(x,y) + C \times Im\{(z-2-i)^{4n}\}$, for any constant $C$ and positive integer $n$.

4. Solve the Dirichlet problem indicated in Fig. 3.

5. (a) Find a solution to the Dirichlet problem indicated in Fig. 4. (Hint: use a combination of $\arg(z-1)$ and $\arg(z+1)$.)

   (b) Show that if $\psi_1(x,y)$ solves the problem in Fig. 4, so does $\psi_1(x,y) + (\text{constant}) \times y$. Explain this indeterminacy in physical terms.

6. The *First Green Formula* states that for any two sufficiently smooth functions $\psi(\mathbf{r})$ and $\phi(\mathbf{r})$,

$$\iiint_D \phi \nabla^2 \psi \, dV + \iiint_D \nabla\phi \cdot \nabla\psi \, dV = \iint_S \phi \frac{\partial \psi}{\partial n} \, dA$$

where the surface $S$ is the boundary of the volume $D$.

   (a) Set $\phi(\mathbf{r}) = 1$ throughout $D$ and suppose that $\psi$ is a *harmonic* function satisfying a pure Neumann problem in $D$. Show that the Neumann data cannot be specified arbitrarily, but must satisfy $\iint_S \frac{\partial \psi}{\partial n} \, dA = 0$.

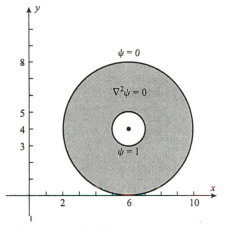

**Figure 3**  *Equilibrium temperature in an annulus*

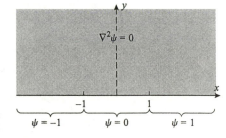

**Figure 4**  *Equilibrium temperature in the upper half-plane*

(b) What is the physical meaning of this result for the case when $\psi$ is the steady state temperature?

7. Using only your physical intuition, sketch the family of isotherms that you would expect to see at equilibrium for slabs with edge temperatures maintained as shown in Fig. 5. Be careful you don't violate the maximum principle.

## 4.7   Schrödinger's Equation

**Summary**   The quantum mechanical description of a point mass moving in a potential $V(\mathbf{r})$ is governed by Schrödinger's equation

$$-\frac{1}{2m}\left(\frac{h}{2\pi}\right)^2 \nabla^2\psi + V(\mathbf{r})\psi = i\frac{h}{2\pi}\frac{\partial\psi}{\partial t}$$

Solutions are uniquely determined by specifying the initial condition $\psi(\mathbf{r},0)$ and the normalization condition

$$\iiint |\psi(\mathbf{r},t)|^2 dx\,dy\,dz = 1,$$

although the latter is sometimes weakened to require only that $\psi$ be finite at infinity. The squared amplitude $|\psi|^2$ is interpreted as the probability density for the location of the particle.

In the beginning of the twentieth century physicists began to come to grips with the main shortcoming of their discipline - its failure to provide a mechanistic explanation of the composition of matter. Although the mechanics of Newton - and its elegant reformulation by Lagrange and Hamilton - modeled terrestrial and astronomical motions accurately, it failed dismally in accounting for atomic phenomena. A completely new description was needed, and eventually quantum mechanics was born.

One of the milestone events of this period was an experiment performed in 1927 by Davisson and Germer. They passed a beam of electrons of known velocity through a slit and observed the pattern that the beam made when impinging on a distant screen. The overall pattern was not what one would expect for a stream of particles; rather, it was a diffraction pattern such as would be produced by monochromatic light *waves*. Nonetheless it was clear (from other experiments) that the pattern was generated by individual *particles* - a point at a time - and not by continuous waves. Thus physicists came to realize that although electrons manifested themselves as particles, their motion was *guided* in some sense by waves; and that instead of trying to adapt Newton's second law to predict the *particle* motions, they should seek an equation for the evolution of the "carrier wave."

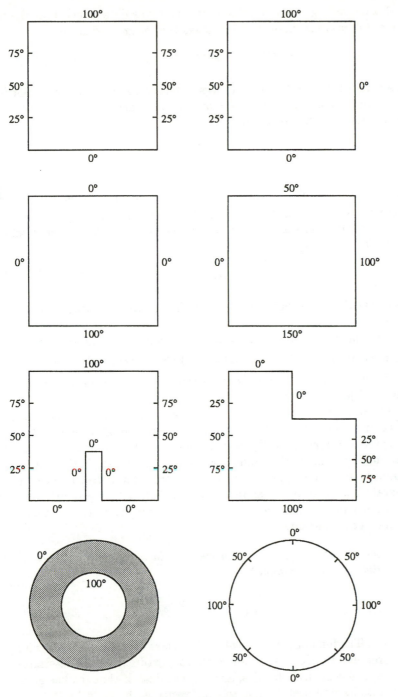

**Figure 5** *Exercise 7*

By measuring the diffraction patterns it was determined that the wave pattern $\psi(x,t)$ for these electrons takes the mathematical form

$$\psi(x,t) \propto e^{i2\pi(x/\lambda-\nu t)} = e^{i(kx-\omega t)} \quad (k=2\pi/\lambda,\ \omega=2\pi\nu). \tag{1}$$

The *wavelength* $\lambda$ of these waves was observed to be inversely proportional to the individual momentum $p=mv$ of each particle, in accordance with an equation attributed to deBroglie:

$$\lambda = \frac{h}{mv} = \frac{h}{p} \tag{2}$$

The *frequency* $\nu$ of the waves was found to be proportional to the electron's energy $E=\frac{1}{2}mv^2 = \frac{p^2}{2m}$, complying with a relation proposed earlier by Planck and Einstein (in explaining the specific heat of solids and the photoelectric effect):

$$\nu = E/h \tag{3}$$

Note that the same constant $h$ - *Planck's constant* - appears in both equations.

Two basic questions then arise: (i) What is the physical meaning of the wave amplitude (1)? (ii) What is the mathematical form of the wave amplitude when the particle's momentum is not constant - i.e. when it moves subject to a force?

The *intensity* of any wave is usually measured by the squared magnitude of its amplitude, $|\psi|^2$. It was in this sense that Davisson and Germer interpreted $\psi(\mathbf{r},t)$: $|\psi(\mathbf{r},t)|^2$ was proportional to the density of electrons that impinged upon the point $\mathbf{r}$ at time $t$. Now an *individual* electron manifests itself at a single point. Thus, in order for a *stream* of electrons to generate this pattern, it was reckoned that $|\psi(\mathbf{r},t)|^2$ is proportional to the probability density that any one electron impinges at $\mathbf{r}$. In other words, the answer to question (i) is

*the probability of finding an electron in a region of space D at time t is proportional to*

$$\iiint_D |\psi(\mathbf{r},t)|^2 dx\,dy\,dz \tag{4}$$

The fundamental probabilistic nature of quantum mechanics represents, of course, a remarkable departure from the deterministic mechanical theories, and a great deal of thought has gone into its interpretation; a good reference is the book by Bohm.

Now to incorporate forces into the wave description, one needs an equation for the evolution of the wave function $\psi(\mathbf{r},t)$. In the one-dimensional, free-electron case, we have the solution indicated in eq. (1). Rewriting (1) as

$$\psi(x, t) \propto e^{i(2\pi/h)[px - Et]},\tag{5}$$

we see that $\psi$ has the form $f(x - \tilde{v}t)$ of a wave propagating to the right at the speed

$$\tilde{v} = E/p = \frac{1}{2}mv^2/mv = v/2\tag{6}$$

Thus from our analysis in Sec. 4.3, we know that *in this case* $\psi$ satisfies the *wave equation* in the form

$$\frac{\partial^2 \psi}{\partial t^2} = \left(\frac{v}{2}\right)^2 \frac{\partial^2 \psi}{\partial x^2}\tag{7}$$

However (7) seems an unlikely candidate for a fundamental physical equation. First of all the appearance of the quantity $(v/2)$, or half the particle velocity, as the speed of propagation is baffling. And second, there is no apparent way to incorporate forces into (7).

A different identity satisfied by (5) follows from the energy-momentum relationship

$$E = p^2/2m.\tag{8}$$

Note that for $\psi$ as in (5) we can construct $E\psi$ and $\frac{p^2}{2m}\psi$ by applying differential operators:

$$E\psi(x, t) \equiv i\frac{h}{2\pi}\frac{\partial \psi}{\partial t},$$

while

$$\frac{p^2}{2m}\psi(x, t) \equiv -\frac{1}{2m}\left(\frac{h}{2\pi}\right)^2 \frac{\partial^2 \psi}{\partial x^2} = \frac{1}{2m}\left(\frac{h}{2\pi i}\frac{\partial}{\partial x}\right)^2 \psi.$$

Therefore we can write the equation

$$E\psi = i\frac{h}{2\pi}\frac{\partial \psi}{\partial t} = \frac{1}{2m}\left(\frac{h}{2\pi i}\frac{\partial}{\partial x}\right)^2 \psi = \frac{p^2}{2m}\psi,\tag{9}$$

*which is obtained from the energy equation (8) by replacing the momentum $p$ with the operator $\frac{h}{2\pi i}\frac{\partial}{\partial x}$ and the energy $E$ with the operator $i\frac{h}{2\pi}\frac{\partial}{\partial t}$, and letting the resulting operator equality act on $\psi$.*

Now the differential equation (9) is a much better candidate for a fundamental law of nature than the old equation (7). It contains no explicit dependence on *a priori* unknown parameters like $v$. Its generalization to three dimensions is clear;

*the vector momentum* $\mathbf{p}$ *is replaced by the operator* $\frac{h}{2\pi i}\nabla$ *and the kinetic energy* $\frac{|\mathbf{p}|^2}{2m} = \frac{\mathbf{p}\cdot\mathbf{p}}{2m}$ by $-\frac{1}{2m}\left(\frac{h}{2\pi}\right)^2\nabla^2$, *resulting in*

$$i\frac{h}{2\pi}\frac{\partial\psi(\mathbf{r},t)}{\partial t} = -\frac{1}{2m}\left(\frac{h}{2\pi}\right)^2\nabla^2\psi(\mathbf{r},t). \tag{10}$$

Furthermore, observe that (10) has a very natural generalization to systems with forces: if $\mathbf{F}$ is derivable from a potential $\mathbf{F} = -\nabla V(r)$, then as we saw in Sec. 4.2, the total energy is given by

$$E = \frac{|\mathbf{p}|^2}{2m} + V(\mathbf{r}) \tag{11}$$

Making the operator replacements in (11), then, we formulate

*the Schrödinger equation of quantum mechanics*:

$$i\frac{h}{2\pi}\frac{\partial\psi}{\partial t} = -\frac{1}{2m}\left(\frac{h}{2\pi}\right)^2\nabla^2\psi + V(r)\psi \tag{12}$$

Schrödinger's equation, with the probabilistic interpretation (4), has proved very successful in explaining natural phenomena on the atomic scale. Additionally, it has been shown to be consistent with Newton's second law on the macroscopic scale. Thus it is accepted as a fundamental law, and must be included - with the wave equation, the heat equation, and Laplace's equation - as a key partial differential equation for our studies.

Note that because (12) describes the evolution of the wave function $\psi$ in time, it must be supplemented with initial conditions; $\psi(\mathbf{r},0)$ must be specified.

Since $|\psi|^2$ is, up to a constant, a probability density function, it is convenient to scale it initially so that

$$\iiint |\psi(\mathbf{r},0)|^2 dx\, dy\, dz = 1 \tag{13}$$

(integrated over all space). This is not always possible; the free electron wave function (1), for example, diverges when inserted into the integral (13).

A nice feature of the Schrödinger equation is that when the "normalization" (13) *is* possible for some time $t_1$, it is automatically satisfied for all subsequent times. We present the argument for the one-dimensional case; the generalization requires vector analysis, and we direct the reader to the exercises.

Rewrite eq. (12) in the form

$$\frac{\partial\psi}{\partial t} = \frac{ih}{4\pi m}\frac{\partial^2\psi}{\partial x^2} - \frac{i2\pi}{h}V(x)\psi \tag{14}$$

The complex conjugate of (14) is

$$\frac{\partial \overline{\psi}}{\partial t} = -\frac{ih}{4\pi m}\frac{\partial^2 \overline{\psi}}{\partial x^2} + \frac{i2\pi}{h}V(x)\overline{\psi} \qquad (15)$$

Now we have

$$\frac{d}{dt}\int_{-\infty}^{\infty}|\psi|^2\,dx = \frac{d}{dt}\int_{-\infty}^{\infty}\overline{\psi}\,\psi\,dx = \int_{-\infty}^{\infty}\left[\overline{\psi}\frac{\partial\psi}{\partial t} + \psi\frac{\partial\overline{\psi}}{\partial t}\right]dx$$

Inserting (14, 15) we observe some cancellation and find

$$\frac{d}{dt}\int_{-\infty}^{\infty}|\psi|^2\,dx = \frac{ih}{4\pi m}\int_{-\infty}^{\infty}\left[\overline{\psi}\frac{\partial^2\psi}{\partial x^2} - \psi\frac{\partial^2\overline{\psi}}{\partial x^2}\right]dx,$$

which we judiciously identify as

$$= \frac{ih}{4\pi m}\int_{-\infty}^{\infty}\frac{\partial}{\partial x}\left[\overline{\psi}\frac{\partial\psi}{\partial x} - \psi\frac{\partial\overline{\psi}}{\partial x}\right]dx$$

$$= \frac{ih}{4\pi m}\left[\overline{\psi}\frac{\partial\psi}{\partial x} - \psi\frac{\partial\overline{\psi}}{\partial x}\right]\Bigg|_{-\infty}^{\infty}$$

$$= 0$$

for (integrable) wave functions approaching zero at infinity. Therefore

$$\frac{d}{dt}\int_{-\infty}^{\infty}|\psi|^2 dx = 0$$

and, consequently, the integral does not change in time.

As these deliberations indicate, there are no transparent boundary conditions to associate with the "carrier wave" $\psi$. Certainly $\psi$ must be *finite* at infinity, for otherwise the interpretation (4) would imply that all matter resides "outside the universe." In our later examples we shall merely require that $\psi(\infty)$ be finite, and we shall analyze (12) for the uniform force field, the square well, the harmonic oscillator, and the hydrogen atom.

Observe that we obtain a *nondimensional* form of (12) by redefining "t" to be "$ht/4m\pi$" and "$V(\mathbf{r})$" to be "$8m\pi^2 V(\mathbf{r})/h^2$":

$$-\nabla^2\psi + V(r)\psi = i\frac{\partial\psi}{\partial t}. \qquad (16)$$

*We shall use this form in all subsequent chapters.*

We close this section by noting that since the kinetic energy $|\mathbf{p}|^2/2m$ is never negative, the conservation of energy principle (Sec. 4.2)

$$E = \frac{|\mathbf{p}|^2}{2m} + V(\mathbf{r}) = constant$$

prevents a classical particle from entering any region in which $V(\mathbf{r})$ exceeds $E$. However, we shall see quantum-mechanical violations of this dictum in some examples; in the classically accessible regions $\psi$ is oscillatory, and in the forbidden zones it decays exponentially. This "tunneling" effect, whose exposition we must defer to Chapter 8, is the key to the operation of many electronic devices which use the Schottky diode.

## Exercises 4.7

1. Derive the relation $\frac{d}{dt}\iiint_{-\infty}^{\infty}|\psi(\mathbf{r},t)|^2 dx\,dy\,dz = 0$ from the three-dimensional Schrödinger equation. (Hint: this requires identities from vector analysis; see Davis and Snider.)

## 4.8  Classification of Second Order Equations

**Summary**  Linear partial differential equations in two dimensions are classified as hyperbolic, parabolic, or elliptic according to the discriminant of their highest-order coefficients. Solutions of hyperbolic equations exhibit finite velocity-of-propagation effects and are uniquely determined by auxiliary conditions analogous to initial and boundary conditions for the wave equation. Solutions of elliptic equations are smooth at interior points and are uniquely determined by boundary conditions analogous to those for Laplace's equation. Parabolic equations exhibit a behavior somewhat intermediate between these two, typified by the heat equation. Such considerations allow the extension of this classification scheme to higher dimensions.

We have seen how the wave, heat, and Laplace equations arise in physical situations, and from this we are able to anticipate many of the qualitative differences between the behavior of their solutions (such as the appropriate auxiliary conditions necessary to determine solutions uniquely). It turns out that these qualitative features and differences persist when the equations are altered in certain ways, and we can classify more general classes of second order partial differential equations as "wave-like, heat-like, or Laplace-like."[5] We shall attempt a rough survey of this aspect of our subject, and we direct the reader to the references (such as Courant and Hilbert or Hellwig) for details.

The basis for the classification scheme lies in the formal similarity between the structure of the two-dimensional versions of the differential equations and the structure of the equations of the conic sections. If we let $y = vt$ in the two-dimensional wave equation, let $y = \gamma t$ in the two-dimensional heat equation, and leave the two-dimensional Laplace equation intact, we

---

[5]A "Schrödinger-like" classification is beyond this text.

derive (exercise 1)

$$\frac{\partial^2 \psi}{\partial x^2} - \frac{\partial^2 \psi}{\partial y^2} = 0 \qquad\qquad \text{(wave)} \qquad (1)$$

$$\frac{\partial^2 \psi}{\partial x^2} - \frac{\partial \psi}{\partial y} = 0 \qquad\qquad \text{(heat)} \qquad (2)$$

$$\frac{\partial^2 \psi}{\partial x^2} + \frac{\partial^2 \psi}{\partial y^2} = 0 \qquad\qquad \text{(Laplace)} \qquad (3)$$

Compare (1, 2, 3) with the conic section equations:

$$x^2 - y^2 = K \qquad\qquad \text{(hyperbola)} \qquad (4)$$
$$x^2 - y = K \qquad\qquad \text{(parabola)} \qquad (5)$$
$$x^2 + y^2 = K \qquad \text{(circle - but more generally, ellipse)} \qquad (6)$$

Because of the similarity in the pattern of signs and exponents, we say that the wave equation is of *hyperbolic* type, the heat equation is of *parabolic* type, and Laplace's equation is of *elliptic* type.

Recall from elementary analytic geometry that the generic conic section

$$ax^2 + 2bxy + cy^2 + dx + ey + f = 0 \qquad (7)$$

can be reduced to one (and only one) of the forms (4, 5, 6) by a linear transformation

$$x' = Ax + By, \; y' = Cx + Dy \qquad (8)$$

followed by a shift $x'' = x' + E, y'' = y' + F$. The particular form (4, 5, 6) achieved depends on the sign of the *discriminant* $b^2 - ac$:
if $b^2 - ac > 0$ the section is a hyperbola;
if $b^2 - ac = 0$ the section is a parabola;
if $b^2 - ac < 0$ the section is an ellipse.
Accordingly, we generalize our terminology and state that the *general linear second order partial differential equation in two variables*

$$a\frac{\partial^2 \psi}{\partial x^2} + 2b\frac{\partial^2 \psi}{\partial x \partial y} + c\frac{\partial^2 \psi}{\partial y^2} + d\frac{\partial \psi}{\partial x} + e\frac{\partial \psi}{\partial y} + f\psi + g = 0 \quad (9)$$

is of *hyperbolic, parabolic,* or *elliptic* type if $b^2 - ac$ is positive, zero, or negative.

It can be shown that the differential equation (9) can be reduced, by a linear change of variables (8), to a form where the left hand side is either (1,2, or 3), according to the sign of the discriminant, and the right hand side contains only derivatives of $\psi$ of order zero or one. If the coefficients $(a, b, c)$ depend on the variables $x$ and $y$, then so does the transformation (8); thus the equation may be of different type at different points in the plane. Note that if the mixed derivative $\partial^2 \psi / \partial x \partial y$ does not appear, classification is immediate; the equation is hyperbolic if the second partials come in with opposite signs, and elliptic if they have the same signs; if one of them is missing then it is parabolic. The *Tricomi* equation

$$y \frac{\partial^2 \psi}{\partial x^2} + \frac{\partial^2 \psi}{\partial y^2} = 0, \tag{10}$$

which describes transonic flow, is elliptic for $y > 0$ and hyperbolic for $y < 0$.

These notions are extended to higher dimensions as follows. The general linear second order partial differential equation in $n$ variables can be written

$$\sum_{i,j=1}^{n} a_{ij} \frac{\partial^2 \psi}{\partial x_i \partial y_j} + \sum_{i=1}^{n} b_i \frac{\partial \psi}{\partial x_i} + c\psi + d = 0 \tag{11}$$

It is said to be *elliptic* at any point where it can be reduced by a linear change of variables to the form

$$\sum_{i=1}^{n} \frac{\partial^2 \psi}{\partial x_i^2} = \text{(zero- or first-order derivative terms)}. \tag{12}$$

It is *hyperbolic* if the form

$$\frac{\partial^2 \psi}{\partial x_n^2} - \sum_{i=1}^{n-1} \frac{\partial^2 \psi}{\partial x_i^2} = \text{(zero- or first-order derivative terms)} \tag{13}$$

can be so achieved; the variable $x_n$ is then called "time-like" and the others $(x_1, x_2, ..., x_{n-1})$ "space-like." *Parabolic* equations result when the form

$$\frac{\partial \psi}{\partial x_n} - \sum_{i=1}^{n-1} \frac{\partial^2 \psi}{\partial x_i^2} = \text{(zero- or first-order derivative terms)} \tag{14}$$

is achieved; again $x_n$ is time-like and the others are space-like. Obviously the three-dimensional wave, heat, and Laplace equation are of the hyperbolic, parabolic, and elliptic type, respectively. More exotic "hypo-elliptic" forms are possible, but occur less frequently in applications.

The importance of this classification scheme is the fact that all equations of a given type have some qualitative features in common. Recall, for example, from Sec. 4.3 that for the wave equation the values of the initial data at a point $x'$ do not influence the values of the solution at $(x, t)$ if $|x - x'|$ exceeds $vt$. This effect is quite general; every hyperbolic equation has a maximum speed of propagation which limits the extent to which the initial data can influence the solution at later times. On the other hand all parabolic equations have infinite speed of propagation, just like the heat equation (Sec. 4.5).

A feature shared by all elliptic equations is *smoothness*. Remember that solutions to the two-dimensional Laplace equation can be visualized as the equilibrium shapes of a membrane stretched across the rim of a drum (Sec. 4.6). Physically we know that the membrane surface will be smooth, even if the rim is jagged. Discontinuities in the boundary data are immediately smoothed out in the interior of the solution region for Laplace's equation - and for all equations of elliptic type (unless the coefficients in the equation itself are discontinuous!). This is also true of parabolic equations. Jagged initial data for the wave equation, however, do not get smoothed; recall (Sec. 4.3) that an initial stretched-string profile $f(x)$ - no matter how jagged - is propagated *intact* (as $f(x - vt)$).

The final issue that we shall discuss is the nature of the auxiliary conditions that are appended to each type of equation. Typically they correspond to what we have seen for the Laplace, heat, and wave equations. The elliptic equation is accompanied by *boundary* conditions, where either $\psi$ or $\frac{\partial \psi}{\partial n}$ or a linear combination of the two (Dirichlet, Neumann, or Robin conditions respectively) is specified *all around* the boundary of the solution region. For a parabolic equation one imposes an *initial* condition, wherein $\psi(x_1, x_2, \cdots, x_{n-1}, x_n)$ is specified on a hyperplane $x_n = 0$ of the time-like variable $x_n$; for a hyperbolic equation one also specifies the derivative $\frac{\partial \psi}{\partial x_n}$ on this hyperplane. In both of these latter cases Dirichlet, Neumann, or Robin conditions are also imposed on the space-like boundaries of the solution region.

Why do we say "typically" in the previous paragraph? It is true that other auxiliary conditions can be imposed mathematically, but they may lead to solutions which have little physical credibility. In this regard one says that a partial differential equation and its auxiliary conditions comprise a *well-posed problem* if the following three conditions are met: (i) a solution to the problem *exists*, (ii) the solution is *unique*, and (iii) the solution depends *continuously* on the auxiliary data. This is to say, if we alter the initial or boundary conditions to a sufficiently small degree, the solutions can not differ drastically.

The next two examples illustrate how imposing the "wrong" auxiliary conditions for an equation leads to ill-posed problems.

**Example 1.** To see why Dirichlet conditions are inappropriate for hyperbolic equations, consider the problem depicted in Fig. 1. If the governing equation were *Laplace's* equation, then the solution $\psi$ could be interpreted as the steady state temperature on a conducting slab whose edges were maintained at zero degrees; and the only possible solution, clearly, would be $\psi(x, y) = 0$. But the hyperbolic equation displayed in the figure has

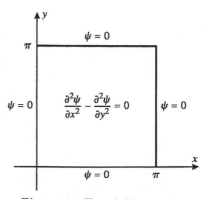

*Figure 1   Hyperbolic equation with Dirichlet conditions*

infinitely many solutions:

$$\psi(x, y) = (constant) \times \sin mx \times \sin my, \tag{15}$$

for any constant and any integer $m$, satisfies all the conditions (exercise 2).

We can concoct a physical interpretation of this behavior. Recall from Section 4.4 our discussion of the forces on a vibrating membrane (Figure 1, Sec. 4.4). We reasoned that the net force on a small rectangular patch would be given by

$$T_x \frac{\partial^2 \psi}{\partial x^2} \Delta x + T_y \frac{\partial^2 \psi}{\partial y^2} \Delta y$$

(equation (1) of that section) where $T_x$ is the tensile force in the $x$ direction and $T_y$ in the $y$ direction. Each tensile force was attributed to a uniform tension $T$ per unit length -

$$T_x = T\Delta y, \ \ T_y = T\Delta x.$$

At equilibrium, then, this force is zero and we get Laplace's equation:

$$T\Delta y \frac{\partial^2 \psi}{\partial x^2} \Delta x + T\Delta x \frac{\partial^2 \psi}{\partial y^2} \Delta y = 0 = \frac{\partial^2 \psi}{\partial x^2} + \frac{\partial^2 \psi}{\partial y^2}.$$

But suppose the membrane had some stiffness and we applied tension $(T)$ in the $x$ direction but *compression* $(-T)$ in the $y$ direction. Then the force balance equation would be hyperbolic:

$$T\Delta y \frac{\partial^2 \psi}{\partial x^2} \Delta x + (-T)\Delta x \frac{\partial^2 \psi}{\partial y^2} \Delta y = 0 = \frac{\partial^2 \psi}{\partial x^2} - \frac{\partial^2 \psi}{\partial y^2},$$

which is the equation in Figure 1. In such a situation we would expect to see the membrane *buckle*, and the presence of the rippled solutions (15) reflect this mechanical instability. ∎

**Example 2.** (*Hadamard's example*) If we try to solve the (elliptic) Laplace equation (3) in the upper half plane ($y \geq 0$) with *initial conditions*

$$\psi(x, 0) = 0, \ \frac{\partial \psi}{\partial y}(x, 0) = \frac{1}{n} \sin nx,$$

we would indeed find the solution

$$\psi(x, y) = (1/n^2) \sin nx \sinh ny$$

(exercise 3). But note that as we take increasingly larger values for $n$, the initial data become infinitesimally small - and yet the values of the solution on the line $y = 1$ become unbounded!

Again we can match this with a physical instability. The wave equation for a string - or, for our current purposes, a wire - was derived in Section 4.3:

$$\frac{\partial^2 \psi}{\partial t^2} = \frac{T}{\rho} \frac{\partial^2 \psi}{\partial x^2}, \tag{16}$$

with tension $T$ and density $\rho$. Now if the wire is under *compression* $(-T)$, and we set $y = t\sqrt{\frac{T}{\rho}}$, then equation (16) becomes Laplace's equation $\frac{\partial^2 \psi}{\partial x^2} + \frac{\partial^2 \psi}{\partial y^2} = 0$. So Hadarmard's solutions reflect the fact that a wire under compression will buckle if its initial shape isn't precisely straight. ∎

There are many other aspects of classification of equations; *systems* of differential equations, for example, can be categorized as to type. At this point, however, we have introduced the reader to the most important differential equations of physics and engineering, and it is time to proceed to the discussion of solution methods.

# Exercises 4.8

1. Describe at which points the following equations are parabolic, hyperbolic, and elliptic:

$$(a) \quad 3\frac{\partial^2 \psi}{\partial x^2} + 4\frac{\partial^2 \psi}{\partial x \partial y} + 2\frac{\partial^2 \psi}{\partial y^2} - \psi = 0$$

$$(b) \quad \frac{\partial^2 \psi}{\partial x^2} + 2x\frac{\partial^2 \psi}{\partial x \partial y} + \frac{\partial^2 \psi}{\partial y^2} + \cos(xy)\,\psi = 0$$

$$(c) \quad e^y\frac{\partial^2 \psi}{\partial x^2} - 2\frac{\partial^2 \psi}{\partial x \partial y} + x\frac{\partial^2 \psi}{\partial y^2} + y^2\frac{\partial \psi}{\partial y} + \psi = 0$$

2. Verify the solution to Example 1.

3. Verify the solution to Example 2.

> For exercises 4-7 you may need to review the chain rule for functions of several variables; see Sec. 1.2.

4. Suppose that $\psi(x, y)$ satisfies Laplace's equation. Perform a change of variables by first elongating the $y$ axis by a factor of 3, and then rotating the new axes through $30^0$. What differential equation does the derived function $\widetilde{\psi}(x', y') = \psi(x(x', y'), y(x', y'))$ satisfy? Verify that it is still elliptic by evaluating the discriminant.

5. Repeat the analysis of exercise 4 for the wave equation (1).

6. Repeat the analysis of exercise 4 for the heat equation (2).

7. Suppose that $\psi(x, y)$ satisfies the wave equation (1). Introduce the change of variables $x' = x + y, y' = x - y$. What differential equation does the derived function $\widetilde{\psi}(x', y') = \psi(x(x', y'), y(x', y'))$ satisfy? Relate this to eq. (3), Sec. 4.3.

# Chapter 5

# The Separation of
# Variables Technique

In this chapter we introduce the most important analytical technique in engineering analysis for solving partial differential equations: separation of variables. Elaboration of the method will consume a large portion of the remainder of the book.

The successful execution of this procedure for a given boundary value problem can be rather lengthy, because three different mathematical procedures are involved:

(i) the use of superposition to decompose a complicated problem into a set of simpler ones;

(ii) the separation of the partial differential equation into a set of ordinary differential equations in an appropriate coordinate system; and

(iii) the construction of eigenfunction expansions (which are generalizations of the Fourier series) which satisfy the boundary conditions.

To help keep all this in perspective we begin the exposition by displaying the final solution formula for a specific problem, and examining its features (Sec. 5.1). Then we come back and begin putting together all the elements that go into the derivation of such formulas. Items (i) and (ii) are covered in this chapter, and Chapter 6 is devoted to the exposition of eigenfunction theory.

Subsequent chapters adapt the basic procedures to more general classes of applications.

## 5.1 Overview of Separation of Variables

**Summary** The solution form generated by the separation of variables procedure for a boundary value problem is partitioned into formulas for subproblems, in each of which all but one boundary condition is homogeneous. The subproblem solution formulas are infinite series, wherein individual terms satisfy the homogeneous equations while the overall sum fulfills the remaining nonhomogeneous condition.

Consider the boundary value problem depicted in Fig. 1. If we interpret the unknown $\psi$ as temperature, we have a two-dimensional slab in which $\psi$ is specified on two edges

$$\psi(0, y) = 0 \tag{1}$$
$$\psi(\pi, y) = y \tag{2}$$

and the heat flux is specified on the remaining edges

$$\frac{\partial \psi(x, 0)}{\partial y} = 0 \tag{3}$$

$$\frac{\partial \psi(x, \pi)}{\partial y} = x(\pi - x) \tag{4}$$

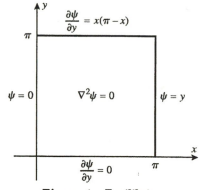

**Figure 1** *Equilibrium temperature specifications*

Thus we have both Dirichlet and Neumann boundary conditions. The underlying partial differential equation,

$$\nabla^2 \psi = 0 \quad (0 < x < \pi, \ 0 < y < \pi) \tag{5}$$

(Laplace's equation), indicates that we are seeking the temperature in the steady state.

The method of separation of variables (which will be discussed at length in later sections) provides the following formula for the solution to eqs. (1) through (5):

$$\psi(x, y) = \frac{x}{2} - \frac{4}{\pi} \sum_{n=0}^{\infty} \frac{\sinh(2n+1)x}{\sinh(2n+1)\pi} \frac{\cos(2n+1)y}{(2n+1)^2}$$
$$+ \frac{8}{\pi} \sum_{n=0}^{\infty} \frac{\sin(2n+1)x}{(2n+1)^4} \frac{\cosh(2n+1)y}{\sinh(2n+1)\pi} \tag{6}$$

A graph of this solution is displayed in Fig. 2, wherein ten terms of each sum have been retained. (Details of the convergence of the series will be covered in Chapter 6.)

Let us examine the formula (6). First observe that each term in the formula is, indeed, a solution of Laplace's equation (5); omitting the coefficients, we have

$$\nabla^2 x = 0,$$

$$\nabla^2 \sinh(2n+1)x \cos(2n+1)y = (2n+1)^2 \sinh(2n+1)x \cos(2n+1)y$$
$$- (2n+1)^2 \sinh(2n+1)x \cos(2n+1)y$$
$$= 0,$$

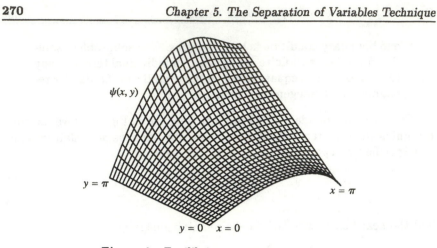

**Figure 2**   *Equilibrium temperature solution*

$$\nabla^2 \sin(2n+1)x \cosh(2n+1)y \;=\; -(2n+1)^2 \sin(2n+1)x \cosh(2n+1)y$$
$$+(2n+1)^2 \sin(2n+1)x \cosh(2n+1)y$$
$$=\; 0,$$

Furthermore because of the factors $x$, $\sinh(2n+1)x$, and $\sin(2n+1)x$, each term equals zero on the left edge (eq. (1)). Similarly, thanks to the factors $\cos(2n+1)y$ and $\cosh(2n+1)y$, whose derivatives are zero at $y = 0$, each term provides zero flux on the bottom edge (eq. (3)). Thus the *homogeneous* equations - (1, 3, and 5) - are satisfied on a term-by-term basis.

Recall (Sec. 1.1) the key feature of linear *homogeneous* equations: any linear combination of solutions is, again, a solution.

Now according to eq. (6) the flux on the upper edge is given by

$$\frac{\partial \psi(x,\pi)}{\partial y} \;=\; (0) + \frac{4}{\pi} \sum_{n=0}^{\infty} \frac{\sinh(2n+1)x}{\sinh(2n+1)\pi} \frac{\sin(2n+1)\pi}{(2n+1)}$$
$$+ \frac{8}{\pi} \sum_{n=0}^{\infty} \frac{\sin(2n+1)x}{(2n+1)^3} \frac{\sinh(2n+1)\pi}{\sinh(2n+1)\pi} \tag{7}$$

Because of the factor $\sin(2n+1)\pi$, every term in the *first* sum is zero, and the second series reduces to

$$\frac{\partial \psi(x,\pi)}{\partial y} = \frac{8}{\pi} \sum_{n=0}^{\infty} \frac{\sin(2n+1)x}{(2n+1)^3} \tag{8}$$

From Fig. 5 of Sec. 3.3 we see that this series is the Fourier sine expansion for the function $x(\pi - x)$. Consequently the Neumann boundary condition (4) is indeed satisfied by our solution formula.

Furthermore, eq. (6) expresses the temperature on the right edge as

$$\psi(\pi, y) \;=\; \frac{\pi}{2} - \frac{4}{\pi}\sum_{n=0}^{\infty}\frac{\sinh(2n+1)\pi}{\sinh(2n+1)\pi}\frac{\cos(2n+1)y}{(2n+1)^2}$$
$$+\frac{8}{\pi}\sum_{n=0}^{\infty}\frac{\sin(2n+1)\pi}{(2n+1)^4}\frac{\cosh(2n+1)y}{\sinh(2n+1)\pi} \tag{9}$$

or, since now the terms of the *second* series are zero,

$$\psi(\pi, y) = \frac{\pi}{2} - \frac{4}{\pi}\sum_{n=0}^{\infty}\frac{\cos(2n+1)y}{(2n+1)^2} \tag{10}$$

According to exercise 4 of Sec. 3.5, this is the Fourier cosine expansion for the function $y$. Therefore the Dirichlet condition (2) is met.

In short, expression (6) is the correct solution to the problem; it satisfies all the required conditions (1) through (5) (and in a very transparent way!).

Let us review some special features of the solution form (6). First notice that our analysis has demonstrated that $\psi(x, y)$ can be partitioned into two pieces $\psi = \psi_1 + \psi_2$ -

$$\psi_1(x, y) = \frac{x}{2} - \frac{4}{\pi}\sum_{n=0}^{\infty}\frac{\sinh(2n+1)x}{\sinh(2n+1)\pi}\frac{\cos(2n+1)y}{(2n+1)^2}$$

and

$$\psi_2(x, y) = \frac{8}{\pi}\sum_{n=0}^{\infty}\frac{\sin(2n+1)x}{(2n+1)^4}\frac{\cosh(2n+1)y}{\sinh(2n+1)\pi}$$

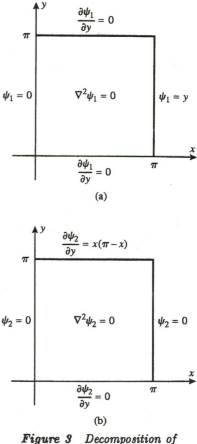

(a)

(b)

***Figure 3*** *Decomposition of solution*

- where the function $\psi_1$ solves a simplified problem, depicted in Fig. 3(a), in which the nonhomogeneous Neumann condition on the top edge has been replaced by its associated homogeneous form; and similarly, $\psi_2$ solves the problem in Fig. 3(b) where the original, nonhomogeneous, Dirichlet condition on the right edge has been "zeroed out."

Thus the display (6) suggests a decomposition of the original problem into subproblems wherein all but one of the governing conditions are homogeneous. *We shall see that this preliminary decomposition into subproblems is an essential step in the separation of variables procedure.*

A second feature of formula (6) is the fact that the *homogeneous* conditions for each subproblem are satisfied by the corresponding series on a *term-by-term* basis, but the *nonhomogeneous* condition entails summing up the series. *In general, the basic strategy of the separation of variables procedure is the compilation of a set of solutions to the homogeneous subset of equations, followed by the assemblage of these into a sum satisfying the nonhomogeneous equation.*

The exercises provide some other illustrations of solution forms resulting from separation of variables. For now, your task is merely to *verify* them. Their *derivation* is the subject of the remainder of this chapter.

# Exercises 5.1

1. Verify that $\psi(x,y) = \psi_1 + \psi_2 + \psi_3 + \psi_4$ solves the Dirichlet problem indicated in Fig. 4, where

$$\psi_1(x,y) = \sin x \sinh y$$
$$\psi_2(x,y) = \sinh 3x \sin 3y$$
$$\psi_3(x,y) = \sin 3x \sinh 3(\pi - y)$$
$$\psi_4(x,y) = \sinh 4(\pi - x) \sin 4y.$$

Which subproblems, involving only one nonhomogeneous boundary condition each, do $\psi_1$, $\psi_2$, $\psi_3$, and $\psi_4$ satisfy individually?

2. Verify that $\psi(x,y) = \psi_1 + \psi_2 + \psi_3$ solves the problem indicated in Fig. 5, where

$$\psi_1(x,y) = -\frac{8}{\pi} \sum_{n=0}^{\infty} \frac{\cosh(2n+1)(\pi-x)}{\sinh(2n+1)\pi} \frac{\sin(2n+1)y}{(2n+1)^4}$$

$$\psi_2(x,y) = \frac{\pi-y}{2} - \frac{4}{\pi} \sum_{n=0}^{\infty} \frac{\cos(2n+1)x}{(2n+1)^2} \frac{\sinh(2n+1)(\pi-y)}{\sinh(2n+1)\pi}$$

$$\psi_3(x,y) = \frac{3\cosh x}{\sinh \pi} \sin y + \frac{2}{3}\frac{\cosh 3x}{\sinh 3\pi} \sin 3y$$

Which subproblems, involving only one nonhomogeneous boundary condition each, do $\psi_1$, $\psi_2$, and $\psi_3$ satisfy individually?

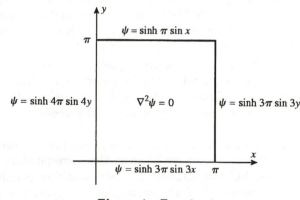

**Figure 4**   *Exercise 1*

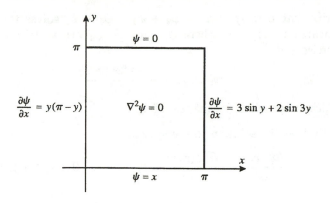

**Figure 5**   *Exercise 2*

3. Verify that $\psi(x,y) = \sum_{n=1}^{\infty} a_n \sin nx \sinh ny$ solves the problem indicated in Fig. 6, where

$$a_n = \frac{2}{\pi \sinh n\pi} \int_0^\pi x^3 \sin nx\, dx.$$

4. Verify that $\psi(x,y) = \psi_1 + \psi_2 + \psi_3 + \psi_4$ solves the problem indicated in Fig. 7, where

$$\psi_1(x,y) = a_0 y + \sum_{n=1}^{\infty} a_n \cos nx \sinh ny$$

$$a_0 = \frac{1}{\pi^2} \int_0^\pi f_1(x)\, dx, \quad a_{n>0} = \frac{2}{\pi \sinh n\pi} \int_0^\pi f_1(x) \cos nx\, dx;$$

$$\psi_2(x,y) = \sum_{n=1}^{\infty} b_n \cosh nx \sin ny$$

$$b_n = \frac{2}{n\pi \sinh n\pi} \int_0^\pi f_2(y) \sin ny\, dy;$$

$$\psi_3(x,y) = c_0(\pi - y) + \sum_{n=1}^{\infty} c_n \cos nx \sinh n(\pi - y)$$

$$c_0 = \frac{1}{\pi^2} \int_0^\pi f_3(x)\, dx, \quad c_{n>0} = \frac{2}{\pi \sinh n\pi} \int_0^\pi f_3(x) \cos nx\, dx;$$

$$\psi_4(x,y) = \sum_{n=1}^{\infty} d_n \cosh n(\pi - x) \sin ny$$

$$d_n = \frac{-2}{n\pi \sinh n\pi} \int_0^\pi f_4(y) \sin ny\, dy;$$

Which subproblems, involving only one nonhomogeneous boundary condition each, do $\psi_1$, $\psi_2$, $\psi_3$, and $\psi_4$ satisfy separately?

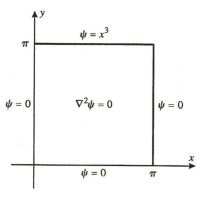

**Figure 6**   *Exercise 3*

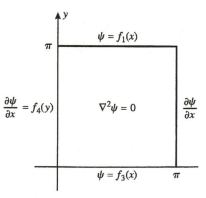

**Figure 7**   *Exercise 4*

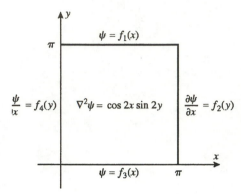

$\psi = f_1(x)$

$\frac{\psi}{x} = f_4(y)$  $\nabla^2\psi = \cos 2x \sin 2y$  $\frac{\partial\psi}{\partial x} = f_2(y)$

$\psi = f_3(x)$

**Figure 8**  **Exercise 5**

5. Verify that $\psi(x,y) = \psi_1 + \psi_2 + \psi_3 + \psi_4 + \psi_5$ solves the problem indicated in Fig. 8, where $\psi_1$ through $\psi_4$ are as in the previous exercise and

$$\psi_5(x,y) = -\frac{\cos 2x \sin 2y}{8}.$$

Which subproblem does $\psi_5$ satisfy?

6. Verify that the solution to the system

$$\frac{\partial\psi(x,t)}{\partial t} = \frac{\partial^2\psi(x,t)}{\partial x^2} \quad (0 < x < \pi, \, 0 < t < \infty),$$

$$\psi(0,t) = \psi(\pi,t) = 0,$$

$$\psi(x,0) = f(x)$$

is given by

$$\psi(x,t) = \sum_{n=1}^{\infty} a_n \sin nx \, e^{-n^2 t},$$

where

$$a_n = \frac{2}{\pi} \int_0^\pi f(x) \sin nx \, dx.$$

The equations describe the evolution of temperature along a rod held at 0 degrees at each end and heated to an initial temperature of $f(x)$.

7. Verify that the solution to the system

$$\frac{\partial^2\psi(x,t)}{\partial t^2} = \frac{\partial^2\psi(x,t)}{\partial x^2} \quad (0 < x < \pi, 0 < t < \infty),$$

$$\psi(0,t) = \psi(\pi,t) = 0,$$

$$\psi(x,0) = f(x), \quad \frac{\partial\psi(x,0)}{\partial t} = g(x)$$

is given by

$$\psi(x,t) = \sum_{n=1}^{\infty} a_n \sin nx \cos nt + \sum_{n=1}^{\infty} b_n \sin nx \sin nt$$

where

$$a_n = \frac{2}{\pi} \int_0^\pi f(x) \sin nx \, dx \quad \text{and} \quad b_n = \frac{2}{n\pi} \int_0^\pi g(x) \sin nx \, dx.$$

The equations describe the motion of a taut string held fixed at each end and released from an initial shape $f(x)$ with velocity $g(x)$.

# 5.2   Separation of Variables: Basic Models

**Summary**   Laplace's equation in two dimensions admits solutions in the separated form $X(x)Y(y)$, where typically one of the factors is a sine or cosine and the other is a hyperbolic sine or cosine. The coefficients and frequencies can be adjusted to make the factors satisfy homogeneous boundary conditions on three sides of a rectangle, while the flexibility of the Fourier expansions enables the fulfillment of a fourth, nonhomogeneous, boundary condition. The more general problem with four non-homogeneous boundary conditions can be decomposed into sub-problems of this form by exploiting the linearity of the equations.

The separation of variables procedure will first be explained for an example involving the computation of the steady state temperature distribution in a planar slab, as in the previous section. However for simplicity we impose only Dirichlet boundary conditions.

Figure 1 depicts a square of thermally conductive material whose dimensions are taken (for mathematical convenience) to be $\pi$ units by $\pi$ units. The points along its top edge are maintained at a temperature of

$$\psi(x,\pi) = \sin x \ \ (0 < x < \pi) \tag{1}$$

(degrees) by external heat sources. The temperature on the right edge is maintained at

$$\psi(\pi,y) = \sin^3 y \ \ (0 < y < \pi), \tag{2}$$

and the other edges are held at

$$\psi(x,0) = x(\pi - x) \ \ (0 < x < \pi), \text{ and} \tag{3}$$

$$\psi(0,y) = f(y) \ \ (0 < y < \pi) \tag{4}$$

respectively; here $f(y)$ is some given function which we choose not to specify at the moment. Inside the square the steady state temperature satisfies Laplace's equation

$$\nabla^2 \psi(x,y) = 0 \ . \tag{5}$$

The differential equation (5), together with the boundary conditions (1) through (4), determines the physical problem.

As a preliminary step we take advantage of the linearity of the equations to simplify the analysis. Suppose we compute the solutions to the following four problems (see Fig. 2):

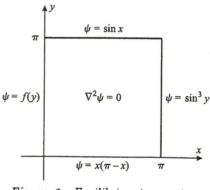

**Figure 1**   *Equilibrium temperature specifications*

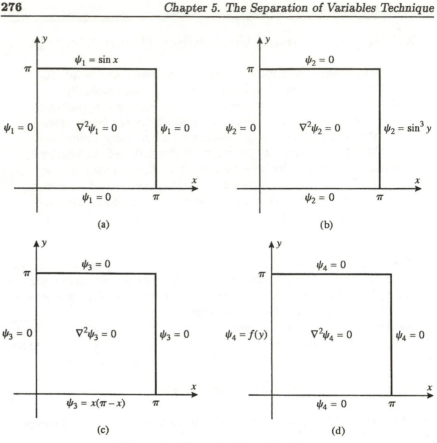

**Figure 2**   *Decomposition of problem*

*Problem* 1. Find $\psi_1(x, y)$ such that

$$\nabla^2\psi_1(x, y) = 0 \text{ inside the square}, \tag{6}$$
$$\psi_1(x, \pi) = \sin x \text{ on the top edge } (0 < x < \pi), \tag{7}$$
$$\psi_1(x, y) = 0 \text{ on the other three sides}. \tag{8}$$

*Problem* 2. Find $\psi_2(x, y)$ such that

$$\nabla^2\psi_2(x, y) = 0 \text{ inside the square}, \tag{9}$$
$$\psi_2(\pi, y) = \sin^3 y \text{ on the right edge } (0 < y < \pi), \tag{10}$$
$$\psi_2(x, y) = 0 \text{ on the other three sides}. \tag{11}$$

*Problem* 3. Find $\psi_3(x, y)$ such that

$$\nabla^2\psi_3(x, y) = 0 \text{ inside the square}, \tag{12}$$
$$\psi_3(x, 0) = x(\pi - x) \text{ on the bottom edge } (0 < x < \pi), \tag{13}$$
$$\psi_3(x, y) = 0 \text{ on the other three sides}. \tag{14}$$

*Problem* 4. Find $\psi_4(x, y)$ such that

$$\nabla^2 \psi_4(x, y) = 0 \text{ inside the square}, \tag{15}$$

$$\psi_4(0, y) = f(y) \text{ on the left edge } (0 < y < \pi), \tag{16}$$

$$\psi_4(x, y) = 0 \text{ on the other three sides}. \tag{17}$$

Then we claim that the sum $\psi(x, y) = \psi_1(x, y) + \psi_2(x, y) + \psi_3(x, y) + \psi_4(x, y)$ solves the original problem. After all, on the top edge

$$\psi(x, \pi) = \sin x + 0 + 0 + 0 = \sin x \ ;$$

on the right edge

$$\psi(\pi, y) = 0 + \sin^3 y + 0 + 0 = \sin^3 y \ ;$$

on the bottom

$$\psi(x, 0) = 0 + 0 + x(\pi - x) + 0 = x(\pi - x) \ ;$$

on the left edge

$$\psi(0, y) = 0 + 0 + 0 + f(y) = f(y) \ ;$$

and inside the square

$$\nabla^2 \psi(x, y) = \nabla^2 \psi_1 + \nabla^2 \psi_2 + \nabla^2 \psi_3 + \nabla^2 \psi_4 = 0 + 0 + 0 + 0 = 0 \ .$$

Thus the sum $\psi_1 + \psi_2 + \psi_3 + \psi_4$ does all that it is supposed to do, that is, solve eqs. (1) through (5).

> By *superposition* we have thus decomposed the original problem into a set of subproblems in each of which only one boundary condition is nonhomogeneous.

Let's go to work on Problems 1 through 4 in turn.

SOLUTION OF PROBLEM 1. The basic "gimmick" of the separation of variables method is to try to find solutions $\psi_1(x, y)$ of the underlying differential equation - Laplace's equation (6) in our case - which can be expressed as a product of factors, *each depending on a single variable*:

$$\psi_1(x, y) = X(x)Y(y) \tag{18}$$

If we insert (18) into (6) we can manipulate the latter into a form where only one variable occurs on each side of the equation:

$$\begin{aligned} \nabla^2 \psi_1(x, y) = 0 \ &= \ \frac{\partial^2}{\partial x^2} X(x)Y(y) + \frac{\partial^2}{\partial y^2} X(x)Y(y) \\ &= \ X''(x)Y(y) + X(x)Y''(y) \ ; \end{aligned}$$

thus

$$\frac{X''(x)}{X(x)} = -\frac{Y''(y)}{Y(y)} \tag{19}$$

and we say that we have *separated* the differential equation.

Consider what equation (19) tells us. Since the left hand side depends only on $x$ and there is no "$x$" on the right hand side, *the left hand side must be constant*; after all, if we changed the $x$ value from $x = x_1$ to $x = x_2$, the right hand side would not be affected. Thus the left hand side would not change either. (That's what equality means!) And if a function of $x$ stays fixed when $x$ changes, then that function is constant.

So the separated equation (19) implies that

$$\frac{X''(x)}{X(x)} = \lambda , \ \ or \ \ X'' - \lambda X = 0 \tag{20}$$

for some (as yet) unspecified constant $\lambda$, known as the *separation constant*. And by the same token $Y(y)$ satisfies

$$Y'' + \lambda Y = 0 . \tag{21}$$

Equation (20) is the *harmonic oscillator* equation and its solutions were analyzed in Example 8 of Sec. 1.1:

$$X(x) = a_1 \cosh \sqrt{\lambda}x + a_2 \sinh \sqrt{\lambda}x \qquad \qquad if \ \lambda > 0 , \tag{22}$$
$$X(x) = b_1 + b_2 x \qquad \qquad if \ \lambda = 0 , \tag{23}$$
$$X(x) = c_1 \cos \sqrt{-\lambda}x + c_2 \sin \sqrt{-\lambda}x \qquad \qquad if \ \lambda < 0 . \tag{24}$$

The corresponding solutions for (21) are

$$Y(y) = d_1 \cos \sqrt{\lambda}y + d_2 \sin \sqrt{\lambda}y \qquad \qquad (\lambda > 0) , \tag{25}$$
$$Y(y) = e_1 + e_2 y \qquad \qquad (\lambda = 0) , \tag{26}$$
$$Y(y) = f_1 \cosh \sqrt{-\lambda}y + f_2 \sinh \sqrt{-\lambda}y \qquad \qquad (\lambda < 0) . \tag{27}$$

To summarize:

> **The function $X(x)Y(y)$ will satisfy Laplace's equation (6) as long as $X$ and $Y$ satisfy (20) and (21), respectively, for any value of the constant $\lambda$. In particular, the functions**
>
> $(a_1 \cosh \sqrt{\lambda}x + a_2 \sinh \sqrt{\lambda}x)(d_1 \cos \sqrt{\lambda}y + d_2 \sin \sqrt{\lambda}y) ,$
> $(b_1 + b_2 x)(e_1 + e_2 y) ,$ and
> $(c_1 \cos \sqrt{-\lambda}x + c_2 \sin \sqrt{-\lambda}x)(f_1 \cosh \sqrt{-\lambda}y + f_2 \sinh \sqrt{-\lambda}y)$
>
> **satisfy (6), whatever the values of the constants ($a_1$ through $f_2$ and $\lambda$).**

Now that we have accumulated a wealth of solutions to Laplace's equation (6), we turn to the matter of manipulating the constants therein so as to satisfy the boundary conditions (7, 8) for Problem 1. Notice that $\psi_1(x, y)$ must be zero for $x = 0$ and $x = \pi$, and for $y = 0$. Expressing this in terms of the separated form,

$$X(0)Y(y) = 0 \ , \ X(\pi)Y(y) = 0, \ X(x)Y(0) = 0 \ , \tag{28}$$

we see immediately that the first two constraints can readily be met by imposing them *on the $X(x)$ factor alone,*

$$X(0) = 0 \ , \ X(\pi) = 0 \ ; \tag{29}$$

and similarly the third constraint is met if

$$Y(0) = 0 \ . \tag{30}$$

Observe that this trick - satisfying the boundary conditions by manipulating a *single* factor - only works because of the *zeros* on the right hand sides of (28). A *nonhomogeneous* condition like $X(0)Y(y) = 5$ could *not* be guaranteed through the adjustment of the factor $X(0)$ alone.

Note also that if we tried to satisfy the first equation in (28) by setting the $Y$ factor equal to zero we would be imposing $Y(y) = 0$, *for all $y$*; the solution $X(x)Y(y)$ that we found would thus be *identically* zero. As we shall see, this will be of no help in solving the overall problem. By using the $X$ factor to satisfy (28), we only constrain the solution at a single point ($x = 0$).

Now eqs. (29) give boundary conditions to go with the (*ordinary*) differential equation $X'' - \lambda X = 0$. If we impose the first condition on the general solution forms (22, 23, 24), we learn that the first coefficient ($a_1, b_1, c_1$) must be zero:

$$
\begin{array}{llllll}
0 &=& a_1 \cosh 0 + a_2 \sinh 0 &=& a_1 \ \text{and} \ X(x) &= a_2 \sinh \sqrt{\lambda} x & (\lambda > 0), \\
0 &=& b_1 + b_2 0 &=& b_1 \ \text{and} \ X(x) &= b_2 x & (\lambda = 0), \\
0 &=& c_1 \cos 0 + c_2 \sin 0 &=& c_1 \ \text{and} \ X(x) &= c_2 \sin \sqrt{-\lambda} x & (\lambda < 0).
\end{array}
$$

To fulfill the second boundary condition, then, we must enforce

$$
\begin{array}{lll}
a_2 \sinh \sqrt{\lambda} \pi = 0 & (\lambda > 0); & \\
b_2 \pi = 0 & (\lambda = 0); & \text{and} \\
c_2 \sin \sqrt{-\lambda} \pi = 0 & (\lambda < 0).
\end{array}
\tag{31}
$$

But if we try to satisfy (31) in a "lazy" way by choosing the second coefficient ($a_2, b_2, c_2$) to be zero also, we will end up with the identically zero solution for $X(x)$. Since we need to find *nontrivial* solutions we must utilize the separation constant $\lambda$, rather than the coefficients, to enforce (31).

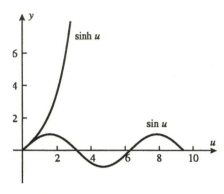

**Figure 3** $\sin u$ and $\sinh u$

Obviously the possibility $\lambda = 0$ can be eliminated, since it dictates $b_2 = 0$ in (31). We can also dismiss $\lambda > 0$ as a possibility, since the sinh function is zero only at the origin (Fig. 3), and this option would force $a_2 = 0$ in (31).

However there are an *infinite number* of *negative* values of $\lambda$ which satisfy (31); namely, the negatives of squares of integers $\lambda = -n^2$:

$$\sin \sqrt{-\lambda}\,\pi = \sin \sqrt{n^2}\,\pi = \sin n\pi = 0 \ .$$

Thus *for $\lambda = -1, -4, -9, -16, ..., -n^2, ...$ we have nontrivial solutions to the system (20, 29) given by the set of functions*

$$
\begin{aligned}
X_1(x) &= \sin x && (\lambda = -1)\\
X_2(x) &= \sin 2x && (\lambda = -4)\\
X_3(x) &= \sin 3x && (\lambda = -9)\\
X_4(x) &= \sin 4x && (\lambda = -16)\\
&\ \ \vdots\\
X_n(x) &= \sin nx && (\lambda = -n^2)\\
&\ \ \vdots
\end{aligned}
\tag{32}
$$

The $y$ function which is associated with $X_n(x) = \sin nx$ through eq. (21) is (from eq. (27))

$$Y_n(y) = f_1 \cosh ny + f_2 \sinh ny \ . \tag{33}$$

According to the boundary condition (30), $Y_n(y)$ must equal zero for $y = 0$. Therefore $f_1$ in (33) must be zero:

$$0 = f_1 \cosh 0 + f_2 \sinh 0 = f_1 \ , \text{ and } Y_n(y) = f_2 \sinh ny \ . \tag{34}$$

Let us once again stop and summarize what has been accomplished. We have concocted an infinite family of solutions to the partial differential equation (6), which also satisfy the three homogeneous boundary conditions (8). These solutions are

$$
\begin{aligned}
\phi_1(x,y) &= \sin x \ \sinh y\\
\phi_2(x,y) &= \sin 2x \ \sinh 2y\\
\phi_3(x,y) &= \sin 3x \ \sinh 3y\\
&\quad\vdots\\
\phi_n(x,y) &= \sin nx \ \sinh ny\\
&\quad\vdots
\end{aligned}
\tag{35}
$$

and any constant multiples of these. (At this point the reader should mentally verify that each function in (35) satisfies eqs. (6) and (8).)

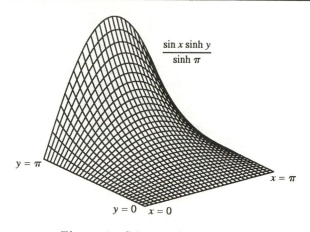

$$\frac{\sin x \sinh y}{\sinh \pi}$$

$y = \pi$

$x = \pi$

$y = 0 \quad x = 0$

*Figure 4    Solution of subproblem 1*

Can we assemble from (35) a solution which also meets the final (non-homogeneous) boundary condition $\psi_1(x, \pi) = \sin x$ (eq. (7))? Of course; it's easy! The very first member of the list (35), multiplied by $1/(\sinh \pi)$, does it:

$$\frac{1}{\sinh \pi} \sin x \, \sinh y \big|_{y=\pi} = \sin x \ .$$

Thus we have completely solved Problem 1 (eqs. (6, 7, 8)); the solution is

$$\psi_1(x, y) = \frac{1}{\sinh \pi} \sin x \sinh y \ . \tag{36}$$

A graph of $\psi_1$ is displayed in Fig. 4.

SOLUTION OF PROBLEM 2. Of course it's not necessary to separate Laplace's equation (9) again; simply by exchanging $x$ and $y$ in the list (35) we obtain a family of solutions to Laplace's equation which are zero on the top, bottom, and left edges:

$$
\begin{aligned}
\phi_1(x, y) &= \sin y \, \sinh x \\
\phi_2(x, y) &= \sin 2y \, \sinh 2x \\
\phi_3(x, y) &= \sin 3y \, \sinh 3x \\
&\vdots \\
\phi_n(x, y) &= \sin ny \, \sinh nx \\
&\vdots
\end{aligned} \tag{37}
$$

Can we assemble from these a solution which equals $\sin^3 y$ on the right edge (eq. (10))? Again it's easy, if we recall the trigonometric identity

$$\sin^3 y = \frac{3}{4} \sin y - \frac{1}{4} \sin 3y \tag{38}$$

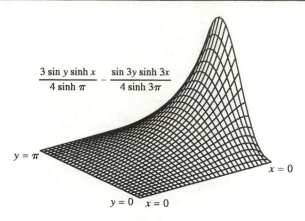

$$\frac{3\sin y \sinh x}{4\sinh \pi} - \frac{\sin 3y \sinh 3x}{4\sinh 3\pi}$$

$y = \pi$

$x = 0$

$y = 0 \quad x = 0$

**Figure 5**   *Solution of subproblem 2*

Extracting the coefficients from (38), we see that the combination

$$\psi_2(x,y) = \frac{3/4}{\sinh \pi}\sin y \sinh x - \frac{1/4}{\sinh 3\pi}\sin 3y \sinh 3x \qquad (39)$$

satisfies eqs. (9, 10, 11) and thus solves Problem 2. See Fig. 5.

SOLUTION OF PROBLEM 3. If $y$ is replaced by $\pi - y$ in (35), the result is a family of solutions to Laplace's equation which are zero on the top, left, and right edges:

$$
\begin{aligned}
\phi_1(x,y) &= \sin x \ \sinh(\pi - y)\\
\phi_2(x,y) &= \sin 2x \ \sinh 2(\pi - y)\\
\phi_3(x,y) &= \sin 3x \ \sinh 3(\pi - y)\\
&\ \ \vdots\\
\phi_n(x,y) &= \sin nx \ \sinh n(\pi - y)\\
&\ \ \vdots
\end{aligned}
\qquad (40)
$$

(The reader should be sure that this point is understood. It is obvious that the functions in (40) are zero on the designated edges, but the claim that they still solve Laplace's equation is due to the fact that this equation does not change form when the transformation

$$y = y(\eta) = \pi - \eta \ , \ \eta = \eta(y) = \pi - y \qquad (41)$$

is made; recall Sec. 2.2. This mathematical transformation is the physical equivalent of measuring "$y$" from the top edge.)

Do we have a trick up our sleeve which will enable us to satisfy the condition $\psi_3(x,0) = x(\pi - x)$ (eq. (13))? Yes we do! The function $x(\pi - x)$

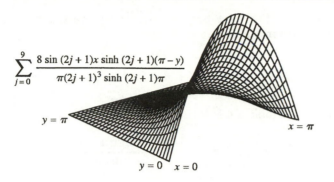

$$\sum_{j=0}^{9} \frac{8 \sin (2j+1)x \sinh (2j+1)(\pi-y)}{\pi(2j+1)^3 \sinh (2j+1)\pi}$$

**Figure 6** *Solution of subproblem 3*

has a *Fourier sine series* expansion which looks like

$$x(\pi - x) = \sum_{n=1}^{\infty} a_n \sin nx \ , \ 0 < x < \pi \ . \tag{42}$$

(In fact the coefficients are given by

$$a_n = \frac{\int_0^\pi x(\pi-x)\sin nx\,dx}{\int_0^\pi \sin^2 nx\,dx} = \left\{ \begin{array}{ll} 0 & , \ n=2j \\ \frac{8}{\pi(2j+1)^3} & , \ n=2j+1; \end{array} \right. \tag{43}$$

see Fig. 5, Sec. 3.3.) Thus using (40, 42, 43) we can assemble a solution to eqs. (12, 13, 14):

$$
\begin{aligned}
\psi_3(x,y) &= \sum_{n=1}^{\infty} a_n \frac{\sin nx \sinh n(\pi-y)}{\sinh n\pi} \\
&= \sum_{j=0}^{\infty} \frac{8}{\pi(2j+1)^3 \sinh(2j+1)\pi}\ \sin(2j+1)x \sinh(2j+1)(\pi-y)
\end{aligned}
$$

$$\tag{44}$$

Fig. 6 depicts the partial sum of the first ten terms of (44).

SOLUTION OF PROBLEM 4. By now the reader should see the general strategy. The family of solutions vanishing on the top, bottom, and right edge is

$$
\begin{aligned}
\phi_1(x,y) &= \sin y \sinh(\pi-x) \\
\phi_2(x,y) &= \sin 2y \sinh 2(\pi-x) \\
\phi_3(x,y) &= \sin 3y \sinh 3(\pi-x) \\
&\ \ \vdots \\
\phi_n(x,y) &= \sin ny \sinh n(\pi-x) \\
&\ \ \vdots
\end{aligned}
$$

$$\tag{45}$$

To assemble a solution $\psi_4(x, y)$ which equals $f(y)$ on the left edge ($x = 0$), first expand $f(y)$ in a sine series:

$$f(y) = \sum_{n=1}^{\infty} a_n \sin nx \ , a_n = \frac{2}{\pi} \int_0^{\pi} f(y) \sin ny \, dy \ . \tag{46}$$

The solution of Problem 4 is then

$$\psi_4(x, y) = \sum_{n=1}^{\infty} a_n \frac{\sin ny \sinh n(\pi - x)}{\sinh n\pi} \tag{47}$$

Assembling the forms (36, 39, 44, 47) completes the solution of the temperature distribution problem! The final answer is

$$
\begin{aligned}
\psi(x, y) \ = \ & \frac{1}{\sinh \pi} \sin x \sinh y \\
+ \ & \frac{3/4}{\sinh \pi} \sin y \sinh x - \frac{1/4}{\sinh 3\pi} \sin 3y \sinh 3x \\
+ \ & \sum_{j=0}^{\infty} \frac{8}{\pi(2j+1)^3 \sinh(2j+1)\pi} \sin(2j+1)x \sinh(2j+1)(\pi - y) \\
+ \ & \sum_{n=1}^{\infty} \frac{2}{\pi} \{ \int_0^{\pi} f(y) \sin ny \, dy \} \frac{\sin ny \sinh n(\pi - x)}{\sinh n\pi} \tag{48}
\end{aligned}
$$

---

That's the method of separation of variables for solving partial differential equations. In future sections we will extend it to more complicated differential equations and boundary conditions, higher dimensions, time dependent equations, and curvilinear coordinates - it's a very powerful technique. But the basic ideas are contained in this example. Let's review the salient features which enabled a successful execution of the solution to this problem.

I. The underlying partial differential equation (5) "separated" when a trial solution form, consisting of factors depending on only one variable each, was inserted (18, 19). Thus solutions to Laplace's equation could be constructed from solutions to the resulting *ordinary* differential equations - each of which contained an unspecified parameter, the separation constant (20, 21). *This is the essence of "separation."*

II. The underlying partial differential equation (5) was *linear*. This came into play twice; first when we decomposed the problem into simpler subproblems, and again when we assembled the solutions to the subproblems by taking linear combinations of the separated solutions (39, 44, 47).

III. The boundary conditions were imposed on constant-coordinate curves, that is on the curves $x = 0, x = \pi, y = 0$, and $y = \pi$. Furthermore the subproblems were defined so that these boundary values were *zero* on all but one of the boundary curves. This enabled us to meet the homogeneous condition along any particular boundary curve simply by imposing it at a single *point* in one of the separated factors (29, 30).

IV. Of the two ODEs that resulted from the separation (20, 21), we *first* analyzed the one $(X)$ which had homogeneous boundary values at *both* ends (29). The differential equation itself had a general solution (22, 23, 24) containing three unspecified constants: the customary coefficients $(a_1, a_2; b_1, b_2;$ or $c_1, c_2)$, and the separation constant $(\lambda)$. In order to obtain nontrivial solutions to the problem we used one of the *coefficients* to meet the boundary condition at one end, but we used the *separation constant* to satisfy the condition at the other end (31). This resulted in an infinite family of solutions, which we later identified as the functions generating the Fourier sine series (32).

V. The homogeneous boundary condition for the *second* separated ODE was satisfied by choice of the coefficients in its general solution (34).

VI. The (final) nonhomogeneous boundary condition for the subproblem was met by assembling all the separated solutions (35) and exploiting the generality of the Fourier sine series expansion (46, 47).

Clearly the tricky step was number IV, where we were lucky not only to find nontrivial solutions to a homogeneous boundary value problem (recall the fourth example in Sec. 1.1), but so many of them that we could expand an *arbitrary* function $(\sin x, \sin^3 y, x[\pi - x], f(y))$ in terms of them!

> The beauty of the method of separation of variables is that step IV will *always* work out this way - the separated second order ordinary differential equation, containing an unspecified parameter and constrained by homogeneous boundary conditions at both ends, will inevitably generate a *complete orthogonal* family of solutions.[1] This aspect of ordinary differential equations is known as the *Sturm-Liouville* theory, and before we take up the subject of partial differential equations in full force we will dedicate a chapter to the development of this important tool.

Let us finish out this section with a few elementary examples demonstrating other boundary conditions. As you study them, refer to the scheme I-VI outlined above and note how each step comes into play.

**Example 1.** The problem indicated in Fig. 7 differs from Problem 1 only in that the Dirichlet condition at the top edge has been replaced by a Neumann condition. Therefore once again we employ a linear combination of the solutions satisfying the other three, homogeneous, conditions (35)

$$\psi(x, y) = \sum_{n=1}^{\infty} b_n \sin nx \sinh ny$$

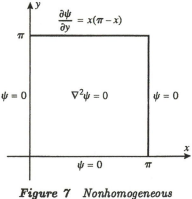

*Figure 7  Nonhomogeneous Neumann condition*

---

[1] "Orthogonality" was defined in Sec. 3.3. "Completeness" means that the family is rich enough to generate any (reasonable) function through its linear combinations.

and impose the nonhomogeneous condition on the top edge. We have

$$
\begin{array}{ccc}
\frac{\partial \psi}{\partial y}(x,\pi) & = & x(\pi - x) \\
\| & & \| \\
\sum_{n=1}^{\infty} n b_n \sin nx \cosh n\pi & = & \sum_{j=0}^{\infty} \frac{8 \sin(2j+1)x}{\pi(2j+1)^3}
\end{array}
$$

(from (42, 43)). Solving for $b_n$, we find that the solution is

$$
\psi(x,y) = \sum_{j=0}^{\infty} \frac{8 \sin(2j+1)x \sinh(2j+1)y}{\pi(2j+1)^4 \cosh(2j+1)\pi}
$$

■

**Example 2.** The problem in Fig. 8 is governed by Laplace's equation (5), three homogeneous Neumann conditions

$$
\frac{\partial \psi}{\partial x}(0,y) = 0, \frac{\partial \psi}{\partial x}(\pi,y) = 0, \frac{\partial \psi}{\partial y}(x,0) = 0, \tag{49}
$$

and a nonhomogeneous Robin condition

$$
\frac{\partial \psi}{\partial y}(x,\pi) = \alpha\{\psi(x,\pi) - T_0\} . \tag{50}
$$

According to the exposition in Sec. 4.5, $\psi$ can be interpreted as the steady state temperature in a slab insulated on three sides, but coupled through "leaky" insulation on the fourth side to the environment, whose temperature equals $T_0$.[2]

Since the previous collection of solutions (35) satisfies homogeneous *Dirichlet* conditions, we must derive a new set of solutions for the Neumann condition (49). Recall that earlier we saw the insertion of the separated form

$$
\psi(x,y) = X(x)Y(y) \tag{51}
$$

into Laplace's equation (5) leads to ordinary differential equations and general solutions of the form

$$
X'' - \lambda X = 0
$$

$$
X(x) = \begin{cases} a_1 \cosh\sqrt{\lambda}x + a_2 \sinh\sqrt{\lambda}x, & \lambda > 0 \\ b_1 + b_2 x, & \lambda = 0 \\ c_1 \cos\sqrt{-\lambda}x + c_2 \sin\sqrt{-\lambda}x, & \lambda < 0 \end{cases} \tag{52}
$$

$$
Y'' + \lambda Y = 0
$$

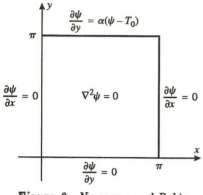

$$
\frac{\partial \psi}{\partial y} = \alpha(\psi - T_0)
$$

$$
\nabla^2 \psi = 0
$$

$$
\frac{\partial \psi}{\partial x} = 0
$$

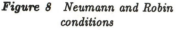

$$
\frac{\partial \psi}{\partial y} = 0
$$

***Figure 8*** ***Neumann and Robin conditions***

---

[2]How good is your physical intuition? Can you predict what the steady state temperature in the slab will be?

$$Y(y) = \begin{cases} d_1 \cos \sqrt{\lambda} y + d_2 \sin \sqrt{\lambda} y, & \lambda > 0 \\ e_1 + e_2 y, & \lambda = 0 \\ f_1 \cosh \sqrt{-\lambda} y + f_2 \sinh \sqrt{-\lambda} y, & \lambda < 0 \end{cases} \tag{53}$$

The first homogeneous condition in (49) is most conveniently met by imposing $X'(0) = 0$, or

$$X'(0) = 0 = \begin{cases} \sqrt{\lambda} a_1 \sinh 0 + \sqrt{\lambda} a_2 \cosh 0 = \sqrt{\lambda} a_2 \\ b_2 \\ -\sqrt{-\lambda} c_1 \sin 0 + \sqrt{-\lambda} c_2 \cos 0 = \sqrt{-\lambda} c_2 \end{cases}$$

$$\text{so} \quad X(x) = \begin{cases} a_1 \cosh \sqrt{\lambda} x, & \lambda > 0 \\ b_1 & \lambda = 0 \\ c_1 \cos \sqrt{-\lambda} x, & \lambda < 0 \end{cases} \tag{54}$$

Imposition of the second condition in (49) implies

$$\begin{array}{rcll} \sqrt{\lambda} a_1 \sinh \sqrt{\lambda} \pi & = & 0 & (\lambda > 0) \\ 0 & = & 0 & (\lambda = 0) \\ -\sqrt{-\lambda} c_1 \sin \sqrt{-\lambda} \pi & = & 0 & (\lambda < 0) \end{array} \tag{55}$$

Since we are not interested in trivial solutions, we cannot set $a_1, b_1$, or $c_1$ equal to zero; we need to satisfy (55) by the selection of $\lambda$.

*The choice $\lambda = 0$ is an acceptable one in this case*, since the second equation in (55) obviously works; the corresponding solution, from eq. (54), is $X(x) = b_1$ (= constant).

No positive values for $\lambda$ yield solutions, because the sinh function in (55) never vanishes. However the sin function vanishes whenever $\sqrt{-\lambda} = n = 1, 2, 3, \ldots$, and the corresponding solution from (54) is $c_1 \cos nx$. So for $\lambda = 0, -1, -4, \ldots, -n^2, \ldots$ we have nontrivial solutions for the $x$ factor in eq. (51) given by constant multiples of the following:

$$\begin{array}{rcll} X_0(x) & = & 1 \ (= \cos 0x) & (\lambda = 0) \\ X_1(x) & = & \cos x & (\lambda = -1) \\ X_2(x) & = & \cos 2x & (\lambda = -4) \\ X_3(x) & = & \cos 3x & (\lambda = -9) \\ X_4(x) & = & \cos 4x & (\lambda = -16) \\ & \vdots & & \\ X_n(x) & = & \cos nx & (\lambda = -n^2) \\ & \vdots & & \end{array} \tag{56}$$

(and we anticipate the utilization of the terms of the Fourier cosine series).

The corresponding $y$ factors, from eq. (53), are $Y_0(y) = e_1 + e_2 y$ (for $\lambda = 0$) and $Y_n(y) = f_1 \cosh ny + f_2 \sinh ny$ ($n = 1, 2, 3, \ldots$). If we impose the third homogeneous boundary condition in (49) on these factors, they reduce to $Y_0(y) = e_1$, $Y_n(y) = f_1 \cosh ny$. Therefore the total collection of solutions to all the homogeneous equations in this problem is given by

$$
\begin{aligned}
\phi_0(x,y) &= 1 \; (= \cos 0x \cosh 0y) \\
\phi_1(x,y) &= \cos x \cosh y \\
\phi_2(x,y) &= \cos 2x \cosh 2y \\
\phi_3(x,y) &= \cos 3x \cosh 3y \\
&\vdots \\
\phi_n(x,y) &= \cos nx \cosh ny \\
&\vdots
\end{aligned}
\tag{57}
$$

Now we seek to satisfy the remaining, nonhomogeneous, boundary condition (50) with a linear combination of the functions in (57):

$$
\psi(x,y) = \sum_{n=0}^{\infty} a_n \cos nx \cosh ny \; .
\tag{58}
$$

Insertion of (58) into (50) yields

$$
\sum_{n=0}^{\infty} n a_n \cos nx \sinh n\pi = \alpha \sum_{n=0}^{\infty} a_n \cos nx \cosh n\pi - \alpha T_0 \; ,
$$

or

$$
\sum_{n=0}^{\infty} [n \sinh n\pi - \alpha \cosh n\pi] a_n \cos nx = -\alpha T_0 \; .
\tag{59}
$$

*Equation (59) has the form of a cosine series* and it will be satisfied if the coefficients on the left, $[n \sinh n\pi - \alpha \cosh n\pi] a_n$, match the coefficients in the expansion of the constant function "$-\alpha T_0$". According to Sec. 3.5, then, we can exploit the orthogonality relations to derive

$$
[0 \sinh 0\pi - \alpha \cosh 0\pi] a_0 = \frac{1}{\pi} \int_0^{\pi} (-\alpha T_0) \, dx = -\alpha T_0 \; (n=0)
$$

$$
[n \sinh n\pi - \alpha \cosh n\pi] a_n = \frac{2}{\pi} \int_0^{\pi} (-\alpha T_0) \cos nx \, dx = 0 \; (n > 0);
$$

i.e., $a_0 = (+)\alpha T_0 / \alpha = T_0$, and $a_n = 0/[n \sinh n\pi - \alpha \cosh n\pi] = 0$. The solution, in accordance with (58), is simply $\psi(x,y) = T_0$. (!) ∎

Did you anticipate this? Physically speaking, one would expect the equilibrium temperature to be uniform if the slab were insulated on *all* sides. Since the insulation is leaky on the top edge, equilibrium is achieved when the interior temperature is uniform and matches that of the environment.

**Example 3.** Let's put nonhomogeneous Robin conditions at both ends of the slab, as in Fig. 9.

The first step is to define the subproblems possessing one nonhomogeneity each. As we have seen, the homogeneous form associated with a Dirichlet condition $\psi = f$ is, of course, $\psi = 0$; and for the Neumann condition $\frac{\partial \psi}{\partial n} = f$, it is $\frac{\partial \psi}{\partial n} = 0$. But the homogeneous equation associated with a Robin condition $\frac{\partial \psi}{\partial n} = \alpha \psi + f$ is $\frac{\partial \psi}{\partial n} = \alpha \psi$. It is *not* $\frac{\partial \psi}{\partial n} = 0$ (!); this is a common error. The operator notation for the Robin condition (recall Sec. 1.1) is $\mathcal{L}\psi = f$, with $\mathcal{L}\psi = \frac{\partial \psi}{\partial n} - \alpha \psi$. Thus the associated homogeneous equation, $\mathcal{L}\psi = 0$, takes the form prescribed.

As a result, the subproblems are as depicted in Fig. 10. Observe how the sum $\psi = \psi_1 + \psi_2$ solves the Robin boundary condition at $y = \pi$:

$$
\begin{aligned}
\frac{\partial \psi}{\partial y} - \alpha \psi &= \frac{\partial(\psi_1 + \psi_2)}{\partial y} - \alpha(\psi_1 + \psi_2) = \frac{\partial \psi_1}{\partial y} - \alpha\psi_1 + \frac{\partial \psi_2}{\partial y} - \alpha\psi_2 \\
&= x + 0 = x \, .
\end{aligned}
$$

(Note in particular that if we had employed, say, $\partial\psi_2/\partial y = 0$ as the associated homogeneous form, we would not have been able to confirm the Robin condition for $(\psi_1 + \psi_2)$.)

To solve the subproblem in Fig. 10(a), we call upon the analysis of the previous example to note that the separated forms

$$
X(x)Y(y) = \begin{cases} \cos nx \, \{f_1 \cosh ny + f_2 \sinh ny\} & (n > 0) \\ (1) \, \{e_1 + e_2 y\} & (n = 0) \end{cases}
$$

already satisfy the differential equation and the Neumann conditions on the left and right. We must therefore choose the coefficients $e_1, e_2$ and $f_1, f_2$ to fulfill the homogeneous Robin condition at $y = 0$:

$$
Y'(0) - \alpha Y(0) = 0 = \begin{cases} n f_2 - \alpha f_1 & (n > 0) \\ e_2 - \alpha e_1 & (n = 0) \end{cases}
$$

Selecting $f_1 = 1$, $f_2 = \alpha/n$, $e_1 = 1$, $e_2 = \alpha$ we arrive at the following collection of solutions, meeting all of the homogeneous conditions of the subproblem:

$$
\begin{aligned}
\phi_0(x, y) &= 1 + \alpha y \\
\phi_1(x, y) &= \cos x \, [\cosh y + \alpha \sinh y] \\
\phi_2(x, y) &= \cos 2x \, [\cosh 2y + (\alpha/2) \sinh 2y] \\
\phi_3(x, y) &= \cos 3x \, [\cosh 3y + (\alpha/3) \sinh 3y] \\
&\vdots \\
\phi_n(x, y) &= \cos nx \, [\cosh ny + (\alpha/n) \sinh ny] \\
&\vdots
\end{aligned}
$$

$$(60)$$

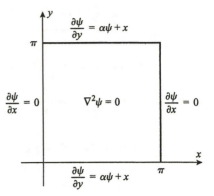

**Figure 9**  *Two Robin conditions*

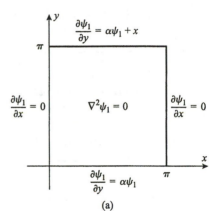

(a)

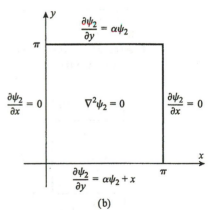

(b)

**Figure 10**  *Decomposition for Fig. 9*

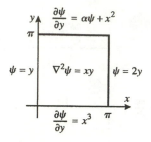

**Figure 11   Example 4**

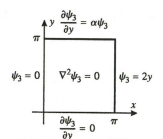

**Figure 12   Subproblem for Example 4**

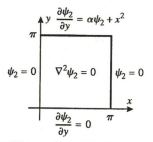

**Figure 13   Subproblem for Example 4**

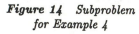

**Figure 14   Subproblem for Example 4**

To satisfy the nonhomogeneous condition on the top edge with a linear combination of these we form

$$\psi(x,y) = a_0(1 + \alpha y) + \sum_{n=1}^{\infty} a_n \cos nx \left[\cosh ny + (\alpha/n) \sinh ny\right] \qquad (61)$$

and insert into the boundary condition to find

$$a_0 \alpha \quad + \quad \sum_{n=1}^{\infty} a_n \cos nx \left[n \sinh n\pi + \alpha \cosh n\pi\right]$$

$$= \quad \alpha a_0(1 + \alpha \pi) + \sum_{n=1}^{\infty} \alpha\, a_n \cos nx \left[\cosh n\pi + (\alpha/n) \sinh n\pi\right] + x$$

or

$$-\alpha^2 a_0 \pi + \sum_{n=1}^{\infty} a_n \cos nx \,(n - \alpha^2/n) \sinh n\pi = x \;. \qquad (62)$$

This has the form of a cosine series expansion of the function $x$, and the coefficient of $\cos nx$ can be extracted by orthogonality:

$$-\alpha^2 a_0 \pi = \frac{1}{\pi} \int_0^{\pi} x\, dx = \frac{\pi}{2} \;,\; \text{ or } \; a_0 = \frac{-1}{2\alpha^2} \;;$$

$$a_n(n - \alpha^2/n) \sinh n\pi = \frac{2}{\pi} \int_0^{\pi} x \cos nx \, dx = \left\{ \begin{array}{ll} 0 & n \text{ even} \\ -4/\pi n^2 & n \text{ odd} \end{array} \right.$$

$$\text{or} \quad a_{n\,(odd)} = \frac{-4}{\pi n^2 (n - \alpha^2/n) \sinh n\pi} \;.$$

The result is the following expression for the solution to the first subproblem:

$$\psi_1(x,y) = \frac{-1}{2\alpha^2}(1 + \alpha y) + \sum_{(odd\ n)} \frac{-4}{\pi} \frac{\cos nx \left[\cosh ny + (\alpha/n) \sinh ny\right]}{n^2(n^2 - \alpha^2/n) \sinh n\pi} \;.$$

The solution to the second subproblem is left as exercise 10.  ∎

If the partial differential equation itself is nonhomogeneous - for example if Laplace's equation $\nabla^2 \psi = 0$ is replaced by *Poisson's* equation $\nabla^2 \psi = \rho(x,y)$ - we first create a subproblem containing this nonhomogeneity but with all boundary conditions homogeneous, and then derive subproblems with single nonhomogeneities on the boundary as before. An example will make this clear.

**Example 4.** The problem depicted in Fig. 11 is decomposed into the sub-problems indicated in Figs. 12 through 16.

Problems of the sort indicated in Fig. 12, where all the boundary conditions are homogeneous but the partial differential equation is nonhomogeneous, are solved using *Green's functions.* They will be discussed in Chapter 8. ∎

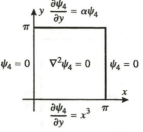

*Figure 15  Subproblem for Example 4*

# Exercises 5.2

1. Use the accumulated set of solutions (35) to solve the problem indicated in Fig. 17 with the boundary condition on the top edge given by

   (a) $\psi(x, \pi) = 4 \sin 3x$   (b) $\psi(x, \pi) = 3 \sin 2x - 2 \sin 3x$
   (c) $\psi(x, \pi) = 0$
   (d) $\psi(x, \pi) = \sin^5 x$ (*Hint*: $\sin^5 x = \frac{5}{8} \sin x - \frac{5}{16} \sin 3x + \frac{1}{16} \sin 5x$)
   (e) $\frac{\partial \psi}{\partial y}(x, \pi) = \sin x$   (f) $\frac{\partial \psi}{\partial y}(x, \pi) = 3 \sin 2x - 2 \sin 3x$

   Verify your answers directly.

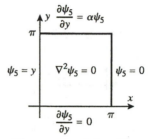

*Figure 16  Subproblem for Example 4*

2. Trace the steps described in the text to find the solution to Laplace's equation in the $\pi$-by-$\pi$ square with the (Neumann) boundary conditions

   $$\frac{\partial \psi}{\partial x}(0, y) = \frac{\partial \psi}{\partial x}(\pi, y) = \frac{\partial \psi}{\partial y}(x, 0) = 0$$

   on the left, right, and lower edges; and on the upper edge,

   (a) $\psi(x, \pi) = \cos x$

   (b) $\psi(x, \pi) = 1$

   (c) $\psi(x, \pi) = 0$

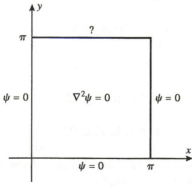

*Figure 17  Exercise 1*

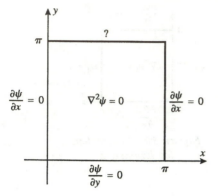

Figure 18   Exercise 2

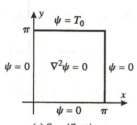

(a) Specifications

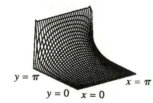

(b) Solution

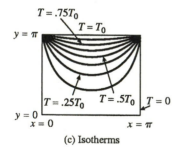

(c) Isotherms

Figure 19   Exercise 4

(d) $\psi(x, \pi) = 3\cos 2x - 2\cos 3x$

(e) $\frac{\partial \psi}{\partial y}(x, \pi) = 3\cos 2x - 2\cos 3x$

(Fig. 18). Verify your answers directly.

3. Use the theory developed in Sec. 1.2 to show that if $\psi(x, y)$ is a solution to Laplace's equation, so is $\psi(a \pm x, b \pm y)$.

4. Derive the representation

$$\psi(x, y) = \frac{4T_0}{\pi}\left\{\frac{\sin x \sinh y}{\sinh \pi} + \frac{\sin 3x \sinh 3y}{3\sinh 3\pi} + \frac{\sin 5x \sinh 5y}{5\sinh 5\pi} + \cdots\right\}$$

to the problem depicted in Fig. 19(a). The solution is illustrated in Fig. 19(b), which was plotted by taking the first 80 terms of the series. The first term looks like Fig. 4 in the text. The *isotherms* (interpreting $\psi$ as temperature) are drawn in Fig. 19(c). Note the discontinuities in temperature at the upper corners. They are dictated by the boundary conditions; the specified values of $\psi$ are incompatible at $(0, \pi)$ and at $(\pi, \pi)$. If $\psi$ represents temperature or voltage, the edges have to be insulated from each other at these corners.

5. Solve the problem depicted in Fig. 20.

6. Solve the problem depicted in Fig. 21.

7. Solve the problem depicted in Fig. 22. (You may need to reread the discussion in Sec. 4.6 on the pure Neumann problem.)

8. Solve the problem depicted in Fig. 23.

9. Solve the problem depicted in Fig. 24.

10. Complete the solution to Example 3.

11. Solve the problem depicted in Fig. 25.

12. Solve the problem depicted in Fig. 26.

13. Solve the problem depicted in Fig. 27.

14. Define the subproblems into which each of the problems in Figs. 28 through 30 must be decomposed.

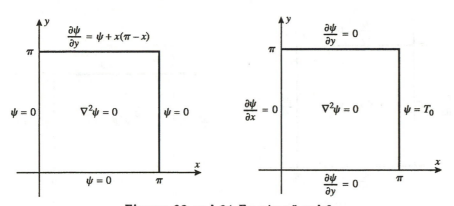

*Figures 20 and 21 Exercises 5 and 6*

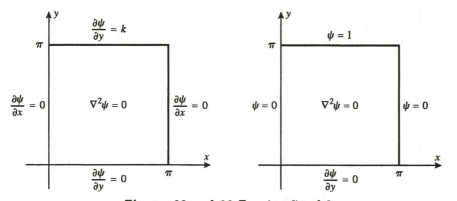

*Figures 22 and 23 Exercises 7 and 8*

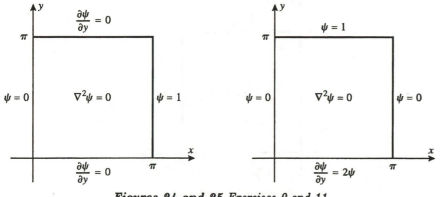

*Figures 24 and 25 Exercises 9 and 11*

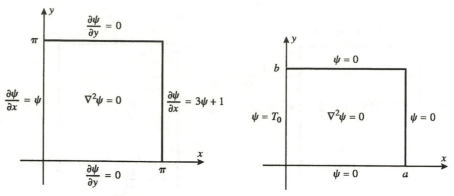

Figures 26 and 27 Exercises 12 and 13

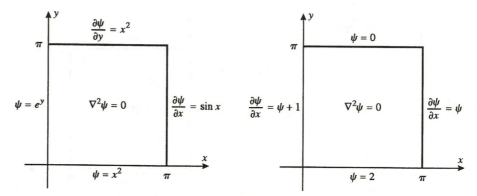

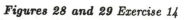

Figures 28 and 29 Exercise 14

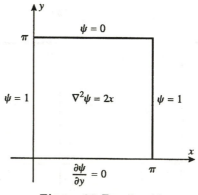

Figure 30 Exercise 14

# 5.3 Separation of the Classical Equations

**Summary** Separating out the time dependence from the wave and heat equations results in a generalization of Laplace's equation known as the Helmholtz equation. The resulting time dependence is either sinusoidal or exponential. The Helmholtz equation separates in cartesian, cylindrical, and spherical coordinates, generating the ordinary differential equations of Bessel and Legendre, as well as the harmonic oscillator and equidimensional equations. Separation of Schrödinger's equation also generates the equations of Airy, Weber, Hermite, Balmer, and Laguerre. Each of these ordinary differential equations can be cast in the form $a_2(x)y'' + a_1(x)y' + a_0(x)y = \lambda g(x)y$, where $\lambda$ is the separation constant for the (generic) variable $x$.

In the preceding sections we saw examples of how separating variables in a partial differential equation can lead to *ordinary* differential equations, which can be exploited to construct a general solution to the original equation. The power of this technique lies in the fact that so many of the classic equations of physics *can* be separated. In this section we shall focus on the separation process, taking note of the types of ordinary differential equations which arise in separating the classical partial differential equations. Chapter 6 will be devoted to characterizing the solutions of these ordinary differential equations. After that we will return to the original partial differential equations, and solve them.

We can achieve some economy of effort by taking advantage of a similarity among the wave equation

$$\nabla^2 \Psi = \frac{\partial^2 \Psi}{\partial t^2} \tag{1}$$

the heat equation

$$\nabla^2 \Psi = \frac{\partial \Psi}{\partial t} \tag{2}$$

and Laplace's equation

$$\nabla^2 \Psi = 0 \tag{3}$$

Many physical systems are governed by these equations, together with boundary conditions imposed on cylindrical or spherical surfaces. Hence it behooves us to investigate the possibility of separating these equations in cylindrical $(\rho, \theta, z)$ and spherical $(r, \theta, \phi)$ coordinates, as well as cartesian.

If we look for separated solutions of the form

$$\Psi = T(t)\psi(...) \tag{4}$$

where $\psi(...)$ is a function only of the spatial variables - $(x, y, z)$, $(\rho, \theta, z)$, or $(r, \theta, \phi)$ - then insertion of (4) into the wave equation (1) produces

$$T\nabla^2\psi = \psi T''$$

and division by $T\psi$ separates out the time dependence:

$$\frac{\nabla^2\psi}{\psi} = \frac{T''}{T} \tag{5}$$

Since the right hand side depends only on $t$ while the left side is independent of $t$, both sides must be constant:

$$\frac{\nabla^2\psi}{\psi} = K \ or \ \nabla^2\psi = K\psi \tag{6}$$

and

$$\frac{T''}{T} = K \ or \ T'' = KT \ . \tag{7}$$

Equation (6) is known as *Helmholtz's equation*, and we shall separate it further. The time equation (7) is simply the harmonic oscillator equation and its solutions are

$$T(t) = \begin{cases} c_1\cosh(\sqrt{K}t) + c_2\sinh(\sqrt{K}t) & \textit{if } K \textit{ is positive,} \\ c_1 + c_2 t & \textit{if } K \textit{ is zero, and} \\ c_1\cos(\sqrt{-K}t) + c_2\sin(\sqrt{-K}t) & \textit{if } K \textit{ is negative.} \end{cases} \tag{8}$$

If the same form $\Psi = T(t)\psi(...)$ is substituted into the heat equation (2), the latter separates into

$$\frac{\nabla^2\psi}{\psi} = \frac{T'}{T}$$

Thus we get

$$\frac{T'}{T} = K \ or \ T' = KT \tag{9}$$

for the time dependence and Helmholtz's equation (6) again for $\psi$. The solutions to (9) are

$$T(t) = ce^{Kt}. \tag{10}$$

Laplace's equation (3) is simply Helmholtz's equation (6) with the parameter $K$ equal to zero. Therefore

*the separated solutions to the wave and heat equations are obtained from solutions to Helmholtz's equation*

$$\nabla^2\psi = K\psi \qquad \textit{(6 repeated)}$$

*by affixing the time factors (8) or (10) respectively. The solutions to Laplace's equation are obtained from solutions to Helmholtz's equation by setting K equal to zero.*

Having disposed of the time factor, we turn to the task of separating Helmholtz's equation in the various coordinate systems.

**Example 1.** The three dimensional Helmholtz equation in cartesian coordinates is

$$\frac{\partial^2 \psi}{\partial x^2} + \frac{\partial^2 \psi}{\partial y^2} + \frac{\partial^2 \psi}{\partial z^2} = K\psi \tag{11}$$

If we attempt to find solutions of the form

$$\psi(x, y, z) = X(x)Y(y)Z(z)$$

then we are led to the condition

$$X''YZ + XY''Z + XYZ'' = KXYZ \ .$$

Dividing by $XYZ$ we can separate, say, the $x$ dependence by arranging as follows:

$$\frac{X''}{X} = -\frac{Y''}{Y} - \frac{Z''}{Z} + K \ .$$

Since the left hand side depends only on $x$ while the right hand side is independent of $x$, both sides must be constant:

$$\frac{X''}{X} = \lambda \text{ or } X'' = \lambda X$$

(the harmonic oscillator equation), and

$$-\frac{Y''}{Y} - \frac{Z''}{Z} = \lambda - K \ . \tag{12}$$

Obviously we can also separate out the y dependence in (12) and conclude

$$Y'' = \mu Y \text{ and } Z'' = \nu Z$$

for any values of the separation constants $\lambda$, $\mu$, and $\nu$ as long as

$$\lambda + \mu + \nu = K$$

(because of (12)).                                                      ∎

**Example 2.** Helmholtz's equation in (two dimensional) polar coordinates takes the form

$$\frac{1}{r}\frac{\partial}{\partial r}\left(r\frac{\partial \psi}{\partial r}\right) + \frac{1}{r^2}\frac{\partial^2 \psi}{\partial \theta^2} = K\psi \tag{13}$$

Inserting the separated trial solution $\psi(r, \theta) = R(r)\Theta(\theta)$ we have

$$\frac{1}{r}(rR')'\Theta + \frac{1}{r^2}R\Theta'' = KR\Theta$$

and a little algebra effects the separation of $\Theta$:

$$\frac{\Theta''}{\Theta} = -\frac{r}{R}(rR')' + r^2 K$$

The $\theta$ equation is the harmonic oscillator equation again

$$\frac{\Theta''}{\Theta} = \lambda \text{ or } \Theta'' = \lambda\Theta$$

When $K = 0$ (Laplace's equation) the radial equation is *equidimensional*:

$$-\frac{r}{R}(rR')' = \lambda \text{ or } r^2 R'' + rR' + \lambda R = 0 .$$

For nonzero $K$ a form of *Bessel's equation* results:

$$-\frac{r^2 R'' + rR'}{R} + r^2 K = \lambda$$

or

$$r^2 R'' + rR' + (-Kr^2 + \lambda)R = 0 . \qquad \blacksquare$$

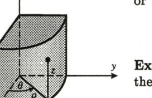

**Figure 1** *Polar
coordinate system*

**Example 3.** Helmholtz's equation in cylindrical coordinates (Fig. 1) takes the form

$$\nabla^2\psi(\rho, \theta, z) = \frac{1}{\rho}\frac{\partial}{\partial\rho}\left(\rho\frac{\partial\psi}{\partial\rho}\right) + \frac{1}{\rho^2}\frac{\partial^2\psi}{\partial\theta^2} + \frac{\partial^2\psi}{\partial z^2} = K\psi \qquad (14)$$

We insert $\psi = R(\rho)\Theta(\theta)Z(z)$ to obtain

$$\frac{(\rho R')'}{\rho}\Theta Z + \frac{R}{\rho^2}\Theta'' Z + R\Theta Z'' = KR\Theta Z \qquad (15)$$

and we isolate $\Theta$,[3]

$$\frac{\Theta''}{\Theta} = -\frac{\rho}{R}(\rho R')' - \rho^2\frac{Z''}{Z} + \rho^2 K = constant = \lambda , \qquad (16)$$

producing the harmonic oscillator equation again:

$$\Theta'' = \lambda\Theta .$$

---

[3]The reader should experiment with (15) to be convinced that the $\rho$ dependence can *not* be separated out at this point, though it will be possible after we have separated $\Theta$! Success at separating variables can depend on the order in which the separations are attempted.

Next we separate out the z dependence from (16):

$$\frac{Z''}{Z} = K - \frac{\lambda}{\rho^2} - \frac{(\rho R')'}{\rho R} = constant = \mu$$

and recover (yet again!)

$$Z'' = \mu Z$$

The radial dependence is then

$$\mu + \frac{(\rho R')'}{\rho R} + \frac{\lambda}{\rho^2} - K = 0$$

or

$$R'' + \frac{1}{\rho}R' + (\mu - K + \frac{\lambda}{\rho^2})R = 0$$

which is equidimensional if $\mu = K$; otherwise it is another form of Bessel's equation.  ■

**Example 4.** Helmholtz's equation in spherical coordinates (Fig. 2) takes the form

$$\nabla^2\psi = \frac{1}{r^2}\frac{\partial}{\partial r}(r^2\frac{\partial\psi}{\partial r}) + \frac{1}{r^2\sin\phi}\frac{\partial}{\partial\phi}(\sin\phi\frac{\partial\psi}{\partial\phi}) + \frac{1}{r^2\sin^2\phi}\frac{\partial^2\psi}{\partial\theta^2} = K\psi \quad (17)$$

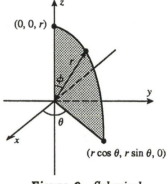

**Figure 2**  *Spherical coordinate system*

The separated ansatz

$$\psi(r,\phi,\theta) = R(r)\Phi(\phi)\Theta(\theta)$$

in (17) results in

$$\frac{(r^2R')'}{r^2}\Phi\Theta + \frac{(\sin\phi\,\Phi')'}{r^2\sin\phi}R\Theta + \frac{\Theta''}{r^2\sin^2\phi}R\Phi = KR\Phi\Theta \qquad (18)$$

We multiply (18) by $r^2\sin^2\phi/R\Phi\Theta$ and manipulate to obtain the familiar equation for $\Theta$:

$$\frac{\Theta''}{\Theta} = -\sin^2\phi\frac{(r^2R')'}{R} - \sin\phi\frac{(\sin\phi\,\Phi')'}{\Phi} + Kr^2\sin^2\phi = constant = \lambda\,.$$

The separation of $\Phi$ then leads to

$$\frac{(\sin\phi\,\Phi')'}{\sin\phi\,\Phi} + \frac{\lambda}{\sin^2\phi} = -\frac{(r^2R')'}{R} + Kr^2 = constant = \mu \qquad (19)$$

or

$$(\sin\phi\,\Phi')' + \frac{\lambda}{\sin\phi}\Phi = \mu\sin\phi\,\Phi\,. \qquad (20)$$

By the methods of Section 1.2 we find that

$$x = \cos\phi \ , \ y(x) = y(\cos\phi) = \Phi(\phi)$$

transforms (20) into the *associated Legendre equation* (exercise 11)

$$(1 - x^2)y'' - 2xy' + \frac{\lambda}{1 - x^2}y = \mu y \ .$$

The remaining radial dependence in (19) satisfies an equidimensional equation if $K = 0$ (Laplace's equation):

$$r^2 R'' + 2rR' + \mu R = 0;$$

and for nonzero $K$ another form of Bessel's equation arises:

$$r^2 R'' + 2rR' + (\mu - Kr^2)R = 0 \ . \tag{21}$$

(In fact solutions to (21) are known as *spherical Bessel functions*.)

Other coordinate systems in which Helmholtz's equation separates are described by Morse and Feshbach. Now we turn to the separation of Schrodinger's equation.

EXAMPLE 5. In appropriate units Schrodinger's equation in one dimension takes the form

$$i\frac{\partial\psi}{\partial t} = -\frac{\partial^2\psi}{\partial x^2} + V(x)\psi \ , \tag{22}$$

where $V(x)$ is the potential energy. The ansatz

$$\psi(x, t) = X(x)T(t)$$

leads to

$$iT'X = -X''T + V(x)XT$$

which separates upon division by $XT$:

$$i\frac{T'}{T} = -\frac{X''}{X} + V(x) \ . \tag{23}$$

Hence both sides of (23) must equal a constant, $\lambda$; i.e.,

$$T' = -i\lambda T \tag{24}$$

and

$$X'' - V(x)X = -\lambda X \ . \tag{25}$$

Equation (24) has solutions $c_1 e^{-i\lambda t}$, but the nature of (25) depends on the potential $V(x)$. Some familiar applications are as follows.

(i) If $V(x)$ is a constant, $V_0$, then (25) is the harmonic oscillator equation again with solutions

$$X(x) = \begin{cases} c_1 \cosh \sqrt{V_0 - \lambda}\, x + c_2 \sinh \sqrt{V_0 - \lambda}\, x & \textit{if } \lambda < V_0, \\ c_1 + c_2 x & \textit{if } \lambda = V_0, \textit{ and} \\ c_1 \cos \sqrt{\lambda - V_0}\, x + c_2 \sin \sqrt{\lambda - V_0}\, x & \textit{if } \lambda > V_0. \end{cases}$$

(ii) For the *quantum mechanical mass-spring oscillator* $V(x)$ equals $x^2$, in appropriate units. The $x$ dependence, eq. (25), then takes a form known historically as *Weber's equation*:

$$X'' + (\lambda - x^2) X = 0 . \tag{26}$$

In Section 1.2 we saw how the change of variables $\Xi(x) = e^{x^2/2} X(x)$ transforms (26) into *Hermite's equation*

$$\Xi'' - 2x\, \Xi' + (\lambda - 1)\, \Xi = 0 .$$

(iii) A uniform force field F results in a potential $V(x) = -Fx$; (25) then takes the form

$$X'' + FxX = -\lambda X , \tag{27}$$

which is easily reduced to *Airy's equation*

$$X'' = \xi X \tag{28}$$

by a linear change of variables $\xi = a + bx$ (exercise 2). ∎

**Example 6.** In appropriate units the Schrödinger equation for the hydrogen atom, in spherical coordinates, is

$$i\frac{\partial \psi}{\partial t} = -\nabla^2 \psi - \frac{q}{r}\psi , \tag{29}$$

where $q$ is a constant characterizing the strength of the electrostatic force. In these coordinates the angular variables separate out just as in Helmholtz's equation, producing the same separation constants $\lambda, \mu$ as in equation (19). The remaining time and radial dependence can be expressed (exercise 7)

$$i\frac{T'}{T} = -\frac{(r^2 R')' + \mu R}{r^2 R} - \frac{q}{r} = constant = K , \tag{30}$$

so the radial factor is governed by *Balmer's equation*

$$(r^2 R')' + \{qr + \mu\}R = -Kr^2 R \tag{31}$$

With a (rather complicated) change of variables, equation (31) transforms into the *associated Laguerre equation*. See exercise 7, Sec. 1.2. ∎

We summarize these computations with a listing of the classical ordinary differential equations, together with the partial differential equations from which they arise. For uniformity of notation the ODEs are all written in a specific format, with $x$ as the independent variable and $y(x)$ as the dependent variable. The author has found this classification very helpful in keeping track of the special functions of applied mathematics.

| Ordinary Differential Equation | Partial Differential Equation | Coordinate System | Variable |
|---|---|---|---|
| Harmonic oscillator $y'' = \lambda y$ | wave, heat, Laplace | cartesian | $x, y, z$ |
| | wave, heat, Laplace | cylindrical | $z$ |
| | wave, heat, Laplace | polar, spherical, cylindrical | $\theta$ |
| | hydrogen atom | spherical | $\theta$ |
| Equidimensional $x^2 y'' + \alpha x y' = \lambda y$ | Laplace | polar, spherical | $r$ |
| | wave, heat, Laplace | cylindrical | $\rho$ |
| Bessel $x^2 y'' + \alpha x y'$ $+ \gamma y$ $= \lambda x^2 y$ | wave, heat | polar, spherical | $r$ |
| | | cylindrical | $\rho$ |
| | Laplace | cylindrical | $\rho$ |
| Legendre $(1 - x^2)y'' - 2xy'$ $+ \beta y/(1 - x^2)$ $= \lambda y$ | wave, heat, Laplace, hydrogen atom | spherical | $\phi$ |
| Airy $y'' + \alpha x y = \lambda y$ | quantum mechanical constant force | one dimensional | $x$ |
| Hermite (Weber) $y'' - 2xy' = \lambda y$ | quantum mechanical harmonic oscillator | one dimensional | $x$ |
| Laguerre (Balmer) $xy'' + (\alpha - x)y' = \lambda y$ | hydrogen atom | spherical | $r$ |

Notice that all these ordinary differential equations conform to the generic format

$$a_2(x)y'' + a_1(x)y' + a_0(x)y = \lambda g(x)y \ ,$$

where $\lambda$ is related to the separation constant for the relevant variable. We shall exploit this commonality in the next chapter to establish that, for selected values of the separation constant, the equation has a family of solutions with all the useful properties of the sine and cosine series.

# Exercises 5.3

1. Carry out the separation of the damped wave equation (telegrapher's equation)

$$\frac{\partial^2 \psi}{\partial t^2} + \nu \frac{\partial \psi}{\partial t} = \frac{\partial^2 \psi}{\partial x^2} + \frac{\partial^2 \psi}{\partial y^2} + \frac{\partial^2 \psi}{\partial z^2}$$

   in three dimensional cartesian coordinates.

2. Find the values of $a$ and $b$ which transform (27) to the form (28).

3. Express the solution to (21) in terms of Bessel functions.

4. Separate the two dimensional wave equation in polar coordinates.

5. Separate the heat equation in cylindrical coordinates.

6. Carry out completely the separation of the heat equation in spherical coordinates.

7. Derive eq. (30).

8. In rarified plasma dynamics the mechanism for the diffusion of ions is the fact that they scatter off of *each other*. In such a case the diffusivity (Sec. 4.5) is proportional to the ion concentration $\psi(x, y, z, t)$ and the diffusion equation becomes nonlinear:

$$\frac{\partial \psi}{\partial t} = (constant) \nabla \cdot (\psi \nabla \psi)$$

   (see the reference by Galeev). Show that a trial solution of the form $\psi(x, y, z, t) = \phi(x, y, z) T(t)$ gives rise to a time behavior

$$T(t) = \frac{-1}{\lambda t + c},$$

   where $c$ and $\lambda$ are constants. Note that this time dependence is inverse power rather than the decaying exponential which characterized the linear diffusion (heat) equation. (Physically speaking, as the density gets lower the number of scattering centers depletes and the diffusion slows down.)

9. The Hamiltonian (kinetic plus potential energy) for a particle in a Coulomb potential, expressed in spherical coordinates, is

$$H(r, \theta, \phi, p_r, p_\theta, p_\phi) = \frac{1}{2m} p_r^2 + \frac{1}{2mr^2 \sin^2 \phi} p_\theta^2 + \frac{1}{2mr^2} p_\phi^2 - \frac{\mu}{r} \quad (32)$$

where $p_r$ is the momentum in the radial direction,

$$p_r = m\dot{r} \ ,$$

$p_\theta$ is the angular momentum about the $z$ axis,

$$p_\theta = [mr^2 \sin^2 \phi]\dot{\theta} \ ,$$

and $p_\phi$ is the canonical angular momentum associated with $\phi$ (see Goldstein for details),

$$p_\phi = mr^2 \dot{\phi} \ .$$

Here $m$ is the mass and $\mu$ is the force constant. The *Hamilton-Jacobi* partial differential results when the momenta are replaced in (32) by the partial derivatives of the "generating function" $S(r, \theta, \phi, t)$ according to

$$p_r \leftarrow \frac{\partial S}{\partial r}, \ \ p_\theta \leftarrow \frac{\partial S}{\partial \theta}, \ \ p_\phi \leftarrow \frac{\partial S}{\partial \phi}, \tag{33}$$

and the result equated with $-\partial S/\partial t$:

$$H\left(r, \theta, \phi, \frac{\partial S}{\partial r}, \frac{\partial S}{\partial \theta}, \frac{\partial S}{\partial \phi}\right) = -\frac{\partial S}{\partial t} \ .$$

Show that this partial differential equation separates into ordinary differential equations if the *additive* (not multiplicative!) ansatz

$$S(r, \theta, \phi, t) = R(r) + \Theta(\theta) + \Phi(\phi) + T(t)$$

is substituted. (For students of mechanics: the separation constants can be interpreted as constants of the motion. Using (33), how many of them can you identify physically?)

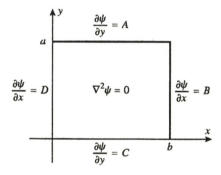

**Figure 3**  *Exercise 10*

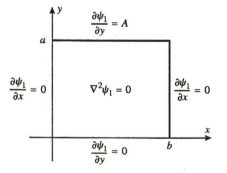

**Figure 4**  *Subproblem for exercise 10*

10. (*The Pure Neumann Problem*) As mentioned in Sec. 4.6, the pure Neumann problem for Laplace's equation contains mathematical and physical subtleties. Even the case of constant flux conditions on the boundary, as depicted in Fig. 3, confounds the basic procedure espoused at the end of the previous section.

(a) Break up the problem in Fig. 3 into subproblems and attempt to solve the subproblem indicated in Fig. 4. Show that the solution is impossible unless $A = 0$, and that even in that case $\psi_1$ is only determined up to a constant.

The trouble with the problem in Fig. 4 can be explained on physical grounds. Solutions to Laplace's equation can be interpreted, as we have seen, as steady-state temperature distributions. But in Fig. 4 the square is insulated on three sides and has a constant heat flux coming in the fourth side; thus the temperature increases without bound, and it can't ever reach a steady state! If $A = 0$ it's insulated on all sides and a uniform steady state temperature is achievable, but since all the equations refer only to *derivatives* of $\psi_1$, the latter is only defined up to a constant. (For example, $\psi_1$ could be either Celsius or Kelvin temperature. Or if $\psi_1$ happens to be electric potential, the reference "ground" level has not been specified in the problem.)

The same holds true for the original problem in Fig. 3. A solution is possible only if the net heat flux in all four sides is zero. Since $\partial\psi/\partial n$ is proportional to heat flux per unit length (in two dimensions), the compatibility condition is

$$Ab + Ba - Cb - Da = 0 \qquad (34)$$

(recall that $\mathbf{n}$ represents the outgoing normal direction). See exercise 6, Sec. 4.6, for mathematical elaboration.

Parts (b), (c), and (d) of this exercise derive a solution prodecure for solving pure Neumann problems.

(b) Analyze the problem in Fig. 3 by separating Laplace's equation with an *additive* ansatz $\psi(x,y) = X(x) + Y(y)$. Derive the conditions

$$X(x) = c_1 + c_2 x + c_3 x^2, \ Y(y) = d_1 + d_2 y + d_3 y^2, c_3 = -d_3 \ . \quad (35)$$

(c) Fit the forms in (35) to the boundary conditions and derive the solution

$$\psi(x,y) = Dx + (B - D)x^2/2b + Cy + (A - C)y^2/2a + K$$

where $K$ is an arbitrary constant. Note that the final condition in (35) is consistent with the compatibility condition (34).

(d) If the boundary conditions in Fig. 3 represent the thermal environment of the square and the compatibility condition (34) is *not* met, then the heat equation

$$\frac{\partial\psi}{\partial t} = \nabla^2\psi$$

has no steady state solution. Use the additive form $\psi(x,y,t) = X(x) + Y(y) + T(t)$ to obtain the solution

$$\psi(x,y,t) = Dx + \frac{B-D}{2b}x^2 + Cy + \frac{A-C}{2a}y^2$$
$$+ \{\frac{B-D}{b} + \frac{A-C}{a}\}t + K \ . \quad (36)$$

(Note that (36) is not a *general* solution, since its initial value $\psi(x, y, 0)$ cannot be adjusted.)

(e) If the boundary conditions in Fig. 3 represent the physical environment of a vibrating drumhead and the compatibility condition (34) is *not* met, then the *wave* equation

$$\frac{\partial^2 \psi}{\partial t^2} = \nabla^2 \psi$$

has no equilibrium solution either. Use the additive form $\psi(x, y, t) = X(x) + Y(y) + T(t)$ to obtain the solution

$$\psi(x, y, t) = Dx + \frac{B - D}{2b} x^2 + Cy + \frac{A - C}{2a} y^2$$
$$+ \{\frac{B - D}{2b} + \frac{A - C}{2a}\} t^2 + K_1 + K_2 t \ . \quad (37)$$

(Note that (37) is not a *general* solution either, since its initial values $\psi(x, y, 0)$, $\partial \psi(x, y, 0)/\partial t$ cannot be adjusted.)

11. Carry out the transformation of eq. (20) into the associated Legendre equation, using the methods of Sec. 1.2 as indicated in the text.

# Chapter 6

# Eigenfunction Expansions

The Fourier expansions have a generalization that is extremely important in engineering applications. This theory, due to Jacques C. F. Sturm and Joseph Liouville, yields new expansions based on functions which reflect the physical properties of the particular system. The new expansions enable us to express the solutions of many *partial* differential equations, which would otherwise be analytically intractible, in a convenient form, just as the sine series enabled us to solve the heat flow problem in Sec. 5.2.

In Chapter 6 we shall survey this theory and introduce the most frequently encountered expansions. The first section previews the basic ideas, illustrating them with a straightforward example. The second section gives a heuristic exposition of the underlying theory. Section 6.3 applies the theory to calculate a few noteworthy expansions, in the context of the physical situations that give rise to them. Section 6.4 discusses the modifications that result when the equation has a singular point. Again, a few of the more familiar expansions are worked out.

Section 6.5 briefly surveys the more rigorous mathematical aspects of the subject (it is optional). The remainder of the chapter is intended to be more archival than expository. The derivations therein generate a catalog (Table 6.3) of expansions that have proved useful in applications. Such an exhaustive tabulation must of necessity be concise and, consequently, user *un*friendly. Therefore as a guide we include Section 6.6, which documents in great detail a contrived (but not unrealistic) application designed to exemplify most of the features of the table.

## 6.1   Introduction

**Summary**   The salient features of the functions giving rise to the Fourier sine series are their completeness, their oscillation property, their orthogonality, and the associated *boundary value problem* $y'' = -n^2 y$, $y(0) = y(\pi) = 0$. The *regular Sturm-*

*Liouville problem,*

$$a_2(x)y'' + a_1(x)y' + a_0(x)y = \lambda g(x)y\,,$$

$$\alpha y(a) + \beta y'(a) = 0\,, \quad \gamma y(b) + \delta y'(b) = 0$$

generalizes the latter and for appropriate values of $\lambda$ its solutions also are complete, oscillate, and are orthogonal with respect to the weight factor

$$w(x) = \frac{g(x)}{a_2(x)}\exp\{\int^x \frac{a_1(\eta)}{a_2(\eta)}\,d\eta\}\,.$$

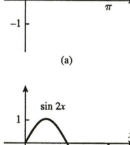

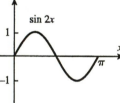

We begin our study of the Sturm-Liouville theory with a review of the relevant properties of the sine series. Recall Sections 3.3 through 3.5.

(i) *Completeness*
The Fourier sine expansion reads

$$f(x) = \sum_{n=1}^{\infty} c_n\phi_n(x) \tag{1}$$

where

$$\phi_n(x) = \sin nx\,.$$

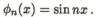

Completeness means that an expansion of the form (1) is possible for any square-integrable $f$ over the interval $0 \le x \le \pi$, when convergence is interpreted in the mean-square sense.

(ii) *Orthogonality*
The coefficients $c_n$ in (1) are easy to find because of the orthogonality property

$$\int_0^\pi \phi_m(x)\phi_n(x)\,dx = 0 \quad \text{if } m \ne n\,; \tag{2}$$

thus multiplying both sides of (1) by $\phi_m(x)$ and integrating yields

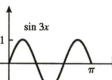

$$c_m = \int_0^\pi f(x)\phi_m(x)\,dx \Big/ \int_0^\pi \phi_m(x)^2\,dx = \frac{2}{\pi}\int_0^\pi f(x)\sin mx\,dx \tag{3}$$

**Figure 1(i)** *Expansion functions for sine series*

(iii) *Oscillation Property*
The expansion functions $\phi_n(x)$ are sketched in Figures 1(i) and 1(ii). Note that the first one, $\sin x$, has no *interior* zeros in the interval, and each subsequent one has one more interior zero than the previous. Thus

the functions must oscillate ever more rapidly for higher $n$. This is somewhat related to the completeness property; in order to synthesize *arbitrary* functions one needs a store of "fast" oscillators, as well as slow ones.

(iv) *Associated Boundary Value Problem*

The functions $\phi_n$ are solutions to the differential equation system

$$y'' = \lambda y, \ y(0) = 0, \ y(\pi) = 0 \qquad (4)$$

with the parameter $\lambda$ equal to $-n^2$.

The reader may recall from Example 4, Sec. 1.1, that boundary value problems are not governed by the same existence-uniqueness theorem as initial value problems, and property (iv) is a manifestation of this fact (the solution is determined only up to a constant multiple). It is helpful to review the analysis conducted in Sec. 5.1.

The general solution to the differential equation $y'' = \lambda y$ is

$$y(x) = \begin{cases} C_1 \cosh \sqrt{\lambda}x + C_2 \sinh \sqrt{\lambda}x & \text{for } \lambda > 0, \\ C_1 + C_2 x & \text{for } \lambda = 0, \\ C_1 \cos \sqrt{-\lambda}x + C_2 \sin \sqrt{-\lambda}x & \text{for } \lambda < 0. \end{cases}$$

Enforcing the boundary condition at $x = 0$ produces

$$C_1 \cdot 1 + C_2 \cdot 0 = C_1 = 0 \ (\textit{regardless of } \lambda).$$

Then the second boundary condition demands either $C_2 = 0$ - and the solution is identically zero - or that $\lambda$ is the negative square of an integer $n$, and the solution is (a constant multiple of) $\sin \sqrt{-(-n^2)}\, x = \sin nx$.

The key to the Sturm-Liouville generalization of the sine series expansions lies in property (iv), the underlying boundary value problem. We begin our exposition by generalizing the differential equation in (4) to the generic form which we encountered when we separated the partial differential equations in Sec. 5.3:

$$a_2(x)y'' + a_1(x)y' + a_0(x)y = \lambda g(x)y, \qquad (5)$$

Here the left hand side can be any second order linear differential operator. However for the moment we assume that (5) has no singular points (Sec. 2.4), so $a_2(x)$ is presumed to be nonzero in the interval of interest. *For simplicity, we assume it positive.* (Of course, we also assume that each coefficient $a_i$ is continuous.) The right hand side has been generalized to permit an additional factor $g(x)$ *which is also presumed to be positive* (and continuous) throughout the interval. Keep in mind that $a_2, a_1, a_0$, and $g$ are *given* coefficients, and that $\lambda$ is still a *parameter.*[1]

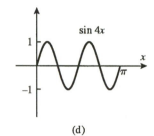

sin 4x

(d)

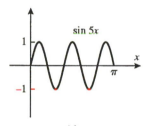

sin 5x

(e)

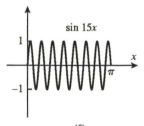

sin 15x

(f)

***Figure 1(ii)***
*Expansion functions for*
*sine series*

---

[1]There is no loss of generality with the positiveness restrictions. If, for example, $a_2$ is negative, multiply the equation by -1. If the coefficient $g$ then becomes negative, redefine $g_{new} = -g_{old}, \ \lambda_{new} = -\lambda_{old}$.

The boundary conditions in (4) are generalized to the forms

$$\alpha y(a) + \beta y'(a) = 0, \ \gamma y(b) + \delta y'(b) = 0. \tag{6}$$

(We presume a and b are both finite.) With proper choice of the constants $\alpha, \beta, \gamma$, and $\delta$, the constraints (6) can require that either $y, y'$, or a linear combination of the two be zero. These are the (one-dimensional) Dirichlet, Neumann, or Robin boundary conditions respectively (Sec. 4.3). We dismiss the meaningless cases $\alpha = \beta = 0, \ \gamma = \delta = 0$.

The zeros on the right hand sides of (6) guarantee that they are *homogeneous* equations so that, whenever $y_1(x)$ and $y_2(x)$ satisfy them, so does every linear combination $C_1 y_1(x) + C_2 y_2(x)$. The significance of this property was emphasized in Secs. 5.1 and 5.2.

Under these assumptions eqs. (5, 6) constitute a *regular Sturm-Liouville problem*. Remarkably, Sturm and Liouville proved that this generalized problem still has all the essential properties (i) through (iv). Specifically, they showed that although (5, 6) usually admits only the trivial solution $y(x) \equiv 0$,

(a) *There is an infinite, decreasing sequence of values for $\lambda$ approaching $-\infty$*

$$\lambda_1 > \lambda_2 > \lambda_3 > \cdots \rightarrow -\infty \tag{7}$$

*for which (5, 6) has nontrivial solutions $\phi_1(x), \phi_2(x), \phi_3(x),\ldots$ respectively.*
(b) *$\phi_n(x)$ has n-1 interior zeroes in the interval (a,b).*[2]
(c) *The $\phi_n$ are "orthogonal with respect to the weight $w(x)$,"* where $w$ is given by

$$w(x) = \frac{g(x)}{a_2(x)} \exp\{\int^x \frac{a_1(\eta)}{a_2(\eta)} \, d\eta\} \tag{8}$$

(the notation indicates an indefinite integral).

"Orthogonality with respect to $w$" means that instead of having

$$\int_a^b \phi_m(x)\phi_n(x) \, dx = 0 \ (m \neq n),$$

we *weight* the points along the interval of integration $[a, b]$ with the factor $w(x)$, producing

$$\int_a^b \phi_m(x)\phi_n(x) \, w(x) \, dx = 0 \ (m \neq n) . \tag{9}$$

---

[2] Sometimes it is natural to enumerate the solutions starting with $n = 0$; in such a case $\phi_n$ has $n$ interior zeros, of course.

(d) *The $\{\phi(x)\}$ form a complete set in the mean-square sense.* This means that any square-integrable function $f(x)$ can be expanded as

$$f(x) = \sum_{n=1}^{\infty} c_n \phi_n(x). \tag{10}$$

Note that to get the coefficient $c_k$ in (10) we multiply both sides by $\phi_k(x)$ **and** $w(x)$ and then integrate to derive

$$c_k = \frac{\int_a^b f(x)\phi_k(x)\,w(x)\,dx}{\int_a^b \phi_k(x)^2 w(x)\,dx}.$$

The mean-square convergence of (10) is also to be interpreted with the weight factor $w$ inserted:

$$\lim_{n\to\infty} \int_a^b |f(x) - \sum_{n=1}^{\infty} c_n \phi_n(x)|^2 w(x)\,dx = 0.$$

Before we see how all this comes about in general it is well to study an illuminating, though somewhat messy, example.

**Example 1.** Figure 2 depicts the steady state temperature distribution inside a portion of a "washer," thermally insulated on the inner and outer curved edges and with specified temperatures on the two straight edges.

In polar coordinates the temperature $\psi(r, \theta)$ is determined by Laplace's equation

$$\frac{1}{r}\frac{\partial}{\partial r}\left(r\frac{\partial \psi}{\partial r}\right) + \frac{1}{r^2}\frac{\partial^2 \psi}{\partial \theta^2} = 0 \tag{11}$$

and the boundary conditions

$$\frac{\partial \psi}{\partial r}(1, \theta) = 0 = \frac{\partial \psi}{\partial r}(2, \theta), \tag{12}$$

$$\psi(r, 0) = 0, \ \psi(r, \theta_1) = g(r). \tag{13}$$

As demonstrated in Sec. 5.3, insertion of the separated trial solution $\psi(r, \theta) = R(r)\Theta(\theta)$ into (11) leads to the ordinary differential equations

$$r^2 R'' + rR' = \lambda R \tag{14}$$

and

$$\Theta'' = -\lambda\Theta. \tag{15}$$

The homogeneous conditions (12) will be met if we take

$$R'(1) = 0, \ R'(2) = 0 \tag{16}$$

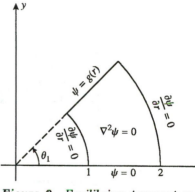

**Figure 2** *Equilibrium temperature a washer section*

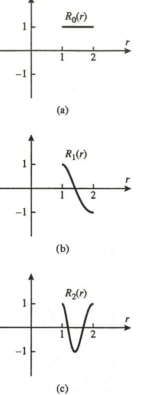

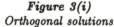

**(a)**

**(b)**

**(c)**

*Figure 3(i)*
*Orthogonal solutions*

as the boundary conditions to go with eq. (14). From the nature of the boundary conditions (flat slopes at the end points) and the oscillation property predicted by the theory, we can anticipate that the orthogonal family of solutions to (14, 16) will have the general shapes indicated in Figs. 3(i) and 3(ii).

In exercise 2 of Sec. 1.1 the general solution to (14) was shown to be

$$R(r) = \begin{cases} C_1 r^{\sqrt{\lambda}} + C_2 r^{-\sqrt{\lambda}} & \text{for } \lambda > 0, \\ C_1 + C_2 \ln r & \text{for } \lambda = 0, \\ C_1 \cos(\sqrt{-\lambda}\ln r) + C_2 \sin(\sqrt{-\lambda}\ln r) & \text{for } \lambda < 0. \end{cases}$$

We impose the first boundary condition, $R'(1) = 0$. For $\lambda > 0$ this gives

$$R'(1) = \sqrt{\lambda}\, C_1 1^{\sqrt{\lambda}-1} - \sqrt{\lambda}\, C_2 1^{-\sqrt{\lambda}-1} = 0,$$

which is met if $C_2 = C_1$. For $\lambda = 0$ this boundary condition demands $C_2 = 0$, and for $\lambda < 0$ we also have $C_2 = 0$. At this point, then, we know

$$R(r) = \begin{cases} C_1 [r^{\sqrt{\lambda}} + r^{-\sqrt{\lambda}}] & \text{for } \lambda > 0, \\ C_1 & \text{for } \lambda = 0, \\ C_1 \cos(\sqrt{-\lambda}\ln r) & \text{for } \lambda < 0. \end{cases}$$

The second boundary condition (16) demands

$$R'(2) = 0 = \begin{cases} C_1 \sqrt{\lambda}\{2^{\sqrt{\lambda}-1} - 2^{-\sqrt{\lambda}-1}\} = \frac{C_1}{2}\sqrt{\lambda}\{2^{\sqrt{\lambda}} - 2^{-\sqrt{\lambda}}\} & \text{for } \lambda > 0, \\ 0 & \text{for } \lambda = 0, \\ -\frac{C_1}{2}\sqrt{-\lambda}\sin(\sqrt{-\lambda}\ln 2) & \text{for } \lambda < 0. \end{cases}$$

The obvious choice - $C_1 = 0$ - leads only to the trivial solution $R(r) = 0$, so instead we manipulate $\lambda$ to enforce this boundary condition. The option $\lambda = 0$ is a feasible possibility, yielding the solution $R(r) = C_1 = \text{constant}$. There are no possibilities for $\lambda > 0$, but negative values of $\lambda$ work if they are chosen so that $\sqrt{-\lambda}\ln 2 = n\pi$ or $\lambda = \lambda_n = -(n\pi/\ln 2)^2$, $n = 1, 2, 3, \ldots$; the corresponding solutions are constant multiples of $R_n(r) = \cos\{n\pi \frac{\ln r}{\ln 2}\}$.

In full agreement with the theory, we have found that *the system* (14, 16) *has nontrivial solutions associated with an infinite discrete set of values of* $\lambda$,

$$\begin{aligned} \lambda_0 &= 0 \\ \lambda_1 &= -(\pi/\ln 2)^2 \approx -20.54 \\ \lambda_2 &= -(2\pi/\ln 2)^2 \approx -82.17 \\ \lambda_3 &= -(3\pi/\ln 2)^2 \approx -184.88 \\ &\vdots \end{aligned} \qquad (17)$$

*converging to $-\infty$; these nontrivial solutions are constant multiples of*

$$R_0(r) = 1$$

$$R_1(r) = \cos \pi \frac{\ln r}{\ln 2}$$

$$R_2(r) = \cos 2\pi \frac{\ln r}{\ln 2}$$

$$R_3(r) = \cos 3\pi \frac{\ln r}{\ln 2}$$

$$\vdots \tag{18}$$

(Obviously it is convenient in this case to begin the numbering with $n = 0$.)

These are the functions depicted in Figures 3(i) and 3(ii), and we confirm that $R_n(r)$ has $n$ interior zeros. In fact if we introduce the change of variable

$$\xi = \pi \frac{\ln r}{\ln 2} \tag{19}$$

we see that the family (18) is merely a transformed version of the functions forming the cosine series $\{1, \cos \xi, \cos 2\xi, \cos 3\xi, ...\}$ on the interval $0 \le \xi \le \pi$.

This identification makes it easy to verify the orthogonality condition. Sturm-Liouville theory predicts that

$$\int_1^2 R_m(r)R_n(r)w(r)\,dr = 0 \ \text{ if } m \ne n, \tag{20}$$

where the weight $w$ is given by (8)

$$w(r) = \frac{1}{r^2} \exp\{\int^r (\frac{\eta}{\eta^2})\,d\eta\} = \frac{\exp(\ln r)}{r^2} = \frac{1}{r}. \tag{21}$$

But in fact, we already *know* the cosines to be orthogonal, in accordance with

$$\int_{\xi=0}^{\xi=\pi} (\cos m\xi)(\cos n\xi)\,d\xi = 0 \ \text{ if } m \ne n. \tag{22}$$

Inserting the change of variable (19) we derive

$$
\begin{aligned}
0 &= \int_{r=1}^{r=2} (\cos m\pi \frac{\ln r}{\ln 2})(\cos n\pi \frac{\ln r}{\ln 2})\frac{d\xi}{dr}\,dr \\
&= \int_{r=1}^{r=2} R_m(r)R_n(r)(\frac{\pi}{r\ln 2})\,dr \\
&= \frac{\pi}{\ln 2} \int_{r=1}^{r=2} R_m(r)R_n(r)(\frac{1}{r})\,dr \tag{23}
\end{aligned}
$$

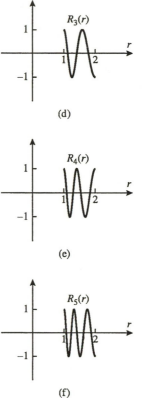

(d)

(e)

(f)

*Figure 3(ii)*
*Orthogonal solutions*

in perfect harmony with (20, 21).

*Completeness* can also be demonstrated through the transformation (19). We know that every (square-integrable) function has a cosine series expansion:

$$F(\xi) = \sum_{n=0}^{\infty} d_n \cos n\xi, \ 0 \le \xi \le \pi \tag{24}$$

Inserting the change of variables (19) we alter (24) to read

$$F(\pi \frac{\ln r}{\ln 2}) = f(r) = \sum_{n=0}^{\infty} d_n \cos n\pi \frac{\ln r}{\ln 2}, \ 1 \le r \le 2. \tag{25}$$

To find the coefficients $d_n$ we exploit the "generalized" orthogonality condition (20); multiplying both sides of the expansion by $\cos m\pi \frac{\ln r}{\ln 2}$ **and also by** $(1/r)$ and integrating, we expose $d_m$:

$$\int_1^2 f(r) \cos m\pi \frac{\ln r}{\ln 2} \left(\frac{1}{r}\right) dr = \sum_{n=0}^{\infty} d_n \int_1^2 \{\cos n\pi \frac{\ln r}{\ln 2} \cos m\pi \frac{\ln r}{\ln 2}\} \frac{dr}{r}$$

$$= d_m \int_1^2 \{\cos m\pi \frac{\ln r}{\ln 2}\}^2 \frac{dr}{r},$$

or

$$d_m = \int_1^2 f(r) \cos m\pi \frac{\ln r}{\ln 2} \frac{dr}{r} \Big/ \int_1^2 \{\cos m\pi \frac{\ln r}{\ln 2}\}^2 \frac{dr}{r}$$

$$= \begin{cases} \dfrac{2}{\ln 2} \int_1^2 f(r) \cos m\pi \dfrac{\ln r}{\ln 2} \dfrac{dr}{r} & m > 0 \\[3mm] \dfrac{1}{\ln 2} \int_1^2 f(r) \dfrac{dr}{r} & m = 0. \end{cases} \tag{26}$$

Returning to the original problem we now consider the general solutions to the $\Theta$ equation (15) for the $\lambda$ values specified in (17):

$$\Theta_n(\theta) = \begin{cases} D_1 + D_2\theta, & n = 0, \\ D_1 \cosh(\frac{n\pi}{\ln 2})\theta + D_2 \sinh(\frac{n\pi}{\ln 2})\theta, & n > 0. \end{cases} \tag{27}$$

The homogeneous boundary condition in (13) requires $\Theta(0) = 0$, so $D_1 = 0$. Thus we have uncovered the following family of functions which solve the homogeneous equations (11), (12), and the first of (13):

$$\phi_0(r,\theta) = \theta \,,$$

$$\phi_1(r,\theta) = \cos\pi\frac{\ln r}{\ln 2}\ \sinh(\frac{\pi}{\ln 2})\theta \,,$$

$$\phi_2(r,\theta) = \cos 2\pi\frac{\ln r}{\ln 2}\ \sinh(\frac{2\pi}{\ln 2})\theta \,,$$

$$\vdots$$

$$\phi_n(r,\theta) = \cos n\pi\frac{\ln r}{\ln 2}\ \sinh(\frac{n\pi}{\ln 2})\theta \,,$$

$$\vdots$$

To satisfy the final, nonhomogeneous boundary condition (13), we take a linear combination of these solutions

$$\psi(r,\theta) = c_0\theta + \sum_{n=1}^{\infty} c_n \cos n\pi\frac{\ln r}{\ln 2}\ \sinh(\frac{n\pi}{\ln 2})\theta \qquad (28)$$

and enforce the boundary condition:

$$\psi(r,\theta_1) = g(r) = c_0\theta_1 + \sum_{n=1}^{\infty} a_n \cos n\pi\frac{\ln r}{\ln 2}\ \sinh(\frac{n\pi}{\ln 2})\theta_1. \qquad (29)$$

Equation (29) has the form of the transformed cosine series (25), so the coefficients are obtained by invoking orthogonality:

$$c_{n>0} = \frac{2}{(\sinh\frac{n\pi\theta_1}{\ln 2})\ln 2}\int_1^2 g(r)\cos n\pi\frac{\ln r}{\ln 2}\frac{dr}{r} \,,$$

$$c_0 = \frac{1}{\theta_1\ln 2}\int_1^2 \frac{g(r)}{r}\,dr \,. \qquad \blacksquare$$

This lengthy example has demonstrated how the differential equation (14) and boundary conditions (16) generate a family of solutions (18) which generalize the Fourier functions in that they are complete, orthogonal, and oscillatory. Admittedly the generalization may not appear spectacular at this point since these solutions are, in fact, only transformed versions of the Fourier functions. But soon we shall exploit the Sturm-Liouville theory to uncover a multitude of new and extremely useful orthogonal families.

# Exercises 6.1

1. Imitate the reasoning in Example 1 to show how the expansion of $f(x)$ in a series of solutions of

$$x^2 y'' + xy' = \lambda y \,, \ y(1) = y(e) = 0$$

corresponds to a transformed sine series. What steady-state temperature problems would utilize this expansion?

**2.** Verify the final form of eq. (26).

**3.** Use the identity $\cos x = (e^{ix} + e^{-ix})/2$ to express the expansion (25) in the form

$$f(r) = \sum_{n=0}^{\infty} \frac{c_n}{2} \{ r^{in\pi/\ln 2} + r^{-in\pi/\ln 2} \},$$

$$c_{n>0} = \frac{1}{\ln 2} \int_1^2 f(r) \{ r^{\frac{in\pi}{\ln 2} - 1} + r^{\frac{-in\pi}{\ln 2} - 1} \} \, dr$$

$$c_0 = \frac{1}{\ln 2} \int_1^2 f(r) \frac{dr}{r}.$$

## 6.2 The Sturm-Liouville Theory

**Summary** When the differential equation for a regular Sturm-Liouville problem is cast in "Sturm-Liouville" form

$$\mathcal{L}y = [p(x)y']' + q(x)y = \lambda w(x)y,$$

then the operator $\mathcal{L}$ is analogous to a self-adjoint matrix, in the sense that $\langle \mathcal{L}f, g \rangle = \langle f, \mathcal{L}g \rangle$ when $f$ and $g$ meet the (homogeneous) boundary conditions. As a result the solutions of the problem for different values of $\lambda$ are orthogonal with respect to the weight $w(x)$: $\langle \phi_i, \phi_j \rangle_w = \|\phi_i\|^2 \delta_{ij}$. To find these "eigenfunctions," one first writes down the general solution to the differential equation as a linear combination of two independent solutions, carrying $\lambda$ as a parameter. The coefficients are chosen so as to fulfill one of the boundary conditions. Then $\lambda$ is selected to meet the other boundary condition: the latter becomes a transcendental equation in $\lambda$ with an infinite number of solutions, accumulating at $-\infty$ if $wp > 0$. The corresponding eigenfunctions are complete and oscillatory.

Alternatively, one may express the two boundary conditions as linear homogeneous algebraic equations for the coefficients. The condition that the determinant vanishes determines the eigenvalues ($\lambda$).

In order to be able to use the Sturm-Liouville expansions with confidence it is advantageous to have some idea of why the theory works. The following analysis, while certainly not rigorous, contains the germ of the essential ideas. The inquisitive reader may consult the references mentioned in Section 6.5 for more details.

The starting point for our discussion is the introduction of an operator notation for the underlying differential equation. Let $\mathcal{M}$ be the second order differential operator

$$\mathcal{M} = a_2(x)\frac{d^2}{dx^2} + a_1(x)\frac{d}{dx} + a_0(x) \tag{1}$$

with the obvious interpretation that $\mathcal{M}$ acts on a function $f(x)$ according to

$$\mathcal{M}f(x) = a_2(x)\frac{d^2 f}{dx^2} + a_1(x)\frac{df}{dx} + a_0(x)f. \tag{2}$$

Then our generalized differential equation from Sec. 6.1, $a_2(x)y'' + a_1(x)y' + a_0(x)y = \lambda g(x)y$, can be abbreviated

$$\mathcal{M}y = \lambda g(x)y. \tag{3}$$

We presume that $a_2(x)$ and $g(x)$ are both positive (redefining $\lambda$ if necessary). The generalized boundary conditions remain as before:

$$\alpha y(a) + \beta y'(a) = 0, \quad \gamma y(b) + \delta y'(b) = 0. \tag{4}$$

Of course the operator $\mathcal{M}$ is *linear*. Thus the differential equation (3), as well as the boundary conditions (4), are *linear homogeneous*; for any constants $C_1$ and $C_2$ and any solutions $y_1$ and $y_2$ the combination $C_1 y_1(x) + C_2 y_2(x)$ is also a solution. In particular, the identically zero function is a solution of (3,4).

All this linear structure suggests a matrix analogy, and it is very profitable to think of $\mathcal{M}$ as a matrix operator in a vector space made up of differentiable functions. With this in mind, the differential equation (3) is analogous (if we omit the factor $g$) to the *eigenvalue* equation for a matrix $A$:

$$Av = \lambda v \tag{5}$$

Accordingly, it is customary to refer to the "privileged" values of the parameter $\lambda_n$ in the Sturm-Liouville theory as *eigenvalues*, while the nomenclature for the associated functions $\phi_n(x)$ is a hybrid word which suggests their eigenvector kinship:

**Definition.** Any function $y(x)$ which satisfies the differential equation (3) and the boundary conditions (4) and which is not identically zero is called an *eigenfunction of* (3, 4). The associated value of $\lambda$ is called an *eigenvalue*.

Note that in extending the eigenvector concept to functions we have introduced two modifications to the basic requirement (5); we *generalize* to allow inclusion of a positive function $g(x)$ in (3), and we *restrict* the eigenfunctions to satisfy the boundary conditions (4) as well as the differential equation (3).

Now let's see why the eigenvalues and eigenfunctions have the Sturm-Liouville properties listed in Sec. 6.1. At the outset we confess that the *completeness* property is very hard to prove, and we direct the reader to the references (recall our discussion of this question with regard to the Fourier series, Section 3.3). To understand the eigenvalue structure and oscillatory behavior, however, we begin by rewriting (3) in the form

$$y'' = \lambda \frac{g(x)}{a_2(x)}y - \frac{a_0(x)}{a_2(x)}y - \frac{a_1(x)}{a_2(x)}y' \tag{6}$$

and restrict our attention to the following special case of (4):

$$y(0) = 0, \quad y(1) = 0. \tag{7}$$

Consider what happens to a function $y(x)$ which starts off from the origin with an initial slope of $+1$ (Fig. 1), and then is constrained to evolve in accordance with (6): Loosely speaking if we take $\lambda$ to be some huge positive number, so that the first term on the right of (6) dominates the other two, then we see that $y''$ has the same sign as $y$ (remember that we have arranged that $g$ and $a_2$ are positive). Since $y$ starts off pointing into the first quadrant, this implies that $y''$ is positive and the graph of $y(x)$ is convex, as in Fig. 2. Qualitatively $y(x)$ acts as though it is *repelled* by the $x$-axis, and clearly the condition $y(1) = 0$ will not be achieved, for large positive $\lambda$.

What happens to this $y(x)$ if $\lambda$ is large *negative*? Then $y''$ has the *opposite* sign as $y$, the curve is *concave* when $y > 0$, and the graph tends to "fight its way" back down to the $x$-axis. Thus we would expect a crossing point $x_1$ to appear eventually, and it should occur very near the origin if $|\lambda|$ is big enough (Fig. 3).

Now we can visualize how the boundary condition $y(1) = 0$ gives rise to the eigenvalues. Refer to Figs. 4(a) through 4(e). We imagine plotting copies of the function $y(x; \lambda)$ as described above, for various values of $\lambda$. When $\lambda$ is large positive, the graphs bend upward. As we decrease $\lambda$, this behavior is curbed and the graphs tend to fight their way back to the $x$-axis; eventually they cross it, and the crossing point migrates to the left. For some critical value $\lambda = \lambda_1$ the crossover occurs at $x = 1$ and *we have our first eigenvalue and eigenfunction.*

Now if we continue to decrease $\lambda$ the graphs go below the $x$-axis and $y$ becomes negative. But then if $\lambda$ is large negative the sign of $y''$ goes positive in (6) and the convexity turns the curves back up towards the axis. Thus we eventually will see a second crossing, migrating to the left, and when it reaches $x = 1$ *we have the second eigenvalue* (Figs. 5(a) through 5(d)).

The pattern should be clear (see Fig. 6). The eigenvalues $\lambda_n$ form a decreasing sequence, and each eigenfunction has one more interior crossing than the previous.

A similar argument explains how the boundary condition $y'(1) = 0$ generates eigenvalues. The first time any of the functions $y(x; \lambda)$ meets *this* condition occurs in Fig. 4(c), which displays the associated eigenfunction

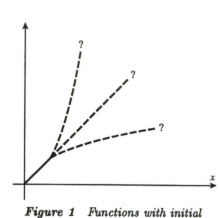

**Figure 1**  *Functions with initial slope = 1*

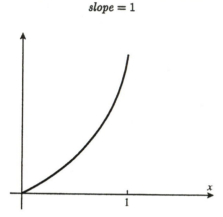

**Figure 2**  *Convex function*

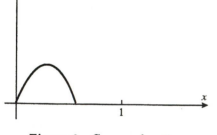

**Figure 3**  *Concave function*

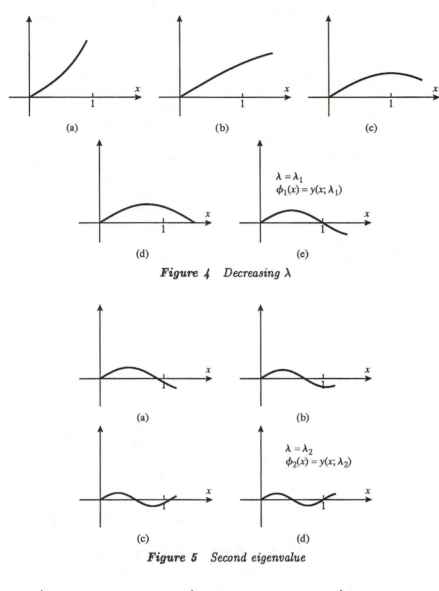

(a)                    (b)                    (c)

(d)                    (e)

$\lambda = \lambda_1$
$\phi_1(x) = y(x; \lambda_1)$

**Figure 4**   *Decreasing* $\lambda$

(a)                    (b)

(c)                    (d)

$\lambda = \lambda_2$
$\phi_2(x) = y(x; \lambda_2)$

**Figure 5**   *Second eigenvalue*

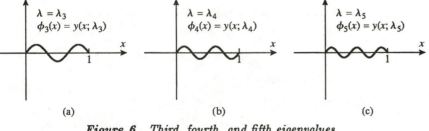

$\lambda = \lambda_3$
$\phi_3(x) = y(x; \lambda_3)$

$\lambda = \lambda_4$
$\phi_4(x) = y(x; \lambda_4)$

$\lambda = \lambda_5$
$\phi_5(x) = y(x; \lambda_5)$

(a)                    (b)                    (c)

**Figure 6**   *Third, fourth, and fifth eigenvalues*

$\phi_1(x)$; $\phi_2(x)$ occurs in Fig. 5(b), and so on. Other aspects of this argument are developed in exercise 1.

(One might object that we have artificially restricted our deliberations to functions which have an initial slope of $+1$ in our derivation, but this is no loss of generality since *by homogeneity* any solution of equations (3, 4) may be scaled up by a constant multiple and remain a solution. *Eigenfunctions, like eigenvectors, are only defined to within a constant factor.*)

Next we analyze the orthogonality property and the emergence of the weight factor $w$ in eq. (8) of Sec. 6.1. The key to this phenomenon is, again, the analogy between the Sturm-Liouville problem (3, 4) and the matrix problem (5) - the *symmetric* matrix eigenvalue problem, in fact, since symmetric matrices have orthogonal eigenvectors. As a preliminary it is useful to review the highlights of the matrix theory, since the Sturm-Liouville derivation proceeds along the same lines.

### HIGHLIGHTS OF THE SYMMETRIC EIGENVALUE PROBLEM

a. The *inner product* $\langle v_1, v_2 \rangle$ of two column vectors $v_1$ and $v_2$ is the scalar $v_1^T v_2$. It is linear in each vector (i.e. "bilinear")

$$\langle c_1 w_1 + c_2 w_2, d_1 v_1 + d_2 v_2 \rangle$$
$$= c_1 d_1 \langle w_1, v_1 \rangle + c_1 d_2 \langle w_1, v_2 \rangle$$
$$+ c_2 d_1 \langle w_2, v_1 \rangle + c_2 d_2 \langle w_2, v_2 \rangle$$

and symmetric: $\langle v_1, v_2 \rangle = \langle v_2, v_1 \rangle$ (for real vectors). $\langle v, v \rangle$ is always positive unless $v = 0$, and it measures the square of the *norm* $\|v\|$ (length) of $v$. In two and three dimensions the inner product can be used to characterize the angle $\theta$ between vectors:

$$\cos \theta = \frac{\langle v_1, v_2 \rangle}{\|v_1\| \, \|v_2\|} \tag{8}$$

because of the Law of Cosines. In higher dimensions we commemorate (8) by calling $v_1$ and $v_2$ *orthogonal* whenever $\langle v_1, v_2 \rangle = 0$ (corresponding to $\theta = 90°$).

b. If the matrix $A$ is symmetric then it can be inserted into the inner product $\langle v_1, v_2 \rangle$ in front of either factor:

$$\langle A v_1, v_2 \rangle = \langle v_1, A v_2 \rangle \tag{9}$$

This follows since

$$\langle A v_1, v_2 \rangle = (A v_1)^T v_2 \;\; = \;\; v_1^T A^T v_2$$
$$= \;\; v_1^T A v_2 = v_1 (A v_2) = \langle v_1, A v_2 \rangle;$$

in fact, (9) *implies* symmetry (exercise 2).

c. If $v_1$ and $v_2$ are *eigenvectors* of $A$ corresponding to different eigenvalues $\lambda_1$ and $\lambda_2$,

$$Av_1 = \lambda_1 v_1, \quad Av_2 = \lambda_2 v_2, \tag{10}$$

then the orthogonality of $v_1$ and $v_2$ is quickly demonstrated by inserting (10) into (9):

$$
\begin{array}{ccc}
\langle Av_1, v_2 \rangle & = & \langle v_1, Av_2 \rangle \\
\| & & \| \\
\langle \lambda_1 v_1, v_2 \rangle & & \langle v_1, \lambda_2 v_2 \rangle \\
\| & & \| \\
\lambda_1 \langle v_1, v_2 \rangle & = & \lambda_2 \langle v_1, v_2 \rangle;
\end{array}
\tag{11}
$$

the final inequality is impossible for $\lambda_1 \neq \lambda_2$ unless the inner product is zero.

Now we shall mimic this derivation for the Sturm-Liouville eigenfunctions by

(a') constructing the analog of the inner product,

(b') fabricating the analog of the symmetric matrix identity (9), and

(c') applying the logic of equations (11).

(a') (*inner product*) The *inner product of two functions $f(x)$ and $g(x)$,* denoted $\langle f, g \rangle$, is defined as

$$\langle f(x), g(x) \rangle = \int_a^b f(x) g(x) \, dx \tag{12}$$

(Recall $a$ and $b$ are the endpoints of the boundary value problem (3, 4). We first used this notation in Sec. 3.3 where it was motivated by the analogy between function tables and column vectors.)

Definition (12) preserves all the essential properties (bilinearity, positivity,[3] symmetry) of the definition in (a) above. The concept of *norm* for a function $f$

$$\|f\| = \sqrt{\langle f, f \rangle} = \left\{ \int_a^b f(x)^2 \, dx \right\}^{1/2} \tag{13}$$

inherited from this definition is, of course, the mean-square integral.

---

[3]See exercise 3 for elaboration of the positivity property.

There is another feature of the definition (12) which is rather nice. We might generalize the formula (8) to define an "angle between functions" via

$$\theta = \arccos \frac{\langle f, g \rangle}{\|f\| \|g\|}, \tag{14}$$

if we could be sure that the ratio in (14) never exceeded 1 in magnitude. And, in fact, the *Schwarz inequality*

$$|\langle f, g \rangle| \leq \|f\| \|g\| \tag{15}$$

(which you will be invited to prove in exercise 5) guarantees this. However, the only real use one makes of this capricious notion is the familiar one — when $\langle f, g \rangle = 0$ we call $f$ and $g$ *orthogonal* (because $\theta = 90°$).

(b') (*symmetry*) The analog of the symmetric matrix identity (9) for the differential operator $\mathcal{M}$ in (1) would be

$$\langle \mathcal{M}f, g \rangle = \langle f, \mathcal{M}g \rangle,$$

Now this equation is FALSE in general. However, Sturm and Liouville showed that it is true under two conditions: when the operator $\mathcal{M}$ takes a certain form, and when the functions $f$ and $g$ satisfy the boundary conditions (4).

DEFINITION. The operator

$$\mathcal{M} = a_2(x) \frac{d^2}{dx^2} + a_1(x) \frac{d}{dx} + a_0(x)$$

is said to be in *Sturm-Liouville* or *self-adjoint* form if

$$a_1(x) = \frac{da_2(x)}{dx}. \tag{16}$$

In other words a *Sturm-Liouville* operator $\mathcal{L}$ acting on a function $f(x)$ can be represented in the form

$$\mathcal{L}f = p(x) \frac{d^2 f}{dx^2} + p'(x) \frac{df}{dx} + q(x)f = [p(x)f']' + q(x)f \tag{17}$$

(We maintain our convention that the coefficient of $f''$ — namely $p(x)$ — is positive in the interval.)

It is an easy matter to convert *any* operator $\mathcal{M}$ to the Sturm-Liouville form, but before we see this let us verify the analog of the symmetric matrix identity (9).

THEOREM If $\mathcal{L}$ is a Sturm-Liouville operator and $f(x)$ and $g(x)$ satisfy the boundary conditions (4) then

$$\langle \mathcal{L}f, g \rangle = \langle f, \mathcal{L}g \rangle. \tag{18}$$

PROOF The important step in shifting the $\mathcal{L}$ from $f(x)$ to $g(x)$ in the inner product integral is the use of integration by parts — twice, in fact (a process for which the Sturm-Liouville form is tailor-made). Observe:

$$\int_a^b [p(x)f']' g\,dx = pf'g|_a^b - \int_a^b [p(x)f']g'\,dx$$

$$= pf'g|_a^b - \int_a^b f'[p(x)g']\,dx$$

$$= pf'g|_a^b - fpg'|_a^b + \int_a^b f[p(x)g']'\,dx$$

Piecing these together we have an intermediate result sometimes called *Green's formula*

$$\langle \mathcal{L}f, g \rangle - \langle f, \mathcal{L}g \rangle = \int_a^b \{[pf']' + qf\}g\,dx$$

$$- \int_a^b f\{[pg']' + qg\}\,dx$$

$$= p\{f'g - fg'\}|_a^b, \qquad (19)$$

valid for *any* $f$ and $g$.

It is clear that if both $f$ and $g$ vanish at $a$ and $b$, the right hand side of (19) is zero; but the same conclusion holds if $f$ and $g$ satisfy *any* form of the boundary conditions (4) (see exercise 6). Thus (18) holds. (See exercise 5 for an alternative derivation, using "Lagrange's identity.")

Now to put the "generic" operator

$$\mathcal{M} = a_2(x)\frac{d^2}{dx^2} + a_1(x)\frac{d}{dx} + a_0(x)$$

into Sturm-Liouville form (17) it is necessary to multiply $\mathcal{M}$ by an *integrating factor* $r(x)$ so that the new coefficients

$$a_2^{new}(x) = r(x)a_2(x)\,, \ a_1^{new}(x) = r(x)a_1(x),$$

satisfy the condition (16). This requires

$$r(x)a_1(x) = \frac{d}{dx}[r(x)a_2(x)]$$

or

$$\frac{r'}{r} = \frac{d\ln r}{dx} = \frac{1}{a_2}[a_1 - a_2'];$$

$$\ln r = \int \frac{a_1}{a_2}\,dx - \int \frac{a_2'}{a_2}\,dx = \int \frac{a_1}{a_2}\,dx - \ln a_2;$$

$$r(x) = \frac{1}{a_2(x)} \exp\left[ \int^x (a_1/a_2)\,dx \right] \tag{20}$$

With this modification the original differential equation (3) takes the Sturm-Liouville form

$$\mathcal{L}y = [py']' + qy = \lambda w y \tag{21}$$

with

$$p(x) = \exp\left\{ \int^x [a_1(x)/a_2(x)]\,dx \right\}, \quad q(x) = p(x)a_0(x)/a_2(x),$$

and

$$w(x) = \frac{g(x)}{a_2(x)} \exp\left\{ \int^x [a_1(x)/a_2(x)]\,dx \right\} = g(x)p(x)/a_2(x).$$

$w(x)$ will turn out to be the weight factor $w$ in eq. (8) of Sec. 6.1.

(c′) (*orthogonality*) Now we are in a position to retrace the logic of equations (11) and derive the orthogonality conditions for the eigenfunctions. If $\phi_1(x)$ and $\phi_2(x)$ are two solutions of (3, 4) corresponding to different eigenvalues $\lambda_1$ and $\lambda_2$, then we have

$$\mathcal{L}\phi_1 = \lambda_1 w \phi_1, \quad \mathcal{L}\phi_2 = \lambda_2 w \phi_2$$

Thus starting from (18) we derive

$$
\begin{array}{ccc}
\langle \mathcal{L}\phi_1, \phi_2 \rangle & = & \langle \phi_1, \mathcal{L}\phi_2 \rangle \\
\| & & \| \\
\langle \lambda_1 w \phi_1, \phi_2 \rangle & & \langle \phi_1, \lambda_2 w \phi_2 \rangle \\
\| & & \| \\
\lambda_1 \langle w \phi_1, \phi_2 \rangle & = & \lambda_2 \langle \phi_1, w \phi_2 \rangle
\end{array}
$$

Since $\lambda_1 \neq \lambda_2$ but $\langle w\phi_1, \phi_2 \rangle = \langle \phi_1, w\phi_2 \rangle$, this latter factor must be zero:

$$\langle w\phi_1, \phi_2 \rangle = \langle \phi_1, w\phi_2 \rangle = \int_a^b \phi_1(x)\phi_2(x)w(x)\,dx = 0.$$

This is the *weighted* orthogonality condition promised in the previous section.

It is convenient to employ a special notation for these weighted inner products. We define

$$\langle f, g \rangle_w = \int_a^b f(x)g(x)w(x)\, dx$$

and, for the associated norm,

$$\|f\|_w = \sqrt{\langle f, f \rangle_w} \tag{22}$$

A compact way of representing the orthogonality condition for the eigenfunctions is

$$\langle \phi_i, \phi_j \rangle_w = \|\phi_i\|_w^2\, \delta_{ij}, \tag{23}$$

where the "Kronecker delta" $\delta_{ij}$ is 1 when $i = j$ and zero otherwise; $\|\phi_i\|_w^2$ is called the "normalization integral."

Based on what we have discovered, we can now outline a procedure for dealing with Sturm-Liouville expansions.

SUMMARY OF THE STURM-LIOUVILLE EXPANSION PROCEDURE

PROBLEM: *to express the function $f(x)$, square-integrable over the interval $[a, b]$, in terms of the solutions to the boundary value problem:*

$$a_2(x)y'' + a_1(x)y' + a_0(x)y = \lambda g(x)y, \tag{3}$$

$$\alpha y(a) + \beta y'(a) = 0, \quad \gamma y(b) + \delta y'(b) = 0, \tag{4}$$

*with $\lambda$ as an auxiliary parameter.* (For generality we'll word this paragraph so that the positivity assumptions on the coefficients $a_2$ and $g$ are unnecessary; but we still presume that the coefficients are continuous and that $a_2$ and $g$ are nonzero.)

SOLUTION PROCEDURE

1. Write down the general solution to (3) as the sum of two independent particular solutions with undetermined coefficients:

$$y = C_1 y_1(x; \lambda) + C_2 y_2(x; \lambda) \tag{24}$$

The constant $\lambda$ will appear as a parameter in the formulas. (In choosing the forms for $y_1$ and $y_2$ it may be helpful to keep in mind that most of the eigenvalues for $\lambda$ will be negative if $a_2$ and $g$ have the same sign, and most will be positive otherwise. See exercise 7 if this is not clear.)

2. Enforce *one* of the boundary conditions (4) by the choice of $C_1$ and $C_2$. In other words, use the relation

$$\alpha[C_1 y_1(a; \lambda) + C_2 y_2(a; \lambda)] + \beta[C_1 y_1'(a; \lambda) + C_2 y_2'(a; \lambda)] = 0$$

to express $C_1$ in terms of $C_2$ or vice versa. (It is easy to see that the resulting solution is a constant multiple of

$$y(x; \lambda) = [\alpha y_2(a; \lambda) + \beta y_2'(a; \lambda)]y_1(x; \lambda)$$
$$- [\alpha y_1(a; \lambda) + \beta y_1'(a; \lambda)]y_2(x; \lambda); \quad (25)$$

see exercise 11.)

3. The remaining boundary condition

$$\gamma y(b; \lambda) + \delta y'(b; \lambda) = 0 \qquad (26)$$

*is regarded as an equation for* $\lambda$. Insert (25) and solve it for the eigenvalues $\{\lambda_n\}$, which will form a sequence going to $\pm\infty$ according to whether $a_2$ and $g$ have opposite or the same signs. The $n$th eigenfunction

$$\phi_n(x) = y(x; \lambda_n)$$

should have $n - 1$ interior zeros (assuming that the eigenvalues are enumerated from $n = 1$). (It is a good idea to plot the first few eigenfunctions to check that none have been overlooked in solving (26).)

4. Compute the weight factor $w$ from

$$w(x) = \left| \frac{g(x)}{a_2(x)} \right| \exp \left\{ \int^x [a_1(x)/a_2(x)] \, dx \right\} \qquad (27)$$

(The absolute value is actually unnecessary, but it is included to keep the "weight" positive.) The eigenfunctions will be orthogonal with respect to $w$:

$$\begin{array}{ccc} \langle \phi_m, \phi_n \rangle_w & = & \|\phi_n\|_w^2 \, \delta_{mn} \\ \| & & \| \\ \int_a^b \phi_m(x)\phi_n(x)w(x) \, dx & & \int_a^b \phi_n^2(x)w(x) \, dx \, \delta_{mn} \end{array}$$

5. The expansion of $f(x)$ is then expressed

$$f(x) = \sum_{n=1}^{\infty} c_n \phi_n(x) \qquad (28)$$

with the coefficients[4] given by

$$c_n = \frac{\langle f, \phi_n \rangle_w}{\|\phi_n\|_w^2} = \frac{\int_a^b f(x)\phi_n(x)w(x)\,dx}{\int_a^b \phi_n(x)^2 w(x)\,dx} \tag{29}$$

**Example 1.** Find the expansion of $f(x) = x$ in terms of the eigenfunctions of the system

$$y'' = \lambda y, \tag{30}$$

$$y(0) = 0, \quad y'(\pi) = 0 \tag{31}$$

SOLUTION. The general solution of (30) is the familiar

$$y(x) = \begin{cases} C_1 \cosh\sqrt{\lambda}x + C_2 \sinh\sqrt{\lambda}x & \text{if } \lambda > 0, \\ C_1 + C_2 x & \text{if } \lambda = 0, \\ C_1 \cos\sqrt{-\lambda}x + C_2 \sin\sqrt{-\lambda}x & \text{if } \lambda < 0. \end{cases}$$

The first boundary condition requires $C_1 = 0$, and $C_2$ then becomes a redundant multiplicative constant (so we can drop it). The second boundary condition states

$$y'(\pi) = 0 = \begin{cases} \sqrt{\lambda}\cosh\sqrt{\lambda}\pi & \text{if } \lambda > 0, \\ 1 & \text{if } \lambda = 0, \\ \sqrt{-\lambda}\cos\sqrt{-\lambda}\pi & \text{if } \lambda < 0, \end{cases}$$

which is only possible for negative values of $\lambda$ satisfying $\sqrt{-\lambda} = \pm\frac{1}{2}, \pm\frac{3}{2}, \cdots, \pm\frac{2n+1}{2}, \cdots$; i.e.,

$$\lambda_n = -\left(\frac{2n+1}{2}\right)^2, \quad n = 0, 1, 2, \ldots \tag{32}$$

The associated solutions to (30, 31) are

$$y(x; \lambda) = \sin\frac{2n+1}{2}x, \quad n = 0, 1, 2, \ldots; \tag{33}$$

see Fig. 7.

The expansion of $f(x)$ reads

$$f(x) = x = \sum_{n=0}^{\infty} c_n \sin\frac{2n+1}{2}x. \tag{34}$$

The formula (27) for the weight is

$$w(x) = \frac{1}{1}e^0 = 1$$

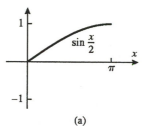

(a)

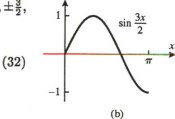

(b)

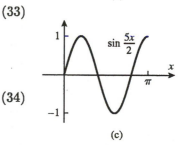

(c)

*Figure 7*
*Eigenfunction for*
*Example 1*

---

[4]In practice these integrals are often too complicated to be evaluated analytically; a numerical scheme such as Simpson's rule must be used. See Appendix A.6.

so the coefficients are given by

$$
\begin{aligned}
c_n &= \frac{\langle x, \sin \frac{2n+1}{2} x \rangle}{\| \sin \frac{2n+1}{2} x \|^2} \\
&= \int_0^\pi x \sin \frac{2n+1}{2} x \, dx \Big/ \int_0^\pi \sin^2 \frac{2n+1}{2} x \, dx \\
&= (-1)^n \frac{8}{(2n+1)^2 \pi}
\end{aligned}
\tag{35}
$$

Most other authors utilize a different formulation of the solution procedure for Sturm-Liouville problems, and the reader must be aware of the alternative approach. Suppose once again the boundary value problem is formulated as

$$
a_2(x)y'' + a_1(x)y' + a_0(x)y = \lambda g(x)y,
$$
$$
\alpha y(a) + \beta y'(a) = 0, \quad \gamma y(b) + \delta y'(b) = 0,
$$

and the general solution of the differential equation is displayed as

$$
C_1 y_1(x; \lambda) + C_2 y_2(x; \lambda).
$$

Then the boundary constraints can be represented as two simultaneous linear equations for $C_1$ and $C_2$:

$$
[\alpha y_1(a; \lambda) + \beta y'(a; \lambda)]C_1 + [\alpha y_2(a; \lambda) + \beta y_2'(a; \lambda)]C_2 = 0
$$
$$
[\gamma y_1(b; \lambda) + \delta y_1'(b; \lambda)]C_1 + [\gamma y_2(b; \lambda) + \delta y_2'(b; \lambda)]C_2 = 0.
\tag{36}
$$

Equations (36) will possess only the trivial solution $C_1 = C_2 = 0$ unless the system is singular, in which case the coefficient determinant must be zero:

$$
\det = \begin{vmatrix} \alpha y_1(a; \lambda) + \beta y_1'(a; \lambda) & \alpha y_2(a; \lambda) + \beta y_2'(a; \lambda) \\ \gamma y_1(b; \lambda) + \delta y_1'(b; \lambda) & \gamma y_2(b; \lambda) + \delta y_2'(b; \lambda) \end{vmatrix} = 0.
\tag{37}
$$

Equation (37) thus determines the eigenvalues $\lambda_n$, and the insertion of $\lambda_n$ into (36) produces (redundant) equations giving $C_1$ in terms of $C_2$ (or vice versa).

Upon comparing this procedure with the one formulated earlier, one sees that eq. (37) replaces eq. (26) as the eigenvalue equation, and (25) results from (either of) (36). Note that the eigenvalues are determined *first* with this scheme, but keep in mind that both eqs. (37) and (26) are *transcendental*: they have an infinite number of solutions.

Although the determinant formulation is more compact, your author has found it to be less insightful and often more confusing in practice. Thus we shall adopt the scheme of eqs. (24) through (27) throughout most of the text.

# EXERCISES 6.2

1. Draw a series of diagrams like Figs. 4 and 5 to illustrate how the eigenfunctions corresponding to the boundary conditions

    (a) $y'(0) = 0$,   $y(1) = 0$

    (b) $y'(0) = 0$,   $y'(1) = 0$

    arise as $\lambda$ ranges from $+\infty$ to $-\infty$.

2. Prove that if eq. (9) holds for every pair of vectors $v_1$ and $v_2$ then $A$ must be symmetric.

3. Concoct a proof of the *positive definiteness* property for the inner product: $\langle f, f \rangle \geq 0$, with equality holding only when $f(x) = 0$ for all $x$. You should presume that f is continuous. (Can you construct a counterexample if the continuity requirement is dropped?)

4. Prove that the Schwarz inequality

$$|\langle f, g \rangle| \leq \|f\| \, \|g\|$$

    holds for any symmetric, bilinear, positive definite inner product[5]. (Hint: start with the observation

$$\langle f + \sigma g, f + \sigma g \rangle \geq 0 \quad \text{for any scalar } \sigma.$$

    Expand this to obtain a quadratic expression (in $\sigma$) *which you know is never negative*. This latter condition, coupled with the classic formula for the roots of any quadratic, yields the Schwarz inequality.)

5. The *adjoint* of the operator $\mathcal{M}$ in (2) is defined by

$$\mathcal{M}_{adj} f(x) = [a_2(x)f]'' - [a_1(x)f]' + a_0(x)f.$$

    (a) Derive *Lagrange's identity*: for any two functions $f$ and $g$

$$g\mathcal{M}f - f\mathcal{M}_{adj}g = \frac{d}{dx}\{a_2 g f' - [a_2 g]' f + a_1 fg\}.$$

    (b) Show that $\mathcal{M}$ is in Sturm-Liouville form if and only if $\mathcal{M} = \mathcal{M}_{adj}$.

    (c) Show that Lagrange's identity implies Green's formula (19) if $\mathcal{M} = \mathcal{M}_{adj}$.

6. (a) Show that the right hand side of (19) equals zero if $f$ and $g$ both satisfy any of the boundary conditions (4).

---

[5]Strictly speaking, an "inner product" has these three properties by definition.

(b) Show that the right hand side of (19) equals zero if $f$ and $g$ both satisfy *periodic boundary conditions*

$$y(a) = y(b), \quad y'(a) = y'(b)$$

and $p(a) = p(b)$.

7. Show how the statement below eq. (24) follows from the fact, demonstrated in the text, that the eigenvalues approach $-\infty$ when $a_2$ and $g$ are both positive. (Hint: See the footnote following eq. (5) in Sec. 6.1.)

8. This exercise shows how the *Gram-Schmidt* process, a tool from matrix analysis, can be used to construct the orthogonal *polynomials* that arise in applications.

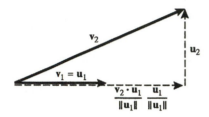

**Figure 8** *Gram-Schmidt construction*

(a) Given two vectors $\mathbf{v}_1$ and $\mathbf{v}_2$, one can construct a pair of *orthogonal* vectors by subtracting, from $\mathbf{v}_2$, its projection along $\mathbf{v}_1$ (Fig. 8). Using the properties of the dot product, show that

$$\mathbf{u}_1 = \mathbf{v}_1$$

and

$$\mathbf{u}_2 = \mathbf{v}_2 - \frac{\mathbf{v}_2 \cdot \mathbf{u}_1}{\mathbf{u}_1 \cdot \mathbf{u}_1}\mathbf{u}_1$$

are orthogonal (or zero; we omit $\mathbf{u}_1$ from the formula for $\mathbf{u}_2$ if $\mathbf{u}_1$ is zero).

(b) Extend this idea; given $\mathbf{v}_1$, $\mathbf{v}_2$, and $\mathbf{v}_3$, define

$$\begin{aligned}
\mathbf{u}_1 &= \mathbf{v}_1 \\
\mathbf{u}_2 &= \mathbf{v}_2 - \frac{\mathbf{v}_2 \cdot \mathbf{u}_1}{\mathbf{u}_1 \cdot \mathbf{u}_1}\mathbf{u}_1
\end{aligned}$$

and

$$\mathbf{u}_3 = \mathbf{v}_3 - \frac{\mathbf{v}_3 \cdot \mathbf{u}_1}{\mathbf{u}_1 \cdot \mathbf{u}_1}\mathbf{u}_1 - \frac{\mathbf{v}_3 \cdot \mathbf{u}_2}{\mathbf{u}_2 \cdot \mathbf{u}_2}\mathbf{u}_2.$$

Show that $\mathbf{u}_1$, $\mathbf{u}_2$, and $\mathbf{u}_3$ are mutually orthogonal or zero (as in a.).

(c) This "Gram-Schmidt" process will also produce orthogonal *functions* when implemented with a symmetric, bilinear, positive definite inner product. Demonstrate this by showing that for any functions $\{f_1(x), f_2(x), f_3(x), f_4(x)\}$, the new functions defined by

$$g_1 = f_1$$

$$g_2 = f_2 - \frac{\langle f_2, g_1 \rangle}{\|g_1\|^2} g_1$$

$$g_3 = f_3 - \frac{\langle f_3, g_1 \rangle}{\|g_1\|^2} g_1 - \frac{\langle f_3, g_2 \rangle}{\|g_2\|^2} g_2$$

$$g_4 = f_4 - \frac{\langle f_4, g_1 \rangle}{\|g_1\|^2} g_1 - \frac{\langle f_4, g_2 \rangle}{\|g_2\|^2} g_2 - \frac{\langle f_4, g_3 \rangle}{\|g_3\|^2} g_3$$

are mutually orthogonal (treating zeros as in part (a)). (Hint: imitate the proof of part (b).)

(d) Start with the functions $\{1, x, x^2, x^3\}$ and the inner product defined by

$$\langle f, g \rangle = \int_{-1}^{1} f(x) g(x)\, dx.$$

Construct the four orthogonal functions according to the Gram-Schmidt prescription. You can check your answer using Table 6.1, because (up to constant factors) these functions are the first four Legendre polynomials! (Hint: Exploit symmetry to simplify the integrals.)

(e) Start with the same functions $\{1, x, x^2, x^3\}$ and the weighted inner product

$$\langle f, g \rangle = \int_{-1}^{1} \frac{f(x) g(x)}{\sqrt{1 - x^2}}\, dx$$

and construct the orthogonal functions. These are proportional to the Chebyshev polynomials of the first kind:

$$T_0(x) = 1$$
$$T_1(x) = x$$
$$T_2(x) = 2x^2 - 1$$
$$T_3(x) = 4x^3 - 3x.$$

(Hint: Again exploit symmetry. The substitution $x = \cos\theta$ will be useful. You should recognize most of the integrals from Fourier series computations.)

**9.** Find the self-adjoint form of the linear second order equations in exercise 4, Sec. 2.3.

**10.** A first step used by many authors in making the arguments in this section more rigorous is the reduction of the general equation (21) to the form

$$\frac{d^2u}{dt^2} = \{\lambda - \frac{g}{u} + (wp)^{-1/4}[(wp)^{1/4}]''\}u$$

through the *Liouville substitution*

$$t = \int^x [w/p]\,dx, \quad u = (wp)^{1/4}y.$$

Verify this reduction, using the methods of Sec. 1.2.

11. Derive (25), the generic form of a solution of (3) which meets the boundary condition (4) at $x = a$.

## 6.3  Regular Sturm-Liouville Expansions

**Summary**  Although Robin boundary conditions, periodic boundary conditions, and Bessel equations lead to eigenvalue problems with unfamiliar solutions, the computation of eigenfunction expansions for all regular Sturm-Liouville problems is a straightforward implementation of the solution procedure outlined in Section 6.2.

According to the scheme outlined in Sec. 6.2, all regular Sturm-Liouville expansions look pretty much alike. For a system of the form

$$a_2(x)y'' + a_1(x)y' + a_0(x)y = \lambda g(x)y, \tag{1}$$

$$\alpha y(a) + \beta y'(a) = 0, \quad \gamma y(b) + \delta y'(b) = 0, \tag{2}$$

one chooses a pair of independent solutions to the differential equation — say, $y_1(x)$ and $y_2(x)$ — and forms a linear combination which meets one of the boundary conditions. Usually this can be accomplished readily by inspection, but the generic form

$$y(x;\lambda) = [\alpha y_2(a;\lambda) + \beta y_2'(a;\lambda)]y_1(x;\lambda) - [\alpha y_1(a;\lambda) + \beta y_1'(a;\lambda)]y_2(x;\lambda) \tag{3}$$

will always satisfy the boundary condition at $x = a$ (recall eq. (25), Sec. 6.2). One then determines the eigenvalues $\lambda_n$ by inserting the formula for $y(x;\lambda)$ into the remaining boundary condition

$$\gamma y(b;\lambda) + \delta y'(b;\lambda) = 0. \tag{4}$$

The expansions then look like

$$f(x) = \sum_{n=1}^{\infty} c_n y(x;\lambda_n) \equiv \sum_{n=1}^{\infty} c_n \phi_n(x), \quad c_n = \frac{\langle f, \phi_n \rangle_w}{\|\phi_n\|_w^2} \tag{5}$$

with $\phi_n(x) = y(x; \lambda_n)$. The various expansions arising in applications differ only in the names of the functions used as solutions $y_1(x; \lambda)$, $y_2(x; \lambda)$. In this section we demonstrate some configurations for Helmholtz's equation which utilize expansions in trigonometric and Bessel functions.

**Example 1.** The equations in Fig. 1 describe the steady state temperature distribution in a rectangular slab. The left and bottom edges are in direct contact with a heat sink maintained at zero degrees, but the right edge is partially shielded from the sink through leaky insulation (giving rise to a Robin boundary condition; $\kappa$ is a given constant). The temperature of the upper edge is maintained at the generic temperature $\psi(x, 2) = g(x)$.

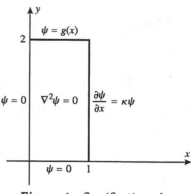

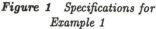

**Figure 1**   *Specifications for Example 1*

As we saw in Sec. 5.3 Laplace's equation separates into harmonic oscillator equations $X'' = \lambda X$ and $Y'' = -\lambda Y$. Since there are homogeneous boundary conditions on both $x = constant$ faces, we address the $x$-dependence first. The general solution to the $X$ equation is

$$X(x) = \begin{cases} C_1 \cosh \sqrt{\lambda}x + C_2 \sinh \sqrt{\lambda}x & \text{if } \lambda > 0, \\ C_1(1) + C_2 x & \text{if } \lambda = 0, \\ C_1 \cos \sqrt{-\lambda}x + C_2 \sin \sqrt{-\lambda}x & \text{if } \lambda < 0, \end{cases}$$

and with $a = 0$ a combination enforcing $\psi(0, y) = X(0)Y(y) = 0$ is

$$X(x; \lambda) = \begin{cases} \sinh \sqrt{\lambda}x & (\lambda > 0), \\ x & (\lambda = 0), \\ \sin \sqrt{-\lambda}x & (\lambda < 0). \end{cases}$$

(In fact strict adherence to the generic form (3) puts minus signs in front of each term.) The Robin boundary condition, which will determine the values of $\lambda$, is

$$\frac{\partial \psi(1, y)}{\partial x} = \kappa \psi(1, y) \text{ or } X'(1)Y(y) = \kappa X(1)Y(y) \text{ or } \kappa X(1) - X'(1) = 0.$$

Hence eq. (4) takes the form

$$0 = \begin{cases} \kappa \sinh \sqrt{\lambda} - \sqrt{\lambda} \cosh \sqrt{\lambda} & (\lambda > 0), \\ \kappa - 1 & (\lambda = 0), \\ \kappa \sin \sqrt{-\lambda} - \sqrt{-\lambda} \cos \sqrt{-\lambda} & (\lambda < 0). \end{cases} \tag{6}$$

Remember that (6) should have an infinite number of solutions $\lambda_n$ clustering at $-\infty$. Most of the solutions, therefore, will come from the third equation in (6). Unless the given constant $\kappa$ happens to equal 1, the middle equation contributes no solutions; for simplicity, we ignore this case (exercise 1).

The other equations in (6) are easier to visualize if we rename the unknown $\lambda$ as $\pm\omega^2$ and rewrite (6) in the form:

$$\begin{array}{ll} \kappa \sinh \omega - \omega \cosh \omega = 0 \text{ or } \omega = \kappa \tanh \omega & (\lambda \doteq \omega^2 > 0) \\ \kappa \sin \omega - \omega \cos \omega = 0 \text{ or } \omega = \kappa \tan \omega & (\lambda = -\omega^2 < 0), \end{array} \tag{7}$$

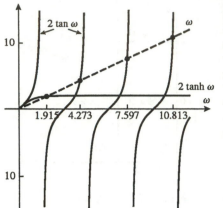

*Figure 2   Eigenvalues for κ = 2*

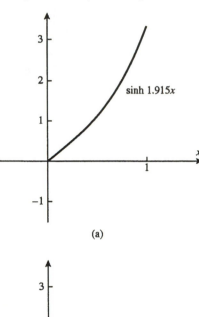

(a)

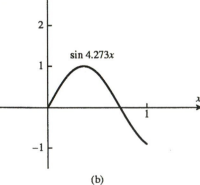

(b)

*Figure 3(i)*

In Fig. 2 we display the graphs of the left and right hand sides of eqs. (7) for the case[6] $\kappa = 2$. The points of intersection correspond to the eigenvalues. The deliberations of Appendix A.5 show how one can find these intersections by visually selecting an approximate solution $\omega_{old}$ and iteratively computing improvements $\omega_{new}$ according to

$$\omega_{new} = 2 \tanh \omega_{old}, \quad \omega_{new} = \tan^{-1}[\omega_{old}/2] \tag{8}$$

(exercise 2).

From these intersections we see that for $\kappa = 2$ there is one positive eigenvalue $\lambda_1 = \omega_1^2 \approx 1.915^2 \approx 3.667$; the rest are negative. A partial list of the eigenvalues and eigenfunctions is as follows:

$$\lambda_1 = \omega_1^2 \approx 1.915^2 \approx 3.6677, \qquad \phi_1(x) = \sinh \sqrt{\lambda_1}x \approx \sinh 1.915x;$$
$$\lambda_2 = -\omega_2^2 \approx -4.273^2 \approx -18.259, \qquad \phi_2(x) = \sin \sqrt{-\lambda_2}x \approx \sin 4.273x;$$
$$\lambda_3 = -\omega_3^2 \approx -7.597^2 \approx -57.714, \qquad \phi_3(x) = \sin \sqrt{-\lambda_3}x \approx \sin 7.597x;$$
$$\lambda_4 = -\omega_4^2 \approx -10.813^2 \approx -116.92, \quad \phi_4(x) = \sin \sqrt{-\lambda_4}x \approx \sin 10.813x;$$
$$\vdots \qquad\qquad\qquad\qquad\qquad \vdots$$

For the parameter value $\kappa = .1$ See Figs. 3(i) and 3(ii). Note that the pattern of oscillations is in accordance with the theory.

the eigenvalues $\lambda = -\omega^2$ are all negative, as demonstrated by Fig. 4. The eigenvalues and eigenfunctions displayed in Fig. 5 are given in this case by

$$\lambda_1 = -\omega_1^2 \approx -1.5044^2 \approx -2.263, \qquad \phi_1(x) = \sin \sqrt{-\lambda_1}x \approx \sin 1.5044x;$$
$$\lambda_2 = -\omega_2^2 \approx -4.6911^2 \approx -22.006, \qquad \phi_2(x) = \sin \sqrt{-\lambda_2}x \approx \sin 4.6911x;$$
$$\lambda_3 = -\omega_3^2 \approx -7.8412^2 \approx -61.485, \qquad \phi_3(x) = \sin \sqrt{-\lambda_3}x \approx \sin 7.8412x;$$
$$\lambda_4 = -\omega_4^2 \approx -10.9864^2 \approx -120.70, \quad \phi_4(x) = \sin \sqrt{-\lambda_4}x \approx \sin 10.9864x;$$
$$\vdots \qquad\qquad\qquad\qquad\qquad \vdots$$

Because the harmonic oscillator equation $X'' = \lambda X$ is already in Sturm-Liouville form, the weight $w(x)$ is 1 and the orthogonality relations for these eigenfunctions become

$$\langle \phi_m, \phi_n \rangle = \int_0^1 \phi_m(x)\phi_n(x)\, dx = \int_0^1 \phi_n(x)^2\, dx\, \delta_{mn} = \|\phi_n\|^2 \delta_{mn}. \tag{9}$$

To complete the analysis of the original problem (Fig. 1) we note that the general solution to the $Y$-equation $Y'' = -\lambda Y$ is given by

$$Y(y) = \begin{cases} D_1 \cos \sqrt{\lambda}y + D_2 \sin \sqrt{\lambda}y & (\lambda > 0), \\ D_1(1) + D_2 y & (\lambda = 0), \\ D_1 \cosh \sqrt{-\lambda}y + D_2 \sinh \sqrt{-\lambda}y & (\lambda < 0), \end{cases}$$

---

[6]In most physical situations $\kappa$ would be negative. We chose $\kappa = 2$ because of the interesting intersection points in Fig. 2. (Compare Fig. 4.)

and the particular combinations satisfying the homogeneous Dirichlet For the parameter value $\kappa = .1$ condition $Y(0) = 0$ on the bottom edge are

$$Y(y) = \begin{cases} \sin\sqrt{\lambda}y & (\lambda > 0), \\ y & (\lambda = 0), \\ \sinh\sqrt{-\lambda}y & (\lambda < 0). \end{cases}$$

Combining these with the corresponding $X$-eigenfunctions we assemble a general format satisfying all the homogeneous conditions of the problem. For $\kappa = 2$ this assemblage looks like

$$\begin{aligned} \psi(x,y) &= c_1 \sinh 1.915x \, \sin 1.915y \\ &+ c_2 \sin 4.273x \, \sinh 4.273y \\ &+ c_3 \sin 7.597x \, \sinh 7.597y \\ &+ c_4 \sin 10.813x \, \sinh 10.813y \\ &+ \cdots \end{aligned} \tag{10}$$

Enforcement of the final, nonhomogeneous boundary condition requires

$$\begin{aligned} g(x) = \psi(x,2) &= c_1 \sinh 1.915x \, \sin(1.915 \times 2) \\ &+ c_2 \sin 4.273x \, \sinh(4.273 \times 2) \\ &+ c_3 \sin 7.597x \, \sinh(7.597 \times 2) \\ &+ c_4 \sin 10.813x \, \sinh(10.813 \times 2) \\ &+ \cdots \end{aligned} \tag{11}$$

Equations (11) reveal the values for the coefficients in (10) when we multiply both sides by a particular eigenfunction and integrate, using (9):

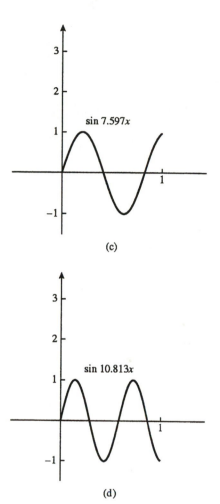

(c)

(d)

*Figure 3(ii)*

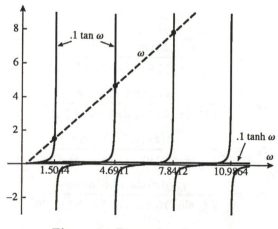

*Figure 4   Eigenvalues for $\kappa = .1$*

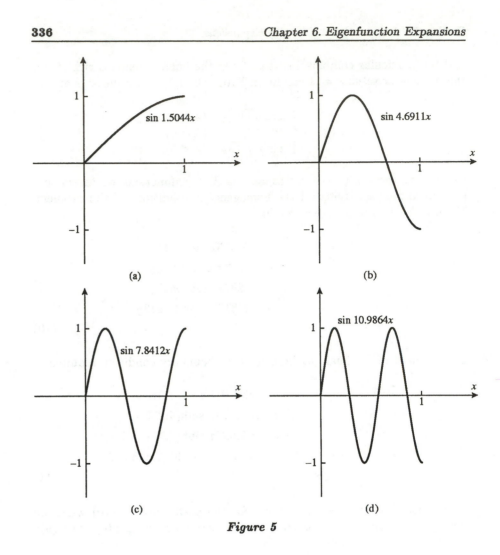

**Figure 5**

$$c_1 = \frac{\int_0^1 g(x) \sinh 1.915x \, dx}{\{\int_0^1 \sinh^2 1.915x \, dx\} \sin 3.830},$$

$$c_2 = \frac{\int_0^1 g(x) \sin 4.273x \, dx}{\{\int_0^1 \sin^2 4.273x \, dx\} \sinh 8.546},$$

$$c_3 = \frac{\int_0^1 g(x) \sin 7.597x \, dx}{\{\int_0^1 \sin^2 7.597x \, dx\} \sinh 15.194},$$

$$c_4 = \frac{\int_0^1 g(x) \sin 10.813x \, dx}{\{\int_0^1 \sin^2 10.813x \, dx\} \sinh 21.626}, \ldots$$

For $\kappa = .1$ the display is a bit more organized since all the eigenvalues are negative:

$$\psi(x, y) = \sum_{n=1}^{\infty} c_n \sin \omega_n x \sinh \omega_n y,$$

$$c_n = \frac{\int_0^1 g(x) \sin \omega_n x\, dx}{\{\int_0^1 \sin^2 \omega_n x\, dx\} \sinh 2\omega_n},$$

$$\omega_n = 1.504,\ 4.691,\ 7.841,\ 10.986,\ \ldots$$

Further simplification of these solution formulas is discussed in exercise 4.

■

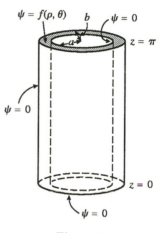

**Example 2.** The analysis of the steady state temperature problem in the heat pipe depicted in Figure 6 relies on Laplace's equation in cylindrical coordinates

$$\nabla^2 \psi(\rho, \theta, z) = \frac{1}{\rho} \frac{\partial}{\partial \rho} (\rho \frac{\partial \psi}{\partial \rho}) + \frac{1}{\rho^2} \frac{\partial^2 \psi}{\partial \theta^2} + \frac{\partial^2 \psi}{\partial z^2} = 0$$

In Sec. 5.3 we saw that separation of this equation leads to the following equations for the $\Theta, z$ and $R$ factors:

$$\Theta'' = \lambda \Theta \tag{12}$$

$$Z'' = \mu Z \tag{13}$$

$$R'' + \frac{1}{\rho} R' + \frac{\lambda}{\rho^2} R = -\mu R \tag{14}$$

**Figure 6**

Let us consider what kind of eigenvalue problem the $\theta$ equation (12) leads to.

What are the boundary conditions for equation (12)? Obviously there are no $\theta$=constant faces on the boundary; the very fact that the "seam" $\theta = 0$ lies inside the pipe material implies that the solution varies smoothly across it. In other words, $\Theta(\theta)$ is a *periodic* function, and its values and slopes must match at 0 and $2\pi$:

$$\Theta(0) = \Theta(2\pi), \quad \Theta'(0) = \Theta'(2\pi). \tag{15}$$

Let's see what new wrinkle this uncovers.

Anticipating as before that the eigenvalues in (12) will be negative, we tentatively write its general solution as

$$C_1 \cos \sqrt{-\lambda}\theta + C_2 \sin \sqrt{-\lambda}\theta. \tag{16}$$

Obviously (15) is satisfied for all $C_1$ and $C_2$ if $\sqrt{-\lambda}$ is an integer, and it is easy to show that any *other* values will yield only the trivial solution $C_1 = C_2 = 0$ (exercise 5). The eigenvalues are thus

$$\lambda_n = -n^2, \quad n = 0, 1, 2, \ldots,$$

and associated with $\lambda_n$ are *two* eigenfunctions

$$\Theta_n^{(1)}(\theta) = \cos n\theta, \quad \Theta_n^{(2)}(\theta) = \sin n\theta, \tag{17}$$

except for $n = 0$ where only the cosine is nontrivial. Of course what we have here is the familiar Fourier series. The weight function is 1 and the expansion formula was derived in Section 3.3:

$$f(\theta) = \sum_{n=0}^{\infty} c_n \cos n\theta + \sum_{n=1}^{\infty} d_n \sin n\theta,$$

$$c_0 = \frac{1}{2\pi} \int_0^{2\pi} f(\theta)\, d\theta, \quad c_{n>0} = \frac{1}{\pi} \int_0^{2\pi} f(\theta) \cos n\theta\, d\theta,$$

$$d_n = \frac{1}{\pi} \int_0^{2\pi} f(\theta) \sin n\theta\, d\theta. \tag{18}$$

∎

Periodic boundary conditions engender a slight exception to the usual Sturm-Liouville theory in that the eigenfunctions are "degenerate" - there can be more than one eigenfunction for each eigenvalue. Such boundary conditions are, as we have seen, a consequence of using *angles* as coordinates. They also arise in solid state physics when one tries to write the Schrödinger equation for an electron in a crystal; see the book by Morrison, Estle, and Lane.

**Example 3.** Next we consider the eigenfunctions that arise from the radial equation

$$R'' + \frac{1}{\rho} R' + \frac{\lambda}{\rho^2} R = -\mu R$$

(eq. (14) repeated) for the heat pipe. The (Dirichlet) boundary conditions indicated in Fig. 6 for $R(\rho)$ are

$$R(a) = 0, \quad R(b) = 0. \tag{19}$$

The values for $\lambda$ in equation (14) were set by the $\theta$ eigenvalue problem in the previous example; they are, therefore, $\lambda_n = -n^2$. Thus the radial equation becomes

$$R'' + \frac{1}{\rho}R' - \frac{n^2}{\rho^2}R = -\mu R. \tag{20}$$

The general solution of this Bessel equation can be written

$$R(\rho; \mu) = \begin{cases} C_1 J_n(\sqrt{\mu}\rho) + C_2 Y_n(\sqrt{\mu}\rho) & (\mu > 0) \\ C_1 \rho^n + C_2 \rho^{-n} & (\mu = 0; \text{ the equidimensional case}) \\ C_1 I_n(\sqrt{-\mu}\rho) + C_2 K_n(\sqrt{-\mu}\rho) & (\mu < 0). \end{cases}$$

(To be even more precise: if $n$ happens to be zero, the solution is $C_1(1) + C_2 \ln \rho$ for $\mu = 0$.) However, according to the solution procedure outlined at the end of the previous section, most of the eigenvalues $\mu$ will be positive (note the minus sign accompanying $\mu$ in (20)), so we tentatively adopt the first form.

The Dirichlet condition corresponds to $\alpha = 1$, $\beta = 0$ in (2), so the solution format (3) reads

$$R(\rho; \mu) = Y_n(\sqrt{\mu}a)J_n(\sqrt{\mu}\rho) - J_n(\sqrt{\mu}a)Y_n(\sqrt{\mu}\rho) \tag{21}$$

and the eigenvalue condition (4) with $\gamma = 1$, $\delta = 0$ is

$$Y_n(\sqrt{\mu}a)J_n(\sqrt{\mu}b) - J_n(\sqrt{\mu}a)Y_n(\sqrt{\mu}b) = 0. \tag{22}$$

The eigenvalues $\mu_p$ which solve (22) can be computed numerically (of course the power series for the Bessel functions have to be encoded). Some of them are listed in Table 2.2. For the case $a = 1$, $b = 2$, $n = 0$ the graph of the left hand side of (22) is displayed in Fig. 7, and the eigenvalues are indicated. The corresponding eigenfunctions

$$R_{n,p}(\rho) = Y_n(\sqrt{\mu_p}a)J_n(\sqrt{\mu_p}\rho) - J_n(\sqrt{\mu_p}a)Y_n(\sqrt{\mu_p}\rho)$$

are depicted for this case in Fig. 8. Note the pattern of the oscillations; our assumption that all $\mu_p$ are positive is vindicated.

It is easy to see that the Sturm-Liouville form of equation (20) is

$$(\rho R')' - \frac{n^2}{\rho}R = -\mu\rho R$$

so we can identify the weight factor as $w(\rho) = \rho$. Then the decomposition of any function $f(\rho)$ into the eigenfunction patterns is given by

$$f(\rho) = \sum_{p=1}^{\infty} c_p R_{n,p}(\rho), \tag{23}$$

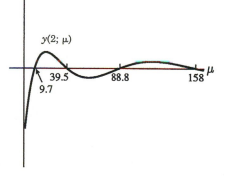

*Figure 7*

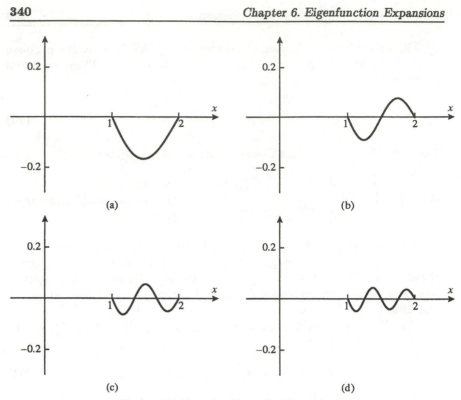

*Figure 8   Eigenfunctions for Example 3*

$$c_p = \frac{\langle f, R_{n,p}\rangle_w}{\|R_{n,p}\|_w^2} = \frac{\int_a^b f(\rho) R_{n,p}(\rho)\rho\, d\rho}{\int_a^b R_{n,p}^2(\rho)\rho\, d\rho}$$

which, with some fussing, can be put in the form

$$c_n = \frac{\pi^2 \mu_p}{2} \frac{J_n^2(\sqrt{\mu_p}b)\int_a^b f(\rho)R_{n,p}(\rho)\rho\, d\rho}{J_n^2(\sqrt{\mu_p}a) - J_n^2(\sqrt{\mu_p}b)} \qquad (24)$$

(exercise 15). Note that (23) provides an *infinite* number of eigenfunction expansions — one for each real value of $n \geq 0$. The eigenvalues themselves depend on $n$ through eq. (22), and a more honest notation would be $\mu_{n,p}$.

Results for other boundary conditions for (20) appear in Table 6.3, entry #4. The final assemblage of the complete solution for three-dimensional problems such as the one depicted in Figure 6 will be discussed in Chapter 7.

∎

**Example 4.** In Section 5.3 we showed that the separation of the Helmholtz equation in spherical coordinates

$$\nabla^2\psi = \frac{1}{r^2}\frac{\partial}{\partial r}\left(r^2\frac{\partial\psi}{\partial r}\right) + \frac{1}{r^2\sin\phi}\frac{\partial}{\partial\phi}\left(\sin\phi\frac{\partial\psi}{\partial\phi}\right) + \frac{1}{r^2\sin^2\phi}\frac{\partial^2\psi}{\partial\theta^2} = K\psi \quad (25)$$

leads to the differential equations

$$\Theta'' = \lambda \Theta \tag{26}$$

$$(\sin \phi \, \Phi')' + \frac{\lambda}{\sin \phi} \Phi = \mu \sin \phi \, \Phi \tag{27}$$

$$r^2 R'' + 2r R' + \mu R = K r^2 R . \tag{28}$$

Equation (26) is, if course, the same eigenfunction problem that we solved in Example 2, so $\lambda_n = -n^2$. In Sections 6.4 and 6.10 we will see that insertion of these $\lambda$'s into the Legendre equation (27) yields eigenvalues for $\mu$ of the form $-\ell(\ell + 1)$, where $\ell$ is a nonnegative integer. For the sake of illustration, then, we simply jump straight to the radial equation (28), which thus takes the form

$$r^2 R'' + 2r R' - \ell(\ell + 1)R = K r^2 R . \tag{29}$$

Equation (29) is a transformed version of the Bessel equation, and its eigenvalues $K$ are typically negative. From the generic Bessel solution ((21), Sec. 1.2), we can show (exercise 6) that its general solution is a linear combination of any two members of the following family of solutions, known as "Spherical Bessel functions": with $\xi = \sqrt{-K} r$,

$j_\ell(\xi) = \sqrt{\pi/2\xi} \, J_{\ell+1/2}(\xi)$    (Spherical Bessel function of the first kind)

$y_\ell(\xi) = \sqrt{\pi/2\xi} \, Y_{\ell+1/2}(\xi)$    (Spherical Bessel function of the second kind)

$h_\ell^{(1)}(\xi) = j_\ell(\xi) + i y_\ell(\xi), \quad h_\ell^{(2)}(\xi) = j_\ell(\xi) - i y_\ell(\xi)$

(Spherical Hankel functions of the first and second kind).

$$\tag{30}$$

As a matter of fact the spherical Bessel functions of integer order can be expressed in terms of elementary functions; recall exercise 3 in Section 2.5. Thus for example

$$j_0(\xi) = \frac{\sin \xi}{\xi}, \quad j_1(\xi) = \frac{\sin \xi}{\xi^2} - \frac{\cos \xi}{\xi},$$

$$j_2(\xi) = \frac{3 - \xi^2}{\xi^3} \sin \xi - \frac{3}{\xi^2} \cos \xi,$$

$$y_0(\xi) = -\frac{\cos \xi}{\xi}, \quad y_1(\xi) = -\frac{\cos \xi}{\xi^2} - \frac{\sin \xi}{\xi},$$

$$y_2(\xi) = \frac{\xi^2 - 3}{\xi^3} \cos \xi - \frac{3}{\xi^2} \sin \xi.$$

See Fig. 9.

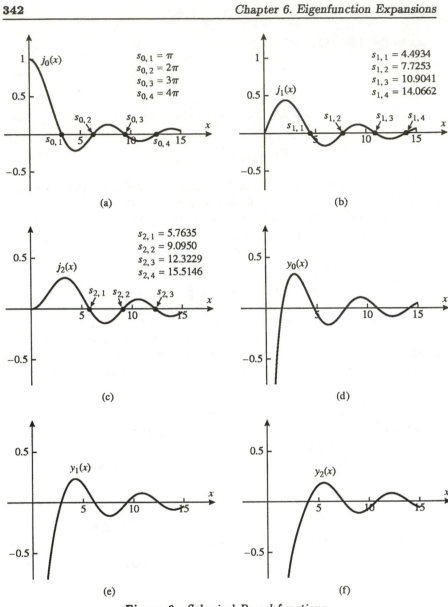

**Figure 9**   *Spherical Bessel functions*

The analysis from here on imitates what we have done before. For Dirichlet conditions the eigenfunctions take the form (3)

$$R_{\ell,n}(r) = y_\ell(\sqrt{-K_{\ell,n}}a)j_\ell(\sqrt{-K_{\ell,n}}r) - j_\ell(\sqrt{-K_{\ell,n}}a)y_\ell(\sqrt{-K_{\ell,n}}r) \quad (31)$$

and by eq. (4) the eigenvalue $K_{\ell,n}$ is the $n$th root of

$$y_\ell(\sqrt{-K}a)j_\ell(\sqrt{-K}b) - j_\ell(\sqrt{-K}a)y_\ell(\sqrt{-K}b) = 0.$$

In (31) we have explicitly acknowledged the dependence of $K$ on $\ell$. Equation (29) is in Sturm-Liouville form, so the weight factor is seen to be $w(r) = r^2$. The eigenfunction expansion takes the form

$$
\begin{aligned}
f(r) &= \sum_{n=1}^{\infty} c_n R_{\ell,n}(r) \\
&= \sum_{n=1}^{\infty} c_n \{ y_\ell(\sqrt{-K_{\ell,n}}a) j_\ell(\sqrt{-K_{\ell,n}}r) - j_\ell(\sqrt{-K_{\ell,n}}a) y_\ell(\sqrt{-K_{\ell,n}}r) \}, \\
c_n &= \frac{\int_a^b f(r) R_{\ell,n}(r) r^2\, dr}{\int_a^b R_{\ell,n}(r)^2 r^2\, dr}.
\end{aligned}
$$

The denominator can be expressed in terms of the usual Bessel functions (exercise 7):

$$
c_n = 2aK_{\ell,n}^2 \frac{J_{\ell+1/2}^2(\sqrt{-K_{\ell,n}}b)}{J_{\ell+1/2}^2(\sqrt{-K_{\ell,n}}a) - J_{\ell+1/2}^2(\sqrt{-K_{\ell,n}}b)} \int_a^b f(r) R_{\ell,n}(r) r^2\, dr.
$$

$$(32)$$

$\blacksquare$

We reiterate that although there are endless sets of eigenfunction expansions resulting from regular Sturm- Liouville problems, they are all handled by the formalism described at the end of Section 6.2. Do not be dismayed by the formidable integrals appearing in the coefficient formulas; accurate *numerical* integration algorithms are readily available. (See the discussion in Appendix A.6.)

**Readers: our final example is extremely tedious. The Lebedev expansions are important in many physical situations (such as edge diffraction anlysis), but they are so unattractive that most textbooks omit them. To keep up your enthusiasm for partial differential equations, your author advises that it might be wise to jump to Section 6.4 at this point, and bookmark the remainder of the present section for future reference.**

**Example 5.** We have seen that separation of Laplace's equation for the equilibrium temperature in the heat pipe of Fig. 6 leads to the following equation for the radial dependence:

$$
R'' + \frac{1}{\rho}R' + \frac{\lambda}{\rho^2}R = -\mu R
$$

(eq. (14) repeated). For the situation in Fig. 6, the boundary conditions on $\Theta$ were homogeneous (periodic), and they dictated the values of $\lambda = -n^2$

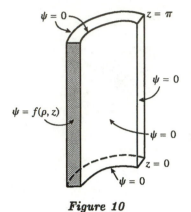

**Figure 10**

in (14); the $\rho$ equation, then, governed the values for $\mu$. For the *quarter heat pipe* shown in Fig. 10, the nonhomogeneity appears on a $\theta$ face and the $z$ faces have homogeneous conditions. Therefore in this case the values of $\mu$ are specified by the $z$ equations

$$Z'' = \mu Z \quad \text{(eq. (13) repeated)}, \quad Z(0) = Z(\pi) = 0,$$

whose eigenvalues we recognize from Section 6.1 as $\mu_s = -s^2$; $s = 1, 2, 3, \cdots$ (sine series). Consequently the radial equation is more appropriately written

$$R'' + \frac{1}{\rho}R' - s^2 R = -\frac{\lambda}{\rho^2}R, \tag{33}$$

and together with the Dirichlet conditions $R(a) = R(b) = 0$ (33) now determines the values of $\lambda$.

For printing convenience we recast the differential equation and boundary conditions in terms of the standard "$y(x)$" notation. In these variables the Sturm-Liouville form of equation (33) is easily seen to be

$$(xy')' - s^2 xy = -\lambda\frac{1}{x}y, \quad y(a) = y(b) = 0. \tag{34}$$

Because of the minus sign on the right, we anticipate that most of the eigenvalues will be positive. So we optimistically take $\lambda = \omega^2$ and invoke the generic Bessel function solution ((21), Sec. 1.2) to express the general solution of the differential equation as

$$C_1 I_{i\omega}(sx) + C_2 K_{i\omega}(sx), \tag{35}$$

with modified Bessel functions. Note that the *order* of the Bessel function is (typically) imaginary in this application.

Does this make sense? The series defining the modified Bessel function of the first kind is given by (Sec. 2.5)

$$I_p(z) = (z/2)^p \sum_{k=0}^{\infty} \frac{(z/2)^{2k}}{k!\Gamma(p+k+1)}$$

and a little algebra (exercise 16) transforms this into

$$I_{i\omega}(sx) = \frac{(sx/2)^{i\omega}}{\Gamma(i\omega+1)} \sum_{k=0}^{\infty} \frac{(sx/2)^{2k}}{k!(i\omega+1)(i\omega+2)\cdots(i\omega+k)}$$

$$= \frac{e^{i\omega\ln(sx/2)}}{\Gamma(i\omega+1)} \sum_{k=0}^{\infty} \frac{(sx/2)^{2k}}{k!(i\omega+1)(i\omega+2)\cdots(i\omega+k)} \tag{36}$$

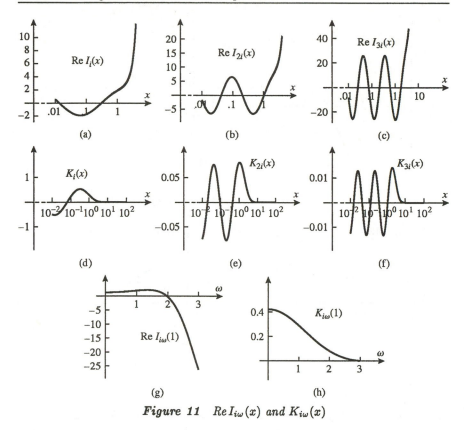

**Figure 11**  $\mathrm{Re}\,I_{i\omega}(x)$ *and* $K_{i\omega}(x)$

(The complex Gamma function has been discussed in exercise 6, Sec. 2.1.)

The series (36) converges to a well-defined complex-valued function of $x$ for all $\omega$, having the property that $I_{i\omega}$ is the complex conjugate of $I_{-i\omega}$ (exercise 16):

$$\overline{I_{i\omega}(sx)} = I_{-i\omega}(sx). \qquad (37)$$

As usual (exercise 8, Sec. 1.1) it follows that the real and imaginary parts of $I_{i\omega}(sx)$ and $K_{i\omega}(sx)$ are also solutions of the differential equation (34). In fact some further algebra reveals that the modified Bessel function of the second kind $K_{i\omega}$ is a constant multiple of the imaginary part of $I_{i\omega}$ (exercise 17). (The real-valued function $K_{i\omega}(sx)$ is known as a *Macdonald function*.)

The real part of the general solution (35), then, can be written as

$$y(x) = C_1 \mathrm{Re}\{I_{i\omega}(sx)\} + C_2 K_{i\omega}(sx).$$

Typical plots of $K_{i\omega}$ and $\mathrm{Re}\{I_{i\omega}\}$ are displayed in Fig. 11.

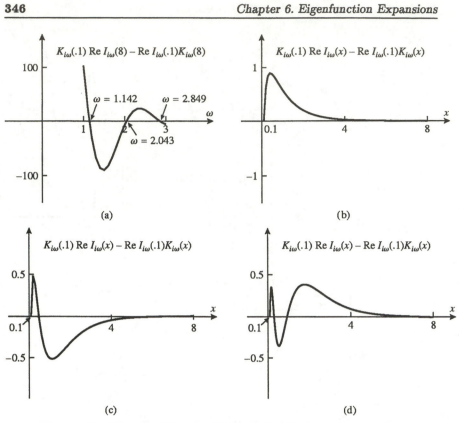

Figure 12    *Eigenfunctions for Example 3: (a) eigenvalue equation;*
(b) $\omega = 1.142$; (c) $\omega = 2.043$; (d) $\omega = 2.849$.

The combination (3) of these solutions meeting the first boundary condition is

$$y(x; \lambda = \omega^2) = K_{i\omega}(sa)\,\mathrm{Re}\{I_{i\omega}(sx)\} - \mathrm{Re}\{I_{i\omega}(sa)\}\,K_{i\omega}(sx)$$

and the eigenvalues $\lambda = \omega^2$ are determined by imposing the second boundary condition (4). Thus, for a Dirichlet condition at $x = b$ the equation

$$K_{i\omega}(sa)\,\mathrm{Re}\{I_{i\omega}(sb)\} - \mathrm{Re}\{I_{i\omega}(sa)\}\,K_{i\omega}(sb) = 0 \qquad (38)$$

determines a discrete set of values $\omega_1 < \omega_2 < \dots$ clustering at infinity, and the associated solution

$$\phi_{s,n}(x) = K_{i\omega_n}(sa)\,\mathrm{Re}\{I_{i\omega_n}(sx)\} - \mathrm{Re}\{I_{i\omega_n}(sa)\}\,K_{i\omega_n}(sx)$$

is the $n$th eigenfunction for the system. As a matter of fact, a little more algebra reveals that the $Re$ operator is unnecessary in this case (exercise 18).

The weight for the orthogonality relation is the coefficient of $\mu y$ in the Sturm-Liouville form (34), so $w(x) = 1/x$. Therefore

$$\int_a^b \phi_{s,n}(x)\phi_{s,m}(x)\frac{1}{x}\,dx = 0 \quad \text{if} \quad n \neq m$$

and the *Lebedev expansion formula* reads

$$f(x) = \sum_{n=1}^{\infty} c_n\{K_{i\omega_n}(sa)I_{i\omega_n}(sx) - I_{i\omega_n}(sa)K_{i\omega_n}(sx)\},$$

$$
\begin{aligned}
c_n &= \frac{\langle f(x), \phi_{s,n}(x)\rangle_w}{\|\phi_{s,n}(x)\|_w^2} \\
&= \frac{\int_a^b f(x)\{K_{i\omega_n}(sa)I_{i\omega_n}(sx) - I_{i\omega_n}(sa)K_{i\omega_n}(sx)\}\,dx/x}{\int_a^b \{K_{i\omega_n}(sa)I_{i\omega_n}(sx) - I_{i\omega_n}(sa)K_{i\omega_n}(sx)\}^2\,dx/x}.
\end{aligned}
$$

Alternatively, (exercise 19)

$$c_n = \frac{2\omega_n K_{i\omega_n}(sb)\int_a^b f(x)\{K_{i\omega_n}(sa)I_{i\omega_n}(sx) - I_{i\omega_n}(sa)K_{i\omega_n}(sx)\}\,dx/x}{K_{i\omega_n}(sa)\frac{\partial}{\partial\omega}\{K_{i\omega_n}(sa)I_{i\omega_n}(sb) - I_{i\omega_n}(sa)K_{i\omega_n}(sb)\}}$$

$$\tag{39}$$

Eigenfunctions for these boundary conditions are displayed in Fig. 12. ∎
More general Lebedev expansions are displayed in entry #5 of Table 6.3.

# Exercises 6.3

**1.** Derive the solution to Example 1 when $\kappa = 1$.

**2.** Using reasonable starting values in eq. (8), confirm the numbers in Fig. 2. What is the next intersection point? What happens if the first equation in (8) is replaced by $\omega_{new} = \tanh^{-1}[\omega_{old}/2]$? Do you know why? (The explanation — the "contracting map principle" — is discussed in Appendix A.5.)

**3.** What equations would you iterate to obtain the values of $\omega_i$ in Example 1 for the case $\kappa = .1$? What are the first four $\omega_i$ if $\kappa = -1$?

**4.** Show that the denominators in the solution display for Example 1 can be simplified, by deriving the identity

$$\int_0^1 \sin^2 \omega_n x\,dx = \frac{1}{2} - 5\cos^2 \omega_n.$$

**5.** Prove that the only values of $\lambda$ which render (16) $2\pi$-periodic are negative squares of integers. (Hint: apply determinant theory to (15).)

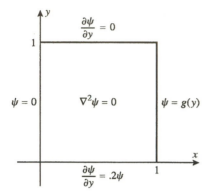

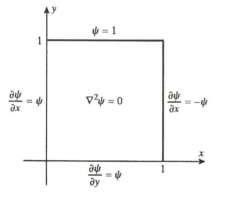

*Figure 13   Exercise 8*

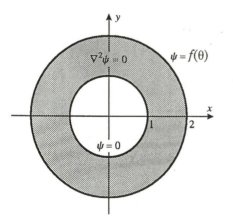

*Figure 14   Exercise 9*

**6.** Derive the solution forms (30).

**7.** Derive the coefficient formula (32). (Hint: with eqs. (30) in mind, compare eq. (31) to eqs. (21, 22) in Example 3. It follows that a spherical Bessel function expansion can be recast as a regular Bessel function expansion. Thus substitute in (23, 24) to derive (32).

**8.** Solve the problem depicted in Fig. 13.

**9.** Solve the problem depicted in Fig. 14.

**10.** Consider the steady-state temperature problem for a washer (an "annulus") depicted in Fig. 15. Show that the solution is given by

$$\psi(r,\theta) = c_0 \ln r + \sum_{n=1}^{\infty} (c_n \cos n\theta + d_n \sin n\theta)(r^n - r^{-n}),$$

$$
\begin{aligned}
c_0 &= \frac{1}{2\pi \ln 2} \int_0^{2\pi} f(\theta)\, d\theta, \\
c_{n>0} &= \frac{1}{\pi(2^n - 2^{-n})} \int_0^{2\pi} f(\theta) \cos n\theta\, d\theta, \\
d_n &= \frac{1}{\pi(2^n - 2^{-n})} \int_0^{2\pi} f(\theta) \sin n\theta\, d\theta.
\end{aligned}
$$

**11.** Solve the problem depicted in Fig. 16.

**12.** Suppose that $\nu$ is a given parameter in the damped harmonic oscillator equation $y'' + \nu y' = \lambda y$, with $\lambda$ as the eigenvalue. Show that with Dirichlet boundary conditions $y(0) = y(1) = 0$, the associated eigenfunction expansion is

$$
\begin{aligned}
f(x) &= \sum_{n=1}^{\infty} a_n e^{-\nu x/2} \sin n\pi x, \\
a_n &= 2 \int_0^1 f(x) e^{+\nu x/2} \sin n\pi x\, dx.
\end{aligned}
$$

**13.** In this exercise we shall extend the analysis of the damped harmonic oscillator equation (exercise 12) to more general boundary conditions, by first applying a mathematical transformation which removes the damping. Consider the Sturm-Liouville problem

$$y'' + \nu y' = \lambda y \tag{40}$$

$$\alpha y(0) + \beta y'(0) = 0, \quad \gamma y(a) + \delta y'(a) = 0. \tag{41}$$

*Figure 15   Exercise 10*

(a) Show that the substitution $y(x) = e^{-\nu x/2} z(x)$ transforms (40,41) to the form

$$z'' = \tilde{\lambda} z, \quad (\tilde{\lambda} = \lambda + \nu^2/4) \tag{42}$$

$$Az(0) + Bz'(0) = 0, \quad Cz(a) + Dz'(a) = 0 \tag{43}$$

where $A = \alpha - \beta\nu/2$, $B = \beta$, $C = \gamma - \delta\nu/2$, $D = \delta$.

(b) By taking the general solution to (42) to be $z(x) = \sin(\omega x + \phi)$ show that the eigenfunctions of (42,43) can be written

$$\phi(x) = \sin(\omega_n x + \arctan[-B\omega_n/A]) \quad (n = 1, 2, 3, \dots)$$

where the eigenvalues $\tilde{\lambda} = -\omega_n^2$ are generated by the solutions to

$$C\sin(\omega a + \arctan[-B\omega/A]) = -D\omega\cos(\omega a + \arctan[-B\omega/A]),$$

and that the function $\phi_0(x) = 1$ is also an eigenfunction if $A = C = 0$.

(c) Show that the eigenfunction expansion corresponding to (42,43) can be written

$$F(x) = \sum_{n=1}^{\infty} c_n \sin(\omega_n x + \arctan[-B\omega_n/A]) + c_0,$$

$$c_n = \frac{2\int_0^a F(x)\sin(\omega_n x + \arctan[-B\omega_n/A])\,dx}{a + \frac{CD}{C^2+D^2\omega_n^2} - \frac{AB}{A^2+B^2\omega_n^2}}$$

for $n > 0$ and if $A = C = 0$ then

$$c_0 = \int_0^a F(x)\,dx/a$$

(but $c_0 = 0$ otherwise).

(d) Let $F(x) = f(x)e^{\nu x/2}$ and manipulate the formulas in (c) to obtain the expansion of $f(x)$ in terms of the eigenfunctions of (40,41). Confirm your answer with the entry #2 in Table 6.3 at the end of this chapter.

14. In this exercise we shall analyze the Sturm-Liouville problem for the *equidimensional* equation by applying a mathematical transformation which reduces it to the damped harmonic oscillator problem of the previous exercise. Consider the system

$$x^2 y'' + \sigma x y' = \lambda y, \quad 0 < a \leq x \leq b < \infty,$$

$$\alpha y(a) + \beta y'(a) = 0, \quad \gamma y(b) + \delta y'(b) = 0. \tag{44}$$

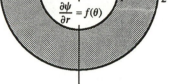

*Figure 16    Exercise 11*

(a) Show that the change of variable $x = ae^t$, $Y(t) = y(ae^t)$ transforms this problem to the following (exercise 2, sec. 2.2):

$$Y'' + (\sigma - 1)Y' = \lambda Y,$$

$$\alpha Y(0) + \frac{\beta}{a}Y'(0) = 0, \quad \gamma Y(\ln \frac{b}{a}) + \frac{\delta}{b}Y'(\ln \frac{b}{a}) = 0. \qquad (45)$$

(b) Use the results of exercise 13 to analyze Eqs. (45), and invert the change of variable to derive the eigenfunctions and expansion for (44) as displayed in Table 6.3, entry #3 at the end of this chapter.

15. Use the Bessel function identities in Table 2.2 to verify the final form of formula (24).

16. Derive eqs. (36) and (37).

17. Verify the identity

$$K_{i\omega}(sx) = \frac{-\pi}{\sinh \omega \pi} \text{Im}\{I_{i\omega}(sx)\} = \frac{\pi i/2}{\sinh \omega \pi}\{I_{i\omega}(sx) - I_{-i\omega}(sx)\}.$$

18. Verify that the expression in eq. (38) equals

$$K_{i\omega_n}(sa)I_{i\omega_n}(sx) - I_{i\omega_n}(sa)K_{i\omega_n}(sx).$$

(Hint: derive the identity

$$iK_{i\nu}(na)\text{Im}\{I_{i\nu}(nx)\} - i\text{Im}\{I_{i\nu}(na)\}K_{i\nu}(nx) = 0$$

and add it to (38)).

19. The following steps will guide you through a derivation of the final form of eq. (39). Refer to the third "normalization integral" in Table 2.2.

(a) Let $\Xi_c(ut) = \alpha_c J_c(ut) + \beta_c H_c^{(1)}(ut)$, $\Upsilon_d(vt) = \gamma_d J_d(vt) + \delta_d H_d^{(1)}(vt)$, where the coefficients in the linear combinations *are not constants* - hence they are subscripted. Argue that the integrated terms in the Table entry must be altered to read

$$\int \left\{(u^2 - v^2)t - \frac{c^2 - d^2}{t}\right\} \Xi_c(ut)\Upsilon_d(vt)\, dt =$$

$$\{u\tilde{\Xi}_{c+1}(ut)\Upsilon_d(vt) - v\Xi_c(ut)\tilde{\Upsilon}_{d+1}(vt)\}t - (c-d)\Xi_c(ut)\Upsilon_d(vt)$$

where

$$\tilde{\Xi}_{c+1}(ut) = \alpha_c J_{c+1}(ut) + \beta_c H_{c+1}^{(1)}(ut)$$

and similarly for $\tilde{\Upsilon}_{d+1}(vt)$.

(b) Now suppose that $u = v = \xi$, that $\Upsilon_d(\xi t) = 0$ at both endpoints of the interval of integration, that $t = x$, and that the corresponding coefficient functions in $\Xi$ and $\Upsilon$ are identical (i.e. $\gamma_d = \alpha_d$, $\delta_d = \beta_d$, so that $\Xi_d(\xi t) = \Upsilon_d(\xi t)$). Show that the normalization integral can be expressed

$$- (c + d) \int_a^b \Upsilon_c(\xi x) \Upsilon_d(\xi x) \frac{dx}{x}$$
$$= -\xi x \frac{\Upsilon_c(\xi x) - \Upsilon_d(\xi x)}{c - d} \tilde{\Upsilon}_{d+1}(\xi x)|_{x=a}^{x=b}. \quad (46)$$

(c) Let $c \to d$ in (46) and derive

$$\int_a^b \Upsilon_d^2(\xi x) \frac{dx}{x} = \frac{\xi x}{2d} \left[ \frac{\partial}{\partial d} \Upsilon_d(\xi x) \right] \tilde{\Upsilon}_{d+1}(\xi x)|_{x=a}^{x=b}. \quad (47)$$

(d) Verify that the eigenfunction for Example 5 is expressed by

$$\phi_{s,n}(x) = K_{i\omega_n}(sa) I_{i\omega_n}(sx) - I_{i\omega_n}(sa) K_{i\omega_n}(sx) =$$
$$- i^{-i\omega} K_{i\omega}(sa) J_{i\omega}(isx) + \frac{\pi i}{2} i^{i\omega} I_{i\omega}(sa) H_{i\omega}^{(1)}(isx) \quad (\omega = \omega_n).$$
$$(48)$$

(e) It follows from (48) that $\phi_{s,n}(x)$ is a suitable instance of $\Upsilon_d(\xi x)$, with $d = i\omega_n$ and $\xi = is$. It also vanishes at the endpoints; in fact the eigenvalue equation (38) states $\Upsilon_{i\omega_n}(isb) = 0$, while $\Upsilon_{i\omega}(isa) = 0$ for *all* $\omega$ (and therefore the contribution of the lower limit in (46) is zero!). Use the identities in Table 2.2 to show that

$$\tilde{\Upsilon}_{i\omega_n+1}(isb) = \frac{\omega_n}{sb} \{ K_{i\omega_n}(sa) I_{i\omega_n}(sb) - I_{i\omega_n}(sa) K_{i\omega_n}(sb) \}$$
$$+ i \{ K_{i\omega_n}(sa) I'_{i\omega_n}(sb) - I_{i\omega_n}(sa) K'_{i\omega_n}(sb) \} \quad (49)$$

(f) Use (38) to reduce (49) to

$$\tilde{\Upsilon}_{i\omega_n+1}(isb) =$$
$$(0) + i \frac{K_{i\omega_n}(sa)}{K_{i\omega_n}(sb)} \{ K_{i\omega_n}(sb) I'_{i\omega_n}(sb) - I_{i\omega_n}(sb) K'_{i\omega_n}(sb) \}$$

(g) Use the Wronskian identity (Table 2.2) to derive

$$\tilde{\Upsilon}_{i\omega_n+1}(isb) = \frac{i}{sb} K_{i\omega_n}(sa) / K_{i\omega_n}(sb).$$

(h) Assemble all this to derive (39).

# 6.4   Singular Sturm-Liouville Expansions

**Summary**   When one or both of the endpoints in a Sturm-Liouville problem goes to $\pm\infty$, or when the coefficient of $y''$ in the differential equation diverges at an endpoint, the problem becomes singular. Usually one then has to forfeit the boundary condition at the singular endpoint, imposing in its stead only the condition of finiteness. The resulting set of eigenvalues can remain discrete, in which case the eigenfunction expansions are infinite series as in the regular case, or the eigenvalues can fill up a continuum, with the expansion summations evolving into integrals. Common examples are:

- the Fourier sine and cosine transforms, where one endpoint lies at infinity,

- the Fourier transform, where both endpoints lie at infinity,

- the Fourier-Bessel series, where the coefficient for the radial dependence diverges at the origin in cylindrical and spherical coordinates, and

- the Legendre series in the azimuthal angle $\phi$, where the coefficient diverges on the $z$-axis.

The Sturm-Liouville problem of finding the eigenfunctions of a differential system

$$a_2(x)y'' + a_1(x)y' + a_0(x)y = \lambda g(x)y \quad (\text{or } \mathcal{M}y = \lambda gy)$$

$$\alpha y(a) + \beta y'(a) = 0, \quad \gamma y(b) + \delta y'(b) = 0$$

becomes quite complicated in general when the coefficient $a_2(x)$ goes to zero at one of the end points ($a$ or $b$), or when the interval $(a,b)$ is infinite. Such situations are called *singular* Sturm-Liouville problems. In this section we shall sketch the derivations of the four cases which occur most commonly in engineering applications. They present enough of the flavor of the subject for the reader to be able to make intelligent use of the exhaustive tabulation of eigenfunction expansions appearing in Table 6.3 at the end of this chapter. For the sake of completeness, heuristic derivations of all the expansions are given in the optional Sections 6.7 through 6.10.

It will become clear, from the diversity of the four basic expansion forms discussed in this section, that any *general* theory of singular Sturm-Liouville problems must be very complex (and beyond the scope of this book). Thus except for a brief excursion in Section 6.5 to unify our findings we shall direct the interested reader to the references for a fuller understanding. Imitating the procedures we use, however, should enable the reader to make

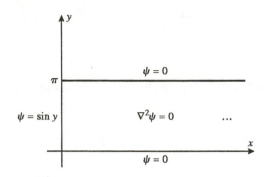

**Figure 1**   *Specifications for Example 1*

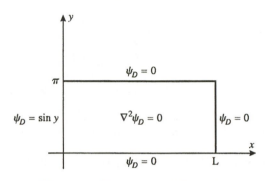

**Figure 2**   *Truncation of Example 1*

a correct analysis of most situations arising in practice (if they are not covered in Table 6.3).

To get a feeling for what to expect from singular Sturm-Liouville problems, we begin by analyzing a few examples entailing the solution of Laplace's equation in an unbounded domain.

**Example 1.** The steady state temperature problem depicted in Fig. 1 calls for solutions to Laplace's equation in a semi-infinite rectangle. Physically speaking, of course, this is an unrealistic idealization, perhaps concocted to simulate a configuration like that in Fig. 2, in the limit where the length $L$ becomes infinite. We employ the subscript "$D$" to reflect the fact that we have arbitrarily imposed a homogeneous *Dirichlet* condition at $x = L$ in Fig. 2.

The solution of the problem in Fig. 2 should be very familiar by now. Laplace's equation separates into the harmonic oscillator equations

$$X'' = \lambda X, \quad Y'' = -\lambda Y \tag{1}$$

and since homogeneous conditions are imposed on the $y = constant$ faces, we are led to the usual eigenfunctions $Y_n(y) = \sin ny$, $\lambda_n = -n^2$ ($n = 1, 2, 3, \dots$). They are mated with the $X$ factors

$$X_n(x) = C_1 \cosh nx + C_2 \sinh nx, \tag{2}$$

and an appropriate combination meeting the Dirichlet condition at $x = L$ is

$$X_n(x) = \sinh nL \cosh nx - \cosh nL \sinh nx = \sinh n(L - x). \qquad (3)$$

The general solution fulfilling all the homogeneous equations is then given by

$$\psi_{Dirichlet}(x, y) = \sum_{n=1}^{\infty} a_n \sin ny \sinh n(L - x),$$

and obviously the nonhomogeneous boundary condition is met with the choice

$$
\begin{aligned}
\psi_{Dirichlet}(x, y) &= \frac{\sin y \sinh(L - x)}{\sinh L} \\
&= \sin y \frac{\sinh L \cosh x - \cosh L \sinh x}{\sinh L} \\
&= \sin y \,[\cosh x - \coth L \sinh x]. \qquad (4)
\end{aligned}
$$

To obtain a solution to the unbounded problem in Fig. 1, we let $L$ go to infinity; thus $\coth L \to 1$ and $\psi_{Dirichlet}$ tends to the solution

$$\psi_{Dirichlet}(x, y) \to \psi(x, y) = \sin y \,[\cosh x - \sinh x] = \sin y \, e^{-x} \qquad (5)$$

which, sure enough, satisfies all of the equations in Fig. 1.

What if we had elected to impose a homogeneous *Neumann* condition at $x = L$ in Fig. 2, instead of the Dirichlet condition? In such a case the appropriate combination of $X$ factors (2) would have been

$$X_n(x) = \cosh nL \cosh nx - \sinh nL \sinh nx = \cosh n(L - x),$$

and instead of (4) we would have the solution

$$
\begin{aligned}
\psi_{Neumann}(x, y) &= \frac{\sin y \cosh(L - x)}{\cosh L} \\
&= \sin y \frac{\cosh L \cosh x - \sinh L \sinh x}{\cosh L} \\
&= \sin y \,[\cosh x - \tanh L \sinh x]. \qquad (6)
\end{aligned}
$$

*But the limiting form of the solution would still agree with* (5):

$$\psi_{Neumann}(x, y) \to \sin y \,[\cosh x - \sinh x] = \psi(x, y).$$

In fact, exercises 1 and 2 show that imposing a variety of boundary conditions at $x = L$ all lead, in the limit, to the same solution (5).

Does this mean that we can ignore the boundary "at $\infty$" for the problem in Fig. 1, confident that $\psi(x, y) = e^{-x} \sin y$ is the only solution?

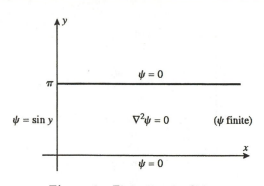

**Figure 3**   *Finiteness condition*

The answer is definitely NO; if we add to $\psi(x,y)$ any constant multiple of the function $\sin y \sinh x$,

$$
\begin{aligned}
\psi(x,y;K) &= \psi(x,y) + K \sin y \sinh x \\
&= e^{-x} \sin y + K \sin y \sinh x, \qquad (7)
\end{aligned}
$$

we obtain an equally valid solution to all the equations!

However, if $K \neq 0$ the solution $\psi(x,y;K)$ *diverges* as $x$ goes to infinity, because of the factor $\sinh x$. This is physically absurd. *The unique member of the solution family $\psi(x,y;K)$ which stays finite as $x \to \infty$ is $\psi(x,y)$:*

$$
\lim_{x \to \infty} \psi(x,y) = \lim_{x \to \infty} e^{-x} \sin y = 0.
$$

So let's go back and rework the problem in Fig. 1 *directly*, imposing only the condition of *finiteness* as $x \to \infty$ (Fig. 3).

We separate the variables and obtain the $Y$ eigenfunctions $\{\sin ny\}$ as before. Next we choose the coefficients in (2)

$$
X_n(x) = C_1 \cosh nx + C_2 \sinh nx
$$

to make $X_n(x)$ *finite* at $\infty$. We take $C_2 = -C_1$, making $X_n(x) = e^{-nx}$. The general solution looks like

$$
\psi(x,y) = \sum_{n=1}^{\infty} c_n e^{-nx} \sin ny
$$

and the particular solution meeting the nonhomogeneous boundary condition is

$$
\psi(x,y) = e^{-x} \sin y
$$

in accordance with (5).  ■

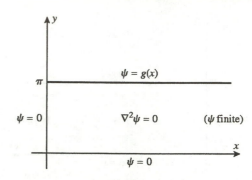

*Figure 4   Specifications for Example 2*

This example provides guidelines for what to expect at singular endpoints in Sturm-Liouville problems; as a rule, *specific* boundary conditions are unnecessary and inappropriate, but the more nebulous *finiteness* condition suffices quite adequately. This notion will be reinforced by further examples in this section. A more thorough discussion of the significance of the finiteness postulate, and its limitations, is given in Sec. 6.5.

**Example 2.** We shift the nonhomogeneous boundary condition from the left edge in Fig. 1 to the top edge in Fig. 4.

For this configuration the procedure of Sec. 5.2 dictates that we begin by analyzing the separated $X$-equation (1) first. The customary solutions, fulfilling the homogeneous Dirichlet condition on the left, are

$$X(x) = \begin{cases} \sinh \sqrt{\lambda}\, x & \text{if } \lambda > 0, \\ x & \text{if } \lambda = 0, \\ \sin \sqrt{-\lambda}\, x & \text{if } \lambda < 0. \end{cases} \tag{8}$$

At this point in a *regular* Sturm-Liouville problem we would now impose the boundary condition on the right to determine the eigenvalues $\lambda$. Since the right end point is singular ($\infty$), this "boundary condition" is

$$X(x) \text{ must be finite as } x \to \infty. \tag{9}$$

Imposing (9) on the solution forms (8) tells us that *every negative value of $\lambda$ is an eigenvalue*, and the eigenfunctions are the sines. The eigenfunction expansion of a function $f(x)$, then, must look something like

$$f(x) = \text{``} \sum_{(\text{all } \lambda < 0)} c_\lambda \sin \sqrt{-\lambda}\, x, \text{''} \quad 0 < x < \infty. \tag{10}$$

But we recognize this form from eqs. (5, 6) of Sec. 3.5; it's the Fourier Sine Transform formula:

$$f(x) = \int_0^\infty S(\omega) \sin \omega x \, d\omega, \tag{11}$$

with $\lambda = -\omega^2$, and the formula for the "expansion coefficient" is

$$S(\omega) = \frac{2}{\pi} \int_0^\infty f(x) \sin \omega x \, dx \qquad (12)$$

Thus our very first encounter with a singular Sturm-Liouville eigenfunction expansion reveals the fragility of the theory of Sec. 6.2. The eigenvalue set is the *continuum* $\infty < \lambda < 0$ in this case; it can't be enumerated $\lambda_1 > \lambda_2 > \lambda_2 > \cdots$. And there is no $n$th eigenfunction with $(n-1)$ interior zeros; all the eigenfunctions $\sin \omega x$ oscillate infinitely often.

*For singular Sturm-Liouville problems it is customary to refer to the set of eigenvalues as the "spectrum", and to reserve the word "eigenvalue" for selected members of the spectrum. This will be discussed in Sec. 6.5. For the moment we shall use the terms "eigenvalue set" and "spectrum" interchangeably.*

The fact that the spectrum here consists of a continuum rather than a discrete set requires that we revise our notion of orthogonality for eigenfunctions. After all, the implementation of the coefficient-extracting procedure of Sec. 6.2 leads to gibberish when applied to the sine transform eq. (11); what sense can be made of

$$\text{``}S(\omega) \, d\omega\text{''} = \frac{\langle f(x), \sin \omega x \rangle}{\| \sin \omega x \|^2} = \frac{\int_0^\infty f(x) \sin \omega x \, dx}{\int_0^\infty | \sin \omega x |^2 \, dx} \, ?$$

The theory of distributions provides the proper machinery for expressing the orthogonality relation. If we insert (12) into (11) and formally reverse the order of integration we derive

$$\begin{aligned} f(x) &= \int_0^\infty \left\{ \int_0^\infty \frac{2}{\pi} f(\xi) \sin \omega \xi \, d\xi \right\} \sin \omega x \, d\omega \\ &= \int_0^\infty \left\{ \int_0^\infty \frac{2}{\pi} \sin \omega \xi \sin \omega x \, d\omega \right\} f(\xi) \, d\xi, \end{aligned}$$

and consequently we identify $\int_0^\infty \frac{2}{\pi} \sin \omega \xi \sin \omega x \, d\omega$ as $\delta(x - \xi)$ (similar to eq. (2), Sec. 3.7). The orthogonality relation for a continuous spectrum thus uses the delta function instead of the Kronecker delta $\delta_{ij}$ (eq. (23), Sec. 6.2), and for the Fourier sine transform it reads

$$\int_0^\infty \sin \omega \xi \sin \omega x \, d\omega = \frac{\pi}{2} \delta(x - \xi)$$

or, with a change of variables to be more consistent with eq. (23) of Sec. 6.2,

$$\int_0^\infty \sin \omega_1 x \, \sin \omega_2 x \, dx = \frac{\pi}{2} \delta(\omega_1 - \omega_2). \qquad (13)$$

Returning to the problem in Fig. 4, we now consider the $Y$-solutions of (1) associated with $X_\omega(x) = \sin \omega x$ - namely,

$$Y_\omega(y) = C_1 \cosh \omega y + C_2 \sinh \omega y \tag{14}$$

- and, selecting the combination in (14) which fulfills the homogeneous Dirichlet condition on the bottom edge, we display the general solution as

$$\psi(x, y) = \int_0^\infty c(\omega) \sin \omega x \sinh \omega y \, d\omega.$$

The final boundary condition requires

$$\psi(x, \pi) = g(x) = \int_0^\infty c(\omega) \sin \omega x \sinh \omega \pi \, d\omega. \tag{15}$$

The "coefficient" $c(\omega)$ is revealed through the orthogonality relation (13) if we multiply both sides of (15) by $\sin \omega' x$ and integrate:

$$
\begin{aligned}
\int_0^\infty g(x) \sin \omega' x \, dx &= \int_0^\infty \left[ \int_0^\infty c(\omega) \sin \omega x \sinh \omega \pi \, d\omega \right] \sin \omega' x \, dx \\
&= \int_0^\infty c(\omega) \sinh \omega \pi \left[ \int_0^\infty \sin \omega x \sin \omega' x \, dx \right] d\omega \\
&= \int_0^\infty c(\omega) \sinh \omega \pi \, \frac{\pi}{2} \delta(\omega - \omega') \, d\omega \\
&= \frac{\pi}{2} c(\omega') \sinh \omega' \pi,
\end{aligned}
$$

or (dropping the prime)

$$c(\omega) = \frac{2}{\pi \sinh \omega \pi} \int_0^\infty g(x) \sin \omega x \, dx.$$

(Equivalently, from (15) we directly identify $c(\omega) \sinh \omega \pi$ as the Fourier sine transform of $g(x)$.)

Reasoning similarly to the above, we recognize the eigenfunctions of the unbounded Neumann problem

$$X'' = \lambda X, \quad X'(0) = 0, \quad X(x) \text{ finite as } x \to \infty$$

as the family $\{\cos \omega x\}$ with $\lambda = -\omega^2$. The expansion is the Fourier cosine transform equations

$$f(x) = \int_0^\infty C(\omega) \cos \omega x \, d\omega, \quad C(\omega) = \frac{2}{\pi} \int_0^\infty f(x) \cos \omega x \, dx,$$

and the orthogonality relation takes the form

$$\int_0^\infty \cos \omega_1 x \cos \omega_2 x \, dx = \frac{\pi}{2} \delta(\omega_1 - \omega_2).$$

The (full) Fourier transform

$$f(x) = \int_{-\infty}^{\infty} F(\omega)e^{i\omega x}\, dx, \quad F(\omega) = \frac{1}{2\pi}\int_{-\infty}^{\infty} f(x)e^{-i\omega x}\, dx$$

identifies the eigenfunctions of the "doubly unbounded" Sturm-Liouville problem

$$X'' = \lambda X, \quad X(x) \text{ finite as } x \to \pm\infty,$$

as the family $\{e^{i\omega x}, -\infty < \omega < \infty\}$ with, as usual, $\lambda = -\omega^2$.                                    ∎

These examples have shown how the eigenvalues of an *ordinary* Sturm-Liouville problem, which form a *discrete* set approaching $-\infty$, can evolve into a *continuum* decreasing to $-\infty$ when the problem becomes singular. This is not a hard and fast rule, however. The eigenvalues of the Bessel and Legendre equations remain discrete in the singular case, as we shall see. And, in fact, in Sec. 6.7 it is shown that the Robin boundary condition for (1) generates a spectrum with both a continuous and a discrete component!

**Example 3.** In Example 3 of Sec. 6.3 we saw that solving Laplace's equation for equilibrium temperature in a cylindrical heat pipe requires knowledge of the eigenfunctions of the Bessel equation

$$R'' + \frac{1}{\rho}R' - \frac{n^2}{\rho^2}R = -\mu R \tag{16}$$

(same as equation (20), Sec. 6.3), with Dirichlet conditions holding on the inner radius $\rho = a$ and the outer radius $\rho = b$ (Fig. 6, Sec. 6.3). Now we consider the pipe to be a *solid* cylinder - mathematically, the "inner radius" $a$ is zero. The point $\rho = 0$ is a singular point for equation (16) (multiply by $\rho^2$), so the appropriate boundary condition is

$$R(\rho) \text{ finite as } \rho \downarrow 0. \tag{17}$$

On the outer wall of the cylinder we shall continue to enforce the Dirichlet condition:

$$R(b) = 0. \tag{18}$$

Previously we identified the general solution to (16) as

$$R(\rho; \lambda) = C_1 J_n(\sqrt{\mu}\rho) + C_2 Y_n(\sqrt{\mu}\rho).$$

Imposing the first boundary condition (17) now requires that $C_2$ equal zero, since Weber's functions $Y_n$ diverge at the origin. Thus the values of $\mu$ must be selected so that the remaining boundary condition, (18), is met:

$$R(b; \mu) = J_n(\sqrt{\mu}b) = 0. \tag{19}$$

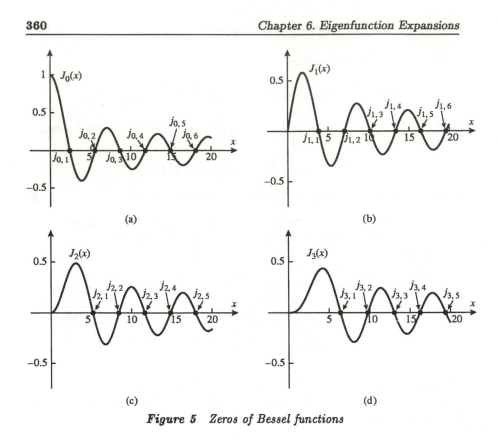

**Figure 5** *Zeros of Bessel functions*

Selected graphs of the function $J_n$ are displayed in Fig. 5. From the figure we see that for each value of $n$ the Bessel function $J_n$ has an infinite number of zeros. Let the $p$th positive zero of $J_n$ be denoted $j_{n,p}$. Then the $p$th eigenvalue $\mu_p$ satisfies

$$\sqrt{\mu_p}\,b = j_{n,p}$$

and the eigenvalues and eigenfunctions for this problem are

$$\mu_p = [j_{n,p}/b]^2, \qquad R_{n,p}(\rho) = J_n(j_{n,p}\rho/b). \tag{20}$$

The zeros of $J_n$ are well known; see Table 2.2. Thus the singular Sturm-Liouville problem (16-18) has a *discrete* set of eigenvalues and eigenfunctions, and the associated expansion is given by

$$f(\rho) = \sum_{p=1}^{\infty} c_p J_n(j_{n,p}\rho/b), \;\; 0 \le \rho \le b \tag{21}$$

We have seen (Sec. 6.3 again) that the weight function for (16) is $w(\rho) = \rho$, so the orthogonality relations for the eigenfunctions corresponding to different eigenvalues — *and hence to different p, but to a common value of*

$n$ — take the form

$$\int_0^b J_n(j_{n,p_1}\rho/b)\, J_n(j_{n,p_2}\rho/b)\, \rho\, d\rho = \int_0^b J_n(j_{n,p_1}\rho/b)^2\, \rho\, d\rho\, \delta_{p_1 p_2}.$$

As a result the coefficients in (21) are given by

$$c_p = \frac{\langle f(\rho), J_n(j_{n,p}\rho/b)\rangle_\rho}{\|J_n(j_{n,p}\rho/b)\|_\rho^2} = \frac{\int_0^b f(\rho) J_n(j_{n,p}\rho/b)\rho\, d\rho}{\int_0^b J_n^2(j_{n,p}\rho/b)\rho\, d\rho} \tag{22}$$

The denominator in (22) can be simplified by means of the identities in Table 2.2; the result is (exercise 3)

$$c_p = \frac{2\int_0^b f(\rho) J_n(j_{n,p}\rho/b)\rho\, d\rho}{b^2 J_{n+1}^2(j_{n,p})} \tag{23}$$

$$= \frac{2\int_0^b f(\rho) J_n(j_{n,p}\rho/b)\rho\, d\rho}{b^2 J_n'^2(j_{n,p})} \tag{24}$$

Equations (21, 22) define the *Fourier-Bessel series* and have been proved to be valid for *any real value* of $n > -1$. Some graphical examples will be displayed in Sec. 6.5. The inner products $\langle f(\rho), J_n(j_{n,p}\rho/b)\rangle_\rho$ are sometimes called *finite Hankel transforms of the first kind.*

EXAMPLE 4. In Sec. 5.3 we saw that both the Helmholtz and Schrodinger equations lead, upon separation in spherical coordinates, to the associated Legendre equation

$$(1-x^2)y'' - 2xy' + \frac{\mu}{1-x^2}y = \lambda y \tag{25}$$

for the cosine of the latitudinal angle, $x = \cos\phi$. In many situations involving this equation $\mu$ is predetermined as the eigenvalue for the $\theta$ (longitudinal) variation,

$$\Theta'' = \mu\Theta\,, \quad \text{periodic boundary conditions}$$

and hence equals $-m^2$, $m = 0, 1, 2, \cdots$. Here we shall merely discuss the case $\mu = 0$, and defer the general case to Sec. 6.10.

The singular points, $x = \pm 1$, of (25) correspond to the north and south poles $\phi = 0, \pi$; as such, we disallow unbounded behavior there and impose the finiteness conditions

$$y(-1),\ y(1)\ \text{finite.} \tag{26}$$

Thus we have a Sturm-Liouville eigenvalue problem *singular at each end.*

In Section 2.4 we considered the Frobenius expansion of the solution to (25) around the singular point $x = -1$, and we found that one solution took the form

$$y_1(x) = \sum_{n=0}^\infty a_n(x+1)^n \tag{27}$$

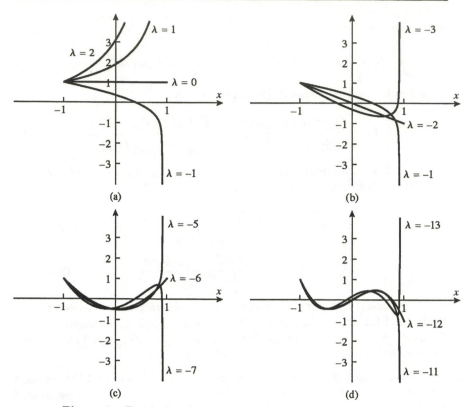

**Figure 6** *Eigenvalue determination for the Legendre equation*

with the coefficients obeying the recurrence relation

$$a_{n+1} = a_n \frac{n(n+1) + \lambda}{2(n+1)^2}, \tag{28}$$

while the second solution had a log term and thus diverged at $x = -1$;

$$
\begin{aligned}
y(x; \lambda) &= C_1 y_1(x) + C_2 y_2(x) \\
&= C_1 \sum_{n=0}^{\infty} a_n (x+1)^n + C_2 \{A \ln(x+1) y_1(x) + \sum_{n=0}^{\infty} b_n (x+1)^n\}.
\end{aligned}
$$

Consequently the first boundary condition in (26) disqualifies the second solution; $C_2 = 0$ and $y(x; \lambda) = \sum_{n=0}^{\infty} a_n (x+1)^n$ .

Now we determine the eigenvalues by demanding finiteness of $y_1(x)$ at $x = +1$ - the *other* boundary condition in (26). In Fig. 6 we display the graphs of $y_1(x)$ for a few selected values of $\lambda$; we arbitrarily set $a_0 = 1$ for this display. The curves indicate that $y_1(x; \lambda)$ diverges to $+\infty$ as $x \to 1$ for positive $\lambda$, but that $\lambda = 0$ yields an eigenvalue since $y_1(x; 0)$ is finite. In fact $y_1(1; 0)$ takes the value 1, and the lack of interior zeros indicates that we have found the *first* eigenfunction. As we decrease $\lambda$ all the curves approach $-\infty$ as $x \to 1$ until we reach $\lambda = -2$, whence $y_1(x; -2)$ is finite and the second eigenfunction results. The next value of $\lambda$ producing a finite $y_1(x; \lambda)$ is $-6$; the next is $-12$; and so on. *The pattern is $\lambda = 0, -2, -6, -12, -20, \ldots$*.

Do you recognize this pattern? Its explanation is revealed by a detailed study of the convergence of the series (27, 28). Recall the ratio test from elementary calculus:

(*Ratio Test*) If the ratio of consecutive terms in the series $\sum_{n=0}^{\infty} s_n$ approaches a limit $L$:

$$\lim_{n \to \infty} s_{n+1}/s_n = L,$$

then the series converges if $|L| < 1$ and diverges if $|L| > 1$.

Now for large values of $n$ the coefficient ratio in (28) tends to the limit

$$\frac{a_{n+1}}{a_n} = \frac{n(n+1)}{2(n+1)^2} + \frac{\lambda}{2(n+1)^2}$$

$$= \frac{1(1 + 1/n)}{2(1 + 1/n)^2} + \frac{\lambda}{2(1 + 1/n)^2} \to \frac{1}{2} + 0.$$

Thus we would expect the ratio of consecutive terms in the series (27) to approach

$$a_{n+1}(x+1)^{n+1}/a_n(x+1)^n \to \frac{1}{2}(x+1) \qquad (29)$$

unless something interrupts this approach to the limit. If $x$ is any number *greater than 1* (or less than $-3$), (29) indicates the terms in the series will start to grow, so the series diverges. Not surprisingly, then, it can be shown that the series also diverges for our critical value, $x$ equal to 1, unless an "interrupt" saves the day.

*The only way to prevent divergence (26) and remain consistent with (28) is to have all the coefficients zero from some point on.* The eigenvalues for $\lambda$, are the values that enforce this condition.

From (28) we see that $a_{\ell+1}$, and all subsequent $a_p$, will equal zero if $\lambda$ takes the value

$$\begin{aligned} \lambda_\ell &= -\ell(\ell+1), \quad \ell = 0, 1, 2, \ldots \\ &= 0, -2, -6, -12, -20, \ldots \end{aligned}$$

These numbers, then, are the eigenvalues, and the eigenfunctions are *polynomials* of degree $\ell$. (In fact we uncovered these polynomial solutions to Legendre's equation in Section 2.3, when we derived the power series expansion formulas around $x = 0$.)

■

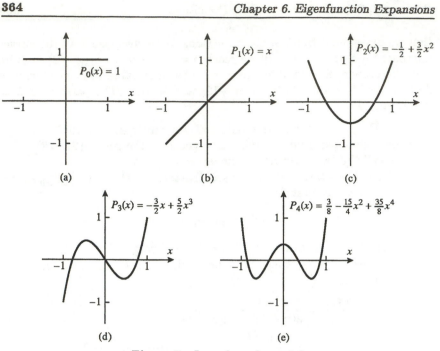

**Figure 7**  *Legendre polynomials*

Historically the *LegendrePolynomial* of degree $\ell$, $P_\ell(x)$, is normalized (through the choice of $a_0$ in the series) so that $P_\ell(1) = 1$. Figure 7 contains sketches of the first few Legendre polynomials and Table 6.1 lists their most important properties. The basic differential equation (25) is already in Sturm-Liouville form, so the weight factor $w(x)$ can be read off immediately; $w(x) = 1$. Thus the Legendre eigenfunction expansion takes the form

$$f(x) = \sum_{\ell=0}^{\infty} c_\ell P_\ell(x), \ -1 \le x \le 1, \ \lambda_\ell = -\ell(\ell+1);$$

$$c_\ell = \frac{\langle f(x), P_\ell(x) \rangle}{\|P_\ell(x)\|^2} = \frac{\int_{-1}^{1} f(x) P_\ell(x)\, dx}{\int_{-1}^{1} P_\ell^2(x)\, dx} = \frac{2\ell+1}{2} \int_{-1}^{1} f(x) P_\ell(x)\, dx$$

(the last equality following from the integral formula in Table 6.1).

When $\mu = -m^2 \ne 0$ the eigenfunctions of (25,26) are called the *associated Legendre functions* $P_\ell^m(x)$. They are not polynomials as a rule (although $P_\ell^0(x) = P_\ell(x)$), but their properties can be derived from the Legendre polynomials; see Sec. 6.10.

Table 6.3 at the end of this chapter lists all the eigenfunction expansions that this author has had occasion to use. Their derivations are long and arduous but they involve no new concepts, and they are documented in the optional Secs. 6.7 through 6.10. The eigenfunction table itself is very general and, of necessity, very terse. Section 6.6 documents a protracted, complicated, and detailed (but not unrealistic) example that exploits the full generality of Table 6.3; it is included for your reference, not for casual reading.

Section 6.5 elaborates on some of the mathematical issues underlying singular Sturm-Liouville expansions. Readers who are eager to proceed with the *applications* of Sturm-Liouville theory to partial differential equations are invited to skip to Chapter 7 at this point.

# Exercises 6.4

1. Solve the problem indicated in Fig. 8 and show that its solution approaches eq. (5) as $L$ goes to infinity.

2. Solve the problem indicated in Fig. 9 and show that its solution approaches eq. (5) as $L$ goes to infinity.

3. Derive the coefficient formulas (23, 24).

4. Solve the problem indicated in Fig. 10.

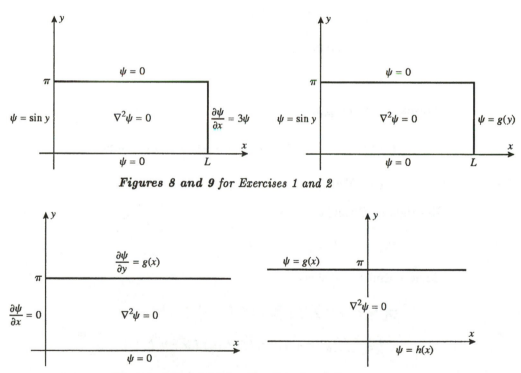

*Figures 8 and 9 for Exercises 1 and 2*

*Figures 10 and 11 for Exercises 4 and 5*

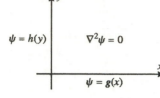

$\psi = h(y)$        $\nabla^2\psi = 0$

$\psi = g(x)$

**Figure 12   Exercise 6**

**5.** Solve the problem indicated in Fig. 11.

**6.** Solve the problem indicated in Fig. 12.

# TABLE 6.1 Legendre Polynomials

Differential equation:

$$(1-x)^2 P_n(x)'' - 2xP_n(x)' = -n(n+1)P_n(x)$$

$$\frac{1}{\sin\phi}\frac{d}{d\phi}\{\sin\phi\,\frac{d}{d\phi}P_n(\cos\phi)\} = -n(n+1)P_n(\cos\phi)$$

Description

$P_n$ is a polynomial of degree $n$ and, up to a constant multiple, is the only solution which is finite at $x = \pm 1$ or $\phi = 0, \pi$.

$P_n(x)$ has $n$ zeros in (-1,1).

$P_{2n}(x)$ is even, $P_{2n}(x)$ is odd.

$P_n(1) = 1$.

Normalization Integral

$$\int_{-1}^{1} P_n(x)\,P_{n'}(x)\,dx = \int_0^\pi P_n(\cos\phi)\,P_{n'}(\cos\phi)\sin\phi\,d\phi$$

$$= \frac{2}{(2n+1)}\delta_{nn'}$$

Rodriguez's Formula

$$P_n(x) = \frac{1}{2^n n!}\{\frac{d}{dx}\}^n(x^2-1)^n$$

Some Useful Expansions

$$\frac{1}{\sqrt{1-2xr+r^2}} = \sum_{n=0}^{\infty} P_n(x)r^n \text{ for } |x| \le 1, |r| < 1$$

$$e^{irx} = \sum_{n=0}^{\infty}(2n+1)i^n j_n(r)P_n(x) \text{ for } r \ge 0, |x| \le 1$$

# 6.5 Mathematical Issues

**Summary**  The convergence properties of the Sturm-Liouville eigenfunction expansions are analogous to those for the Fourier series; the expansions are optimal in a weighted least-squares sense, and they converge uniformly for twice-continuously differentiable target functions if the latter satisfy the same boundary conditions as the eigenfunctions. Finiteness is the strongest boundary condition that can be imposed at a singular end point in the common "limit point" cases, but there are exceptional problems where no solution remains finite, or a variety of boundary conditions can be imposed, at a singular endpoint. Any eigenfunction corresponding to an element of the point spectrum is square integrable and possesses a finite number of zeros.

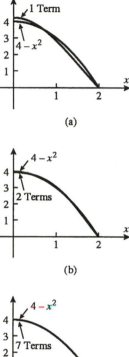

In this section we are going to survey the mathematical details concerning the nature of the Sturm-Liouville eigenfunction expansions. Obviously we have deferred discussion of such matters because we didn't want to interrupt our exposition of the formal computation of these series.[7]

The question of convergence is dismissed quickly: in practically all cases it replicates the Fourier series theory exactly. Thus for integrable functions eigenfunction expansions converge and are "best fits" in a *weighted* $(\rho(x))$ least squares sense; at jump discontinuities they converge to the midpoint; and they converge uniformly for twice-continuously differentiable functions, if the target functions satisfy the same boundary conditions as the eigenfunctions.

**Example 1.** Figure 1 displays one-term, two-term, and seven-term sums of the Fourier-Bessel expansion for the quadratic function $4 - x^2$:

$$4 - x^2 = \sum_{p=1}^{\infty} c_p J_0(j_{0,p}x/2), \quad 0 \le x \le 2.$$

The convergence looks good at all points. However if we try the same function with the series

$$4 - x^2 = \sum_{p=1}^{\infty} d_p J_2(j_{2,p}x/2), \quad 0 \le x \le 2,$$

the fact that $J_2(0) = 0$ hinders the approximation. In Fig. 2 we can see the series straining against this constraint as it strives to fit the target function.

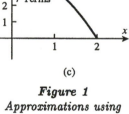

*Figure 1*
*Approximations using*
$J_0(j_{0,p}x/2)$

---

[7]The eigenfunction expansions bearing the names of Hermite, Airy, Laguerre, and square-well arise from problems in quantum theory, and are deferred to Sec. 7.4; they are, however, listed in Table 6.3.

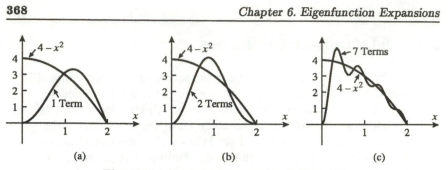

**Figure 2**   *Approximations using* $J_2(j_{2,p}x/2)$

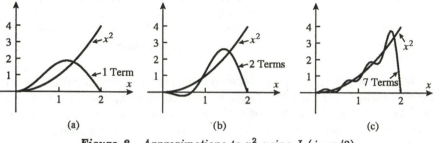

**Figure 3**   *Approximations to* $x^2$ *using* $J_2(j_{2,p}x/2)$

In Fig. 3 we use the same eigenfunctions but change the target function to $x^2$:

$$x^2 = \sum_{p=1}^{\infty} e_p J_2(j_{2,p}x/2), \quad 0 \le x \le 2.$$

Now the mismatch of boundary conditions at $x = 2$ is the limiting factor. If, instead, we use a Dini series

$$x^2 = \sum_{p=1}^{\infty} f_p J_2(j'_{2,p}x/2), \quad 0 \le x \le 2,$$

then Fig. 4 demonstrates improved convergence, but still the eigenfunctions have to depart from the target function at the right end in order to flatten out their slopes.      ■

> When the target function $f$ has *known* discontinuities or mismatches at the end points, one way of getting good numerical approximations is to subtract off a "simple" function possessing the same type of discontinuity and then expand the difference. This is discussed in Section 8.4.

Now let us turn to some of the complications engendered by the limiting processes for *singular* Sturm-Liouville systems.

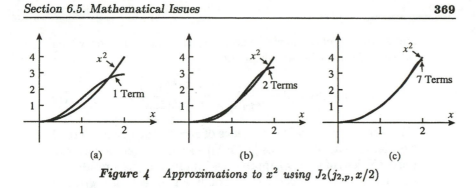

**Figure 4**  *Approximations to $x^2$ using $J_2(j_{2,p}, x/2)$*

## A. Limit points versus limit circles

In Section 6.4 we encountered situations wherein the general solution of the Sturm-Liouville differential equation was expressible as a linear combination of two particular solutions - one of which diverged as the singular end point was approached, and one of which remained finite. We invariably dismissed the divergent solution on the basis that "its behavior would spoil the convergence of the expansions." A more cogent argument to this effect might go as follows.

Let the system under consideration be given by

$$[p(x)y']' + q(x)y' = \lambda \rho(x)y \tag{1}$$

on the interval $a < x \le b$, where $a$ is the singular endpoint and the boundary condition

$$\gamma y(b) + \delta y'(b) = 0 \tag{2}$$

is imposed at $b$. We analyze this problem by concocting another boundary condition at $x = \tilde{a} > a$:

$$[p(x)y']' + q(x)y = \lambda \rho(x)y, \tag{3}$$

$$\gamma y(b) + \delta y'(b) = 0, \tag{4}$$

$$\alpha y(\tilde{a}) + \beta y'(\tilde{a}) = 0 \tag{5}$$

($\alpha$ and $\beta$ constants, not both zero), and considering the behavior of the eigenfunctions of this *regular* Sturm-Liouville problem as $\tilde{a} \to a$.

Suppose $y_1(x; \lambda)$ is a particular solution of (3) which remains finite as $x \to a$, and $y_2(x; \lambda)$ is an independent solution diverging at $a$. The general solution to (3) is given by

$$C_1 y_1(x; \lambda) + C_2 y_2(x; \lambda) \tag{6}$$

and simple algebra shows (exercise 1) that any solution which satisfies (5) must be a multiple of

$$y(x; \lambda) = y_1(x; \lambda) - \frac{\alpha\, y_1(\tilde{a}; \lambda) + \beta\, y_1'(\tilde{a}; \lambda)}{\alpha\, y_2(\tilde{a}; \lambda) + \beta\, y_2'(\tilde{a}; \lambda)}\, y_2(x; \lambda). \quad (7)$$

(We ignore the unlikely occurrence of zero in the denominator.) Now let $\tilde{a} \to a$. Then since $|y_2(\tilde{a}; \lambda)| \to \infty$, the denominator in the coefficient of $y_2(x; \lambda)$ approaches infinity and, as a result, the solution $y(x; \lambda)$ approaches the nondiverging solution $y_1(x; \lambda)$ for every $x$ to the right of $a$. In this manner we justify "throwing out" the diverging solutions for the singular case.

This argument is much more compelling as regards the dismissal of the divergent solutions, but it invites two very intriguing criticisms.

(i) There might be a possibility of avoiding the "evaporation" of the coefficient of $y_2(x; \lambda)$ in (7) by letting $\alpha$ and $\beta$ *vary with* $\tilde{a}$ in such a way as to keep the denominator finite. In other words one could conceivably alter the boundary conditions imposed at $\tilde{a}$, while $\tilde{a}$ was approaching $a$, so that in this limit the solution of (3, 5) took the form (6) with $C_2$ nonzero. In such a case the eigenfunctions of the (singular) problem *would*, obviously, diverge at $a$!

Now if this happened, the eigenfunction expansions clearly couldn't converge *pointwise* on $a \leq x \leq b$. Only *mean-square* convergence would be possible, and this would require that the eigenfunctions themselves, though divergent at $a$, would have to be square-integrable.

Suffice it to say that there are known examples of this phenomenon.[8]; the Bessel equation of order $\nu$ on the interval $[0, b]$ is one, if $-1 < \nu < 1$ In other words expansions different from the Fourier-Bessel series derived in Section 6.4 are possible. However their applications seem to be limited.

(ii) When the above anomaly does not occur, then it must be the case that the *same* solution $y_1(x; \lambda)$ is obtained as the limit of $y(x; \lambda)$ in (7) *regardless of the choice of boundary conditions (i.e., choice of $\alpha$ and $\beta$) that we concoct for the (regular) sub-interval Sturm-Liouville problems!* As a matter of fact nothing in our argument even guarantees that $y_1(x; \lambda)$ will obey (5) at the point $a$. The boundary condition at $x = \tilde{a}$ is "lost" in the limit, and the only thing we can expect of the eigenfunctions is finiteness in this case.

In general when one analyzes singular Sturm-Liouville systems as the limit of regular systems there is a natural division of the examples into two categories known, for reasons we won't go into here, as the "limit point" case and the "limit circle" case.

---

[8]See the reference by Levitan and Sargsjan.

In the limit point case, which occurs most commonly in applications, it is *impossible* to enforce a boundary condition of the form (5) at $a$. Furthermore at most *one* solution of the differential equation (up to a constant multiple, as always!) remains square-integrable with respect to the weight $\rho$. Observation (ii) above pertains to the limit point case.

In the limit circle case, the eigenfunctions **do** continue to maintain boundary conditions as the limit $\tilde{a} = a$ is achieved. All solutions to the differential equation remain square-integrable in the limit, and it is *necessary* to impose a boundary condition at $x = a$ in order to single out (constant multiples of) a particular solution as an eigenfunction. Thus the behavior of the singular Sturm Liouville problem is much like the regular problem, in the limit circle case. The Bessel equation of order $-1 < \nu < 1$, on the interval $0 \leq x \leq b$, exemplifies this case.[9]

## B. Point spectrum versus continuous spectrum

In Section 6.4 we saw that the Fourier sine transform and its associated expansion,

$$f(x) = \int_0^\infty S(\omega) \sin \omega x \, d\omega, \quad S(\omega) = \frac{2}{\pi} \int_0^\infty f(x) \sin \omega x \, dx \qquad (8)$$

arose as the limit of a Sturm-Liouville eigenfunction expansion

$$f(x) = \sum_{n=1}^\infty c_n \sin n\pi x / b \qquad (9)$$

as the endpoint $b$ went to infinity (and the problem became singular). The set of eigenvalues of (9),

$$\lambda_n = -n^2 \pi^2 / b^2,$$

evolved into a *continuum*, filling up the negative half-line.

On the other hand the Fourier-Bessel expansion

$$f(x) = \sum_{p=1}^\infty c_p J_\nu(j_{\nu,p} x/b), \quad 0 \leq x \leq b \qquad (10)$$

arose from a singular Sturm-Liouville problem for which the eigenvalue set remained *discrete*:

$$\lambda_p = -[j_{\nu,p}/b]^2.$$

---

[9]In rough terms, the points on the "limit circle" each correspond to distinct boundary conditions that can be imposed. If the limit circle shrinks to a "limit point", then there is no flexibility in the choice of boundary conditions. See the references by Coddington and Levinson or Levitan and Sargsjan.

We shall even uncover a situation (Example 1, Sec. 6.7) wherein the eigenvalue set included the negative half-line *plus* an isolated positive number.

It is traditional to designate what we have called the "eigenvalue set" as the *spectrum* of the Sturm-Liouville problem, in the singular case. The nomenclature *eigenvalue* is reserved for the isolated points in the spectrum; thus the eigenvalues belong to the "point spectrum" or "discrete spectrum", and the rest of the spectrum is the "continuous spectrum". Clearly for regular Sturm-Liouville problems the entire spectrum is discrete, and thus the "eigenvalue" terminology is appropriate.

Some general theorems are known which relate the continuous and point spectrum to the coefficients in the underlying differential equation.[10] Here we shall limit ourselves to the following observation, which you can verify by scanning the entries in Table 6.3.

> Eigenfunctions with eigenvalues in the point spectrum are square integrable (with respect to the weight $w(x)$) and possess only a finite number of zeros in the interval.

This statement is beautifully illustrated by entry 10 in Table 6.3.

## Exercises 6.5

**1.** Derive eq. (7).

## 6.6   An Example for Reference

**Summary**   Table 6.3 is a very exhaustive listing of the eigenfunction expansions that arise in applications. In order to maintain generality while keeping the size of the table within bounds, it is necessary to employ terse notational conventions which may render the use of the table somewhat unwieldy. Therefore this section documents a single protracted and detailed, but not unrealistic, example that is designed to demonstrate all the complications that can arise in its use.

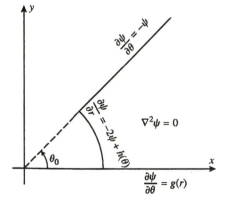

**Figure 1**

The steady state temperature configuration in Fig. 1 is governed by two Robin conditions and a Neumann condition. (The Robin condition on the inner arc ($r = 1$) is nonstandard; since $\partial\psi/\partial r$ is the *inward* normal derivative, one would expect the coefficient "$-2$" to be positive from the thermodynamic considerations of Sec. 4.5. For this example, however, we wish to demonstrate the full generality of Table 6.3.)

----

[10]See the reference by Levitan and Sargsjan.

The basic procedure of Sec. 5.2 dictates the decomposition into sub-problems as indicated in Figs. 2 and 3. Laplace's equation, as we have seen, separates in polar coordinates into the harmonic oscillator equation and the equidimensional equation,

$$\Theta'' = \lambda\Theta \quad \text{and} \quad r^2 R'' + rR' = -\lambda R. \tag{1}$$

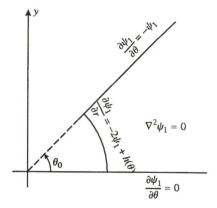

*Figure 2*

SOLUTION OF PROBLEM 1. For Problem 1 the nonhomogeneity is on a constant-$r$ face, so we first address the Sturm-Liouville problem defined by the $\Theta$ equations:

$$\Theta'' = \lambda\Theta; \quad \Theta'(0) = 0, \quad \Theta(\theta_0) + \Theta'(\theta_0) = 0. \tag{2}$$

The reader will surely be able to find the eigenfunctions for (31) directly, but we shall use entry #2 in Table 6.3, which reads

$$y'' + \nu y' = \lambda y; \quad \alpha y(0) + \beta y'(0) = 0, \ \gamma y(L) + \delta y'(L) = 0. \tag{3}$$

Equations (2) and (3) match for $\nu = \alpha = 0$, $\beta = \gamma = \delta = 1$, $L = \theta_0$. The eigenvalues are therefore given by $\lambda_p = -\omega_p^2 - \nu^2/4 = -\omega_p^2$, where $\omega_p$ is the $p$th solution ("possibly imaginary") to

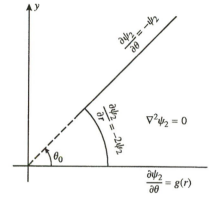

*Figure 3*

$$(\delta\nu/2 - \gamma)\sin\left[\omega L + \arctan\left(-\frac{\beta}{\alpha - \beta\nu/2}\omega\right)\right]$$

$$= \omega\delta\cos\left[\omega L + \arctan\left(-\frac{\beta}{\alpha - \beta\nu/2}\omega\right)\right]$$

or, since $\arctan(-\frac{1}{0-0}) = \arctan(-\infty) = -\frac{\pi}{2}$,

$$(0 - 1)\sin(\omega\theta_0 - \frac{\pi}{2}) = \omega\,(1)\cos(\omega\theta_0 - \frac{\pi}{2}),$$

or finally

$$-\cos\omega\theta_0 = -\omega\sin\omega\theta_0, \quad \omega = \cot\omega\theta_0. \tag{4}$$

There is no "exceptional" eigenvalue since $\alpha - \beta\nu/2 = 0$ but $\gamma - \delta\nu/2 = 1 \neq 0$.

An "imaginary" solution $\omega = i\tau$ would satisfy

$$i\tau = \cot i\tau\theta_0 = i\coth\tau\theta_0 \quad \text{or} \quad \tau = -\coth\tau\theta_0 \tag{5}$$

(recall $\cos ix = \cosh x$, $\sin ix = i\sinh x$). Equation (5) clearly has no solutions; its left and right members have opposite signs.

The eigenfunctions are given by

$$\Theta_p(\theta) = e^{-\nu\theta/2}\sin\left[\omega_p\theta + \arctan\left(-\frac{\beta}{\alpha - \beta\nu/2}\omega_p\right)\right] = \cos\omega_p\theta,$$

and the orthogonality relations read

$$\int_0^L \Theta_p(\theta)\Theta_{p'}(\theta)e^{\nu\theta}\,d\theta = \frac{1}{2}\left\{L + \frac{\gamma\delta}{\gamma^2 + \omega_p^2\delta^2} - \frac{\alpha\beta}{\alpha^2 + \omega_p^2\beta^2}\right\}\delta_{pp'},$$

or

$$\int_0^{\theta_0} \cos\omega_p\theta\cos\omega_{p'}\theta\,d\theta = \frac{1}{2}\left\{\theta_0 + \frac{1}{1+\omega_p^2}\right\}\delta_{pp'}. \tag{6}$$

The corresponding solutions of the equation (1) for the radial factor $R(r)$ with $\lambda = -\omega_p^2$ are

$$R_p(r) = D_1 r^{\omega_p} + D_2 r^{-\omega_p}.$$

The combination which is finite at infinity is $D_2 r^{-\omega_p}$. Therefore the general solution of Problem 1 is given by

$$\psi_1(r,\theta) = \sum_{p=1}^\infty c_p r^{-\omega_p}\cos\omega_p\theta,$$

and the nonhomogeneous boundary condition reads

$$\begin{aligned}
\frac{\partial\psi_1(1,\theta)}{\partial r} \quad &+\quad 2\psi_1(1,\Theta) = h(\theta)\\
&= \sum_{p=1}^\infty c_p(-\omega_p)\,1^{-\omega_p-1}\cos\omega_p\theta + 2\sum_{p=1}^\infty c_p\,1^{-\omega_p}\cos\omega_p\theta\\
&= \sum_{p=1}^\infty c_p(2-\omega_p)\cos\omega_p\theta.
\end{aligned}$$

The coefficients are derived from (6):

$$\begin{aligned}
\int_0^{\theta_0} h(\theta)\cos\omega_{p'}\theta\,d\theta &= \sum_{p=1}^\infty c_p(2-\omega_p)\frac{1}{2}\left\{\theta_0 + \frac{1}{1+\omega_p^2}\right\}\delta_{pp'}\\
&= c_{p'}(2-\omega_{p'})\frac{1}{2}\left\{\theta_0 + \frac{1}{1+\omega_{p'}^2}\right\}.
\end{aligned}$$

or (dropping the prime)

$$c_p = \frac{2}{2-\omega_p}\int_0^{\theta_0} h(\theta)\cos\omega_p\theta\,d\theta \bigg/ \left\{\theta_0 + \frac{1}{1+\omega_p^2}\right\}. \tag{7}$$

SOLUTION OF PROBLEM 2. For Problem 2 the radial equations determine the Sturm-Liouville problem

$$r^2 R'' + rR' = -\lambda R;\quad 2R(1) + R'(1) = 0,\ R(\infty)\ \text{finite}.$$

This matches up with entry #14 in Table 6.3:

$$\rho^2 R'' + \sigma\rho R' = \lambda R; \quad \alpha R(a) + \beta R'(a) = 0, \ R(\rho)\rho^{(\sigma-1)/2} \text{ finite as } \rho \to \infty,$$

with $\sigma = a = b = 1$, $\alpha = 2$, and "$\lambda$"$= -\lambda$. The eigenvalues are the elements of the continuum

$$\text{``}\lambda\text{''} = -\omega^2 - (\sigma-1)^2/4 = -\omega^2, \ 0 \le \omega < \infty,$$

and also the isolated point

$$\text{``}\tilde{\lambda}\text{''} = [a\alpha/\beta]^2 - a\alpha(\sigma-1)/\beta = 4,$$

since $\beta = 1 \ne 0$ and $\alpha a/\beta = 2 > (\sigma-1)/2 = 0$. The corresponding eigenfunctions are

$$\begin{aligned}
R_\omega(r) &= \rho^{(1-\sigma)/2} \sin\left\{\omega \ln \rho/a - \arctan \frac{2\beta\omega}{2\alpha a + \beta(1-\sigma)}\right\} \\
&= r^0 \sin\left\{\omega \ln r/1 - \arctan \frac{2(1)\omega}{2(1)(1) + 1(0)}\right\} \\
&= \sin\{\omega \ln r - \arctan\omega\},
\end{aligned}$$

and

$$\tilde{R}(r) = r^{-a\alpha/\beta} = r^{-2}.$$

The weight function $w(r) = r^{\sigma-2} = r^{-1}$. The orthogonality relations read

$$\int_a^\infty R_\omega(r)R_{\omega'}(r)r^{\sigma-2}\,dr = \int_1^\infty R_\omega(r)R_{\omega'}(r)\,dr/r = \frac{\pi}{2}\delta(\omega - \omega'), \quad (8)$$

$$\int_a^\infty R_\omega(r)r^{-a\alpha/\beta}r^{\sigma-2}\,dr = \int_1^\infty R_\omega(r)r^{-2}\,dr/r = 0, \quad (9)$$

and

$$\int_a^\infty r^{-2a\alpha/\beta}r^{\sigma-2}\,dr = a^{\sigma-1-2a\alpha/\beta}/[2a\alpha/\beta - \sigma + 1] \text{ or } \int_1^\infty r^{-4}\,dr/r = \frac{1}{4}. \tag{10}$$

The solutions to $\theta'' = \lambda\theta$ which accompany these radial eigenfunctions are

$$\Theta_\omega(\theta) = D_1 \cosh\omega\theta + D_2 \sinh\omega\theta \quad \text{for } \lambda = \omega^2,$$

$$\tilde{\Theta}(\theta) = D_1 \cos 2\theta + D_2 \sin 2\theta \qquad \text{for } \tilde{\lambda} = -4.$$

The homogeneous boundary condition at $\theta = \theta_0$ requires

$$\omega D_1 \sinh\omega\theta_0 + \omega D_2 \cosh\omega\theta_0 = -D_1 \cosh\omega\theta_0 - D_2 \sinh\omega\theta_0 \quad (\lambda = \omega^2),$$

$$-2D_1 \sin 2\theta_0 + 2D_2 \cos 2\theta_0 = -D_1 \cos 2\theta_0 - D_2 \sin 2\theta_0 \quad (\tilde{\lambda} = -4),$$

and we fulfill these by choosing

$$D_1 = [\omega \cosh \omega\theta_0 + \sinh \omega\theta_0] \text{ and } D_2 = -[\omega \sinh \omega\theta_0 + \cosh \omega\theta_0] \quad (\lambda = \omega^2),$$

$$D_1 = [2 \cos 2\theta_0 + \sin 2\theta_0] \text{ and } D_2 = -[-2 \sin 2\theta_0 + \cos 2\theta_0] \quad (\tilde{\lambda} = -4).$$

This results in the compact forms (exercise 4)

$$\begin{aligned}
\Theta_\omega(\theta) &= \omega \cosh \omega(\theta - \theta_0) + \sinh \omega(\theta - \theta_0) \quad (\lambda = -\omega^2), \\
\tilde{\Theta}(\theta) &= 2 \cos 2(\theta - \theta_0) + \sin 2(\theta - \theta_0).
\end{aligned} \tag{11}$$

The assembled general solution to Problem 2 is

$$\begin{aligned}
\psi_2(r, \theta) &= d_0 \tilde{\Theta}(\theta) \tilde{R}(r) + \int_0^\infty d(\omega) \Theta_\omega(\theta) R_\omega(r) \, d\omega \\
&= d_0 [2 \cos 2(\theta - \theta_0) + \sin 2(\theta - \theta_0)] r^{-2} +
\end{aligned}$$

$$\int_0^\infty d(\omega) [\omega \cosh \omega(\theta - \theta_0) + \sinh \omega(\theta - \theta_0)] \sin\{\omega \ln r - \arctan \omega\} \, d\omega.$$

It must meet the nonhomogeneous boundary condition

$$\begin{aligned}
\frac{\partial \psi_2(r, 0)}{\partial \theta} &= g(r) \\
&= d_0 [-4 \sin(-2\theta_0) + 2 \cos(-2\theta_0)] r^{-2} +
\end{aligned}$$

$$\int_0^\infty d(\omega) [\omega^2 \sinh(-\omega\theta_0) + \omega \cosh(-\omega\theta_0)] R_\omega(r) \, d\omega. \tag{12}$$

According to (9, 10) we can isolate $d_0$ by multiplying both members of (12) by $r^{-2}$ and by the weight factor $r^{-1}$ and integrating:

$$\int_1^\infty g(r) r^{-2} \, dr/r = d_0 [-4 \sin(-2\theta_0) + 2 \cos(-2\theta_0)] \int_1^\infty r^{-4} \, dr/r +$$

$$\int_0^\infty d(\omega) [\omega^2 \sinh(-\omega\theta_0) + \omega \cosh(-\omega\theta_0)] \left\{ \int_1^\infty R_\omega(r) r^{-2} \, dr/r \right\} \, d\omega$$

$$= d_0 [4 \sin 2\theta_0 + 2 \cos 2\theta_0]/4 + (0);$$

hence

$$d_0 = \frac{4}{4 \sin 2\theta_0 + 2 \cos 2\theta_0} \int_1^\infty g(r) r^{-3} \, dr. \tag{13}$$

Similarly (8, 9) indicate that $b(\omega)$ will be isolated if we multiply both members of (12) by $R_{\omega'}(r)$ and by the weight factor $r^{-1}$ and integrate:

$$\int_1^\infty g(r)R_{\omega'}(r)\,dr/r = d_0[4\sin 2\theta_0 + 2\cos 2\theta_0]\int_1^\infty r^{-2}R_{\omega'}(r)\,dr/r +$$

$$\int_0^\infty d(\omega)[\omega^2\sinh(-\omega\theta_0) + \omega\cosh(-\omega\theta_0)]\left\{\int_1^\infty R_\omega(r)R_{\omega'}(r)\,dr/r\right\}\,d\omega$$

$$= (0) +$$

$$\int_0^\infty d(\omega)[-\omega^2\sinh\omega\theta_0 + \omega\cosh\omega\theta_0]\left\{\frac{\pi}{2}\delta(\omega - \omega')\right\}\,d\omega$$

$$= d(\omega')[-\omega'^2\sinh\omega'\theta_0 + \omega'\cosh\omega'\theta_0]\frac{\pi}{2}.$$

Hence

$$d(\omega) = \frac{2}{\pi[-\omega^2\sinh\omega\theta_0 + \omega\cosh\omega\theta_0]}\int_1^\infty g(r)R_\omega(r)\,dr/r. \qquad (14)$$

The complete solution to the problem in Fig. 1 is thus given by

$$\psi(r,\theta) = \sum_{p=1}^\infty c_p r^{-\omega_p}\cos\omega_p\theta + d_0[2\cos 2(\theta - \theta_0) + \sin 2(\theta - \theta_0)]r^{-2} +$$

$$\int_0^\infty d(\omega)[\omega\cosh\omega(\theta - \theta_0) + \sinh\omega(\theta - \theta_0)]\sin\{\omega\ln r - \arctan\omega\}\,d\omega,$$

with the definitions (7, 13, 14).

# Exercises 6.6

**1.** Derive the solution forms (11). Then rederive these forms from the original differential equation $\Theta'' = \lambda\Theta$ by first introducing the change of variable $\theta' = \theta_0 - \theta$ (which measures angles from the other side of the sector).

# 6.7   Derivation of Expansions for the Constant Coefficient Equation

**Summary**   The spectrum for the harmonic oscillator equation with Robin boundary conditions on a semi-infinite interval may consist of a continuum and an isolated point. The eigenfunctions for the damped harmonic oscillator equation can be extracted from the undamped case by a change of variable.

In Sec. 6.3 we have seen that the eigenfunctions for the harmonic oscillator equation

$$y'' = \lambda y \tag{1}$$

give rise to the Fourier sine transform for the homogeneous Dirichlet condition on the semiinfinite interval, the Fourier cosine transform for the homogeneous Neumann condition on the semiinfinite interval, and the (full) Fourier transform for the doubly infinite interval. The next example shows how the homogeneous Robin boundary condition on the semiinfinite interval can generate an eigenfunction expansion whose spectrum consists of a continuum and an isolated point.

**Example 1.** We consider the differential equation (1) together with the boundary conditions

$$\alpha y(0) + \beta y'(0) = 0, \quad y(\infty) \quad \text{finite.} \tag{2}$$

For $\lambda > 0$ the general solution, as we have seen, is the sum of two exponentials –

$$y(x; \lambda) = C_1 e^{\sqrt{\lambda}x} + C_2 e^{-\sqrt{\lambda}x}$$

– one of which is disqualified by the finiteness condition at $\infty$. Insertion of $e^{-\sqrt{\lambda}x}$ into (2) results in

$$\alpha - \beta\sqrt{\lambda} = 0,$$

*which has the positive solution*

$$\lambda = (\alpha/\beta)^2 \tag{3}$$

*if* $\alpha/\beta > 0$. In such a case, then, we get an isolated positive eigenvalue (3) with the associated eigenfunction $e^{-\alpha x/\beta}$.

Zero is not an eigenvalue because the general solution $c_1 + c_2 x$ cannot meet both boundary conditions (unless $\alpha = 0$, which reduces to the Neumann condition).

For negative values of $\lambda$ it is easiest to analyze equation (2) if we write the general solution in the form

$$y(x) = C \sin\{\sqrt{-\lambda}x - \phi\} \tag{4}$$

(recall exercise 3, Sec. 2.1), in which case (2) demands

$$\tan \phi = \frac{\beta \sqrt{-\lambda}}{\alpha} \tag{5}$$

(exercise 3). Then the functions

$$y(x; \lambda) = \sin\{\sqrt{-\lambda}x - \arctan \frac{\beta \sqrt{-\lambda}}{\alpha}\} \tag{6}$$

all satisfy the finiteness condition at $\infty$ for $\lambda < 0$, and thus are eigenfunctions. Therefore the expansion for boundary condition (2) takes the curious form

$$f(x) = Ae^{-\alpha x/\beta} + \int_0^\infty G(\omega) \sin\{\omega x - \arctan \frac{\beta \omega}{\alpha}\} \, d\omega \tag{7}$$

with $\lambda = (\alpha/\beta)^2$ in the first term, $\lambda = -\omega^2$ in the integral.

The evaluation of the coefficient $A$ proceeds without difficulty. $A = 0$ if $\alpha/\beta \leq 0$; otherwise, since the usual orthogonality argument shows

$$\int_0^\infty e^{-\alpha x/\beta} \sin\{\omega x - \arctan \frac{\beta \omega}{\alpha}\} \, dx = 0,$$

we have

$$A = \langle f, e^{-\alpha x/\beta} \rangle / \|e^{-\alpha x/\beta}\|^2 = \frac{2\alpha}{\beta} \int_0^\infty f(x) e^{-\alpha x/\beta} \, dx.$$

However the formula for $G(\omega)$ is more subtle (because the inner products $\int_0^\infty y(x; \lambda) y(x; \lambda') \, dx$ diverge). Exercise 1 carefully traces the evolution of the expansions for the *regular* Sturm-Liouville problem

$$y'' = \lambda y, \quad \alpha y(0) + \beta y'(0) = 0, \quad y(a) = 0 \tag{8}$$

as $a$ goes to $\infty$ to derive

$$G(\omega) = \frac{2}{\pi} \int_0^\infty f(x) \sin\{\omega x - \arctan \frac{\beta \omega}{\alpha}\} \, dx. \tag{9}$$

∎

**Example 2.** Now we proceed to derive the eigenfunctions for the *damped harmonic oscillator* equation

$$y'' + \nu y' = \lambda y \tag{10}$$

on the interval $0 \leq x < \infty$. We take the general boundary condition at $x = 0$

$$\alpha y(0) + \beta y'(0) = 0, \tag{11}$$

and, for reasons to be seen, we temporarily defer statement of the finiteness condition at $\infty$.

The substitution $y(x) = e^{-\nu x/2}z(x)$ reduces eqs. (10, 11) to the form analyzed in the previous example. We have

$$y = e^{-\nu x/2}z; \quad y' = -\frac{\nu}{2}e^{-\nu x/2}z + e^{-\nu x/2}z';$$

$$y'' = \frac{\nu^2}{4}z - \nu e^{-\nu x/2}z' + e^{-\nu x/2}z''.$$

Substitution into (10) results in

$$\frac{\nu^2}{4}z - \nu e^{-\nu x/2}z' + e^{-\nu x/2}z'' - \left(\frac{\nu^2}{2}\right)e^{-\nu x/2}z + \nu e^{-\nu x/2}z' = \lambda e^{-\nu x/2}z$$

or

$$z'' = (\lambda + \nu^2/4)z = \tilde{\lambda}z \tag{12}$$

(identical with eq. (1)). The boundary condition (11) evolves into

$$\alpha z(0) + \beta\{-\left(\frac{\nu}{2}\right)z(0) + z'(0)\} = [\alpha - \frac{\beta\nu}{2}]z(0) + \beta z'(0) = 0, \tag{13}$$

which is in the form of Eq. (2). From the previous example, then, we can read off the transformed eigenvalues $\tilde{\lambda}$, the transformed eigenfunctions, and the appropriate boundary condition at $\infty$ – namely, $z(x)$ *should be bounded as $x \to \infty$*. In terms of the given data (10, 11), these become

$$\lambda = -\omega^2 - \frac{\nu^2}{4} \quad (0 \le \omega < \infty)$$

$$\text{and also } [(\alpha - \frac{\beta\nu}{2})/\beta]^2 - \frac{\nu^2}{4} \text{ if } (\alpha - \frac{\beta\nu}{2})/\beta > 0;$$

$$\phi_\omega(x) = e^{-\nu x/2}\sin[\omega x + \arctan\left(-\frac{\beta}{\alpha - \beta\nu/2}\omega\right)]$$

$$\text{and also } e^{-\nu x/2}e^{-(\alpha-\beta\nu/2)x/\beta} = e^{-\alpha x/\beta} \quad \text{if } (\alpha - \frac{\beta\nu}{2})/\beta > 0.$$

The finiteness condition now reads

$$y(x)e^{+\nu x/2} \quad \text{bounded as} x \to \infty.$$

To derive the expansion theorem for these eigenfunctions we replace $\alpha$ by $\alpha - \beta\nu/2$ in (7), multiply throughout by $e^{-\nu x/2}$, and reinterpret $f(x)e^{-\nu x/2} = \tilde{f}(x)$ as the function being expanded:

$$\begin{aligned}
\tilde{f}(x) &= f(x)e^{-\nu x/2} = Ae^{-\nu x/2}e^{-(\alpha-\beta\nu/2)x/\beta} \\
&\quad + \int_0^\infty G(\omega)e^{-\nu x/2}\sin\{\omega x - \arctan\left(-\frac{\beta}{\alpha - \beta\nu/2}\omega\right)\}\,d\omega \\
&= Ae^{-\alpha x/\beta} + \int_0^\infty G(\omega)\phi_\omega(x)\,d\omega, \tag{14}
\end{aligned}$$

$$
\begin{aligned}
G(\omega) &= \frac{2}{\pi} \int_0^\infty f(x) \sin\{\omega x - \arctan\left(-\frac{\beta}{\alpha - \beta\nu/2}\omega\right)\} \, dx \\
&= \frac{2}{\pi} \int_0^\infty \tilde{f}(x) e^{\nu x/2} \sin\{\omega x - \arctan\left(\frac{\beta}{\alpha - \beta\nu/2}\omega\right)\} \, dx \\
&= \frac{2}{\pi} \int_0^\infty \tilde{f}(x) \phi_\omega(x) e^{(+)\nu x} \, dx, \tag{15}
\end{aligned}
$$

and, if $\left(\alpha - \frac{\beta\nu}{2}\right)/\beta > 0$,

$$
\begin{aligned}
A &= \frac{\beta}{2\alpha - \beta\nu} \int_0^\infty f(x) e^{-(\alpha - \beta\nu/2)x/\beta} \, dx \\
&= \frac{\beta}{2\alpha - \beta\nu} \int_0^\infty \tilde{f}(x) e^{-\alpha x/b} e^{(+)\nu x} \, dx \tag{16}
\end{aligned}
$$

(otherwise $A = 0$). Note how the weight factor $e^{\nu x}$ crops up, as it must, in (15, 16).   ∎

The eigenfunction development for eq. (10) on the interval $(-\infty, \infty)$ is assigned as exercise 2. All of these expansions are tabulated in Table 6.3.

## Exercises 6.7

1. In this exercise we shall trace the evolution of the eigenfunction expansions of the regular Sturm-Liouville problem (8) as $a \to \infty$. Assume $\alpha \neq 0, \beta \neq 0$.

   (a) Show that the solutions to the differential equation and the second boundary condition can be written as (constant multiples of) $\sinh \omega(x - a)$ if $\lambda = \omega^2 > 0$, $\sin \omega(x - a)$ if $\lambda = -\omega^2 < 0$, and $(x - a)$ if $\lambda = 0$.

   (b) Show that the Robin boundary condition, which determines the eigenvalues in this formulation, is expressed

   $$
   \omega = \frac{\alpha}{\beta} \tanh \omega a \quad \text{if} \quad \lambda = \omega^2 > 0, \tag{17}
   $$

   $$
   \omega = \frac{\alpha}{\beta} \tan \omega a \quad \text{if} \quad \lambda = -\omega^2 < 0, \tag{18}
   $$

   and no solution for $\lambda = 0$.

   (c) By considering graphs of $\tanh \omega a$ for increasing values of $a$, show that if $\alpha/\beta > 0$ eq. (17) has one solution which approaches $\omega = \alpha/\beta$, while it has no solutions otherwise. Show furthermore that the eigenfunction $\sinh \omega(x - a)$, normalized by dividing by $-e^{a\omega}/2$, approaches $e^{-\alpha x/\beta}$ in this circumstance.

(d) By considering graphs of $\tan \omega a$ for increasing values of $a$, show that the solutions of (18) "fill up" the interval $(0, \infty)$, and are spaced roughly $\pi/a$ apart. Show furthermore that when $\omega$ satisfies (18) the eigenfunction $\sin \omega(x - a)$ can be written as $\sin[\omega x - \arctan(\beta\omega/\alpha)]$.

(e) Show that

$$\int_0^a \sin^2 \omega_n(x - a)\, dx = \frac{1}{2}\{a - \alpha\beta/(\alpha^2 + \omega_n^2\beta^2)\}$$

when $\omega_n$ satisfies (18). (Hint: use Table 6.3, entry # 2).

(f) If the eigenfunction expansion for the regular Sturm–Liouville problem (8) is written in the form

$$f(x) = g_0 \sinh \omega_0(x - a) + \sum_{n=1}^{\infty} \Delta\omega_n g_n \sin[\omega_n x - \arctan(\beta\omega_n/\alpha)]$$

$$(19)$$

with $\Delta\omega_n = \omega_{n+1} - \omega_n$ show that the formula for $g_n$ is given by

$$g_n = \frac{2\int_0^a f(x)\sin[\omega_n x - \arctan(\beta\omega_n/\alpha)]dx}{[a - \alpha\beta/(\alpha^2 + \omega_n^2\beta^2)]\Delta\omega_n}. \qquad (20)$$

(g) With the observation $\Delta\omega_n \approx \pi/a$ for large $a$ argue that (19, 20) tend to the forms

$$f(x) = Ae^{-\alpha x/\beta} + \int_0^{\infty} G(\omega)\sin[\omega x - \arctan(\beta\omega/\alpha)]\, d\omega,$$

$$G(\omega) = \frac{2}{\pi}\int_0^{\infty} f(x)\sin[\omega x - \arctan(\beta\omega/\alpha)]\, dx,$$

and $A$ is as described in Example 1.

2. Use the procedure of Example 2 to analyze the singular Sturm–Liouville problem

$$y'' + \nu y' = \lambda y, \quad y(x)e^{\nu x/2} \quad \text{finite as } x \to \pm\infty.$$

Compare your results with entry #12 of Table 6.3.

3. Derive eq. (5).

# 6.8 Derivation of Expansions for the Equidimensional Equation

**Summary** The singular problems for the equidimensional equation result from letting the lower boundary go to zero and/or the upper boundary to $\infty$. Since the equidimensional equation is reducible by a logarithmic change of variables to a constant coefficient equation, it is not surprising that its eigenfunction expansions, the Mellin transforms, are logarithmic sine\cosine transforms.

In Sec. 5.3 we saw how the equidimensional equation

$$x^2 y'' + \sigma x y' = \lambda y \tag{1}$$

governs the radial factor $R(r)$ when Laplace's equation is separated in polar or cylindrical coordinates ($\sigma = 1$) and spherical coordinates ($\sigma = 2$). The *regular* Sturm-Liouville problem generated by the boundary conditions

$$y'(1) = 0, \qquad y'(2) = 0$$

was analyzed for $\sigma = 1$ in Example 1, Sec. 6.1. Now we consider the singular cases, with the left end point at the origin, or the right end point at infinity, or both.

**Example 1.** First we consider the problem

$$x^2 y'' + x y' = \lambda y, \quad 0 < x \leq a, \tag{2}$$

with the boundary condition at $x = a$ taken in the form

$$y(a) = 0. \tag{3}$$

We defer the statement of the finiteness condition at $x = 0$. We employ a change of variable which reduces the equidimensional equation to the constant coefficient equation (exercise 3, Sec. 1.2). Let

$$x = ae^{-t}, \ t = -\ln x/a; \ 0 < x \leq a, \ 0 \leq t < \infty; \ y(ae^{-t}) = Y(t). \tag{4}$$

Then by the formulas in Sec. 1.2,

$$y' = \frac{Y'}{-x}, \quad y'' = \frac{Y''}{x^2} + \frac{Y'}{x^2},$$

so the transformed problem is

$$x^2 \{\frac{Y''}{x^2} + \frac{Y'}{x^2}\} + x\frac{Y'}{-x} = Y'' = \lambda Y, \quad Y(0) = 0; \ 0 \leq t < \infty. \tag{5}$$

These equations were revealed in Example 2 of Sec. 6.4 to generate the Fourier Sine Transform. Its eigenvalues are $\lambda = -\omega^2$, $0 \le \omega < \infty$; its eigenfunctions are $\sin \omega t$; its second boundary condition is boundedness of $Y(t)$ at $\infty$; and the expansion theorem reads

$$f(t) = \int_0^\infty S(\omega) \sin \omega t \, d\omega, \ S(\omega) = \frac{2}{\pi} \int_0^\infty f(t) \sin \omega t \, dt. \qquad (6)$$

Inverting the substitution (4), then, we append the boundary condition

$$y(0) \ \text{finite (more precisely, } y(x) \text{ bounded as } x \ge 0) \qquad (7)$$

to (2, 3) and conclude that *its* eigenvalues are, again, $\lambda = -\omega^2$, $0 \le \omega < \infty$, and its eigenfunctions are $-\sin\left[\omega \ln x/a\right]$. The expansion theorem reads

$$\tilde{f}(x) = f(-\ln x/a) = \int_0^\infty S(\omega) \sin\left[-\omega \ln x/a\right] d\omega, \ 0 < x \le a, \qquad (8)$$

$$
\begin{aligned}
S(\omega) &= \frac{2}{\pi} \int_{t=0 \ (x=a)}^{t=\infty \ (x=0)} f(-\ln x/a) \sin\left[-\omega \ln x/a\right] d[-\ln x/a] \\
&= \frac{2}{\pi} \int_0^\infty \tilde{f}(x) \sin\left[-\omega \ln x/a\right] \frac{dx}{x}. \qquad (9)
\end{aligned}
$$

(Clearly the nuisance minus sign can be dropped.) Note how the weight factor $1/x$ crops up, as it must, in (9).

It is sometimes convenient to express the integrals (8, 9) in terms of the functions $x^{\pm i\omega}$. Observe that $S(\omega)$ can be written

$$
\begin{aligned}
S(\omega) &= -\frac{2}{\pi} \frac{1}{2i} \int_0^a \tilde{f}(x) \{e^{i\omega \ln x/a} - e^{-i\omega \ln x/a}\} \frac{dx}{x} \\
&= \frac{i}{\pi} \int_0^a \tilde{f}(x) \{(x/a)^{i\omega} - (x/a)^{-i\omega}\} \frac{dx}{x} \\
&= \frac{a^{-i\omega} i}{\pi} \int_0^a \tilde{f}(x) \{x^{i\omega-1} - a^{2i\omega}/x^{i\omega+1}\} dx.
\end{aligned}
$$

The *finite Mellin transform of the first kind* of $f(x)$ is defined to be

$$m_1[s; a, f(x)] = \int_0^a \tilde{f}(x) \{x^{s-1} - a^{2s}/x^{s+1}\} dx.$$

Therefore we have

$$S(\omega) = \frac{a^{-i\omega} i}{\pi} m_1[i\omega; a, f(x)].$$

Moreover since $S(\omega)$ is obviously an odd function, eq. (8,9) can be written

$$
\begin{aligned}
f(x) &= \frac{1}{2i}\int_{-\infty}^{\infty} S(\omega)e^{-i\omega\ln(x/a)}\,d\omega \\
&= \frac{1}{2i}\int_{-\infty}^{\infty}\frac{a^{-i\omega}i}{\pi}m_1[i\omega;a,f(x)]x^{-i\omega}a^{i\omega}\,d\omega \\
&= \frac{1}{2i}\int_{-\infty}^{\infty}x^{-i\omega}m_1[i\omega;a,f(x)]\,d\omega,
\end{aligned}
$$

which can be interpreted as an *inverse Mellin transform.*                    ∎

The finite Mellin transform of the first kind is, at bottom, a logarithmic sine transform, and it provides the eigenfunction expansion for the equidimensional equation (2) with the boundary condition (3). As one might expect, the finite Mellin transform of the second kind is a logarithmic *cosine* transform generated by the corresponding Sturm-Liouville problem with the homogeneous *Neumann* condition at the right end point (exercise 1):

$$
x^2 y'' + xy' = \lambda y, \quad y(0)\,\text{finite}, \quad y'(a)=0. \tag{10}
$$

**Example 2.** The (infinite) Mellin transform is generated by the differential equation (2) on the interval $(0,\infty)$. It is reduced to a constant-coefficient system by the change of variables

$$
x = e^{-t}, \quad t = -\ln x.
$$

We obtain the transformed differential equation $Y'' = \lambda Y$ again, with $t$ in $(-\infty,\infty)$. Thus we recover the Fourier Integral expansion:

$$
f(t) = \int_{-\infty}^{\infty} F(\omega)e^{i\omega t}\,d\omega, \quad F(\omega) = \frac{1}{2\pi}\int_{-\infty}^{\infty} f(t)e^{-i\omega t}\,dt
$$

($\lambda = -\omega^2$, eigenfunctions $= e^{i\omega t}$). Inverting the change of variables gives us the eigenfunctions $e^{-i\omega\ln x}$ in the original system, and the corresponding expansion reads

$$
f(-\ln x) = \tilde{f}(x) = \int_{-\infty}^{\infty} F(\omega)e^{-i\omega\ln x}\,d\omega,
$$

$$
\begin{aligned}
F(\omega) &= \frac{1}{2\pi}\int_{t=-\infty\,(x=\infty)}^{t=\infty\,(x=0)}\tilde{f}(x)e^{i\omega\ln x}\,d(-\ln x) \\
&= \frac{1}{2\pi}\int_{x=0}^{x=\infty}\frac{\tilde{f}(x)}{x}e^{i\omega\ln x}\,dx.
\end{aligned}
$$

These equations are often rewritten as

$$\tilde{f}(x) = \frac{1}{2\pi} \int_{-\infty}^{\infty} M[i\omega; \tilde{f}] x^{-i\omega} d\omega$$

where the *Mellin transform* is defined by

$$M[s; f] = \int_{0}^{\infty} f(x) x^{s-1} dx. \qquad\blacksquare$$

The extensions of these computations to Robin boundary conditions, cases where $\sigma \neq 1$, and the infinite interval $[a, \infty)$ are all handled by the same procedure: use a change of variable of the form (4) to transform to a constant-coefficient system, and exploit the Fourier representations. We leave the details to the exercises. Sneddon's book and Naylor's papers are authoritative references for the Mellin transforms (but watch out for misprints).

# Exercises 6.8

1. Work out the eigenfunction expansion for eqs. (10); derive the Mellin transform of the second kind in Table 6.3, entry #13.

2. Work out the eigenfunction expansion for eq. (2) with a Robin boundary condition at $a$. See Table 6.3, entry #13.

3. Work out the eigenfunction expansion for the "exterior" problem where the differential equation in (2) holds for $a < x < \infty$ and $y(a) = 0$. Use the change of variable $x = ae^t$. What is the boundary condition at infinity? See Table 6.3, entry #14 (recall we used this expansion in Example 5, Sec. 6.5).

4. Generalize exercise 3 for a Robin boundary condition at $a$.

5. Work out the eigenfunction expansion for eq. (1) on $(0, a)$ with

    (a) a Dirichlet boundary condition at $a$;

    (b) a Neumann boundary condition at $a$;

    (c) a Robin boundary condition at $a$.

    See Table 6.3, entry #13.

6. Work out the eigenfunction expansion for eq. (1) on $(a, \infty)$ with

    (a) a Dirichlet boundary condition at $a$;

    (b) a Neumann boundary condition at $a$;

    (c) a Robin boundary condition at $a$.

See Table 6.3, entry #14.

**7.** Work out the eigenfunction expansion for eq. (1) on $(0, \infty)$. See Table 6.3, entry # 15.

# 6.9 Derivations of Expansions for Bessel's Equation

**Summary** The various expansions engendered by singular Sturm-Liouville problems for Bessel's equation include, besides the Fourier-Bessel series of Sec. 6.4, the Dini series, the Hankel transform, Weber's integral theorem, Macdonald's series, and the Kontorovich-Lebedev transforms.

In Sec. 6.3 we saw that Bessel's equation, in the form

$$x^2 y'' + xy' = \mu y + \lambda x^2 y, \tag{1}$$

gives rise to *two* regular Sturm-Liouville eigenfunction expansions, according to whether $\mu$ or $\lambda$ is regarded as the eigenvalue. Now we address the singular cases: either the left end point is zero or the right end point is infinity, or both. Also there may be homogeneous Dirichlet, Neumann, or Robin boundary conditions at the non-singular point (when there is one). We shall present analyses of six typical problems and direct the reader to the exercises, the papers of Naylor, the textbook of Sneddon, or simply to Table 6.3 for the other cases.

**Example 1.** *(Dini Series)* With $\mu = \nu^2$ in (1) the singular Sturm-Liouville problem

$$x^2 y'' + xy' - \nu^2 y = \lambda x^2 y; \quad y(0) \quad \text{finite}, \quad y'(b) = 0 \tag{2}$$

is identical to the equations generating the Fourier-Bessel series in Sec. 6.4, except for the Neumann boundary condition at $x = b$. The analysis of the eigenfunctions is perfectly analogous: we take

$$C_1 J_\nu(\sqrt{-\lambda}x) + C_2 Y_\nu(\sqrt{-\lambda}x) \tag{3}$$

as the general solution, disqualify $Y_\nu$ by the finiteness condition, and identify the eigenvalues through the Neumann condition $J'_\nu(\sqrt{-\lambda}b) = 0$.

The nomenclature of Fig. 1 designates the $p$th positive zero of $J'_\nu(x)$ as $j'_{\nu,p}$, whence we identify the eigenvalues as $\lambda = -j'^2_{\nu,p}/b^2$, the eigenfunctions as $J_\nu(j'_{\nu,p}x/b)$, and the *Dini expansion* as

$$f(x) = \sum_{p=1}^{\infty} c_p J_\nu(j'_{\nu,p}x/b), \quad c_p = \frac{\int_0^b f(x) J_\nu(j'_{\nu,p}x/b)x\,dx}{\int_0^b J_\nu^2(j'_{\nu,p}x/b)x\,dx}. \tag{4}$$

Selected values of $j'_{\nu,p}$ are listed in Table 2.2.

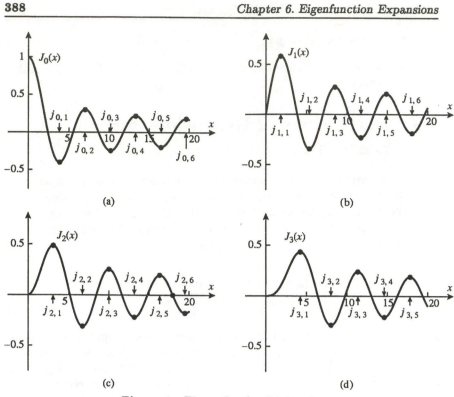

**Figure 1**  *Eigenvalue for Dini series*

*Equation* (4) *is wrong if* $\nu = 0$! You can see this by examining the first eigenfunction ($\nu = 0$, $p = 1$). (This merely requires rescaling the first graph in Fig. 1.) Recall that the general theory predicts that the first eigenfunction can have *no* interior zeros. Thus $J_0(j'_{0,1}x/b)$ must be the *second* eigenfunction, and we still must find the first. (Note from the graphs that $J_\nu(j'_{0,1}x/b)$ has no interior zeros for $\nu > 0$.)

To recover the missing eigenfunction, we set $\nu = 0$ in (2) and observe that $\lambda = 0$ is an eigenvalue, generating the eigenfunction $\Phi_0(x) = $ constant $= 1$ (say). Thus the expression (4) must be modified for $\nu = 0$ to read

$$f(x) = c_0 + \sum_{p=1}^{\infty} c_p J_0(j'_{0,p}x/b),$$

$$c_0 = \frac{\int_0^b f(x)(1)x\,dx}{\int_0^b 1^2 x\,dx} = \frac{2}{b^2}\int_0^b f(x)x\,dx; \quad c_{p>0} \text{ as in (4)}. \quad \blacksquare$$

Although Fourier published the Fourier-Bessel series in 1822, the "missing" term in the related Dini series went unnoticed until 1880 – and was not computed correctly until 1904! Happily,

our heuristic oscillation argument exposed it immediately.[11] Abramowitz
and Stegun's handbook compensates for this by calling 0 the
*first* root of $J_0'(x)$ and renumbering $j_{0,p}'$ as $j_{0,p+1}'$. Graphs of
some Dini series expansions appear in Sec. 6.5.

**Example 2.** *(The Hankel Transform)* If we let $b \to \infty$ in the preceding
example we get a Sturm-Liouville problem that is singular at *both* ends.
The general solution of the differential equation is still expressed by the
Bessel functions (3), and the finiteness condition at $x = 0$ again enforces
$c_2 = 0$. As $x \to \infty$ the functions $y(x; \lambda) = J_\nu(\sqrt{-\lambda}x)$ remain bounded for
all *negative* $\lambda$ (they go to zero, in fact; see Sec. 2.5). For $\lambda > 0$, however,
the $J_\nu$'s become modified Bessel functions $I_\nu$ and diverge exponentially at
$\infty$ (remember, also, the $K_\nu$'s diverge at 0). Thus *every negative number*
$\lambda = -s^2$ gives rise to an eigenfunction bounded at both ends and we expect
the Fourier-Bessel expansion of Sec. 6.4 to evolve into

$$f(x) = \int_0^\infty F(s)J_\nu(sx)ds, \quad 0 < x < \infty, \; \lambda = -s^2. \tag{5}$$

$F(s)$ is known as the *Hankel Transform* of $f(x)$. To derive a formula
for $F(s)$ we imitate the process used to derive the Fourier transform from
the Fourier series. Thus we rewrite the Fourier-Bessel series formulas ((21,
22) in Sec. 6.4) as

$$s_p = j_{\nu,p}/b, \quad \Delta s_p = (j_{\nu,p+1} - j_{\nu,p})/b, \tag{6}$$

$$f(x) = \sum_{p=1}^\infty F(s_p)J_\nu(s_px)\Delta s_p, \quad 0 \le x \le b \tag{7}$$

$$F(s_p = j_{\nu,p}/b) = a_p/\Delta s_p = \frac{2\int_0^b f(x)J_\nu(s_px)x\,dx}{b^2 J_\nu'^2(j_{\nu,p})(j_{\nu,p+1} - j_{\nu,p})/b}. \tag{8}$$

As $b \to \infty$, $\Delta s_p \to 0$ and the formula (7) clearly approaches (5).

Tracing the evolution of (8) takes some care. First we let $s$ be fixed
but arbitrary, and we choose a sequence of values of $b$ approaching infinity
through the relationship

$$b = b_p = sj_{\nu,p}, \quad p = 1, 2, 3, \ldots.$$

For such values eq. (8) evolves according to

$$\lim_{b \to \infty} F(s) = \lim_{p \to \infty} \frac{2\int_0^{b_p} f(x)J_\nu(sx)x\,dx}{b - p^2 J_\nu'^2(j_{\nu,p})(j_{\nu,p+1} - j_{\nu,p})/b - p}. \tag{9}$$

---

[11]Born too late!

To estimate the denominator, we first note that the asymptotic formula for Bessel functions for large arguments (Table 2.2) -

$$J_\nu(x) \approx (2/\pi x)^{1/2} \cos\{x - \nu\pi/2 - \pi/4\}, \quad x \gg 1 \tag{10}$$

- indicates that for large $x$ the separation between the zeros of $J_\nu$ approaches the separation between the zeros of the cosine (i.e., $\pi$). Thus $j_{\nu,p+1} - j_{\nu,p} \approx \pi$. To the same order of accuracy

$$\begin{aligned} J_\nu'(x) &\approx -(1/2\pi x^3)^{1/2} \cos\{x - \nu\pi/2 - \pi/4\} \\ &\quad -(2/\pi x)^{1/2} \sin\{x - \nu\pi/2 - \pi/4\} \\ &\approx -(2/\pi x)^{1/2} \sin\{x - \nu\pi/2 - \pi/4\}. \end{aligned}$$

Since the sine of an argument is $\pm 1$ when its cosine is zero, for large $p$

$$|J_\nu'(j_{\nu,p})| \approx (2/\pi j_{\nu,p})^{1/2}. \tag{11}$$

Insertion of this information into (9) reveals the formula for the Hankel transform:

$$F(s) = s \int_0^\infty f(x) J_\nu(sx) x \, dx \tag{12}$$

(exercise 1).                                                                 ∎

**Example 3.** *(Weber's Integral Theorem)* Now we shall analyze the Bessel equation (2) in an interval which is singular only on the right:

$$y(a) = 0, \quad y(\infty) \quad \text{finite}.$$

Our point of departure will be the analysis of the regular Sturm-Liouville problem in Example 2, Sec. 6.3. For the interval $(a, b)$ we found the eigenfunctions to be

$$\phi_{\nu,p}(x) = Y_\nu(s_p a) J_\nu(s_p x) - J_\nu(s_p a) Y_\nu(s_p x)$$

where the eigenvalues $\lambda_p = -s_p^2$ solve

$$\phi_\nu(b; s) = Y_\nu(sa) J_\nu(sb) - J_\nu(sa) Y_\nu(sb) = 0.$$

For the expansion formula we employ the form from Sec. 6.3

$$f(x) = \sum_{p=1}^\infty c_p \phi_{\nu,p}(x)$$

$$c_p = \frac{2 s_p^2 \int_a^b f(x) \phi_{\nu,p}(x) x \, dx}{b^2 \phi_{\nu,p}'^2(b) - a^2 \phi_{\nu,p}'^2(a)}. \tag{13}$$

Tracing the progression of these formulas as $b \to \infty$ entails some slight modification of the strategy of the previous example. For large $b$ we have

$$
\begin{aligned}
\phi_\nu(b; s) &= Y_\nu(sa)J_\nu(sb) - J_\nu(sa)Y_\nu(sb) \\
&\approx \left[\frac{2}{\pi sb}\right]^{1/2} \{Y_\nu(sa)\cos(sb - \frac{\nu\pi}{2} - \frac{\pi}{4}) \\
&\quad - J_\nu(sa)\sin(sb - \frac{\nu\pi}{2} - \frac{\pi}{4})\} \\
&\approx \left[\frac{2}{\pi sb}\right]^{1/2} \{J_\nu^2(sa) + Y_\nu^2(sa)\}^{1/2} \cos(sb - \frac{\nu\pi}{2} - \frac{\pi}{4} + \theta)
\end{aligned}
$$

where $\theta = \arctan\{J_\nu(sa)/Y_\nu(sa)\}$. Thus the zeros $s_p$ of $\phi_\nu(b; s)$ satisfy

$$ s_{p+1} - s_p \approx \pi/b $$

asymptotically and at any of these zeros

$$ |\phi_\nu'(b; s_p)| = |\phi_{\nu,p}'(b)| \approx s_p \left[\frac{2}{\pi s_p b}\right]^{1/2} \{J_\nu^2(s_p a) + Y_\nu^2(s_p a)\}^{1/2}. \qquad (14) $$

We rewrite the expansion formula as

$$ f(x) = \sum_{p=1}^{\infty} (c_p/\Delta s_p)\phi_{\nu,p}(x; s_p)\Delta s_p \qquad (15) $$

$$ (c_p/\Delta s_p) = \frac{2s_p^2 \int_a^b f(x)\phi_{\nu,p}(x)x\,dx}{[b^2 \phi_{\nu,p}'^2(b) - a^2 \phi_{\nu,p}'^2(a)]\Delta s_p}. \qquad (16) $$

Noting that $\Delta s_p \approx \pi/b$ approaches zero as $b \to \infty$ we substitute (14, 16) into (15) and take the limit to find (exercise 2)

$$ f(x) = \int_0^\infty F(\mu)\{Y_\nu(sa)J_\nu(sx) - J_\nu(sa)Y_\nu(sx)\}ds $$

where

$$ F(s) = \frac{s \int_a^\infty f(x)\{Y_\nu(sa)J_\nu(sx) - J_\nu(sa)Y_\nu(sx)\}x\,dx}{[Y_\nu^2(sa) + J_\nu^2(sa)]^{1/2}} \qquad (17) $$

(a form of *Weber's formula*). ■

**Example 4.** *(Macdonald Series)* Now we turn to the form of Bessel's equation (1) where $\mu$ is the eigenvalue and $\lambda = s^2$. The problem to be considered is defined by the equations

$$ x^2 y'' + xy' - s^2 y = \mu y; \quad y(a) = 0, \quad y(\infty) \quad \text{finite.} \qquad (18) $$

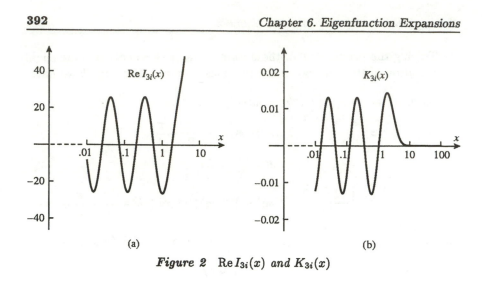

**Figure 2**   $\operatorname{Re} I_{3i}(x)$ and $K_{3i}(x)$

As we saw in Example 5, Sec. 6.3, the general solution to (18) can be written

$$y(x) = C_1 \operatorname{Re} I_{i\omega}(sx) + C_2 K_{i\omega}(sx). \tag{19}$$

Figure 2 indicates that as $x$ approaches infinity $K_{i\omega}(sx) \to 0$ but $\operatorname{Re} I_{i\omega}(sx)$ diverges.

This is confirmed by the approximations from Table 2.2 for large $z$:

$$K_l(z) \approx (\pi/2z)^{1/2} e^{-z}, \quad I_l \approx e^z / (2\pi z)^{1/2}$$

(which are valid for complex $l$).

Therefore the finiteness condition at $x = b = \infty$ disenfranchises the $I_{i\omega}$ term in (19); the eigenfunctions are the Macdonald functions $K_{i\omega}(sx)$ and the equation determining the eigenvalues $\lambda_p = -\omega_p^2$ is

$$K_{i\omega_p}(sa) = 0 \tag{20}$$

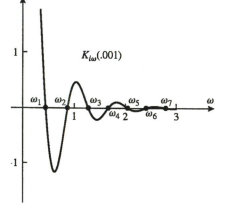

**Figure 3**   *Eigenvalues for the Mcdonald series*

Graphs depicting the dependence of $K_{i\omega}(sa)$ on the parameter $\omega$ are shown in Fig. 3. The eigenvalues are discrete and the expansion reads (exercise 3)

$$f(x) = \sum_{p=1}^{\infty} c_p K_{i\omega_p}(sx), \quad a < x < \infty$$

$$c_p = \frac{\int_a^\infty f(x) K_{i\omega_p}(sx)\, dx/x}{\int_a^\infty \{K_{i\omega_p}(sx)\}^2\, dx/x} = \frac{\int_a^\infty f(x) K_{i\omega_p}(sx)\, dx/x}{\frac{sa}{2\omega_p} K_{i\omega_p+1}(sa) \frac{\partial}{\partial\omega} K_{i\omega}(sa)|_{\omega_p}}. \tag{21}$$

■

**Example 5.** *(Kontorovich-Lebedev Transform)* If we let $a$ approach zero in the previous example we obtain a Sturm-Liouville problem which is singular at both ends. To divine the limiting form of the expansion (21) we again imitate the analysis leading to the Hankel transform (Example 2). First we note from the "Identities" section in Table 2.2 that

$$K_{i\omega_p+1}(sa) = i\omega_p K_{i\omega_p}(sa)/sa - K'_{i\omega_p}(sa) = -K'_{i\omega_p}(sa) \qquad (22)$$

(recall (20)). Furthermore, from the "Wronskians" section in the table we have

$$K'_{i\omega_p}(sa) = -\frac{1}{saI_{i\omega_p}(sa)} + K_{i\omega_p}(sa)I'_{i\omega_p}(sa) = -\frac{1}{saI_{i\omega_p}(sa)}; \qquad (23)$$

consequently

$$K_{i\omega_p+1}(sa) = \frac{1}{saI_{i\omega_p}(sa)}.$$

Now observe from the Identities that

$$\frac{\partial}{\partial\omega}K_{i\omega}(sa) = -i\pi\cot(i\omega\pi)K_{i\omega}(sa)$$

$$+\frac{i\pi}{2}\frac{1}{i\sinh\omega\pi}\{-I_{-i\omega}(sa)\ln(sa/2) - I_{i\omega}(sa)\ln(sa/2) + \mathcal{O}(1)\}$$

where we note that $(sa/2)^{\pm i\omega} = e^{\pm i\omega\ln(sa/2)}$ is bounded, and thus dominated by $\ln(sa/2)$, as $a \to 0$. In Sec. 6.3 we showed that $I_{-i\omega}(sa)$ is the conjugate of $I_{i\omega}(sa)$; and since $\operatorname{Im}I_{i\omega_p}(sa) = 0$ (eq. (20)), we have $I_{-i\omega}(sa) = I_{i\omega}(sa)$ at $\omega = \omega_p$. Therefore

$$K_{i\omega_p+1}(sa)\frac{\partial}{\partial\omega}K_{i\omega}(sa)|_{\omega_p} = \frac{1}{saI_{i\omega_p}(sa)}\frac{(-\pi/2)}{\sinh\omega_p\pi}\{2I_{i\omega_p}(sa)\ln(sa/2)+\mathcal{O}(1)\}$$

Finally we observe that for small $a$, the oscillatory behavior of

$$K_{i\omega}(sa) \approx \frac{-\pi}{\sinh\omega\pi}\operatorname{Im}\frac{(sa/2)^{i\omega}}{\Gamma(i\omega+1)} = \frac{-\pi}{\sinh\omega\pi}\operatorname{Im}\frac{e^{i\omega\ln(sa/2)}}{\Gamma(i\omega+1)},$$

as a function of $\omega$, is governed principally by the rapidly varying sinusoidal factor of period $2\pi/|\ln(sa/2)| = -2\pi/\ln(sa/2)$ (see Fig. 3). Since a sinusoid has two zeros per period, the zeros become dense in the interval $(0,\infty)$ and are roughly separated by

$$\Delta\omega_p = \omega_{p+1} - \omega_p \approx -\pi/\ln(sa/2).$$

Insertion of all these relations into (21) renders the expansion in the form (exercise 4)

$$f(x) =$$

$$\sum_{p=1}^{\infty} F(\omega_p; s)K_{i\omega_p}(sx)\frac{2\omega_p}{sa}\frac{saI_{i\omega_p}(sa)\sinh\omega_p\pi}{(-\pi)[I_{i\omega_p}(sa)\ln(sa/2) + \mathcal{O}(1)]}$$

$$\times\frac{\Delta\omega_p}{(-\pi)/\log(sa/2)}$$

with

$$F(\omega_p; s) = \int_a^\infty f(x) K_{i\omega_p}(sx) \, dx/x. \tag{24}$$

Letting $a \to 0$ we introduce the *Kontorovich-Lebedev* transform of $f(x)$,

$$F(\omega; s) = \int_0^\infty f(x) K_{i\omega}(sx) dx/x \quad (\omega \geq 0), \tag{25}$$

and observe that the expansion evolves into an inverse transform

$$f(x) = \frac{2}{\pi^2} \int_0^\infty F(\omega; s) K_{i\omega}(sx) \, \omega \sinh \omega\pi \, d\omega. \tag{26}$$

∎

**Example 6.** *(Finite Lebedev Transform)* Now we consider the Bessel equation in the same form as the previous two examples, with the singular boundary condition on the left:

$$x^2 y'' + xy' - s^2 y = \mu y; \quad y(0) \quad \text{finite}, \quad y(b) = 0. \tag{27}$$

In Example 5 of Sec. 6.3 we analyzed this problem with a *regular* boundary condition on the left: $y(a) = 0$ $(a > 0)$. With minor modification (exercise 5) those results can be summarized as follows. The solution of Bessel's equation satisfying the boundary condition at $b$ is the (real) function

$$K_{i\omega}(sb) I_{i\omega}(sx) - I_{i\omega}(sb) K_{i\omega}(sx), \tag{28}$$

the eigenvalues $\lambda = -\omega^2$ are determined by

$$K_{i\omega_p}(sb) I_{i\omega_p}(sa) - I_{i\omega_p}(sb) K_{i\omega_p}(sa) = 0, \tag{29}$$

and the expansion is given by

$$f(x) = \sum_{p=1}^\infty c_p \{ K_{i\omega_p}(sb) I_{i\omega_p}(sx) - I_{i\omega_p}(sb) K_{i\omega_p}(sx) \},$$

$$c_p = \frac{-2\omega_p K_{i\omega_p}(sa) \int_a^b f(x) \{ K_{i\omega_p}(sb) I_{i\omega_p}(sx) - I_{i\omega_p}(sb) K_{i\omega_p}(sx) \} \, dx/x}{K_{i\omega_p}(sb) \frac{\partial}{\partial\omega} \{ K_{i\omega_p}(sb) I_{i\omega_p}(sa) - I_{i\omega_p}(sb) K_{i\omega_p}(sa) \}}. \tag{30}$$

For small $a$ the left hand side of equation (29) determining the eigenvalues looks like

$$K_{i\omega}(sb)(sa/2)^{i\omega}/(i\omega)! - I_{i\omega}(sb)(i\omega - 1)!(sa/2)^{-i\omega}/2. \tag{31}$$

Since $\ln(sa/2) \to -\infty$ as $a \downarrow 0$, the rapidly oscillating factors $(sa/2)^{\pm i\omega} = e^{\pm i\nu \ln(sa/2)}$ determine the zeros of (31), as a function of $\omega$. The period of these factors is $-2\pi/\ln(sa/2)$, and since there are two zeros per period we estimate the distance between the roots of (31) to equal

$$\omega_{p+1} - \omega_p = \Delta\omega \approx \frac{-\pi}{\ln(sa/2)}.$$

Note also that (29) implies

$$\frac{K_{i\omega_p}(sa)}{K_{i\omega_p}(sb)} = \frac{I_{i\omega_p}(sa)}{I_{i\omega_p}(sb)} = \frac{\overline{I_{i\omega_p}(sa)}}{\overline{I_{i\omega_p}(sb)}} \qquad (32)$$

(the final equality holding since the left hand side is real).

Now let's estimate the derivative in the denominator of (30). As $a$ approaches zero the terms containing the divergent factor $\ln(sa/2)$ will dominate, so

$$\frac{\partial}{\partial\omega} I_{i\omega}(sa) \approx i I_{i\omega}(sa) \ln(sa/2)$$

and

$$\frac{\partial}{\partial\omega} K_{i\omega}(sa) \approx \frac{i\pi}{2} \csc i\omega\pi \{-I_{-i\omega}(sa) - I_{i\omega}(sa)\} \ln(sa/2)$$

$$\approx \frac{-\pi}{\sinh\omega\pi} \operatorname{Re} I_{i\omega}(sa) \ln(sa/2).$$

As a result we have

$$\frac{\partial}{\partial\omega} \{K_{i\omega_p}(sb) I_{i\omega_p}(sa) - I_{i\omega_p}(sb) K_{i\omega_p}(sa)\}$$

$$\approx \{i K_{i\omega_p}(sb) I_{i\omega_p}(sa) - I_{i\omega_p}(sb) \frac{-\pi}{\sinh\omega\pi} \operatorname{Re} I_{i\omega_p}(sa)\} \ln(sa/2)$$

$$\approx \{i I_{i\omega_p}(sb) K_{i\omega_p}(sa) + I_{i\omega_p}(sb) \frac{\pi}{\sinh\omega\pi} \operatorname{Re} I_{i\omega_p}(sa)\} \ln(sa/2)$$

(by eq. (29))

$$\approx I_{i\omega_p}(sb) \frac{\pi}{\sinh\omega_p\pi} \{\operatorname{Re} I_{i\omega_p}(sa) - i \operatorname{Im} I_{i\omega_p}(sa)\} \ln(sa/2)$$

(by exercise 17, Sec. 6.3)

$$\approx I_{i\omega_p}(sb) \frac{\pi}{\sinh\omega_p\pi} \overline{I_{i\omega_p}(sa)} \ln(sa/2).$$

Now we rewrite the expansion eqs. (30) with the substitutions derived above:

$$f(x) = \sum_{p=1}^{\infty} \frac{c_p}{\Delta\omega} \{K_{i\omega_p}(sb) I_{i\omega_p}(sx) - I_{i\omega_p}(sb) K_{i\omega_p}(sx)\} \Delta\omega,$$

$$\frac{c_p}{\Delta\omega} \approx \frac{-2\omega_p \overline{I_{i\omega_p}(sa)} \int_a^b f(x) \{K_{i\omega_p}(sb) I_{i\omega_p}(sx) - I_{i\omega_p}(sb) K_{i\omega_p}(sx)\} \, dx/x}{\frac{-\pi}{\ln(sa/2)} \overline{I_{i\omega_p}(sb)} I_{i\omega_p}(sb) \frac{\pi}{\sinh\omega_p\pi} \overline{I_{i\omega_p}(sa)} \ln(sa/2)}$$

and obtain the *finite Lebedev transform* as $a \downarrow 0$:

$$f(x) = \int_0^\infty F(\omega)\{K_{i\omega}(sb)I_{i\omega}(sx) - I_{i\omega}(sb)K_{i\omega}(sx)\}\,d\omega,$$

$$F(\omega) = \frac{2\omega \sinh \omega \pi \int_0^b f(x)\{K_{i\omega}(sb)I_{i\omega}(sx) - I_{i\omega}(sb)K_{i\omega}(sx)\}\,dx/x}{\pi^2 |I_{i\omega}(sb)|^2}. \quad \blacksquare$$

**Example 7.** *Spherical Bessel functions* (Example 4, Sec. 6.3) satisfy the differential equation

$$y'' + \frac{2}{x}y' - \frac{l(l+1)}{x^2}y = \lambda y. \tag{33}$$

(Recall the factor $-l(l+1)$ is the eigenvalue for the polar angle $\phi$. See Example 4, Sec. 6.4.) Equation (33) generates a singular Sturm-Liouville problem if the boundary conditions are applied at 0 or $\infty$. For boundary conditions of the form

$$y(0) \quad \text{finite}, \quad y(b) = 0 \tag{34}$$

finiteness disqualifies all solutions except (multiples of) $j_l(\sqrt{-\lambda}x)$; and if $s_{l,p}$ denotes the $p$th positive zero of $j_l$ then the eigenfunctions are $j_{l,p}(s_{l,p}x/b)$ (see Fig. 9, Sec. 6.3).

The associated expansions can be deduced from the regular Fourier-Bessel function expansions as in Example 3, Sec. 6.4. Thus, for example, to find an expansion for $f(x)$ in $0 \le x \le b$ with eigenfunctions vanishing at $b$, we start by writing

$$f(x) = \sum_{p=1}^\infty c_p j_l(s_{l,p}x/b), \quad 0 \le x \le b, \quad \lambda_{l,p} = -s_{l,p}^2/b^2.$$

Using eqs. (26), Sec. 6.3, we change this to

$$f(x) = \sum_{p=1}^\infty c_p \sqrt{b\pi/2s_{l,p}x}\, J_{l+1/2}(s_{l,p}x/b),$$

or

$$f(x)\sqrt{x} = \sum_{p=1}^\infty c_p \sqrt{b\pi/2s_{l,p}}\, J_{l+1/2}(s_{l,p}x/b). \tag{35}$$

Note $J_{l+1/2}(s_{l,p}) = \sqrt{2s_{l,p}/\pi}\,j_l(s_{l,p}) = 0$; thus (35) has the form of a Fourier-Bessel expansion, and we can extract the formula for the coefficients from eqs. (21–24) of Sec. 6.4:

$$c_p\sqrt{b\pi/2s_{l,p}} = \frac{2\int_0^b f(x)\sqrt{x}\, J_{l+1/2}(s_{l,p}x/b)x\,dx}{b^2 J_{l+3/2}^2(s_{l,p})}$$

or

$$c_p = \frac{2 \int_0^b f(x) j_l(s_{l,p} x/b) x^2 dx}{b^3 j_{l+1}^2(s_{l,p})}. \tag{36}$$

■

Expansions for the boundary condition $y'(b) = 0$ can also be computed from the generalized Dini series (Table 6.3 entry 18). If the interval $(0, b)$ extends to infinity the *Spherical Hankel Transform* can be derived from the (usual) Hankel Transform of Example 2 (exercise 6).

## Exercises 6.9

**1.** Verify the statement leading to formula (12).

**2.** Verify formula (17).

**3.** Verify formula (21).

**4.** Derive eqs. (24).

**5.** Derive the representations (28–30) from the analysis of Example 3, Sec. 6.3.

**6.** Derive the spherical Hankel transform equations in Table 6.3, entry 26, from the Hankel transform in Example 2.

## 6.10   The Associated Legendre Equation

**Summary**   When Helmholtz's and Schrödinger's equations are separated in spherical coordinates, the associated Legendre equation governs the behavior of the latitudinal angle factor. Its eigenfunctions, the associated Legendre functions, generalize the Legendre polynomials and possess similar computational characteristics.

In Sec. 5.3 we have seen that when either Helmholtz's or Schrödinger's equation is separated in spherical coordinates the behavior of the latitudinal angle factor $\Phi(\phi)$ is governed by the equation

$$(\sin \phi \, \Phi')' + \frac{\mu}{\sin \phi} \Phi = \lambda \sin \phi \, \Phi. \tag{1}$$

(Here we have altered the notation slightly.) In most applications $\mu$ is the eigenvalue for the longitudinal ($\theta$) angle factor in the separated Laplacian, and periodicity demands $\mu = -m^2$ where $m$ is an integer (see Section 6.3). By the methods in Sec. 1.2 one can show that the substitution

$$x = \cos \phi, \quad y(x) = y(\cos \phi) = \Phi(\phi)$$

(and $\mu = -m^2$) transforms (1) into the associated Legendre equation

$$(1 - x^2)y'' - 2xy' - \frac{m^2}{1 - x^2}y = \lambda y \qquad (2)$$

(exercise 1). The boundary conditions for (2) are finiteness at the north and south poles, where $x = \pm 1$.

There is a neat trick which enables us to analyze (2) easily. Some lengthy algebra reveals that the *Associated Legendre function*

$$P_\ell^m(x) = (-1)^m (1 - x^2)^{m/2} \frac{d^m}{dx^m} P_\ell(x), \quad \ell \geq m \geq 0, \qquad (3)$$

satisfies[12] (2), with $\lambda = -\ell(\ell + 1)$. Here $P_\ell(x)$ is the Legendre polynomial introduced in Sec. 6.4. Taking this for granted for the moment, we argue that these functions must then be the eigenfunctions for this problem — or more precisely, the $p$th eigenfunction must be

$$X_p(x) = P_{m+p}^m(x), \quad p = 0, 1, 2, \ldots \qquad (4)$$

and the $p$th eigenvalue equals

$$\lambda_p = -(m + p)(m + p + 1). \qquad (5)$$

After all,

(i) (we have promised to show that) $X_p$ satisfies the differential equation (2) with $\lambda = \lambda_p$;

(ii) $X_p$ is finite at $x = \pm 1$ — it's $(-1)^m (1 - x^2)^{m/2}$ times a polynomial;

(iii) since the Legendre polynomial $P_{m+p}(x)$ has $(m + p)$ zeros in (-1, 1), its $m$th derivative must have $p$ zeros therein (see Fig. 1 and exercise 2).

Thus the associated Legendre functions comprise the complete set of eigenfunctions for this problem and the appropriate eigenfunction expansion is

$$f(x) = \sum_{p=0}^{\infty} c_p P_{m+p}^m(x) \quad \text{or} \quad \sum_{\ell=m}^{\infty} d_\ell P_\ell^m(x), \qquad (6)$$

the second (reindexed) form being more common in practice. Like the Fourier-Bessel series, there is an expansion available for each value of $m$;

---

[12]Some authors omit the factor $(-1)^m$. Here we have chosen to go along with the tables of Abramowitz and Stegun. (Incidentally, the latter contains errors on this point; equations 8.6.6, 8.6.7, and Fig. 8.4 therein are inconsistent.)

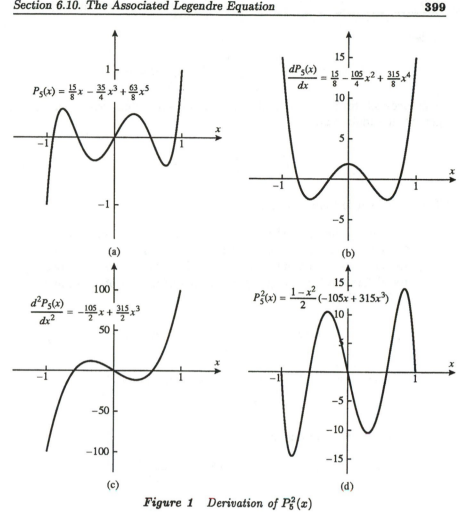

$$P_5(x) = \frac{15}{8}x - \frac{35}{4}x^3 + \frac{63}{8}x^5$$

(a)

$$\frac{dP_5(x)}{dx} = \frac{15}{8} - \frac{105}{4}x^2 + \frac{315}{8}x^4$$

(b)

$$\frac{d^2P_5(x)}{dx^2} = -\frac{105}{2}x + \frac{315}{2}x^3$$

(c)

$$P_5^2(x) = \frac{1-x^2}{2}(-105x + 315x^3)$$

(d)

**Figure 1**   *Derivation of $P_5^2(x)$*

when $m = 0$ the Legendre polynomial expansion results. The weight function $w(x) = 1$ as before, and the formula for the coefficients is

$$
\begin{aligned}
d_\ell &= \frac{\langle f(x), P_\ell^m(x) \rangle}{\| P_\ell^m \|^2} = \frac{\int_{-1}^{1} f(x) P_\ell^m(x)\, dx}{\int_{-1}^{1} P_\ell^m(x)^2\, dx} \\[2mm]
&= \frac{(2\ell + 1)(\ell - m)!}{2(\ell + m)!} \int_{-1}^{1} f(x) P_\ell^m(x)\, dx \qquad (7)
\end{aligned}
$$

where we have employed Table 6.2 to simplify the denominator.

When the parameter $m$ is negative the formula (3) is nonsensical. However, the Rodriguez formula in Table 6.2 does, in fact, continue to define solutions to the basic differential equation (2). We omit detailed consideration of this complication and simply state that *when $m$ is less than zero but greater than or equal to $-\ell$, the associated Legendre function is taken* *to be*

$$P_\ell^m(x) = (-1)^m \frac{(\ell+m)!}{(\ell-m)!} P_\ell^{-m}(x), \quad 0 < -m \le \ell. \qquad (8)$$

This necessitates a slight modification of the expansion expression (6); in general it should read

$$f(x) = \sum_{\ell=|m|}^{\infty} d_\ell P_\ell^m(x). \qquad (9)$$

The coefficient formula (7) remains intact.

Here is the proof that the associated Legendre function defined by (3) satisfies the differential equation (2). For brevity we replace the differentiation symbol $d/dx$ by $D$. The combinatorial formula for taking higher order derivatives of a product is *Leibniz's formula*:

$$D^n(fg) = \sum_{j=0}^{n} \frac{n!}{j!(n-j)!} D^j f \, D^{n-j} g \qquad (10)$$

Now use this to differentiate the Legendre equation

$$(1-x^2)y'' - 2xy' = \lambda y$$

$m$ times. For the first term we obtain

$$\begin{aligned}
D^m\{(1-x^2)y''\} &= (1-x^2)D^{m+2}y + \frac{m!}{1!(m-1)!}(-2x)D^{m+1}y \\
&\quad + \frac{m!}{2!(m-2)!}(-2)D^m y \\
&= (1-x^2)Y'' - 2mxY' - m(m-1)Y
\end{aligned}$$

where $Y = D^m y$. Continuing, we find

$$D^m\{-2xy'\} = -2xD^{m+1}y + m(-2)D^m y = -2xY' - 2mY$$

$$D^m\{\lambda y\} = \lambda Y$$

Thus $Y$ satisfies the equation

$$(1-x^2)Y'' - 2(m+1)xY' - m(m+1)Y = \lambda Y, \qquad (11)$$

which is very close to the associated Legendre equation! We need to get rid of the factor $(m+1)$. The trick is to set

$$Z(x) = (1-x^2)^k Y(x) = (1-x^2)^k D^m y.$$

We'll adjust the value of $k$ later. We have

$$
\begin{aligned}
Z' &= (1-x^2)^k Y' + k(-2x)(1-x^2)^{k-1} Y \\
&= (1-x^2)^k Y' - 2kx(1-x^2)^{k-1} Y \\
Z'' &= (1-x^2)^k Y'' - 4kx(1-x^2)^{k-1} Y' \\
&\quad - 2k(1-x^2)^{k-1} Y + 4x^2 k(k-1)(1-x^2)^{k-2} Y.
\end{aligned}
$$

Next force $Z$ to satisfy the associated Legendre equation.

$$
\begin{aligned}
(1-x^2)Z'' - 2xZ' &= (1-x^2)^k \{ (1-x^2)Y'' - 4kxY' - 2kY \\
&\quad + \frac{4x^2 k(k-1)}{1-x^2} Y \} \\
&\quad + (1-x^2)^k \{ -2xY' + \frac{4kx^2}{(1-x^2)} Y \} \\
&= (1-x^2)^k \{ (1-x^2)Y'' - 2x(2k+1)Y' \\
&\quad - 2kY + \frac{4k^2 x^2}{(1-x^2)} Y \}.
\end{aligned}
$$

If we set $k = m/2$ the terms in $Y''$ and $Y'$ can be replaced using (11). Reintroducing $Z$ we then have

$$
\begin{aligned}
(1-x^2)Z'' - 2xZ' &= m(m+1)Z + \lambda Z - 2\frac{m}{2} Z \\
&\quad + \frac{4x^2 (m/2)^2}{(1-x^2)} Z \\
&= \lambda Z + m^2 Z + \frac{m^2 x^2}{1-x^2} Z \\
&= \lambda Z + \frac{m^2}{1-x^2} Z,
\end{aligned}
$$

which coincides with the associated Legendre equation. Now if $\lambda$ takes the value $-\ell(\ell+1)$, then $Y$ is $D^m y = D^m P_\ell(x)$ and

$$
Z(x) = (1-x^2)^{m/2} Y(x) = (1-x^2)^{m/2} D^m P_\ell(x)
$$

is, up to the factor $(-1)^m$, the associated Legendre function.

In practice we find that the associated Legendre functions are not widely used by themselves; they are usually wedded with the Fourier functions in the variable $\theta$ and the resulting combinations are the *Spherical Harmonics*. They will be discussed in Section 7.1.

## Exercises 6.10

**1.** Demonstrate the equivalence of eqs. (1) and (2).

**2.** Draw a sketch to demonstrate the fact that between any two zeros of a differentiable function there must occur a zero of the derivative (*Rolle's theorem*). (Why must the function be differentiable?) Use this to show that $P_n^m(x)$ has $(n-m)$ zeros in (-1, 1) (Fig. 1).

# TABLE 6.2 Associated Legendre Functions

Differential equation: for $n \geq |m|$,

$$(1-x)^2 P_n^m(x)'' - 2x P_n^m(x)' - \frac{m^2}{1-x^2} P_n^m(x) = -n(n+1) P_n^m(x)$$

$$\frac{1}{\sin\phi} \frac{d}{d\phi} \{\sin\phi \frac{d}{d\phi} P_n^m(\cos\phi)\} - \frac{m^2}{\sin^2\phi} P_n^m(\cos\phi) = -n(n+1) P_n^m(\cos\phi)$$

Description

$P_n^m$ is, up to a constant multiple, the only solution which is finite at $x = \pm 1$ or $\phi = 0, \pi$.

$P_n^m(x)$ is odd if $n+m$ is odd, even if $n+m$ is even.

$P_n^m(\pm 1) = 0$ for $m \geq 1$.

$|m|$ must be $\leq n$.

$P_n^0$ is the Legendre polynomial of degree $n$.

Normalization Integral

$$\int_{-1}^{1} P_n^m(x) P_{n'}^m(x) \, dx = \int_0^\pi P_n^m(\cos\phi) P_{n'}^m(\cos\phi) \sin\phi \, d\phi$$

$$= \frac{2(n+m)!}{(2n+1)(n-m)!} \delta_{nn'}$$

Rodriguez's Formula

$$P_n^m(x) = (-1)^m (1-x^2)^{m/2} \{\frac{d}{dx}\}^m P_n(x)$$

$$= (-1)^m \frac{1}{2^n n!} (1-x^2)^{m/2} \{\frac{d}{dx}\}^{m+n} (x^2-1)^n$$

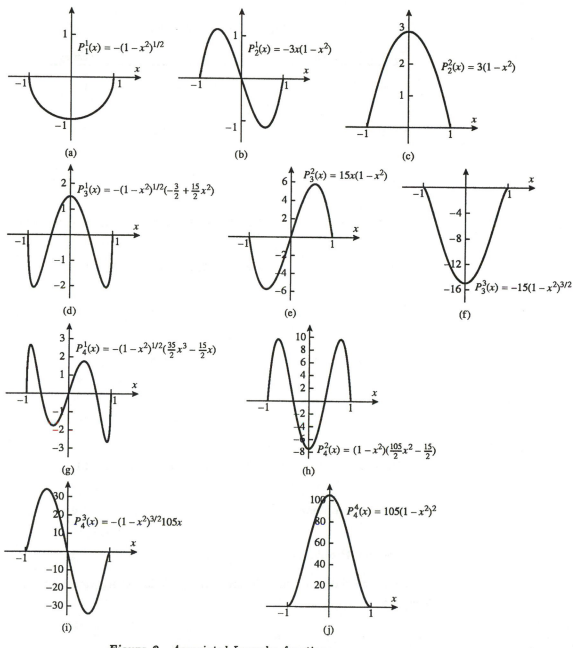

**Figure 2**  *Associated Legendre functions*

# TABLE 6.3 Eigenfunction Expansions

## Guide to the Expansions

| Type of Equation | | Entry |
|---|---|---|
| Generic | | 1 |
| Constant Coefficient | $y'' + \nu y' = \lambda y$ | 2, 7-12 |
| Equidimensional | $x^2 y'' + \sigma x y' = \lambda y$ | 3, 13-15 |
| Bessel | $x^2 y'' + \sigma x y' = \lambda_1 y + \lambda_2 x^2 y$ | 4, 5, 6, 16-26 |
| Legendre | $(1 - x^2)y'' - 2xy' - \frac{m^2}{1-x^2} y = \lambda y$ | 27, 28 |
| or | $(\sin \phi \, \Phi')' - \frac{m^2}{\sin \phi} \Phi = \lambda \sin \phi \, \Phi$ | |
| Spherical Harmonics | $\nabla^2 R(r) \, Y_{\ell m}(\phi, \theta)$ $= \frac{(r^2 R')' - \ell(\ell+1)R}{r^2} Y_{\ell m}(\phi, \theta)$ | 29 |
| Hermite | $y'' - x^2 y = \lambda y$ | 30 |
| Chebyshev | $(1 - x^2)y'' - \sigma x y' = \lambda y$ | 31 |
| **Regular Problems** | | 1-6 |
| **Singular Problems** | | 7-31 |

Note: Section 6.6 demonstrates the use of this table in solving partial differential equations.

## PART ONE. **Regular Sturm-Liouville Problems**

### 1. <u>GENERIC EIGENFUNCTION EXPANSION</u>

Differential Equation

$$a_2(x)\, y'' + a_1(x)\, y' + a_0(x)\, y = \lambda\, g(x)\, y(x), \ a < x < b$$

Boundary Conditions

$$\alpha y(a) + \beta y'(a) = 0, \quad \gamma y(b) + \delta y'(b) = 0$$

Let $y_1(x; \lambda)$ and $y_2(x, \lambda)$ be independent solutions to the differential equation.

Eigenvalues

$\lambda_p$ is the $p$th root of the equation

$$[\alpha y_2(a; \lambda) + \beta y_2'(a; \lambda)][\gamma y_1(b; \lambda) + \delta y_1'(b; \lambda)]$$

$$- [\alpha y_1(a; \lambda) + \beta y_1'(a; \lambda)][\gamma y_2(b; \lambda) + \delta y_2'(b; \lambda)] = 0$$

Eigenfunctions

$$\phi_p(x) = [\alpha y_2(a; \lambda_p) + \beta y_2'(a; \lambda_p)] y_1(x; \lambda_p)$$

$$- [\alpha y_1(a; \lambda_p) + \beta y_1'(a; \lambda_p)] y_2(x; \lambda_p)$$

Weight Function $\quad w(x) = \dfrac{g(x)}{a_2(x)} \exp\{\displaystyle\int^x [a_1(x)/a_2(x)]\, dx\}$

Orthogonality Integrals

$$\int_a^b \phi_p(x)\, \phi_{p'}(x)\, w(x)\, dx = 0 \quad \text{if } p \neq p'$$

Expansion

$$f(x) = \sum_{p=1}^{\infty} c_p\, \phi_p(x)$$

$$c_p = \frac{\int_a^b f(x)\, \phi_p(x)\, w(x)\, dx}{\int_a^b \phi_p^2(x)\, w(x)\, dx}$$

Reference: Sec. 6.2 (exercise 10)

## 2. CONSTANT COEFFICIENT EQUATIONS (Damped Harmonic Oscillator; Sine, Cosine Series, Fourier Series)

Differential Equation

$$y'' + \nu y' = \lambda y, \; 0 < x < L$$

Boundary Conditions (see below for Periodic Boundary Conditions)

$$\alpha y(0) + \beta y'(0) = 0, \quad \gamma y(L) + \delta y'(L) = 0$$

Eigenvalues

$\lambda_p = -\omega_p^2 - \nu^2/4$, where $\omega_p$ is the $p$th solution (possibly imaginary) to
$$(\delta\nu/2 - \gamma) \sin[\omega L + \arctan(-\tfrac{\beta}{\alpha - \beta\nu/2}\, \omega)] =$$
$$\omega\delta \, \cos[\omega L + \arctan(-\tfrac{\beta}{\alpha - \beta\nu/2}\, \omega)]$$

also $\lambda_0 = -\nu^2/4$, if $\alpha - \beta\nu/2 = \gamma - \delta\nu/2 = 0$
or $\lambda_0 = 0$, if $\nu = 0$ and $L = \frac{\beta\gamma - \alpha\delta}{\alpha\gamma}$.

Eigenfunctions

$$\phi_p(x) = e^{-\nu x/2} \sin[\omega_p x + \arctan(-\tfrac{\beta}{\alpha - \beta\nu/2}\, \omega_p)]$$

also $\phi_0(x) = e^{-\nu x/2}$, if $\alpha - \beta\nu/2 = \gamma - \delta\nu/2 = 0$
or $\tilde{\phi}_0(x) = \beta - \alpha x$, if $\nu = 0$ and $L = \frac{\beta\gamma - \alpha\delta}{\alpha\gamma}$.

Weight Function     $w(x) = e^{\nu x}$

Orthogonality Integral

$$\int_0^L \phi_0^2(x)\, e^{\nu x}\, dx = L$$

$$\int_0^L \tilde{\phi}_0^2(x)\, dx = \beta^2 L - 2\alpha\beta\frac{L^2}{2} + \alpha^2\frac{L^3}{3}$$

$$\int_0^L \phi_p(x)\, \phi_{p'}(x)\, e^{\nu x}\, dx = \tfrac{1}{2}\left\{ L + \frac{\gamma\delta}{\gamma^2 + \omega_p^2\delta^2} - \frac{\alpha\beta}{\alpha^2 + \omega_p^2\beta^2} \right\} \delta_{pp'} \text{ otherwise}$$

Expansion

$$f(x) = \sum_{p=1}^{\infty} c_p\, \phi_p(x) + c_0\, \phi_0(x) + \tilde{c}_0\, \tilde{\phi}_0(x)$$

$$c_p = 2 \int_0^L f(x)\, \phi_p(x)\, e^{\nu x}\, dx \Big/ \Big\{ L + \frac{\gamma\delta}{\gamma^2 + \omega_p^2 \delta^2} - \frac{\alpha\beta}{\alpha^2 + \omega_p^2 \beta^2} \Big\} \quad (p \neq 0)$$

$$c_0 = \begin{cases} \int_0^L f(x)\, \phi_0(x)\, e^{\nu x}\, dx / L & \text{if } \alpha - \beta\nu/2 = \gamma - \delta\nu/2 = 0 \\ \\ 0 & \text{otherwise} \end{cases}$$

$$\tilde{c}_0 = \begin{cases} \int_0^L f(x)\, \tilde{\phi}_0(x)\, dx \Big/ \Big\{ \beta^2 L - 2\alpha\beta \frac{L^2}{2} + \alpha^2 \frac{L^3}{3} \Big\} & \text{if } \nu = 0 \text{ and } L = \frac{\beta\gamma - \alpha\delta}{\alpha\gamma} \\ \\ 0 & \text{otherwise} \end{cases}$$

Reference: Sec. 6.3 (exercise 13)

Fourier Sine Series: Take $\nu = \beta = \delta = 0$, $\alpha = \gamma = 1$.

Fourier Cosine Series: Take $\nu = \alpha = \gamma = 0$, $\beta = \delta = 1$.

*Periodic Boundary Conditions: Fourier Series*

Differential Equation

$$y'' = \lambda y, \quad -L < x < L$$

Boundary Conditions

$$y(-L) = y(L), \quad y'(-L) = y'(L)$$

Eigenvalues

$$\lambda_p = -(p\pi/L)^2, \quad p = 0, \pm 1, \pm 2, \ldots$$

Eigenfunctions

$$\phi_0(x) = 1, \quad \phi_p(x) = \left\{ \begin{array}{c} \cos \frac{p\pi x}{L} \\ \sin \frac{p\pi x}{L} \end{array} \right\}, \quad p = 1, 2, 3, \ldots$$

or $\phi_p(x) = e^{ip\pi x/L}$; $p = 0, \pm 1, \pm 2, \ldots$

Weight Function $\quad w(x) = 1$

Orthogonality Integrals

$$\int_{-L}^{L} \cos \frac{p\pi x}{L} \sin p'\pi x / L \, dx = 0,$$

$$\int_{-L}^{L} \sin \frac{p\pi x}{L} \, \sin p'\pi x/L \, dx = L \, \delta_{pp'},$$

$$\int_{-L}^{L} \cos \frac{p\pi x}{L} \, \cos p'\pi x/L \, dx = \begin{cases} 2L & \text{if } p = p' = 0 \\ L \, \delta_{pp'}, & \text{otherwise} \end{cases}$$

$$\int_{-L}^{L} e^{ip\pi x/L} \, e^{-ip'\pi x/L} \, dx = 2L \, \delta_{pp'}$$

**Expansion (Real Form)**

$$f(x) = \sum_{p=0}^{\infty} c_p \cos \frac{p\pi x}{L} + \sum_{p=1}^{\infty} d_p \sin \frac{p\pi x}{L}$$

$$c_0 = \frac{1}{2L} \int_{-L}^{+L} f(x) \, dx, \ c_p = \frac{1}{L} \int_{-L}^{+L} f(x) \cos \frac{p\pi x}{L} \, dx \ (p > 0),$$

$$d_p = \frac{1}{L} \int_{-L}^{+L} f(x) \sin \frac{p\pi x}{L} \, dx, \ p > 0$$

**Expansion (Complex Form)**

$$f(x) = \sum_{p=-\infty}^{\infty} g_p \, e^{ip\pi x/L};$$

$$g_p = \frac{1}{2L} \int_{-L}^{L} f(x) \, e^{-ip\pi x/L} \, dx$$

Reference: Chap. 3; Sec. 6.3 (Example 2)

---

## 3. EQUIDIMENSIONAL EQUATIONS

**Differential Equation**

$$x^2 y'' + \sigma x y' = \lambda y, \ 0 < a < x < b$$

**Boundary Conditions**

$$\alpha y(a) + \beta y'(a) = 0, \quad \gamma y(b) + \delta y'(b) = 0$$

**Eigenvalues**

$\lambda_p = -\omega_p^2 - (\sigma - 1)^2/4$ , where $\omega_p$ is the $p$th solution (possible imaginary) to
$$[\delta\sigma - \delta - 2b\gamma] \sin[\omega \ln \tfrac{b}{a} + \arctan \tfrac{2\beta\omega}{\beta\sigma-\beta-2a\alpha}] = 2\omega\delta \cos[\omega \ln \tfrac{b}{a} + \arctan \tfrac{2\beta\omega}{\beta\sigma-\beta-2a\alpha}]$$

also $\lambda_0 = -(\sigma - 1)^2/4$, if $\alpha - \beta\frac{\sigma-1}{2} = \gamma - \delta\frac{\sigma-1}{2} = 0$

Eigenfunctions

$$\phi_p(x) = x^{(1-\sigma)/2} \sin[\omega_p \ln \tfrac{x}{a} + \arctan(\tfrac{2\beta}{\beta\sigma - \beta - 2a\alpha}\, \omega_p)]$$

also $\phi_0(x) = x^{(1-\sigma)/2}$, if $\alpha - \beta\frac{\sigma-1}{2} = \gamma - \delta\frac{\sigma-1}{2} = 0$

Weight Function    $w(x) = x^{\sigma-2}$

Orthogonality Integral

$$\int_a^b \phi_p(x)\,\phi_{p'}(x)\,x^{\sigma-2}\,dx = \begin{cases} \ln\frac{b}{a} & \text{if } p = p' = 0; \\[2mm] & \text{otherwise,} \\[1mm] \frac{1}{2}\left\{\ln\frac{b}{a} + \dfrac{b\gamma\delta}{b^2\gamma^2 + \omega_p^2\delta^2} - \dfrac{a\alpha\beta}{a^2\alpha^2 + \omega_p^2\beta^2}\right\}\delta_{pp'} \end{cases}$$

Expansion

$$f(x) = \sum_{p=1}^\infty c_p\phi_p(x) + c_0\phi_0(x);$$

$$c_{p\neq0} = 2\int_a^b f(x)\,\phi_p(x)\,x^{\sigma-2}\,dx \Big/ \{\ln\tfrac{b}{a} + \dfrac{b\gamma\delta}{b^2\gamma^2 + \omega_p^2\delta^2} - \dfrac{a\alpha\beta}{a^2\alpha^2 + \omega_p^2\beta^2}\}$$

$$c_0 = \begin{cases} \int_a^b f(x)\,x^{(\sigma-3)/2}\,dx \big/ \ln(b/a) & \text{if } \alpha - \beta\frac{\sigma-1}{2} = \gamma - \delta\frac{\sigma-1}{2} = 0 \\[2mm] 0, & \text{otherwise} \end{cases}$$

Reference: Sec. 6.3 (exercise 14)

## 4. REGULAR BESSEL FUNCTION SERIES

Differential Equation

$$x^2 y'' + x y' - \nu^2 y = \lambda x^2 y, \; 0 < a < x < b$$

*Case One: Dirichlet-Dirichlet Conditions*

Boundary Conditions

$$y(a) = 0, \; y(b) = 0$$

Eigenvalues

$$\lambda_p = -\mu_{\nu,p}^2, \text{ where } \mu = \mu_{\nu,p} \text{ is the } p\text{th positive root of}$$
$$Y_\nu(\mu a)\,J_\nu(\mu b) - J_\nu(\mu a)\,Y_\nu(\mu b) = 0$$

Eigenfunctions

$$\phi_p(x) = Y_\nu(\mu_{\nu,p}a)\,J_\nu(\mu_{\nu,p}x) - J_\nu(\mu_{\nu,p}a)\,Y_\nu(\mu_{\nu,p}x)$$

Weight Function    $w(x) = x$

Orthogonality Integral

$$\int_a^b \phi_{\nu,p}(x)\,\phi_{\nu,p'}(x)\,x\,dx = \frac{2}{\pi^2 \mu_{\nu,p}^2}\left\{ \frac{J_\nu^2(\mu_{\nu,p}a)}{J_\nu^2(\mu_{\nu,p}b)} - 1\right\}\delta_{pp'}$$

Expansion

$$f(x) = \sum_{p=1}^{\infty} c_p\phi_{\nu,p}(x)$$

$$c_p = \frac{\pi^2}{2}\,\frac{\mu_{\nu,p}^2 J_\nu^2(\mu_{\nu,p}b)\int_a^b f(x)\phi_{\nu,p}(x)x\,dx}{J_\nu^2(\mu_{\nu,p}a) - J_\nu^2(\mu_{\nu,p}b)}$$

*Case Two: Neumann-Dirichlet Conditions*

Boundary Conditions

$$y'(a) = 0,\; y(b) = 0;\; 0 < a < x < b \text{ or } 0 < b < x < a)$$

Eigenvalues

$\lambda_p = -\mu_{\nu,p}^2$, where $\mu = \mu_{\nu,p}$ is the $p$th positive root of
$Y_\nu'(\mu a)\,J_\nu(\mu b) - J_\nu'(\mu a)\,Y_\nu(\nu b) = 0$
and $J_\nu(z)$ and $Y_\nu(z)$ are the Bessel functions of the
first and second kind, respectively, of order $\nu$

Eigenfunction

$$\phi_p(x) = Y_\nu'(\mu_{\nu,p}a)\,J_\nu(\mu_{\nu,p}x) - J_\nu'(\mu_{\nu,p}a)\,Y_\nu(\mu_{\nu,p}x)$$

Weight Function    $w(x) = x$

Orthogonality Integral

$$\int_a^b \phi_{\nu,p}(x)\,\phi_{\nu,p'}(x)\,x\,dx =$$

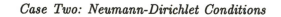

$$\frac{2}{\pi^2 \mu_{\nu,p}^2}\left\{ \frac{{J_\nu'}^2(\mu_{\nu,p}a)}{J_\nu^2(\mu_{\nu,p}b)} - 1 + \frac{\nu^2}{\mu_{\nu,p}^2 a^2}\right\}\delta_{pp'}$$

Expansion

$$f(x) = \sum_{p=1}^{\infty} c_p \phi_{\nu,p}(x)$$

$$c_p = \frac{\pi^2}{2} \frac{\mu_{\nu,p}^2 J_\nu^2(\mu_{\nu,p}b) \int_a^b f(x)\phi_{\nu,p}(x)x\,dx}{J_\nu'^2(\mu_{\nu,p}a) - J_\nu^2(\mu_{\nu,p}b)\{1 - \nu^2/\mu_{\nu,p}^2 a^2\}}$$

*Case Three: Neumann-Neumann Conditions*

Boundary Conditions

$$y'(a) = 0, \; y'(b) = 0$$

Eigenvalues

$\lambda_p = -\mu_{\nu,p}^2$, where $\mu = \mu_{\nu,p}$ is the $p$th positive root of
$$Y_\nu'(\mu a)\, J_\nu'(\mu b) - J_\nu'(\mu a)\, Y_\nu'(\mu b) = 0$$

Eigenfunctions

$$\phi_p(x) = Y_\nu'(\mu_{\nu,p}a)\, J_\nu(\mu_{\nu,p}x) - J_\nu'(\mu_{\nu,p}a)\, Y_\nu(\mu_{\nu,p}x)$$

Weight Function $\quad w(x) = x$

Orthogonality Integral

$$\int_a^b \phi_{\nu,p}(x)\,\phi_{\nu,p'}(x)\,x\,dx =$$

$$\frac{2}{\pi^2 \mu_{\nu,p}^2}\left\{ \left[\frac{J_\nu'(\mu_{\nu,p}a)}{J_\nu'(\mu_{\nu,p}b)}\right]^2 \left[1 - \frac{\nu^2}{\mu_{\nu,p}^2 b^2}\right] - 1 + \frac{\nu^2}{\mu_{\nu,p}^2 a^2}\right\}\delta_{pp'}$$

Expansion

$$f(x) = \sum_{p=1}^{\infty} c_p \phi_{\nu,p}(x)$$

$$c_p = \frac{\pi^2}{2} \frac{\mu_{\nu,p}^2 J_\nu'^2(\mu_{\nu,p}b) \int_a^b f(x)\phi_{\nu,p}(x)x\,dx}{\{1 - \nu^2/\mu_{\nu,p}^2 b^2\}J_\nu'^2(\mu_{\nu,p}a) - \{1 - \nu^2/\mu_{\nu,p}^2 a^2\}J_\nu'^2(\mu_{\nu,p}b)}$$

Note: for Robin boundary conditions see Cinelli's paper.

Note: the inner products in the coefficients $c_p$ are commonly known as finite Hankel transforms.

Reference: Sec. 6.3 (Example 3), Table 2.3, Sneddon, Cinelli

## 5. LEBEDEV EXPANSION

Differential Equation

$$x^2 y'' + xy' - \mu^2 x^2 y = \lambda y, \ 0 < a < x < b$$

Boundary Conditions

$$\alpha y(a) + \beta y'(a) = 0, \ \ \gamma y(b) + \delta y'(b) = 0$$

Eigenvalues

$\lambda_p = -\nu_p^2$, where $\nu = \nu_p$ is the $p$th root of
$$\gamma \, \phi(b; \mu, \nu) + \delta \, \tfrac{\partial}{\partial b} \phi(b; \mu, \nu) = 0,$$
and $\phi(x; \mu, \nu)$ is defined below

Eigenfunctions

$$\phi(x; \mu, \nu_p) = \alpha [K_{i\nu_p}(\mu a) \, I_{i\nu_p}(\mu x) - I_{i\nu_p}(\mu a) \, K_{i\nu_p}(\mu x)]$$

$$+ \beta \mu [K'_{i\nu_p}(\mu a) I_{i\nu_p}(\mu x) - I'_{i\nu_p}(\mu a) K_{i\nu_p}(\mu x)]$$

where $I_{i\nu}$ and $K_{i\nu}$ are the modified Bessel functions of the first and second kind, respectively, of order $i\nu$

Weight Function    $w(x) = 1/x$

Orthogonality Integral

$$\int_a^b \phi(x; \mu, \nu_p) \, \phi(x; \mu, \nu_{p'}) \, \tfrac{dx}{x} =$$

$$\frac{\tfrac{\partial}{\partial \nu}[\gamma \, \phi(b; \mu, \nu_p) + \delta \, \tfrac{\partial}{\partial b} \phi(b : \mu, \nu_p)][\alpha \, K_{i\nu_p}(\mu a) + \beta \mu \, K'_{i\nu_p}(\mu a)]}{2\nu_p[\gamma \, K_{i\nu_p}(\mu b) + \delta \mu \, K'_{i\nu_p}(\mu b)]} \delta_{pp'}$$

Expansion

$$f(x) = \sum_{p=1}^{\infty} c_p \, \phi(x; \mu, \nu_p)$$

$$c_p = \frac{2\nu_p[\gamma \, K_{i\nu_p}(\mu b) + \delta \mu \, K'_{i\nu_p}(\mu b)]}{\tfrac{\partial}{\partial \nu}[\gamma \, \phi(b; \mu, \nu_p) + \delta \, \tfrac{\partial}{\partial b} \phi(b; \mu, \nu_p)][\alpha \, K_{i\nu_p}(\mu a) + \beta \mu \, K'_{i\nu_p}(\mu a)]}$$

Reference: Sec. 6.3 (Example 5); Sec. 6.6; Naylor (1966)

## 6. SPHERICAL BESSEL FUNCTION SERIES

Differential Equation

$$x^2 y'' + 2xy' - \ell(\ell+1)y = \lambda x^2 y, \ 0 < a < x < b$$

Boundary Conditions

$$y(a) = 0, \ \ y(b) = 0$$

Eigenvalues

$$\lambda_p = -s_{\ell,p}^2, \text{ where } s = s_{\ell,p} \text{ is the } p\text{th positive root of}$$
$$y_\ell(sa)\, j_\ell(sb) - j_\ell(sa)\, y_\ell(sb) = 0$$

Eigenfunctions

$$\phi_{\ell,p}(x) = y_\ell(s_{\ell,p}a)\, j_\ell(s_{\ell,p}x) - j_\ell(s_{\ell,p}a)\, y_\ell(s_{\ell,p}x)$$

Weight function $\quad w(x) = x^2$

Orthogonality Integral

$$\int_a^b \phi_{\ell,p}(x)\, \phi_{\ell,p'}(x)\, x^2\, dx = \frac{J_{\ell+1/2}^2(s_{\ell,p}a) - J_{\ell+1/2}^2(s_{\ell,p}b)}{2\, a\, s_{\ell,p}^4\, J_{\ell+1/2}^2(s_{\ell,p}b)}\, \delta_{pp'}$$

Expansion

$$f(x) = \sum_{p=1}^\infty c_p\, \phi_{\ell,p}(x)$$

$$c_p = 2\, a\, s_{\ell,p}^4\, \frac{J_{\ell+1/2}^2(s_{\ell,p}b)}{J_{\ell+1/2}^2(s_{\ell,p}a) - J_{\ell+1/2}^2(s_{\ell,p}b)} \int_a^b f(x)\, \phi_{\ell,p}(x)\, x^2\, dx$$

Reference: Sec. 6.3 (Example 4)

(For other boundary conditions relate the general solution to the Bessel functions and use the corresponding series.)

# Part Two. **Singular Sturm-Liouville Problems**

## 7. FOURIER INTEGRAL

### Differential Equation

$$y'' = \lambda y, \quad -\infty < x < \infty$$

### Boundary Conditions

$$y(\pm\infty) \text{ finite}$$

### Eigenvalues

$$\lambda_\omega = -\omega^2; \quad -\infty < \omega < \infty$$

### Eigenfunctions

$$\phi_\omega(x) = e^{i\omega x}$$

### Weight Function $\quad w(x) = 1$

### Orthogonality Integral

$$\int_{-\infty}^{\infty} e^{i(\omega-\omega')x}\, dx = 2\pi\, \delta\left(\omega - \omega'\right)$$

### Expansion

$$f(x) = \int_{-\infty}^{\infty} F(\omega)\, e^{i\omega x}\, d\omega, \quad -\infty < x < \infty;$$

$$F(\omega) = \tfrac{1}{2\pi} \int_{-\infty}^{\infty} f(x)\, e^{-i\omega x}\, dx$$

Reference: Sec. 3.6, Sec. 3.7, Sec. 6.4 (Example 2)

---

## 8. FOURIER SINE INTEGRAL

### Differential Equation

$$y'' = \lambda y, \, 0 < x < \infty$$

### Boundary Conditions

$$y(0) = 0, \;\; y(\infty) \text{ finite}$$

Eigenvalues

$$\lambda_\omega = -\omega^2; \quad 0 \leq \omega < \infty$$

Eigenfunctions

$$\phi_\omega(x) = \sin \omega x$$

Weight Function $\quad w(x) = 1$

Orthogonality Integral

$$\int_0^\infty \sin \omega x \, \sin \omega' x \, dx = \tfrac{\pi}{2} \delta(\omega - \omega') \quad (0 \leq \omega, \, \omega' < \infty)$$

Expansion

$$f(x) = \int_0^\infty S(\omega) \, \sin \omega x \, d\omega;$$

$$S(\omega) = \tfrac{2}{\pi} \int_0^\infty f(x) \, \sin \omega x \, dx$$

Reference: Sec. 3.6, Sec. 6.4 (Example 2)

---

## 9. FOURIER COSINE INTEGRAL

Differential Equation

$$y'' = \lambda y, \; 0 < x < \infty$$

Boundary Conditions

$$y'(0) = 0, \quad y(\infty) \text{ finite}$$

Eigenvalues

$$\lambda = -\omega^2; \quad 0 \leq \omega < \infty$$

Eigenfunctions

$$y_\omega(x) = \cos \omega x$$

Weight Function $\quad w(x) = 1$

Orthogonality Integral

$$\int_0^\infty \cos \omega x \, \cos \omega' x \, dx = \tfrac{\pi}{2}\,\delta(\omega - \omega') \quad (0 \le \omega, \, \omega' < \infty)$$

Expansion

$$f(x) = \int_0^\infty C(\omega) \, \cos \omega x \, d\omega;$$

$$C(\omega) = \tfrac{2}{\pi} \int_0^\infty f(x) \, \cos \omega x \, dx$$

Reference: Sec. 3.6, Sec. 6.4 (Example 2)

---

## 10. FOURIER INTEGRAL WITH ROBIN BOUNDARY CONDITION

Differential Equation

$$y'' = \lambda y, \; 0 < x < \infty$$

Boundary Conditions

$$\alpha y(0) + \beta y'(0) = 0, \quad y(\infty) \text{ finite}$$

Eigenvalues

$$\lambda = -\omega^2, \; 0 \le \omega < \infty; \text{ also } \lambda = (\alpha/\beta)^2 \text{ if } \beta \neq 0 \text{ and } \alpha/\beta > 0$$

Eigenfunctions

$$\phi_\omega(x) = \sin\{\omega x - \arctan \tfrac{\beta \omega}{\alpha}\}; \text{ also } e^{-\alpha x/\beta} \text{ if } \beta \neq 0 \text{ and } \alpha/\beta > 0$$

Weight Function    $w(x) = 1$

Orthogonality Integral

$$\int_0^\infty \phi_\omega(x) \, \phi_{\omega'}(x) \, dx = \tfrac{\pi}{2}\,\delta(\omega - \omega')$$

$$\int_0^\infty \phi_\omega(x) \, e^{-\alpha x/\beta} \, dx = 0 \text{ if } \beta \neq 0 \text{ and } \alpha/\beta > 0$$

$$\int_0^\infty e^{-2\alpha x/\beta} \, dx = \tfrac{\beta}{2\alpha} \text{ if } \beta \neq 0 \text{ and } \alpha/\beta > 0$$

Expansion

$$f(x) = A \, e^{-\alpha x/\beta} + \int_0^\infty G(\omega) \, \sin\{\omega x - \arctan \tfrac{\beta \omega}{\alpha}\} \, d\omega$$

$$G(\omega) = \tfrac{2}{\pi} \int_0^\infty f(x) \sin\{\omega x - \arctan \tfrac{\beta\omega}{\alpha}\}\, dx$$

$$A = \begin{cases} \tfrac{2\alpha}{\beta} \int_0^\infty f(x)\, e^{-\alpha x/\beta}\, dx & \text{if } \beta \neq 0 \text{ and } \alpha/\beta > 0, \\[2mm] 0 & \text{otherwise} \end{cases}$$

Reference: Sec. 6.7 (Example 1)

---

## 11. CONSTANT COEFFICIENT ON $[0, \infty)$

**Differential Equation**

$$y'' + \nu y' = \lambda y, \ 0 < x < \infty$$

**Boundary Conditions**

$$\alpha y(0) + \beta y'(0) = 0, \quad y(x)\, e^{\nu x/2} \text{ finite at } \infty$$

**Eigenvalues**

$$\lambda = -\omega^2 - \nu^2/4, \quad 0 \leq \omega < \infty;$$

also $(\alpha/\beta)^2 - \alpha\nu/\beta$ if $\beta \neq 0$ and $\alpha/\beta > \nu/2$

**Eigenfunctions**

$$\phi_\omega(x) = e^{-\nu x/2} \sin\{\omega x - \arctan \tfrac{\beta\omega}{\alpha - \beta\nu/2}\};$$

also $e^{-\alpha x/\beta}$ if $\beta \neq 0$ and $\alpha/\beta > \nu/2$

**Weight Function** $\quad w(x) = e^{\nu x}$

**Orthogonality Integral**

$$\int_0^\infty \phi_\omega(x)\, \phi_{\omega'}\, e^{\nu x}\, dx = \tfrac{\pi}{2}\, \delta(\omega - \omega')$$

$$\int_0^\infty \phi_\omega(x)\, e^{-\alpha x/\beta}\, e^{\nu x}\, dx = 0 \text{ if } \beta \neq 0 \text{ and } \alpha/\beta > \nu/2$$

$$\int_0^\infty e^{-2\alpha x/\beta}\, e^{\nu x}\, dx = \tfrac{\beta}{2\alpha - \beta\nu} \text{ if } \beta \neq 0 \text{ and } \alpha/\beta > \nu/2$$

**Expansion**

$$f(x) = A\, e^{-\alpha x/\beta} + \int_0^\infty G(\omega)\, e^{-\nu x/2} \sin\{\omega x - \arctan \tfrac{\beta\omega}{\alpha}\}\, d\omega$$

$$G(\omega) = \tfrac{2}{\pi} \int_0^\infty f(x)\, e^{\nu x/2} \sin\{\omega x - \arctan \tfrac{\beta \omega}{\alpha}\}\, dx$$

$$A = \begin{cases} \dfrac{2\alpha - \beta\nu}{\beta} \int_0^\infty f(x)\, e^{-\alpha x/\beta}\, e^{\nu x}\, dx & \text{if } \beta \neq 0 \text{ and } \alpha/\beta > \nu/2 \\[2mm] 0 & \text{otherwise} \end{cases}$$

Reference: Sec. 6.7 (exercise 2)

---

## 12. <u>CONSTANT COEFFICIENT EQUATION ON $(-\infty, \infty)$</u>

Differential Equation

$$y'' + \nu y' = \lambda y, \quad -\infty < x < \infty$$

Boundary Conditions

$$y(x)\, e^{\nu x/2} \text{ finite at } \pm\infty$$

Eigenvalues

$$\lambda = -\omega^2 - \nu^2/4, \quad -\infty < \omega < \infty$$

Eigenfunctions

$$\phi_\omega(x) = e^{-\nu x/2} e^{i\omega x}$$

Weight Function $\quad w(x) = e^{\nu x}$

Orthogonality Integral

$$\int_{-\infty}^\infty \phi_\omega(x)\, \overline{\phi_{\omega'}(x)}\, e^{\nu x}\, dx = 2\pi\, \delta(\omega - \omega')$$

Expansion

$$f(x) = \int_{-\infty}^\infty G(\omega)\, e^{-\nu x/2}\, e^{i\omega x}\, d\omega$$

$$G(\omega) = \tfrac{1}{2\pi} \int_{-\infty}^\infty f(x)\, e^{\nu x/2}\, e^{-i\omega x}\, dx$$

Reference: Sec. 6.7

## 13. EQUIDIMENSIONAL EQUATION ON $(0, a]$ (FINITE MELLIN TRANSFORM)

**Differential Equation**

$$x^2 y'' + \sigma x y' = \lambda y, \ 0 < x < a$$

**Boundary Condition**

$$\alpha y(a) + \beta y'(a) = 0, \quad y(x)\, x^{(\sigma-1)/2} \text{ finite as } x \downarrow 0$$

**Eigenvalues**

$$\lambda = -\omega^2 - (1-\sigma)^2/4, \quad 0 \le \omega < \infty$$

$$\text{also } [a\alpha/\beta]^2 + a\alpha(1-\sigma)/\beta, \text{ if } \beta \ne 0 \text{ and } \alpha a/\beta < \tfrac{\sigma-1}{2}$$

**Eigenfunctions**

$$\phi_\omega(x) = x^{(1-\sigma)/2} \sin\{\omega \ln x/a - \arctan \tfrac{2\beta\omega}{2\alpha a + \beta(1-\sigma)}\}$$

$$\text{also } x^{-a\alpha/\beta} \text{ if } \beta \ne 0 \text{ and } \alpha a/\beta < \tfrac{\sigma-1}{2}$$

**Weight Function**    $w(x) = x^{\sigma-2}$

**Orthogonality Integral**

$$\int_0^a \phi_\omega(x)\, \phi_{\omega'}(x)\, x^{\sigma-2}\, dx = \tfrac{\pi}{2} \delta(\omega - \omega')$$

$$\int_0^a \phi_\omega(x)\, x^{-a\alpha/\beta}\, x^{\sigma-2}\, dx = 0 \text{ if } \beta \ne 0 \text{ and } \alpha a/\beta < \tfrac{\sigma-1}{2}$$

$$\int_0^a x^{-2a\alpha/\beta}\, x^{\sigma-2}\, dx = a^{\sigma-1-2a\alpha/\beta}/(\sigma - 1 - 2a\alpha/\beta) \text{ if } \beta \ne 0 \text{ and } \alpha a/\beta < \tfrac{\sigma-1}{2}$$

**Expansion**

$$f(x) = A\, x^{-a\alpha/\beta} + \int_0^\infty G(\omega)\, \phi_\omega(x)\, d\omega$$

$$G(\omega) = \tfrac{2}{\pi} \int_0^a f(x)\, \phi_\omega(x)\, x^{\sigma-2}\, dx$$

$$A = \begin{cases} \int_0^a f(x)\, x^{-a\alpha/\beta}\, x^{\sigma-2}\, dx\, [\sigma - 1 - 2a\alpha/\beta]/a^{\sigma-1-2a\alpha/\beta} \\ \qquad \text{if } \beta \ne 0 \text{ and } \alpha a/\beta < \tfrac{\sigma-1}{2}; \\ \\ 0 \qquad \text{otherwise} \end{cases}$$

**Reference:** Sec. 6.8

*Finite Mellin Transform of the First Kind*: Take $\sigma = \alpha = 1$, $\beta = 0$;

$$f(x) = \tfrac{1}{2\pi} \int_{-\infty}^{\infty} x^{-i\omega} m_1(i\omega; a) \, d\omega$$

$$m_1(s; a) = \int_0^a f(x) \{ x^{s-1} - a^{2s}/x^{s+1} \} \, dx$$

*Finite Mellin Transform of the Second Kind*: Take $\sigma = \beta = 1$, $\alpha = 0$;

$$f(x) = \tfrac{1}{2\pi} \int_{-\infty}^{\infty} x^{-i\omega} m_2(i\omega; a) \, d\omega$$

$$m_2(s; a) = \int_0^a f(x) \{ x^{s-1} + a^{2s}/x^{s+1} \} \, dx$$

---

## 14. EQUIDIMENSIONAL EQUATION ON $[a, \infty)$

Differential Equation

$$x^2 y'' + \sigma x y' = \lambda y, \; a < x < \infty$$

Boundary Condition

$$\alpha y(a) + \beta y'(a) = 0, \quad y(x) \, x^{(\sigma-1)/2} \text{ finite as } x \to \infty$$

Eigenvalues

$$\lambda = -\omega^2 - (\sigma - 1)^2/4, \quad 0 \le \omega < \infty$$

$$\text{also } [a\alpha/\beta]^2 - a\alpha(\sigma - 1)/\beta, \text{ if } \beta \ne 0 \text{ and } \alpha a/\beta > \tfrac{\sigma-1}{2}$$

Eigenfunctions

$$\phi_\omega(x) = x^{(1-\sigma)/2} \sin\{ \omega \ln x/a - \arctan \tfrac{2\beta\omega}{2\alpha a + \beta(1-\sigma)} \}$$

$$\text{also } x^{-a\alpha/\beta} \text{ if } \beta \ne 0 \text{ and } \alpha a/\beta > \tfrac{\sigma-1}{2}$$

Weight Function     $w(x) = x^{\sigma-2}$

Orthogonality Integral

$$\int_a^\infty \phi_\omega(x) \, \phi_{\omega'}(x) \, x^{\sigma-2} \, dx = \tfrac{\pi}{2} \delta(\omega - \omega')$$

$$\int_a^\infty \phi_\omega(x) \, x^{-a\alpha/\beta} \, x^{\sigma-2} \, dx = 0 \text{ if } \alpha a/\beta > \tfrac{\sigma-1}{2}$$

$$\int_a^\infty x^{-2a\alpha/\beta} \, x^{\sigma-2} \, dx = a^{\sigma-1-2a\alpha/\beta}/(2a\alpha/\beta - \sigma + 1)$$
$$\text{if } \alpha a/\beta > \tfrac{\sigma-1}{2}$$

Expansion

$$f(x) = A\,x^{-a\alpha/\beta} + \int_0^\infty G(\omega)\,\phi_\omega(x)\,d\omega$$

$$G(\omega) = \tfrac{2}{\pi} \int_a^\infty f(x)\,\phi_\omega(x)\,x^{\sigma-2}\,dx$$

$$A = \begin{cases} \int_a^\infty f(x)\,x^{-a\alpha/\beta}\,x^{\sigma-2}\,dx\,(2a\alpha/\beta - \sigma + 1)/a^{\sigma-1-2a\alpha/\beta} \\ \quad \text{if } \alpha a/\beta > \frac{\sigma-1}{2} \\ 0 \qquad \text{otherwise} \end{cases}$$

Reference: Sec. 6.8

---

## 15. <u>EQUIDIMENSIONAL EQUATION ON $(0,\infty)$ (MELLIN TRANSFORM)</u>

Differential Equation

$$x^2 y'' + \sigma x y' = \lambda y, \; 0 < x < \infty$$

Boundary Condition

$$y(x)\,x^{(\sigma-1)/2} \text{ finite at } 0, \infty$$

Eigenvalues

$$\lambda_\omega = -\omega^2 - (\sigma-1)^2/4; \quad -\infty < \omega < \infty$$

Eigenfunctions

$$\phi_\omega(x) = x^{(1-\sigma)/2} e^{-i\omega \ln x} = x^{(1-\sigma)/2}\,x^{-i\omega}$$

Weight Function $\quad w(x) = x^{\sigma-2}$

Orthogonality Integral

$$\int_0^\infty \phi_\omega(x)\,\overline{\phi_{\omega'}(x)}\,x^{\sigma-2}\,dx = 2\pi\,\delta(\omega - \omega')$$

Expansion

$$f(x) = \int_{-\infty}^\infty G(\omega)\,\phi_\omega(x)\,d\omega$$

$$G(\omega) = \tfrac{1}{2\pi} \int_0^\infty f(x)\,\overline{\phi_\omega(x)}\,x^{\sigma-2}\,dx$$

Reference: Sec. 6.8

Mellin Transform: Take $\sigma = 1$;

$$f(x) = \tfrac{1}{2\pi} \int_{-\infty}^{\infty} M(i\omega)\, x^{-i\omega}\, d\omega$$

$$M(s) = \int_0^{\infty} f(x)\, x^{s-1}\, dx$$

---

## 16. FOURIER-BESSEL SERIES

Differential Equation

$$x^2 y'' + x y' - \nu^2 y = \lambda x^2 y, \; 0 < x < b$$

Boundary Conditions

$$y(0) \text{ finite}, \; y(b) = 0$$

Eigenvalues

$$\lambda_p = -j_{\nu,p}^2/b^2; \quad p = 1, 2, 3, \ldots$$
$$\text{where } j_{\nu,p} \text{ is the } p\text{th positive zero of } J_\nu(z)$$

Eigenfunctions

$$\phi_p(x) = J_\nu(j_{\nu,p} x/b)$$

Weight Function      $w(x) = x$

Orthogonality Integral

$$\int_0^b J_\nu(j_{\nu,p} x/b)\, J_\nu(j_{\nu,p'}\, x/b)\, x\, dx = \frac{b^2 J_{\nu+1}^2(j_{\nu,p})}{2}\, \delta_{pp'}$$

Expansion

$$f(x) = \sum_{p=1}^{\infty} c_p J_\nu(j_{\nu,p} x/b), \quad (\nu \geq -1/2)$$

$$c_p = \frac{2 \int_0^b f(x) J_\nu(j_{\nu,p} x/b)\, x\, dx}{b^2 J_{\nu+1}^2(j_{\nu,p})}$$
$$\text{(The integral is the finite}$$
$$\text{Hankel transform of the}$$
$$\text{first kind.)}$$

Reference: Sec. 6.4 (Example 3), Table 2.3

## 17. <u>DINI SERIES</u>

Differential Equation

$$x^2 y'' + xy' - \nu^2 y = \lambda x^2 y, \ 0 < x < b$$

Boundary Conditions

$$y(0) \text{ finite}, \ y'(b) = 0$$

Eigenvalues

$$\lambda_p = -(j'_{\nu,p}/b)^2$$

where $j'_{\nu,p}$ is the $p$th positive zero of $J'_\nu(z)$;

also $\lambda_0 = 0$, if $\nu = 0$

Eigenfunctions

$$\phi_p(x) = J_\nu(j'_{\nu,p} x/b)$$

also $\phi_0(x) = 1$, if $\nu = 0$

Weight Function $\quad w(x) = x$

Orthogonality Integral

$$\int_0^b J_\nu(j'_{\nu,p} x/b) \, J_\nu(j'_{\nu,p'} x/b) \, x \, dx = \tfrac{b^2}{2} J_\nu^2(j'_{\nu,p})[1 - \nu^2/j'^2_{\nu,p}] \, \delta_{pp'}$$

$$\int_0^b (1) \, J_0(j'_{0,p} x/b) \, x \, dx = 0, \quad p > 0$$

Expansion

$$f(x) = \sum_{p=1}^\infty c_p J_\nu(j'_{\nu,p} x/b) + c_0, \quad (\nu \geq -1/2)$$

$$c_0 = 2 \int_0^b f(x) \, x \, dx/b^2 \text{ if } \nu = 0, \ a_0 = 0 \text{ otherwise;}$$

$$c_{p>0} = \frac{2 \int_0^b f(x) \, J_\nu(j'_{\nu,p} x/b) \, x \, dx}{b^2 J_\nu^2(j'_{\nu,p})\{1 - \nu^2/j'^2_{\nu,p}\}}$$

Reference: Sec. 6.9 (Example 1), Table 2.3

## 18. GENERALIZED DINI SERIES

Differential Equation

$$x^2 y'' + x y' - \nu^2 y = \lambda x^2 y, \ 0 < x < b$$

Boundary Conditions

$$y(0) \text{ finite}, \ \gamma y(b) + \delta y'(b) = 0$$

Eigenvalues and Eigenfunctions

$$\lambda_p = -(\mu_{\nu,p}/b)^2, \ \phi_p(x) = J_\nu(\mu_{\nu,p} x/b)$$

where $\mu_{\nu,p}$ is the $p$th positive zero of
$(b\gamma/\delta) J_\nu(z) + z J_\nu'(z)$;

also if $\nu = -b\gamma/\delta$ then $\lambda_0 = 0$ is an
eigenvalue and $\phi_0(x) = x^\nu$ is an eigenfunction;

if $\nu < -b\gamma/\delta$ then $\lambda_0 = (\mu_{\nu,0}/b)^2$ is an eigenvalue and $\phi_0(x) = I_\nu(\mu_{\nu,0} x/b)$
is an eigenfunction, where $\{\pm i\mu_{\nu,0}\}$ are the
imaginary zeros of $(b\gamma/\delta) J_\nu(z) + z J_\nu'(z)$

Weight Function    $w(x) = x$

Orthogonality Integral

$$\int_0^b J_\nu(\mu_{\nu,p} x/b) \, J_\nu(\mu_{\nu,p'} x/b) \, x \, dx =$$

$$(b^2/2)\{J_\nu^2(\mu_{\nu,p})[1 - \nu^2/\mu_{\nu,p}^2] + J_\nu'^2(\mu_{\nu,p})\} \delta_{pp'}$$

(the normalizations for the exceptional eigenfunctions
when $\nu \leq -b\gamma/\delta$ can be derived from the expansion below)

Expansion

$$f(x) = \sum_{p=1}^\infty c_p J_\nu(\mu_{\nu,p} x/b) + \phi_0(x), \quad (\nu \geq -1/2)$$

$$\phi_0(x) = \begin{cases} 2(\nu+1) x^\nu \int_0^b f(\xi) \, \xi^{\nu+1} \, d\xi / b^{2\nu+2} & \text{if } \nu = -\frac{\beta\gamma}{\delta}, \\[2ex] \dfrac{2 I_\nu(\mu_{\nu,0} x/b) \int_0^b f(\xi) \, I_\nu(\mu_{\nu,0}\xi/b) \, \xi \, d\xi}{b^2\{(1 + \nu^2/\mu_{\nu,0}^2) I_\nu^2(\mu_{\nu,0}) - I_\nu'^2(\mu_{\nu,0})\}} & \text{if } \nu < -\frac{\beta\gamma}{\delta}, \\[2ex] 0 & \text{otherwise} \end{cases}$$

$$c_{p>0} = \frac{2\int_0^b f(x)\, J_\nu(\mu_{\nu,p} x/b) x\, dx}{b^2\{J_\nu^2(\mu_{\nu,p})[1 - \nu^2/\mu_{\nu,p}^2] + J'^2_\nu(\mu_{\nu,p})\}}$$

The integral is the finite
Hankel transform of
the second kind.)

Reference: Watson

## 19. HANKEL TRANSFORM

Differential Equation

$$x^2 y'' + xy' - \nu^2 y = \lambda x^2 y,\ 0 < x < \infty$$

Boundary Conditions

$$y(0)\ \text{finite},\ y(\infty)\ \text{finite}$$

Eigenvalues

$$\lambda_s = -s^2,\quad 0 \le s < \infty$$

Eigenfunctions

$$\phi_s(x) = J_\nu(sx)$$

Weight Function $\quad w(x) = x$

Orthogonality Integral

$$\int_0^\infty J_\nu(sx)\, J_\nu(s'x)\, x\, dx = \delta(s - s')/s$$

Expansion

$$f(x) = \int_0^\infty F(s)\, J_\nu(sx)\, ds \quad (\nu \ge -1/2)$$

$$F(s) = s \int_0^\infty f(x)\, J_\nu(sx)\, x\, dx$$

Reference: Sec. 6.9 (Example 2)

## 20. <u>WEBER'S INTEGRAL FORMULA</u>

Differential Equation

$$x^2 y'' + xy' - \nu^2 y' = \lambda x^2 y, \ 0 < a < x < \infty$$

Boundary Conditions

$$y(a) = 0, \ \ y(\infty) \text{ finite}$$

Eigenvalues

$$\lambda_\mu = -\mu^2, \ \ 0 \le \mu < \infty$$

Eigenfunctions

$$\phi_{\nu,\mu}(x) = Y_\nu(\mu a) J_\nu(\mu x) - J_\nu(\mu a) Y_\nu(\mu x)$$

Weight Function $\quad w(x) = x$

Orthogonality Integral

$$\int_0^\infty \phi_{\nu,\mu}(x) \phi_{\nu,\mu'} \, x \, dx = \{[Y_\nu^2(\mu a) + J_\nu^2(\mu a)]^{1/2}/\mu\} \, \delta(\mu - \mu')$$

Expansion

$$f(x) = \int_0^\infty F(\mu)\{Y_\nu(\mu a) J_\nu(\mu x) - J_\nu(\mu a) Y_\nu(\mu x)\} d\mu$$

$$F(\mu) = \frac{\mu \int_a^\infty f(x)\{Y_\nu(\mu a) J_\nu(\mu x) - J_\nu(\mu a) Y_\nu(\mu x)\} x \, dx}{[Y_\nu^2(\mu a) + J_\nu^2(\mu a)]^{1/2}}$$

Reference: Sec. 6.9 (Example 3)

---

## 21. <u>MACDONALD SERIES</u>

Differential Equation

$$x^2 y'' + xy' - \mu^2 x^2 y = \lambda y, \ 0 < a < x < \infty$$

Boundary Conditions

$$y(a) = 0, \ \ y(\infty) \text{ finite}$$

Eigenvalues

$$\lambda_p = -\nu_p^2, \text{ where } \nu_p \text{ is the } p\text{th root of } K_{i\nu}(\mu a) = 0$$

Eigenfunctions

$$\phi_p(x) = K_{i\nu_p}(\mu x)$$

Weight Function $\quad w(x) = 1/x$

Orthogonality Integral

$$\int_a^\infty K_{i\nu_p}(\mu x) \, K_{i\nu_{p'}}(\mu x) \, dx/x = \frac{\mu a}{2\nu_p} K_{i\nu_p+1}(\mu a) \frac{\partial}{\partial \nu} K_{i\nu}(\mu a)|_{\nu=\nu_p}$$

Expansion

$$f(x) = \sum_{p=1}^\infty c_p \, K_{i\nu_p}(\mu x)$$

$$c_p = \frac{\int_a^\infty f(x) \, K_{i\nu_p}(\mu x) \, dx/x}{\frac{\mu a}{2\nu_p} K_{i\nu_p+1}(\mu a) \frac{\partial}{\partial \nu} K_{i\nu}(\mu a)|_{\nu=\nu_p}}$$

Reference: Sec. 6.9 (Example 4)

---

## 22. KONTOROVICH-LEBEDEV TRANSFORM

Differential Equation

$$x^2 y'' + x y' - \mu^2 x^2 y = \lambda y, \ 0 < x < \infty$$

Boundary Conditions

$$y(x) \text{ finite at } 0 \text{ and } \infty$$

Eigenvalues

$$\lambda = -\nu^2, \ 0 \le \nu < \infty$$

Eigenfunctions

$$\phi(x) = K_{i\nu}(\mu x)$$

Weight Function $\quad w(x) = 1/x$

Orthogonality Integral

$$\int_0^\infty K_{i\nu}(\mu x)\, K_{i\nu'}(\mu x)\, dx/x = \frac{\pi^2}{2\nu \sinh \nu\pi}\, \delta(\nu - \nu')$$

Expansion Theorem

$$f(x) = \int_0^\infty F(\nu)\, K_{i\nu}(\mu x)\, d\nu$$

$$F(\nu) = \frac{2}{\pi^2} \int_0^\infty f(x)\, K_{i\nu}(\mu x)\, \nu \sinh \nu\pi\, dx/x$$

Reference: Sec. 6.9 (Example 5), Sneddon

---

## 23. <u>FINITE LEBEDEV TRANSFORM</u> (Dirichlet condition)

Differential Equation

$$x^2 y'' + x y' - \mu^2 x^2 y = \lambda y,\ 0 < x < b$$

Boundary Conditions

$$y(0)\ \text{finite},\ y(b) = 0$$

Eigenvalues

$$\lambda = -\nu^2,\ \ 0 \le \nu < \infty$$

Eigenfunctions

$$\phi_{\nu,\mu}(x) = K_{i\nu}(\mu b) I_{i\nu}(\mu x) - I_{i\nu}(\mu b) K_{i\nu}(\mu x)$$

Weight Function     $w(x) = 1/x$

Orthogonality Integral

$$\int_0^b \phi_{\nu,\mu}(x)\, \phi_{\nu',\mu}(x)\, \frac{dx}{x} = \frac{\pi^2}{2\nu \sinh \nu\pi} |I_{i\nu}(\mu b)|^2\, \delta(\nu - \nu')$$

Expansion

$$f(x) = \int_0^\infty F(\nu)\{K_{i\nu}(\mu b) I_{i\nu}(\mu x) - I_{i\nu}(\mu b) K_{i\nu}(\mu x)\}\, d\nu$$

$$F(\nu) = \frac{2\nu \sinh \nu\pi \int_0^b f(x)\{K_{i\nu}(\mu b) I_{i\nu}(\mu x) - I_{i\nu}(\mu b) K_{i\nu}(\mu x)\}\frac{dx}{x}}{\pi^2\, |I_{i\nu}(\mu b)|^2}$$

Reference: Sec. 6.9 (Example 6), Naylor (1963)

---

## 24. FINITE LEBEDEV TRANSFORM (Neumann condition)

### Differential Equation

$$x^2 y'' + xy' - \mu^2 x^2 y = \lambda y, \; 0 < x < b$$

### Boundary Conditions

$$y(0) \text{ finite}, \; y'(b) = 0$$

### Eigenvalues

$$\lambda = -\nu^2, \quad 0 \le \nu < \infty$$

### Eigenfunctions

$$\phi_{\nu,\mu}(x) = I'_{i\nu}(\mu b) K_{i\nu}(\mu x) - K'_{i\nu}(\mu b) I_{i\nu}(\mu x)$$

### Weight Function $\quad w(x) = 1/x$

### Orthogonality Integral

$$\int_0^b \phi_{\nu,\mu}(x)\, \phi_{\nu',\mu}(x)\, \frac{dx}{x} = \frac{\pi^2}{2\nu \sinh \nu\pi} |I'_{i\nu}(\mu b)|^2 \, \delta(\nu - \nu')$$

### Expansion

$$f(x) = \int_0^\infty F(\nu) \left\{ I'_{i\nu}(\mu b) K_{i\nu}(\mu x) - K'_{i\nu}(\mu b) I_{i\nu}(\mu x) \right\} d\nu$$

$$F(\nu) = \int_0^b f(x) \frac{2\nu \sinh \nu\pi F(\nu)\{ I'_{i\nu}(\mu b) K_{i\nu}(\mu x) - K'_{i\nu}(\mu b) I_{i\nu}(\mu x) \}}{\pi^2 |I'_{i\nu}(\mu b)|^2} \frac{dx}{x}$$

Reference: Naylor (1963)

---

## 25. SPHERICAL BESSEL FUNCTION SERIES

### Differential Equation

$$x^2 y'' + 2xy' - \ell(\ell+1)y = \lambda x^2 y, \; 0 < x < b$$

Boundary Conditions

$$y(b) = 0, \quad y(x)\, x^{1/2} \text{ finite at } 0$$

Eigenvalues

$$\lambda_p = -(s_{\ell,p}/b)^2, \text{ where } s_{\ell,p} \text{ is the } p\text{th positive zero of } j_\ell(z)$$

Eigenfunctions

$$\phi_p(x) = j_\ell(s_{\ell,p}x/b)$$

Weight Function    $w(x) = x^2$

Orthogonality Integral

$$\int_0^b j_\ell(s_{\ell,p}x/b)\, j_\ell(s_{\ell,p'}x/b)\, x^2\, dx = \frac{b^3}{2} j_{\ell+1}^2(s_{\ell,p})\, \delta_{pp'}$$

Expansion

$$f(x) = \sum_{p=1}^{\infty} c_p\, j_\ell(s_{\ell,p}x/b), \quad (\ell \geq -1)$$

$$c_p = \frac{2 \int_0^b f(x)\, j_\ell(s_{\ell,p}x/b)\, x^2\, dx}{b^3 j_{\ell+1}^2(s_{\ell,p})}$$

(For other boundary conditions relate $j_\ell$ to the Bessel functions and use the corresponding generalized Dini series.)

Reference: Sec. 6.9 (Example 7)

---

## 26. SPHERICAL HANKEL TRANSFORM

Differential Equation

$$x^2 y'' + 2xy' - \ell(\ell+1)y = \lambda x^2 y, \ 0 < x < \infty$$

Boundary Conditions

$$y(x)\, x^{1/2} \text{ finite at } 0 \text{ and } \infty$$

Eigenvalues

$$\lambda_s = -s^2, \quad 0 \le s < \infty$$

Eigenfunctions

$$\phi_s(x) = j_\ell(sx)$$

Weight Function $\quad w(x) = x^2$

Orthogonality Integral

$$\int_0^\infty j_\ell(sx)\, j_\ell(s'x)\, x^2\, dx = \tfrac{\pi}{2s^2}\, \delta(s - s')$$

Expansion

$$f(x) = \int_0^\infty F(s)\, j_\ell(sx)\, ds, \quad (\ell \ge -1)$$

$$F(s) = \tfrac{2}{\pi}s^2 \int_0^\infty f(x)\, j_\ell(sx)\, x^2\, dx$$

Reference: Sec. 6.9 (exercise 6)

---

## 27. LEGENDRE SERIES

$$x = \cos\phi, \; y(x) = y(\cos\phi) = \Phi(\phi)$$

| *x*-formulation | *φ*-formulation |
|---|---|

Differential Equation

$$(1 - x^2)y'' - 2xy' = \lambda y, \; -1 < x < 1 \quad (\sin\phi\, \Phi')' = \lambda \sin\phi\, \Phi, \; 0 < \phi < \pi$$

Boundary Conditions

$y(\pm 1)$ finite $\qquad\qquad\qquad\qquad \Phi(0),\; \Phi(\pi)$ finite

Eigenvalues

$$\lambda_\ell = -\ell(\ell + 1), \quad \ell = 0, 1, 2, \ldots$$

Eigenfunctions

$\Phi_\ell(x) = P_\ell(x) \qquad\qquad\qquad\qquad \Phi_\ell(\phi) = P_\ell(\cos\phi)$

*(Legendre Polynomial)*

Weight Function

$$w(x) = 1 \qquad\qquad\qquad w(\phi) = \sin\phi$$

**Orthogonality Integral**

$$\int_{-1}^{1} P_\ell(x)\, P_{\ell'}(x)\, dx \;=\; \frac{2}{2\ell+1}\,\delta_{\ell\ell'}$$
$$= \int_0^\pi P_\ell(\cos\phi)\, P_{\ell'}(\cos\phi)\, \sin\phi\, d\phi$$

**Expansion**

$$f(x) = \sum_{\ell=0}^\infty c_\ell\, P_\ell(x) \qquad g(\phi) = \sum_{\ell=0}^\infty d_\ell\, P_\ell(\cos\phi)$$

$$c_\ell = \frac{2\ell+1}{2} \int_{-1}^{1} f(x)\, P_\ell(x)\, dx \quad d_\ell = \frac{2\ell+1}{2} \int_0^\pi g(\phi)\, P_\ell(\cos\phi)\, \sin\phi\, d\phi$$

**Reference: Sec. 6.4 (Example 4)**

---

## 28. ASSOCIATED LEGENDRE SERIES

$$x = \cos\phi,\; y(x) = y(\cos\phi) = \Phi(\phi)$$

| _x_-formulation | _φ_-formulation |
|---|---|

**Differential Equation** $(m = 0, \pm 1, \pm 2, \dots)$

$$(1-x^2)y'' - 2xy' - \frac{m^2}{1-x^2}y = \lambda y \qquad (\sin\phi\, \Phi')' - \frac{m^2}{\sin\phi}\Phi = \lambda \sin\phi\, \Phi$$
$$-1 < x < 1 \qquad\qquad\qquad\qquad 0 < \phi < \pi$$

**Boundary Conditions**

$$y(\pm 1) \text{ finite} \qquad\qquad\qquad \Phi(0),\; \Phi(\pi) \text{ finite}$$

**Eigenvalues**

$$\lambda_\ell = -\ell(\ell+1), \quad \ell = |m|, |m|+1, |m|+2, \dots$$

**Eigenfunctions**

$$\phi_\ell(x) = P_\ell^m(x) \qquad\qquad\qquad \phi_\ell(\phi) = P_\ell^m(\cos\phi)$$
$$(\textit{Associated Legendre function})$$

**Weight Function**

$$w(x) = 1 \qquad\qquad\qquad\qquad w(\phi) = \sin\phi$$

Orthogonality Integral

$$\int_{-1}^{1} P_\ell^m(x) \, P_{\ell'}^m(x) \, dx = \frac{2(\ell+m)!}{(2\ell+1)(\ell-m)!} \, \delta_{\ell\ell'}$$

$$= \int_0^\pi P_\ell^m(\cos\phi) \, P_{\ell'}^m(\cos\phi) \, \sin\phi \, d\phi$$

Expansion

$$f(x) = \sum_{\ell=|m|}^{\infty} c_\ell \, P_\ell^m(x) \qquad\qquad g(\phi) = \sum_{\ell=|m|}^{\infty} d_\ell \, P_\ell^m(\cos\phi)$$

$$c_\ell = \frac{(2\ell+1)(\ell-m)!}{2(\ell+m)!} \int_{-1}^{1} f(x) \, P_\ell^m(x) \, dx$$

$$d_\ell = \frac{(2\ell+1)(\ell-m)!}{2(\ell+m)!} \int_0^\pi g(\phi) \, P_\ell^m(\cos\phi) \, \sin\phi \, d\phi$$

Reference: Sec. 6.10

---

## 29. SPHERICAL HARMONICS SERIES

Differential Equation

$$\nabla^2 R(r) \, Y_{\ell m}(\phi,\theta) = \frac{(r^2 R')' - \ell(\ell+1)R}{r^2} \, Y_{\ell m}(\phi,\theta)$$

Boundary Conditions

$Y_{\ell m}$ finite at $\phi = 0, \pi$ and $2\pi$-periodic in $\theta$

Eigenfunction

$$Y_{\ell m}(\phi,\theta) = \sqrt{\frac{(2\ell+1)(\ell-m)!}{4\pi(\ell+m)!}} \, P_\ell^m(\cos\phi) e^{im\theta}$$

$Y_{\ell m}$ is the spherical harmonic of order $\ell$, $m$; $R$ is any function

Weight Function $\quad w(\phi,\theta) = \sin\phi$

Orthogonality Integral

$$\int_0^\pi \int_0^{2\pi} Y_{\ell m}(\phi,\theta) \, \overline{Y_{\ell' m'}(\phi,\theta)} \, \sin\phi \, d\theta \, d\phi = \delta_{\ell\ell'} \, \delta_{mm'}$$

Expansion

$$f(\phi,\theta) = \sum_{\ell=0}^{\infty} \sum_{m=-\ell}^{\ell} c_{\ell m} \, Y_{\ell m}(\phi,\theta);$$

$$c_{\ell m} = \int_0^\pi \int_0^{2\pi} f(\phi, \theta) \, \overline{Y_{\ell m}(\phi, \theta)} \, \sin\phi \, d\theta \, d\phi$$

Reference: Sec. 7.1 (Example 4)

---

## 30. HERMITE SERIES

Differential Equation

$$y'' - x^2 y = \lambda y, \quad -\infty < x < \infty$$

Boundary Conditions

$$y(\pm\infty) \text{ finite}$$

Eigenvalues

$$\lambda_n = -(2n + 1), \quad n = 0, 1, 2, \ldots$$

Eigenfunctions

$$\phi_n(x) = e^{-x^2/2} H_n(x), \text{ where } H_n(x) \text{ is the Hermite polynomial}$$
$$\text{of order } n$$

Weight Function     $w(x) = 1$

Orthogonality Integral

$$\int_{-\infty}^\infty e^{-x^2} H_n(x) \, H_{n'}(x) \, dx = 2^n \sqrt{\pi} \, n! \, \delta_{nn'}$$

Expansion

$$f(x) = \sum_{n=0}^\infty c_n \, e^{-x^2/2} \, H_n(x)$$

$$c_n = \frac{\int_{-\infty}^\infty f(x) \, H_n(x) \, e^{-x^2/2} \, dx}{2^n \sqrt{\pi} n!}$$

Reference: Sec. 7.4 (Example 3)

## 31. CHEBYSHEV SERIES

*First Kind*

### Differential Equation

$$(1 - x^2)y'' - xy' = \lambda y, \quad -1 < x < 1$$

### Boundary Value Problem

$y(\pm 1)$ finite

### Eigenvalues

$$\lambda_p = -p^2, \quad p = 0, 1, 2, \ldots$$

### Eigenfunctions

$$\phi_p(x) = T_p(x) = \cos(p \cos^{-1} x)$$
$$(\text{Chebyshev polynomial of the first kind})$$

### Weight Function $\quad w(x) = (1 - x^2)^{-1/2}$

### Orthogonality Integral

$$\int_{-1}^{1} T_p(x) \, T_{p'}(x)(1 - x^2)^{-1/2} \, dx = \begin{cases} \pi, & p = p' = 0 \\ \pi/2 \, \delta_{pp'} & \text{otherwise} \end{cases}$$

### Expansion

$$f(x) = \sum_{p=0}^{\infty} a_p \, T_p(x), \quad -1 < x < 1;$$

$$c_0 = \int_{-1}^{1} f(x)(1 - x^2)^{-1/2} \, dx/\pi,$$

$$c_{n>0} = 2 \int_{-1}^{1} f(x) \, T_n(x)(1 - x^2)^{-1/2} \, dx/\pi.$$

---

*Second Kind*

### Differential Equation

$$(1 - x^2)y'' - 3xy' = \lambda y, \quad -1 < x < 1$$

### Boundary Conditions

$y(\pm 1)$ finite

Eigenvalues

$$\lambda_p = -p\,(p+2), \quad p = 0, 1, 2, \ldots$$

Eigenfunctions

$$\phi_p(x) = U_p(x) = \frac{\sin[(p+1)\cos^{-1} x]}{\sqrt{1 - x^2}}$$
$$(\textit{Chebyshev polynomial of the second kind})$$

Weight Function    $w(x) = (1 - x^2)^{1/2}$

Orthogonality Integral

$$\int_{-1}^{1} T_p(x)\, T_{p'}(x)(1 - x^2)^{1/2}\, dx = \pi/2\, \delta_{pp'}$$

Expansion

$$f(x) = \sum_{p=0}^{\infty} c_p\, U_p(x)$$

$$c_n = 2 \int_{-1}^{1} f(x)\, U_n(x)\,(1 - x^2)^{1/2}\, dx/\pi.$$

Reference: Abramowitz and Stegun

# Chapter 7

# Applications of Eigenfunctions to Partial Differential Equations

In Sec. 5.2 we demonstrated how the technique of separation of variables could be used to solve some of the partial differential equations (PDEs) of engineering. The present chapter is devoted to an extensive compilation of examples of the method, reflecting its stature as the dominant technique in engineering analysis for such problems.

The main features of the method were listed in Sec. 5.2. Although we shall see, in the course of this chapter, that some of the "rules" become modified in certain situations, the basic strategy that they dictate is a sound and unambiguous implementation of the method of separation of variables. We restate this strategy here.

1. The problem to be solved specifies a linear homogeneous partial differential equation which must hold in a given domain, together with conditions imposed on the solution at the boundary of the domain. These boundary conditions dictate the value of certain linear combinations of the unknown function and its derivative in the direction normal to the boundary. The domain itself is a "curvilinear rectangle" whose sides are constant-coordinate curves for the appropriate curvilinear coordinates $(x, y, r, \theta, \dots)$.

2. Linearity is exploited to decompose the problem into a superposition of subproblems, each of which involve *homogeneous* boundary conditions on all but one of the boundary curves.

3. A trial solution, expressed as a product of factors each of which depends on only one independent variable (curvilinear coordinate), is inserted into the partial differential equation. If the result can be manipulated into a set

of *ordinary* differential equations (ODEs), each containing an unspecified parameter known as the "separation constant," the problem is separable. This separation step has been demonstrated in Sec. 5.3 for all the equations to be studied.

4. The first variable to be resolved is the one with homogeneous boundary conditions imposed at *both* ends. These boundary conditions are carried over to the ODE for that coordinate. The result is a Sturm-Liouville problem, which is treated by the procedure covered in Chapter 6 - to wit:

> 4a. The general solution of the ODE is expressed as a linear combination of two particular solutions, with unspecified constant coefficients.

> 4b. The (homogeneous) boundary condition at one end is enforced by choice of one of the coefficients.

> 4c. The (homogeneous) boundary condition at the other end is enforced by choice of the separation constant.

> 4d. The acceptable values of the separation constant constitute an infinite set of numbers called the eigenvalues. Associated with each eigenvalue is a solution to the ODE known as the eigenfunction. The eigenfunctions constitute a complete orthogonal set.

5. The ODE involving the coordinate with a nonhomogeneous boundary condition is then addressed. First a general solution to the ODE is constructed, as in step 4a. The *homogeneous* boundary condition is enforced by choice of the coefficients.

6. The result of steps 1 through 5 is an infinite family of solutions to the PDE, each of which satisfies the homogeneous boundary conditions. The final, nonhomogeneous, boundary condition is then fulfilled by forming the appropriate superposition of these solutions, with coefficients prescribed by the orthogonality property.

Steps 1 through 6 describe the fundamental algorithm for initiating our study, but we reiterate that some of these rules will be extended and modified as we treat more exotic examples. The time variation in particular requires special treatment, because *boundary* conditions in this variable are inappropriate (except in the context of control theory); *initial* conditions are imposed instead. Also the procedure must be adapted to higher dimensions. But these adjustments are minor, and adherence to the fundamental program will lead to a successful execution of many solutions in physics and engineering.

# 7.1  Eigenfunction Expansions in Higher Dimensions

**Summary**   The extension of the basic procedure, outlined in the introduction, to higher dimensional problems entails multivariate eigenfunction expansions - the spherical harmonics, for example. The orthogonality relations, however, render this a straightforward process.

The separation-of-variables approach to multidimensional problems is a straightforward extension of the procedure that we have been following:

(1) decomposing the problem into subproblems in which only one boundary condition is nonhomogeneous,

(2) separating the equation,

(3) analyzing the homogeneous one-dimensional problems,

(4) analyzing the nonhomogeneous problem,

(5) assembling the solution, and

(6) using orthogonality to get the coefficients.

Note the *plural* - "problems" - in step 3. A *three*-dimensional problem with a nonhomogeneous boundary condition on only one face leads, upon separation, to *two* Sturm-Liouville problems. Consider the following example.

**Example 1.** The equilibrium temperature $\psi$ inside a box is governed by Laplace's equation

$$\nabla^2 \psi(x,y,z) = \frac{\partial^2 \psi}{\partial x^2} + \frac{\partial^2 \psi}{\partial y^2} + \frac{\partial^2 \psi}{\partial z^2} = 0.$$

We take the box to have edge lengths $a$, $b$, and $c$. The boundary conditions

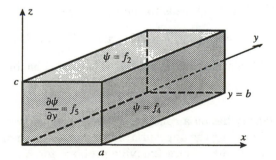

**Figure 1**   *Specifications for Example 1*

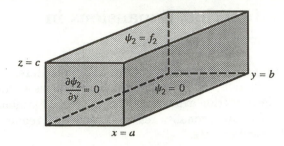

**Figure 2**  *Subproblem for Example 1*

are expressed (Fig. 1)

$$\begin{aligned}
\psi(x,y,0) &= f_1(x,y), \quad \psi(x,y,c) = f_2(x,y), \\
\psi(0,y,z) &= f_3(y,z), \quad \psi(a,y,z) = f_4(y,z), \\
\frac{\partial \psi(x,0,z)}{\partial y} &= f_5(x,z), \quad \frac{\partial \psi(x,b,z)}{\partial y} = f_6(x,z).
\end{aligned} \tag{1}$$

Clearly there are six subproblems for this situation, of which the following is typical: find $\psi_2$ satisfying Laplace's equation

$$\nabla^2 \psi_2(x,y,z) = 0 \tag{2}$$

and the boundary conditions

$$\begin{aligned}
\psi_2(x,y,c) &= f_2(x,y), \\
\psi_2(x,y,0) &= 0, \; \psi_2(0,y,z) = 0, \; \psi_2(a,y,z) = 0 \\
\frac{\partial \psi_2(x,0,z)}{\partial y} &= 0, \; \frac{\partial \psi_2(x,b,z)}{\partial y} = 0.
\end{aligned} \tag{3}$$

See Fig. 2. We relegate the formulation and solution of the remaining subproblems to exercise 1. Note that constant-coordinate *curves* in two dimensions become constant-coordinate *surfaces* in three dimensions. We proceed with (2,3) and omit the subscripts for convenience.

The separation of Laplace's equation in three-dimensional cartesian coordinates was demonstrated in Sec. 5.3. The substitution of $\psi = X(x)Y(y)Z(z)$ into (2) leads to the conditions

$$X'' = \lambda X, \quad Y'' = \mu Y, \quad \text{and} \quad Z'' = \nu Z, \tag{4}$$

for any values of the separation constants $\lambda$, $\mu$, and $\nu$ as long as

$$\lambda + \mu + \nu = 0. \tag{5}$$

The nonhomogeneity lies on a constant-$z$ face, so we address the $X$ and $Y$ equations first. Coupling the appropriate homogeneous conditions (3) with (4) we obtain the familiar Sturm-Liouville problem for $X$

$$X'' = \lambda X, \quad X(0) = X(a) = 0,$$

whose solutions are the terms of the sine series (Table 6.3 #2)

$$X_m(x) = \sin \frac{m\pi}{a} x, \quad \lambda_m = -\left(\frac{m\pi}{a}\right)^2, \quad m = 1, 2, 3, \ldots \tag{6}$$

The $Y$ equation from (4) and its corresponding boundary condition from (3) imply

$$Y'' = \mu Y, \quad Y'(0) = Y'(b) = 0,$$

leading to the cosine series

$$Y_n(y) = \cos \frac{n\pi}{b} y, \quad \mu_n = -\left(\frac{n\pi}{b}\right)^2, \quad n = 0, 1, 2, 3, \ldots \tag{7}$$

Note that $\mu$ is entirely independent of $\lambda$, and that the $y$-eigenfunctions are enumerated with a different index ($n$) from the $x$-eigenfunctions ($m$).

The parameter $\nu$ in the equation (4) for $Z$ is fixed by condition (5), so insertion of the eigenvalue expressions in (6, 7) dictates

$$Z'' = \left(\frac{m^2 \pi^2}{a^2} + \frac{n^2 \pi^2}{b^2}\right) Z. \tag{8}$$

The homogeneous boundary condition from (3) implies

$$Z(0) = 0, \tag{9}$$

and the solution to (8, 9) is $Z = (constant) \sinh \sqrt{\frac{m^2\pi^2}{a^2} + \frac{n^2\pi^2}{b^2}} z$. Thus the general solution to our problem, satisfying all the *homogeneous* conditions, is

$$\begin{aligned}
\psi_2(x, y, z) &= \sum_{m=1}^{\infty} \sum_{n=0}^{\infty} a_{mn} \sin\left(\frac{m\pi}{a} x\right) \cos\left(\frac{n\pi}{b} y\right) \sinh \sqrt{\frac{m^2\pi^2}{a^2} + \frac{n^2\pi^2}{b^2}} z \\
&= a_{1,0} \sin \frac{\pi x}{a} \sinh \frac{\pi z}{a} \\
&\quad + a_{2,0} \sin \frac{2\pi x}{a} \sinh \frac{2\pi z}{a} \\
&\quad + a_{1,1} \sin \frac{\pi}{a} x \cos \frac{\pi}{b} y \sinh \sqrt{\frac{\pi^2}{a^2} + \frac{\pi^2}{b^2}} z \\
&\quad + a_{3,0} \sin \frac{3\pi x}{a} \sinh \frac{3\pi z}{a} \\
&\quad + a_{2,1} \sin \frac{2\pi}{a} x \cos \frac{\pi}{b} y \sinh \sqrt{\frac{4\pi^2}{a^2} + \frac{\pi^2}{b^2}} z \\
&\quad + a_{1,2} \sin \frac{\pi}{a} x \cos \frac{2\pi}{b} y \sinh \sqrt{\frac{\pi^2}{a^2} + \frac{4\pi^2}{b^2}} z + \cdots
\end{aligned}$$

To meet the final nonhomogeneous condition in (3) we must have

$$f_2(x, y) = \sum_{m=1}^{\infty} \sum_{n=0}^{\infty} a_{mn} \sin\left(\frac{m\pi}{a} x\right) \cos\left(\frac{n\pi}{b} y\right) \sinh \sqrt{\frac{m^2\pi^2}{a^2} + \frac{n^2\pi^2}{b^2}} c. \tag{10}$$

Note that (10) is a *multivariate* eigenfunction expansion; the target function, $f_2(x, y)$, depends on $x$ *and* $y$. The successful conclusion of our analysis, then, rests on two considerations: ($a$) Does the theory guarantee that such an expansion is valid? ($b$) How do we compute the coefficients $a_{mn}$?

Let's answer the easy question (b) first. The key, as in the univariate case, is the orthogonality of the eigenfunctions. Assuming that (10) holds and converges fast enough to justify termwise integration, we multiply each side by *both* $\sin p\pi x/a$ *and* $\cos q\pi y/b$ and integrate:

$$
\int_0^a \int_0^b f_2(x, y) \sin \frac{p\pi}{a}x \cos \frac{q\pi}{b}y \, dy \, dx
$$

$$
= \sum_{m=1}^\infty \sum_{n=0}^\infty a_{mn} \sinh \sqrt{\frac{m^2\pi^2}{a^2} + \frac{n^2\pi^2}{b^2}}\, c
$$

$$
\times \int_0^a \sin \frac{m\pi}{a}x \sin \frac{p\pi}{a}x \, dx \int_0^b \cos \frac{n\pi}{b}y \cos \frac{q\pi}{b}y \, dy
$$

$$
= \sum_{m=1}^\infty \sum_{n=0}^\infty a_{mn} \sinh \sqrt{\frac{m^2\pi^2}{a^2} + \frac{n^2\pi^2}{b^2}}\, c
$$

$$
\times \int_0^a \sin^2 \frac{p\pi}{a}x \, dx \int_0^b \cos^2 \frac{q\pi}{b}y \, dy \, \delta_{mp}\, \delta_{nq}
$$

$$
= a_{pq} \sinh \sqrt{\frac{p^2\pi^2}{a^2} + \frac{q^2\pi^2}{b^2}}\, c
$$

$$
\times \int_0^a \sin^2 \frac{p\pi}{a}x \, dx \int_0^b \cos^2 \frac{q\pi}{b}y \, dy
$$

or

$$
a_{p,0} = \frac{2}{ab} \int_0^a \int_0^b f_2(x, y) \sin \frac{p\pi}{a}x \, dy \, dx \Big/ \sinh \frac{p\pi}{a}c,
$$

and for $q \geq 1$

$$
a_{p,q} = \frac{4}{ab} \int_0^a \int_0^b f_2(x, y) \sin \frac{p\pi}{a}x \cos \frac{q\pi}{b}y \, dy \, dx \Big/ \sinh \sqrt{\frac{p^2\pi^2}{a^2} + \frac{q^2\pi^2}{b^2}}\, c.
$$

So the evaluation of the coefficients is almost as easy in the multivariate case as it was in Chapter 6. But question (a) remains: how can we conclude, from the completeness of the eigenfunctions in the *univariate* case, that expansions of *multivariate* functions like (10) are valid? The argument goes as follows. First we conceptually regard one of the variables ($x$) as frozen. To emphasize this, we rewrite $x$ as $x_0$. This makes $f_2(x_0, y)$ a function of only one variable ($y$), and consequently we can expand it in a cosine series:

$$
f_2(x_0, y) = \sum_{n=0}^\infty C_n \cos \frac{n\pi y}{b};
$$

$$C_0 = \frac{1}{b} \int_0^b f_2(x_0, y)\, dy,$$

$$C_{n>0} = \frac{2}{b} \int_0^b f_2(x_0, y) \cos \frac{n\pi y}{b}\, dy. \tag{11}$$

But we must acknowledge that if we change the value $x_0$, the function $f_2(x_0, y)$ will change, and its cosine expansion will thus also change. *In other words, the values of the coefficients $C_n$ depend on the value chosen for $x_2$; they are functions of $x_0$.* Thus the expansion (11) is more honestly written

$$f_2(x, y) = \sum_{n=0}^{\infty} C_n(x) \cos \frac{n\pi y}{b} = C_0(x) + C_1(x) \cos \frac{\pi y}{b} + \cdots;$$

$$C_0(x) = \frac{1}{b} \int_0^b f_2(x, y)\, dy,$$

$$C_{n>0}(x) = \frac{2}{b} \int_0^b f(x, y) \cos \frac{n\pi y}{b}\, dy, \tag{12}$$

where the subscript on $x_0$ has been dropped since we no longer regard $x$ as fixed.

Now each coefficient $C_n(x)$, as a function of $x$, has a sine series expansion. For instance $C_0(x)$ has the expansion

$$C_0(x) = \sum_{m=1}^{\infty} c_{m0} \sin \frac{m\pi x}{a}, \quad c_{m0} = \frac{2}{a} \int_0^a C_0(x) \sin \frac{m\pi x}{a}\, dx,$$

and in general

$$C_n(x) = \sum_{m=1}^{\infty} c_{mn} \sin \frac{m\pi x}{a},$$

$$c_{mn} = \frac{2}{a} \int_0^a C_n(x) \sin \frac{m\pi x}{a}\, dx. \tag{13}$$

Notice the necessity of the two subscripts on $c_{mn}$; $m$ indexes the terms in the sum (13), while $n$ designates which particular coefficient in (12) we are working on.

Assembling (12) and (13) we arrive at the multivariate expansion

$$f_2(x, y) = \sum_{m=1}^{\infty} \sum_{n=0}^{\infty} c_{mn} \sin \frac{m\pi x}{a} \cos \frac{n\pi y}{b}, \quad 0 \le x \le a,\ 0 \le y \le b, \tag{14}$$

$$c_{m0} = \frac{2}{ab} \int_0^a \int_0^b f_2(x, y) \sin \frac{m\pi x}{a}\, dy\, dx,$$

$$c_{mn} = \frac{4}{ab} \int_0^a \int_0^b f_2(x, y)$$

$$\times \sin \frac{m\pi x}{a} \cos \frac{n\pi y}{b}\, dy\, dx, \quad (n > 0). \tag{15}$$

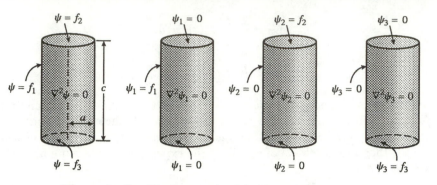

***Figure 3*** *Specifications and subproblems for Example 2*

Equation (14) justifies the form (10), and (15) verifies the equivalence of the coefficients: $c_{mn}$ equals $a_{mn}$ times $\sinh\sqrt{\frac{m^2\pi^2}{a^2} + \frac{n^2\pi^2}{b^2}}c$.  ∎

Another aspect of multivariate eigenfunction expansions is illustrated in the next example.

**Example 2.** The analysis of the equilibrium temperature inside a cylinder of radius $a$ and height $c$, with boundary conditions depicted in Fig. 3, is best carried out in cylindrical coordinates:

$$\nabla^2\psi(\rho,\theta,z) = \frac{1}{\rho}\frac{\partial}{\partial\rho}\left(\rho\frac{\partial\psi}{\partial\rho}\right) + \frac{1}{\rho^2}\frac{\partial^2\psi}{\partial\theta^2} + \frac{\partial^2\psi}{\partial z^2} = 0,$$

$$\begin{aligned}
\psi(a,\theta,z) &= f_1(\theta,z) \text{ on the wall,}\\
\psi(a,\theta,c) &= f_2(\rho,\theta) \text{ on the top, and}\\
\psi(\rho,\theta,0) &= f_3(\rho,\theta) \text{ on the bottom.}
\end{aligned}$$

Our first step is to break the problem up into subproblems:

PROBLEM 1. $\nabla^2\psi_1 = 0$, $\psi_1 = f_1$ on the wall, $\psi_1 = 0$ on the top and bottom.

PROBLEM 2. $\nabla^2\psi_2 = 0$, $\psi_2 = f_2$ on the top, $\psi_2 = 0$ on the wall and bottom.

PROBLEM 3. $\nabla^2\psi_3 = 0$, $\psi_3 = f_3$ on the bottom, $\psi_3 = 0$ on the wall and top.

SOLUTION OF PROBLEM 1. The deliberations of Sec. 5.3 showed that the insertion of the form $\psi_1 = R(\rho)\Theta(\theta)Z(z)$ into Laplace's equation results in the separated equations

$$\Theta'' = \lambda\Theta, \quad Z'' = \mu Z, \quad \text{and} \quad \rho^2 R'' + \rho R' = -\mu\rho^2 R - \lambda R. \tag{16}$$

Inasmuch as there is no physical boundary on the $\theta$ "faces," we append periodic boundary conditions to the equation $\theta'' = \lambda\theta$ and obtain eigenfunctions corresponding to the Fourier series (Example 2, Sec. 6.3 or Table

6.3 entry #2):

$$\Theta_n(\theta) = a_n \cos n\theta + b_n \sin n\theta, \ \ \lambda_n = -n^2, \ \ n = 0, 1, 2, \ldots \quad (17)$$

Since the nonhomogeneity in this subproblem lies on the $\rho = constant$ face we next look for eigenfunctions in $z$. The $Z$ equation in (16), with the boundary conditions $Z(0) = Z(c) = 0$, generates the sine series:

$$Z_m(z) = \sin \frac{m\pi}{c} z, \quad \mu_m = -\left(\frac{m\pi}{c}\right)^2.$$

Substitution of $\lambda_n$ and $\mu_m$ into the radial equation in (16) produces

$$\rho^2 R'' + \rho R' + \left(-\frac{m^2\pi^2}{c^2}\rho^2 - n^2\right) R = 0,$$

whose general solution is

$$C_1 I_n\left(\frac{m\pi}{c}\rho\right) + C_2 K_n\left(\frac{m\pi}{c}\rho\right).$$

Since $K_n$ diverges for $\rho = 0$ the finiteness postulate demands $C_2 = 0$.

The assembled general solution to Problem 1 is thus

$$\psi_1(\rho, \theta, z) \ = \ \sum_{n=0}^{\infty}\sum_{m=1}^{\infty} a_{nm} \cos n\theta \sin \frac{m\pi z}{c} I_n\left(\frac{m\pi}{c}\rho\right)$$

$$+ \sum_{n=1}^{\infty}\sum_{m=1}^{\infty} b_{nm} \sin n\theta \sin \frac{m\pi z}{c} I_n\left(\frac{m\pi}{c}\rho\right).$$

The nonhomogeneous boundary condition at the wall will be met if

$$f_1(\theta, z) \ = \ \sum_{n=0}^{\infty}\sum_{m=1}^{\infty} a_{nm} \cos n\theta \sin \frac{m\pi z}{c} I_n\left(\frac{m\pi}{c}a\right)$$

$$+ \sum_{n=1}^{\infty}\sum_{m=1}^{\infty} b_{nm} \sin n\theta \sin \frac{m\pi z}{c} I_n\left(\frac{m\pi}{c}a\right). \quad (18)$$

We isolate coefficients by multiplying each side of (18) by $(\cos n\theta \sin m\pi z/c)$ or $(\sin n\theta \sin m\pi z/c)$ and integrating; orthogonality yields

$$a_{0m} = \frac{2}{2\pi c} \int_0^{2\pi} \int_0^c f_1(\theta, z) \sin \frac{m\pi z}{c} \, dz \, d\theta / I_0\left(\frac{m\pi}{c}a\right),$$

$$a_{n>0,m} = \frac{2}{\pi c} \int_0^{2\pi} \int_0^c f_1(\theta, z) \cos n\theta \sin \frac{m\pi z}{c} \, dz \, d\theta / I_n\left(\frac{m\pi}{c}a\right),$$

and

$$b_{n>0,m} = \frac{2}{\pi c} \int_0^{2\pi} \int_0^c f_1(\theta, z) \sin n\theta \sin \frac{m\pi z}{c} \, dz \, d\theta / I_n\left(\frac{m\pi}{c}a\right).$$

SOLUTION OF PROBLEM 2. We separate out the $\rho$ dependence as before and obtain the Fourier series (17) again. Since the $\rho$ equation has homogeneous boundary conditions for Problem 2, we insert $\lambda_n$ into the radial equation (16) and consider the Sturm-Liouville problem

$$\rho^2 R'' + \rho R' - n^2 R = -\mu \rho^2 R, \quad R(0) \text{ finite}, \quad R(a) = 0. \tag{19}$$

In Sec. 6.4, Example 3 (see also Table 6.3 #16), we showed that the eigenfunctions of (19) are

$$R_p^{(n)}(\rho) = J_n(j_{n,p}\rho/a), \quad \mu_p^{(n)} = j_{n,p}^2/a^2, \quad p = 1, 2, 3, \ldots$$

where $j_{n,p}$ is the $p$th positive zero of $J_n(x)$. They generate the Fourier Bessel series.

The separated $z$ equation now becomes

$$Z'' = (j_{n,p}^2/a^2)z,$$

with $Z(0) = 0$ for this subproblem. Hence $Z_{n,p}(z) = \sinh(j_{n,p}z/a)$ and the general solution is

$$\psi_2(\rho, \theta, z) = \sum_{n=0}^{\infty} \sum_{p=1}^{\infty} a_{np} \cos n\theta \, J_n(j_{n,p}\rho/a) \sinh(j_{n,p}z/a)$$

$$+ \sum_{n=0}^{\infty} \sum_{p=1}^{\infty} b_{np} \sin n\theta \, J_n(j_{n,p}\rho/a) \sinh(j_{n,p}z/a).$$

The nonhomogeneous boundary condition is

$$f_2(\rho, \theta) = \sum_{n=0}^{\infty} \sum_{p=1}^{\infty} a_{np} \cos n\theta \, J_n(j_{n,p}\rho/a) \sinh(j_{n,p}c/a)$$

$$+ \sum_{n=0}^{\infty} \sum_{p=1}^{\infty} b_{np} \sin n\theta \, J_n(j_{n,p}\rho/a) \sinh(j_{n,p}c/a). \tag{20}$$

Expansion (20) has an interesting feature. Suppose we analyze it in the manner of eq. (10). For *fixed* $\rho$, $f_2(\rho, \theta)$ has a Fourier series expansion

$$f_2(\rho, \theta) = \sum_{n=0}^{\infty} C_n(\rho) \cos n\theta + \sum_{n=1}^{\infty} D_n(\rho) \sin n\theta,$$

and if we expand the coefficients $C_n(\rho)$ and $D_n(\rho)$ as Fourier-Bessel series

$$C_n(\rho) = \sum_{p=1}^{\infty} c_{np} J_n(j_{n,p}\rho/a), \quad D_n(\rho) = \sum_{p=1}^{\infty} d_{np} J_n(j_{n,p}\rho/a),$$

we obtain the form (20). Observe that the Fourier-Bessel expansions of the $n$th coefficients are in terms of Bessel functions of order $n$; $C_1$ and $D_1$ are expanded in terms of $J_1$, $C_2$ and $D_2$ in terms of $J_2$, etc. Each coefficient pair is expanded in terms of a different eigenfunction family!

Now the orthogonality relation for the Fourier-Bessel eigenfunctions (Table 6.3 #16) only guarantees that

$$\int_0^a J_m(j_{m,q}\rho/a)\, J_n(j_{n,p}\rho/a)\, \rho\, d\rho = 0 \text{ for } p \neq q,$$

*when the subscripts $m$ and $n$ match.* Does this foil the extraction of the coefficients in (20)? No; the trigonometric factors save the day. Observe that

$$\int_0^a \int_0^{2\pi} \sin m\theta\, J_m(j_{m,q}\rho/a)\, \sin n\theta\, J_n(j_{n,p}\rho/a)\, \rho\, d\theta\, d\rho$$

will be zero for $m \neq n$, because of the $\theta$ integral.

Therefore the coefficients in (20) are isolated by multiplying by $[\cos m\theta\, J_m(j_{m,q}\rho/a)]$ or $[\sin m\theta\, J_m(j_{m,q}\rho/a)]$, and by the weight function $\rho$, and integrating (exercise 2):

$$a_{op} = \frac{2}{2\pi a^2 J_1^2(j_{o,p})\sinh(j_{o,p}c/a)} \int_0^a \int_0^{2\pi} f_2(\rho,\theta)\, d\theta\, J_0(j_{0,p}\rho/a)\, \rho\, d\rho,$$

$$a_{n>o,p} = \frac{2}{\pi a^2 J_{n+1}^2(j_{n,p})\sinh(j_{n,p}c/a)} \int_0^a \int_0^{2\pi} f_2(\rho,\theta)\, \cos n\theta$$
$$\times\, J_n(j_{n,p}\rho/a)\, \rho\, d\theta\, d\rho,$$

$$b_{np} = \frac{2}{\pi a^2 J_{n+1}^2(j_{n,p})\sinh(j_{n,p}c/a)} \int_0^a \int_0^{2\pi} f_2(\rho,\theta)\, \sin n\theta$$
$$\times\, J_n(j_{n,p}\rho/a)\, \rho\, d\theta\, d\rho. \tag{21}$$

PROBLEM 3 is simply Problem 2 with the cylinder held upside down. Thus the solution $\psi$ is obtained by replacing $z$ with $c - z$ in the corresponding formulas. The overall solution $\psi = \psi_1 + \psi_2 + \psi_3$.  ∎

**Example 3.** Now suppose the cylinder of the previous example is sliced so as to produce a wedge spanned by an angle $\theta_1$, as in Fig. 4. For simplicity we take homogeneous Dirichlet conditions on all walls except the side $\theta = \theta_1$, where the potential is assigned specific boundary values

$$\psi(\rho,\theta_1,z) = f(\rho,z). \tag{22}$$

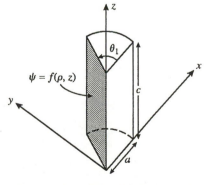

**Figure 4**   *Wedge for Example 3*

(Thus no subproblem decomposition is necessary.)

Since there are homogeneous boundary conditions on each face $z = 0$, $z = c$ we employ the $z$ eigenfunctions $\sin \frac{m\pi z}{c}$ with $\mu_m = -\left(\frac{m\pi}{c}\right)^2$ as before. The radial dependence is analyzed next. From (16) we have

$$\rho^2 R'' + \rho R' - \left(\frac{m\pi}{c}\right)^2 \rho^2 R = -\lambda R. \tag{23}$$

The boundary conditions for (23) are

$$R(a) = 0, \quad R(0) \text{ finite}. \tag{24}$$

According to Table 6.3 #23, eqs. (23), (24) give rise to the finite Lebedev transform with the eigenfunctions

$$K_{i\nu}(m\pi a/c)I_{i\nu}(m\pi\rho/c) - I_{i\nu}(m\pi a/c)K_{i\nu}(m\pi\rho/c)$$

and the continuum of eigenvalues

$$\lambda = -\nu^2, \quad 0 \leq \nu < \infty.$$

The $\theta$ dependence in eq. (16) is thus governed by $\theta'' = \nu^2\theta$ with $\theta(0) = 0$ as a boundary condition. Hence

$$\theta_\nu(\theta) = \sinh \nu\theta.$$

In assembling the eigenfunction expansion for the solution we acknowledge the discrete index $m$ and the continuous index $\nu$ with the coefficient form $g_m(\nu)$:

$$\psi(\rho,\theta,z) = \sum_{m=1}^{\infty} \int_0^\infty g_m(\nu) \sin \frac{m\pi z}{c} \sinh \nu\theta \left\{ K_{i\nu}\left(\frac{m\pi a}{c}\right) I_{i\nu}\left(\frac{m\pi\rho}{c}\right) \right.$$
$$\left. - I_{i\nu}\left(\frac{m\pi a}{c}\right) K_{i\nu}\left(\frac{m\pi\rho}{c}\right) \right\} d\nu.$$

To determine $g_m(\nu)$ from the nonhomogeneous boundary condition (22) -

$$f(\rho,z) = \sum_{m=1}^{\infty} \int_0^\infty g_m(\nu) \sin \frac{m\pi z}{c} \sinh \nu\theta_1 \left\{ K_{i\nu}\left(\frac{m\pi a}{c}\right) I_{i\nu}\left(\frac{m\pi\rho}{c}\right) \right.$$
$$\left. - I_{i\nu}\left(\frac{m\pi a}{c}\right) K_{i\nu}\left(\frac{m\pi\rho}{c}\right) \right\} d\nu \tag{25}$$

- we multiply each side of (25) by the product

$$\sin \frac{m'\pi z}{c} \left\{ K_{i\nu'}\left(\frac{m'\pi a}{c}\right) I_{i\nu'}\left(\frac{m'\pi\rho}{c}\right) - I_{i\nu'}\left(\frac{m'\pi a}{c}\right) K_{i\nu'}\left(\frac{m'\pi a}{c}\right) \right\}$$

and by the weight $1/\rho$ and integrate. The orthogonality relations imply (exercise 5)

$$
\int_o^a \int_0^c f(\rho, z) \sin \frac{m'\pi z}{c} \left\{ K_{i\nu'}\left(\frac{m'\pi a}{c}\right) I_{i\nu'}\left(\frac{m'\pi\rho}{c}\right) \right.
$$
$$
\left. - I_{i\nu'}\left(\frac{m'\pi a}{c}\right) K_{i\nu'}\left(\frac{m'\pi\rho}{c}\right) \right\} dz \frac{d\rho}{\rho}
$$
$$
= \frac{c\pi^2}{4\nu'} g_{m'}(\nu') \frac{\sinh \nu'\theta_1}{\sinh \nu'\pi} \left| I_{i\nu'}\left(\frac{m'\pi a}{c}\right) \right|^2,
$$

or (dropping the primes)

$$
g_m(\nu) = \frac{4\nu}{c\pi^2} \frac{\sinh \nu\pi}{\sinh \nu\theta_1} \frac{1}{|I_{i\nu}(m\pi a/c)|^2} \int_0^a \int_0^c f(\rho, z) \sin \frac{m\pi z}{c} \times
$$
$$
\left\{ K_{i\nu}\left(\frac{m\pi a}{c}\right) I_{i\nu}\left(\frac{m\pi\rho}{c}\right) - I_{i\nu}\left(\frac{m\pi a}{c}\right) K_{i\nu}\left(\frac{m\pi\rho}{c}\right) \right\} dz \frac{d\rho}{\rho}. \quad (26)
$$
∎

**Example 4.** The electrostatic potential $\psi$ inside a sphere of radius $b$ is best analyzed in a spherical coordinate system Fig. 5.

We take the Dirichlet boundary condition

$$
\psi(b, \phi, \theta) = f(\phi, \theta) \quad (27)
$$

on the surface of the sphere. In Sec. 5.3 we saw that the separated form $\psi(r, \phi, \theta) = R(r)\Phi(\phi)\Theta(\theta)$ in Laplace's equation results in

$$
\Theta'' = \lambda\Theta, \quad (\sin\phi\,\Phi')' + \frac{\lambda}{\sin\phi}\Phi = \mu\sin\phi\,\Phi, \quad r^2 R'' + 2rR' + \mu R = 0. \quad (28)
$$

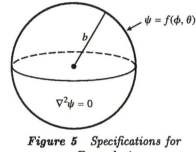

**Figure 5** *Specifications for Example 4*

As usual, we get the Fourier series for $\theta$; however, for spherical problems it is traditional to use the complex exponential form (recall Sec. 3.3)

$$
\Theta_m(\theta) = e^{im\theta}, \quad \lambda_m = -m^2, \quad m = 0, \pm 1, \pm 2, \ldots
$$

In Sec. 6.10 (see also Table 6.3 #28) it is shown that the $\phi$ equation is the Associated Legendre equation, and the eigenfunctions which remain finite at $\phi = 0$ and $\pi$ (the "north and south poles") are the Associated Legendre Functions

$$
\Phi_{ml}(\phi) = P_l^m(\cos\phi), \quad \mu_{ml} = -l(l+1), \quad l = |m|, |m|+1, |m|+2, \ldots
$$

The remaining radial dependence in (28) satisfies an equidimensional equation,

$$
r^2 R'' + 2rR' - l(l+1)R = 0,
$$

whose general solution is easily seen to be

$$
R_l(r) = C_1 r^l + C_2 r^{-l-1},
$$

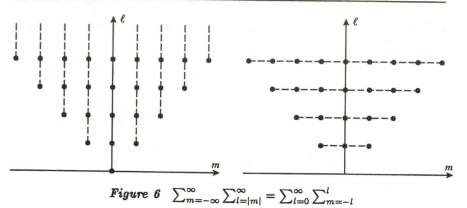

*Figure 6* $\sum_{m=-\infty}^{\infty}\sum_{l=|m|}^{\infty} = \sum_{l=0}^{\infty}\sum_{m=-l}^{l}$

and we take $C_2 = 0$ to ensure finiteness at the origin.

The assembled general solution is thus

$$\psi(r,\phi,\theta) = \sum_{m=-\infty}^{\infty}\sum_{l=|m|}^{\infty} a_{lm} r^l P_l^m(\cos\phi)\, e^{im\theta}. \tag{29}$$

At $r = b$, the Dirichlet condition (27) becomes

$$f(\phi,\theta) = \sum_{m=-\infty}^{\infty}\sum_{l=|m|}^{\infty} a_{lm} b^l P_l^m(\cos\phi) e^{im\theta}.$$

From the orthogonality relations one then extracts (exercise 6)

$$a_{lm} = \frac{(2l+1)(l-m)}{4\pi(l+m)!}\int_{\phi=0}^{\pi}\int_{\theta=0}^{2\pi} f(\phi,\theta)\, P_l^m(\cos\phi)\, e^{-im\theta}\sin\phi\, d\theta\, d\phi / b^l. \tag{30}$$

The points in the $m, l$ plane that are involved in the summation (29) are diagramed in Fig. 6. From this picture it is easy to see that the order of summation can be reversed by writing

$$\psi(r,\phi,\theta) = \sum_{l=0}^{\infty}\sum_{m=-l}^{l} a_{lm} r^l P_l^m(\cos\phi) e^{im\theta}.$$

It has become traditional to combine the two angular factors in (29) according to the following definition:

$$Y_{lm}(\phi,\theta) = \sqrt{\frac{(2l+1)(l-m)!}{4\pi(l+m)!}}\, P_l^m(\cos\phi) e^{im\theta}, \quad -l \le m \le l. \tag{31}$$

These functions are called *spherical harmonics* and are used extensively in radiation theory and atomic physics. They are tabulated in Table 7.1. Because of the choice of coefficient in (31),

**the orthogonality relation is quite convenient:**

$$\int_0^\pi \int_0^{2\pi} Y_{lm}(\phi,\theta)\overline{Y_{l'm'}(\phi,\theta)} \sin\phi \, d\theta d\phi = \begin{cases} 1 & \text{if } l = l' \text{ and } m = m', \\ 0 & \text{otherwise.} \end{cases} \quad (32)$$

The solution of our problem then takes a very transparent form (exercise 6):

$$\psi(r,\phi,\theta) = \sum_{l=0}^\infty \sum_{m=-l}^l c_{lm} \, r^l \, Y_{lm}(\phi,\theta),$$

$$c_{lm} = \int_0^\pi \int_0^{2\pi} f(\phi,\theta) \, \overline{Y_{lm}(\phi,\theta)} \, \sin\phi \, d\theta \, d\phi / b^l. \quad (33)$$

It is worth recording for future reference that in the course of this derivation we have verified a statement made in Table 6.3 # 29: namely, the spherical harmonic $Y_{lm}(\phi,\theta)$ satisfies the identity

$$\nabla^2 R(r)Y_{lm}(\phi,\theta) = \frac{(r^2 R')' - l(l+1)R}{r^2} Y_{lm} \quad (34)$$

for *any* function $R$ of the radial coordinate whatever; see exercise 8.

# EXERCISES 7.1

1. Work out the solutions to the remaining subproblems in the Example 1.

2. Verify eqs. (21).

3. Rework the Example 2 with the Dirichlet boundary condition on the wall replaced by a Neumann boundary condition.

4. Rework Example 3 with a nonhomogeneous Neumann condition at $\rho = a$ and nonhomogeneous Dirichlet conditions on the other two faces.

5. Verify eqs. (26).

6. Verify eqs. (30), (32), (33).

7. Expand $\sin^3 \phi \sin^2 \theta$ in terms of spherical harmonics.

8. Use the definition (31) to verify identity (34).

9. Rework Example 4 with the Dirichlet boundary condition replaced by a Neumann boundary condition. What is the restriction on the latter? (Hint: recall the discussion of the Neumann problem in Sec. 4.6.)

# TABLE 7.1 The Spherical Harmonics

$$Y_{l,m}(\phi,\theta) = \sqrt{\frac{(2l+1)(l-m)!}{4\pi(l+m)!}}\, P_{l,m}(\cos\phi)\, e^{im\theta}$$

Note: $Y_{l,-m}(\phi,\theta) = (-1)^m \overline{Y_{l,m}(\phi,\theta)}$.

| $l$ \\ $m$ | 0 | 1 |
|---|---|---|
| 0 | $\frac{1}{4\pi}$ | |
| 1 | $\sqrt{\frac{3}{4\pi}}\cos\phi$ | $-\sqrt{\frac{3}{8\pi}}\sin\phi\, e^{i\theta}$ |
| 2 | $\sqrt{\frac{5}{4\pi}}\left(\frac{3}{2}\cos^2\phi - \frac{1}{2}\right)$ | $-\sqrt{\frac{15}{8\pi}}\sin\phi\cos\phi\, e^{i\theta}$ |
| 3 | $\sqrt{\frac{7}{4\pi}}\left(\frac{5}{2}\cos^3\phi - \frac{3}{2}\cos\phi\right)$ | $-\sqrt{\frac{21}{4\pi}}\sin\phi\left(\frac{5}{4}\cos^2\phi - \frac{1}{4}\right)e^{i\theta}$ |

| $l$ \\ $m$ | 2 | 3 |
|---|---|---|
| 2 | $\frac{1}{4}\sqrt{\frac{15}{2\pi}}\sin^2\phi\, e^{i2\theta}$ | |
| 3 | $\frac{1}{4}\sqrt{\frac{105}{2\pi}}\sin^2\phi\cos\phi\, e^{i2\theta}$ | $-\frac{1}{4}\sqrt{\frac{35}{4\pi}}\sin^3\phi\, e^{i3\theta}$ |

ADDITION THEOREM: If $\psi$ is the angle between the directions $(\phi_1,\theta_1)$ and $(\phi_2,\theta_2)$, so that

$$\cos\psi = \cos\phi_1\cos\phi_2 + \sin\phi_1\sin\phi_2\cos(\theta_1 - \theta_2),$$

then

$$P_l(\cos\psi) = \frac{4\pi}{2l+1} \sum_{m=-l}^{l} \overline{Y_{l,m}(\phi_1,\theta_1)}\, Y_{l,m}(\phi_2,\theta_2).$$

# 7.2 Time-Dependent Problems: The Heat Equation

*"Time is nature's way of keeping things from happening all at once."* (A. Hynek)

**Summary** Because the time variation of mechanical systems does not engender Sturm-Liouville boundary value problems, time-dependent problems with time-independent boundary conditions are first decomposed into stationary solutions and transient solutions. For the latter the separated time equation is treated last. The time factors for the heat equation are exponential decays, with time constants dependent on the eigenvalues of the spatial equations.

The separation of variables procedure requires some adjusting before it can be applied to time-dependent problems. The reason for this is that the time variation is not governed by a Sturm-Liouville boundary value problem, and as a result we are not able to employ time eigenfunctions in the solution. To illustrate, consider the following example.

**Example 1.** A typical problem for the heat equation is depicted in Fig. 1. The temperature pattern $\bar{\Psi}(x,y,t)$ in a (two-dimensional) slab is initially given by

$$\Psi(x,y,t=0) = f(x,y); \tag{1}$$

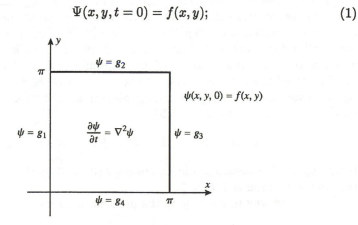

**Figure 1** *Specifications for Example 1*

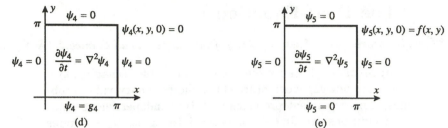

**Figure 2** *Proposed decomposition for Example 1*

it evolves according to the heat equation

$$\frac{\partial \Psi}{\partial t} = \nabla^2 \Psi; \tag{2}$$

but its boundary values are constrained by the heat sources along the edges according to the Dirichlet conditions

$$\begin{aligned}
\Psi(0,y,t) &= g_1(y); \quad \Psi(x,\pi,t) = g_2(x); \\
\Psi(\pi,y,t) &= g_3(y); \quad \Psi(x,0,t) = g_4(x).
\end{aligned} \tag{3}$$

The partial differential equation (2), then, is accompanied by five nonhomogeneous auxiliary conditions (1) and (3).

Probably the decomposition illustrated in Fig. 2 suggests itself to the reader as the natural extension of the methodology expounded previously. However *the superposition depicted is not workable!* Consider, for instance, the solution for Problem 1. One would be inclined to work out the eigenfunctions in $y$ and $t$ first, and out of them assemble a general solution which could be fitted to the nonhomogeneous boundary condition on the left edge. But the $t$ variable does not generate eigenfunctions; the separated equations for the heat equation are (Sec. 5.3)

$$X'' = \lambda X, \quad Y'' = \mu Y, \quad T' = KT; \quad K = \lambda + \mu. \tag{4}$$

If we append the homogeneous condition $T(0) = 0$ to the time equation, the unique solution is $T(t) = 0$.

The *proper* way to decompose the problem is to let

$$\Psi(x,y,t) = \Psi_0(x,y,t) + \Psi_\infty(x,y). \tag{5}$$

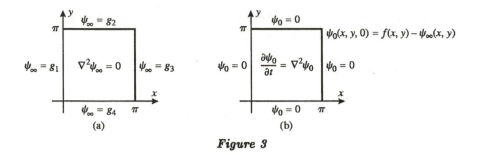

**Figure 3**

Here $\Psi_\infty(x,y)$ is a time-independent, or stationary, solution of the heat equation –

$$\nabla^2 \Psi_\infty(x,y) = -\frac{\partial \Psi_\infty}{\partial t} = 0$$

– which satisfies the boundary conditions on the edges:

$$\Psi_\infty(0,y) = g_1(y); \;\; \Psi_\infty(x,\pi) = g_2(x); \;\; \Psi_\infty(\pi,y) = g_3(y); \;\; \Psi_\infty(x,0) = g_4(x); \tag{6}$$

and $\Psi_0(x,y,t)$ satisfies the (full) heat equation with *homogeneous* boundary conditions

$$\Psi_0(0,y) = 0; \;\; \Psi_0(x,\pi) = 0; \;\; \Psi_0(\pi,y) = 0; \;\; \Psi_0(x,0) = 0, \tag{7}$$

and an initial condition obtained by combining (1) and (5):

$$\Psi_0(x,y,t=0) = f(x,y) - \Psi_\infty(x,y). \tag{8}$$

The reader should mentally verify that the combination $\Psi_0 + \Psi_\infty$ satisfies (1), (2), and (3). See Fig. 3.

The subscript "∞" is appropriate because we shall see that in most physical situations $\Psi(x,y,t)$ converges to $\Psi_\infty(x,y)$ as $t \to \infty$. Thus $\Psi_\infty(x,y)$ corresponds to the "steady state" solution, and $\Psi_0(x,y,t)$ to the "transient" solution.

Note that the calculation of $\Psi_\infty$, itself, calls for a decomposition into four subproblems. Note also that if the boundary conditions vary in time, there can be no time-independent "stationary" solution. *This decomposition is only valid when the boundary conditions are time-independent.* The procedure for time-dependent boundary conditions will be described in Sec. 8.2.

The separation of variables methodology *can* be applied to the subproblems of Fig. 3. Since the calculation of $\Psi_\infty$ is a time–independent problem, it can be accomplished by the procedures of Chapter 6. In exercise 1 the reader is invited to derive the formula

$$\Psi_\infty(x,y) = \sum_{n=1}^{\infty} a_n \sin nx \sinh ny + \sum_{n=1}^{\infty} b_n \sin nx \sinh n(\pi - y)$$

$$+ \sum_{n=1}^{\infty} c_n \sinh nx \sin ny + \sum_{n=1}^{\infty} d_n \sinh n(\pi - x) \sin ny,$$

$$a_n = \frac{2}{\pi \sinh n\pi} \int_0^{\pi} g_2(x) \sin nx \, dx, \quad \text{etc.} \tag{9}$$

With $\Psi_\infty$ known, the necessary adjustment (8) in the initial condition is made and $\Psi_0$ is found by

*a.* computing the eigenfunctions in the $x$ and $y$ variables:

$$X_m(x) = \sin mx, \quad \lambda_m = -m^2, \quad Y_n(y) = \sin ny, \quad \mu_n = -n^2;$$

*b.* solving the time equation:

$$T' = KT = (-m^2 - n^2)T, \quad T(t) = (const)e^{-(m^2+n^2)t};$$

*c.* assembling the general solution:

$$\Psi_0(x,y,t) = \sum_{m=1}^{\infty} \sum_{n=1}^{\infty} A_{mn} \sin nx \sin ny \, e^{-(m^2+n^2)t}; \quad \text{and}$$

*d.* selecting the coefficients to meet the initial condition (8), which is the nonhomogeneous condition for this subproblem:

$$\Psi_0(x,y,0) = f(x,y) - \Psi_\infty(x,y) = \sum_{m=1}^{\infty} \sum_{n=1}^{\infty} A_{mn} \sin mx \sin ny,$$

$$A_{mn} = \frac{4}{\pi^2} \int_0^{\pi} \int_0^{\pi} [f(x,y) - \Psi_\infty(x,y)] \sin mx \sin ny \, dx \, dy.$$

Notice that $\Psi_0(x,y,t) \to 0$ as $t \to \infty$ because of the decaying exponentials; $\Psi_0$ is indeed transient, and $\Psi$ approaches $\Psi_\infty$, the steady state.

EXAMPLE 2. The one dimensional heat equation

$$\frac{\partial \Psi}{\partial t} = \frac{\partial^2 \Psi}{\partial x^2}, \quad \Psi = \Psi(x,t), \tag{10}$$

describes temperature distribution in a thin rod (Fig. 4). We append boundary conditions corresponding to constant-temperature sources at the ends of the rod

$$\Psi(0,t) = 0, \quad \Psi(\pi,t) = 5, \tag{11}$$

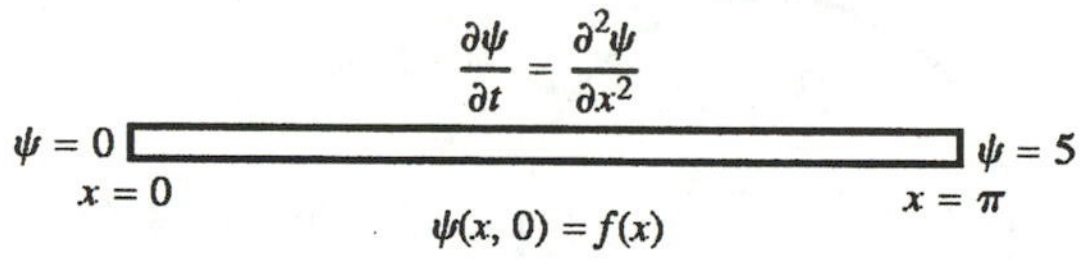

$$\frac{\partial \psi}{\partial t} = \frac{\partial^2 \psi}{\partial x^2}$$

$\psi = 0$ [================================] $\psi = 5$
$x = 0$ $\qquad\qquad\qquad\qquad\qquad\qquad\qquad$ $x = \pi$

$$\psi(x, 0) = f(x)$$

***Figure 4*** *Specifications for Example 2*

and prescribe the initial temperature

$$\Psi(x, 0) = f(x). \tag{12}$$

The stationary solution $\Psi_\infty$ solves the simple ODE system

$$\Psi_\infty{}'' = 0, \quad \Psi_\infty(0) = 0, \quad \Psi_\infty(\pi) = 5; \tag{13}$$

thus $\Psi_\infty(x) = 5x/\pi$. Consequently the transient solution $\Psi$ satisfies

$$\frac{\partial \Psi_0}{\partial t} = \frac{\partial^2 \Psi_0}{\partial x^2}, \quad \Psi_0(0, t) = \Psi_0(\pi, t) = 0, \quad \Psi_0(x, 0) = f(x) - \frac{5x}{\pi}. \tag{14}$$

Separation of the heat equation yields

$$X'' = \lambda X, \quad T' = \lambda T, \tag{15}$$

so the $x$ eigenfunctions are $X_n(x) = \sin nx$, $\lambda_n = -n^2$; the associated time dependence is thus given by $T_n(t) = e^{-n^2 t}$, and the general solution

$$\Psi_0(x, t) = \sum_{n=1}^{\infty} a_n \sin nx \, e^{-n^2 t} \tag{16}$$

satisfies all the homogeneous equations in (14). To meet the nonhomogeneous initial condition in (14), we invoke orthogonality as before and find

$$a_n = \frac{2}{\pi} \int_0^\pi [f(x) - 5x/\pi] \sin nx \, dx. \tag{17}$$

The sum $\Psi_0 + \Psi_\infty$ solves the problem.                                   ∎

A good way of interpreting these eigenfunction formulas is the following: the system decomposes the initial state $[f(x) - 5x/\pi]$ into "eigenmodes" ($\sin nx$) through eq. (17), assigns to each mode an exponential time factor ($e^{-n^2 t}$), and then lets the solution evolve. The higher modes for the heated rod are short lived. The first mode, $\sin x$, persists the longest as the steady state $5x/\pi$ is approached.

**Example 3.** Heat flow inside a solid sphere is described by the heat equation for the temperature $\Psi$ in spherical coordinates:

$$\frac{\partial \Psi}{\partial t} = \nabla^2 \Psi = \frac{1}{r^2} \frac{\partial}{\partial r} \left( r^2 \frac{\partial \psi}{\partial r} \right) + \frac{1}{r^2 \sin \phi} \frac{\partial}{\partial \phi} \left( \sin \phi \frac{\partial \Psi}{\partial \phi} \right)$$
$$+ \frac{1}{r^2 \sin^2 \phi} \frac{\partial^2 \Psi}{\partial \theta^2}. \tag{18}$$

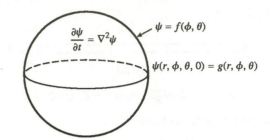

***Figure 5*** *Specifications for Example 3*

See Fig. 5. If the temperature is specified on the surface we have the Dirichlet boundary condition

$$\Psi = f(\phi, \theta) \quad \text{when} \quad r = b; \tag{19}$$

and for initial conditions we take

$$\Psi = g(r, \phi, \theta) \quad \text{at} \quad t = 0. \tag{20}$$

The stationary solution $\Psi_\infty(r, \phi, \theta)$ satisfies

$$\nabla^2 \Psi_\infty = 0, \quad \Psi_\infty = f(\phi, \theta) \quad \text{at} \quad r = b, \tag{21}$$

and in Sec. 7.1 we found the solution of (21) to be

$$\Psi_\infty(r, \phi, \theta) = \sum_{l=0}^{\infty} \sum_{m=-l}^{l} a_{lm} r^l Y_{lm}(\phi, \theta),$$

where

$$a_{lm} = b^{-l} \int_0^\pi \int_0^{2\pi} f(\phi, \theta) \, \overline{Y_{lm}(\phi, \theta)} \sin \phi \, d\theta \, d\phi.$$

The transient solution $\Psi_0$ thus obeys

$$\frac{\partial \Psi_0}{\partial t} = \nabla^2 \Psi_0, \quad \Psi_0(r, \phi, \theta, 0) = g(r, \phi, \theta) - \Psi_\infty(r, \phi, \theta),$$
$$\Psi_0(b, \phi, \theta, t) = 0. \tag{22}$$

According to Sec. 5.3 the separation of the heat equation in spherical coordinates gives rise to the following ordinary differential equations:

the time equation $\quad T' = KT$

the $\theta$ equation $\quad \Theta'' = \lambda \Theta$

the associated Legendre equation $\quad (\sin \phi \, \Phi')' + \frac{\lambda}{\sin \phi} \Phi = \mu \sin \phi \, \Phi$

and the radial equation $\quad r^2 R'' + 2r R' + (\mu - Kr^2) R = 0.$

Exactly as in Sec. 7.1 the periodicity and finiteness conditions in $\theta$ and $\phi$ lead to the spherical harmonics with the eigenvalues

$$\Theta(\theta)\Phi(\phi) = Y_{lm}(\phi,\theta), \quad \lambda = -m^2, \quad \mu = -l(l+1).$$

The radial dependence then generates the Sturm-Liouville problem

$$r^2 R'' + 2rR' - l(l+1)R = Kr^2 R, \quad R(0) \text{ finite}, \quad R(b) = 0,$$

with $K$ as the eigenvalue. This system was discussed in Example 7 of Sec. 6.9 (see also Table 6.3 #25), where the eigenfunctions are identified as the spherical Bessel functions

$$R_{l,p}(r) = j_l(s_{l,p}r/b),$$

and the eigenvalues for $K$ are

$$K_{l,p} = -s_{l,p}^2/b^2;$$

here $s_{l,p}$ is the $p$th positive zero of $j_l(x)$.

The solution of the time equation with these eigenvalues is

$$T_{l,p}(t) = e^{-s_{l,p}^2 t/b^2}.$$

Therefore the general solution is expressed

$$
\begin{aligned}
\Psi_0(r,\phi,\theta,t) &= \sum_{l=0}^{\infty}\sum_{m=-l}^{l}\sum_{p=1}^{\infty} a_{lmp} \\
&\quad \times j_l(s_{l,p}r/b)Y_{lm}(\phi,\theta)e^{-s_{l,p}^2 t/b^2},
\end{aligned}
\tag{23}
$$

and fitting the initial condition (22) requires

$$a_{lmp} = \frac{2\int_0^b \int_0^\pi \int_0^{2\pi} F(r,\phi,\theta)\,\overline{Y_{lm}(\phi,\theta)}\,j_l(s_{l,p}r/b)\,r^2\sin\phi\,d\theta\,d\phi\,dr}{b^3 j_{l+1}^2(s_{l,p})}, \tag{24}$$

where $F(r,\phi,\theta) = g(r,\phi,\theta) - \Psi_\infty(r,\phi,\theta)$ (exercise 7). ∎

**Example 4.** An important step in the manufacture of semiconductors for electronic circuits is the diffusion of impurity ions into a semiconducting medium such as silicon (see Streetman). For *gaseous deposition* processes, a fixed impurity concentration level is continuously maintained at the surface while the impurities diffuse into the interior (Fig. 6). On the other hand when *implantation procedures* are used, a sharply peaked concentration of impurities is established at a shallow depth initially, and it is allowed to diffuse into the semiconductor without replenishment (Fig. 7). It is important for circuit designers to be able to predict the impurity concentration

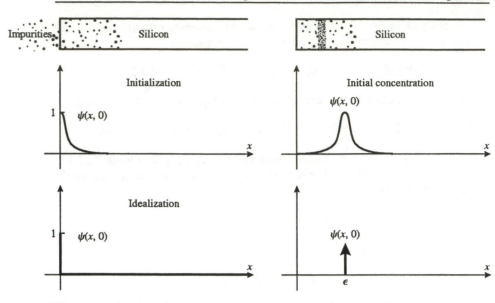

**Figure 6**   *Gaseous deposition*          **Figure 7**   *Implantation*

profile. We shall derive the solutions for idealized conditions: the flow is assumed to be one dimensional and the semiconductor to be semiinfinite. Furthermore we model the initial impurity concentration for the implantation process as a delta function centered at $x = \varepsilon$; for gaseous diffusion the impurity concentration is initially zero except for the end $x = 0$, where it is fixed at some value which we take to be unity. See Figs. 6 and 7.

As discussed in Sec. 4.5, the diffusion of impurities is described by the same physical law (Fick's law) as the flow of heat. Thus the impurity concentration $\Psi(x,t)$ satisfies the heat equation (10). The initial condition for the implantation process is

$$\Psi(x,0) = \delta(x - \varepsilon). \tag{25}$$

Since there is no replenishment, the flux is zero on the surface -

$$\frac{\partial \Psi}{\partial x}(0,t) = 0; \tag{26}$$

and of course $\Psi$ is finite at $x = \infty$:

$$\Psi(\infty,t) \quad \text{finite}. \tag{27}$$

Both spatial boundary conditions (26, 27) are homogeneous, and the steady-state solution of $\Psi''_{\infty}(x) = 0, \Psi'_{\infty}(0) = 0, \Psi_{\infty}(\infty)$ *bounded* is indeterminant:

$$\Psi_{\infty}(x) = constant = K.$$

The separation of eq. (10) was given by eq. (15). For the boundary conditions (26), (27) we identify the eigenfunctions under entry #9 in Table

6.3, the Fourier Cosine Integral. Attaching the time factors we have the solution expansion for the transient solution

$$\Psi_0(x,t) = \int_0^\infty C(\omega) \cos \omega x \, e^{-\omega^2 t} d\omega. \tag{28}$$

Incorporating the steady-state solution into the initial condition (25) for $\Psi_0$, we have

$$\delta(x - \varepsilon) - K = \int_0^\infty C(\omega) \cos \omega x \, d\omega.$$

To find $C(\omega)$ we multiply by $\cos \omega' t$ and integrate over $x$, employing the orthogonality condition in the Table to derive

$$\int_0^\infty [\delta(x - \varepsilon) - K] \cos \omega' x \, dx = \int_0^\infty \int_0^\infty C(\omega) \cos \omega x \cos \omega' x \, dx \, d\omega;$$

$$\cos \omega' \varepsilon - \frac{\pi}{2} K \, \delta(\omega') = \frac{\pi}{2} \int_0^\infty C(\omega) \, \delta(\omega - \omega') \, d\omega = \frac{\pi}{2} C(\omega').$$

Hence

$$\Psi_0(x,t) = \int_0^\infty C(\omega) \cos \omega x \, e^{-\omega^2 t} d\omega = \frac{2}{\pi} \int_0^\infty \cos \omega \varepsilon \cos \omega x \, e^{-\omega^2 t} d\omega - K$$

and

$$\Psi = \Psi_0 + \Psi_\infty = \frac{2}{\pi} \int_0^\infty \cos \omega \varepsilon \cos \omega x \, e^{-\omega^2 t} d\omega. \tag{29}$$

Using analytic function theory one can show that

$$\int_0^\infty e^{-u^2} \cos 2bu \, du = \frac{\sqrt{\pi}}{2} e^{-b^2} \tag{30}$$

(see, for example, Saff and Snider). Thus with appropriate substitutions (29) can be reduced to (exercise 11)

$$\Psi(x,t) = \frac{1}{2\sqrt{\pi t}} \left\{ \exp[-(x - \varepsilon)^2 / 4t] + \exp[-(x + \varepsilon)^2 / 4t] \right\}. \tag{31}$$

As $\varepsilon$ approaches zero the solution converges to

$$\Psi(x,t) = \frac{1}{\sqrt{\pi t}} \exp[-x^2 / 4t], \tag{32}$$

which is proportional to the *Gaussian distribution* $e^{-u^2/2}/\sqrt{2\pi}$ for fixed $t$.

For gaseous deposition the continuous replenishment at $x = 0$ imposes the *nonhomogeneous* boundary condition

$$\Psi(0,t) = 1, \tag{33}$$

together with the finiteness condition (27). The initial condition is

$$\Psi(x, 0) = 0 \quad \text{(for } x > 0). \tag{34}$$

For the stationary solution we have

$$\Psi_\infty{}'' = 0, \quad \Psi_\infty(0) = 1, \quad \Psi_\infty(\infty) \text{ finite,} \tag{35}$$

whose solution is easily seen to be $\Psi_\infty(x) = 1$. The transient subproblem then becomes

$$\frac{\partial \Psi_0}{\partial t} = \frac{\partial^2 \Psi_0}{\partial x^2}, \quad \Psi_0(0, t) = 0, \quad \Psi_0(\infty, t) \text{ finite}, \quad \Psi_0(x, 0) = -1. \tag{36}$$

This problem separates as before, with the boundary conditions dictating the Fourier Sine Integral from Table 6.3 #8. Thus

$$\Psi_0(x, t) = \int_0^\infty S(\omega) \sin \omega x \, e^{-\omega^2 t} d\omega. \tag{37}$$

The initial condition requires

$$-1 = \int_0^\infty S(\omega) \sin \omega x \, d\omega,$$

so according to the Table

$$S(\omega) = (-1)\frac{2}{\pi} \int_0^\infty \sin \omega x \, d\omega. \tag{38}$$

The integral in (38) diverges, but according to exercise 2 Sec. 3.7 it has a distribution interpretation as $1/\omega$; thus $S(\omega) = -2/\pi\omega$ and

$$\Psi_0(x, t) = -\frac{2}{\pi} \int_0^\infty \frac{\sin \omega x}{\omega} e^{-\omega^2 t} d\omega. \tag{39}$$

The integral in (39) can be expressed in terms of the area under the Gaussian distribution. Observe that

$$\frac{\sin \omega x}{\omega} = \int_0^x \cos \omega \xi \, d\xi \tag{40}$$

and that insertion of (40) into (39) and reversal of the order of integration (which can be justified mathematically) results in

$$\Psi_0(x, t) = -\frac{2}{\pi} \int_0^x \int_0^\infty \cos \omega \xi \, e^{-\omega^2 t} d\omega \, d\xi. \tag{41}$$

Identity (30), with appropriate substitutions (exercise 10), reduces (41) to

$$\Psi_0(x, t) = -\frac{-2}{\sqrt{\pi}} \int_0^{x/2\sqrt{t}} e^{-u^2} du \equiv -\text{Erf}\left[\frac{x}{2\sqrt{\pi}}\right],$$

the negative of the so-called *Error Function*. The net concentration $\Psi = \Psi_\infty + \Psi_0$ is known as the *Complementary Error Function*

$$\Psi(x,t) = 1 - \text{Erf}\left[\frac{x}{2\sqrt{\pi}}\right] = \text{Erfc}\left[\frac{x}{2\sqrt{t}}\right]. \qquad (42)$$

■

# Exercises 7.2

1. Verify the formula for $\Psi_\infty$ in Example 1.

2. Solve the heat equation in a rectangular box with dimensions $(a, b, c)$. Take the initial temperature to be $g(x, y, z)$ and take the boundary conditions to be $\Psi(0, y, z, t) = 1$, $\Psi = 0$ on all other faces.

3. Solve the heat equation in a cylinder of unit height and radius. Take the initial temperature to be $g(r, \theta, z)$, with $\Psi = 1$ on the top, $\Psi = 0$ on the bottom, and $\Psi = 1$ on the side wall.

4. Solve the problem depicted in Fig. 1 with the boundary condition on the right edge replaced by $\partial\Psi/\partial x = .1\Psi + g_3$ (recall Example 1, Sec. 6.3).

5. Verify eq. (24).

6. Solve the heat equation for the temperature in a rod subjected to the conditions

$$\Psi(0, t) = 1, \quad \Psi(1, t) = 2, \quad \Psi(x, 0) = 1 + x.$$

7. Solve the heat equation for the temperature in a rod subjected to the conditions

$$\Psi(0, t) = 1, \quad \frac{\partial\Psi}{\partial x}\left(\frac{\pi}{2}, t\right) = 0, \quad \Psi(x, 0) = f(x).$$

8. The following problem has no physical significance whatever, but it provides good practice for our separation of variables procedure:

$$(1 - x^2)\frac{\partial^2\Psi}{\partial x^2} - x\frac{\partial\Psi}{\partial x} = \frac{\partial\Psi}{\partial t}, \quad -1 < x < 1,$$

$$\frac{\partial\Psi}{\partial x} \text{ finite at } x = \pm 1, \quad \Psi(x, 0) = 1 - x^2.$$

Solve this problem completely. (Hint: use Table 6.3.)

9. Use trig identities, algebraic substitutions, and identity (30) to reduce eq. (29) to the form (32).

**10.** Use algebraic substitutions and identity (30) to reduce eq. (41) to the form (42).

**11.** At the end of Sec. 4.5 we adopted the convention that the *diffusivity*, $\gamma$ in eq. (3) of that section, would be unity. Review Examples 1 through 4 in the present section and restore this parameter in the solution formulas.

## 7.3   Initial Value Problems for The Wave Equation

**Summary**   Solving the wave equation entails extracting the time-independent subproblem satisfying the nonhomogeneous boundary conditions, then assembling a general transient solution and fitting the (compensated) initial values of $\Psi$ and $\partial\Psi/\partial t$. The separated time equation is the harmonic oscillator equation, with the frequencies of oscillation of the individual modes determined by the eigenvalues of the separated spatial equations. The modes themselves are standing waves, and the set of allowed frequencies dictates many of the important characteristics of the physical system.

The application of separation of variables to the wave equation is almost identical to that for the heat equation; the only essential difference is that the wave equation carries *two*, rather than one, initial conditions. But the basic steps - extraction of the stationary solution and assemblage of eigenmodes for the transient - are the same. (**We presume herein that the boundary conditions are time-independent; the general case is studied in Sec. 8.2.**)

**Example 1.** The wave equation in one dimension

$$\frac{\partial^2\Psi}{\partial t^2} = \frac{\partial^2\Psi}{\partial x^2},$$

is best visualized as describing vibrations of a taut string (Fig. 1).

We shall assume that the string is held fixed at both ends with the right end elevated -

$$\Psi(0,t) = 0, \quad \Psi(a,t) = 1,$$

and that the initial displacement and velocity along the string are specified by the equations

$$\Psi(x,0) = f(x), \quad \frac{\partial\Psi}{\partial t}(x,0) = g(x).$$

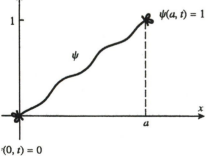

**Figure 1**   *Taut vibrating string*

The stationary solution $\Psi_\infty(x)$ satisfies the system

$$\Psi_\infty'' = 0, \quad \Psi_\infty(0) = 0, \quad \Psi(a) = 1;$$

thus we see immediately that

$$\Psi_\infty(x) = x/a. \tag{1}$$

Consequently the transient solution $\Psi_0(x,t)$ must solve

$$\frac{\partial^2 \Psi_0}{\partial t^2} = \frac{\partial^2 \Psi_0}{\partial x^2}, \quad \Psi_0(0,t) = \Psi_0(a,t) = 0, \tag{2}$$

$$\Psi_0(x,0) = f(x) - \frac{x}{a}, \quad \frac{\partial \Psi_0}{\partial t}(x,0) = g(x). \tag{3}$$

(Note that because $\partial \Psi_\infty / \partial t = 0$, no correction needs to be made to the initial velocity $g(x)$.) As demonstrated in Sec. 5.3, the ordinary differential equations resulting from separating the one-dimensional wave equation are

$$X'' = \lambda X, \quad T'' = \lambda T. \tag{4}$$

The first of these, with the homogeneous boundary conditions (2), generates the sine-series eigenfunctions:

$$X_n(x) = \sin n\pi x/a, \quad \lambda_n = -n^2\pi^2/a^2 \quad (n = 1,2,3,...).$$

The associated time factors, then, are the solutions to the second of eqs. (4) with $\lambda = \lambda_n$:

$$T_n(t) = C_1 \cos n\pi t/a + C_2 \sin n\pi t/a,$$

and the eigenmode expansion of $\Psi_0$ takes the form

$$\Psi_0(x,t) = \sum_{n=1}^{\infty} [c_n \cos n\pi t/a + d_n \sin n\pi t/a] \sin \frac{n\pi x}{a}. \tag{5}$$

To fit the initial conditions (3) we apply orthogonality to the relations

$$\Psi_0(x,0) = f(x) - \frac{x}{a} = \sum_{n=1}^{\infty} c_n \sin \frac{n\pi x}{a},$$

$$\frac{\partial \Psi_0}{\partial t}(x,0) = g(x) = \sum_{n=1}^{\infty} \frac{n\pi}{a} d_n \sin \frac{n\pi x}{a},$$

and obtain

$$c_n = \frac{2}{a} \int_0^a [f(x) - \frac{x}{a}] \sin \frac{n\pi x}{a} dx, \quad d_n = \frac{2}{n\pi} \int_0^a g(x) \sin \frac{n\pi x}{a} dx.$$

The overall solution is then $\Psi_0(x,t) + \Psi_\infty(x)$.

We wish to consider the significance of the individual eigenmodes in the solution formula (5). To be specific, let us address the case when the string is deformed into some shape $f(x)$ and released *from rest*, so that initial velocities $g(x)$ are zero. We also assume that the boundary conditions are homogeneous (so that we can omit $\Psi_\infty$). The solution then is given by

$$\Psi(x,t) = \sum_{n=1}^\infty c_n \cos\frac{n\pi t}{a} \sin\frac{n\pi x}{a},$$

$$c_n = \frac{2}{a}\int_0^1 f(x)\sin\frac{n\pi x}{a}\,dx. \tag{6}$$

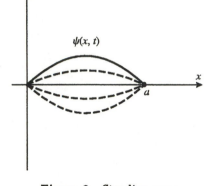

**Figure 2   Standing wave**

Notice that if the initial displacement $f(x)$ were proportional to, say, $\sin \pi x/a$ then by orthogonality the only term that would survive in (6) would be the one where $n = 1$; i.e., the solution would be some multiple of $\sin \pi x/a \cos \pi t/a$. This solution is called a "standing" wave (exercise 5, Sec. 4.3) and it is illustrated in Fig. 2. It is one-half wavelength long and vibrates at a frequency of $\pi/a$ radians per unit time.

Thus by choosing the initial configuration of the string to have the same shape as one of the individual terms in the solution, we can activate only that mode. The string equations, in effect, break up the initial shape of the string into these modes and assign each mode a vibrational frequency (which is a multiple of $\pi/a$ radians per second). ∎

This phenomenon is exemplified by the plucking of a guitar string. With the pick the instrumentalist bends the string into a deformed shape and then releases it from rest. If the string is plucked near the center, as in Fig. 3, the "fundamental" mode ($n = 1$) shown in Fig. 2 is strongly activated. If the string is plucked nearer the end - as in Fig. 4, where the pick is held at

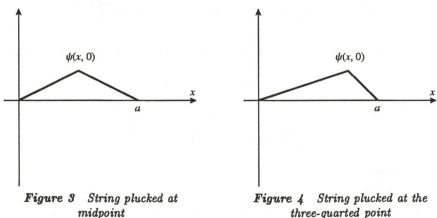

**Figure 3   String plucked at midpoint**

**Figure 4   String plucked at the three-quarted point**

the 3/4 point - the higher modes are more strongly activated and the guitar sound has a "twangy" quality. See exercise 1.

The different modes of vibration of a guitar string are distinguishable to the human ear by the frequency of the sound they generate; different frequencies are discerned as different "pitches." The *fundamental* is the frequency of the lowest mode, and integer multiples of the fundamental frequency are called "harmonics." A cello and a trombone sound different even if they are playing the same fundamental - $F^{\#}$, say - because of the difference in the intensities of the harmonics they generate.

Many engineering devices operate better in some modes than others. For example, certain modes propagate down an optical fiber or a waveguide with less attenuation than the others. In such cases engineers design the systems to suppress the undesirable modes, either by shaping the initial excitation or by introducing devices (like resistor cards in a waveguide) which damp out certain modes preferentially.

The fundamental mode on a guitar string can be suppressed by holding the finger lightly in contact with the midpoint of the string (the twelfth fret on the guitar), creating a stationary point or "node" there. According to eq. (6) the pitch of the note then doubles, producing an *octave*. This style of fingering, called "playing harmonics," is used frequently in musical performance.

*Note that the separation-of-variable computations for analyzing a system operating in a single mode are considerably reduced, since the final expansion step is unnecessary.*

An important characteristic of any engineering device is the set of frequencies supported by its eigenmodes - its "eigenfrequencies." According to eq. (6) the eigenfrequencies of the vibrating string are the harmonics

$$\omega_n = \frac{n\pi}{a}, \quad n = 1, 2, 3, \dots \tag{7}$$

To focus our study on the eigenmodes, we shall impose homogeneous boundary conditions (thereby obviating $\Psi_\infty$) in the subsequent examples.

**Example 2.** Sound waves inside a resonant rectangular cavity (Fig. 5) are governed by the three dimensional wave equation

$$\frac{\partial^2 \Psi}{\partial t^2} = \nabla^2 \Psi,$$

where $\Psi(x, y, z, t)$ represents the incremental pressure. If the box has rigid walls the boundary conditions are homogeneous Neumann conditions

$$\frac{\partial \Psi}{\partial x} = 0 \quad \text{at} \quad x = 0 \quad \text{and} \quad x = a,$$

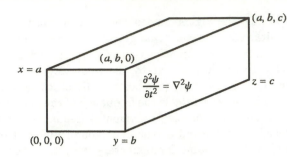

***Figure 5*** *Rectangular cavity*

$$\frac{\partial \Psi}{\partial y} = 0 \quad \text{at} \quad y = 0 \quad \text{and} \quad y = b,$$

and

$$\frac{\partial \Psi}{\partial z} = 0 \quad \text{at} \quad z = 0 \quad \text{and} \quad z = c. \tag{8}$$

To make the problem complete we impose the initial conditions

$$\Psi = 0 \text{ at } t = 0, \quad \frac{\partial \Psi}{\partial t} = g(x, y, z) \text{ at } t = 0. \tag{9}$$

The separation of the wave equation in three-dimensional cartesian coordinates is derived in Sec. 5.3; the single-variable factors satisfy

$$\frac{T''}{T} = \frac{X''}{X} + \frac{Y''}{Y} + \frac{Z''}{Z}. \tag{10}$$

Separation of the three spacial coordinates leads to the Sturm-Liouville problems

$$\frac{X''}{X} = \lambda, \qquad X'(0) = X'(a) = 0;$$

$$\frac{Y''}{Y} = \mu, \qquad Y'(0) = Y'(b) = 0;$$

$$\frac{Z''}{Z} = \nu, \qquad Z'(0) = Z'(c) = 0; \tag{11}$$

and the remaining ordinary differential equation is

$$\frac{T''}{T} = \lambda + \mu + \nu \tag{12}$$

The Sturm-Liouville problems all generate cosine series with eigenfunction-eigenvalues given by

$$\begin{aligned}
X_m(x) &= \cos \frac{m\pi}{a} x, \quad \lambda_m = -\left(\frac{m\pi}{a}\right)^2, \ m = 0, 1, 2, \ldots; \\
Y_n(y) &= \cos \frac{n\pi}{b} y, \quad \mu_n = -\left(\frac{n\pi}{b}\right)^2, \ n = 0, 1, 2, \ldots; \\
Z_p(z) &= \cos \frac{p\pi}{c} z, \quad \nu_p = -\left(\frac{p\pi}{c}\right)^2, \ p = 0, 1, 2, \ldots.
\end{aligned}$$

Equation (12) has the solution

$$T_{mnp}(t) = C_1 \cos \sqrt{\left(\frac{m\pi}{a}\right)^2 + \left(\frac{n\pi}{b}\right)^2 + \left(\frac{p\pi}{c}\right)^2}\, t$$

$$+ C_2 \sin \sqrt{\left(\frac{m\pi}{a}\right)^2 + \left(\frac{n\pi}{b}\right)^2 + \left(\frac{p\pi}{c}\right)^2}\, t \qquad (13)$$

(or $C_1 + C_2 t$ if $m = n = p = 0$) and we can enforce the first (homogeneous) initial condition (9) by setting $C_1 = 0$. The assembled general solution for $\Psi(x, y, z, t)$ is

$$\sum_{m=0}^{\infty} \sum_{n=0}^{\infty} \sum_{p=0}^{\infty} B_{mnp} \cos \frac{m\pi}{a} x \cos \frac{n\pi}{b} y \cos \frac{p\pi}{c} z$$

$$\times \left\{ \sin \sqrt{\left(\frac{m\pi}{a}\right)^2 + \left(\frac{n\pi}{b}\right)^2 + \left(\frac{p\pi}{c}\right)^2}\, t + B_{000} t \right\},$$

$$(mpn) \neq (0, 0, 0), \qquad (14)$$

and orthogonality yields the expressions for the coefficients in terms of the initial data $g(x, y, z)$ (eq. (9)):

$$B_{mnp} = \frac{2^3 \displaystyle\int_0^a \int_0^b \int_0^c g(x, y, z) \cos \frac{m\pi}{a} x \cos \frac{n\pi}{b} y \cos \frac{p\pi}{c} z \, dx\, dy\, dz}{abc \sqrt{\left(\frac{m\pi}{a}\right)^2 + \left(\frac{n\pi}{b}\right)^2 + \left(\frac{p\pi}{c}\right)^2}}, \qquad (15)$$

for $m, n, p > 0$; for $B_{0np}$, $B_{m0p}$, or $B_{mn0}$ use one-half of (15); for $B_{00p}$, $B_{0n0}$, or $B_{m00}$ use one-fourth; and for $B_{000}$ use $\int_0^a \int_0^b \int_0^c g(x, y, z)\, dz\, dy\, dx / abc$ (exercise 2). (The $B_{000}$ mode is not a true sound wave.)

Our interpretation of (14) is that any initial disturbance inside the cavity is decomposed into the basic threefold family of eigenmodes through eq. (15), and each mode vibrates as a standing wave at the frequency $\omega$ assigned to that mode by eq. (13):

$$\omega_{mnp} = \sqrt{\left(\frac{m\pi}{a}\right)^2 + \left(\frac{n\pi}{b}\right)^2 + \left(\frac{p\pi}{c}\right)^2}. \qquad (16)$$

Only these discrete frequencies of vibration are permitted inside the cavity.

■

The allowed frequencies are displayed in Fig. 6 for cavities of various dimensions. Allowed frequencies for the one dimensional vibrating string (and for some subsequent electromagnetic examples) are also shown for comparison. Note the difference between the patterns. The string frequencies form a *harmonic* series, i.e. they are all multiples of a *fundamental* frequency.

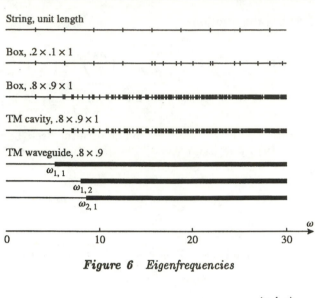

**Figure 6** *Eigenfrequencies*

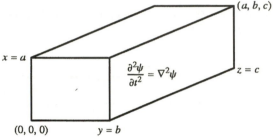

**Figure 7** *Rectangular cavity*

The cavity frequencies, on the other hand, are incommensurable, and they even overlap occasionally. However the lower frequencies of a *long thin* cavity are nearly regular.

Musically the *harmonic* patterns are pleasing to the ear, while the superposition of incommensurable frequencies sounds noisy. Organ pipes and wind instruments are designed to be long and thin because of this feature.

**Example 3.** Transverse magnetic ("TM") waves inside a resonant rectangular cavity are governed by a set of equations very similar to those for the sound waves of Example 2. The complete electromagnetic field is determined once the longitudinal component of the electric field **E** is known; see the references by Jackson, Kraus, or Panofsky and Phillips. For this example we take the $z$ component as longitudinal in Fig. 7, and we denote $E_z$ by $\Psi(x, y, z, t)$.

As a consequence of Maxwell's equations it can be shown (exercise 2, Sec. 4.4) that $\Psi$ satisfies the wave equation

$$\frac{\partial^2 \Psi}{\partial t^2} = \nabla^2 \Psi \tag{17}$$

inside the cavity, while for perfectly conducting walls the boundary conditions are slightly different from those for sound waves:

$$\Psi = 0 \qquad \text{at } x = 0 \text{ and } x = a,$$
$$\Psi = 0 \qquad \text{at } y = 0 \text{ and } y = b,$$

and

$$\frac{\partial \Psi}{\partial z} = 0 \quad \text{at } z = 0 \text{ and } z = c. \tag{18}$$

For simplicity we take the same initial conditions as before, eqs. (9). Clearly separation of the spatial coordinates (in the fashion of eqs. (11) now leads to *sine* series for the $x$ and $y$ dependences and the cosine series for $z$:

$$X_m(x) = \sin\frac{m\pi}{a}x, \quad \lambda_m = -\left(\frac{m\pi}{a}\right)^2, \quad m = 1, 2, \dots;$$

$$Y_n(y) = \sin\frac{n\pi}{b}x, \quad \mu_n = -\left(\frac{n\pi}{b}\right)^2, \quad n = 1, 2, \dots;$$

$$Z_p(z) = \cos\frac{p\pi}{c}x, \quad \nu_m = -\left(\frac{p\pi}{c}\right)^2, \quad p = 0, 1, 2, \dots.$$

The time factor remains the same, so the solution for initial conditions like eqs. (9) is

$$\Psi(x, y, z, t) = \sum_{m=1}^{\infty}\sum_{n=1}^{\infty}\sum_{p=0}^{\infty} B_{mnp} \sin\frac{m\pi}{a}x \sin\frac{n\pi}{b}y \cos\frac{p\pi}{c}z$$

$$\times \sin\sqrt{\left(\frac{m\pi}{a}\right)^2 + \left(\frac{n\pi}{b}\right)^2 + \left(\frac{p\pi}{c}\right)^2}\, t,$$

$$B_{mnp} = \frac{2^3 \int_0^a \int_0^b \int_0^c g(x, y, z) \sin\frac{m\pi}{a}x \sin\frac{n\pi}{b}y \cos\frac{p\pi}{c}z\, dz\, dy\, dx}{abc\sqrt{\left(\frac{m\pi}{a}\right)^2 + \left(\frac{n\pi}{b}\right)^2 + \left(\frac{p\pi}{c}\right)^2}}, \tag{19}$$

for $p > 0$; for $B_{mn0}$ use one-half of (19).

(The $m = 0$, $n = 0$ modes are not supported by transverse magnetic waves; observe their omission in Fig. 6.) ∎

**Example 4.** The resonant cavity of Example 3 becomes a "waveguide" if it is extended indefinitely in the $z$ direction (Fig. 8).

There are two different ways of analyzing the behavior of a waveguide. One can ask either

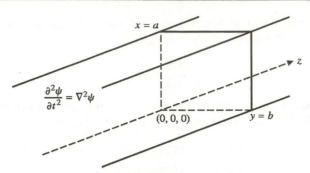

*Figure 8   Rectangular waveguide*

1. how does an electromagnetic field existing inside the guide at $t = 0$ evolve in time, or

2. how does an electromagnetic disturbance at one end of the guide propagate down the tube?

Clearly these two questions involve distinct initial-boundary conditions, and the mathematical description of each solution will be different. Here we shall use time domain methodology to analyze question 1. Question 2 involves a boundary value problem with time dependent conditions imposed on a spatial boundary, and we defer its treatment until Sec. 8.2. At that time we shall compare the physical significance of the two descriptions.

In either case the function $E_z = \Psi(x, y, z, t)$ satisfies the wave equation

$$\frac{\partial^2 \Psi}{\partial t^2} = \nabla^2 \Psi,$$

and the boundary conditions for perfectly conducting walls

$$\Psi = 0 \quad \text{at} \quad x = 0, \quad x = a, \quad y = 0, \quad \text{and} \quad y = b.$$

We take the waveguide to be doubly infinite as in Fig. 8, so that[1] finiteness conditions prevail at $z = \pm\infty$:

$$\Psi(x, y, \infty, t) \text{ finite}, \quad \Psi(x, y, -\infty, t) \text{ finite}.$$

For initial conditions we again assume $\Psi = 0$, $\partial\Psi/\partial t = g(x, y, z)$ (eqs. (9)).

The separated Sturm-Liouville problems for $x$ and $y$ yield the same sine functions as in the previous example, but the $z$ equation is now

$$\frac{Z''}{Z} = \nu, \quad Z(\pm\infty) \quad \text{finite}, \tag{20}$$

(compare (11)). Equation (20) generates the Fourier integral, with a continuum of eigenvalues (Sec. 6.4 and Table 6.3 #7):

$$Z_k(z) = e^{ikz}, \quad \lambda_k = -k^2, \quad -\infty < k < \infty.$$

---

[1]The analysis of the waveguide closed at one end is very similar; see exercise 5.

Thus the time dependence for these modes (recall (12)) is given by

$$T_{mnk}(t) = \sin \sqrt{\left(\frac{m\pi}{a}\right)^2 + \left(\frac{n\pi}{b}\right)^2 + k^2}\, t.$$

The solution is

$$\Psi(x,y,z,t) = \sum_{m=1}^{\infty} \sum_{n=1}^{\infty} \int_{-\infty}^{\infty} A_{mn}(k) \sin \frac{m\pi}{a}x \sin \frac{n\pi}{b}y\, e^{ikz}$$

$$\times \sin \sqrt{\left(\frac{m\pi}{a}\right)^2 + \left(\frac{n\pi}{b}\right)^2 + k^2}\, t\, dk, \tag{21}$$

with the coefficients given by

$$A_{mn}(k) = \frac{2 \int_0^a \int_0^b \int_{-\infty}^{\infty} g(x,y,z) \sin \frac{m\pi}{a}x \sin \frac{n\pi}{b}y\, e^{-ikz}\, dz\, dy\, dx}{\pi ab \sqrt{\left(\frac{m\pi}{a}\right)^2 + \left(\frac{n\pi}{b}\right)^2 + k^2}}. \tag{22}$$

Like the resonant cavity, the waveguide breaks up the initial disturbance into modes through eq. (22), and creates standing waves (21) by assigning each mode a frequency of vibration in accordance with

$$\omega_{mnk} = \sqrt{\left(\frac{m\pi}{a}\right)^2 + \left(\frac{n\pi}{b}\right)^2 + k^2}$$
$$(m = 1, 2, ...; \ n = 1, 2, ...; \ -\infty < k < \infty). \tag{23}$$

The waveguide differs from the cavity in that the waveguide eigenfrequencies are characterized by two - instead of three - *discrete* parameters, and a *continuous* parameter ($k$). This leads to a denser frequency graph as depicted in Fig. 6. *The continuum waveguide eigenfrequencies can be regarded as the limiting form of the discrete cavity eigenfrequencies, as the length of the cavity increases.*

Because $k$ can take *any* value the eigenfrequency overlapping for the waveguide is much more severe than for the cavity, and it is customary in displaying the waveguide frequencies to segregate them into "waveguide modes," corresponding to distinct values of the discrete subscripts $m$ and $n$. See Fig. 6. Note that strictly speaking a "waveguide mode" is a *family* of modes; the value of $k$ must also be specified to define a single mode. However, the $X(x)Y(y)$ factors for any mode are independent of $k$.

From (23) we see that each waveguide mode supports electromagnetic vibrations for a continuum of frequencies bounded below by the "cut-off" frequency for that mode, given by

$$\omega_{mn} = \sqrt{\left(\frac{m\pi}{a}\right)^2 + \left(\frac{n\pi}{b}\right)^2}. \tag{24}$$

In particular, then, *the waveguide does not support any transverse magnetic standing waves at frequencies below*

$$\omega_{11} = \sqrt{\left(\frac{\pi}{a}\right)^2 + \left(\frac{\pi}{b}\right)^2}.$$

■

An alternative description of the waveguide modes is revealed by writing the time factor in eq. (21) in exponential notation:

$$
\begin{aligned}
\Psi &= \sum_{m=1}^{\infty} \sum_{n=1}^{\infty} \int_{-\infty}^{\infty} d_{mn} \sin \frac{m\pi}{a}x \sin \frac{n\pi}{b}y \, e^{ikz} \\
&\quad \times \frac{\left[\exp(i\omega_{mnk}t) - \exp(-i\omega_{mnk}t)\right] dk}{2i} \\
&= \frac{1}{2i} \sum_{m=1}^{\infty} \sum_{n=1}^{\infty} \int_{-\infty}^{\infty} d_{mn}(k) \sin \frac{m\pi}{a}x \\
&\quad \times \sin \frac{n\pi}{b}y \, \exp[i(kz + \omega_{mnk})t)] \, dk \\
&\quad - \frac{1}{2i} \sum_{m=1}^{\infty} \sum_{n=1}^{\infty} \int_{-\infty}^{\infty} d_{mn}(k) \sin \frac{m\pi}{a}x \\
&\quad \times \sin \frac{n\pi}{b}y \, \exp[i(kz - \omega_{mnk}t)] \, dk.
\end{aligned}
\tag{25}
$$

Recall from Sec. 4.3 that any function of the composite variable $(z - vt)$ can be visualized as a waveform propagating in the $z$ direction at the speed $v$. Since $\exp[i(kz \pm \omega_{mnk}t)] = \exp[ik(z \pm \omega_{mnk}t/k)]$ is of this form, for fixed $x$ and $y$ the integrands in (25) are (complex) sinusoidal waves traveling along the length of the tube; the wave for the $(m, n, k)$th mode propagates at a speed given by

$$|v| = \omega_{mnk}/|k| = \sqrt{1 + \omega_{mn}^2/k^2}. \tag{26}$$

The waves in the first sum in (25) propagate to the right (in Fig. 8) for negative $k$ and to the left for positive $k$; the opposite is true for the second sum.

**Example 5.** Now we are going to remove the walls of the waveguide and address solutions of the wave equation in "free space." Since there are no boundaries, we impose finiteness conditions at $\pm\infty$ in each direction. Equations (9) will again serve to define the initial conditions. The separated equations (11), (12) become modified to

$$\frac{X''}{X} = \lambda, \quad \frac{Y''}{Y} = \mu, \quad \frac{Z''}{Z} = \nu;$$

$$X(\pm\infty), \ Y(\pm\infty), \ Z(\pm\infty) \text{ bounded}; \quad \frac{T''}{T} = \lambda + \mu + \nu.$$

According to Table 6.3 #7 the eigenfunction expansions are all Fourier integrals. Thus we write the individual factors as

$$
\begin{aligned}
X_{k_1}(x) &= e^{ik_1 x}, \quad \lambda = -k_1^2, \ -\infty < k_1 < \infty; \\
Y_{k_2}(x) &= e^{ik_2 y}, \quad \mu = -k_2^2, \ -\infty < k_2 < \infty; \\
Z_{k_3}(x) &= e^{ik_3 y}, \quad \nu = -k_3^2, \ -\infty < k_3 < \infty; \\
T_{k_1 k_2 k_3}(t) &= \sin \sqrt{k_1^2 + k_2^2 + k_3^2}\, t.
\end{aligned}
$$

The general solution is expressed

$$
\begin{aligned}
\Psi(x, y, z, t) = \int_{-\infty}^{\infty} \int_{-\infty}^{\infty} \int_{-\infty}^{\infty} & A(k_1, k_2, k_3) e^{ik_1 x}\, e^{ik_2 y}\, e^{ik_3 z} \\
& \times \sin \sqrt{k_1^2 + k_2^2 + k_3^2}\, t \, dk_1 \, dk_2 \, dk_3
\end{aligned}
\tag{27}
$$

with $A$ determined by the initial data (9):

$$
\begin{aligned}
A(k_1, k_2, k_3) = \frac{1}{8\pi^3} \int_{-\infty}^{\infty} \int_{-\infty}^{\infty} \int_{-\infty}^{\infty} & \frac{g(x, y, z)}{\sqrt{k_1^2 + k_2^2 + k_3^2}} \\
& \times e^{-i(k_1 x + k_2 y + k_3 z)} dx \, dy \, dz.
\end{aligned}
\tag{28}
$$

Equation (28) decomposes the initial field into the sinusoidal eigenmodes, and eq. (27) assigns the sinusoidal time factor with frequency $\omega_{k_1 k_2 k_3} = \sqrt{k_1^2 + k_2^2 + k_3^2}$ to the $(k_1, k_2, k_3)$th mode.

If we define the *wave vector* $\mathbf{K}$ to be $k_1 \mathbf{i} + k_2 \mathbf{j} + k_3 \mathbf{k}$ then the general solution can be more compactly expressed with vector notation:

$$\Psi(\mathbf{r}, t) = \int_{-\infty}^{\infty} \int_{-\infty}^{\infty} \int_{-\infty}^{\infty} A(\mathbf{K})\, e^{i\mathbf{K} \cdot \mathbf{r}} \sin |\mathbf{K}| t \, dk_1 \, dk_2 \, dk_3$$

where, as usual, $\mathbf{r} = x\mathbf{i} + y\mathbf{j} + z\mathbf{k}$. Further insight is offered by rewriting the time factor as an exponential:

$$
\begin{aligned}
\Psi(\mathbf{r}, t) = & \int_{-\infty}^{\infty} \int_{-\infty}^{\infty} \int_{-\infty}^{\infty} \frac{A(\mathbf{K})}{2i} e^{i(\mathbf{K} \cdot \mathbf{r} + \omega t)} dk_1 \, dk_2 \, dk_3 \\
& - \int_{-\infty}^{\infty} \int_{-\infty}^{\infty} \int_{-\infty}^{\infty} \frac{A(\mathbf{K})}{2i} e^{i(\mathbf{K} \cdot \mathbf{r} - \omega t)} dk_1 \, dk_2 \, dk_3,
\end{aligned}
\tag{29}
$$

where $\omega = |\mathbf{K}|$. The individual eigenmode $e^{i(\mathbf{K}\cdot\mathbf{r}\pm\omega t)}$ in (29) is called a *plane wave*, because at any fixed time $t$ it is constant in planes perpendicular to the wave vector $\mathbf{K}$ (recall $\mathbf{K}\cdot\mathbf{r}$ depends only on the component of $\mathbf{r}$ along $\mathbf{K}$).  ∎

The nature of the eigenmodes in (29) is revealed if we consider the case where $\mathbf{K}$ points along the $z$ axis so that $\mathbf{K} = |\mathbf{K}|\mathbf{k} = \omega\mathbf{k}$. The mode function takes the form

$$e^{i(\mathbf{K}\cdot\mathbf{r}\pm\omega t)} = e^{i(|\mathbf{K}|z\pm\omega t)} = e^{i\omega(z\pm t)},$$

which is a (complex) sine wave, independent of $x$ and $y$, and propagating in the $z$ direction at unit speed ($\omega/|\mathbf{K}| = 1$ in these units[2]). More generally, the eigenmode $e^{i(\mathbf{K}\cdot\mathbf{r}+\omega t)}$ propagates in the $-\mathbf{K}$ direction at unit speed, since a time shift $\Delta t$ can be compensated with a shift in $\mathbf{r}$ by an amount $\Delta t$ in the $-\mathbf{K}$ direction.

Thus eq. (29) represents $\Psi(\mathbf{r}, \mathbf{t})$ as a superposition of plane waves, propagating in all directions. The representation (27), on the other hand, expresses $\Psi$ in terms of standing waves.

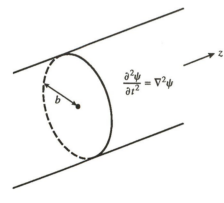

**Figure 9**  *Cylindrical waveguide*

**Example 6.** The equations governing transverse magnetic waves propagating down a *cylindrical* waveguide (Fig. 9) are identical with those for the rectangular guide, except for the idiosyncracies of the cylindrical coordinate system:

$$\frac{\partial^2 \Psi}{\partial t^2} = \nabla^2 \Psi,$$

$\Psi(b, \theta, z, t) = 0$ (for a perfectly conducting wall),

$\Psi(\rho, \theta, z, t)$ finite at $\rho = 0$, $z = \infty$, and $z = -\infty$.

For initial conditions we take a cylindrical coordinate version of (29):

$$\Psi(\rho, \theta, z, 0) = 0, \quad \frac{\partial \Psi}{\partial t}(\rho, \theta, z, 0) = g(\rho, \theta, z). \tag{30}$$

Section 5.3 demonstrates that the separation of the wave equation in cylindrical coordinates results in the ordinary differential equations

$$\frac{\theta''}{\theta} = \lambda, \quad \frac{(\rho R')'}{\rho R} + \frac{\lambda}{\rho^2} = \mu, \quad \frac{Z''}{Z} = \nu, \quad \frac{T''}{T} = \mu + \nu. \tag{31}$$

The periodicity condition thus generates the usual Fourier series for $\Theta$:

$$\Theta_n(\theta) = a_n \cos n\theta + b_n \sin n\theta, \quad \lambda_n = -n^2, \quad n = 0, 1, 2, \dots. \tag{32}$$

---

[2]Recall that at the end of Sec. 4.4 we adopted the convention that the speed of propagation, $v$ in eq. (32) of that section, would be unity. In exercise 15 you will be invited to restore the parameter $v$ in these solution formulas.

The conditions $R(0)$ finite, $R(b) = 0$ result in the Fourier-Bessel series (Table 6.3 #16) with eigenfunctions

$$R_p^{(n)}(\rho) = J_n(j_{n,p}\rho/b), \quad \mu_p^{(n)} = -j_{n,p}^2/b^2, \quad p = 1, 2, 3 \ldots. \quad (33)$$

And for the $z$ dependence we get the Fourier transform

$$Z_k(z) = e^{ikz}, \quad \nu_k = -k^2, \quad -\infty < k < \infty. \quad (34)$$

Finally we have for the time factors

$$\frac{T''}{T} = -j_{n,p}^2/b^2 - k^2,$$

and thanks to the initial conditions (30), the solutions are

$$T_{n,p,k} = \sin \omega_{npk} t, \quad \omega_{npk} = \sqrt{j_{n,p}^2/b^2 + k^2}. \quad (35)$$

Initial disturbances are decomposed into these modes and assigned frequencies according to (35). Like the rectangular waveguide, the eigenfrequencies are indexed by two discrete parameters $(n, p)$ and one continuous parameter $(k)$. The cutoff frequency for the $(n, p)$th waveguide mode is

$$\omega_{np} = j_{n,p}/b.$$

We leave the derivation of the expansion formulas as exercise 6.  ∎

**Example 7.** For the *coaxial* waveguide operated in a transverse magnetic mode[3] (Fig. 10) the finiteness condition at the origin is replaced by vanishing of the electric field at the inner wall: the equations become

$$\frac{\partial^2 \Psi}{\partial t^2} = \nabla^2 \Psi,$$

$$\Psi(b, \theta, z, t) = \Psi(a, \theta, z, t) = 0$$

$$\Psi(\rho, \theta, z, t) \text{ finite at } z = \pm\infty,$$

$$\Psi(\rho, \theta, z, 0) = 0, \quad \frac{\partial \Psi}{\partial t}(\rho, \theta, z, 0) = g(\rho, \theta, z).$$

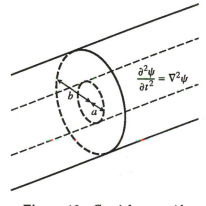

**Figure 10**   *Coaxial waveguide*

The only difference between this and the previous example is the inner boundary condition on the radial factor:

$$\frac{(\rho R')'}{\rho R} - \frac{n^2}{\rho^2} = \text{constant} = -\mu, \quad R(a) = R(b) = 0.$$

---

[3]Coaxial waveguides are customarily operated in a transverse electromagnetic mode, where both the electric and magnetic fields are transverse to the $z$ axis. However the transverse magnetic modes can be present. See the references by Kraus and Carver; Ramo, Whinnery, and van Duzer.

This Sturm-Liouville problem was analyzed in Sec. 6.3, and the eigenfunctions were found to be (see also Table 6.3 #4)

$$R_{np}(\rho; \mu_{np}) = Y_n(\sqrt{\mu_{np}}a)J_n(\sqrt{\mu_{np}}\rho) - J_n(\sqrt{\mu_{np}}a)Y_n(\sqrt{\mu_{np}}\rho),$$

where the eigenvalues $\mu_{np}$ are the roots of $R_{np}(b; \mu) = 0$ (Figs. 7, 8 of Sec. 6.3). Exercise 7 demonstrates that the eigenfrequencies are the numbers

$$\omega_{npk} = \sqrt{\mu_{n,p} + k^2}, \tag{36}$$

with the cutoff frequency for the $(n, p)th$ waveguide mode given by

$$\omega_{np} = \sqrt{\mu_{np}}. \qquad\blacksquare$$

**Example 8.** Sound waves inside a resonant spherical cavity with a "soft" wall are governed by the wave equation in spherical coordinates:

$$\frac{\partial^2 \Psi}{\partial t^2} = \nabla^2 \Psi = \frac{1}{r^2}\frac{\partial}{\partial r}\left(r^2 \frac{\partial \Psi}{\partial r}\right) + \frac{1}{r^2 \sin\phi}\frac{\partial}{\partial \phi}\left(\sin\phi \frac{\partial \Psi}{\partial \phi}\right) + \frac{1}{r^2 \sin^2\phi}\frac{\partial^2 \Psi}{\partial \theta^2} \tag{37}$$

The boundary condition for a soft wall is

$$\Psi(b, \phi, \theta, t) = 0. \tag{38}$$

The spherical version of our initial conditions (9) reads

$$\Psi = 0, \quad \frac{\partial \Psi}{\partial t} = g(r, \phi, \theta) \quad \text{at} \quad t = 0. \tag{39}$$

The separation of eq. (37) is almost exactly like that for the spherical heat equation in the previous section (Example 4); however the equation for the time factor $T' = KT$ is replaced by (see Sec. 5.3) $T'' = KT$, where $K$ is the eigenvalue for the radial equation

$$r^2 R'' + 2rR' - l(l+1)R = Kr^2 R. \tag{40}$$

Imitating the derivation in Sec. 6.9 (see also Table 6.3 #17) we obtain the spatial eigenfunctions

$$j_l(s_{l,p}r/b)Y_{lm}(\phi, \theta), \quad K_{l,p} = -s_{l,p}^2/b^2,$$

where $s_{l,p}$ is the $p$th positive zero of $j_l(x)$. The time factor is thus independent of $m$:

$$T_{lmp}(t) = \sin(s_{l,p}t/b),$$

and the assembled general solution is

$$\Psi(r, \phi, \theta, t) = \sum_{l=0}^{\infty} \sum_{m=-l}^{l} \sum_{p=1}^{\infty} B_{lmp}\, j_l(s_{l,p}r/b)Y_{lm}(\phi, \theta)\sin(s_{l,p}t/b),$$

where (exercise 8)

$$B_{lmp} = \frac{2 \int_0^b \int_0^\pi \int_0^{2\pi} g(r,\phi,\theta) \overline{Y_{lm}(\phi,\theta)} \, j_l(s_{l,p} r/b) \, r^2 \sin\phi \, d\theta \, d\phi \, dr}{b^2 j_{l+1}^2(s_{l,p}) s_{l,p}}. \quad (41)$$

The modal frequencies supported by this cavity are given by the numbers

$$\omega_{lmp} = s_{l,p}/b \quad (l = 0,1,2,\ldots; -l \le m \le l; \ p = 1,2,3,\ldots).$$

(There is considerable overlapping of the eigenfrequencies since $\omega_{lmp}$ is independent of $m$.) ∎

**Example 9.** To obtain another decomposition (alternative to Example 5, that is) of the solutions to the wave equation in free space we model the latter as an infinite sphere. Thus we replace the boundary condition (38) in the previous example by a finiteness condition at infinity. Equation (40) then generates the Spherical Hankel Transform, according to Table 6.3 #26, with eigenfunctions $j_l(sr)$ and a continuum of eigenvalues $K = -s^2$ ($0 \le s < \infty$). The general solution for the initial conditions (39) takes the form

$$\Psi(r,\theta,\phi,t) = \sum_{l=0}^\infty \sum_{m=-l}^l \int_{s=0}^\infty B_{lm}(s) j_l(sr) Y_{lm}(\phi,\theta) \sin st \, ds \quad (42)$$

with coefficients given by

$$B_{lm}(s) = \frac{2}{\pi} s \int_0^\infty \int_0^\pi \int_0^{2\pi} g(r,\theta,\phi) \, j_l(sr) \overline{Y_{lm}(\phi,\theta)} \, r^2 \sin\phi \, d\theta \, d\phi \, dr \quad (43)$$

(exercise 10). ∎

Equation (42) expresses $\Psi$ in terms of standing waves. The traveling-wave version of the solution can be written using *spherical Hankel functions*. These are defined by the equations

$$h_l^{(1)}(\xi) = j_l(\xi) + i y_l(\xi), \quad h_l^{(2)}(\xi) = j_l(\xi) - i y_l(\xi).$$

Thus we derive

$$j_l(sr) = \{h_l^{(1)}(sr) + h_l^{(2)}(sr)\}/2 \quad (44)$$

and the asymptotic approximations for large $r$:

$$h_l^{(1)}(sr) \approx i^{-l-1} e^{isr}/sr, \quad h_l^{(2)}(sr) \approx i^{-l-1} e^{-isr}/sr \quad (45)$$

(exercise 11).

Using (44) and the exponential form of the time factors $\sin st = [e^{ist} - e^{-ist}]/2i$ we judiciously reorganize (42) as

$$\begin{aligned}
\Psi(r,\theta,\phi,t) &= \sum_{l=0}^\infty \sum_{m=-l}^l \int_{s=0}^\infty \frac{B_{lm}(s)}{4i} Y_{lm}(\phi,\theta) \\
&\times \{h_l^{(1)}(sr)e^{ist} - h_l^{(2)}(sr)e^{-ist} \\
&- h_l^{(1)}(sr)e^{-ist} + h_l^{(2)}(sr)e^{ist}\} ds. \quad (46)
\end{aligned}$$

Note that for large $r$ the approximate form of

$$h_l^{(1)}(sr)e^{-ist} \approx i^{-l-1}e^{is(r-t)}/sr$$

is a function of $(r - t)$, divided by $r$. Therefore it can be interpreted as a traveling wave propagating outward (at unit speed) in the **radial** direction *with magnitude diminishing as* $1/r$. The term $h_l^{(2)}(sr)e^{ist}$ is also an outgoing radial wave, and the terms $h_l^{(1)}(sr)e^{ist}$ and $h_l^{(2)}(sr)e^{-ist}$ are incoming radial waves, all "fading" like $1/r$.

Therefore in a certain sense the expression (39) represents $\Psi$ as a superposition of traveling radial waves.[4] Note, however, that this interpretation is flawed. The individual incoming and outgoing waves are divergent at the origin (one can show

$$|h_l^{(1,2)}(sr)| \approx \frac{(2l)!}{2^l l!}(sr)^{-l-1}$$

for small $r$). Thus an incoming wave cannot exist alone; it needs an outgoing companion to cancel the infinity at the origin (as in eq. (44)).

In Examples 5 and 9 we have exhibited two very different decompositions of the free-space wave equation solutions. A cylindrical coordinate version appears in exercise 13.

# Exercises 7.3

1. Work out the eigenmode expansion coefficients for the picked guitar string (see Figs. 4). How does the intensity of the fourth harmonic vary with the location of the point of application of the pick? The relevant sine series expansion can be extracted from Fig. 5, Sec. 3.3.

2. Verify formula (15) and the modifications stated directly after it.

3. Solve the wave equation in a rectangle with dimensions $(a, b)$ and homogeneous Dirichlet conditions on the edges. What are the eigenfrequencies?

4. Solve the wave equation in an infinite rectangular strip with homogeneous Dirichlet conditions on the edges. Compare the eigenfrequencies with those in exercise 3.

5. Work out the eigenfunctions for transverse magnetic waves in a waveguide which is *semiinfinite* ($z \geq 0$); the boundary condition on the face $z = 0$ is $\partial \Psi / \partial z = 0$.

---

[4]The waves are not, however, spherically uniform, because the factors $Y_{lm}(\phi, \theta)$ vary around the sphere (except for $Y_{00}$: see Table 7.1).

6. Work out the expansion formulas for the complete solution to the initial value problem (30) for the cylindrical waveguide.

7. Work out the general solution for transverse magnetic waves in a coaxial waveguide and verify eq. (36).

8. Verify eq. (41).

9. Consider the one-dimensional wave equation with damping:

$$\frac{\partial^2 \Psi}{\partial t^2} + \eta \frac{\partial \Psi}{\partial t} = \frac{\partial^2 \Psi}{\partial x^2}. \tag{47}$$

Show that if the damping coefficient $\eta < 2\pi$, then the general solution to (47) with homogeneous Dirichlet endpoint conditions at $x = 0, 1$ is

$$\Psi(x,t) = \sum_{n=1}^{\infty} \sin n\pi x \, e^{-\eta t/2} [a_n \sin \sqrt{n^2 \pi^2 - \eta^2/4} \, t$$
$$+ \quad b_n \cos \sqrt{n^2 \pi^2 - \eta^2/4} \, t].$$

What are the formulas for $a_n, b_n$ if the initial conditions are

$$\Psi(x,0) = f(x), \quad \frac{\partial \Psi}{\partial t}(x,0) = g(x)?$$

How do these formulas change if, say, $6\pi < \eta < 8\pi$?

10. Derive eqs. (42, 43).

11. Derive eqs. (44, 45) (see Table 2.2).

12. Rework Examples 2 through 9 with the initial conditions $\Psi = f$, $\partial \Psi / \partial t = 0$ at $t = 0$ instead of eq. (9). Identify the traveling waves.

13. (a) Imagine free space as an infinitely large cylinder, and derive the following expansion for solutions to the wave equation with initial conditions given by (30):

$$\Psi(\rho, \theta, z, t) =$$
$$\int_0^\infty \int_{-\infty}^\infty \sum_{n=0}^\infty [a_n(k,s) \cos n\theta + b_n(k,s) \sin n\theta]$$
$$\times e^{ikz} J_n(s\rho) \sin \sqrt{s^2 + k^2} \, t \, dk \, ds. \tag{48}$$

Express the coefficients $a$ and $b$ in terms of the initial data.

(b) The substitution (Table 2.2) $J_n(s\rho) = \{H_n^{(1)}(s\rho) + H_n^{(2)}(s\rho)\}/2$ and the exponential form of the time factors leads to a traveling-wave representation. For large $\rho$ we have

$$e^{ikz} H_n^{(1)}(s\rho) \, e^{-i\sqrt{s^2+k^2} t} \approx \frac{1-i}{\sqrt{\pi s \rho}} (-i)^n e^{i(kz+s\rho-\sqrt{s^2+k^2} t)},$$

which can be interpreted as a wave propagating out from the $z$ axis in the direction $k\mathbf{k} + s\mathbf{u}_\rho$, where $\mathbf{u}_\rho$ is a unit vector in the direction of $\rho$. Work out the traveling-wave version of (48).

14. Work out the modal frequencies for the spherical cavity (Example 8) with rigid walls ($\partial \Psi / \partial r = 0$ at $r = b$).

15. At the end of Sec. 4.4 we adopted the convention that the speed of propagation, $v$ in eq. (32) of that section, would be unity. Review the examples in the present section and restore this parameter in the solution formulas.

## 7.4 The Schrödinger Equation

**Summary** In classical mechanics, a particle cannot cross a barrier where the potential energy is higher than its total energy. However, the Schrödinger equation does permit such "tunneling;" its solutions are oscillatory in regions where the kinetic energy is positive, and nonoscillatory (but finite) in the forbidden regions. Square-well potentials give rise to wave functions which are piecewise sinusoidal or exponential. Constant force fields generate Airy function solutions. Mass-spring oscillator systems (the "quantum mechanical harmonic oscillator") has solutions which are Hermite polynomials tempered with Gaussians. The three-dimensional hydrogen atom wave functions are spherical harmonics times associated Laguerre polynomials times exponentials. The constant force field has a continuous spectrum; each of the others is discrete (in part, at least).

As we discussed in Sec. 4.7 the Schrödinger equation for a mechanical system is formed by writing down the Hamiltonian $H$ - which is the sum of the kinetic energy and potential energy of a particle - and replacing the kinetic energy therein with the operator - $h^2 \nabla^2 / 2m$, and then equating $H\Psi$ with $(ih)$ times the time derivative of the wave function $\Psi$. For one dimensional problems we have $\Psi = \Psi(x, t)$, and units can be chosen so that the Schrödinger equation reads

$$i\frac{\partial \Psi}{\partial t} = H\Psi = -\frac{\partial^2 \Psi}{\partial x^2} + V(x)\Psi, \tag{1}$$

with $V(x)$ as the potential energy.

The separated form $\Psi(x, t) = X(x)T(t)$ in (1) produces

$$i\frac{T'}{T} = \frac{HX}{X} = \frac{-X'' + V(x)X}{X} = \text{constant} = E \tag{2}$$

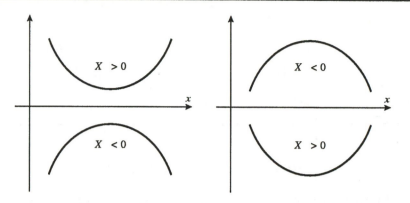

**Figure 1** *Convexity of interpretation of $X''(x)$*

where we have chosen to call the separation constant $E$ since it is an eigenvalue of the Hamiltonian, or *energy*, operator. The formulation of the Sturm-Liouville differential equation corresponding to (2) is

$$X'' - V(x)X = -EX, \tag{3}$$

so the separation constant $E$ is the negative of the eigenvalue $\lambda$ in the nomenclature of Sec. 6.2. Obviously the time factor associated with $X(x; E)$ via eq. (2) is

$$T(t) = e^{-iEt}. \tag{4}$$

If (3) is rewritten

$$X'' = -[E - V(x)]X \tag{5}$$

then we can predict the qualitative nature of $X$ by convexity arguments, similar to those in Sec. 6.2, based on the sign of $[E - V(x)]$. When the total energy $E$ is less than $V(x)$ then the sign of $X''$ is the same as that of $X$ and the graph of $X$ bends away from the axis (like the exponential functions; see Fig. 1(a)). Classically such points would be inaccessible to the particle, since the kinetic energy would have to be negative there.

When $E > V(x)$ the graph of $X$ bends towards the axis, and the solution is wavelike (Fig. 1(b)). In fact it oscillates faster when $E - V$ is greater. Thus faster oscillations correspond classically to higher kinetic energy. We shall observe this behavior in the examples.

**Example 1.** The potential function for the attractive square well is given by

$$V(x) = \begin{cases} 0 & \text{for } |x| < 1 \\ U & (= \text{constant} > 0) \text{ for } |x| > 1. \end{cases}$$

See Fig. 2.

The analysis of this problem is basically very simple, but the algebraic bookkeeping can be a nightmare unless it is organized systematically.

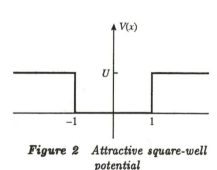

**Figure 2** *Attractive square-well potential*

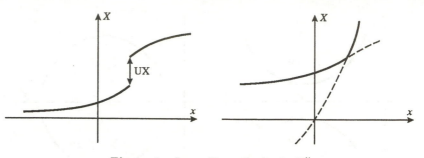

*Figure 3 Jump discontinuity in $X''$*

The separated differential equation, with the finiteness conditions, takes the form

$$X'' = -EX, \ -1 < x < 1; \ X'' = (U - E)X, \ |x| > 1; \ X \text{ finite at } \pm\infty. \tag{6}$$

First we must decide how to interpret the discontinuities in the differential equation. Equations (6) dictate that at $x = \pm 1, X''$ must undergo a *jump* discontinuity of size $UX$. Now Fig. 3 demonstrates that a jump discontinuity in $X''$ arises when $X'$ possesses a corner; *note, however, that $X'$ - as well as $X$ - is still continuous.* Therefore we attack this problem by solving the differential equation in the three regions $x < -1, -1 < x < 1$, and $x > 1$ separately, and then matching the expressions for $X$ and $X'$ at these points.

> *We do not interpret the continuity conditions at $\pm 1$ as boundary conditions*; the boundaries - which will determine the eigenvalues $\lambda$ - are at $\pm\infty$. The points $x = \pm 1$ are interior points where the differential equation happens to have discontinuous coefficients.[5]

Anticipating that most if not all of the eigenvalues will be negative, and hence $E > 0$, we begin by displaying a pair of independent solutions *in the central region* in the form

$$X_1(x; E) = \cos\sqrt{E}x, \ X_2(x; E) = \sin\sqrt{E}x, \ -1 < x < 1. \tag{7}$$

If the central solution $X_1$ is extended into the right hand region $x > 1$, clearly it must be expressible in terms of the general solution for that region. For convenience in applying the continuity conditions we express this general solution (for $x > 1$) in terms of the shifted variable $(x - 1)$:

$$X_1(x; E) = a_1 \cos\sqrt{E - U}(x - 1) + a_2 \sin\sqrt{E - U}(x - 1). \tag{8}$$

---

[5]Many other books *do* regard continuity conditions as boundary conditions. By distinguishing between the two, this author feels that the determination of the eigenvalues becomes more transparent.

Now it is easy to see that (8) will match $\cos\sqrt{E}x$ in value and first derivative at $x = 1$ if

$$a_1 = \cos\sqrt{E}, \quad a_2 = -\frac{\sqrt{E}}{\sqrt{E-U}}\sin\sqrt{E}. \tag{9}$$

Following this same strategy for the other discontinuity, and for the solution $X_2$, will lead us to the *general* solution to the differential equation in (6). Before we write out this general solution it is convenient to introduce a more compact notation. We define

$$k = \sqrt{E}, \quad k' = \sqrt{E-U}. \tag{10}$$

The general solution then can be written (exercise 1)

$$X(x;E) = c_1 X_1(x;X) + c_2 X_2(x;E),$$

where $X_1(x;E)$ is displayed as

$$\underline{X_1(x;E)}$$

| $[\cos k][\cos k'(x+1)]$ $+$ $(k/k')[\sin k][\sin k'(x+1)]$ | $\cos kx$ | $[\cos k][\cos k'(x-1)]$ $-$ $(k/k')[\sin k][\sin k'(x-1)]$ |
|---|---|---|
| $(x < -1)$ | $(-1 < x < 1)$ | $(x > 1)$ |

$$\tag{11}$$

Following the same reasoning we display $X_2(x;E)$ as

$$\underline{X_2(x;E)}$$

| $-[\sin k][\cos k'(x+1)]$ $+$ $(k/k')[\cos k][\sin k'(x+1)]$ | $\sin kx$ | $[\sin k][\cos k'(x-1)]$ $+$ $(k/k')[\cos k][\sin k'(x-1)]$ |
|---|---|---|
| $(x < -1)$ | $(-1 < x < 1)$ | $(x > 1)$ |

$$\tag{12}$$

Observe that $X_1$ is an even function, and $X_2$ is odd (exercise 1).

It is immediately clear from (10, 11, 12) that if $E$ is greater than $U$ then $k$ and $k'$ are real, and both $X_1$ and $X_2$ satisfy the finiteness condition at $\pm\infty$.[6] Thus the energy levels ($E$) include the continuum from $U$ to $\infty$.

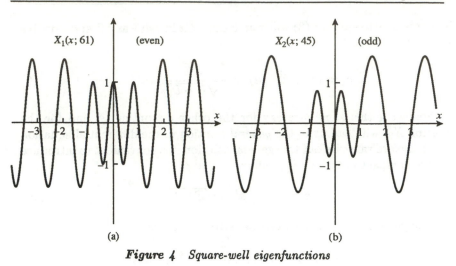

**Figure 4** *Square-well eigenfunctions*

The particular number $U$, itself, is not an energy level in general (exercise 2).

Typical eigenfunctions are illustrated in Fig. 4 for $U = 36$. Note from eq. (10) that $k > k'$; the solutions oscillate faster in the central region. This could be anticipated from our earlier discussion since, classically, the particle has more kinetic energy inside the well.

To study the possibility of other energy levels less than $U$ (or eigenvalues greater than $-U$), we break up the analysis into two cases:

**Case 1**: $0 < E < U$ (and thus $k' = iK'$ with $0 < K' < \sqrt{U}$, and $0 < k < \sqrt{U}$);

**Case 2**: $E \leq 0$ (and thus $k' = iK'$ with $K' \geq \sqrt{U}, k = iK$ with $K \geq 0$).

*Case 1* $(0 < E < U)$. The solution forms (11, 12) remain valid if we keep in mind the relations $\cos i\theta = \cosh \theta, \sin i\theta = i \sin \theta$. In terms of real functions the even and odd solutions become (exercise 1)

$$\underline{X_1(x; E)}$$

| $[\cos k][\cosh K'(x + 1)]$ $+$ $(k/K')[\sin K][\sinh K'(x + 1)]$ | $\cos kx$ | $[\cos k][\cosh K'(x - 1)]$ $-$ $(k/K')[\sin K][\sinh K'(x - 1)]$ |
|:---:|:---:|:---:|
| $(x < -1)$ | $(-1 < x < 1)$ | $(x > 1)$ |

$$\tag{13}$$

---

[6]This is another example - like the Fourier series of Sec. 6.3 (Example 5) - of a Sturm-Liouville problem having *two* independent eigenfunctions associated with the same eigenvalue.

$$X_2(x; E)$$

| $\begin{array}{c} -[\sin k][\cosh K'(x+1)] \\ + \\ (k/K')[\cos k][\sinh K'(x+1)] \end{array}$ | $\sin kx$ | $\begin{array}{c} [\sin k][\cosh K'(x-1)] \\ + \\ (k/K')[\cos k][\sinh K'(x-1)] \end{array}$ |
|:---:|:---:|:---:|
| $(x < -1)$ | $(-1 < x < 1)$ | $(x > 1)$ |

$$\tag{14}$$

According to our basic Sturm-Liouville procedure, we should next determine $c_1$ and $c_2$ so that $c_1 X_1 + c_2 X_2$ is finite at $-\infty$; the equation for the eigenvalues will appear when we subsequently impose this condition at $+\infty$. Since

$$\cosh K'(x+1) = \frac{e^{K'(x+1)} + e^{-K'(x+1)}}{2}, \tag{15}$$

$$\sinh K'(x+1) = \frac{e^{K'(x+1)} - e^{-K'(x+1)}}{2}, \tag{16}$$

the "offending" term that must be canceled at $-\infty$ is $e^{-K'(x+1)}$. The coefficient of this term in $c_1 X_1 + c_2 X_2$, for $x < -1$, is read from (13, 14) to be

$$c_1[\cos k - \frac{k}{K'} \sin k] + c_2[-\sin k - \frac{k}{K'} \cos k]. \tag{17}$$

Hence we ensure finiteness at $-\infty$ by selecting values for $c_1$ and $c_2$ making (17) zero:

$$c_1 = [\sin k + \frac{k}{K'} \cos k] C, \ \ c_2 = [\cos k - \frac{k}{K'} \sin k] C \tag{18}$$

for some constant $C$.

At $+\infty$ the offending term in (16) is $e^{K'(x-1)}$, and its coefficient for $x > 1$ is read from (13, 14):

$$c_1[\cos k - \frac{k}{K'} \sin k] + c_2[\sin k + \frac{k}{K'} \cos k]. \tag{19}$$

Forcing (19) to vanish, then, should yield the eigenvalues or energy levels in this range. Inserting the expressions (18) we find the equation for the energy levels to be

$$2C^2[\sin k + \frac{k}{K'} \cos k][\cos k - \frac{k}{K'} \sin k] = 0 \ \ \ (= 2c_1 c_2). \tag{20}$$

Notice from eqs. (10) that

$$K' = \sqrt{U - k^2}. \tag{21}$$

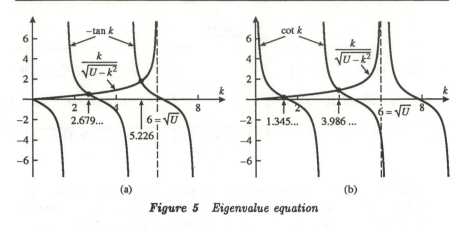

***Figure 5*** *Eigenvalue equation*

Therefore (20) has solutions if either the first factor vanishes

$$\frac{k}{\sqrt{U - k^2}} = -\tan k \tag{22}$$

(in which case $c_1 = 0$ and the eigenfunction is the *odd* function $X_2(x; E)$), or the second factor vanishes

$$\frac{k}{\sqrt{U - k^2}} = \cot k \tag{23}$$

(yielding the *even* eigenfunction $X_1(x; E)$ since $c_2 = 0$). The intersections of the graphs in Fig. 5 reveal that eqs. (22, 23) *do* possess some solutions (notice in particular from the nature of Fig. 5(b) that there will *always* be at least one intersection - and hence one even eigenfunction - for any $U > 0$). The values of $k$ have to be obtained numerically from eqs. (22, 23), and each solution $k$ generates an isolated energy level via (recall (10))

$$E = k^2. \tag{24}$$

For $U = 36$ there are two even and two odd eigenfunctions, and they are displayed in Fig. 6.

These solutions correspond to particles with total energy between 0 and $U$. Classically such particles would be confined to the well $|x| < 1$, but quantum mechanically they have a finite probability of leaking out.

*Case 2* ($E < 0$). Again we use the solution forms (11, 12) with the substitutions $k' = iK', k = iK$. The even solution (13) remains real, but the odd solution $X_2(x; E)$ becomes pure imaginary. Thus we set $\tilde{X}_2(x; E) = X_2(x; E)/i$ to obtain the forms (exercise 1)

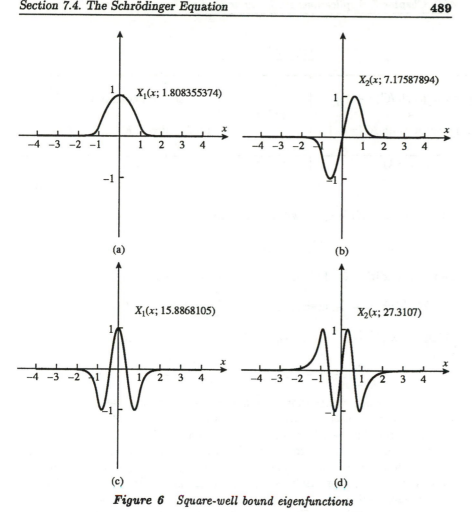

**Figure 6** *Square-well bound eigenfunctions*

$$\underline{X_1(x; E)}$$

| $[\cosh K][\cosh K'(x+1)]$ $-$ $(K/K')[\sinh K][\sinh K'(x+1)]$ | $\cosh Kx$ | $[\cosh K][\cosh K'(x-1)]$ $+$ $(K/K')[\sinh K][\sinh K'(x-1)]$ |
|---|---|---|
| $(x < -1)$ | $(-1 < x < 1)$ | $(x > 1)$ |

$$(25)$$

$$\tilde{X}_2(x; E)$$

| $-[\sinh K][\cosh K'(x+1)]$<br>$+$<br>$(K/K')[\cosh K][\sinh K'(x+1)]$ | $\sinh Kx$ | $[\sinh K][\cosh K'(x-1)]$<br>$+$<br>$(K/K')[\cosh K][\sinh K'(x-1)]$ |
|---|---|---|
| $(x < -1)$ | $(-1 < x < 1)$ | $(x > 1)$ |

$$\tag{26}$$

The solution $c_1 X_1 + c_2 X_2$ will be finite at $-\infty$ if (exercise 1)

$$c_1 = [\sinh K + \frac{K}{K'}\cosh K]C, \quad c_2 = [\cosh K + \frac{K}{K'}\sinh K]C,$$

and the condition for finiteness at $+\infty$ then becomes

$$2C^2[\sinh K + \frac{K}{K'}\cosh K][\cosh K + \frac{K}{K'}\sinh K] = 0 \quad (= 2c_1 c_2). \tag{27}$$

With $K' = \sqrt{U + K^2}$ (recall eq. (10)) we see that (27) has solutions whenever

$$\frac{K}{\sqrt{U + K^2}} = -\tanh K \quad \text{or} \quad \frac{K}{\sqrt{U + K^2}} = -\coth K.$$

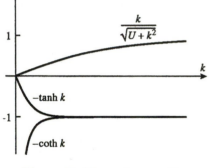

**Figure 7** *Eigenvalue equation*

But Fig. 7 demonstrates that these equations are inconsistent; *there are no negative energy levels* (or positive eigenvalues).

To summarize: we have seen that the energy levels of the attractive quantum mechanical square well model comprise *all* numbers greater than $U$ and a few isolated values between 0 and $U$. As $U$ gets bigger (and the well gets deeper) more isolated energy levels arise. In fact the *infinitely deep square well* model has a countably infinite set of energy levels; it is analyzed in exercise 3. Exercise 4 shows that the *repulsive* square well ($U < 0$) has a continuum of energy levels greater than $U$ and no others; typical eigenfunctions are displayed in Fig. 8 for $U = -36$.

The eigenfunction expansions for this problem are tedious and seldom necessary in practice. We omit them. ∎

**Example 2.** The potential function for a uniform constant force field $F$ is $V(x) = -Fx$. Classically this force field carries *all* particles "downhill" (to the right for $F > 0$), and a particle of energy $E$ can only move upstream to the point $x_0 = -E/F$, where it runs out of kinetic energy.

The separated equation (3) takes the form

$$X'' + FxX = -EX, \quad X(\pm\infty) \text{ finite.} \tag{28}$$

In exercise 6 you will be invited to show that the general solution to (28) can be expressed in terms of Airy functions (exercise 2, Sec. 2.3) as

$$X(x; E) = c_1 Ai\{-(E + Fx)/F^{2/3}\} + c_2 Bi\{-(E + Fx)/F^{2/3}\}. \tag{29}$$

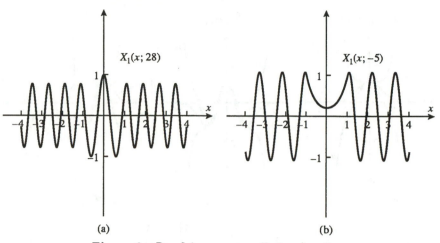

(a)                                            (b)

**Figure 8**  *Repulsive square-well eigenfunctions*

From Fig. 1 of Sec. 2.3 we see that $Bi\{-(E+Fx)/F^{2/3}\}$ will diverge at $+\infty$ (or $-\infty$ if $F$ happens to be negative). Thus we must take $c_2$ to be zero. But then $Ai\{-(E+Fx)/F^{2/3}\}$ is finite at $\pm\infty$ regardless of the value of $E$!

As a consequence *every real number* is an eigenvalue for this problem, and the eigenfunctions are the Airy functions of the first kind; see Fig. 9. As noted, classically the particle would be confined to the region of positive kinetic energy $x > -E/F$, gaining kinetic energy as it moves to the right. Quantum mechanically, then, its oscillation rate increases to the right, while there is a finite probability of its leaking in to the region $x < -E/F$.  ∎

**Example 3.** The quantum mechanical harmonic oscillator problem refers to a mass-spring system as in Sec. 3.2, but described by the Schrödinger equation. The potential energy can be dimensionalized to take the form $V(x) = x^2$. Classically, then, a particle of energy $E$ can venture no further than $x = \pm\sqrt{E}$ from the origin.

The separated equation (3) takes a form known as *Weber's equation*:

$$X'' - x^2 X = -EX, \quad X \text{ finite at } \pm\infty. \tag{30}$$

Entry 30 in Table 6.3 tabulates the eigenfunctions for eq. (30). We sketch the derivation here; see exercise 7 for details.

The substitutions

$$X(x) = e^{-x^2/2} z(x), \quad E = 1 - \mu \tag{31}$$

reduce (30) to the *Hermite differential equation* for the factor $z(x)$:

$$z'' - 2xz' = \mu z. \tag{32}$$

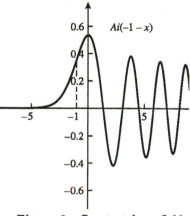

**Figure 9**  *Constant force field eigenfunction*

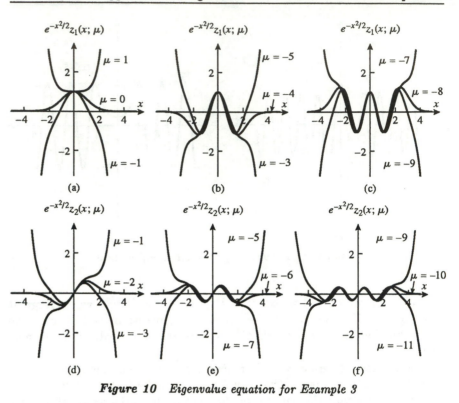

**Figure 10**   *Eigenvalue equation for Example 3*

The solutions of (32) can be organized into power series of the form

$$z_1(x) = \sum_{m=0}^{\infty} a_{2m} x^{2m}, \quad z_2(x) = \sum_{m=0}^{\infty} a_{2m+1} x^{2m+1}, \tag{33}$$

where the coefficients of the even function $z_1$ satisfy

$$a_{2j+2} = a_{2j} \frac{4j + \mu}{(2j+1)(2j+2)} \tag{34}$$

and those of the odd function $z_1$ satisfy

$$a_{2j+3} = a_{2j+1} \frac{4j + 2 + \mu}{(2j+2)(2j+3)}. \tag{35}$$

The nature of the eigenvalues is revealed in Fig. 10, where $X_1(x) = e^{-x^2/2} z_1(x)$ and $X_2(x) = e^{-x^2/2} z_2(x)$ are plotted for a selection of values of $\mu$. Clearly the values $\mu = 0, -2, -4, -6, -8, \ldots$ produce the eigenfunctions (i.e., solutions finite at infinity).

A rough explanation of the behavior depicted in these diagrams is as follows.

For large $j$, the recursion relations (34) for the even function $z_1$ approach the form

$$a_{2j+2} \approx a_{2j}/j, \tag{36}$$

which is characteristic of the series for $e^{x^2}$;

$$e^{x^2} = 1 + \frac{x^2}{1!} + \frac{x^4}{2!} + \frac{x^6}{3!} + \frac{x^8}{4!} + \cdots .$$

This would indicate that, if the approach to the limit is not interrupted, the solution $z_1$ will diverge at infinity roughly like $e^{x^2}$ and the *original* function $X_1 = e^{-x^2/2}z_1$ will diverge like $e^{x^2/2}$.

As in the case of the Legendre polynomials, the only way we can prevent this divergence is by choosing $\mu$ so as to truncate the series (34). The choice

$$\mu_p = -4p, \quad n = 0, 1, 2, \ldots \tag{37}$$

dictates that the associated function $z_1$ will be a *polynomial* of degree $p$, and thus $X_1 = e^{-x^2/2}z_1$ will go to zero at infinity.

Similarly the choice $\mu_p = -4p - 2$ produces odd eigenfunctions approaching zero at infinity.

Historically the polynomial solutions to (32) have been chosen so that the term of highest degree is $(2x)^p$; these functions are known as the *Hermite polynomials* and their properties are listed at the end of this section in Table 7.2.

$$H_p(x) = 2^p x^p + \text{(lower order terms)}.$$

The eigenfunctions of the original problem (30), then, are

$$\phi_p(x) = e^{-x^2/2}H_p(x), \quad p = 0, 1, 2, \ldots$$

with corresponding energy levels

$$E_p = 2p + 1$$

increasing to $\infty$. The weight function for the Weber equation (30) is $w(x) = 1$, so the eigenfunction expansions take the form

$$f(x) = \sum_{n=0}^{\infty} c_p e^{-x^2/2} H_p(x), \quad -\infty < x < \infty, \lambda_p = -2p - 1 \quad (= -E_p),$$

with the coefficients given by

$$c_p = \frac{< f(x), H_p(x)e^{-x^2/2} >}{\|H_p(x)e^{-x^2/2}\|^2} = \frac{\int_{-\infty}^{\infty} f(x)H_p(x)e^{-x^2/2}dx}{\int_{-\infty}^{\infty} H_p^2(x)e^{-x^2}dx}$$
$$= \frac{\int_{-\infty}^{\infty} f(x)H_p(x)e^{-x^2/2}dx}{2^p \pi^{1/2} p!}.$$

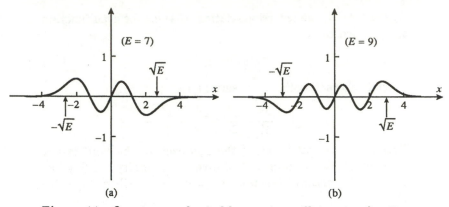

***Figure 11*** *Quantum-mechanical harmonic oscillator eigenfunctions*

Graphs of the eigenfunctions in Fig. 11 illustrate the transitions between wavelike and exponential-like behavior at the classical turning points, and the more rapid oscillatory behavior in regions of higher kinetic energy (i.e., nearer the origin). ∎

**Example 4.** The Schrödinger equation for a particle moving in an attractive inverse square central field (such as the hydrogen atom) is

$$i\frac{\partial \Psi}{\partial t} = -\nabla^2 \Psi - \frac{q}{r}\Psi \tag{38}$$

($q$ is a constant characterizing the strength of the field.) The boundary conditions are finiteness at $r = 0$ and $\infty$ and at $\phi = 0$ and $\pi$, and $2\pi$-periodicity in $\theta$.

The separation of (38) was exhibited in Sec. 5.3. However, it is instructive to observe how the spherical harmonic identity

$$\nabla^2 R(r)Y_{lm}(\phi,\theta) = \frac{(r^2 R')' - l(l+1)R}{r^2}Y_{lm}(\phi,\theta) \tag{39}$$

(Table 6.3 # 29) leads directly to the radial equation for separated solutions of the form

$$\Psi = R(r)Y_{lm}(\phi,\theta)T(t). \tag{40}$$

Substitution of (39, 40) into (38) produces

$$iT'RY_{lm} = -\frac{(r^2 R')' - l(l+1)R}{r^2}Y_{lm}T - \frac{q}{r}RY_{lm}T$$

and as a result we conclude

$$i\frac{T'}{T} = -\frac{(r^2 R')' - l(l+1)R}{r^2 R} - \frac{q}{r} = \text{constant} = E.$$

Thus the time factor is as in (4) and the radial dependence is governed by Balmer's equation

$$(r^2 R')' + \{qr - l(l+1)\}R = -Er^2 R, \quad R \text{ finite at } 0 \text{ and } \infty. \tag{41}$$

From the theory we know that most if not all of the energy levels $E$ (the negatives of the eigenvalues) are positive. In fact laborious computations show that *every* positive number is an energy level, whether $q$ is positive or negative. The positive-energy eigenfunctions are known in physics as "scattered states," and we refer the reader to Messiah's book for the details of these solutions. Here we confine our attention to the "bound" states, corresponding to positive $q$ and negative $E$. They are historically more significant, since they explained the spectral lines observed by pre-quantum experimental physicists.

Since we are assuming $E < 0$ we take the *positive* square root $\sqrt{-E}$ and make the following substitutions:

$$\xi = 2\sqrt{-E}r,$$

$$R(r = \xi/2\sqrt{-E}) = \xi^l e^{-\xi/2} Z(\xi),$$

$$k = 2l + 1$$

$$\mu = l + 1 - \frac{q}{2\sqrt{-E}}. \tag{42}$$

Then (41) transforms into the *Associated Laguerre Equation*

$$\xi \Xi'' + (k + 1 - \xi)\Xi' = \mu\Xi; \tag{43}$$

(see exercise 8 for details of these and other computations).

The indicial equation for (43) indicates that its general solution can be written

$$\Xi(\xi) = \sum_{j=0}^{\infty} a_j \xi^j + \xi^{-k} \sum_{j=0}^{\infty} b_j \xi^j$$

and since $k = 2l+1 > 0$ in our application the second solution is disqualified by the finiteness condition at the origin (even after multiplication by $\xi^l$, as prescribed in (42)).

The analysis of the eigenvalue equation, which is determined by the condition at infinity, is facilitated by the observation that if $Y(\xi)$ satisfies the *Laguerre equation*

$$\xi Y'' + (1 - \xi)Y' = \mu Y \tag{44}$$

then $\Xi = Y^{(k)}(\xi)$ satisfies the associated Laguerre equation with $\mu$ replaced by $\mu + k$. Thus we focus on (44).

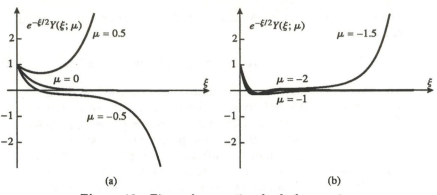

**Figure 12**   Eigenvalue equation for hydrogen atom

The recurrence relation for the regular (non-logarithmic) solution of (44) is

$$a_{j+1} = a_j \frac{j + \mu}{(j+1)(j+k+1)}. \tag{45}$$

The graphs of $e^{-\xi/2}$ times these solutions (recall (42)), for $k = 1$ ($l = 0$), are shown for a selection of values of $\mu$ in Fig. 12. Eigenfunctions are seen to occur for $\mu = 0, -1, -2, \ldots$. This is explained, briefly, similarly to the Hermite functions: it can be shown that unless the series generated by (45) truncates, the solutions to (44) diverge roughly like $e^\xi$ at infinity when $\mu$ is real. (Note, in fact, that if $\mu = k + 1$ the solution generated by (45) *is* $e^\xi$.) This, in turn, implies divergence of the solutions to our original equation (41).

From (45), then, we see that the values of $\mu$ which truncate the series are the nonpositive integers

$$\mu = -p, \quad p = 0, 1, 2, \ldots \tag{46}$$

and the associated solutions to (44), normalized by the condition $Y(0) = 1$, are known as the *Laguerre polynomials*:

$$\begin{aligned}
Y(\xi; \mu = -p) &= L_p(\xi) \\
&= 1 - \frac{p!}{1!(p-1)!}(\xi/1!) + \frac{p!}{2!(p-2)!}(\xi^2/2!) - \cdots \\
&\quad + (-1)^p(\xi^p/p!)
\end{aligned}$$

(see Table 7.3). The derived solutions to (43), properly signed, are the *Associated Laguerre polynomials*:

$$L_p^k(\xi) = (-1)^k \{d/dx\}^k L_{p+k}(\xi).$$

Restoring the substitutions (42) we are led to a family of eigenfunctions for the hydrogen atom given by

$$R_{pl}(r) = (2\sqrt{-E_p}\,r)^l \exp\{-\sqrt{-E_p}\,r\} L_p^{2l+1}(2\sqrt{-E_p}\,r)$$

with corresponding energy levels

$$E_p = -q^2/4(p+l+1)^2. \tag{47}$$

Since the eigenvalues $\lambda_p = -E_p$ decrease to zero, and not to $-\infty$, we know we have not constructed *all* the eigenfunctions for this problem. The positive energy "scattered states" can be analyzed by allowing $\mu$ to become complex in (44). (Note that $\mu$ is *not* an eigenvalue, nor are the Laguerre polynomials eigenfunctions, since - without the exponential factors - the latter diverge at infinity.) The additional eigenfunctions are known to specialists as *Regular Spherical Coulomb Functions* and their energy levels, as we said earlier, fill up the positive real line.

The eigenfunctions and energy levels are usually indexed in atomic physics by $\{n, l, m\}$ instead of $\{p, l, m\}$, where $n = p + l + 1 = \{l + 1, l + 2, \dots\}$. Spectroscopists had predicted the pattern (47) of energy levels before quantum mechanics was formulated, and the above derivation was hailed as a resounding confirmation of the correctness of Schrödinger's equation. Traditionally $n$ is known as the principal quantum number, $l$ as the orbital quantum number, and $m$ as the magnetic quantum number. With this indexing the energy level (47) depends only on the principal quantum number. ∎

# Exercises 7.4

1. (a) Verify displays (11 through 14, 25, and 26).

   (b) Show that (11) defines an even, and (12) an odd, function.

2. Show that $E = U$ is not an energy level of eqs. (6) unless $U$ happens to take one of the values $n^2\pi^2$ or $(n + 1/2)^2\pi^2$ ($n$ an integer).

3. The infinitely deep square well potential is obtained by letting $U \to \infty$ in the example in the text. Thus the possibility $E > U$ no longer exists.

   (a) (Case 1) What are the limiting values of $k$ at the points of intersection of the graphs in Fig. 5 as $U \to \infty$?

   (b) Show that the eigenfunctions defined by eqs. (13, 14) approach zero for $|x| > 1$, as $U \to \infty$.

   (c) From part (b) we conclude that the infinite square well problem can be described by the equations

   $$X'' = -EX, \quad X(-1) = X(1) = 0.$$

   Work out the energy levels and eigenfunctions for this Sturm-Liouville system, and confirm your answer in part (a). Are there any negative energy levels?

4. Work out the eigenfunctions for the repulsive square well potential (i.e. take $U < 0$ in the example in the text). Show that Fig. 8(a) depicts the typical eigenfunction if $E > 0$, whereas Fig. 8(b) describes $U < E < 0$. Show that there are no energy levels less than $U$. Interpret.

5. Find the isolated eigenvalues for the attractive square well potential if $U = 1$; if $U = 50$.

6. Derive eq. (29).

7. Verify the equations used in the derivation of the Hermite polynomial expansion.

8. Verify the equations used in the derivation of the Laguerre polynomial expansion.

# TABLE 7.2 Hermite Polynomials

Differential equation:

$$H_n''(x) - 2xH_n'(x) = -2nH_n(x)$$

Description

$H_n(x)$ is the only solution, up to a constant multiple, which does not grow exponentially at infinity.

$H_n(-x) = (-1)^n H_n(x)$.

$H_n(x)$ has $n$ real zeros.

Normalization Integral

$$\int_{-\infty}^{\infty} H_{n_1}(x) H_{n_2}(x) e^{-x^2}\, dx = 2^n \pi^{1/2} n_1!\, \delta_{n_1 n_2}$$

Formula

$$H_n(x) = (-1)^n e^{x^2} \{d/dx\}^n e^{-x^2}$$

See Figure 13.

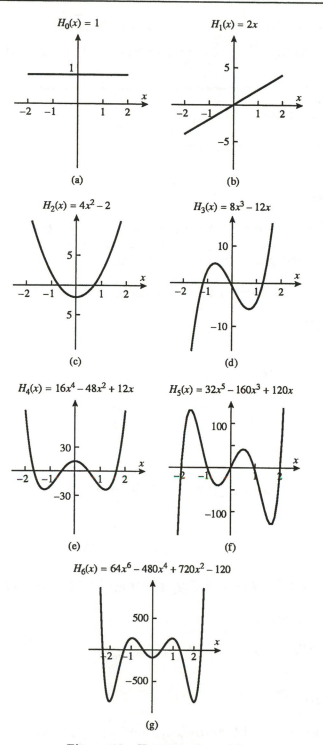

$H_0(x) = 1$

(a)

$H_1(x) = 2x$

(b)

$H_2(x) = 4x^2 - 2$

(c)

$H_3(x) = 8x^3 - 12x$

(d)

$H_4(x) = 16x^4 - 48x^2 + 12x$

(e)

$H_5(x) = 32x^5 - 160x^3 + 120x$

(f)

$H_6(x) = 64x^6 - 480x^4 + 720x^2 - 120$

(g)

**Figure 13    Hermite polynomials**

# TABLE 7.3 Laguerre Polynomials

Differential equation:

$$xL_n''(x) + (1-x)L_n'(x) = -nL_n(x)$$

Description

> The Laguerre polynomial $L_n(x)$ is the only solution,
> up to a constant multiple, which is finite at zero
> and grows slower than $e^x$ at infinity.

> $L_n(x)$ has degree $n$ and $n$ positive zeros.

> $L_n(0) = 1$.

Normalization Integral

$$\int_0^\infty L_{n_1}(x)L_{n_2}(x)e^{-x}\,dx = \delta_{n_1 n_2}$$

Formula

$$L_n(x) = \sum_{j=0}^n \frac{n!}{(n-j)!\,j!^2}(-x)^j = \frac{e^x}{n!}\{d/dx\}^n(x^n e^{-x})$$

# Associated Laguerre Polynomials

Differential equation:

$$xL_n^{k''}(x) + (k+1-x)L_n^{k'}(x) = -nL_n(x)$$

Description

> The associated Laguerre polynomial $L_n^k(x)$ is the only solution,
> up to a constant multiple, which is finite at zero

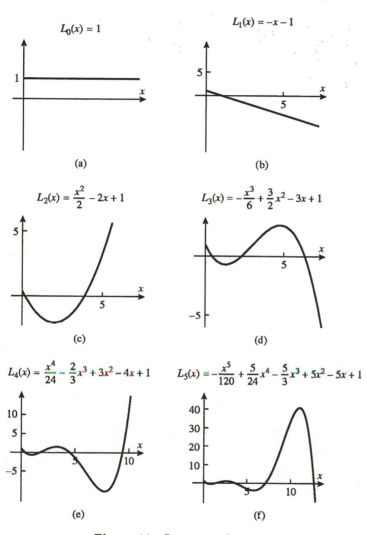

**Figure 14**   *Laguerre polynomials*

and grows slower than $e^x$ at infinity.

$L_n^k(x)$ has degree $n$ and $n$ positive zeros.

Normalization Integral

$$\int_0^\infty L_{n_1}^k(x)L_{n_2}^k(x)e^{-x}x^k\,dx = \tfrac{(n+k)!}{n!}\delta_{n_1 n_2}$$

Formula

$$L_n^k(x) = (-1)^k\{d/dx\}^k L_{n+k}(x)$$

See Figure 14.

---

## 7.5   Shortcuts

**Summary**   The basic similarities in the separated solution forms for the Helmholtz and related equations in cartesian, cylindrical, and spherical coordinates enable the deployment of compact universal forms applicable to most situations.

Now that the basic tenets of the separation of variables technique have been thoroughly expounded, we are ready to look for short cuts in its implementation. In the course of surveying the applications of the separation of variables technique we have had occasion to observe the repeated occurrence of some of the eigenfunction families. The terms of the various types of Fourier series, for example, arise again and again. This makes feasible our task: **the tabulation of generic forms of separated solutions of the basic Helmholtz equation in rectangular, cylindrical, and spherical coordinates.**

Recall from Section 5.3 that the Helmholtz equation

$$\nabla^2\psi = \Lambda\psi \tag{1}$$

provides solutions to

- Laplace's equation $\nabla^2\psi = 0$ if $\Lambda = 0$,

- the heat equation $\frac{\partial\Psi}{\partial t} = \nabla^2\Psi$ in the form $\Psi = \psi e^{\Lambda t}$, and

- the wave equation $\frac{\partial^2\Psi}{\partial t^2} = \nabla^2\Psi$ in the form $\Psi = \psi[d_1\cos\sqrt{-\Lambda}\,t + d_2\sin\sqrt{-\Lambda}\,t]$ or, if $\Lambda = 0$, $\Psi = \psi[d_1 + d_2 t]$.

In fact, the form of (1) suggests that we regard $\Lambda$ as an eigenvalue for the *partial* differential operator $\nabla^2$.

**Example 1.** In the several examples in Chapters 5 and 6 devoted to the two dimensional Laplace equation, we have observed solutions in the forms

$$\cos nx \cosh ny, \quad \sinh \frac{2n-1}{2}x \sin \frac{2n-1}{2}y,$$

$$\sin \mu_n x \sinh \mu_n y \quad (\mu_n = 1.5044, \ 4.6911, \ 7.8412, \ldots). \tag{2}$$

In fact in Sec. 5.2 we pointed out that for *any* value of the constants the functions

$$(a_1 \cosh \sqrt{\lambda}x + a_2 \sinh \sqrt{\lambda}x)(d_1 \cos \sqrt{\lambda}y + d_2 \sin \sqrt{\lambda}y) \tag{3}$$

$$(b_1 \cos \sqrt{-\lambda}x + b_2 \sin \sqrt{-\lambda}x)(e_1 \cosh \sqrt{-\lambda}y + e_2 \sinh \sqrt{-\lambda}y) \tag{4}$$

$$(c_1 + c_2 x)(f_1 + f_2 y) \tag{5}$$

were solutions. For brevity we combine forms (3, 4) by writing

$$(a_1 \cos \mu x + a_2 \sin \mu x)(b_1 \cos \nu y + b_2 \sin \nu y) \tag{6}$$

with the proviso that

$$\mu^2 + \nu^2 = 0, \tag{7}$$

so that one or the other of $(\mu, \nu)$ is imaginary, rendering the corresponding cosine or sine as cosh or sinh.

As confirmation, recall that we have seen any number of times that the separated form of $\nabla^2 X(x)Y(y) = 0$ is

$$\frac{X''}{X} = -\frac{Y''}{Y}, \tag{8}$$

and the form (6) with condition (7) fulfills (8). ∎

**Example 2.** Let's use the format (3–5) to solve the problem depicted in Fig. 1. Because there are homogeneous boundary conditions on both $y$ ends, we need to choose the form for which the $y$ factors are eigenfunctions. Since eigenfunctions are oscillatory (recall Sec. 6.1), we select

$$\psi_\nu(x,y) = (a_1 \cosh \nu x + a_2 \sinh \nu x)(b_1 \cos \nu y + b_2 \sin \nu y) \tag{9}$$

The homogeneous boundary conditions are

$$\psi_\nu(x,0) = \psi_\nu(x,1) = \psi_\nu(0,y) = 0.$$

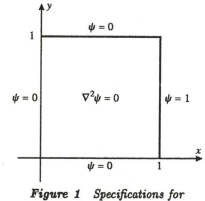

**Figure 1** *Specifications for Example 2*

To make $\psi_\nu$ zero at $y = 0$ we take $b_1 = 0$. The vanishing of $\psi_\nu$ at $y = 1$ must be accomplished through the choice of $\nu$; thus we set $\nu = n\pi, n = 1, 2, 3 \dots$. Now $\psi_\nu$ will be zero on the left if $a_1 = 0$. All that's left to do is to assemble the solutions we have found –

$$\psi(x, y) = \sum_{n=1}^{\infty} c_n \sinh n\pi x \sin n\pi y, \tag{10}$$

and (with the observation that we have a complete orthogonal family - the sine series) enforce the final boundary condition:

$$\psi(1, y) = 1 = \sum_{n=1}^{\infty} c_n \sinh n\pi \sin n\pi y$$

implies

$$c_n = \frac{1}{\sinh n\pi} \frac{\int_0^1 (1) \sin n\pi y \, dy}{\int_0^1 \sin^2 n\pi y \, dy} = \begin{cases} 0, & n \text{ even} \\ \frac{4}{n\pi \sinh n\pi}, & n \text{ odd}. \end{cases} \qquad \blacksquare$$

Shortcuts invariably carry hazards, and the initial reasoning in the following example contains a subtle error. Once it is manifested, however, the cure is clear.

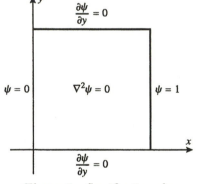

**Figure 2**  *Specifications for Example 3*

**Example 3.** For the situation depicted in Fig. 2, we follow the same reasoning as above and choose form (9). The bottom and top homogeneous Neumann conditions on $y$ are now accommodated by the choice $b_2 = 0, \nu = n\pi, n = 0, 1, 2, \dots$. The left condition is met as before if $a_1 = 0$, and the assembled solution reads

$$\psi(x, y) = \sum_{n=0}^{\infty} c_n \sinh n\pi x \cos n\pi y. \tag{11}$$

(Do you see the flaw yet?) The final boundary condition is

$$\psi(1, y) = 1 = \sum_{n=0}^{\infty} c_n \sinh n\pi \cos n\pi y \tag{12}$$

and by orthogonality we would expect (12) to be satisfied if

$$c_n = \frac{1}{\sinh n\pi} \frac{\int_0^1 (1) \cos n\pi y \, dy}{\int_0^1 \cos^2 n\pi y \, dy}. \tag{13}$$

Alas! Equation (13) says $c_n = 0$ for $n > 0$, but

$$c_0 = \frac{1}{\sinh 0} \frac{\int_0^1 (1) \cos 0 \, dy}{\int_0^1 \cos^2 0 \, dy} = \frac{1}{0} (!).$$

What went wrong?

The difficulty lies in a minor detail we overlooked in adopting (9). If $\mu = \nu = 0$, the general solutions of the separated equations (6) are *not* given by the form (9), but rather by (5)

$$(c_1 + c_2 x)(f_1 + f_2 y). \tag{14}$$

Now, the homogeneous boundary conditions for the configuration in Fig. 2, applied to (14), require $f_2 = 0, c_1 = 0$. So when this correction is written into the assembled general solution, eq. (11) should read

$$\psi(x, y) = c_0 x + \sum_{n=1}^{\infty} c_n \sinh n\pi x \, \cos n\pi y \tag{15}$$

(or, to maintain the cosine series format

$$= c_0 x \cos 0y + \sum_{n=1}^{\infty} c_n \sinh n\pi x \, \cos n\pi y).$$

The condition $\psi(1, y) = 1$ in (15) then gives

$$c_{n>0} = 0, \quad c_0 = \frac{1}{1} \frac{\int_0^1 1 \cos 0 \, dy}{\int_0^1 \cos^2 0 \, dy} = 1$$

instead of (13). In fact the solution is simply $\psi(x, y) = x$. ∎

By the same token one easily sees that the separated solutions of the wave equation in one (spatial) dimension are

$$(a_1 + a_2 x)(b_1 + b_2 t) \quad \text{and}$$

$$(a_1 \cos \mu x + a_2 \sin \mu x)(b_1 \cos \mu t + b_2 \sin \mu t),$$

since the separated form of $\partial^2 \psi / \partial t^2 = \partial^2 \psi / \partial x^2$ is

$$\frac{X''}{X} = \frac{T''}{T}.$$

Table 7.4 at the end of this section lists the separated solution sets for the common partial differential equations in the three standard cordinate systems. Readers are invited to verify the entries.

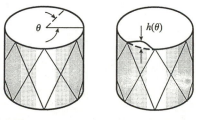

**Example 4.** A drumhead is an example of a stretched thin membrane; as such, it vibrates in accordance with the wave equation. Let us use Table 7.4 to analyze the motion of the drumhead of radius $a$ depicted in Fig. 3.

*Figure 3   Drumhead with warped rim*

At equilibrium the displacement $\psi(r, \theta, t)$ will be zero, if the rim is not warped. More generally, however, the rim itself will have a displacement $h(\theta)$ and we have to solve the initial-boundary-value problem

$$\frac{\partial^2 \psi}{\partial t^2} = \nabla^2 \psi, \ \psi(a, \theta, t) = h(\theta), \ \psi(r, \theta, 0) = f(r, \theta), \frac{\partial \psi}{\partial t}(r, \theta, 0) = g(r, \theta).$$
(16)

First we establish the steady state solution:

$$\nabla^2 \psi_\infty = 0, \ \psi_\infty(a, \theta) = h(\theta).$$
(17)

From Table 7.4 we immediately read off the general form of the solutions to Laplace's equation in two dimensional polar coordinates which are periodic in $\theta$ and finite at $r = 0$:

$$\psi_\infty(r, \theta) = \sum_{n=0}^\infty c_n r^n \cos n\theta + \sum_{n=1}^\infty d_n r^n \sin n\theta.$$
(18)

(Since (18) contains the full Fourier series we are confident that no solutions have been overlooked.) The choice of the constants in (18) that enforce the boundary condition is dictated by the usual formulas

$$c_0 = \frac{1}{2\pi} \int_0^{2\pi} h(\theta) d\theta, \ c_{n>0} = \frac{a^{-n}}{\pi} \int_0^{2\pi} h(\theta) \cos n\theta d\theta,$$

$$d_{n>0} = \frac{a^{-n}}{\pi} \int_0^{2\pi} h(\theta) \sin n\theta d\theta.$$

Now we have to solve for the vibratory modes:

$$\frac{\partial^2 \psi_0}{\partial t^2} = \nabla^2 \psi_0, \psi_0(a, \theta, t) = 0, \ \psi_0(r, \theta, 0) = f(r, \theta) - \psi_\infty(r, \theta),$$

$$\frac{\partial \psi_0}{\partial t}(r, \theta, 0) = g(r, \theta).$$

We need solutions from Table 7.4 of the time-domain *wave* equation which are periodic in $\theta$ and finite at $r = 0$. Also they must vanish at $r = a$. This dictates the form

$$
\begin{aligned}
\psi_0(r, \theta, t) \ = \ & \sum_{n=0}^\infty \sum_{p=1}^\infty a_{np} \cos n\theta \, J_n(j_{n,p} r/a) \cos j_{n,p} t/a \\
& + \sum_{n=1}^\infty \sum_{p=1}^\infty b_{np} \sin n\theta \, J_n(j_{n,p} r/a) \cos j_{n,p} t/a \\
& + \sum_{n=0}^\infty \sum_{p=1}^\infty c_{np} \cos n\theta \, J_n(j_{n,p} r/a) \sin j_{n,p} t/a \\
& + \sum_{n=1}^\infty \sum_{p=1}^\infty d_{np} \sin n\theta \, J_n(j_{n,p} r/a) \sin j_{n,p} t/a
\end{aligned}
$$
(19)

where $j_{n,p}$ is the $p$th zero of $J_n$ (Sec. 6.4 Example 3; see also Table 6.3 #16). Note that we have segregated the cosine and sine time factors; this facilitates the enforcement of the initial conditions. First we set $t = 0$ in (19);

$$f(r,\theta) - \psi_\infty(r,\theta) = \sum_{n=0}^\infty \sum_{p=1}^\infty a_{np} \cos n\theta\, J_n(j_{n,p}r/a)$$
$$+ \sum_{n=1}^\infty \sum_{p=1}^\infty b_{np} \sin n\theta\, J_n(j_{n,p}r/a).$$

By orthogonality we derive

$$b_{mp} = \frac{2}{\pi a^2 J_{m+1}^2(j_{m,p})} \int_0^a \int_0^{2\pi} [f(r,\theta) - \phi_\infty(r,\theta)] \sin m\theta\, J_m(j_{m,p}r/a) r\, d\theta\, dr.$$

$$a_{m>0,p} = \frac{2}{\pi a^2 J_{m+1}^2(j_{m,p})} \int_0^a \int_0^{2\pi} [f(r,\theta) - \phi_\infty(r,\theta)] \cos m\theta\, J_m(j_{m,p}r/a) r\, d\theta\, dr,$$

$$a_{0p} = \frac{1}{\pi a^2 J_1^2(j_{0,p})} \int_0^a \int_0^{2\pi} [f(r,\theta) - \phi_\infty(r,\theta)] J_0(j_{0,p}r/a) r\, d\theta\, dr.$$

Next we differentiate (19) with respect to time and set $t = 0$ to obtain the initial values of $\partial \psi / \partial t$:

$$d_{mp} = \frac{2}{\pi a j_{m,p} J_{m+1}^2(j_{m,p})} \int_0^a \int_0^{2\pi} g(r,\theta) \sin m\theta\, J_m(j_{m,p}r/a) r\, d\theta\, dr.$$

$$c_{m>0,p} = \frac{2}{\pi a j_{m,p} J_{m+1}^2(j_{m,p})} \int_0^a \int_0^{2\pi} g(r,\theta) \cos m\theta\, J_m(j_{m,p}r/a) r\, d\theta\, dr,$$

$$c_{0p} = \frac{1}{\pi a j_{0,p} J_1^2(j_{0,p})} \int_0^a \int_0^{2\pi} g(r,\theta)\, J_0(j_{0,p}r/a) r\, d\theta\, dr.$$

And of course $\psi = \psi_1 + \psi_\infty$. ∎

**Example 5.** The *Chladne plate* is a thin slice of metal, usually square or round, which also vibrates in accordance with the two dimensional wave equation when stroked with a violin bow along the edge; see Fig. 4.

If we consider a circular Chladne plate of radius 1, we again cull solutions from Table 7.4 which are periodic in $\theta$ and finite at $r = 0$;

$$(a_1 \cos n\theta + a_2 \sin n\theta)\, J_n(kr)\, (d_1 \cos kt + d_2 \sin kt),\ n = 0, 1, 2, \ldots.$$

Although one bows the outer edge of the Chladne plate to excite the vibrations, high speed photography has revealed that most of the time the plate vibrates freely; the bow only forces it occasionally, to refresh its energy. Thus we take the outer edge to be unconstrained. Structural mechanics tells us that this gives rise to the edge condition $\partial \phi / \partial r = 0$, so the values

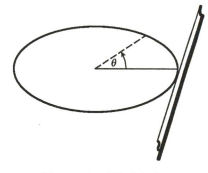

*Figure 4   Chladne plate*

of $k$ are proportional to the zeros of the derivative of $J_n$ and the radial expansion is the Dini series (Sec. 6.9; see also Table 6.3 #17).

Imposing the initial conditions

$$\psi(r,\theta,t) = f(r,\theta), \ \frac{\partial \psi}{\partial t} = g(r,\theta) \quad \text{at} \quad t = 0,$$

we invite the reader in exercise 1 to verify the solution form

$$
\begin{aligned}
\psi(r,\theta,t) &= \sum_{n=0}^{\infty}\sum_{p=1}^{\infty} \cos n\theta \, J_n(j'_{n,p}r)\,(a_{np}\cos j'_{np}t + b_{np}\sin j'_{np}t) \\
&+ \sum_{n=1}^{\infty}\sum_{p=1}^{\infty} \sin n\theta \, J_n(j'_{n,p}r)\,(c_{np}\cos j'_{np}t + d_{np}\sin j'_{np}t) \\
&+ a + bt.
\end{aligned}
\tag{20}
$$

The initial condition on $\psi$ is met through the choice

$$a = \frac{1}{\pi}\int_0^1\int_0^{2\pi} f(r,\theta)\,r\,d\theta\,dr,$$

$$a_{0p} = \frac{1}{\pi J_0^2(j'_{0p})}\int_0^1\int_0^{2\pi} f(r,\theta)J_0(j'_{0p}r)\,r\,d\theta\,dr,$$

$$a_{np} = \frac{2}{\pi J_n^2(j'_{np})[1 - n^2/j'^2_{np}]}\int_0^1\int_0^{2\pi} f(r,\theta)\cos n\theta \, J_n(j'_{n,p}\,r)r\,d\theta\,dr,$$

$$c_{np} = \frac{2}{\pi J_n^2(j'_{np})[1 - n^2/j'^2_{np}]}\int_0^1\int_0^{2\pi} f(r,\theta)\sin n\theta \, J_n(j'_{n,p}\,r)r\,d\theta\,dr;$$

and the initial condition on $\partial\psi/\partial t$ is met if

$$b = \frac{1}{\pi}\int_0^1\int_0^{2\pi} g(r,\theta)\,r\,d\theta\,dr,$$

$$b_{0p} = \frac{1}{\pi j'_{0p}J_0^2(j'_{0p})}\int_0^1\int_0^{2\pi} f(r,\theta)J_0(j'_{0p}r)\,r\,d\theta\,dr,$$

$$b_{np} = \frac{2}{\pi j'_{np}J_n^2(j'_{np})[1 - n^2/j'^2_{np}]}\int_0^1\int_0^{2\pi} f(r,\theta)\cos n\theta \, J_n(j'_{n,p}\,r)\,r\,d\theta\,dr,$$

$$d_{np} = \frac{2}{\pi j'_{np}J_n^2(j'_{np})[1 - n^2/j'^2_{np}]}\int_0^1\int_0^{2\pi} f(r,\theta)\sin n\theta \, J_n(j'_{n,p}\,r)r\,d\theta\,dr.$$

The nonoscillatory terms $a + bt$ are necessitated by the exceptional term in the Dini series discussed in Sec. 6.9. This so-called "rigid body mode" accounts for any translational motion of the plate. (After all, with the outer edge unconstrained there is nothing to keep the plate from floating away!)

■

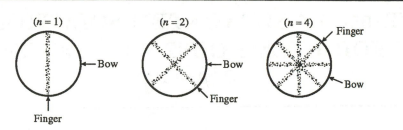

**Figure 5** *Chladne patterns*

The vibratory modes of the Chladne plate can be displayed easily with a classroom demonstration by sprinkling ground pepper on the plate before bowing it. The pepper will be thrown off at points of high-amplitude vibration and will congregate at the "nodes", or quiescent points. By placing your fingertip on the edge of the plate while bowing, you suppress all modes except those having a line of nodes at your finger, and a maximum at the point of bowing.

If we designate the latter as the line $\theta = 0$ and the former as $\theta = \theta_1$, then the $\sin n\theta$ terms in (20) are suppressed (since they have *nodes* at $\theta = 0$); similarly all $\cos n\theta$ terms are suppressed (by the finger) unless $n$ takes a value making $\cos n\theta_1 = 0$ (and usually the lowest such mode dominates the vibration). This has a remarkable effect, because then *extra* nodal lines appear at the *other* angles $\theta_i$ for which $\cos n\theta_i = 0$! The drawings in Fig. 5 illustrate this effect.

# Exercises 7.5

**1.** Verify eq. (20).

**2.** Why does the Chladne plate sound so awful? (Hint: look at the values of $j'_{n,p}$ in Table 2.2 and reread the comment on organ pipes in Sec. 7.3.)

**3.** Use Table 7.4 to re-solve some of the examples worked out previously in the text.

**4.** Concoct a Sturm-Liouville problem for which the first eigenfunction is $1+x$. (Hint: use $y'' = \lambda y$ with Robin boundary conditions at $x = 0$ and 1.)

# TABLE 7.4 SEPARATED SOLUTIONS FOR HELMHOLTZ'S EQUATION $\nabla^2 \psi = \Lambda \psi$

*Please note:*

1. Solutions of *Laplace's equation* $\nabla^2 \psi = 0$ are obtained by setting $\Lambda = 0$.

2. Time domain solutions of the *heat equation* $\frac{\partial \psi}{\partial t} = \nabla^2 \psi$ are obtained by appending the factor $e^{\Lambda t}$.

3. Time domain solutions of the *wave equation* $\frac{\partial^2 \psi}{\partial t^2} = \nabla^2 \psi$ are obtained by appending the factor $[d_1 \cos \sqrt{-\Lambda}\, t + d_2 \sin \sqrt{-\Lambda}\, t]$ or, if $\Lambda = 0$, $[d_1 + d_2 t]$.

4. For time domain solutions, the values of $\Lambda$ (eigenvalues clustering at $-\infty$) are to be determined by the spatial boundary conditions, and all spatial factors are oscillatory (eigenfunctions). For frequency-domain (Fourier) and $s$-plane (Laplace) analysis and for Laplace's equation $\Lambda$ is regarded as given and all but one of the spatial factors are oscillatory (see Sec. 8.2).

5. If $\nu = 0$ replace factors $[c_1 \cos \nu u + c_2 \sin \nu u]$ by $[c_1 + c_2 u]$, for $u = \theta, x, y, z,$ or $t$.

6. If $\kappa = 0$ replace factors $[c_1 h_\ell^{(1)}(\kappa r) + c_2 h_\ell^{(2)}(\kappa r)]$ by $[c_1 r^\ell + c_2 r^{-\ell-1}]$.

7. If $\kappa = 0$, replace factors $[c_1 I_\nu(\kappa \rho) + c_2 I_{-\nu}(\kappa \rho)]$ by $[c_1 \rho^\nu + c_2 \rho^{-\nu}]$.

8. If $\nu = 0$ replace factors $[c_1 \rho^\nu + c_2 \rho^{-\nu}]$ by $[c_1 + c_2 \ln \rho]$.

9. Use $Y_n$ or $K_n$ instead of $J_{-n}$ or $I_{-n}$, respectively, when $n$ is an integer.

10. If $\kappa = 0$, replace factors $[c_1 \operatorname{Re} I_{i\nu}(\kappa \rho) + c_2 K_{i\nu}(\kappa \rho)]$ by $[c_1 \cos(\nu \ln \rho) + c_2 \sin(\nu \ln \rho)]$.

11. If $\nu = 0$ replace factors $[c_1 \cos(\nu \ln \rho) + c_2 \sin(\nu \ln \rho)]$ by $[c_1 + c_2 \ln \rho]$.

---

I. *RECTANGULAR COORDINATES*:

$$\frac{\partial^2 \psi}{\partial x^2} + \frac{\partial^2 \psi}{\partial y^2} + \frac{\partial^2 \psi}{\partial z^2} = \Lambda \psi$$

*y, z factors oscillatory*:

$$\psi = [a_1 \cosh \kappa x + a_2 \sinh \kappa x]\,[b_1 \cos \mu y + b_2 \sin \mu y]$$
$$\times [c_1 \cos \nu z + c_2 \sin \nu z]$$

where $\kappa^2 = \mu^2 + \nu^2 + \Lambda$.

*all factors oscillatory*:

$$\psi = [a_1 \cos \kappa x + a_2 \sin \kappa x]\,[b_1 \cos \mu y + b_2 \sin \mu y]$$
$$\times [c_1 \cos \nu z + c_2 \sin \nu z]$$

where $\Lambda = -\kappa^2 - \mu^2 - \nu^2$.

(For two dimensions, set $\nu = 0$; for one dimension, set $\nu = \mu = 0$.)

---

## II. *SPHERICAL COORDINATES*:

$$\frac{1}{r^2}\frac{\partial}{\partial r}\left(r^2\frac{\partial \psi}{\partial r}\right) + \frac{1}{r^2 \sin \phi}\frac{\partial}{\partial \phi}\left(\sin \phi \frac{\partial \psi}{\partial \phi}\right) + \frac{1}{r^2 \sin^2 \phi}\frac{\partial^2 \psi}{\partial \theta^2} = \Lambda \psi$$

*$\phi, \theta$ factors oscillatory* (periodic in $\theta$ and bounded at $\phi = 0, \pi$):

$$\psi = [a_1 h_\ell^{(1)}(\kappa r) + a_2 h_\ell^{(2)}(\kappa r)]Y_{\ell m}(\phi, \theta)$$

where $\kappa^2 = -\Lambda$.

*all factors oscillatory* (periodic in $\theta$ and bounded at $\phi = 0, \pi$):

$$\psi = [a_1 j_\ell(\kappa r) + a_2 y_\ell(\kappa r)]Y_{\ell m}(\phi, \theta)$$

where $\Lambda = -\kappa^2$.

---

## III. *CYLINDRICAL COORDINATES*:

$$\frac{1}{\rho}\frac{\partial}{\partial \rho}\left(\rho\frac{\partial \psi}{\partial \rho}\right) + \frac{1}{\rho^2}\frac{\partial^2 \psi}{\partial \theta^2} + \frac{\partial^2 \psi}{\partial z^2} = \Lambda \psi$$

*$\theta$ and $\rho$ factors oscillatory*:

$$\psi = [a_1 \cos \nu\theta + a_2 \sin \nu\theta]\,[b_1 J_\nu(\kappa\rho) + b_2 J_{-\nu}(\kappa\rho)]$$
$$\times [c_1 \cosh \mu z + c_2 \sinh \mu z]$$

where $\mu^2 = \Lambda + \kappa^2$.

*θ and z factors oscillatory*:

$$\psi = [a_1 \cos \nu\theta + a_2 \sin \nu\theta] \, [b_1 I_\nu(\kappa\rho) + b_2 I_{-\nu}(\kappa\rho)]$$
$$\times \, [c_1 \cos \mu z + c_2 \sin \mu z]$$

where $\kappa^2 = \Lambda + \mu^2$.

*ρ and z factors oscillatory*:

$$\psi = [a_1 \cosh \nu\theta + a_2 \sinh \nu\theta] \, [b_1 \operatorname{Re} I_{i\nu}(\kappa\rho) + b_2 K_{i\nu}(\kappa\rho)]$$
$$\times \, [c_1 \cos \mu z + c_2 \sin \mu z]$$

where $\kappa^2 = \Lambda + \mu^2$.

*all factors oscillatory*:

$$\psi = [a_1 \cos \nu\theta + a_2 \sin \nu\theta] \, [b_1 J_\nu(\kappa\rho) + b_2 J_{-\nu}(\kappa\rho)]$$
$$\times \, [c_1 \cos \mu z + c_2 \sin \mu z]$$

where $\Lambda = -\kappa^2 - \mu^2$.

(For two dimensional polar coordinates set $\mu = 0$ and replace $\rho$ by $r$.)

# Chapter 8

# Green's Functions and Transform Methods

We have seen how to break up a problem involving a linear homogeneous differential equation and nonhomogeneous boundary conditions, such as depicted in Fig. 1, into subproblems wherein the equation remains homogeneous and all but one of the boundary conditions are homogeneous also (Figs. 2(i) and 2(ii)).

Now we are ready to deal with the remaining issue - a nonhomogeneity in the equation itself. Thus consider the problem depicted in Fig. 3. This situation can be decomposed into four subproblems like those in Fig. 2, *plus an additional subproblem where the only nonhomogeneity is in the differential equation*; see Figs. 4(i)–4(iii).

To deal with the new subproblem, then, *we need a technique for solving a* **nonhomogeneous** *differential equation subject to* **homogeneous** *boundary conditions*. This is provided by the Green's function method.

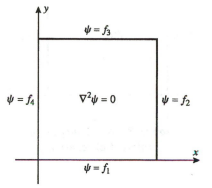

**Figure 1** *Typical homogeneous differential equation*

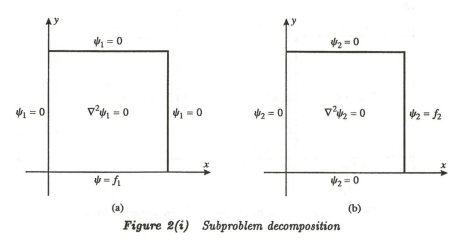

**Figure 2(i)** *Subproblem decomposition*

513

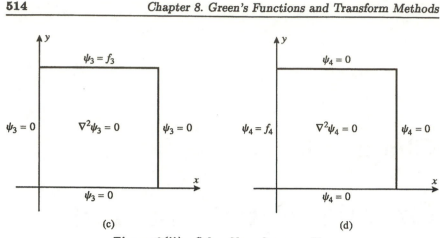

*Figure 2(ii)   Subproblem decomposition*

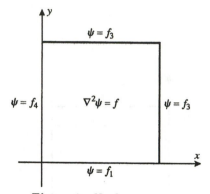

*Figure 3   Nonhomogeneous differential equation*

The overall strategy for analyzing the effect of the nonhomogeneity - or "source term" $f$ - in the differential equation was introduced in Sec. 1.4. We regard the nonhomogeneity as a superposition of impulses (recall Fig. 2, Sec. 1.4); we work out the "Green's function," which is the system's response to a single impulse; and finally we add these impulse responses together to assemble the particular solution.

Section 8.1 demonstrates how the Green's function can be expressed in terms of the eigenfunctions of the differential equation. The procedure is general and straightforward for time-independent problems and blends in perfectly with the overall methodology of this book.

However the absence of time eigenfunctions, as we have pointed out, undermines the direct application of much of the theory developed so far. In Section 8.2 we remedy this shortcoming by showing how the Fourier or Laplace transform can be judiciously employed to remove the time dependence. This tactic - coupled with the eigenfunction expansions for the

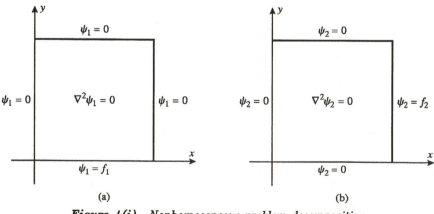

*Figure 4(i)   Nonhomogeneous problem decomposition*

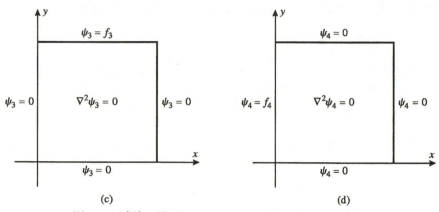

**Figure 4(ii)**   *Nonhomogeneous problem decomposition*

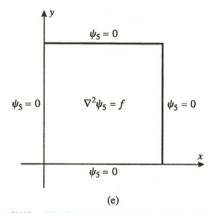

**Figure 4(iii)**   *Nonhomogeneous problem decomposition*

Green's functions, the decompositions and the time independent procedures discussed in Chapters 5 through 7 - constitutes a complete implementation of the separation of variables methodology.

We close the chapter with a look at some refinements. Closed-form expressions for Green's functions are very difficult to obtain, but occasionally they turn up in the theorems of potential theory and contour integration. Section 8.3 lists a compilation of known Green's functions for particular differential operators, together with some applications. The derivations are omitted.

Section 8.4 shows how Green's functions can be used to enhance the accuracy of the eigenfunction expansions of solutions to differential equations, for situations when the convergence is slowed by the appearance of discontinuities in the boundary data.

Section 8.5 discusses the implementation and physical interpretation of the transform methodology when applied to systems which radiate.

# 8.1   Expansions for Green's Functions

**Summary**   The solution of a nonhomogeneous differential equation with homogeneous boundary conditions can be reduced to the solution of the same system with the nonhomogeneity replaced by an impulse. This solution, called the Green's function, is readily computed in terms of the eigenfunctions of the system.

Section 1.4 demonstrated how to reduce the solution of a linear nonhomogeneous ordinary differential equation

$$\mathcal{L}y = a_2(x)y'' + a_1(x)y' + a_0(x)y = f(x) \tag{1}$$

to the impulse response or Green's function, which is a solution $G(x; x')$ of the simpler equation

$$\mathcal{L}G = a_2(x)\frac{\partial^2 G(x; x')}{\partial x^2} + a_1(x)\frac{\partial G(x; x')}{\partial x} + a_0(x)G(x; x') = \delta(x - x'). \tag{2}$$

Here $x'$ denotes the point where the impulse is applied. A solution to (2) is given in terms of $G$ by

$$y(x) = \int G(x; x')f(x')\, dx', \tag{3}$$

integrated over the interval of interest.

A convenient way of remembering why (3) works is to substitute it directly into (1). Since $\mathcal{L}$ operates only on the variable $x$, we can bring it inside the integral sign:

$$\mathcal{L}y = \mathcal{L}\int G(x; x')f(x')\, dx' \;\; = \;\; \int \mathcal{L}G(x; x')f(x')\, dx'$$

$$= \int \delta(x - x')f(x')\, dx' = f(x)$$

In Sec. 1.4 we also showed how to piece together a Green's function out of solutions to the associated homogeneous equation. Now we'll show how to construct a Green's function for (1) out of eigenfunctions. This construction has two advantages: it provides solutions which automatically satisfy the same homogeneous boundary conditions as the eigenfunctions, and it readily extends to partial differential equations.

For the theoretical discussion we shall presume that $\mathcal{L}$ has been put in Sturm-Liouville form, but we'll correct for this at the conclusion.

Thus suppose that $\{\phi_n(x), n = 1, 2, 3, \dots\}$ is a complete set of eigenfunctions of the Sturm-Liouville operator $\mathcal{L}$, satisfying prescribed homogeneous boundary conditions:

$$\mathcal{L}\phi_n(x) = \lambda_n w(x)\phi_n(x), \quad \alpha\phi_n(a) + \beta\phi_n'(a) = \gamma\phi_n(b) + \delta\phi_n'(b) = 0 \tag{4}$$

($w(x)$ is the weight function). Then to expand $G(x; x')$ (for fixed $x'$) in these eigenfunctions

$$G(x; x') = \sum_{n=1}^{\infty} a_n \phi_n(x) \tag{5}$$

we simply enforce eq. (2):

$$\mathcal{L}G(x; x') = \sum_{n=1}^{\infty} a_n \mathcal{L}\phi_n(x) = \sum_{n=1}^{\infty} a_n \lambda_n \rho(x) \phi_n(x) = \delta(x - x'). \tag{6}$$

The coefficients are easily found by using orthogonality: we multiply both sides by $\phi_m$ and integrate, isolating $a_n$:

$$a_m = \frac{\int_a^b \delta(x - x') \phi_m(x)\, dx}{\int_a^b \phi_m^2(x) w(x)\, dx\, \lambda_m} = \frac{\phi_m(x')}{\|\phi_m\|_w^2\, \lambda_m}. \tag{7}$$

That's it!

**Theorem.** *A Green's function for (1) is given by*

$$G(x; x') = \sum_{n=1}^{\infty} \frac{1}{\|\phi_n\|_w^2\, \lambda_n} \phi_n(x') \phi_n(x), \tag{8}$$

*where the functions $\phi_n(x)$ are a complete orthogonal set satisfying the Sturm-Liouville system (4).*

A solution to $\mathcal{L}y(x) = f(x)$ is thus given by

$$y(x) = \int_a^b G(x; x') f(x')\, dx'. \tag{9}$$

Note that the solution $y(x)$ defined by (9) satisfies the same boundary conditions as do the eigenfunctions (since it is a linear combination of them). Therefore we have

$$\alpha y(a) + \beta y'(a) = 0, \ \gamma y(b) + V \delta y'(b) = 0.$$

**Example 1.** Consider the system

$$y'' + y = 3 \sin 2\pi x, \ \ y(0) = 0, y(1) = 0. \tag{10}$$

We need the eigenfunctions for the operator $\mathcal{L}\phi = \phi'' + \phi$, or

$$\phi'' + \phi = \lambda \phi, \ \phi(0) = 0, \phi(1) = 0. \tag{11}$$

Recasting the differential equation as $\phi'' = (\lambda - 1)\phi$, we recognize (11) as a disguised version of the Sturm-Liouville problem defining the sine series eigenfunctions $\phi_n(x) = \sin n\pi x$ -

$$\phi_n'' = -n^2 \pi^2 \phi_n, \ \phi_n(0) = \phi_n(1) = 0. \tag{12}$$

Identifying $-n^2\pi^2$ as $(\lambda - 1)$, we have $\lambda_n = 1 - n^2\pi^2$ as the eigenvalues of (11). The weight $w(x) = 1$ and the expansion (8) reads

$$G(x; x') = \sum_{n=1}^{\infty} \frac{\sin n\pi x \sin n\pi x'}{\| \sin n\pi x \|^2 \lambda_n} = \sum_{n=1}^{\infty} \frac{2}{1 - n^2\pi^2} \sin n\pi x \sin n\pi x'.$$

Finally, the solution to the system (10) is

$$y(x) = \int_0^1 G(x; x') \, 3 \sin 2\pi x' \, dx', \tag{13}$$

which simplifies in this case to

$$y(x) = \frac{3}{1 - 4\pi^2} \sin 2\pi x \tag{14}$$

(exercise 1). ∎

The extension of these notions to nonself-adjoint operators and to higher dimensions is fairly straightforward. Suppose the differential equation to be solved is expressed as

$$\mathcal{M}\psi(x, y, z) = f(x, y, z). \tag{15}$$

Here $\mathcal{M}$ is a partial differential operator. The boundary conditions associated with (15) are presumed to be homogeneous.

It may be helpful to have a specific operator in mind. Suppose, for example, that $\mathcal{M}$ is the Laplacian operator and we have to solve

$$\nabla^2 \psi(x, y, z) = f(x, y, z)$$

inside a unit cube $0 \le x, y, z \le 1$ with homogeneous Dirichlet boundary conditions. The eigenfunction equation for $\nabla^2$ is, of course, Helmholtz's equation

$$\nabla^2 \Phi(x, y, z) = \Lambda \Phi,$$

and its eigenfunctions in this case consist of the familiar sine factors:

$$\nabla^2 [\sin m\pi x \, \sin n\pi y \, \sin p\pi z]$$
$$= -(m^2 + n^2 + p^2)\pi^2 [\sin m\pi x \, \sin n\pi y \, \sin p\pi z] \tag{16}$$

(Table 7.4). They are orthogonal in the unit cube with respect to the weight factor $w = 1$.

The Green's function that we seek is a function $G(x,y,z;x',y',z')$ which, as a function of $(x,y,z)$, satisfies the homogeneous boundary conditions and also the equation

$$\mathcal{M}G(x,y,z;x',y',z') = \delta(x - x')\delta(y - y')\delta(z - z') \qquad (17)$$

For then the function

$$\psi(x,y,z) = \iiint f(x',y',z')\,G(x,y,z;x',y',z')\,dx'\,dy'\,dz' \qquad (18)$$

will satisfy the homogeneous boundary conditions and the differential equation (15):

$$
\begin{aligned}
\mathcal{M}\psi(x,y,z) &= \mathcal{M}\iiint f(x',y',z')\,G(x,y,z;x',y',z')\,dx'\,dy'\,dz' \\
&= \iiint f(x',y',z')\mathcal{M}G(x,y,z;x',y',z')\,dx'\,dy'\,dz'
\end{aligned}
$$

(since $\mathcal{M}$ only operates on $x,y,z$)

$$
\begin{aligned}
&= \iiint f(x',y',z')\delta(x - x')\delta(y - y')\delta(z - z')\,dx'\,dy'\,dz' \\
&= f(x,y,z).
\end{aligned}
$$

Our goal is to construct $G$ out of the (multidimensional) eigenfunctions of $\mathcal{M}$, which satisfy an eigenvalue equation (containing, possibly, an extra factor $g$[1]):

$$\mathcal{M}\Phi_{mnp}(x,y,z) = \Lambda_{mnp}g(x,y,z)\Phi_{mnp}(x,y,z). \qquad (19)$$

These eigenfunctions are orthogonal with respect to the weight factor $w(x,y,z)$, which will be distinct from $g(x,y,z)$ if $\mathcal{M}$ is not self-adjoint.

If we postulate the expansion of the Green's function to be

$$G(x,y,z;x',y',z') = \sum_m \sum_n \sum_p a_{mnp}\Phi_{mnp}(x,y,z), \qquad (20)$$

then the coefficients $a_{mnp}$ can be determined by insertion into the equation (17):

$$
\begin{aligned}
\mathcal{M}G &= \sum_m \sum_n \sum_p a_{mnp}\mathcal{M}\Phi_{mnp}(x,y,z) \\
&= \sum_m \sum_n \sum_p a_{mnp}\Lambda_{mnp}\,g(x,y,z)\,\Phi_{mnp}(x,y,z) \\
&= \delta(x - x')\delta(y - y')\delta(z - z'). \qquad (21)
\end{aligned}
$$

---

[1]Recall (Sec. 6.2) that a factor $g$ appears in the eigenvalue equation for some differential operators.

Multiplication of (21) by $\Phi_{rst}(x, y, z)w(x, y, z)/g(x, y, z)$ and integrating produces

$$\sum_m \sum_n \sum_p a_{mnp}\Lambda_{mnp} \iiint \Phi_{mnp}(x, y, z)\,\Phi_{rst}(x, y, z)\,w(x, y, z)\,dx\,dy\,dz$$

$$= \iiint \delta(x - x')\delta(y - y')\delta(z - z')\,\Phi_{rst}(x, y, z)\,]\tfrac{w(x,y,z)}{g(x,y,z)}dx\,dy\,dz$$

which, by orthogonality, reduces to

$$a_{rst}\Lambda_{rst} \iiint w(x, y, z)\Phi_{rst}^2(x, y, z)\,dx\,dy\,dz$$

$$= \Phi_{rst}(x', y'z')w(x', y', z')/g(x', y', z'). \tag{22}$$

Insertion of (22) into (20) gives the desired expansion. For archival purposes we state the result in full generality.

**Theorem.** *The Green's function for (15), satisfying the same homogeneous boundary conditions as the eigenfunctions* $\Phi_{mnp}(x, y, z)$ *in (19), is given by*

$$G(x, y, z; x', y', z')$$

$$= \sum_m \sum_n \sum_p \frac{w(x', y', z')/g(x', y', z')}{\Lambda_{mnp}\|\Phi_{mnp}(x, y, z)\|_w^2}$$

$$\times \Phi_{mnp}(x, y, z)\Phi_{mnp}(x', y', z'). \tag{23}$$

**Example 2.** Consider Poisson's equation

$$\nabla^2\Psi(x, y) = x^2 y$$

in the square depicted in Fig. 1 with homogeneous Dirichlet conditions on all sides. ($\Psi$ can be interpreted as the two-dimensional electrostatic potential due to a charge density $-x^2 y$ inside a square with grounded walls.)

We need to select from Table 7.4 an eigenfunction family for the Laplacian - in other words, solutions to Helmholtz's equation. We find

$$\nabla^2\Phi = \nabla^2(a_1 \cos \mu x + a_2 \sin \mu x)(b_1 \cos \nu y + b_2 \sin \nu y) = -(\mu^2 + \nu^2)\Phi. \tag{24}$$

Thus, choosing the constants to satisfy the boundary conditions we have

$$\phi_{mn}(x, y) = \sin mx \sin ny, \quad \lambda_{mn} = -m^2 - n^2; \quad m, n = 1, 2, 3, \ldots.$$

The weights are all unity, the factor $g$ in eq. (19) is unity, and (23) becomes

$$G(x, y; x'y') = \sum_{m=1}^\infty \sum_{n=1}^\infty -\frac{1}{(m^2 + n^2)(\pi/2)^2} \sin mx \sin ny \sin mx' \sin ny'.$$

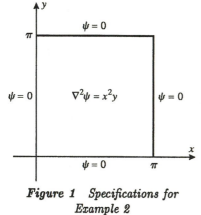

*Figure 1   Specifications for Example 2*

Therefore the solution to the problem is

$$\Psi(x,y) = \int_0^\pi \int_0^\pi G(x,y;x'y')\, x'^2 y'\, dx'\, dy'$$

$$= -\sum_{m=1}^\infty \sum_{n=1}^\infty \frac{4}{\pi^2(m^2+n^2)} \sin mx \sin ny$$

$$\times \int_0^\pi x'^2 \sin mx'\, dx' \int_0^\pi y' \sin ny'\, dy'$$

We omit the evaluation of the integrals. ∎

**Example 3.** For the equation

$$\mathcal{M}y = x^2 y'' + xy' - y = x^3, \ y(0) \text{ finite}, \ y(1) = 0 \tag{25}$$

we employ the eigensystem generating the Fourier-Bessel series (Table 6.3 #16):

$$x^2 y_n'' + xy_n' - y_n = -j_{1,n}^2 x^2 y_n, \ y_n(0) \text{ finite}, \ y_n(1) = 0; \ y_n(x) = J_1(j_{1,n}x). \tag{26}$$

The weight $w$ for this equation is $x$, but the factor $g$ in (26) is $x^2$ (since (26) is not in Sturm-Liouville form). Formula (23) reads

$$G(x;x') = \sum_{n=1}^\infty \frac{x'/x'^2}{-j_{1,n}^2 \|J(j_{1,n}x)\|_{(\rho=x)}^2} J_1(j_{1,n}x) J_1(j_{1,n}x')$$

$$= \sum_{n=1}^\infty \frac{2/x'}{J_2^2(j_{1,n})} J_1(j_{1,n}x) J_1(j_{1,n}x')$$

and the solution to (25) is

$$y(x) = \int_0^1 G(x;x')\, x'^3\, dx'$$

$$= \sum_{n=1}^\infty J_1(j_{1,n}x) \frac{2}{J_2^2(j_{1,n})} \int_0^1 x'^2\, J_1(j_{1,n}x')\, dx'. \quad ∎$$

For curvilinear two- and three-dimensional configurations it is convenient to modify the fundamental differential equation (21) defining the Green's function. Let us demonstrate the situation with a specific differential operator. Consider the Laplacian in spherical coordinates:

$$\nabla^2 \psi = \frac{1}{r^2} \frac{\partial}{\partial r}\left(r^2 \frac{\partial \psi}{\partial r}\right) + \frac{1}{r^2 \sin\phi} \frac{\partial}{\partial \phi}\left(\sin\phi \frac{\partial \psi}{\partial \phi}\right) + \frac{1}{r^2 \sin^2\phi} \frac{\partial^2 \psi}{\partial \theta^2}.$$

Now the natural extension of (21) might appear to be

$$\nabla^2 G(r,\theta,\phi;r',\theta',\phi') = \delta(r-r')\delta(\theta-\theta')\delta(\phi-\phi') \quad \textbf{(wrong)}. \tag{27}$$

The difficulty here is that the right-hand side of (27) does not represent a *unit* impulse when integrated over *volume*. Recall that the differential element of volume in spherical coordinates is given by $r^2 \sin\phi\, d\theta\, d\phi\, dr$ (see Davis and Snider), so that the total intensity of the impulse in (27) is

$$\int_0^\infty \int_0^\pi \int_0^{2\pi} \delta(r-r')\,\delta(\theta-\theta')\,\delta(\phi-\phi')\, r^2 \sin\phi\, d\theta\, d\phi\, dr = r'^2 \sin\phi',$$

instead of unity!

Clearly a *unit* impulse would have to compensate for the (*Jacobian*) "scale factor" $r^2 \sin\phi$ in the spherical volume element, so the proper generalization of (21) is the following:

$$\nabla^2 G(r,\theta,\phi;r',\theta',\phi') = \frac{\delta(r-r')\delta(\theta-\theta')\delta(\phi-\phi')}{r^2 \sin\phi}. \tag{28}$$

Now the right-hand side truly represents a *unit* impulse in three dimensions, and it is consistent with $\delta(x-x')\delta(y-y')\delta(z-z')$ in (21) in the sense that its volume integral is unity.

By the same token, in cylindrical coordinates where the volume element is $\rho\, d\rho\, d\theta\, dz$, one should use

$$\nabla^2 G(\rho,\theta,z;\rho',\theta',z') = \frac{\delta(\rho-\rho')\delta(\theta-\theta')\delta(z-z')}{\rho}, \tag{29}$$

and in polar coodinates,

$$\nabla^2 G(r,\theta,r',\theta') = \frac{\delta(r-r')\delta(\theta-\theta')}{r} \tag{30}$$

Naturally the introduction of a scale factor into eq. (21) alters the formula (23) for the Green's function. We leave it as a simple task (exercise 8) for the reader to modify the derivation to show the following.

**Theorem.** *Suppose the factor $g$ appears in the eigenfunction equation for the operator $\mathcal{M}$:*

$$\mathcal{M}\Phi_{mnp}(s,t,u) = \Lambda_{mnp}g(s,t,u)\Phi_{mnp}(s,t,u).$$

*Suppose further that the scale factor $h$ appears in the equation defining the Green's function for $\mathcal{M}$:*

$$\mathcal{M}G(s,t,u;s',t',u') = h(s,t,u)\delta(s-s')\delta(t-t')\delta(u-u').$$

*Then the Green's function is given by*

$$G(s,t,u;s',t',u') =$$
$$\sum_m \sum_n \sum_p \frac{w(s',t',u')h(s',t',u')/g(s',t',u')}{\Lambda_{mnp}\|\Phi_{mnp}(s,t,u)\|_w^2}$$
$$\times \Phi_{mnp}(s,t,u)\,\Phi_{mnp}(s',t',u') \tag{31}$$

**Example 4.** To solve

$$\nabla^2 \psi(r, \phi, \theta) = (\tan r\phi)(\sec \theta) \tag{32}$$

inside the unit sphere with the boundary conditions depicted in Fig. 2, we expand the Green's function in terms of the eigenfunctions of the Laplacian in spherical coordinates. For homogeneous Dirichlet conditions on a unit sphere these are spherical Bessel functions times spherical harmonics (Table 6.3 #25, 29):

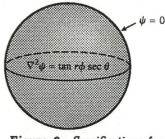

**Figure 2** *Specifications for Example 4*

$$\nabla^2 j_\ell(s_{\ell,p}r)Y_{\ell m}(\phi, \theta) = -s_{\ell,p}^2 \, j_\ell(s_{\ell,p}r)\, Y_{\ell m}(\phi, \theta). \tag{33}$$

According to the table they are orthogonal with respect to the weight $w(r, \theta, \phi) = r^2 \sin \phi$:

$$\int_0^1 \int_0^\pi \int_0^{2\pi} j_\ell(s_{\ell,p}r)Y_{\ell m}(\phi, \theta)j_{\ell'}(s_{\ell',p'}r)\overline{Y_{\ell'm'}(\phi, \theta)}r^2 \sin\phi \, d\theta \, d\phi \, dr$$

$$= [j_{\ell+1}^2(s_{\ell,p})/2]\delta_{\ell\ell'}\delta_{mm'}\delta_{pp'}. \tag{34}$$

To implement (31) we note that the factor $g$ in the eigenfunction equation (33) is unity. Furthermore the scale factor $h(r, \theta, \phi)$ for spherical coordinates, displayed in (28), happens to be the reciprocal of the weight factor $w(r, \theta, \phi)$. The derivation in exercise 8 reveals that the complex conjugation operation on $Y_{\ell m}$ in the orthogonality relation (34) resurfaces as a minor modification in formula (31), resulting in the expression

$$G(r, \phi, \theta; r', \phi', \theta') = -\sum_{p=1}^\infty \sum_{\ell=0}^\infty \sum_{m=-\ell}^\ell \frac{2}{s_{\ell,p}^2 \, j_{\ell+1}^2(s_{\ell,p})} j_\ell(s_{\ell,p}r)\, Y_{\ell m}(\phi, \theta)$$

$$\times j_\ell(s_{\ell,p}r')\, \overline{Y_{\ell m}(\phi', \theta')} \tag{35}$$

The solution to (32) is thus expressed

$$\psi(r, \phi, \theta) = \int_0^1 \int_0^\pi \int_0^{2\pi} G(r, \phi, \theta; r', \phi', \theta') \tan(r'\phi') \sec\theta'$$

$$\times r'^2 \sin\phi' \, d\theta' \, d\phi' \, dr'. \tag{36}$$

∎

If the differential equation is time dependent, the procedure described above cannot be implemented directly (since there are no eigenfunctions for the time variable). We defer discussion of this case to Sec. 8.2.

What happens if one of the eigenvalues of the differential operator $\mathcal{M}$ is zero? Certainly this appears to foil the mathematics because of this zero arising in the denominator of the expansion (23). In a sense, however, the problem is not with the mathematics, but with the physics - *the solution to the differential equation (15) and the (homogeneous) boundary conditions is not unique.* Any multiple of the eigenfunction corresponding to $\Lambda = 0$ can be added to a solution of (15), and the sum will be another solution. Thus there are not enough conditions imposed on the physical model to determine its response unambiguously. Exercises 5 and 6 elaborate on this phenomenon.

# Exercises 8.1

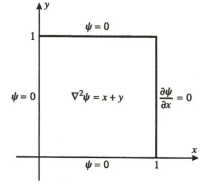

**Figure 3    Exercise 3**

1. (a) Verify that eq. (14) follows from (13).

   (b) Verify that eq. (14) solves the system (10).

2. Solve by Green's functions: $y'' - y = x$, $y(0) = 0$, $y(1) = 0$.

3. Express the solution to the problem in Fig. 3 by Green's functions.

4. An alternative way of deriving a formula for the solution to (15) in terms of the eigenfunctions (19) of $\mathcal{M}$ is to expand both the solution $\psi$ and the function $f/g$ ($g$ defined by (19)) in terms of the eigenfunctions:

$$\frac{f(x_1, x_2)}{g(x_1, x_2)} = \sum_{n_1, n_2} c_{n_1, n_2} \Phi_{n_1, n_2}(x_1, x_2), \quad c_{n_1, n_2} = \frac{< f/g, \Phi_{n_1, n_2} >_w}{\|\Phi_{n_1, n_2}\|_w^2};$$

(37)

$$\psi(x_1, x_2) = \sum_{n_1, n_2} a_{n_1, n_2} \Phi_{n_1, n_2}(x_1, x_2), \quad a_{n_1, n_2} \text{ to be determined.}$$

(38)

(a) Show that insertion of (37, 38) into (15) yields the formula

$$a_{n_1, n_2} = c_{n_1, n_2} / \Lambda_{n_1, n_2} \tag{39}$$

if none of the eigenvalues are zero.

(b) Verify that expressions (37-39) and expressions (18, 23) ultimately produce the same formula for the solution $\psi$.

(c) Now assume that one of the eigenvalues for $\mathcal{M}$ is zero; specifically, let $\Lambda_{00} = 0$. By comparing the expansions that you derived in part (a), prove that the equation $\mathcal{M}\psi = f$ has *no solution* unless $f/g$ is orthogonal to the eigenfunction $\Phi_{00}$ -

$$< f/g, \Phi_{00} >_w = 0 \tag{40}$$

- and that when (40) holds, $\mathcal{M}\psi = f$ has *an infinite number of solutions* because any constant multiple of $\Phi_{00}$ can be added to $\psi$. (It is understood that the homogeneous boundary conditions are imposed.)

This result is sometimes called the "Fredholm Alternative."

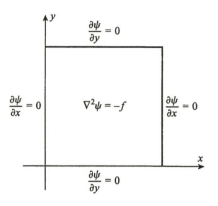

*Figure 4    Exercise 5*

5. The problem defined by Fig. 4 depicts the steady state temperature distribution in a slab insulated on all sides and heated by an external source/sink. The function $f(x, y)$ describes the amount of heat applied at $(x, y)$ per unit time, per unit area; the net heat applied (per unit time) is $\iint f(x, y)\, dx\, dy$ (integrated over the slab). Physically we know that unless this net heat is zero, no steady state can be achieved. Work out the eigenfunctions for this problem and demonstrate that this conclusion is born out by the Fredholm Alternative. What is the physical significance of the indeterminacy in the solution? (Recall discussion of the pure Neumann problem in Sec. 4.6.)

6. Consider the generic differential equation $\mathcal{M}\psi = f$ in a region with nonhomogeneous boundary conditions imposed around its sides. Suppose the Green's function for the corresponding equation with homogeneous boundary conditions is $G(x, y, z; x', y', z')$, *and also suppose that you have knowledge of a function* $\chi(x, y, z)$ *which satisfies the boundary conditions* (but not the differential equation). Show that

$$\psi(x, y, z) = \chi(x, y, z) +$$
$$\iiint G(x, y, z; x', y', z')\, [f - \mathcal{M}\chi(x', y', z')]\, dx'\, dy'\, dz'$$

7. Derive eqs. (31, 35).

## 8.2    Transform Methods

**Summary**  When a time dependent differential equation or boundary condition is Fourier- or Laplace-transformed, the time variable is replaced by the parameter $\omega$ or $s$. With this parameter treated as a constant, the transformed problem is formally time independent and can be analyzed by the methodology of the preceding sections.

The absence of time eigenfunctions, as we have seen, undermines the direct application of our separation of variables techniques to time dependent problems. We have finessed this difficulty for the wave and heat equations by first meeting all the spatial nonhomogeneous boundary conditions with a time-independent "stationary" solution. Then in the "transient" subproblem the treatment of the time factors is deferred until the end. This is

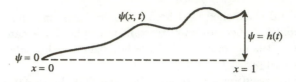

*Figure 1   Time-dependent boundary condition*

feasible precisely because all the spatial boundary conditions in the transient subproblem are homogeneous.

Unfortunately, our artifice cannot be applied in all cases. Consider, for example, the situation of a vibrating string with one end fixed, and the other jiggled up and down (see Fig. 1). The motion is governed by the wave equation

$$\frac{\partial^2 \Psi}{\partial t^2} = \frac{\partial^2 \Psi}{\partial x^2},$$

the initial conditions

$$\Psi(x,0) = f(x), \; \frac{\partial \Psi}{\partial t} = g(x),$$

and the edge conditions

$$\Psi(0,t) = 0, \; \Psi(1,t) = h(t). \tag{1}$$

Normally we would create a transient subproblem with homogeneous edge conditions by first extracting the solution to the steady state subproblem

$$\frac{\partial^2 \Psi_\infty}{\partial x^2} = 0; \; \Psi_\infty(0) = 0, \; \Psi_\infty(1) = h(t). \tag{2}$$

However eqs. (2) clearly do *not* define a time independent problem! In fact their solution is $\Psi_\infty(x,t) = xh(t)$, and since this does not (likely) solve the wave equation, it is of no help in pursuing the solution to eqs. (1).

In general the time domain procedure is ineffective in solving time dependent partial differential equations with time dependent boundary conditions, because the "stationary" subproblem which is supposed to satisfy the spatial boundary conditions simply is *not* stationary - it, too, depends on time.

Therefore we need another approach for time dependent problems. *This is provided by the Fourier and Laplace transforms; by exchanging the time variable for a parameter in the frequency domain or the s plane, we render the resulting equations time independent, so that they can be solved by the methods studied previously.*

Some examples will make the procedure clear. Recall (Sec. 3.9) that the Fourier transform has the more direct physical interpretation in terms of steady-state responses, while the Laplace transform has the advantage of exhibiting the effects of the initial configuration.

**Example 1.** We will analyze eqs. (1), for the vibrating string driven at one end, by employing the Laplace transform. We transform the time dependence of $\Psi(x, t)$:

$$\Phi(x, s) = \int_0^\infty e^{-st} \Psi(x, t)\, dt \tag{3}$$

That is, we temporarily regard $x$ as a fixed parameter and take the Laplace transform of the time function $\Psi(x, t)$; there is a different transform for each value of $x$, and the notation $\Phi(x, s)$ reflects this.

By the usual rules the transform of the second time derivative of $\Psi$ is given by

$$
\begin{aligned}
\int_0^\infty e^{-st} \frac{\partial^2 \Psi}{\partial t^2}(x, t)\, dt &= s^2 \Phi(x, s) - s\Psi(x, 0) - \frac{\partial \Psi}{\partial t}(x, 0) \\
&= s^2 \Phi(x, s) - sf(x) - g(x),
\end{aligned} \tag{4}
$$

while the spatial derivatives simply come out of the integral:

$$\int_0^\infty e^{-st} \frac{\partial^2 \Psi}{\partial x^2}(x, t)\, dt = \frac{\partial^2}{\partial x^2} \int_0^\infty e^{-st} \Psi(x, t)\, dt = \frac{\partial^2 \Phi}{\partial x^2}(x, s). \tag{5}$$

Thus the transform $\Phi$ of $\Psi$ obeys the system derived by combining (1, 3, 4, and 5):

$$\frac{\partial^2 \Phi}{\partial x^2} = s^2 \Phi - sf(x) - g(x),$$

$$\Phi(0, s) = 0,$$

$$\Phi(1, s) = \int_0^\infty e^{-st} h(t)\, dt = H(s) \quad \text{(the Laplace transform of } h(t)\text{)}. \tag{6}$$

Since the time dependence is gone from the system (6), for fixed $s$ we have an ordinary nonhomogeneous differential equation with boundary conditions. It can be solved using the methods of Sec. 1.4 (exercise 1):

$$
\begin{aligned}
\Phi(x, s) = {}& -\sinh sx \int_0^x [sf(\xi) + g(\xi)] \cosh s\xi\, d\xi \\
& + \cosh sx \int_0^x [sf(\xi) + g(\xi)] \sinh s\xi\, d\xi + A\frac{\sinh sx}{\sinh s}, \tag{7}
\end{aligned}
$$

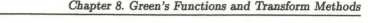

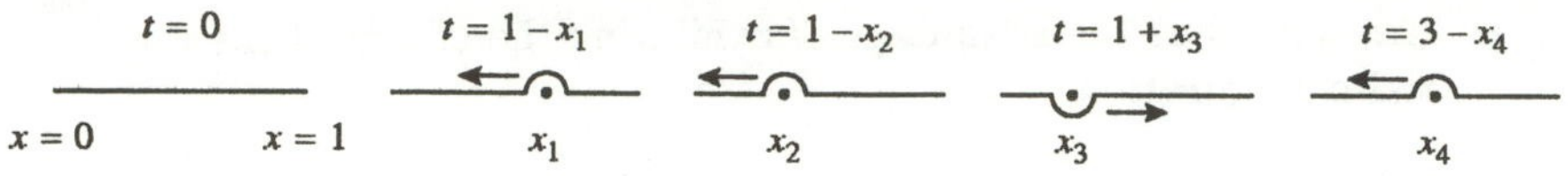

$t = 0$        $t = 1 - x_1$      $t = 1 - x_2$      $t = 1 + x_3$      $t = 3 - x_4$

$x = 0$        $x = 1$        $x_1$        $x_2$        $x_3$        $x_4$

*Figure 2*   *Snapshots of string*

where $A$ is given by

$$A = H(s) + \sinh s \int_0^1 [s f(\xi) + g(\xi)] \cosh s\xi \, d\xi$$

$$- \cosh s \int_0^1 [s f(\xi) + g(\xi)] \sinh s\xi \, d\xi. \tag{8}$$

The solution of eqs. (1) is thus given by the inverse transform of (7) (the Bromwich integral)

$$\Psi(x, t) = \frac{e^{at}}{2\pi} \int_{-\infty}^{\infty} \Phi(x, a + i\omega) e^{i\omega t} \, d\omega$$

for $a$ sufficiently large (recall Sec. 3.8).        ∎

We observe the two somewhat unsatisfactory features of the Laplace transform method - the formidability of evaluating the inverse transform, and the difficulty of physically visualizing the solution from its transformed format. For this example, however, we can interpret some features of the solution if the initial state is quiescent: $f(x) = g(x) = 0$. In such a case we use the geometric series (Sec. 2.2) to write (7) as

$$\Phi(x, s) = H(s) \frac{\sinh sx}{\sinh s} = H(s) \left\{ \frac{e^{sx} - e^{-sx}}{e^s - e^{-s}} \right\}$$

$$= H(s) \left\{ e^{-s(1-x)} - e^{-s(1+x)} \right\} \frac{1}{1 - e^{-2s}}$$

$$= H(s) e^{-s(1-x)} \left\{ 1 + e^{-2s} + e^{-4s} + e^{-6s} + \cdots \right\}$$

$$\quad - H(s) e^{-s(1+x)} \left\{ 1 + e^{-2s} + e^{-4s} + e^{-6s} + \cdots \right\}$$

$$= H(s) \left\{ e^{-s(1-x)} - e^{-s(1+x)} + e^{-s(1-x+2)} \right.$$

$$\left. - e^{-s(1+x+2)} + e^{-s(1-x+4)} - e^{-s(1+x+4)} + \cdots \right\} \tag{9}$$

Since the exponential factors all correspond to time delays when the inverse transform is taken (Table 3.2), the first term of (9) says that the motion of the right end $h(t)$ is propagated to the point $x$ with a time delay equal to the distance $(1 - x)$ from $x$ to the right end (see Fig. 2)[2]. The second term says

---

[2] Recall that the speed of propagation in these units in unity (end of Sec. 4.4).

the motion $-h(t)$ arrives after a delay of $1+x$; this corresponds to $h(t)$ traveling the length of the string (unity) and back to $x$, having been negated by the reflection at the (fixed) left end. The original motion $h(t)$ then recurs 2 units after its previous appearance (third term); it has made a round trip, reflecting off both ends. Next the reflected motion recurs (fourth term); and so on. (Recall the discussion in Sec. 4.3.)

**Example 2.** Now we reconsider the motion of the vibrating string, driven at one end, by using Fourier analysis. The Fourier transform of the time dependence of the solution is given by

$$\Phi(x,\omega) = \frac{1}{2\pi} \int_{-\infty}^{\infty} \Psi(x,t)e^{-i\omega t}dt.$$

(We trust that use of the same symbol - $\Phi$ - for both the Laplace and Fourier transforms will cause no confusion, since the presence of the parameter $s$ or $\omega$ identifies the type of transform.) The inverse transform represents $\Psi(x,t)$ as a combination of sinusoids:

$$\Psi(x,t) = \int_{-\infty}^{\infty} \Phi(x,\omega)e^{i\omega t}d\omega. \tag{10}$$

From (10) we are reminded that the operator $\partial/\partial t$ is replaced in the transformed domain by $(i\omega)$ (Sec. 3.6).

As discussed in Sec. 3.9, for frequency analysis we drop the initial conditions from the formulation (1) (and presume that $h(t)$ is given for all time):

$$\frac{\partial^2 \Psi}{\partial t^2} = \frac{\partial^2 \Psi}{\partial x^2}, \ \Psi(0,t) = 0, \ \Psi(1,t) = h(t). \tag{11}$$

The Fourier transform of (11) then reads

$$\frac{\partial^2 \Phi(x,\omega)}{\partial x^2} = -\omega^2 \Phi(x,\omega); \ \Phi(0,\omega) = 0, \ \Phi(1,\omega) = H(\omega), \tag{12}$$

where we have taken the transform of the boundary function $h(t)$ to be

$$H(\omega) = \frac{1}{2\pi} \int_{-\infty}^{\infty} h(t)e^{-i\omega t}dt. \tag{13}$$

The general solution to the (ordinary) differential equation (12) is

$$\Phi(x,\omega) = C_1 \cos\omega x + C_2 \sin\omega x, \tag{14}$$

and the first boundary condition dictates $C_1 = 0$. The second boundary condition becomes

$$C_2 \sin\omega = H(\omega), \tag{15}$$

giving rise to the (modal) form

$$\Phi(x,\omega) = \frac{\sin \omega x}{\sin \omega} H(\omega) \qquad (16)$$

and the corresponding solution

$$\Psi(x,t) = \int_{-\infty}^{\infty} \frac{\sin \omega x}{\sin \omega} H(\omega) e^{i\omega t} d\omega. \qquad (17)$$

∎

The obvious deficiency in formula (16) for the cases when $\omega = n\pi$, $(n = 0, \pm 1, \pm 2, \dots)$ serves as an alert to resonance phenomena. Recall that in Sec. 7.3 we found that the functions $\sin n\pi x \, e^{in\pi t}$ are the solutions to the *unforced* problem $[h(t) = 0]$ associated with (11), so one would expect that the response to forces at those frequencies would be singular (recall Sec. 3.2). In exercise 6 we guide the interested reader through a reinterpretation of our analysis using the distribution calculus of Sec. 1.3, and we recover mathematically correct solutions for $\omega = n\pi$. They are *physically* unrealistic, however, since damping mechanisms must be incorporated in any model operated near resonances.

The Laplace and Fourier transforms need not be restricted to problems with nonhomogeneous time dependent spatial boundary conditions; they can be used for any situation involving the wave or heat equations. The reader may prefer to avoid them if the boundary conditions are time independent, however, simply because the inverse transform can be rather obscure. Consider the following.

**Example 3.** The temperature profile in a thin rod insulated at both ends satisfies the system

$$\frac{\partial \Psi}{\partial t} = \frac{\partial^2 \Psi}{\partial x^2}; \ \frac{\partial \Psi}{\partial x}(0,t) = 0, \ \frac{\partial \Psi}{\partial x}(\pi,t) = 0, \ \Psi(x,0) = f(x). \qquad (18)$$

Its solution by the "time domain" procedure is easily computed to be

$$\Psi(x,t) = \sum_{n=0}^{\infty} a_n \cos nx \, e^{-n^2 t}, \quad a_n = \frac{2}{\pi} \int_0^{\pi} f(x) \cos nx \, dx \qquad (19)$$

(for $a_0$ take one-half of the expression; see exercise 2).

The system of equations determining the Laplace transform $\Phi(x,s)$ is obtained from (18) in the manner of Example 1:

$$\frac{\partial^2 \Phi}{\partial x^2} = s\Phi - f(x), \ \frac{\partial \Phi}{\partial x}(0,s) = 0, \ \frac{\partial \Phi}{\partial x}(\pi,s) = 0.$$

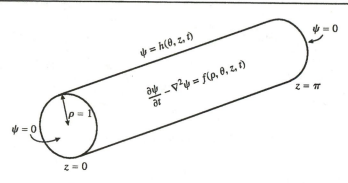

**Figure 3** *Specifications for Example 4*

Its solution is (exercise 3)

$$\Phi(x,s) = \frac{1}{\sqrt{s}} \int_0^x f(\xi) \sinh \sqrt{s}(\xi - x) \, d\xi + A \cosh \sqrt{s}x \tag{20}$$

where $A$ is given by

$$A = \frac{\int_0^\pi f(\xi) \cosh \sqrt{s}(\xi - \pi) \, d\xi}{\sqrt{s} \sinh \sqrt{s}\pi}$$

In principle, the solution is completed by putting (20) into the Bromwich integral. However the characteristics of the solution are far more transparent from the form derived using the time domain procedure (19). And while it is true that one can show that the inverse transform of (20) agrees with eq. (19), the additional labor is considerable.[3] ∎

The reader may have noticed in Examples 1 and 3 that the Laplace transform of the original, homogeneous, differential equation was rendered *non*homogeneous by the initial conditions; thus Green's functions had to be invoked to complete the solution. The next example illustrates the employment of transforms and Green's functions in a higher dimensional problem.

**Example 4.** The temperature distribution inside the cylinder in Fig. 3, with time dependent internal heating sources described by $f(r,\theta,z,t)$, is governed by the equation

$$\frac{\partial \Psi}{\partial t} - \nabla^2 \Psi = f, \tag{21}$$

---

[3]We *can*, however, read off the asymptotic form of the solution from (20) by applying the final value theorem (Sec. 3.8):

$$\lim_{t \to \infty} \Psi(x,t) = \lim_{s \to 0} s\Phi(x,s) = \frac{1}{\pi} \int_0^\pi f(\xi) \, d\xi.$$

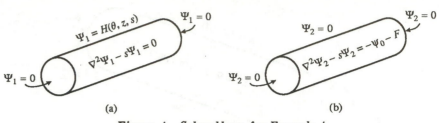

**Figure 4**   *Subproblems for Example 4*

the initial condition

$$\Psi(\rho, \theta, z, t = 0) = g(\rho, \theta, z),$$

and the time-dependent Dirichlet boundary conditions

$$\Psi(\rho, \theta, z = 0, t) = \Psi(\rho, \theta, z = \pi, t) = 0, \ \Psi(\rho = 1, \theta, z, t) = h(\theta, z, t). \quad (22)$$

To get rid of the time dependencies we Laplace-transform the time variable in (21, 22) to obtain

$$s\Phi(\rho, \theta, z, s) - g(\rho, \theta, z) - \nabla^2 \Phi(\rho, \theta, z, s) = F(\rho, \theta, z, s)$$

or

$$\nabla^2 \Phi(\rho, \theta, z, s) - s\Phi(\rho, \theta, z, s) = -g(\rho, \theta, z) - F(\rho, \theta, z, s) \quad (23)$$

and

$$\Phi(\rho, \theta, z = 0, s) = \Phi(\rho, \theta, z = \pi, s) = 0, \ \Phi(\rho = 1, \theta, z, s) = H(\theta, z, s), \quad (24)$$

where $\Phi(\rho, \theta, z, s), F(\rho, \theta, z, s)$, and $H(\theta, z, s)$ are the transforms of

$$\Psi(\rho, \theta, z, t), \ f(\rho, \theta, z, t), \ \text{and} \ h(\theta, z, t)$$

respectively.

Because the system (23, 24) has two nonhomogeneities we define two subproblems, depicted in Fig. 4.

PROBLEM 1. The *homogeneous* differential equation

$$\nabla^2 \Phi_1(\rho, \theta, z, s) - s\Phi_1(\rho, \theta, z, s) = 0 \quad (25)$$

is subjected to the *nonhomogeneous* boundary conditions

$$\Phi_1(\rho, \theta, z = 0, s) = \Phi_1(\rho, \theta, z = \pi, s) = 0, \ \Phi_1(\rho = 1, \theta, z, s) = H(\theta, z, s). \quad (26)$$

PROBLEM 2. The *nonhomogeneous* differential equation

$$\nabla^2 \Phi_2(\rho, \theta, z, s) - s\Phi_2(\rho, \theta, z, s) = -g(\rho, \theta, z) - F(\rho, \theta, z, s) \quad (27)$$

is subjected to the *homogeneous* boundary conditions

$$\Phi_2(\rho,\theta,z=0,s) = \Phi_2(\rho,\theta,z=\pi,s) = 0, \ \Phi_2(\rho=1,\theta,z,s) = 0. \quad (28)$$

To solve problem 1 we observe that eq. (25) is Helmholtz's equation with the parameter $s$ playing the role of "$\Lambda$" in Table 7.4. Because the nonhomogeneity is on the radial ($\rho$) boundary, we seek separated solutions with the $\theta$ and $z$ factors as eigenfunctions. The table thus provides us with the format

$$[a_1 \cos \nu\theta + a_2 \sin \nu\theta]\,[b_1 I_\nu(\kappa\rho) + b_2 I_{-\nu}(\kappa\rho)]\,[c_1 \cos \mu z + c_2 \sin \mu z]$$

where $\kappa^2 = s + \mu^2$. For periodicity in $\theta$ we take $\nu = n = 0,1,2,\ldots$, and the homogeneous end conditions dictate $c_1 = 0, \mu = m = 1,2,3,\ldots$. We replace $I_{-n}(\kappa\rho)$ by $K_n(\kappa\rho)$, but then dismiss it thanks to the finiteness condition at $\rho = 0$. Acknowledging the exceptional term for $n = 0$ we have as our general solution

$$\Phi_1(\rho,\theta,z,s) = \sum_{m=0}^{\infty} a_{m,0}\, I_0(\sqrt{s+m^2}\rho)\, \sin mz$$

$$+ \sum_{n=1}^{\infty} [a_{mn} \cos n\theta + b_{mn} \sin n\theta]\, I_n(\sqrt{s+m^2}\rho)\, \sin mz. \quad (29)$$

The fitting of the final boundary condition in (26) is left as an exercise.

The Green's function for problem 2 has to be expanded in terms of the eigenfunctions of the operator $(\nabla^2 - s)$; that is, they are solutions to

$$\nabla^2\Phi - s\Phi = \Lambda\Phi \ \text{or}\ \nabla^2\Phi = (\Lambda + s)\Phi. \quad (30)$$

Thus we turn to Table 7.4 for solutions to Helmholtz's equation with the eigenvalue "$\Lambda$" therein identified as *our* $(\Lambda + s)$. Since all three factors must be oscillatory we read off the separated solutions as

$$[a_1 \cos \nu\theta + a_2 \sin \nu\theta]\,[b_1 J_\nu(\kappa\rho) + b_2 J_{-\nu}(\kappa\rho)]\,[c_1 \cos \mu z + c_2 \sin \mu z]$$

with the identification $\Lambda + s = -\kappa^2 - \mu^2$. Imposing the homogeneous Dirichlet condition at $\rho = 1$ we mold the eigenfunctions out of these solutions:

$$\begin{aligned}
\Phi_{0pm} &= J_0(j_{0,p}\rho)\, \sin mz, \\
\Phi_{npm}^{(1)} &= \cos n\theta\, J_n(j_{n,p}\rho)\, \sin mz, \ \text{and} \\
\Phi_{npm}^{(2)} &= \sin n\theta\, J_n(j_{n,p}\rho)\, \sin mz
\end{aligned} \quad (31)$$

(see Table 6.3 #16) with $\Lambda_{npm} = -j_{n,p}^2/a^2 - m^2\pi^2/c^2 - s; \ n = 1,2,\ldots; \ p = 1,2,3,\ldots; \ m = 1,2,3,\ldots$.

To implement eq. (31) of Sec. 8.1 for the Green's function expansion, we note that $g = 1, h = 1/\rho$ for the cylindrical coordinate system, and the

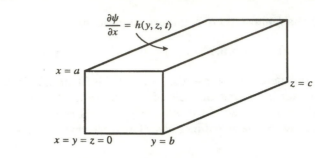

***Figure 5***   *Specifications for Example 5*

weight $w = \rho$ for the eigenfunctions (31); thus

$$
\begin{aligned}
G(\rho, \theta, z; \rho', \theta', z') \;=\; & \sum_{p=1}^{\infty} \sum_{m=1}^{\infty} \frac{1}{-(j_{0,p}^2 + m^2 + s)\|\Phi_{0pm}\|^2} \\
& \times J_0(j_{0,p}\rho) \sin mz \, J_0(j_{0,p}\rho') \sin mz' \\[2ex]
+ \; & \sum_{n=1}^{\infty} \sum_{p=1}^{\infty} \sum_{m=1}^{\infty} \frac{1}{-(j_{n,p}^2 + m^2 + s)\|\Phi_{npm}\|^2} \\
& \times \cos n\theta \, J_n(j_{n,p}\rho) \sin mz \, \cos n\theta' \, J_n(j_{n,p}\rho') \sin mz' \\[2ex]
+ \; & \sum_{n=1}^{\infty} \sum_{p=1}^{\infty} \sum_{m=1}^{\infty} \frac{1}{-(j_{n,p}^2 + m^2 + s)\|\Phi_{npm}\|^2} \\
& \times \sin n\theta \, J_n(j_{n,p}\rho) \sin mz \, \sin n\theta' \, J_n(j_{n,p}\rho') \sin mz'
\end{aligned}
$$

and the solution to problem 2 is

$$
\begin{aligned}
\Phi_2(\rho, \theta, z, s) \;=\; & -\int_0^{2\pi} \int_0^1 \int_0^{\pi} G(\rho, \theta, z; \rho', \theta', z') \\
& \times [g(\rho', \theta', z') + F(\rho', \theta', z', s)] \, \rho' \, dz' \, d\rho' \, d\theta'.
\end{aligned}
$$

The temperature $\Psi(\rho, \theta, z)$, then, is the inverse Laplace transform of $\Phi_1 + \Phi_2$.                                                                                          ∎

Example 4 certainly demonstrates that the use of Green's functions in higher dimensional Laplace-transformed problems is often cumbersome. If the differential equation is homogeneous, they can be avoided by employing a different subproblem decomposition. An example will illustrate.

**Example 5.** Let us reconsider Example 2 of Sec. 7.3, describing sound waves in a resonant rectangular cavity (Fig. 5). Now, however, we suppose the top "lid" of the box is vibrated so that the boundary condition at $x = a$ specifies the normal derivative of the incremental pressure to be some (given) time-dependent function

$$\frac{\partial \Psi}{\partial x} = h(y, z, t) \text{ at } x = a. \tag{32}$$

On the other five walls we take the usual homogeneous Neumann boundary conditions. The three dimensional (homogeneous) wave equation

$$\frac{\partial^2 \Psi}{\partial t^2} = \nabla^2 \Psi \tag{33}$$

governs the motion, and for initial conditions we take

$$\Psi(x, y, z, 0) = f(x, y, z), \frac{\partial \Psi}{\partial t}(x, y, z, 0) = g(x, y, z). \tag{34}$$

Now if we form the Laplace transform of (33) we have

$$s^2 \Phi(x, y, z, s) - s f(x, y, z) - g(x, y, z) = \nabla^2 \Phi(x, y, z, s)$$

(recall eq. (4)), or

$$\nabla^2 \Phi - s^2 \Phi = -sf - g. \tag{35}$$

Equation (35) is nonhomogeneous and requires Green's functions for its solution. We finesse this computation by going to a subproblem decomposition.

The three nonhomogeneities in this problem are due to the functions $h(y, z, t)$, $f(x, y, z)$, and $g(x, y, z)$. Therefore we decompose as follows:

$$\Psi(x, y, z, t) = \Psi_1(x, y, z, t) + \Psi_2(x, y, z, t) + \Psi_3(x, y, z, t).$$

SUBPROBLEM 1: Find $\Psi_1$ satisfying the wave equation (33) and

$$\frac{\partial \Psi_1}{\partial x} = h(y, z, t) \text{ at } x = a, \ \frac{\partial \Psi_1}{\partial n} = 0 \text{ on the other walls,} \tag{36}$$

$$\Psi_1(x, y, z, 0) = \frac{\partial \Psi_1}{\partial t}(x, y, z, 0) = 0. \tag{37}$$

SUBPROBLEM 2: Find $\Psi_2$ satisfying the wave equation (33) and

$$\frac{\partial \Psi_2}{\partial n} = 0 \text{ on all walls,}$$

$$\Psi_2(x, y, z, 0) = f(x, y, z), \frac{\partial \Psi_2}{\partial t}(x, y, z, 0) = 0.$$

SUBPROBLEM 3: Find $\Psi_3$ satisfying the wave equation (33) and

$$\frac{\partial \Psi_3}{\partial n} = 0 \text{ on all walls,}$$

$$\Psi_3(x, y, z, 0) = 0, \ \frac{\partial \Psi_3}{\partial t}(x, y, z, 0) = g(x, y, z).$$

Subproblem 3 was solved in Example 2, Sec. 7.3. As we noted above eq. (14) in that section, the initial disturbance $g$ was decomposed into the eigenmodes

$$\cos \frac{m\pi}{a} x \cos \frac{n\pi}{b} y \cos \frac{p\pi}{c} z$$

and the time factors

$$\sin \sqrt{(\frac{m\pi}{a})^2 + (\frac{n\pi}{b})^2 + (\frac{p\pi}{c})^2} t \tag{38}$$

were appended. Subproblem 2 is solved similarly (with cos replacing sin in (38)).

Now because of the homogeneous initial conditions (37) for Subproblem 1, eq. (35) for the Laplace transform of the wave equation is replaced by the homogeneous Helmholtz equation

$$\nabla^2 \Phi_1(x, y, z, s) - s^2 \Phi_1(x, y, z, s) = 0. \tag{39}$$

The boundary conditions for (39) are

$$\frac{\partial \Phi_1}{\partial x} = H(y, z, s) \text{ for } x = a, \ \frac{\partial \Phi_1}{\partial n} = 0 \text{ on all other walls}, \tag{40}$$

where $H(y, z, s)$ is the Laplace transform of $h(y, z, t)$.

From Table 7.4 we see that the separated solutions of the Helmholtz equation have the form

$$[a_1 \cosh \kappa x + a_2 \sinh \kappa x] [b_1 \cos \mu y + b_2 \sin \mu y] [c_1 \cos \nu z + c_2 \sin \nu z]$$

where $\kappa^2 = \mu^2 + \nu^2 + s$. The homogeneous Neumann conditions on each of the $y$ and $z$ faces, and on the face at $x = 0$, reduce this to

$$\cosh \sqrt{s^2 + (\frac{n\pi}{b})^2 + (\frac{p\pi}{c})^2} x \cos \frac{n\pi}{b} y \cos \frac{p\pi}{c} z$$

and the assembled solution reads

$$\Psi_1(x, y, z, s) = \sum_{n=0}^{\infty} \sum_{p=0}^{\infty} a_{np}$$

$$\cdot \cosh \sqrt{s^2 + (\frac{n\pi}{b})^2 + (\frac{p\pi}{c})^2} x \cos \frac{n\pi}{b} y \cos \frac{p\pi}{c} z \tag{41}$$

The nonhomogeneous boundary condition (40) is met by setting

$$a_{np} = \frac{4 \int_0^b \int_0^c H(y, z, s) \cos \frac{n\pi}{b} y \cos \frac{p\pi}{c} z \, dz \, dy}{bc \sqrt{s^2 + (\frac{n\pi}{b})^2 + (\frac{p\pi}{c})^2} \sinh \sqrt{s^2 + (\frac{n\pi}{b})^2 + (\frac{p\pi}{c})^2} a} \tag{42}$$

(take one-half of (42) for $a_{0p}$ and $a_{n0}$; take one-quarter for $a_{00}$).

Finally $\Psi_1(x, y, z, t)$ is represented by applying the Bromwich integral to (41). ∎

Before concluding this section we must point out that the Fourier and Laplace transforms are only useful in general when the coefficients in the partial differential equation do not depend explicitly on time. Thus we have seen that if we transform

$$\frac{\partial^2 \Psi}{\partial t^2} = \frac{\partial^2 \Psi}{\partial x^2}$$

we get the Helmholtz equation (24). But the transform of, say,

$$\frac{\partial^2 \Psi}{\partial t^2} = t\nabla^2 \Psi$$

involves transforming a product, and the *convolution* comes into play (Sec. 3.9, exercise 9). Usually this complication foils our procedure.

So we must confess that time dependent partial differential equations with time dependent coefficients, coupled with time dependent spatial boundary conditions, are beyond the scope of this text. Fortunately they are rare in engineering applications (other than control theory).

The applications of transform methods to radiation problems wil be studied in Sec. 8.5.

# Exercises 8.2

**1.** Verify - or, better, derive - the solution formulas (7, 8).

**2.** Derive eqs. (19).

**3.** Derive eq. (20).

**4.** Re-solve Example 3, Sec. 7.3, using the Laplace transform.

**5.** Find the Laplace transform of the solution to the problem

$$\frac{\partial \Psi}{\partial t} = \frac{\partial^2 \Psi}{\partial x^2}, \ \Psi(0,t) = 0, \ \Psi(1,t) = 5, \ \Psi(x,0) = f(x).$$

**6.** In this exercise you are going to rework Example 2 with the distribution calculus. It is obvious from (16) that the frequencies $\omega = n\pi$ ($n = 0, \pm 1, \pm 2, \ldots$) are troublesome, so take the forcing function $h(t)$ to have the form

$$h(t) = \int_{-\infty}^{\infty} H(\omega)e^{i\omega t}d\omega + Be^{iN\pi t} + A$$

$$= \int_{-\infty}^{\infty} [H(\omega) + B\delta(\omega - N\pi) + A\delta(\omega)]e^{i\omega t}\, d\omega$$

where $H(\omega)$ is an ordinary function. Similarly write the solution as

$$\Psi(x,t) = \int_{-\infty}^{\infty} X(x;\omega)e^{i\omega t}\,d\omega.$$

(a) Take the Fourier transform of the differential equation and boundary conditions (11) to derive

$$\frac{\partial^2 X}{\partial x^2} = -\omega^2 X, \quad X(0;\omega) = 0,$$
$$X(1;\omega) = H(\omega) + B\delta(\omega - N\pi) + A\delta(\omega). \quad (43)$$

(b) Show that the solution to the differential equation and the first boundary condition in (43) is

$$X(x;\omega) = c\sin\omega x \qquad (44)$$

for any $c$, possibly depending on $\omega$: $c = c(\omega)$.

(c) The second boundary condition in (43) imposes the relation

$$c(\omega)\sin\omega = H(\omega) + B\delta(\omega - N\pi) + A\delta(\omega) \qquad (45)$$

The distribution solution of this equation for $c(\omega)$ is tricky because of the zeros of $\sin\omega$. In exercise 8, Sec. 1.3, we derived

$$\sin\omega\,\delta(\omega - n\pi) = 0 \quad \text{and} \quad \sin\omega\,\delta'(\omega - n\pi) = (-1)^{n+1}\delta(\omega - n\pi) \qquad (46)$$

Use eqs. (46) to show that the solution to (45) is given by

$$c(\omega) = \frac{H(\omega)}{\sin\omega} - (-1)^N B\delta'(\omega - N\pi) - A\delta'(\omega)$$
$$+ \sum_{n=-\infty}^{\infty} d_n\delta(\omega - n\pi) \qquad (47)$$

where the coefficients $d_n$ are arbitrary.

(d) Assemble the Fourier integral for $\Psi(x,t)$:

$$\Psi(x,t) = \int_{-\infty}^{\infty} c(\omega)\sin\omega x\,e^{i\omega t}\,d\omega$$
$$= \int_{-\infty}^{\infty} \frac{H(\omega)}{\sin\omega}\sin\omega x\,e^{i\omega t}\,d\omega$$
$$+ (-1)^N B[x\cos N\pi x\,e^{iN\pi t} + it\sin N\pi x\,e^{iN\pi t}]$$
$$+ Ax + \sum_{n=-\infty}^{\infty} d_n\sin n\pi x\,e^{in\pi t} \qquad (48)$$

(e) Verify directly that (48) solves the differential equation and boundary conditions (11).

**7.** Fit the coefficients in (29) to the boundary conditions (26).

**8.** Work out the Green's function for the wave equation inside the unit sphere, with homogeneous Dirichlet conditions and with the time dependence removed by Fourier transformation.

**9.** (*For advanced students*) Use residue theory and the Bromwich integral to verify the consistency of eqs. (19) and (20). (Appendix B may be helpful.)

## 8.3   Closed-Form Green's Functions

**Summary**   Green's functions in closed form are known for certain equations and geometries. For the heat and wave equations in free space, the time dependence can be handled explicitly. The derivations do not involve eigenfunctions, but the functions are tabulated herein for reference.

The methods of Sec. 8.1 are very general and yield eigenfunction expansions of the Green's functions for any problem that is amenable to the separation of variables technique. Closed-form expressions for Green's functions are also known for some simple geometries, and we compile a listing in this section. The derivations will not be given, inasmuch as they usually lie outside the customary domain of differential equation techniques.

The best-known Green's function is derived from the identity

$$\nabla^2 \frac{1}{\sqrt{x^2 + y^2 + z^2}} = \nabla^2 \frac{1}{|\mathbf{R}|} = -4\pi \, \delta(x) \, \delta(y) \, \delta(z) \tag{1}$$

where $\mathbf{R} = x\mathbf{i} + y\mathbf{j} + z\mathbf{k}$. It is not difficult (though somewhat laborious; see exercise 1) to show that $\nabla^2(1/|\mathbf{R}|)$ is zero for $\mathbf{R} \neq \mathbf{0}$, but the fact that the singularity gives rise to the indicated combination of delta functions requires some vector analysis; see Davis and Snider, Jackson, or Hellwig. To fit (1) into the Green's function format, we "relativize" $\mathbf{R} = x\mathbf{i} + y\mathbf{j} + z\mathbf{k}$ with respect to $\mathbf{R}' = x'\mathbf{i} + y'\mathbf{j} + z'\mathbf{k}$ and define

$$G_0(x, y, z; x', y', z') = \frac{-1}{4\pi \sqrt{(x - x')^2 + (y - y')^2 + (z - z')^2}} = \frac{-1}{4\pi |\mathbf{R} - \mathbf{R}'|} \tag{2}$$

so that (1) implies

$$\nabla^2 G_0(x, y, z; x', y', z') = \delta(x - x') \, \delta(y - y') \, \delta(z - z'). \tag{3}$$

This $G_0$ is then a Green's function for the Laplacian operator, satisfying finiteness conditions at infinity.

**Example 1.** The solution to the three dimensional Poisson equation

$$\nabla^2 \psi(x,y,z) = f(x,y,z) \tag{4}$$

which goes to zero at infinity is given by

$$\psi(x,y,z) = \frac{-1}{4\pi} \iiint\limits_{-\infty}^{+\infty} \frac{f(x',y',z')}{\sqrt{(x-x')^2 + (y-y')^2 + (z-z')^2}} \, dx' \, dy' \, dz' . \tag{5}$$

(In electrostatics $f$ is proportional to the charge density and $\psi$ is the voltage; then (4) is a form of *Gauss' Law* and (5) is the *Coulomb potential.*)   ■

A more versatile Green's function can be obtained by adding a solution $\gamma(x,y,z)$ of the (homogeneous) *Laplace* equation to the function $-1/4\pi|\mathbf{R} - \mathbf{R}'|$; the sum $\tilde{G} = -1/4\pi|\mathbf{R} - \mathbf{R}'| + \gamma$ still satisfies the basic relationship

$$\nabla^2 \tilde{G} = \nabla^2 \left( \frac{-1}{4\pi|\mathbf{R} - \mathbf{R}'|} + \gamma \right) = \delta(x - x') \, \delta(y - y') \, \delta(z - z'),$$

but $\gamma$ may be adjusted so that homogeneous boundary conditions are satisfied by $\tilde{G}$ on some specified *finite* surface.

For instance if $z' > 0$ the function

$$\gamma_1(x,y,z;x',y',z') = \frac{1}{4\pi\sqrt{(x-x')^2 + (y-y')^2 + (z+z')^2}}$$

satisfies Laplace's equation in the region $z > 0$ (since its singularity is at $(x',y',-z')$), and the combination

$$
\begin{aligned}
G_1(\mathbf{R};\mathbf{R}') \;=\; & \frac{-1}{4\pi\sqrt{(x-x')^2 + (y-y')^2 + (z-z')^2}} \\
& + \frac{1}{4\pi\sqrt{(x-x')^2 + (y-y')^2 + (z+z')^2}} \tag{6}
\end{aligned}
$$

obviously satisfies homogeneous Dirichlet conditions on the plane $z = 0$:

$$G_1(x,y,0;x',y',z') = 0.$$

Thus $G_1$ is a Green's function for Poisson's equation in the half-space $z > 0$ with Dirichlet boundary conditions.

**Example 2.** The solution to the Dirichlet problem in the half-space

$$\nabla^2 \psi(x,y,z) = f(x,y,z) \quad (z > 0), \quad \psi(x,y,0) = 0$$

is given by

$$\psi(x,y,z) = \int_0^\infty \int_{-\infty}^\infty \int_{-\infty}^\infty G_1(x,y,z;x',y',z')f(x',y',z')dx'\,dy'\,dz' \ . \quad (7)$$

∎

In the references by Jackson and Hellwig it is shown that the function

$$\gamma_2(\mathbf{R};\mathbf{R}') = \frac{a}{4\pi|\mathbf{R}'|}\ \left|\mathbf{R} - \frac{a^2}{|\mathbf{R}'|^2}\mathbf{R}'\right|^{-1}, \quad (8)$$

added to $G_0$ in (2), produces a Green's function which equals zero on the sphere $|\mathbf{R}| = a$ (see exercise 2). Thus the function $G_2 = G_0 + \gamma$ can be used to solve Poison's equation either inside *or outside* a sphere, with Dirichlet boundary conditions.

**Example 3.** The solution to the Dirichlet problem

$$\nabla^2\psi(r,\theta,\phi) = f(r,\theta,\phi) \quad (r > a),$$

$$\psi(a,\theta,\phi) = 0, \quad \psi(r,\theta,\phi) \to 0 \text{ as } r \to \infty$$

is given by

$$\begin{aligned}
\psi(r,\theta,\phi) &= \int_a^\infty \int_0^\pi \int_0^{2\pi} G_2(r,\theta,\phi;r',\theta',\phi') \\
&\quad \times f(r',\theta',\phi')\,r'^2 \sin\phi'\,d\phi'\,d\theta'\,dr'
\end{aligned} \quad (9)$$

∎

Recall that in spherical coordinates the functions $G_0$, $G_1$, $G_2$ all satisfy

$$\nabla^2 G(\mathbf{R};\mathbf{R}') = \delta(r-r')\,\delta(\theta-\theta')\,\delta(\phi-\phi')/r^2\sin\phi,$$

since the right hand side describes a *unit* impulse (Sec. 8.1). Note also the volume element $r'^2 \sin\phi'\,d\phi'\,d\theta'\,dr'$ in the solution representation (9).

For the two-dimensional Laplacian operator the search for Green's functions is motivated by the identity (Hellwig)

$$\begin{aligned}
\nabla^2\left[\frac{1}{2\pi}\ln|\mathbf{R}-\mathbf{R}'|\right] &= \nabla^2\frac{1}{2\pi}\ln\sqrt{(x-x')^2 + (y-y')^2 + (z-z')^2} \\
&= \delta(x-x')\,\delta(y-y').
\end{aligned}$$

However because $(1/2\pi)\ln|\mathbf{R}-\mathbf{R}'|$ does not approach zero at infinity, it is seldom employed directly as a Green's function. The combination

$$G_3(\mathbf{R};\mathbf{R}') = \frac{1}{2\pi}\left\{\ln|\mathbf{R}-\mathbf{R}'| + \ln\frac{a}{|\mathbf{R}'|} - \ln\left|\mathbf{R} - \frac{a^2}{|\mathbf{R}'|^2}\mathbf{R}'\right|\right\} \quad (10)$$

satisfies

$$\nabla^2 G_3(\mathbf{R}; \mathbf{R}') = \delta(x - x')\,\delta(y - y'), \quad G_3(\mathbf{R}; \mathbf{R}') = 0 \ \text{for}\ |\mathbf{R}| = a. \quad \blacksquare$$

**Example 4.** The solution to the Dirichlet problem

$$\nabla^2 \psi(r, \theta) = f(r, \theta) \quad (r < a), \quad \psi(a, \theta) = 0$$

is given by

$$\psi(r, \theta) = \int_0^a \int_0^{2\pi} G_3(r, \theta; r', \theta') f(r', \theta')\, r'\, d\theta'\, dr' \qquad \blacksquare$$

*Conformal mapping* can be applied to $G_3$ to find Green's functions for other two dimensional configurations; see Saff and Snider or Churchill.

A Green's function for Helmholtz's equation $\nabla^2 \psi = \Lambda \psi$ in three dimensions is a solution to

$$\nabla^2 G_4(\mathbf{R}; \mathbf{R}', \Lambda) - \Lambda G_4(\mathbf{R}; \mathbf{R}', \Lambda) = \delta(x - x')\,\delta(y - y')\,\delta(z - z'). \quad (11)$$

Such a function is given by (Arfken or Davis and Snider)

$$G_4(\mathbf{R}; \mathbf{R}', \Lambda) = -\frac{e^{\pm\sqrt{\Lambda}|\mathbf{R} - \mathbf{R}'|}}{4\pi|\mathbf{R} - \mathbf{R}'|} \qquad (12)$$

where the $\pm$ sign is dictated by the conditions at infinity. Some examples will serve as clarification.

**Example 5.** The initial value problem for the three dimensional nonhomogeneous heat equation

$$\frac{\partial \Psi}{\partial t} - \nabla^2 \Psi = f(x, y, z, t),$$

$$\Psi(x, y, z, t = 0) = g(x, y, z), \quad \Psi \to 0 \ \text{as}\ |\mathbf{R}| \to \infty$$

is Laplace-transformed to read

$$s\Phi - \nabla^2 \Phi = F(x, y, z; s) - g(x, y, z), \quad \Phi \to 0 \ \text{as}\ |\mathbf{R}| \to \infty. \quad (13)$$

The left hand side of eq. (13) has the format of (11) with $\Lambda = s$ (and an overall change of sign). With $\Lambda = s$ in (12) and the $\pm$ sign chosen so that $G_4 \to 0$ at $\infty$, we obtain as the solution to (13)

$$\Phi(x, y, z; s) = (+)\iiint\limits_{-\infty}^{+\infty} \frac{e^{-\sqrt{s}|\mathbf{R} - \mathbf{R}'|}}{4\pi|\mathbf{R} - \mathbf{R}'|}$$

$$\times [F(x', y', z'; s) + g(x', y', z')]\, dx'\, dy'\, dz'$$

and the inverse transform yields the solution $\Psi(x, y, z, t)$. ∎

**Example 6.** To find solutions to the three dimensional nonhomogeneous wave equation

$$\frac{\partial^2 \Psi}{\partial t^2} - \nabla^2 \Psi = f(x, y, z, t) \tag{14}$$

we can Fourier-transform the time variable to obtain

$$-\omega^2 \Phi(x, y, z; \omega) - \nabla^2 \Phi(x, y, z; \omega) = F(x, y, z; \omega). \tag{15}$$

Equation (15) has the format of (11) with $\Lambda = -\omega^2$ (and, again, an overall change of sign). We tentatively employ (12) to write the solution to (15) as

$$\Phi(x, y, z; \omega) = (+) \iiint\limits_{-\infty}^{+\infty} \frac{e^{\pm i\omega|\mathbf{R}-\mathbf{R}'|}}{4\pi|\mathbf{R}-\mathbf{R}'|} F(x', y', z'; \omega) \, dx' \, dy' \, dz'. \tag{16}$$

The inverse transform of (16) will take the form

$$
\begin{aligned}
\Psi(x, y, z, t) &= \int_{-\infty}^{\infty} \psi(x, y, z; \omega) e^{i\omega t} \, d\omega \\
&= (+) \iiint\limits_{-\infty}^{\infty} \frac{e^{\pm i\omega|\mathbf{R}-\mathbf{R}'|+i\omega t}}{4\pi|\mathbf{R}-\mathbf{R}'|} \\
&\quad \times F(x', y', z'; \omega) \, dx' \, dy' \, dz'.
\end{aligned} \tag{17}
$$

*Either* choice of sign in the factor $e^{i\omega(t\pm|\mathbf{R}-\mathbf{R}'|)}$ yields a mathematically acceptable solution to (14); the (+) sign yields incoming waves and the (-) sign yields outgoing waves.[4] ∎

By linearity one can easily see that if

$$-\frac{e^{+i\omega|\mathbf{R}-\mathbf{R}'|}}{4\pi|\mathbf{R}-\mathbf{R}'|} \quad \text{and} \quad -\frac{e^{-i\omega|\mathbf{R}-\mathbf{R}'|}}{4\pi|\mathbf{R}-\mathbf{R}'|}$$

both satisfy eq. (11), then so must the combination

$$-\frac{1}{4\pi|\mathbf{R}-\mathbf{R}'|} \left[ \frac{e^{+i\omega|\mathbf{R}-\mathbf{R}'|}}{2} + \frac{e^{-i\omega|\mathbf{R}-\mathbf{R}'|}}{2} \right] = -\frac{\cos(\omega|\mathbf{R}-\mathbf{R}'|)}{4\pi|\mathbf{R}-\mathbf{R}'|}.$$

Utilization of this Green's function leads to standing-wave solutions to (15):

$$
\begin{aligned}
\Psi(x, y, z, t) &= \int_{-\infty}^{\infty} \Phi(x, y, z; \omega) e^{i\omega t} \, d\omega \\
&= (+) \iiint\limits_{-\infty}^{\infty} \frac{\cos(\omega|\mathbf{R}-\mathbf{R}'|)}{4\pi|\mathbf{R}-\mathbf{R}'|} \\
&\quad \times F(x', y', z'; \omega) \, dx' \, dy' \, dz'.
\end{aligned}
$$

---

[4] Section 8.5 contains further elaboration on this point.

Two-dimensional Green's functions for Helmholtz's equations are derived in Hildebrand's book. With $\Lambda = s$, for the Laplace-transformed heat equation $-(1/2\pi)K_0(\sqrt{s}|\mathbf{R} - \mathbf{R}'|)$ is the Green's function which goes to zero at infinity:

$$[\nabla^2 - s]\,\frac{-1}{2\pi}K_0(\sqrt{s}|\mathbf{R} - \mathbf{R}'|) = \delta(x - x')\,\delta(y - y'). \tag{18}$$

For the Fourier-transformed wave equation, where $\Lambda = -\omega^2$, the Green's function corresponding to outgoing waves is $(i/4)H_0^{(2)}(\omega|\mathbf{R} - \mathbf{R}'|)$:

$$[\nabla^2 + \omega^2]\,\frac{i}{4}H_0^{(2)}(\omega|\mathbf{R} - \mathbf{R}'|) = \delta(x - x')\,\delta(y - y'). \tag{19}$$

See exercise 3.

Green's functions for the Helmholtz equation in one dimension are given by $\frac{-1}{2(\pm\sqrt{\Lambda})}e^{\pm\sqrt{\Lambda}|x-x'|}$ with the sign determined appropriately. For the Laplace-transformed heat equation we use

$$\left[\frac{\partial^2}{\partial x^2} - s\right]\frac{1}{2\sqrt{s}}e^{-\sqrt{s}|x-x'|} = \delta(x - x'). \tag{20}$$

For the Fourier-transformed wave equation the form

$$\left[\frac{\partial^2}{\partial x^2} + \omega^2\right]\frac{i}{2\omega}e^{-i\omega|x-x'|} = \delta(x - x') \tag{21}$$

yields waves travelling to the right. See exercise 4.

A Green's function for the *time domain* analysis[5] of the one dimensional heat equation would be a solution to

$$\frac{\partial G_5}{\partial t} - \frac{\partial^2 G_5}{\partial x^2} = \delta(x - x')\,\delta(t - t'). \tag{22}$$

Hildebrand shows that such a function, approaching zero at $x = \pm\infty$, is given for $t > 0$ by

$$G_5(x, t; x', t') = \begin{cases} \frac{1}{\sqrt{4\pi(t-t')}}e^{-(x-x')^2/4(t-t')}, & 0 < t' < t \\ 0, & \text{otherwise} \end{cases} \tag{23}$$

**Example 7.** A solution to

$$\frac{\partial \Psi}{\partial t} - \frac{\partial^2 \Psi}{\partial x^2} = f(x, t), \;\; t \geq 0; \;\; \Psi \text{ finite at } x = \pm\infty$$

---

[5]Green's functions for time domain analysis are sometimes known as *Riemann functions* in the mathematical literature.

is given by

$$\Psi(x,t) = \int_0^t \int_{-\infty}^{\infty} \frac{e^{-(x-x')^2/4(t-t')}}{\sqrt{4\pi(t-t')}} f(x',t')\, dx'\, dt' \, .$$ ∎

Similarly the time domain analysis of the nonhomogeneous one dimensional wave equation requires solutions to

$$\frac{\partial^2 G_6}{\partial t^2} - \frac{\partial^2 G_6}{\partial x^2} = \delta(x-x')\,\delta(t-t'). \tag{24}$$

Such a function, finite at $x = \pm\infty$, is exhibited in Friedman's book to be

$$G_6(x,t;,x',t') = \begin{cases} \frac{1}{2}, & t' < t - |x-x'| \\ 0, & \text{otherwise} \end{cases} \tag{25}$$

for *all* $t$, $-\infty < t < \infty$.

**Example 8.** A solution to

$$\frac{\partial^2 \Psi}{\partial t^2} - \frac{\partial^2 \Psi}{\partial x^2} = f(x,t), \quad \Psi \text{ finite at } x = \pm\infty$$

is computed, using (25), to be

$$\Psi(x,t) = \int_{t'=-\infty}^{t} \int_{x'=x-(t-t')}^{x+(t-t')} \frac{f(x',t')}{2}\, dx'\, dt' \tag{26}$$

(exercise 5). Recall this formula from exercise 3, Sec. 4.3. ∎

Equation (26) demonstrates, once again, the finite velocity of propagation inherent in the wave equation (Sec. 4.3); the values of $f(x',t')$ for points $x'$ beyond $(t-t')$ do not influence the solution at $(x,t)$.

In Courant and Hilbert's second volume it is shown that the Green's function for the initial-value problem for the wave equation in two dimensions

$$\frac{\partial^2 G_7}{\partial t^2} - \nabla^2 G_7 = \delta(x-x')\,\delta(y-y')\,\delta(t-t') \tag{27}$$

is given by

$$G_7(\mathbf{R};\mathbf{R}',t;t') = \begin{cases} \dfrac{1/2\pi}{\sqrt{(t-t')^2 - |\mathbf{R}-\mathbf{R}'|^2}}, & 0 < t' < t \text{ and } |\mathbf{R}-\mathbf{R}'| < t - t' \\ 0, & \text{otherwise} \end{cases} \tag{28}$$

Thus a solution to the two-dimensional problem

$$\frac{\partial^2 \Psi}{\partial t^2} - \nabla^2 \Psi = f(\mathbf{R},t), \quad \Psi(\mathbf{R},0) = \frac{\partial \Psi}{\partial t}(\mathbf{R},0) = 0 \tag{29}$$

can be written

$$\Psi(\mathbf{R}, t) = \int_{t'=0}^{t} \int_{|\mathbf{R}-\mathbf{R}'|<t-t'} \frac{f(\mathbf{R}', t')}{2\pi \sqrt{(t-t')^2 - |\mathbf{R} - \mathbf{R}'|^2}} \, dx' \, dy' \, dt' . \quad (30)$$

Friedman also shows that a three dimensional solution to

$$\frac{\partial^2 G_7}{\partial t^2} - \nabla^2 G_7 = \delta(x - x') \, \delta(y - y') \, \delta(z - z') \, \delta(t - t') \quad (31)$$

is given by

$$G_7(\mathbf{R}; \mathbf{R}', t; t') = \begin{cases} \frac{\delta(t-t'-|\mathbf{R}-\mathbf{R}'|)}{4\pi|\mathbf{R}-\mathbf{R}'|}, & t > t' \\ 0, & \text{otherwise} \end{cases} \quad (32)$$

The resulting solution to the problem

$$\frac{\partial^2 \Psi}{\partial t^2} - \nabla^2 \Psi = f(\mathbf{R}, t) \quad (33)$$

is

$$\Psi(\mathbf{R}, t) = \iiint_{-\infty}^{+\infty} \int_{t'=-\infty}^{\infty} \frac{\delta(t - t' - |\mathbf{R} - \mathbf{R}'|)}{4\pi|\mathbf{R} - \mathbf{R}'|} f(\mathbf{R}', t') \, dt' \, dx' \, dy' \, dz'$$

$$= \iiint_{-\infty}^{+\infty} \frac{f(\mathbf{R}', t - |\mathbf{R} - \mathbf{R}'|)}{4\pi|\mathbf{R} - \mathbf{R}'|} \, dx' \, dy' \, dz' . \quad (34)$$

Because (34) has the appearance of (5) with an altered time parameter, the former is sometimes known as the *retarded potential*. Note that in three dimensions only the values of $f$ at points $(\mathbf{R}', t')$ which are *exactly* at a distance $|\mathbf{R} - \mathbf{R}'| = t - t'$ contribute to $\Psi(\mathbf{R}, t)$. In one and two dimensions all points *within* this distance contribute (eqs. (26, 30)). Courant and Hilbert contains an interesting discussion relating this phenomenon to the Huyghens construction.

# Exercises 8.3

1. Verify that $\nabla^2 \frac{1}{|\mathbf{R}-\mathbf{R}'|} = 0$ for $\mathbf{R} \neq \mathbf{R}'$.

2. Use exercise 1 to show that $\nabla^2[G_0 + \gamma_2] = 0$ for $|\mathbf{R}|, |\mathbf{R}'| < a$, and that $G_0 = -\gamma_2$ when $|\mathbf{R}| = a$ (eqs. (2, 8)).

3. (a) Use (18) to express a solution to the problem

$$\frac{\partial \Psi}{\partial t} - \frac{\partial^2 \Psi}{\partial x^2} - \frac{\partial^2 \Psi}{\partial y^2} = \sin xy , \quad \Psi(x, y, 0) = \cos xy$$

in the Laplace domain.

(b) Use (19) to express a solution to

$$\frac{\partial^2 \Psi}{\partial t^2} - \frac{\partial^2 \Psi}{\partial x^2} - \frac{\partial^2 \Psi}{\partial y^2} = \sin xy$$

in the Fourier domain.

4.  (a) Use (20) to express a solution to the problem

$$\frac{\partial \Psi}{\partial t} - \frac{\partial^2 \Psi}{\partial x^2} = \sin x, \quad \Psi(x,0) = \cos x$$

in the Laplace domain.

(b) Use (21) to express a solution to

$$\frac{\partial^2 \Psi}{\partial t^2} - \frac{\partial^2 \Psi}{\partial x^2} = \sin x$$

in the Fourier domain.

(c) Use the procedure of Sec. 1.4 to derive the Green's functions (20, 21).

5. Derive the representation (26) from the Green's function (25).

## 8.4  Discontinuities

**Summary**  The eigenfunction expansions of solutions to problems with discontinuities converge slowly. Accuracy can be improved if a known function, exhibiting the same kind of discontinuity, is first subtracted from the solution; the "smoothed" remainder can then be modeled satisfactorily with eigenfunctions. The flexibility afforded by Green's functions render this a feasible procedure.

Consider the problems indicated in Figs. 1-3. The axes therein have been rotated to clarify the features of the solutions.

Figure 1 could describe the electrostatic potential inside a two dimensional rectangle, three of whose sides are grounded while the potential $\psi = (2 - x)$ volts is applied to the fourth. Clearly the upper and right edges must be insulated from each other by a small piece of dielectric. In Fig. 2 we can imagine a conducting rod, whose temperature profile is initially heated to $\psi = 10$ degrees, suddenly thrust (at $t = 0$) into an environment where its ends are in contact with ice. Fig. 3 might describe a guitar string struck with a mallet just away from the end, or "bridge".

Of course these configurations are only idealizations. Any *real* string would break before it would permit itself to be deformed into the discontinuous shape

$$\psi(x) = \begin{cases} e^{-x^2} & \text{for } 0 < x \leq L \\ 0, & x = 0 \end{cases} \tag{1}$$

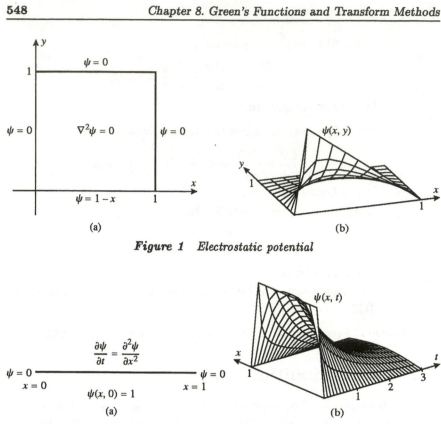

**Figure 1**  *Electrostatic potential*

**Figure 2**  *Heated conducting rod*

(although a "slinky" can approximate this situation rather well!). And no real dielectric of "infinitesimal" thickness could prevent an electrical short between the bottom and left edges in Fig. 1. However since these situations can be approximated closely in practice, it is reasonable to seek mathematical solutions for them.

We are going to discuss these solutions on three levels:

**(i)** the qualitative nature of the mathematical solutions;

**(ii)** the validity and *practical* utility of the straightforward separation of variables procedure;

**(iii)** an artifice for enhancing the accuracy of the solution representations.

The general mathematical theory (recall Sec. 4.8) states that *elliptic* partial differential equations — such as Laplace's equation — and *parabolic* partial differential equations — such as the heat equation — typically have solutions which, although they may change rapidly, are continuously differentiable to all orders in the interior of the solution region[6]. In other words

---

[6]Of course if the *coefficients* in the equation have discontinuities the solutions will

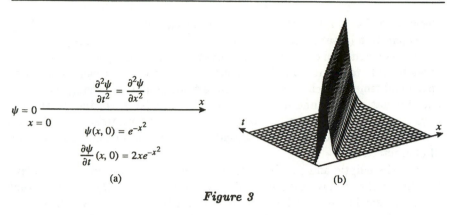

$$\psi = 0 \xrightarrow{\quad \dfrac{\partial^2 \psi}{\partial t^2} = \dfrac{\partial^2 \psi}{\partial x^2} \quad} x$$

$x = 0$

$$\psi(x, 0) = e^{-x^2}$$

$$\frac{\partial \psi}{\partial t}(x, 0) = 2xe^{-x^2}$$

(a)          (b)

**Figure 3**

the discontinuities on the boundary are smoothed out immediately in the interior, as demonstrated in Figs. 1 and 2. The interior values approach the boundary values where the latter are continuous, while at the points of discontinuity the interior values waver between the limiting boundary values on either side of the discontinuity. Shortly we will present an example which illustrates this behavior more specifically.

On the other hand, for *hyperbolic* partial differential equations - such as the wave equation - the discontinuities propagate into the interior of the (space-time) solution region. The solution expressions themselves have to be treated as "distributions", involving the generalized functions of Sec. 1.5. If a slinky is released from the initial position described in Fig. 3, the kink will propagate down the spring as shown with practically no distortion until it is reflected at the other end. It follows a *shock* wave front and does not get smoothed.

Remarkably, the series solutions given by the separation of variables procedure *remain valid* in these circumstances! They converge to the solution at every interior point for elliptic and parabolic problems, and they converge to the solution at all of its points of continuity for hyperbolic problems. However this conscientious performance is only realized if *all* the terms in the sums are retained. Truncating the series (which, of course, is a practical necessity) leads to the type of difficulty illustrated in Figs. 2 and 3 of Sec. 6.5; one can't hope to model discontinuous functional behavior accurately, using smooth eigenfunctions.

The trick that is employed to compute solutions more accurately in these cases is to employ (yet another!) decomposition $\psi = \psi_1 + \psi_2$ into subproblems. First we construct an auxiliary function $\psi_1$ which contains the same *discontinuities* as the actual solution $\psi$. Then we solve for the difference $\psi_2$:

$$\psi_2 = \psi - \psi_1 \tag{2}$$

inherit them. We have seen this phenomenon in the analysis of the quantum mechanical square well potential (Sec. 7.4).

Because $\psi_2$ has no discontinuities (they get canceled by the subtraction), the separation of variables eigenfunction series fits it quite nicely.

The construction of $\psi_1$ is rather difficult in the hyperbolic case, since one must be able to figure how the discontinuous "shocks" propagate through space and time - which, it would seem, is tantamount to solving the differential equation! Fortunately the shock fronts evolve according to simpler laws than does the smooth part of the solution,[7] and we refer the dedicated reader to Courant and Hilbert's second volume for details on this "method of characteristics."

For the elliptic and parabolic cases all that is required of the auxiliary function is that it have a discontinuity *on the boundary only*, matching that of the assigned boundary data. In other words if the boundary values jump by, say, 5 units at a corner, then the auxiliary function should also jump by 5 units at the corner; if the boundary function has a jump in its first derivative at the midpoint of an edge, then the auxiliary function must have the same jump. But $\psi_1$ need have no other special properties than this.

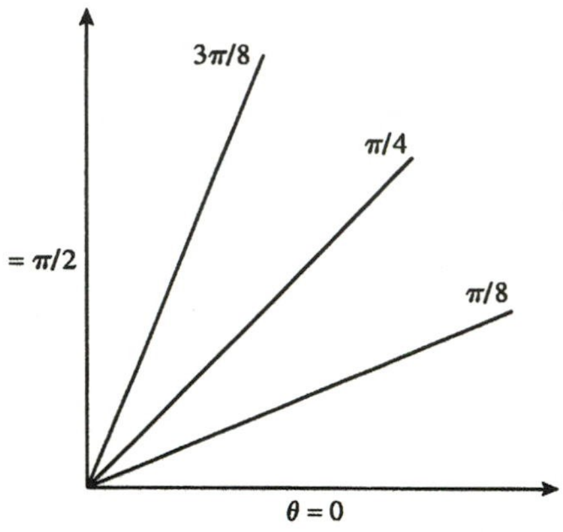

**Example 1.** A well-known example of a function with a corner discontinuity is the polar angle $\theta$ for points in the first quadrant; it jumps by $\pi/2$ at the origin and is infinitely often differentiable elsewhere (Fig. 4). Let's see how $\theta = \arctan(y/x)$ aids in the solution of the problem depicted in Fig. 1.

Since $\theta$ jumps *up* by $\pi/2$ as we move from the $x$ axis to the $y$ axis, while we need a function which jumps *down* by 1 unit, we set

$$\psi_1(x,y) = -\frac{2}{\pi}\theta = -\frac{2}{\pi}\arctan\frac{y}{x}. \tag{3}$$

Since $\nabla^2\psi = 0$, the differential equation determining the remainder $\psi_2(x,y) = \psi - \psi_1$ becomes (see Fig. 5)

$$\nabla^2\psi_2(x,y) = -\nabla^2\psi_1(x,y) = \frac{2}{\pi}\nabla^2\arctan\frac{y}{x} \tag{4}$$

with the (continuous) boundary conditions

$$\psi_2(x,0) = 1-x, \quad \psi_2(x,1) = \frac{2}{\pi}\arctan\frac{1}{x},$$

$$\psi_2(0,y) = 1, \quad \psi_2(1,y) = \frac{2}{\pi}\arctan y. \tag{5}$$

As luck would have it the Laplacian of the function $\theta = \arctan(y/x)$ is identically zero! (Recall Sec. 4.6.) Therefore eq. (4) is simply Laplace's equation

$$\nabla^2\psi_2 = 0 \tag{6}$$

*Figure 4   Corner discontinuity*

---

[7]For electromagnetic waves the shocks follow the "rays" of geometric optics.

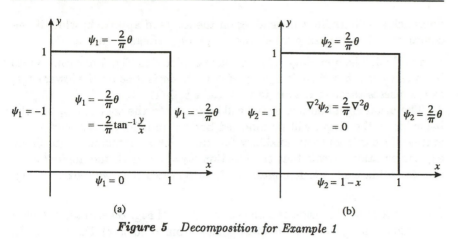

**Figure 5**   *Decomposition for Example 1*

and eqs. (5, 6) have the straightforward solution (exercise 2)

$$
\begin{aligned}
\psi_2(x,y) \;=\;& 2\sum_{n=1}^{\infty}\left\{\int_0^1 (1-x)\sin n\pi x\,dx\right\}\frac{\sin n\pi x\,\sinh n\pi(1-y)}{\sinh n\pi} \\
&+ 2\sum_{n=1}^{\infty}\left\{\int_0^1 \left[\frac{2}{\pi}\arctan\frac{1}{x}\right]\sin n\pi x\,dx\right\}\frac{\sin n\pi x\,\sinh n\pi y}{\sinh n\pi} \\
&+ 2\sum_{n=1}^{\infty}\left\{\int_0^1 (1)\sin n\pi y\,dy\right\}\frac{\sin n\pi y\,\sinh n\pi(1-x)}{\sinh n\pi} \\
&+ 2\sum_{n=1}^{\infty}\left\{\int_0^1 \left[\frac{2}{\pi}\arctan y\right]\sin n\pi y\,dy\right\}\frac{\sin n\pi y\,\sinh n\pi x}{\sinh n\pi} \quad (7)
\end{aligned}
$$

The overall solution is then

$$
\psi(x,y)=\psi_2(x,y)-\frac{2}{\pi}\arctan\frac{y}{x}. \qquad (8)
$$

∎

Of course we were lucky in this example to have selected an auxiliary function, $\psi_1$, which satisfied Laplace's equation. In general we will be left with a *non*homogeneous partial differential equation (3), with a known right hand side, and we will have to use Green's functions to obtain the solution.

Expression (8) tells us a lot about the solution to this problem around the discontinuous corner. The function $\psi_2$, being a solution to a smooth subproblem, approaches its limiting value - one - near the corner. Thus the "singular" behavior of $\psi$ is modeled by $\psi_1=-2\theta/\pi$ there. And $\psi_1$ is easy to visualize — see Fig. 6.

We see that the equipotentials $\psi=constant$ fan out radially from the corner. If one approaches the corner along a radial line, the values of $\psi$

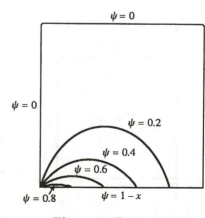

**Figure 6**   *Equipotentials*

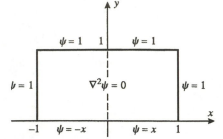

**Figure 7**  *Specifications for Example 2*

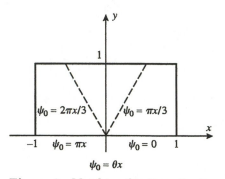

**Figure 8**  *Matching the discontinuity in Example 2*

approach a definite limit depending on the angle of approach; while if one comes in on a "drunken weave," the values of $\psi$ range from 0 to -1.

**Example 2.** The boundary value problem depicted in Fig. 7 has *continuous* boundary data, but there is a jump of -2 in its *derivative* on the lower edge, as one passes through the origin from right to left.

The usual eigenfunction series will converge to the solution $\psi$, but the behavior at the origin will be modeled better, and the convergence accelerated, if we introduce an auxiliary function with a matching jump. Some experimentation reveals that the function $\psi_0 = \theta x$, with the angle $\theta$ as in the preceding example, has *most* of the properties that we seek (see Fig. 8):

**(i)** $\theta x$ is continuous inside and on the rectangle of Fig. 8 because, although along the lower edge $\theta$ itself jumps from 0 to $\pi$ at the origin, $\theta x$ changes from $\theta x = 0$ to $\theta x = \pi x$ and the transition is continuous at $x = 0$;

**(ii)** on the other hand the $x$ derivative of $\theta x$ along the lower edge jumps by $\pi$ as we cross the origin from right to left;

**(iii)** $\theta x$ is continuously differentiable away from the origin.

Thus the function $\psi_1 = -(2/\pi)\theta x$ matches the jump in the derivative of the solution $\psi$ on the boundary and, using Green's functions, we can accurately calculate the "excess" $\psi_2 = \psi - \psi_1$ by solving the system

$$\nabla^2\psi_2 = -\nabla^2\psi_1 = (2/\pi)\nabla^2(\theta x), \quad \psi_2 = \psi + (2/\pi)\theta x \quad \text{on the boundary.}$$

(See exercise 6.)                                                                      ∎

Clearly the construction of an explicit auxiliary function possessing specified boundary properties may require considerable ingenuity on the part of the analyst. For reasons of expository continuity we have ignored boundary "mismatches" in most of this book. The interested reader may want to review the examples in Chapters 6 and 7 and contemplate the implementation of this technique therein.

# Exercises 8.4

**1.** Show that $\nabla^2 \arctan \frac{y}{x} = 0$ (in the first and fourth quadrants).

**2.** Derive eq. (7).

**3.** Formulate the auxiliary function and the smooth, possibly nonhomogeneous, subproblems for the problem illustrated in Fig. 2.

**4.** (a) Express the decomposition of the problem illustrated in Fig. 9 into auxiliary functions and smooth problems.

   (b) Solve the problems formulated in (a). Don't work out integrals.

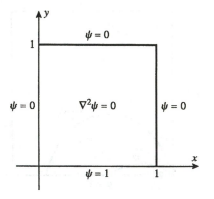

**Figure 9**  *Exercise 4*

5. Express the complete decomposition of the problem depicted in Fig. 10 into auxiliary functions and smooth, possibly nonhomogeneous, subproblem(s).

6. Carry out the complete solution of Example 2.

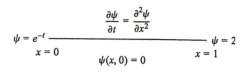

$$\frac{\partial \psi}{\partial t} = \frac{\partial^2 \psi}{\partial x^2}$$

$\psi = e^{-t}$ at $x = 0$, $\quad \psi(x,0) = 0$, $\quad \psi = 2$ at $x = 1$

**Figure 10**   *Exercise 5*

# 8.5   Radiation Problems

**Summary**   The solutions to the wave equation produced by localized time-dependent excitations are studied by both Laplace and Fourier transforms. In keeping with the deliberations of Chapter 3, the latter provides direct meaningful information about sinusoidal solutions, while the former, supplemented with analytic function theory to enable the estimate of the inverse transform, provides information on the transient features of the solution. The influence of the remote past in the Fourier approach necessitates another boundary-condition postulate, the Sommerfeld radiation condition. It is seen that when an electromagnetic wave mode is driven at a frequency below its cutoff frequency, its solution is attenuated, while traveling waves result for frequencies above cutoff.

In this section we further demonstrate the strength of the Laplace and Fourier transforms with some tough boundary value problems for the wave equation. In the situations analyzed in Sec. 7.3 we presumed that we were given the initial ($t = 0$) values of the field $\Psi$ and $\partial\Psi/\partial t$, *for all values of* $(x, y, z)$; and our task was to predict how these values would evolve in time. Now we take the initial state to be quiescent, but we are presented with the time history of a *localized disturbance* and our job is to describe how it propagates out to infinity. This is the *radiation problem*. We will analyze examples of *guided* radiation inside a rectangular waveguide and along a fiber optic cable, and of radiation into three dimensions from a spherical source. A good reference for the underlying physics is the book by Jones.

The localized excitation is specified as a time dependent boundary condition; hence the *transform* techniques are required. As in Sec. 3.9, the Fourier transform yields answers in a format which is suitable for the analysis of steady state responses to sinusoidal excitations (even if the inverse transform cannot be evaluated in closed form). The interpretation of the Laplace transform is seldom so transparent; however, using integration methods from analytic function theory one can often extract from the Bromwich integral a description of the *transient* features of the propagation - features which are not available from Fourier analysis. Therefore for the first two examples we expound both the Fourier and Laplace descriptions, directing the more advanced readers to Appendix B for the derivation of the extra information afforded by the latter.

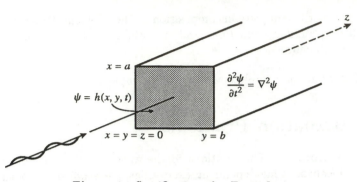

*Figure 1  Specifications for Example 1*

The examples also demonstrate that the finiteness boundary condition at infinity is accommodated quite comfortably in the Laplace description, while Fourier analysis requires a further postulate - the "Sommerfeld condition."

First we reconsider the rectangular waveguide, which was analyzed as an *initial* value problem in Example 4 of Sec. 7.3. Here we treat the more common situation where electromagnetic waves are launched into the end of the tube (Fig. 1).

**Example 1.** Recall (Sec. 7.3) that the propagation of transverse magnetic (TM) waves is governed by the wave equation

$$\frac{\partial^2 \Psi}{\partial t^2} = \nabla^2 \Psi \tag{1}$$

and the idealized boundary conditions expressing perfect conductivity of the side walls

$$\Psi = 0 \quad \text{at} \quad x = 0 \quad \text{and} \quad x = a,$$

$$\Psi = 0 \quad \text{at} \quad y = 0 \quad \text{and} \quad y = b. \tag{2}$$

If the disturbance is launched into the end $z = 0$, we have a time dependent boundary condition which we take to be

$$\Psi(x, y, 0, t) = h(x, y, t). \tag{3}$$

Finiteness conditions prevail at the far end:

$$\Psi \quad \text{finite at} \quad z = \infty; \tag{4}$$

and, since the initial value problem for the waveguide was solved in Sec. 7.3, now we presume

$$\Psi = \frac{\partial \Psi}{\partial t} = 0 \quad \text{at} \quad t = 0.$$

We start by Laplace-transforming the time dependence:

$$\Phi(x, y, z, s) = \int_0^\infty e^{-st}\Psi(x, y, z, t)\, dt. \tag{5}$$

As in Sec. 8.2, the quiescent initial conditions reduce the transformed wave equation to the *Helmholtz equation* (or "reduced wave equation")

$$\nabla^2\Phi = s^2\Phi, \tag{6}$$

with the spatial boundary conditions inherited by $\Phi$:

$$\Phi = 0 \quad \text{at} \quad x = 0 \quad \text{and} \quad x = a,$$

$$\Phi = 0 \quad \text{at} \quad y = 0 \quad \text{and} \quad y = b,$$

$$\Phi \quad \text{finite at} \quad z = \infty, \tag{7}$$

and

$$\Phi(x, y, 0, s) = H(x, y, s) = \int_0^\infty e^{-st}h(x, y, t)\, dt. \tag{8}$$

The solution format for (6), with eigenfunctions in the $X$ and $Y$ factors, is given in Table 7.3 as

$$[a_1 \cos \mu x + a_2 \sin \mu x]\,[b_1 \cos \nu y + b_2 \sin \nu y]\,[c_1 \cosh \kappa z + c_2 \sinh \kappa z] \tag{9}$$

where $\kappa^2 = \mu^2 + \nu^2 + s^2$. The homogeneous boundary conditions render this as

$$\sin \frac{m\pi}{a}x\, \sin \frac{n\pi}{b}y\, [c_1 \cosh \sqrt{s^2 + \omega_{mn}^2}\,z + c_2 \sinh \sqrt{s^2 + \omega_{mn}^2}\,z], \tag{10}$$

where

$$\omega_{mn} = \sqrt{\left(\frac{m\pi}{a}\right)^2 + \left(\frac{n\pi}{b}\right)^2} \tag{11}$$

(Recall the interpretation of $\omega_{mn}$ as the "cut-off" frequency for the $(m, n)$th waveguide mode in Sec. 7.3.) Since $s$ is regarded as a positive real number in the Laplace transform, the finiteness condition at $z = \infty$ requires $c_1 = -c_2$ and reduces (10) to

$$\sin \frac{m\pi}{a}x\, \sin \frac{n\pi}{b}y\, \exp\{-z\sqrt{s^2 + \omega_{mn}^2}\}.$$

Thus the general solution is

$$\Phi(x, y, z, s) = \sum_{m=1}^{\infty}\sum_{n=1}^{\infty} a_{mn} \sin \frac{m\pi}{a}x\, \sin \frac{n\pi}{b}y\, \exp\{-z\sqrt{s^2 + \omega_{mn}^2}\}. \tag{12}$$

It will match the edge disturbance (8) at $z = 0$ if

$$H(x, y, s) = \sum_{m=1}^{\infty} \sum_{n=1}^{\infty} a_{mn} \sin \frac{m\pi}{a} x \sin \frac{n\pi}{b} y,$$

i.e. if

$$a_{mn} = a_{mn}(s) = \frac{4}{ab} \int_0^a \int_0^b H(x, y, s) \sin \frac{m\pi}{a} x \sin \frac{n\pi}{b} y \, dy \, dx. \qquad (13)$$

The waveguide response is then obtained (in principle) by inserting (13) into (12) and taking the inverse Laplace transform (Bromwich integral).

It is enlightening to see what happens if one tries to excite a *single* eigenmode of the waveguide, by properly shaping the edge disturbance $h(x, y, t)$. Thus let us pick integers $p, q$ and set

$$h(x, y, t) = \sin \frac{p\pi}{a} x \sin \frac{q\pi}{b} y \sin \Omega t$$

where $\Omega$ is one of the frequencies supported by the $(p, q)$ mode family - i.e.

$$\Omega^2 > \omega_{pq}^2 = \left( \frac{p\pi}{a} \right)^2 + \left( \frac{q\pi}{b} \right)^2.$$

(Recall that the general solution derived in Sec. 7.3 was

$$\sum_{m=1}^{\infty} \sum_{n=1}^{\infty} \int_{-\infty}^{\infty} A_{mn}(k) \sin \frac{m\pi}{a} x \sin \frac{n\pi}{b} y \, e^{ikz}$$

$$\times \sin \sqrt{\left( \frac{m\pi}{a} \right)^2 + \left( \frac{n\pi}{b} \right)^2 + k^2} \, t \, dk,$$

for the doubly infinite guide. Thus all modes with a cross-section pattern like $\sin \frac{m\pi}{a} x \sin \frac{n\pi}{b} y$ vibrate with a frequency greater than $\omega_{mn}$.) By eq. (8) we have

$$H(x, y, s) = \sin \frac{p\pi}{a} x \sin \frac{q\pi}{b} y \frac{\Omega}{\Omega^2 + s^2}$$

and the sums in (12) collapse to

$$\Phi(x, y, z, s) = \sin \frac{p\pi}{a} x \sin \frac{q\pi}{b} y \frac{\Omega}{\Omega^2 + s^2} \exp\{-z \sqrt{s^2 + \omega_{pq}^2}\} \qquad (14)$$

The Bromwich integral of (14) is very difficult to evaluate exactly, but in Appendix B we employ analytic function methods to derive

$$\Psi(x, y, z, t) = \begin{cases} 0, & z > t \\ \sin[\Omega t - \sqrt{\Omega^2 - \omega_{pq}^2} z] + \mathcal{O}(1/t), & z < t \end{cases} \qquad (15)$$

times $\sin \frac{p\pi}{a}x \sin \frac{q\pi}{b}y$. For $z < t$ this describes a traveling sine wave moving to the right with the speed

$$v_0 = \Omega / \sqrt{\Omega^2 - \omega_{pq}^2} \tag{16}$$

plus a start-up transient which eventually dies out (like $1/t$). However (15) also says that the leading edge of the disturbance ($t = z$) propagates with the slower velocity

$$v_1 = 1.$$

We shall elaborate on the physical meaning of these two "wave velocities" in Sec. 9.3.

What happens if we pick $\Omega$ less than $\omega_{pq}$ — i.e. if we try to force a wave down the tube with a frequency *below* the cutoff frequency of that mode family? Formula (14) still holds but the computations in the Appendix show that in this case

$$\Psi(x, y, z, t) = \begin{cases} 0, & t < z \\ \sin \Omega t \exp\{-z\sqrt{\omega_{pq}^2 - \Omega^2}\} + \mathcal{O}(1/t), & t > z \end{cases} \tag{17}$$

times $\sin \frac{p\pi}{a}x \sin \frac{q\pi}{b}y$. Again we see the wave front propagating at unit velocity, but no traveling wave is propagated; instead a standing wave pattern is set up whose amplitude attenuates to the right. These nonpropagating patterns are called *evanescent waves*. They do not arise as natural waveguide modes, since there was no evidence of them in the derivation in Sec. 7.3. They have to be "driven into the waveguide" through the boundary conditions (or internal nonhomogeneities - see exercise 8). ∎

Let's go back and re-analyze this problem in the frequency domain. (Therefore we ignore the initial conditions and assume the boundary disturbance $h(x, y, t)$ has been active for all time.) Fourier analysis expresses the solution as a superposition of sinusoids of the form

$$\Psi(x, y, z, t) = \int_{-\infty}^{\infty} \chi(x, y, z, \omega) e^{i\omega t} \, d\omega.$$

(Since we have used $\Phi$ for the Laplace transform, we employ $\chi$ here.) Recall that time derivatives are replaced by factors of $(i\omega)$ in the Fourier description, so the wave equation (1) reduces to Helmholtz's equation (6) with $s^2$ replaced by $-\omega^2$

$$\nabla^2 \chi(x, y, z, \omega) = -\omega^2 \chi(x, y, z, \omega) \tag{18}$$

Separation of $\chi$ into $X(x)Y(y)Z(z)$ leads to the same $X, Y$ eigenfunctions as before (9), but the minus sign in (18) calls for a broader general solution format for $Z$:

$$Z(z) = \begin{cases} A_1 e^{i\sqrt{\omega^2 - (\frac{m\pi}{a})^2 - (\frac{n\pi}{b})^2}z} \\ \quad + A_2 e^{-i\sqrt{\omega^2 - (\frac{m\pi}{a})^2 - (\frac{n\pi}{b})^2}z} & \text{if} \quad \omega^2 > (\frac{m\pi}{a})^2 + (\frac{n\pi}{b})^2 \\ B_1 z + B_2 & \text{if} \quad \omega^2 = (\frac{m\pi}{a})^2 + (\frac{n\pi}{b})^2 \\ C_1 e^{\sqrt{(\frac{m\pi}{a})^2 + (\frac{n\pi}{b})^2 - \omega^2}z} \\ \quad + C_2 e^{-\sqrt{(\frac{m\pi}{a})^2 + (\frac{n\pi}{b})^2 - \omega^2}z} & \text{if} \quad \omega^2 < (\frac{m\pi}{a})^2 + (\frac{n\pi}{b})^2 \end{cases}$$

$$(19)$$

Interpreting the Fourier transform as the response to excitation at a specific frequency, we see in (19) a confirmation of the qualitative features predicted by (15, 17) - the sinusoidal waveforms for $\omega > \omega_{mn} = \sqrt{(\frac{m\pi}{a})^2 + (\frac{n\pi}{b})^2}$ and the exponentials for $\omega < \omega_{mn}$.

The boundary condition of finiteness at $z = +\infty$ demands that $B_1 = 0$ and $C_1 = 0$, so the evanescent waves produced by subcutoff frequencies are again in evidence. But how do we decide about $A_1$ and $A_2$?

If we reattach the time factor $e^{i\omega t}$ to the first form in (19) we find

$$A_1 e^{i\{\sqrt{\omega^2 - (\frac{m\pi}{a})^2 - (\frac{n\pi}{b})^2}z + \omega t\}} + A_2 e^{-i\{\sqrt{\omega^2 - (\frac{m\pi}{a})^2 - (\frac{n\pi}{b})^2}z - \omega t\}}.$$

Thus $A_2$ is the coefficient of a wave moving to the right in Fig. 1 at speed

$$v_0 = \omega \Big/ \sqrt{\omega^2 - (\frac{m\pi}{a})^2 - (\frac{n\pi}{b})^2}. \qquad (20)$$

It is an *outgoing* wave in our configuration. Similarly $A_1$ gives rise to an *incoming* wave, moving to the left. Surely the outgoing wave is the appropriate response to the disturbance $h(x, y, t)$ at the end of the tube (3). But what is the story behind the incoming wave? And why didn't it appear in the Laplace transform?

This is best understood by recalling that the Fourier description is tailored to represent a system for all time, from minus infinity to plus infinity. Its "initial conditions" correspond to the system's status at $t = -\infty$; recall the remarks at the end of Sec. 3.9. Now since the waveguide extends from $z = 0$ to $z = \infty$, if there were some disturbance in the tube at "$t = -\infty$" then by any *finite* time $t$ its outgoing components would have propagated past every finite point $z$, but its incoming components would keep arriving (there being no damping mechanism). This possibility has to be accommodated by the Fourier description, and the indeterminate coefficient $A_1$ is the consequence.

In the Laplace description we prescribed *quiescent* initial conditions throughout the waveguide at $t = 0$. This had the effect of zeroing out such "built-in" waves, and no indeterminacy occurred in the computations.

So, unless there is some compelling reason to account for the incoming wave components[8] in the Fourier description, we usually dismiss them by convention. That is, we impose the *Sommerfeld radiation* condition, which disenfranchises the incoming waves. Consequently we take $A_1 = 0$.

*Note that the Sommerfeld condition, like finiteness, is a homogeneous boundary condition.* If $\Psi_1$ and $\Psi_2$ each have no incoming components, neither does $a_1\Psi_1 + a_2\Psi_2$. (See exercise 2.)

The general response to a sinusoidal edge disturbance oscillating at the frequency $\omega$, satisfying all the *homogeneous* boundary conditions including the radiation condition, can be written

$$
\begin{aligned}
\chi(x, y, z, \omega)e^{i\omega t} \; = \; & \sum_m \sum_n a_{mn} \sin\frac{m\pi}{a}x \, \sin\frac{n\pi}{b}y \\
& \times e^{-i\{\sqrt{\omega^2 - (\frac{m\pi}{a})^2 - (\frac{n\pi}{b})^2}\, z - \omega t\}} \\[2mm]
+ \; & \sum_m \sum_n a_{mn} \sin\frac{m\pi}{a}x \, \sin\frac{n\pi}{b}y \\
& \times e^{-\sqrt{(\frac{m\pi}{a})^2 + (\frac{n\pi}{b})^2 - \omega^2}\, z}\, e^{i\omega t} \\[2mm]
+ \; & a_{MN} \sin\frac{M\pi}{a}x \, \sin\frac{N\pi}{b}y \, e^{i\omega t}.
\end{aligned}
\tag{21}
$$

Here the first sum extends over $m, n$ satisfying $\omega_{mn} < \omega$; accordingly, its terms are sinusoidal waves moving to the right. The second sum, for which $\omega_{mn} > \omega$, generates evanescent waveforms which attenuate exponentially to the right. The final term is only present if $\omega$ happens to equal a particular $\omega_{MN}$, and it produces a standing wave. The boundary condition for $\chi$ when $z = 0$ is obtained by Fourier-transforming $h(x, y, t)$ in (3) and can be expressed in the form

$$
\chi(x, y, 0, \omega) = \eta(x, y, \omega) = \frac{1}{2\pi}\int_{-\infty}^{\infty} h(x, y, t)\, e^{-i\omega t}\, dt.
\tag{22}
$$

Letting $z = 0$ in (21), we gather the summations and enforce (22):

$$
\eta(x, y, \omega) = \sum_{m=1}^{\infty} \sum_{n=1}^{\infty} a_{mn} \sin\frac{m\pi}{a}x \, \sin\frac{n\pi}{b}y.
$$

---

[8]The analysis of *scattering* phenomena requires inclusion of the incoming wave; see Messiah and Jackson.

Orthogonality readily identifies the coefficients:

$$a_{mn} = \frac{4}{ab} \int_0^a \int_0^b \eta(x, y, \omega) \sin \frac{m\pi}{a} x \, \sin \frac{n\pi}{b} y \, dy \, dx . \tag{23}$$

**Let us compare the description (21) with the traveling-wave description for the initial value problem analyzed in Section 7.3 (see eq. (24) therein).**

| Initial Value Formulation | Boundary Value Formulation |
|---|---|
| The *given* nonhomogeneous data are the initial values:<br><br>$$\Psi = f(x, y, z)$$<br><br>$$\frac{\partial \Psi}{\partial t} = g(x, y, z) \quad \text{at} \quad t = 0$$ | The *given* nonhomogeneous data are the boundary values:<br><br>$$\Psi = h(x, y, t) \quad \text{at} \quad z = 0.$$ |
| The eigenfunctions for the X and Y factors are the same. ||
| We Fourier-transform the dependence in $z$. The $Z$ factor is $e^{ikz}$, where the coefficient of $z$ - the wave number $k$ - can take any real value.<br>Then we append the $T$ factor $e^{i\omega t}$, with the frequency $\omega$ chosen according to<br><br>$$\omega = \sqrt{\left(\frac{m\pi}{a}\right)^2 + \left(\frac{n\pi}{b}\right)^2 + k^2} \tag{24}$$<br><br>The frequency $\omega$ is always real (regardless of $k$). | We Fourier-transform the dependence in $t$. The $T$ factor is $e^{i\omega t}$, where the coefficient of $t$ - the frequency $\omega$ - can take any real value. Then we append the $Z$ factor $e^{ikz}$, with the wave number $k$ chosen according to<br><br>$$k = \pm\sqrt{\omega^2 - \left(\frac{m\pi}{a}\right)^2 - \left(\frac{n\pi}{b}\right)^2} \tag{25}$$<br><br>The wave number $k$ will be real for high $\omega$ (with $k < 0$ for waves moving to the right) or imaginary for low $\omega$ (with $ik < 0$ for waves attenuated to the right). |

Equations (24) and (25) are equivalent; they both say that the time frequency $\omega$ and the wave number $k$ are related by the so-called "dispersion law"[9]

$$\omega^2 = \left(\frac{m\pi}{a}\right)^2 + \left(\frac{n\pi}{b}\right)^2 + k^2 . \tag{26}$$

---

[9]Dispersion laws can be very helpful in analyzing the underlying physics; see Sec. 9.3.

*Note that the dispersion law can be derived with either the initial-value or boundary-value formulation.*

This example displays both the beauty and shortcomings of the Fourier frequency domain approach. The fact that the $m, n$ mode propagates waves down the tube for driving frequencies above cutoff while it attenuates the lower frequencies is derived from Fourier analysis with far less labor than by the Laplace method. The wave speed $v_0$ (20) is obtained as well (compare (16)). However no information about the startup transient, nor the speed $v_1$ of the initial wavefront, is available. These transient effects are best exhibited by the Laplace transform.

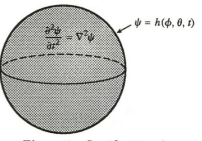

**Figure 2** *Specifications for Example 2*

**Example 2.** Now we consider the problem of radiation emitted by a pulsating sphere (Fig. 2). The differential pressure $\Psi(r, \phi, \theta, t)$ is described by the wave equation in spherical coordinates. The pulsations create a pressure wave at the outer radius of the sphere $r = b$, which we express as a time dependent boundary condition

$$\Psi(b, \phi, \theta, t) = h(\phi, \theta, t).$$

We begin with the Laplace transform approach. Transforming the time dependence we get

$$\Phi(r, \phi, \theta, s) = \int_0^\infty e^{-st} \Psi(r, \phi, \theta, t)\, dt,$$

$$H(\phi, \theta, s) = \int_0^\infty e^{-st} h(\phi, \theta, t)\, dt.$$

The transformed wave equation with quiescent initial conditions is the Helmholtz equation, which together with the transformed boundary condition produces the system

$$\nabla^2 \Phi = s^2 \Phi, \quad \Phi(b, \phi, \theta, s) = H(\phi, \theta, s). \tag{27}$$

With the nonhomogeneous condition imposed on a constant-$r$ face, we will have oscillatory eigenfunctions in $\theta$ and $\phi$; consequently Table 7.4 states that the general solution has the form

$$R(r) = c_1 h_\ell^{(1)}(isr) + c_2 h_\ell^{(2)}(isr).$$

The spherical Hankel functions were discussed in Example 9 of Sec. 7.3. For large $r$, we have

$$h_\ell^{(1)}(isr) \approx i^{-\ell-2} e^{-sr}/sr, \quad h_\ell^{(2)}(isr) \approx i^{-\ell-2} e^{sr}/sr$$

(eqs. (45), Sec. 7.3), and the requirement of finiteness at $r = \infty$ disenfranchises $h_\ell^{(2)}$, so $c_2 = 0$. Our assembled solution is thus

$$\Phi(r,\phi,\theta,s) = \sum_{\ell=0}^{\infty} \sum_{m=-\ell}^{\ell} a_{\ell m}(s) h_\ell^{(1)}(isr) Y_{\ell m}(\phi,\theta) \tag{28}$$

and we can satisfy the condition at $r = b$ (27) by using orthogonality to find the coefficients $a_{\ell m}$.

To excite a particular $p, q$ mode at the frequency $\Omega$ we select the boundary condition

$$h(\phi,\theta,t) = Y_{pq}(\phi,\theta)\sin\Omega t, \quad H(\phi,\theta,s) = Y_{pq}(\phi,\theta)\frac{\Omega}{\Omega^2 + s^2} \tag{29}$$

All the coefficients but one in the expansion (28) then vanish and we have

$$\Phi(r,\phi,\theta,s) = \frac{\Omega}{\Omega^2 + s^2} \frac{h_p^{(1)}(isr)}{h_p^{(1)}(isb)} Y_{pq}(\phi,\theta). \tag{30}$$

In Appendix B we show that the inverse transform of (30) is given by

$$\Psi(r,\phi,\theta,t) = \begin{cases} 0, & \text{for} \quad t < r - b \\ \{\dfrac{h_p^{(2)}(\Omega r)}{2i\, h_p^{(2)}(\Omega b)} e^{i\Omega t} \\ \quad - \dfrac{h_p^{(1)}(\Omega r)}{2i\, h_p^{(1)}(\Omega b)} e^{-i\Omega t}\} Y_{pq}(\phi,\theta), & \text{for} \quad t > r - b \end{cases} \tag{31}$$

Note that since (eqs. (45), Sec. 7.3 again)

$$h_p^{(1)}(\Omega r) \approx (-i)^{p+1}\frac{e^{i\Omega r}}{\Omega r}, \quad h_p^{(2)}(\Omega r) \approx i^{p+1}\frac{e^{-i\Omega r}}{\Omega r} \quad \text{for large } r, \tag{32}$$

both terms in (31) have time dependence for large $r$ of the form $f(t - r)$, and thus represent *outgoing* waves (attenuated by $r^{-1}$). Their velocity is unity, which in this case is the same as the speed of the leading edge of the disturbance (according to (31)).

The frequency domain analysis of this problem proceeds as follows. We write $\Psi(r,\phi,\theta,t) = e^{i\omega t}\chi(r,\phi,\theta,\omega)$ and for $\nabla^2\chi = -\omega^2\chi$ Table 7.4 yields

$$[c_1 h_\ell^{(1)}(\omega r) + c_2 h_\ell^{(2)}(\omega r)]Y_{\ell m}(\phi,\theta).$$

In (32) we have the asymptotic approximations for these forms. Now when we append the factor $e^{i\omega t}$ we see that $h_\ell^{(1)}(\omega r)$ gives rise to an incoming,

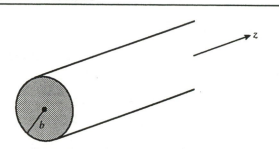

***Figure 3***    *Geometry for Example 3*

and $h_\ell^{(2)}(\omega r)$ an outgoing, wave. The Sommerfeld radiation condition thus calls for $c_1 = 0$, and our general solution is

$$\chi(r, \phi, \theta) = \sum_{\ell=0}^{\infty} \sum_{m=-\ell}^{\ell} a_{\ell m} Y_{\ell m}(\phi, \theta) h_\ell^{(2)}(\omega r) e^{i\omega t} \qquad (33)$$

This is readily fitted to the boundary condition at $r = b$.     ∎

**Example 3.** The propagation of light waves in optical fibers is more complicated than ordinary waveguide propagation because the perfectly conducting walls are replaced by dielectrics. For details of the physics we refer the reader to Keiser's text. Here for simplicity we shall demonstrate the analysis for transverse magnetic (TM) waves propagated along a fiber without protective cladding, but the basic mathematical procedure will work for the other modes as well. Be warned - propagation along fibers is complicated, and this research-level calculation will be the longest example in the book!

As is the case for the usual waveguides, all components of the TM wave can be determined once the longitudinal component of the electric field $E$ is known. Thus for the geometry depicted in Fig. 3 we let $\Psi(\rho, \theta, z, t)$ equal $E_z$.

The analysis of the fiber-air interface through Maxwell's equations is rather complicated, but (as is shown in the references) it leads to the following results:

1. Any TM wave must be axially symmetric; $\Psi$ does not vary with $\theta$.

2. On either side of the interface $\Psi$ satisfies the wave equation, but the speed of light is slower in the fiber than in air. Thus we introduce the *index of refraction $\eta$ ($\eta > 1$)*:

$$\frac{\partial^2 \Psi}{\partial t^2} = \frac{1}{\eta^2} \nabla^2 \Psi, \quad 0 \le \rho < b; \quad \frac{\partial^2 \Psi}{\partial t^2} = \nabla^2 \Psi, \quad \rho > b \qquad (34)$$

3. The tangential components of both the electric and magnetic fields are continuous across the interface. The resulting conditions on $\Psi$

will be introduced below.[10]

If we introduce the function $\eta(\rho)$ which expresses the index of refraction in the two regions,

$$\eta(\rho) = \begin{cases} 1, & \text{for } \rho > b \\ \eta, & \text{for } \rho < b, \end{cases}$$

then the pair of wave equations (34) can be analyzed simultaneously:

$$\frac{\partial^2 \Psi}{\partial t^2} = \frac{1}{\eta(\rho)^2} \nabla^2 \Psi = \frac{1}{\eta(\rho)^2} \left\{ \frac{\partial^2 \Psi}{\partial \rho^2} + \frac{1}{r} \frac{\partial \Psi}{\partial \rho} + \frac{\partial^2 \Psi}{\partial z^2} \right\} \tag{35}$$

(recall there is no $\theta$ dependence).

We shall demonstrate only the frequency domain approach for analyzing this problem. We excite the fiber with oscillations at the end $z = 0$:

$$\Psi(\rho, 0, t) = h(\rho)e^{i\omega t} \tag{36}$$

and postulate the separated form

$$\Psi(\rho, z, t) = e^{i\omega t} R(\rho) Z(z). \tag{37}$$

Inserting (37) into (35) and separating leads to

$$-\frac{Z''}{Z} = \frac{1}{R} \left\{ R'' + \frac{R'}{\rho} \right\} + \eta(\rho)^2 \omega^2 = \text{constant} = \lambda. \tag{38}$$

Because the nonhomogeneity is imposed on a constant-$z$ face, the radial direction is to be analyzed first. The eigenvalue equation for $R$ is

$$R'' + \frac{1}{\rho} R' + \eta(\rho)^2 \omega^2 R = \lambda R,$$

which is a Bessel equation whose solutions can be expressed[11]

$$R(\rho) = \begin{cases} \begin{aligned} & c_1 J_0(\sqrt{\eta^2 \omega^2 - \lambda}\rho) \\ & + c_2 Y_0(\sqrt{\eta^2 \omega^2 - \lambda}\rho), \quad \rho < b \\ & d_1 J_0(\sqrt{\omega^2 - \lambda}\rho) \\ & + d_2 Y_0(\sqrt{\omega^2 - \lambda}\rho), \quad \rho > b \end{aligned} \end{cases} \text{if } \lambda < \omega^2, \tag{39}$$

---

[10]Recalling a similar situation with discontinuities in the quantum mechanical square well problem (Example 1, Sec. 7.4), one might expect that $\Psi$ and $\partial \Psi / \partial \rho$ must be continuous, but as we shall see this is not true. The explanation is that although *Maxwell's* equations are supposed to hold *everywhere* - including the interface - the wave equation itself is a consequence of Maxwell's equations *only in the homogeneous portion of the media*. Maxwell's equations themselves demand only the continuity of the *tangential* components of the fields.

[11]The exceptional cases $\lambda = \omega^2$, $\lambda = \eta^2 \omega^2$ are dealt with in exercise 6.

or

$$R(\rho) = \left\{ \begin{array}{ll} c_1' J_0(\sqrt{\eta^2\omega^2 - \lambda}\rho) \\ \quad + c_2' Y_0(\sqrt{\eta^2\omega^2 - \lambda}\rho), & \rho < b \\ d_1' I_0(\sqrt{\lambda - \omega^2}\rho) \\ \quad + d_2' K_0(\sqrt{\lambda - \omega^2}\rho), & \rho > b \end{array} \right\} \text{if} \quad \omega^2 < \lambda < \eta^2\omega^2, \quad (40)$$

or

$$R(\rho) = \left\{ \begin{array}{ll} c_1'' I_0(\sqrt{\lambda - \eta^2\omega^2}\rho) \\ \quad + c_2'' K_0(\sqrt{\lambda - \eta^2\omega^2}\rho), & \rho < b \\ d_1'' I_0(\sqrt{\lambda - \omega^2}\rho) \\ \quad + d_2'' K_0(\sqrt{\lambda - \omega^2}\rho), & \rho > b \end{array} \right\} \text{if} \quad \eta^2\omega^2 < \lambda. \quad (41)$$

The best way to proceed from here is to follow the steps of the straightforward Sturm-Liouville methodology discussed in Chapter 6. We must distinguish the *boundary* conditions at $\rho = 0$ and $\rho = \infty$ from the *interface* conditions at $\rho = b$; the latter are frequently called "boundary" conditions also, but they are not *auxiliary* conditions imposed on the differential equations (of Maxwell); they follow *from* the equations. We have not truly found the "general solution" to the differential equations until these interface conditions have been met. The eigenvalues are then determined by the (*bona fide*) boundary conditions at $\rho = 0$ and $\infty$.

So, following the steps outlined in Sec. 6.2, we must first write down the general solution to the differential equation as a linear combination of two independent solutions, with undetermined coefficients. Equations (39-41) do not suffice, because they have *four* coefficients each. We must eliminate the $d$'s (expressing them in terms of the $c$'s) by invoking the interface conditions.

Since $\Psi = E_z$ is a component of the electric field tangent to the interface, it must be continuous. Hence

$$\lim_{\rho \uparrow b} R(\rho) = \lim_{\rho \downarrow b} R(\rho). \quad (42)$$

This is one equation relating the $d$'s and $c$'s. For instance, in the range $\lambda < \omega^2$ eq. (42) reads

$$c_1 J_0(\sqrt{\eta^2\omega^2 - \lambda}b) + c_2 Y_0(\sqrt{\eta^2\omega^2 - \lambda}b)$$
$$= d_1 J_0(\sqrt{\omega^2 - \lambda}b) + d_2 Y_0(\sqrt{\omega^2 - \lambda}b).$$

We refer the reader to Keiser for the derivation of the other interface condition; it follows from Maxwell's equations by requiring continuity of the $\theta$ component of the magnetic field:

$$\lim_{\rho \uparrow b} \frac{\eta^2}{\eta^2\omega^2 - \lambda} R'(\rho) = \lim_{\rho \downarrow b} \frac{1}{\omega^2 - \lambda} R'(\rho). \quad (43)$$

For reasons to be given below, we will not execute the solution of (42, 43) for the $d$'s in terms of the $c$'s at this point. They are simple linear algebraic equations and *we shall proceed as if they had been solved.*

The next step in the Sturm-Liouville procedure is to eliminate one of the $c$'s by imposing *one* of the homogeneous boundary conditions - finiteness at $\rho = 0$ and $\infty$. This is easy, since the function accompanying $c_2$ (or $c_2'$, or $c_2''$) in each case is infinite at $\rho = 0$. Hence $c_2 = c_2' = c_2'' = 0$ and we can now take the overall multiplicative constant $c_1$ ($c_1'$, $c_1''$) to be unity. (This will simplify conditions (42, 43), which is one reason we weren't eager to implement them earlier!)

The final step is the determination of the eigenvalues $\lambda$ through the application of the other homogeneous boundary condition (finiteness at $r = \infty$).

Since the Bessel functions $J_0$ and $Y_0$ are both finite at infinity, from eq. (39) we see that *all* numbers $\lambda$ below $\omega^2$ are eigenvalues (regardless of how the expressions for $d_1$ and $d_2$ turn out). But eqs. (40, 41) give finite answers at $r = \infty$ only if the coefficients $d_1'$, $d_1'''$ of the Bessel function $K_0$ are zero (Sec. 2.5). Thus in order to uncover the eigenvalues in these ranges we have to carry out the solutions of the interface conditions (42, 43) for $d_1'$ and $d_2'''$ and set the expressions equal to zero. This computation is expedited by the use of matrices.

For $\omega^2 < \lambda < \eta^2 \omega^2$ the interface conditions are expressed (remember that $c_2' = 0$, $c_1' = 1$)

$$\begin{bmatrix} I_0(\sqrt{\lambda - \omega^2}b) & K_0(\sqrt{\lambda - \omega^2}b) \\ \frac{-1}{\sqrt{\lambda - \omega^2}} I_0'(\sqrt{\lambda - \omega^2}b) & \frac{-1}{\sqrt{\lambda - \omega^2}} K_0'(\sqrt{\lambda - \omega^2}b) \end{bmatrix} \begin{bmatrix} d_1' \\ d_2' \end{bmatrix}$$

$$= \begin{bmatrix} J_0(\sqrt{\eta^2 \omega^2 - \lambda}b) \\ \frac{\eta^2 J_0'(\sqrt{\eta^2 \omega^2 - \lambda}b)}{\sqrt{\eta^2 \omega^2 - \lambda}} \end{bmatrix}$$

To make $d_1'$ vanish we can simply set the numerator in Cramer's rule for $d_1'$ equal to zero:

$$\det \begin{vmatrix} J_0(\sqrt{\eta^2 \omega^2 - \lambda}b) & K_0(\sqrt{\lambda - \omega^2}b) \\ \frac{\eta^2 J_0'(\sqrt{\eta^2 \omega^2 - \lambda}b)}{\sqrt{\eta^2 \omega^2 - \lambda}} & \frac{-1}{\sqrt{\lambda - \omega^2}} K_0'(\sqrt{\lambda - \omega^2}b) \end{vmatrix} = 0$$

or

$$\frac{J_0'(\sqrt{\eta^2 \omega^2 - \lambda}b)}{J_0(\sqrt{\eta^2 \omega^2 - \lambda}b)} \sqrt{\eta^2 \omega^2 - \lambda} = -\frac{1}{\eta^2} \frac{K_0'(\sqrt{\lambda - \omega^2}b)}{K_0(\sqrt{\lambda - \omega^2}b)} \sqrt{\lambda - \omega^2}. \tag{44}$$

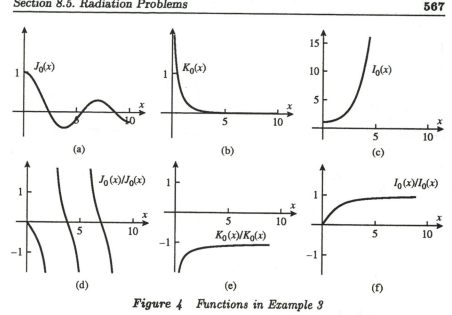

(a)      (b)      (c)

(d)      (e)      (f)

**Figure 4**   *Functions in Example 3*

Figure 4 exhibits the relevant functions in eq. (44). The graphs of each side of the equation, as functions of $\lambda$, are displayed in Fig. 5. The eigenvalues are located at the points of intersection of the curves.

It is clear from the figure that the number of intersections will equal the number of positive vertical asymptotes for the "$J_0$-function", which equals the number of zeros of $J_0(\sqrt{\eta^2\omega^2 - \lambda}\, b)$ in the interval from $\lambda = \omega^2$ to $\lambda = \eta^2\omega^2$ (or, equivalently, the number of zeros of $J_0(x)$ between $x = 0$ and $x = \sqrt{\eta^2\omega^2 - \omega^2}\, b = \sqrt{\eta^2 - 1}\,\omega b$). As $\omega$ gets smaller this interval decreases and encompasses fewer of these zeros. Note in particular that if $\sqrt{\eta^2\omega^2 - \omega^2}\, b$ is *less* than $j_{0,1}$, the first positive zero of $J_0$, there will be *no* eigenvalues. If $\omega$ is increased so that this quantity is between $j_{0,1}$ and $j_{0,2}$, one eigenvalue will result. And a new eigenvalue "kicks in" each time $\sqrt{\eta^2\omega^2 - \omega^2}\, b$ exceeds a $j_{0,p}$ (Fig. 6). The $p$th mode can only be excited if the frequency $\omega$ exceeds the cutoff value

$$\omega_p = j_{0,p}/b\sqrt{\eta^2 - 1}. \tag{45}$$

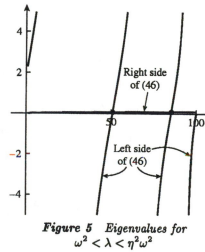

**Figure 5**   *Eigenvalues for* $\omega^2 < \lambda < \eta^2\omega^2$

For $\lambda > \eta^2\omega^2$ the condition that $d_1''$ vanish leads to (exercise 5)

$$\frac{I_0'(\sqrt{\lambda - \eta^2\omega^2}\, b)}{I_0(\sqrt{\lambda - \eta^2\omega^2}\, b)} = \frac{1}{\eta^2}\sqrt{\frac{\lambda - \omega^2}{\lambda - \eta^2\omega^2}}\,\frac{K_0'(\sqrt{\lambda - \omega^2}\, b)}{K_0(\sqrt{\lambda - \omega^2}\, b)} \tag{46}$$

but, as Fig. 4 demonstrates, $I_0'/I_0$ is positive and $K_0'/K_0$ is negative; thus there are no eigenvalues in this range.

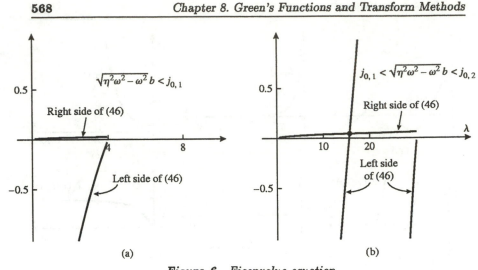

**Figure 6**   *Eigenvalue equation*

The $z$ dependence of eq. (38) is given by

$$Z(z) = a_1 e^{i\sqrt{\lambda}z} + a_2 e^{-i\sqrt{\lambda}z}.$$

If $\lambda < 0$ the condition of finiteness at $z = \infty$ rules out the term $a_2 e^{-i\sqrt{\lambda}z} = a_1 e^{\sqrt{-\lambda}z}$, so we set

$$Z_{\lambda<0}(z) = e^{-i\sqrt{-\lambda}z}.$$

For $\lambda > 0$ we invoke the Sommerfeld radiation condition to eliminate the incoming wave engendered by $a_1$; thus

$$Z_{\lambda>0}(z) = e^{-i\sqrt{\lambda}z}.$$

(For $\lambda = 0$ the solution is $c_1 + c_2 z$, with $c_2$ disenfranchised by the condition at $\infty$; thus *either* form above suffices for this case.)

Our assembled general solution at this point looks like

$$
\begin{aligned}
\Psi(\rho, z, t) = \chi(\rho, z)e^{i\omega t} \quad = \quad & \int_{\lambda=-\infty}^{0} c(\lambda) R_\lambda(\rho) e^{-\sqrt{-\lambda}z} e^{i\omega t} \, d\lambda \\
+ \quad & \int_{\lambda=0}^{\omega^2} c(\lambda) R_\lambda(\rho) e^{-i\sqrt{\lambda}z} e^{i\omega t} \, d\lambda \\
+ \quad & \sum_{\omega^2 < \lambda_p < \eta^2 \omega^2} c(\lambda_p) R_{\lambda_p}(\rho) e^{-i\sqrt{\lambda}z} e^{i\omega t} \quad (47)
\end{aligned}
$$

The function $c(\lambda)$ is the expansion coefficient which will be used to meet the boundary condition at $z = 0$ (36). The function $R_\lambda(p)$ is the radial solution (39, 40) with the appropriate expressions for $d_1$ and $d_2$ (determined by eqs.

(42, 43); see exercise 6) inserted. The final term in (47) results from the isolated eigenvalues, if there are any, in the range $\omega^2 < \lambda < \eta^2\omega^2$; $R_{\lambda_p}(\rho)$ is the corresponding eigenfunction:

$$R_p(\rho) = \begin{cases} J_0(\sqrt{\eta^2\omega^2 - \lambda_p}\,\rho), & \rho < b \\ K_0(\sqrt{\lambda_p - \omega^2}\,\rho)\,J_0(\sqrt{\eta^2\omega^2 - \lambda_p}\,b)/K_0(\sqrt{\lambda_p - \omega^2}\,b), & \rho > b \end{cases}$$

In applications of fiber optics for communications only the modes in the final term of eq. (47) are of interest. The exponentially decaying $z$ dependence of the first term indicates that no waves are propagated along the fiber. The second term does give wave propagation, but a closer look at Maxwell's equations (see Keiser) reveals that these modes carry off finite amounts of electromagnetic energy to infinity in the radial direction (the functions $J_0(\sqrt{\omega^2 - \lambda}\,\rho)$, $Y_0(\sqrt{\omega^2 - \lambda}\,\rho)$ don't decrease fast enough). These "radiation modes" require continuous replenishment from the source at $z = 0$. Thus the usable "guided modes" can only be excited at high frequencies, exceeding the cutoff numbers given in eq. (45). ∎

# Exercises 8.5

1. Rework Example 1 with Neumann conditions imposed at the edge of the waveguide: $\frac{\partial\Psi}{\partial z}$ is given at $x = 0$. Carry out the *distribution* solution to part (a) (as in exercise 6, Sec. 8.2). Show that the Neumann conditions - but not the Dirichlet conditions - allow some indeterminant mixing in of the standing wave solutions of Sec. 7.3.

2. Some texts quote a more rigorous form of the Sommerfeld radiation condition for spherical coordinate waveforms in the frequency domain $\Psi(r, \theta, \phi, t) = \chi(r, \theta, \phi)e^{i\omega t}$; namely,

$$\frac{\partial\Psi}{\partial r} - i\omega\Psi = \mathcal{O}(r^{-2}) \quad \text{as} \quad r \to \infty.$$

Show that the outgoing waves $e^{i\omega(r-t)}$, $h_\ell^{(2)}(\omega r)e^{i\omega t}$, and $h_\ell^{(1)}(\omega r)e^{-i\omega t}$ satisfy this form of the radiation condition, while the incoming waves $e^{i\omega(r+t)}$, $h_\ell^{(1)}(\omega r)e^{i\omega t}$, and $h_\ell^{(2)}(\omega r)e^{-i\omega t}$ do not.

3. Analyze the cylindrical waveguide in the manner of Example 1.

4. Supplement eqs. (39-41) with the solution forms for $\lambda = \omega^2$ and $\lambda = \eta^2\omega^2$.

5. Derive eq. (46).

6. Work out the expression for the function $R_\lambda(\rho)$ in eq. (47).

7. Consider the problem of radiation emanating off of the $z = 0$ plane into the "free half-space" $z > 0$. Express the field $\Psi$ in terms of traveling plane waves (Example 5, Sec. 7.3). Construct an example of a boundary condition at $z = 0$ which generates only the plane wave propagating in the $+\mathbf{k}$ direction ($e^{i(k_3 z - \omega t)}$). Are there boundary conditions which generate "surface waves" propagating only in the $+\mathbf{i}$ direction? How about the oblique direction $\mathbf{i} + \mathbf{k}$?

8. Consider the doubly infinite rectangular waveguide excited *internally* by a source term in the wave equation

$$\frac{\partial^2 \Psi}{\partial t^2} = \nabla^2 \Psi + f(x, y, z, t).$$

Analyze this problem in the frequency domain and identify the traveling and evanescent waves.

# Chapter 9

# Perturbations, Small Waves, and Dispersion

*It is recommended that the reader review Appendix A before embarking on the study of this chapter.*

Perturbation theory is a simple procedure for estimating the solution to a difficult problem when the solution of a nearby, simpler, problem is known. As an introduction to the method, we shall begin by considering an algebraic equation which can be solved *exactly*, and then see how perturbation methods generate approximate solutions to the equation. The successive corrections are obtained by solving *linear* problems - hence the alternative nomenclature, "linearization". The remainder of the chapter is devoted to the application of the perturbation technique to problems involving transcendental equations, differential equations, matrices, eigenfunctions, and integral equations. The wave oscillations which result when a physical system is perturbed can often be used to understand the system better, so we conclude with a survey of this methodology.

## 9.1 Perturbation Methods for Algebraic Equations

**Summary** The perturbation method for solving a complicated algebraic equation employs linearization to simplify the computation of the Taylor expansion of the solution, expanded around solutions to simpler equations. The method is quite robust, but it is foiled when the equation has multiple roots or when a "singular" perturbation changes the degree of the equation.

Consider the problem of solving the quadratic equation

$$x^2 - 5x + c = 0 \tag{1}$$

when the coefficient $c$ is known to be "close" to 6; for definiteness, let's suppose $c = 6.01$. If $c$ were exactly 6, the solutions could easily be determined by factoring:

$$x^2 - 5x + 6 = (x - 2)(x - 3),$$

so 2 and 3 are the roots.

Let us advertise the fact that $c$ is close to 6 by writing

$$c = 6 + \epsilon$$

(bearing in mind the usual mathematical convention that $\epsilon$ is a small number). Then the *exact* solution of eq. (1),

$$x^2 - 5x + 6 + \epsilon = 0, \tag{2}$$

can be expressed by the quadratic formula

$$x^{(1)} = \frac{5 - \sqrt{5^2 - 4 \times 1 \times (6 + \epsilon)}}{2} = \frac{5 - \sqrt{1 - 4\epsilon}}{2} \tag{3}$$

$$x^{(2)} = \frac{5 + \sqrt{5^2 - 4 \times 1 \times (6 + \epsilon)}}{2} = \frac{5 + \sqrt{1 - 4\epsilon}}{2} \tag{4}$$

Obviously $x^{(1)}$ is the root "near" to 2.

Now here's the important conceptual leap. Although we have in mind that $\epsilon$ equals .01, it can be regarded as a *variable*, which we set equal to .01 when we are done. Thus we *imbed* the problem $x^2 - 5x + 6.01 = 0$ in a family of problems $x^2 - 5x + 6 + \epsilon = 0$, for $0 \leq \epsilon \leq 1$ (say).

With this intepretation formula (3) identifies $x^{(1)}$ as a *function of $\epsilon$* - valid for a *range* of $\epsilon$ - and, as such, it has a Taylor expansion around $\epsilon = 0$. In fact, a little computation shows

$$\begin{aligned} x^{(1)} &= x^{(1)}(\epsilon) = x^{(1)}(0) + \epsilon \frac{dx^{(1)}}{d\epsilon}(0) + \frac{\epsilon^2}{2!} \frac{d^2 x^{(1)}}{d\epsilon^2}(0) + \frac{\epsilon^3}{3!} \frac{d^3 x^{(1)}}{d\epsilon^3}(0) + \cdots \\ &= 2 + \epsilon + \epsilon^2 + 2\epsilon^3 + \cdots \end{aligned} \tag{5}$$

(exercise 2). Thus we generate the approximations (recall Sec. 2.2)

$$x^{(1)} = 2 + \mathcal{O}(\epsilon) \tag{6}$$
$$x^{(1)} = 2 + \epsilon + \mathcal{O}(\epsilon^2) \ (= 2.01 + \mathcal{O}(\epsilon^2)) \tag{7}$$
$$x^{(1)} = 2 + \epsilon + \epsilon^2 + \mathcal{O}(\epsilon^3) \ (= 2.0101 + \mathcal{O}(\epsilon^3)) \tag{8}$$

*As a matter of fact, the exact solution is $x^{(1)} = 2.010102052\ldots$ .*

Perturbation theory is a procedure which enables the computation of the approximations (6-8) *without first solving the problem exactly*, as we did in (3, 4).

The perturbation technique is based on the observation that the first order approximation (7) can be formally obtained from (5) by neglecting all powers of $\epsilon$ greater than one — and the second order approximation (8), by neglecting powers higher than two. Now *this operating strategy can be implemented from the outset, by inserting the formal series*

$$x^{(1)} = x_0 + x_1\epsilon + x_2\epsilon^2 + \cdots \tag{9}$$

*into the original equation and dropping the higher powers of $\epsilon$.*

In other words we insert (9) into (2),

$$(x_0 + x_1\epsilon + x_2\epsilon^2 + \cdots)^2 - 5(x_0 + x_1\epsilon + x_2\epsilon^2 + \cdots) + 6 + \epsilon = 0,$$

and expand the terms with the convention that $\epsilon^2 = \epsilon^3 = \cdots = 0$;

$$x_0^2 + 2x_0x_1\epsilon + (0) - 5x_0 - 5x_1\epsilon - (0) + 6 + \epsilon = 0. \tag{10}$$

If we are seeking the solution near 2 - i.e. the solution which is 2 when $\epsilon = 0$ - then clearly $x_0 = 2$. The zeroth order terms in (10) cancel and we are left with

$$2 \times 2x_1\epsilon - 5x_1\epsilon + \epsilon = 0, \text{ or } x_1 = 1.$$

We have obtained the approximation $x^{(1)} \approx 2 + \epsilon = 2.01$, in agreement with (7), without solving the quadratic equation!

This process exemplifies *first order perturbation theory*. For second order perturbation theory we insert (9) into (2) and adopt the convention that $\epsilon^3 = \epsilon^4 = \cdots = 0$:

$$x_0^2 + 2x_0x_1\epsilon + (2x_0x_2 + x_1^2)\epsilon^2 - 5x_0 - 5x_1\epsilon - 5x_2\epsilon^2 + 6 + \epsilon = 0$$

We already know that the choice $x_0 = 2, x_1 = 1$ makes the zeroth and first order terms drop out. The second order terms become

$$(2 \times 2x_2 + 1)\epsilon^2 - 5x_2\epsilon^2 = 0,$$

or

$$x_2 = 1,$$

and we recover the second order approximation, in accordance with (8).

**Example 1.** Solve

$$e^x + x = 1.02 \tag{11}$$

SOLUTION Observing that $x_0 = 0$ solves the nearby problem $e^x + x = 1$, we regard $\epsilon = .02$ as a perturbation and propose the solution form

$$x = x_0 + x_1\epsilon + x_2\epsilon^2 + \cdots \tag{12}$$

for substitution into

$$e^x + x = 1 + \epsilon. \tag{13}$$

The zeroth order computation, of course, is

$$e^{x_0} + x_0 = 1$$

with solution $x_0 = 0$.

To first order we have

$$e^{x_0 + x_1 \epsilon} + x_0 + x_1 \epsilon = 1 + \epsilon.$$

or

$$e^{0 + x_1 \epsilon} + 0 + x_1 \epsilon = 1 + \epsilon.$$

Using the Taylor series for the exponential ($e^{x_1 \epsilon} = 1 + x_1 \epsilon + x_1^2 \epsilon^2 / 2! + \cdots$) but setting $\epsilon^2 = 0$ we find

$$1 + x_1 \epsilon + x_1 \epsilon = 1 + \epsilon$$

or

$$x_1 = .5.$$

Thus, by (12)

$$x \approx 0 + .5 \times .02 = .01$$

is the first order approximation.

Keeping second order terms in (13) we find

$$e^{.5\epsilon + x_2 \epsilon^2} + .5\epsilon + x_2 \epsilon^2 = 1 + \epsilon \tag{14}$$

With

$$
\begin{aligned}
e^{.5\epsilon + x_2 \epsilon^2} &= 1 + (.5\epsilon + x_2 \epsilon^2) + (.5\epsilon + x_2 \epsilon^2)^2 / 2! + (.5\epsilon + x_2 \epsilon^2)^3 / 3! + \cdots \\
&= 1 + .5\epsilon + (x_2 + .5^2 / 2!)\epsilon^2 + \mathcal{O}(\epsilon^3)
\end{aligned}
$$

the second order terms in (14) are

$$x_2 \epsilon^2 + .125\epsilon^2 + x_2 \epsilon^2 = 0,$$

or

$$x_2 = -.0625.$$

The second order approximation is

$$x \approx 0 + .5 \times .02 - .0625 \times .02^2 = .009975.$$

*In fact high-precision computation shows*

$$e^{.009975} + .009975 \approx 1.0199999 \qquad \blacksquare$$

Perturbation theory doesn't always "work", because some problems have answers which are not differentiable with respect to the parameter $\epsilon$. For example, consider the quadratic equation

$$x^2 - (2 + \epsilon)x + 1 = 0.$$

Clearly for $\epsilon$ near zero the solutions should be near $x_0 = 1$ (a double root). The exact solutions are

$$x^{(1)} = \frac{2 + \epsilon + \sqrt{(2+\epsilon^2) - 4 \times 1 \times 1}}{2} = \frac{2 + \epsilon + \sqrt{4\epsilon + \epsilon^2}}{2}$$

$$x^{(2)} = \frac{2 + \epsilon - \sqrt{(2+\epsilon^2) - 4 \times 1 \times 1}}{2} = \frac{2 + \epsilon - \sqrt{4\epsilon + \epsilon^2}}{2},$$

and the derivative of, say, $x^{(1)}$ with respect to $\epsilon$ is

$$\frac{dx^{(1)}}{d\epsilon} = \frac{1}{2} + \frac{2 + \epsilon}{2\sqrt{4\epsilon + \epsilon^2}}. \qquad (15)$$

But (15) becomes infinite as $\epsilon$ approaches zero! Thus there is no Taylor expansion for this root near $\epsilon = 0$.

Would the perturbation approach expose this irregularity? Yes it would! Insertion of (9), with $x_0 = 1$, into (15) produces to first order

$$x_0^2 + 2x_0 x_1 \epsilon - 2x_0 - 2x_1 \epsilon - \epsilon x_0 + 1 = 0$$

or

$$x_1 = \frac{x_0}{2x_0 - 2} = \frac{1}{0} \ (!)$$

Thus perturbation theory conspicuously fails and we say the problem is "ill-conditioned". *Polynomials with multiple roots are always ill-conditioned with respect to perturbations of their coefficients* (exercise 3).

Perturbation theory also has applications in matrix computations. We present two illustrations.

**Example 2.** Suppose the inverse of a matrix $A$ is known. If $A$ is altered slightly - to $A + \epsilon B$ - then the inverse should be expressible as a perturbation of $A^{-1}$:

$$(A + \epsilon B)^{-1} = A^{-1} + \epsilon C + \mathcal{O}(\epsilon^2).$$

To find $C$ we substitute into the equation

$$(A + \epsilon B)(A^{-1} + \epsilon C) = I + \mathcal{O}(\epsilon^2)$$

and observe the first-order terms:

$$\epsilon BA^{-1} + \epsilon AC = 0 \text{ or } C = -A^{-1}BA^{-1}.$$

The approximation for $(A + \epsilon B)^{-1}$ is thus

$$(A + \epsilon B)^{-1} = A^{-1} - \epsilon A^{-1}BA^{-1} + \mathcal{O}(\epsilon^2)$$

For instance if we take $A + \epsilon B$ to be

$$\begin{bmatrix} 2.1 & .1 & .2 \\ .1 & -1.1 & 0 \\ 0 & 0 & 1 \end{bmatrix} = \begin{bmatrix} 2 & 0 & 0 \\ 0 & -1 & 0 \\ 0 & 0 & 1 \end{bmatrix} + (.1)\begin{bmatrix} 1 & 1 & 2 \\ 1 & -1 & 0 \\ 0 & 0 & 0 \end{bmatrix} = A + \epsilon B,$$

its approximate inverse is given by

$$\begin{bmatrix} .5 & 0 & 0 \\ 0 & -1 & 0 \\ 0 & 0 & 1 \end{bmatrix} - (.1)\begin{bmatrix} .5 & 0 & 0 \\ 0 & -1 & 0 \\ 0 & 0 & 1 \end{bmatrix}\begin{bmatrix} 1 & 1 & 2 \\ 1 & -1 & 0 \\ 0 & 0 & 0 \end{bmatrix}\begin{bmatrix} .5 & 0 & 0 \\ 0 & -1 & 0 \\ 0 & 0 & 1 \end{bmatrix}$$

$$= \begin{bmatrix} .475 & .05 & -.1 \\ .05 & -.9 & 0 \\ 0 & 0 & 1 \end{bmatrix}$$

*In fact the exact inverse, rounded to four decimals, is*

$$\begin{bmatrix} .4741 & .0431 & -.0948 \\ .0431 & -.9052 & -.0086 \\ 0 & 0 & 1 \end{bmatrix} \qquad \blacksquare$$

**Example 3.** A real symmetric matrix has real eigenvalues and a complete orthogonal set of eigenvectors. Let $A$ be this matrix, $\{\lambda_1, \lambda_2, \dots, \lambda_n\}$ its eigenvalues, and $\{v_1, v_2, \dots, v_n\}$ the corresponding eigenvectors:

$$Av_i = \lambda_i v_i, \quad v_i^T v_j = 0 \text{ for } i \neq j. \tag{16}$$

Suppose $A$ is altered to become $A' = A + \epsilon B$. Then each eigenvalue and eigenvector is perturbed. Let us consider approximations for the perturbed eigenvalue $\lambda_1'$ and corresponding eigenvector $v_1'$.

Since $\lambda_1'$ is assumed to be a perturbation of $\lambda_1$ we propose the form

$$\lambda_1' = \lambda_1 + \epsilon \mu_1 + \mathcal{O}(\epsilon^2)$$

for the new eigenvalue. The corresponding new eigenvector $v_1'$ presumably differs from $v_1$ by $\mathcal{O}(\epsilon)$. If we express the first-order correction in terms of the unperturbed eigenvectors

$$v_1' = v_1 + \epsilon \sum_{i=1}^{n} a_i v_i + \mathcal{O}(\epsilon^2)$$

then our task is to find formulas for $\mu_1$ and the coefficients $a_i$.

We work out the first-order expansion for the equation $A'\mathbf{v}_1' = \lambda_1'\mathbf{v}_1'$:

$$A\mathbf{v}_1 + \epsilon B\mathbf{v}_1 \quad + \quad \epsilon \sum_{i=1}^{n} a_i A\mathbf{v}_i = \lambda_1\mathbf{v}_1 + \epsilon\mu_1\mathbf{v}_1 + \epsilon \sum_{i=1}^{n} a_i\lambda_1\mathbf{v}_i$$

$$\|$$

$$\lambda_1\mathbf{v}_1 + \epsilon B\mathbf{v}_1 \quad + \quad \epsilon \sum_{i=1}^{n} a_i\lambda_i\mathbf{v}_i = \lambda_1\mathbf{v}_1 + \epsilon\mu_1\mathbf{v}_1 + \epsilon \sum_{i=1}^{n} a_i\lambda_1\mathbf{v}_i$$

(where we have used (16)). Exploiting the orthogonality of the $\mathbf{v}_i$, we premultiply each side of the last equality by $\mathbf{v}_1^T$ to obtain the formula for the correction to the eigenvalue:

$$\mu_1 = \mathbf{v}_1^T B\mathbf{v}_1 / \mathbf{v}_1^T\mathbf{v}_1, \tag{17}$$

while premultiplication by $\mathbf{v}_i^T$ isolates $a_i$:

$$a_i = \mathbf{v}_i^T B\mathbf{v}_1 / [\lambda_1 - \lambda_i]\mathbf{v}_i^T\mathbf{v}_i, \ i \neq 1. \tag{18}$$

Note that $a_1$ cancels out of eq. (17). It is determined by the requirement that the lengths of $\mathbf{v}_1'$ and $\mathbf{v}_1$ be equal, to first order. By orthogonality,

$$\mathbf{v}_1'^T\mathbf{v}_1' = (1 + \epsilon a_1)^2\mathbf{v}_1^T\mathbf{v}_1 + \epsilon^2 \sum_{i=2}^{n} a_i^2\mathbf{v}_i^T\mathbf{v}_i + \mathcal{O}(\epsilon^3)$$

$$= \mathbf{v}_1^T\mathbf{v}_1 + 2\epsilon a_1\mathbf{v}_1^T\mathbf{v}_1 + \mathcal{O}(\epsilon^2) \ ;$$

hence $\mathbf{v}_1'^T\mathbf{v}_1' = \mathbf{v}_1^T\mathbf{v}_1$ to first order if

$$a_1 = 0.$$

For the matrix in the previous example the unperturbed eigenvalues are 2, -1, and +1:

$$\begin{bmatrix} 2 & 0 & 0 \\ 0 & -1 & 0 \\ 0 & 0 & 1 \end{bmatrix} \begin{bmatrix} 1 \\ 0 \\ 0 \end{bmatrix} = (2) \begin{bmatrix} 1 \\ 0 \\ 0 \end{bmatrix},$$

$$\begin{bmatrix} 2 & 0 & 0 \\ 0 & -1 & 0 \\ 0 & 0 & 1 \end{bmatrix} \begin{bmatrix} 0 \\ 1 \\ 0 \end{bmatrix} = (-1) \begin{bmatrix} 0 \\ 1 \\ 0 \end{bmatrix},$$

$$\begin{bmatrix} 2 & 0 & 0 \\ 0 & -1 & 0 \\ 0 & 0 & 1 \end{bmatrix} \begin{bmatrix} 0 \\ 0 \\ 1 \end{bmatrix} = (1) \begin{bmatrix} 0 \\ 0 \\ 1 \end{bmatrix}.$$

For the eigenvalue 2, Eqs. (17, 18) give

$$\mu_1 = [\,1\ 0\ 0\,] \begin{bmatrix} 1 & 1 & 2 \\ 1 & -1 & 0 \\ 0 & 0 & 0 \end{bmatrix} \begin{bmatrix} 1 \\ 0 \\ 0 \end{bmatrix} \div (1) = 1,$$

$$a_2 = [\,0\ 1\ 0\,] \begin{bmatrix} 1 & 1 & 2 \\ 1 & -1 & 0 \\ 0 & 0 & 0 \end{bmatrix} \begin{bmatrix} 1 \\ 0 \\ 0 \end{bmatrix} \div (1)(2+1) = 1/3,$$

$$a_3 = [\,0\ 0\ 1\,] \begin{bmatrix} 1 & 1 & 2 \\ 1 & -1 & 0 \\ 0 & 0 & 0 \end{bmatrix} \begin{bmatrix} 1 \\ 0 \\ 0 \end{bmatrix} \div (1)(2-1) = 0.$$

Thus $\lambda_1' \approx 2.1, \mathbf{v}_1' \approx [1\ \ \frac{1}{30}\ \ 0]^T$. *In fact electronic calculations show* $\lambda_1' = 2.103\cdots, \mathbf{v}_1' = [1\ \ .0312\cdots\ \ 0]^T$.     ■

The denominator in formula (18) makes it clear that standard perturbation theory fails for multiple eigenvalues, and can be inaccurate for closely-spaced eigenvalues. Perturbing the eigenvalues of a *non*symmetric matrix is somewhat more complicated, and we direct the reader to Stewart's text for details.

Before we close this section we present one more illustration of a possible shortcoming of the perturbation theory approach.

**Example 4.** Suppose we try to use perturbation theory to solve the equation

$$\epsilon x^2 + x + .25 = 0. \tag{19}$$

Inserting the solution form (12) and retaining only first-order terms we derive

$$\epsilon x_0^2 + x_0 + \epsilon x_1 + .25 = 0.$$

The unperturbed solution is $x_0 = -.25$, and the first-order correction is

$$x_1 = -x_0^2 = -.0625,$$

so

$$x \approx -.25 - .0625\epsilon. \tag{20}$$

This seems all well and good - (20) satisfies (19) very accurately for small $\epsilon$ (for $\epsilon = .1$, insertion of $x = -.25625$ into the left hand side of (19) produces

.000316 $\cdots$ ). But the perturbation procedure provides no hint that (19) has not one but *two* solutions! For $\epsilon = .1$ they are

$$x_\pm = \frac{-1 \pm \sqrt{12 - 4\epsilon \times .25}}{2\epsilon}; x_+ = -.2558\cdots, \ x_- = -9.7434\cdots;$$

As $\epsilon \to 0$ one of the solutions converges to $x_0 = -.25$, and the other wanders off to infinity.

The perturbation term $\epsilon x^2$ in (19) altered the character of the equation drastically - from linear to quadratic. It illustrates what is known as a *singular perturbation*, and we direct the reader to the specialized literature (Smith, Bellman) for further discussion of this case.     ■

Why are the perturbation calculations so easy, when the original problem may be quite complex (as for eq. (11))? The reason is that the equation for the unknown coefficient is always *linear*. After all, in first order perturbation theory the unknown coefficient $x_1$ is always accompanied by the factor $\epsilon$, and since all terms involving $\epsilon^2$ are dropped no higher powers of $x_1$ can occur. Similarly, for second order theory the unknown coefficient $x_2$ is accompanied by $\epsilon^2$ - but we neglect $\epsilon^3$, so higher powers of $x_2$ don't arise. For this reason perturbation theory is sometimes called "linearization."

# Exercises 9.1

1. Approximate the root near -1 for the equations

   (a) $x^3 - x^2 - x + .99 = 0$ ;

   (b) $.99x^3 - x^2 - x + 1 = 0$ .

2. Verify eq. (5).

3. Analyze why the statement made in the text about multiple roots is true. (Hint: suppose $p(x) = a_0 + a_1 x + a_2 x^2 + \cdots + a_n x^n = 0$ for $x = x_0$. Use the chain rule to express $\partial x / \partial a_j$ in terms of $p'(x)$ and $\partial p(x) / \partial a_j$. What happens as $x \to x_0$?

4. Extend the perturbation theory idea to two variables and find an approximate solution near $(0, 0)$ to the system

$$x^2 + y^2 + x = .01,$$
$$xy - x - y = 0.$$

Test the accuracy of your solution.

5. Suppose you must analyze an equation subject to *two* perturbations; to be specific, consider the equation

$$x^3 + (1 + \epsilon_1)x^2 + (1 + \epsilon_2)x + 1 = 0$$

to be solved near $x = -1$. Develop the perturbation approach for this case. Show, in particular, that the perturbations can be analyzed one at a time to achieve first order accuracy; "coupling" effects don't show up until second order. Test the accuracy of your solutions for $\epsilon_1 = .01, \epsilon_2 = .02$.

6. Test the perturbation formula for the inverse with some other matrices.

7. Work out second-order perturbation theory for the matrix inverse. Try out some test cases. (When $A$ is the identity matrix, the series resulting from 2nd, 3rd, ... , order perturbation theory is known as the *Neumann series*.)

8. Use perturbation theory to estimate the other eigenvalues and eigenvectors for Example 3.

9. Prove the following statement: if $\mathbf{v}$ differs from an eigenvector of a symmetric matrix $A$ by $\mathcal{O}(\epsilon)$, then $\mathbf{v}^T A \mathbf{v} / \mathbf{v}^T \mathbf{v}$ differs from the eigenvalue by $\mathcal{O}(\epsilon^2)$. This ratio is known as the *Rayleigh quotient*.

10. (*Perturbation theory for Sturm-Liouville problems*) The perturbation theory for the symmetric-matrix eigenvalue problem can be formally carried over to regular Sturm-Liouville problems. In the notation of Section 6.2 let $\mathcal{L}$ denote a second-order differential operator in Sturm-Liouville form with weight function $w(x)$, eigenvalues $\lambda_n$, and eigenfunctions $\phi_n$:

$$\mathcal{L}\phi_n = \lambda_n w \phi_n$$

(assume homogeneous boundary conditions).

Suppose that $\mathcal{L}$ is perturbed by the addition of a function $\epsilon h(x)$. For simplicity we assume that $h$ is simply a scalar function and does not contain differential operators. Thus the weight function $w$ remains the same and the new eigenvalues and eigenfunctions, denoted $\lambda_n + \epsilon \mu_n$ and $\psi_n$ respectively, satisfy

$$[\mathcal{L} + \epsilon h]\psi_n = (\lambda_n + \epsilon \mu_n) w \psi_n. \tag{21}$$

We seek formulas for $\mu_n$ and $\psi_n$.

(a) Since the original eigenfunctions form a complete orthogonal set, we can expand the new eigenfunctions in terms of the old.

$$\psi_n(x) = \sum_{m=1}^{\infty} c_m \phi_m(x), \ \ c_m = \langle \psi_n, \phi_m \rangle_w / \|\phi_n\|_w^2. \tag{22}$$

But since $\psi_n$ differs from $\phi_n$ by $\mathcal{O}(\epsilon)$ (presumably) we can express (22) as

$$\psi_n = a_n \phi_n + \epsilon \sum_{\substack{m=1 \\ (m \neq n)}}^{\infty} a_m \phi_m.$$

Use orthogonality to show that

$$\|\psi_n\|^2 = |a_n|^2 \|\phi_n\|^2 + \mathcal{O}(\epsilon^2), \tag{23}$$

and since $\|\psi_n\|$ and $\|\phi_n\|$ should agree to first order, conclude that $a_n = 1$.

(b) Combine eqs. (21) and (23) and obtain, to first order in $\epsilon$,

$$\lambda_n e \phi_n + \epsilon \sum_{\substack{m=1 \\ (m \neq n)}}^{\infty} a_m \lambda_m w \phi_m + \epsilon h \phi_n$$

$$= \lambda_n w \phi_n + \epsilon \sum_{\substack{m=1 \\ (m \neq n)}}^{\infty} a_m \lambda_n w \phi_m + \epsilon \mu_n w \phi_n. \tag{24}$$

(c) Multiply both sides of (24) with $\phi_m$ and integrate to derive the formula for the perturbation of the eigenvalue $\lambda_n$:

$$\mu_n = \int_a^b h(x) \phi_n^2(x) dx \, / \, \|\phi_n\|_w^2.$$

(d) Multiply both sides of (23) with $\phi_m$ and integrate to derive the formula for the coefficient $a_m$:

$$a_m = \int_a^b \phi_m(x) h(x) \phi_n(x) dx \, / \, (\lambda_m - \lambda_n) \|\phi_m\|_w^2. \tag{25}$$

Clearly (25) reveals that first-order perturbation theory is invalid when $\lambda_n$ is a multiple eigenvalue, and inaccurate when the eigenvalues are closely spaced. Thus for a continuous spectrum advanced methods must be used. We refer the reader to Friedrichs's work for further details.

# 9.2 Perturbation Methods for Differential Equations

**Summary** Linearization can be used to compute an approximation to a solution of a perturbed differential equation by using Green's functions for the unperturbed equation. The approximation, for an initial value problem, is valid for small perturbations and moderate excursions of the independent variable, unless a resonance of the unperturbed equation is excited or the perturbation affects the order of the equation. The perturbation can lie in the equation itself, or in the auxiliary conditions.

The idea of expanding an unknown in powers of a small parameter can be extended to *functions*, enabling us to find approximate solutions to analytically intractible differential equations. We simply take the format of equation (17) of the previous section and adapt it to functions:

$$y(x; \varepsilon) = y_0(x) + \varepsilon y_1(x) + \varepsilon^2 y_2(x) + \cdots \tag{1}$$

Then we proceed as before - neglecting $\varepsilon$ for zeroth-order perturbation theory, neglecting $\varepsilon^2$ for first-order corrections, neglecting $\varepsilon^3$ for second-order theory, etc. (one seldom carries perturbation theory beyond second order for differential equations). An example will illustrate the basic procedure.

**Example 1.** The nonlinear system

$$y' + y + \varepsilon y^2 = 0, \quad y(0) = 1 \tag{2}$$

will be used to demonstrate the ideas. Inserting the expression (1) and retaining second-order terms we find

$$y_0' + \varepsilon y_1' + \varepsilon^2 y_2' + y_0 + \varepsilon y_1 + \varepsilon^2 y_2 + \varepsilon y_0^2 + \varepsilon^2 2 y_0 y_1 + \mathcal{O}(\varepsilon^3) = 0 \tag{3}$$

*The initial condition also must be addressed:*

$$y_0(0) + \varepsilon y_1(0) + \varepsilon^2 y_2(0) + \mathcal{O}(\varepsilon^3) = 1. \tag{4}$$

The zeroth-order equations, obtained by setting $\varepsilon = 0$ in $(3, 4)$, are

$$y_0' + y_0 = 0, \quad y_0(0) = 1;$$

and their solution is

$$y_0(x) = e^{-x}. \tag{5}$$

To first order $(3, 4)$ require

$$\varepsilon y_1' + \varepsilon y_1 + \varepsilon y_0^2 = 0 \text{ or } y_1' + y_1 = -y_0^2 = -e^{-2x}, \quad y_1(0) = 0. \tag{6}$$

(Note how the initial condition is derived from (4). Note also that the zeroth order solution becomes a *nonhomogeneity* for the first order equations.) The solution is found by the methods of Sec. 1.1 to be (exercise 1)

$$y_1(x) = e^{-2x} - e^{-x}. \tag{7}$$

The second-order correction satisfies

$$\varepsilon^2\{y_2' + y_2 + 2y_0y_1\} = 0 \quad \text{or} \quad y_2'' + y_2 = -2y_0y_1 = -2e^{-x}[e^{-2x} - e^{-x}],$$
$$y_2(0) = 0, \tag{8}$$

and its solution is

$$y_2(x) = e^{-3x} - 2e^{-2x} + e^{-x}. \tag{9}$$

The overall solution, to second order, is then

$$y(x) = e^{-x} + \varepsilon(e^{-2x} - e^{-x}) + \varepsilon^2(e^{-3x} - 2e^{-2x} + e^{-x}) + \mathcal{O}(\varepsilon^3). \tag{10}$$

As a matter of fact, the original problem (2) can be solved in closed form in this case. The expansion of the exact solution,

$$y(x) = e^{-x}/[1 + \varepsilon(1 - e^{-x})], \tag{11}$$

in powers of $\varepsilon$ agrees with (10); see exercise 1. Plots of (10) and (11) in Fig. 1 demonstrate the accuracy of the perturbation approach for $\varepsilon = .6$. ∎

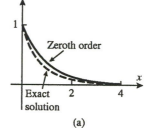

(a)

**Example 2.** The *Duffing* equation (recall Exercise 6, Sec. 3.2) can be interpreted as describing oscillatory motion generated by a spring with a nonlinear spring constant $(1 + \varepsilon y^2)$:

$$y'' + y + \varepsilon y^3 = 0, \quad y(0) = 1, \quad y'(0) = 0. \tag{12}$$

The initial conditions indicate that the spring is stretched and then released from rest. To first order the perturbation equations read as follows:

$$y_0'' + \varepsilon y_1'' + y_0 + \varepsilon y_1 + \varepsilon y_0^3 = 0,$$
$$y_0(0) + \varepsilon y_1(0) = 1, \quad y_0'(0) + \varepsilon y_1'(0) = 0. \tag{13}$$

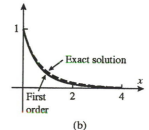

(b)

The zeroth-order problem

$$y_0'' + y_0 = 0, \quad y_0(0) = 1, \quad y_0'(0) = 0 \tag{14}$$

has the solution

$$y_0(x) = \cos x. \tag{15}$$

The equations for the first-order perturbation then constitute a nonhomogeneous linear second order differential equation

$$\varepsilon y_1'' + \varepsilon y_1 + \varepsilon y_0^3 = 0, \quad \text{or}$$
$$y_1'' + y_1 = -y_0^3 = -\cos^3 x = -\frac{\cos 3x + 3\cos x}{4} \tag{16}$$

with homogeneous initial conditions:

$$y_1(0) = 0, \quad y_1'(0) = 0. \tag{17}$$

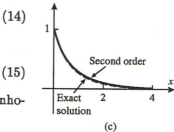

(c)

***Figure 1*** *Solutions for Example 1*

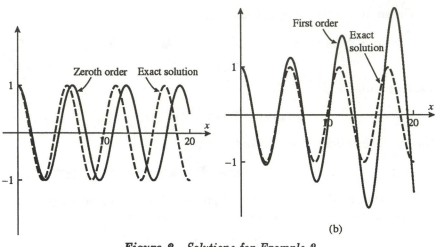

**Figure 2**   *Solutions for Example 2*

They are solved by using the appropriate Green's function. The solution (exercise 2) is found to be

$$y_1(x) = \frac{\cos 3x}{32} - \frac{3x \sin x}{8} - \frac{\cos x}{32}. \tag{18}$$

The function

$$y_0(x) + \varepsilon y_1(x) = \cos x + \varepsilon \left\{ \frac{\cos 3x}{32} - \frac{3x \sin x}{8} - \frac{\cos x}{32} \right\} \tag{19}$$

is plotted in Fig. 2 together with a computer-generated solution of (12) for $\varepsilon = .3$. Notice that although the perturbation correction improves the fit over the first oscillation, its accuracy deteriorates for larger $x$. The reason for this is revealed in expression (18). We have been working under the presumption that the term $\varepsilon y_1(x)$ is small, simply because $\varepsilon$ is small. However $y_1(x)$ grows without bound for large $x$, because of the middle term in (18) - a result of the *resonant* forcing term $\cos x$ in (16) (recall Section 3.2). Thus the legitimacy of the perturbation approximation in this case is only valid for a limited range of $x$.                                      ∎

Problems like the Duffing equation, for which the (presumedly small) perturbation terms grow without bound, frequently arise in the mathematical modeling of oscillatory systems. The unbounded terms are known as "secular" perturbations, and advanced techniques have been devised for describing the solutions more accurately. We discuss *Linstedt's* method in exercise 3. See the texts of Smith and Bellman for further reference.

**Example 3.** The (linear) equations

$$y'' + (1 + \varepsilon x)y = 0, \quad y(0) = 1, \quad y'(0) = 0 \tag{20}$$

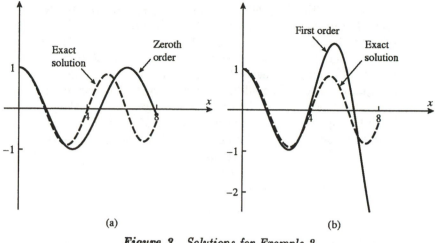

**Figure 3**    *Solutions for Example 3*

describe the motion of a unit mass on a spring whose spring constant $(1+\varepsilon x)$ stiffens with time $(x)$.

To first order the perturbation equations read as follows:

$$y_0'' + \varepsilon y_1'' + y_0 + \varepsilon y_1 + \varepsilon x y_0 = 0,$$
$$y_0(0) + \varepsilon y_1(0) = 1, \quad y_0'(0) + \varepsilon y_1'(0) = 0. \tag{21}$$

The unperturbed solution is identical with that for the Duffing spring in the previous example:

$$y_0(x) = \cos x. \tag{22}$$

The first-order correction is again described by a nonhomogeneous linear differential equation with homogeneous initial conditions:

$$y_1'' + y_1 + x y_0 = 0, \quad \text{or} \quad y_1'' + y_1 = -x y_0 = -x \cos x,$$
$$y_1(0) = 0, \quad y_1'(0) = 0.$$

And the Green's function provides the solution (exercise 4):

$$y_1(x) = -\frac{x^2 \sin x + x \cos x + \sin x}{4}. \tag{23}$$

The complete first-order solution, $y_0 + \varepsilon y_1$, is plotted in Fig. 3 for $\varepsilon = .2$ along with

$$\pi Bi'(-\varepsilon^{-2/3})\, Ai(-[1+\varepsilon x]\varepsilon^{-2/3}) - \pi Ai'(-\varepsilon^{-2/3})\, Bi(-[1+\varepsilon x]\varepsilon^{-2/3}) \tag{24}$$

which, as exercise 5 shows, expresses the exact solution in terms of Airy functions. Again the perturbation is secular and the improvements offered by perturbation theory eventually deteriorate.    ∎

Now recall the *singular* perturbation demonstrated in Example 4 from Section 9.1; when the quadratic term $\varepsilon x^2$ was added to the *linear* algebraic equation $x + .25 = 0$, it altered the character of the equation drastically - an extra solution was created. The singular perturbation has its analog in differential equations, too. The addition of a perturbation containing a higher derivative than the original equation changes the *order* of the equation - and thus the number of initial conditions required to specify the solution. We briefly demonstrate the phenomenon with a simple example.

**Example 4.** Consider the system

$$\varepsilon y'' + y' + .25 y = 0; \quad y(0) = 0, \quad y'(0) = 1. \tag{25}$$

The *exact* solution to this (constant coefficient) equation is (exercise 6)

$$y(x) = \frac{\varepsilon}{\sqrt{1-\varepsilon}}[\exp(\omega_+ x) - \exp(\omega_- x)]; \quad \omega_\pm = \frac{-1 \pm \sqrt{1-\varepsilon}}{2\varepsilon}. \tag{26}$$

If we analyze (25) with perturbation theory the calculations go as follows. To zeroth order,

$$y_0' + .25 y_0 = 0, \quad y_0(0) = 0, \quad y_0'(0) = 1; \tag{27}$$

The differential equation and the first initial condition imply that $y_0(x) \equiv 0$ - *and thus we have no flexibility to meet the condition on* $y'(0)$.

Since $y_0(x) \equiv 0$, the first-order correction satisfies

$$\varepsilon y_0'' + \varepsilon y_1' + .25\varepsilon y_1 = 0 \quad \text{or} \quad y_1' + y_1 = -y_0'' = 0, \quad y_1(0) = 0 \tag{28}$$

so $y_1(x) \equiv 0$ also. (In fact it is not difficult to see that the perturbation solution is identically zero to all orders.)

The graphs of the exact solution and the perturbation computation in Fig. 4 demonstrate the typical behavior of problems with singular perturbations. For $x$ significantly large the agreement is good, but there is a "boundary layer" near $x = 0$ where the unsatisfied initial condition $y'(0) = 1$ causes considerable mismatch. Again the reader is referred to the specialized literature for more on this subject.  ∎

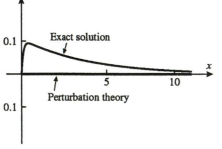

*Figure 4   Solutions for Example 4*

The application of perturbation theory to *partial* differential equations is demonstrated in the final two examples.

**Example 5.** The equation for steady-state temperature distributions in a nonhomogeneous medium is

$$\nabla \cdot (k\nabla \psi) = 0 \tag{29}$$

where $k$ is the thermal conductivity. Usually we have taken $k$ to be constant (uniform) so that (29) reduces to Laplace's equation. For nonuniform conductivities, however, the correct equation is seen to be

$$k\nabla^2\psi + (\nabla k)\cdot\nabla\psi = 0 \quad\text{or}\quad \nabla^2\psi + (\nabla k/k)\cdot\nabla\psi = 0. \tag{30}$$

Let us regard the second term as a perturbation and find the approximate solution of (30) in, say, a rectangle with boundary conditions (Fig. 5)

$$\psi(x,0) = 0, \quad \psi(0,y) = 0, \quad \psi(x,\pi) = 0, \quad \psi(\pi,y) = 1. \tag{31}$$

We treat the gradient $\nabla k$ as the *vector* expansion parameter $\varepsilon$ for the analysis. The equation (30) is rewritten as

$$\nabla^2\psi + \frac{\varepsilon}{k}\cdot\nabla\psi = 0,$$

and we propose the solution form

$$\psi(x,y) = \psi_0(x,y) + |\varepsilon|\psi_1(x,y) + \mathcal{O}(\varepsilon^2), \quad |\varepsilon| = |\nabla k|. \tag{32}$$

To zeroth order we have

$$\nabla^2\psi_0 = 0, \quad \psi_0(x,0) = 0, \quad \psi_0(0,y) = 0, \quad \psi_0(x,\pi) = 0, \quad \psi_0(\pi,y) = 1 \tag{33}$$

which was solved in exercise 4, Sec. 5.2:

$$\psi_0(x,y) = \frac{4}{\pi}\sum_{n=0}^{\infty}\frac{\sinh(2n+1)x\sin(2n+1)y}{(2n+1)\sinh(2n+1)\pi}. \tag{34}$$

The first-order correction then satisfies

$$|\varepsilon|\nabla^2\psi_1 + \frac{\varepsilon\cdot\nabla\psi_0}{k} = 0 \quad\text{or}\quad \nabla^2\psi_1 = -\frac{\nabla k\cdot\nabla\psi_0}{|\nabla k|\,k}, \tag{35}$$

$$\psi_1 = 0 \text{ on all four sides.} \tag{36}$$

This last system is a nonhomogeneous differential equation with homogeneous boundary conditions and is readily solved by the Green's function method:

$$\psi_1(x,y) = -\iint G(x,y;x',y')\frac{\nabla k\cdot\nabla\psi_0}{|\nabla k|\,k}\,dx'\,dy'\,. \tag{37}$$

The reader is invited to execute this formula in exercise 7. ∎

Perturbation theory can even be used to treat *geometric* effects, such as partial differential equations with conditions imposed on warped boundaries, where ordinarily separation of variables would seem inapplicable! The final example illustrates the procedure.

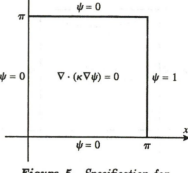

**Figure 5** *Specification for Example 5*

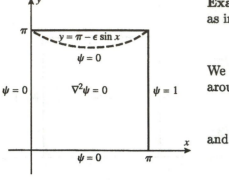

**Figure 6**

**Example 6.** Suppose the upper edge of the square in Example 5 is warped as in Fig. 6, so that it is described by the equation

$$y_{top}(x) = \pi - \varepsilon \sin x. \tag{38}$$

We then seek the equilibrium temperature with the heat sources arranged around the edges as before (but this time with *uniform* conductivity!):

$$\nabla^2 \psi = 0; \quad \psi(x,0) = \psi(0,y) = 0, \quad \psi(\pi, y) = 1, \tag{39}$$

and

$$\psi(x, \pi - \varepsilon \sin x) = 0. \tag{40}$$

Using the Taylor expansion we can write the last condition as

$$\psi(x, \pi) - \varepsilon \sin x \frac{\partial \psi}{\partial y}(x, \pi) + \mathcal{O}(\varepsilon^2) = 0. \tag{41}$$

Now insertion of $\psi(x,y) = \psi_0(x,y) + \varepsilon \psi_1(x,y) + \mathcal{O}(\varepsilon^2)$ into (39, 41) yields the identical zeroth-order system as in the previous example

$$\nabla^2 \psi_0 = 0, \quad \psi_0(x,0) = 0, \quad \psi_0(0,y) = 0, \quad \psi_0(x,\pi) = 0, \quad \psi_0(\pi,y) = 1 \tag{42}$$

with the boundary conditions imposed *on the unperturbed edges!* Thus

$$\psi_0(x,y) = \frac{4}{\pi} \sum_{n=0}^{\infty} \frac{\sinh(2n+1)x \sin(2n+1)y}{(2n+1)\sinh(2n+1)\pi}. \tag{43}$$

The first-order correction equations also address the unperturbed edge:

$$\nabla^2 \psi_1 = 0; \quad \psi_1(x,0) = \psi_1(0,y) = 0, \quad \psi_1(\pi,y) = 0, \text{ and}$$
$$\psi_1(x,\pi) = -\sin x \frac{\partial \psi_0}{\partial y}(x,\pi). \tag{44}$$

Thus equations (44) are solved in the same manner as the zeroth-order equations:

$$\psi_1(x,y) = \sum_{n=1}^{\infty} a_n \sin nx \sinh ny,$$

$$a_n = \frac{2}{\pi} \int_0^{\pi} \left\{ -\sin x \frac{\partial \psi_0}{\partial y}(x,\pi) \right\} \sin nx \, dx \Big/ \sinh n\pi \tag{45}$$

(see exercise 8).       ■

    This final example suggests the power of the perturbation approach. Many apparently intractable problems in engineering have been successfully analyzed by cleverly recasting them as perturbations of nearby, solvable, configurations.

# Exercises 9.2

1. (a) Derive the solution forms (5), (7), and (9).

   (b) Derive the solution (11).

   (c) Show that the Taylor expansion of (11) agrees with (10).

2. Derive the expression (18) for the perturbation of the Duffing equation.

3. (*Lindstedt's method*) Upon contemplating Fig. 2 one might speculate that the source of the secular perturbation is that the formalism of classical perturbation theory is unable to account for the shift in the oscillation *frequency* (interpreting $x$ as *time*) induced by the nonlinearity. With the Linstedt procedure we give ourselves some slack by allowing a reparametrization of the *independent variable*

$$x = f(\xi), \tag{46}$$

and seek a perturbation expansion for the dependent variable $y$ as a function of $\xi$, *choosing $f(\xi)$ so that the expansion contains no secular terms*. Thus we set

$$x = f(\xi) = \xi(1 + \varepsilon c_1 + \varepsilon^2 c_2 + \cdots) = \xi + \varepsilon c_1 \xi + \mathcal{O}(\varepsilon^2). \tag{47}$$

   (a) Show that by the rules for changing variables introduced in Sec. 1.2, the Duffing system (12) transforms to

$$\frac{d^2 y}{d\xi^2} + y + \varepsilon(2c_1 y + y^3) = \mathcal{O}(\varepsilon^2), \quad y(0) = 1, \quad \frac{dy}{d\xi}(0) = 0. \tag{48}$$

   (b) With $y(\xi) = y_0(\xi) + \varepsilon y_1(\xi) + \mathcal{O}(\varepsilon^2)$ the zeroth-order solution to (48) is clearly $y_0(\xi) = \cos\xi$. Derive the equation for $y_1(\xi)$,

$$\frac{d^2 y_1}{d\xi^2} + y_1 = -\left(2c_1 + \frac{3}{4}\right)\cos\xi - \frac{1}{4}\cos 3\xi. \tag{49}$$

   (c) Explain why the choice $c_1 = -3/8$ will eliminate any secular terms from the solution to (49).

   (d) Assemble these ideas to derive the approximation

$$y(x) \approx \cos\frac{x}{1 - 3\varepsilon/8} + \frac{\varepsilon}{32}\left\{\cos\frac{3x}{1 - 3\varepsilon/8} - \cos\frac{x}{1 - 3\varepsilon/8}\right\}. \tag{50}$$

   to order $\mathcal{O}(\varepsilon^2)$.

   (e) Compare the graph of (50) with that of the computer-generated solution in Fig. 2.

**4.** Derive the perturbation solution (23) for the aging spring.

**5.** Derive the exact solution (24) for the aging spring.

**6.** Derive the exact solution (26) for the singular perturbation example.

**7.** Carry out the solution formula (37).

**8.** Carry out the solution formula (45).

**9.** Compute the second-order perturbation expansions for

$$y' = \varepsilon y^2, \quad y(0) = 1.$$

Compare with the exact solution. (We analyzed this example for $\varepsilon = 1$ in Sec. 2.3 and in exercise 9 of Sec. 1.1.)

**10.** Work out the first-order perturbation expansion for

　　a. $y'' + \varepsilon y^2 = 0$, $y(0) = 0$, $y'(0) = 1$.
　　b. $y'' + \varepsilon y^2 = x$, $y(0) = 0$, $y'(0) = 1$.
　　c. $y'' + \varepsilon y'^2 + y = 0$, $y(0) = 1$ , $y'(0) = 0$.
　　d. $y'' + \varepsilon \sin y = 0$, $y(0) = 0$, $y'(0) = 1$.

**11.** Consider the damping term $(\varepsilon y')$ as a perturbation effect in the equation for the harmonic oscillator

$$y'' + \varepsilon y' + y = 0, \quad y(0) = 0, \quad y'(0) = 1.$$

Compare the exact solution with the perturbation approximation.

**12.** Use perturbation theory to approximate the steady state temperature distribution in the square of Fig. 6 if the *right* edge – instead of the top edge – is warped:

$$x = \pi - \varepsilon \sin y.$$

**13.** Use perturbation theory to approximate the steady state temperature distribution in the square of Fig. 6 if the Dirichlet condition on the warped edge is replaced by the Neumann condition

$$\frac{\partial \psi}{\partial n} = f(x).$$

**14.** (*A perturbation technique for integral equations*) Suppose the unknown function $y(x)$ satisfies an *integral equation* of the form

$$y(x) = f(x) + \varepsilon \int_a^b K(x,t)\, y(t)\, dt, \tag{51}$$

where $f(x)$ and $K(x,y)$ are given. Equation (51) is known as a *Fredholm equation of the second kind*; if the upper limit "$b$" is replaced by the variable "$x$" it becomes a *Volterra equation of the second kind* (see Arfken's text).

(a) Show that substitution of the expression (1) into (51) and matching the powers of $\varepsilon$ result in the *Neumann series*

$$y_0(x) = f(x),$$

$$y_1(x) = \int_a^b K(x,t)\, y_0(t)\, dt,$$

$$y_2(x) = \int_a^b K(x,t)\, y_1(t)\, dt, \dots . \tag{52}$$

(b) The Neumann series can be expected to converge for a Fredholm equation if the parameter $\varepsilon$ is small enough. Carry out 4 terms of the expansion for the equation

$$y(x) = x + .5 \int_{-1}^1 (t-x)\, y(t)\, dt.$$

Verify that the exact solution is $y(x) = (3x+1)/4$, and compare with your approximation.

(c) The Neumann series for a Volterra equation can be expected to converge if either $\varepsilon$ is small or if the range of integration $[a, x]$ is sufficiently short. Carry out 3 terms of the expansion for the equation

$$y(x) = 1 - 2 \int_0^x t\, y(t)\, dt.$$

Verify that the exact solution is $y(x) = e^{-x^2}$, and compare with your approximation.

(d) Show that if the *Kernel* $K(x,t)$ in a Volterra equation is independent of $x$, the integral equation can be reformulated as a first order differential equation with initial conditions. Re-solve part (c) by this procedure.

## 9.3  Dispersion Laws and Wave Velocities

**Summary**  Coupled systems of partial differential equations are usually very difficult to solve completely, but sometimes the dispersion law - the relation between the wave number and frequency for oscillatory perturbations - can be derived by separating variables. Many physical properties of the system can be extracted from this formula: the phase, signal-front, and group velocities, in particular.

Maxwell's equations (Sec. 4.4) comprise a familiar example of a system of coupled partial differential equations. There are six unknowns (the scalar components of $\mathbf{E}$ and $\mathbf{B}$) and six (scalar) equations. In exercise 2 of Sec. 4.4 it was shown how one could, under certain circumstances, decouple the system by deriving a single partial differential equation - the wave equation - for each component of the electromagnetic field. This leads to the successful solution of such problems as waveguide radiation, by the separation of variables procedure. In general, though, when a system is so complex that it cannot be reduced to a single equation, the eigenfunction expansion techniques may be thwarted.

Perturbations of a given configuration, however, frequently yield to the analysis, and the results provide significant information about the physical system. We illustrate by studying some plasma oscillations. The text by Chen is an excellent reference for the background physics.

**Example 1.** A plasma can be described roughly as a fluid of ionized particles whose motion is subject to the laws of both fluid mechanics and electromagnetics. We consider a one-dimensional plasma with an electron mass density $\rho(x,t)$ (mass per unit volume). This gives rise to an electron charge density $-q\rho$, where $q$ is the electron's charge-to-mass ration $(e/m)$. The unperturbed state is assumed to be quiescent: the density $\rho_0$ is uniform, and overall charge neutrality is maintained by a background of positive ions, assumed to be so massive that their charge density $n = +q\rho$ does not vary. The unperturbed electron velocity $\mathbf{v}_0$ and electric field $\mathbf{E}_0$ are zero. The magnetic field is neglected.

The fluid force which drives the electron motion is the pressure gradient $(-\nabla p)$; but because the electrons are charged, they are also subject to the electrical force $-q\rho\mathbf{E}$ per unit volume. Consequently Newton's second law as stated in eq. (10) of Sec. 4.4 has to be extended:

$$\rho\frac{d\mathbf{v}}{dt} = -\nabla p - q\rho\mathbf{E}.$$

In one dimension with the convective derivative and the speed of sound[1] $v_s$ inserted (Sec. 4.4), this becomes

$$\rho\left[\frac{\partial v}{\partial t} + v\cdot\frac{\partial v}{\partial x}\right] = v_s^2\frac{\partial\rho}{\partial x} - q\rho E. \tag{1}$$

The conservation of mass is expressed by eq. (11) in Sec. 4.4:

$$\frac{\partial\rho}{\partial t} = -\frac{\partial\rho v}{\partial x} \tag{2}$$

---

[1] Our derivation will show that $v_s = \sqrt{dp/d\rho}$ is the speed at which sound waves would propagate in the pure electron gas if the electrons were uncharged and the ions did not participate in the motion. See Chen for details.

Maxwell's equation $\nabla \cdot (\epsilon_0 \mathbf{E}) = (net\,charge\,density)$ takes the one-dimensional form

$$\epsilon_0 \frac{\partial E}{\partial x} = n - q\rho, \tag{3}$$

where $\epsilon_0$ is the permittivity of free space.

Equations (1-3) constitute a (nonlinear) coupled system of partial differential equations for the unknowns $\rho, v$, and $E$. Note that the unperturbed values $\rho_0 = constant = n/q, v_0 = 0, E_0 = 0$ satisfy (1-3).

Now let's derive the first order perturbation equations. We omit the perturbation parameter - denoted $\epsilon$ in the previous sections - to avoid confusion with $\epsilon_0$. Setting

$$\rho = \rho_0 + \rho_1, \; v = v_0 + v_1 = v_1, \; E = E_0 + E_1 = E_1$$

and neglecting products of the perturbations, we reduce eq. (1) to

$$\rho_0 \frac{\partial v_1}{\partial t} = -v_s^2 \frac{\partial \rho_1}{\partial x} - q\rho_0 E_1. \tag{4}$$

Equaion (2) becomes

$$\frac{\partial \rho_1}{\partial t} = -\rho_0 \frac{\partial v_1}{\partial x}, \tag{5}$$

and (3) yields

$$\epsilon_0 \frac{\partial E_1}{\partial x} = -q\rho_1. \tag{6}$$

We envision the perturbations as being set up by some type of boundary oscillation, so this time-dependent system shall be analyzed in the frequency domain. Thus for each frequency $\omega$ the contributions take the form

$$\rho_1(x,t) = P_1(x,\omega)e^{i\omega t}, v_1(x,t) = V_1(x,\omega)e^{i\omega t}, E_1(x,t) = \mathcal{E}_1(x,\omega)e^{i\omega t} \tag{7}$$

and eqs. (4-6) become

$$i\rho_0\omega V_1 = -v_s^2 \frac{\partial P_1}{\partial x} - q\rho_0 \mathcal{E}_1, \; i\omega P_1 = -\rho_0 \frac{\partial V_1}{\partial x}, \; \epsilon_0 \frac{\partial \mathcal{E}_1}{\partial x} = -qP_1. \tag{8}$$

These are coupled *ordinary* differential equations. Bearing in mind the success of the eigenfunction expansions, however, let's try to find solutions with a common shape factor $X(x,\omega)$:

$$P_1(x,\omega) = A(\omega)X(x,\omega), \; V_1(x,\omega) = B(\omega)X(x,\omega), \; \mathcal{E}_1(x,\omega) = C(\omega)X(x,\omega) \tag{9}$$

Insertion of (9) into (8) yields

$$i\omega\rho_0 BX = -v_s^2 AX' - q\rho_0 CX, \; i\omega AX = -\rho_0 BX', \; \epsilon_0 CX' = -qAX$$

or, after division by $X$,

$$v_s^2 \frac{X'}{X} A + i\omega\rho_0 B + q\rho_0 C = 0,$$

$$i\omega A + \rho_0 \frac{X'}{X} B = 0,$$

$$qA + \epsilon_0 \frac{X'}{X} C = 0. \tag{10}$$

Clearly eqs. (10) imply $X'/X$ is independent of $x$ - say, $X'/X = \mu$ - and the equations can be expressed in matrix form as

$$\begin{bmatrix} v_s^2 \mu & i\omega\rho_0 & q\rho_0 \\ i\omega & \rho_0\mu & 0 \\ q & 0 & \epsilon_0\mu \end{bmatrix} \begin{bmatrix} A \\ B \\ C \end{bmatrix} = \begin{bmatrix} 0 \\ 0 \\ 0 \end{bmatrix}.$$

This homogeneous system admits nontrivial solutions only if the determinant vanishes:

$$v_s^2 \rho_0 \epsilon_0 \mu^3 + \omega^2 \epsilon_0 \rho_0 \mu - q^2 \rho_0^2 \mu = 0 \tag{11}$$

One solution of (11) is $\mu = X'/X = 0$, generating the perturbation $A = 0$, $B/C = iq/\omega$, or (constant multiples of)

$$\rho_1(x,t) = 0, \; v_1(x,t) = \frac{iq}{\omega} e^{i\omega t}, \; \mathcal{E}_1(x,t) = e^{i\omega t} \tag{12}$$

The functions in (12) are independent of $x$; they describe oscillations of the plasma as a rigid body, and have little interest.

The other solutions of (11) are

$$\mu = \mu(\omega) = \pm \sqrt{\frac{q^2 \rho_0 - \epsilon_0 \omega^2}{\epsilon_0 v_s^2}} \tag{13}$$

with the corresponding perturbations given by (constant multiples of)

$$\rho_1(x,t) = e^{i\omega t} e^{\mu x}, \; v_1(x,t) = -\frac{i\omega}{\mu\rho_0} e^{i\omega t} e^{\mu x},$$

$$E_1(x,t) = -\frac{q}{\mu\epsilon_0} e^{i\omega t} e^{\mu x} \tag{14}$$

Thus if a disturbance is created in the plasma density at some point (say, $x = 0$) in this (idealized one-dimensional) plasma, then we can Fourier-analyze the disturbance and assemble the combination of functions in (13,

14) which matches this boundary condition. We shall not go into the specific physics of the boundary condition.

Note that for the frequency components with $\omega^2 > q^2\rho_0/\epsilon_0$, $\mu$ in (13) is imaginary. It is convenient, then, to take $\mu = -ik$ and write (13, 14) as

$$k = k(\omega) = \pm\frac{\sqrt{\epsilon_0\omega^2 - q^2\rho_0}}{\epsilon_0 v_s^2}, \tag{15}$$

$$\rho_0(x,t) = e^{i(\omega t - kx)}, v_1(x,t) = -\frac{i\omega}{\mu\rho_0}e^{i(\omega t - kx)},$$

$$E_1(x,t) = -\frac{q}{\mu\epsilon_0}e^{i(\omega t - kx)} \tag{16}$$

We recognize the functions in (16) as traveling waves (known in plasma physics as *Bohm-Gross waves*). The minus sign was affixed to the "$-k$" so that the waves migrate to the right, fulfilling the Sommerfeld condition for $x > 0$, when $\omega$ and $k$ are positive. (Since either $\omega$ or $k$ can be negative, this luxury sacrifices no loss of generality.)

When $\omega^2 < q^2\rho_0/\epsilon_0$, we take $\mu = -ik < 0$ in (14) for finiteness at $x = \infty$, and obtain waves attenuated to the right (exactly analogous to our experience with waveguides, Sec. 8.5).                                          ∎

Suppose that instead of treating the *boundary* value problem, we consider the *initial* value problem where the disturbance is given at $t = 0$ for all $x$. Then we can Fourier–analyze the $x$ dependence first and subsequently derive the time factors. You are invited to carry out this program in exercise 1, where you will find that the solutions take the same form as (16), but with the parameter $k$ occurring as the Fourier "spatial frequency" or *wave number*; $k$ thus takes all real values from $-\infty$ to $\infty$, and $\omega$ then is expressed in terms of $k$ by

$$\omega = \omega(k) = \pm\frac{\sqrt{\epsilon_0 v_s^2 k^2 + q^2\rho_0}}{\epsilon_0} \tag{17}$$

Just as we saw in our waveguide analysis (Sec. 8.5), the relations (15, 17) between $\omega$ and $k$ are identical - whether derived through boundary-value or initial-value analysis - and can be expressed generically as

$$\omega^2 = \frac{q^2\rho_0}{\epsilon_0} + v_s^2 k^2. \tag{18}$$

A relation of this type between frequency and wave number is called a "dispersion law."

Dispersion laws are very convenient tools for analyzing physical systems, as we shall see in the remainder of this section. We reiterate that dispersion laws are equally applicable to initial value problems and boundary value problems. For the former we regard the wave number $k$ as a (real) Fourier expansion parameter, and use the law to assign time factors $e^{i\omega t}$ to the eigenfunctions $e^{-ikx}$; for the latter the frequency $\omega$ is the (real) expansion parameter, and the law determines the spacial factor $e^{-ikx}$.

In this book we have consistently expressed the $X(x)$ eigenfunctions as $e^{ikx}$ rather than $e^{-ikx}$, so we would be inclined to write the solution formula in terms of $e^{i(\omega t + kx)}$. Of course since $k$ runs from $-\infty$ to $\infty$ in the Fourier expansion, it makes no difference which way we write it. But most analysts are accustomed to thinking in terms of waves running to the *right*, while $e^{i(\omega t + kx)}$ appears to be running to the *left* unless we specify $k < 0$ (or, more precisely, unless $k$ has the opposite sign of $\omega$). So in the present context we use $e^{-ikx}$. The Fourier expansion thus takes the (modified) form

$$f(x) = \int_{-\infty}^{\infty} F(k)e^{-ikx}\,dk, \ F(k) = \frac{1}{2\pi}\int_{-\infty}^{\infty} f(x)e^{ikx}\,dx.$$

A second feature of dispersion laws is that they can be used to determine certain physical properties by experimental measurements. For example if we create oscillations at frequency $\omega$ at two points in the plasma, we can use interferometry to determine the distance between the maxima of the standing waves (Sec. 7.3 Example 3) that result. From the form $e^{-ikx}$ we see that this distance is the wavelength $2\pi/k$.[2] Thus we can plot $\omega$ versus $k$ and, by comparison with the dispersion law (18), determine the plasma density or the sound speed.

An extremely important aspect of dispersion laws is the ease with which they can be derived. As the analysis of Example 1 shows, all we need to do is to seek trial solutions of the partial differential equations of the form $(constant) \times e^{i(\omega t - kx)}$, and determine the relation between $\omega$ and $k$ which permits nontrivial solutions.

**Example 2.** Maxwell's equations for the electromagnetic field inside a uniform media with conductivity $\sigma$ are

$$\nabla \cdot \mathbf{E} = 0, \ \nabla \cdot \mathbf{B} = 0, \tag{19}$$

$$\nabla \times \mathbf{E} = -\frac{\partial \mathbf{B}}{\partial t}, \ \nabla \times \mathbf{B} = \mu\epsilon\frac{\partial \mathbf{E}}{\partial t} + \mu\sigma\mathbf{E}. \tag{20}$$

To find a dispersion law we seek solutions of the form

$$\mathbf{E} = \tilde{\mathbf{E}}e^{i(\omega t - kx)}, \ \mathbf{B} = \tilde{\mathbf{B}}e^{i(\omega t - kx)}, \tag{21}$$

---

[2]Or the half-wavelength $\pi/k$ if, as is usual, we can only measure the amplitude of the response.

where $\tilde{\mathbf{E}}$, $\tilde{\mathbf{B}}$ are constant vectors. Insertion of (21) into (19, 20) reveals that

$$ik\tilde{E}_x = 0, \; ik\tilde{B}_x = 0, \tag{22}$$

$$-ik\tilde{E}_y + i\omega\tilde{B}_z = 0, \; (i\omega\mu\epsilon + \mu\sigma)\tilde{E}_y - ik\tilde{B}_z = 0 \tag{23}$$

$$ik\tilde{E}_z + i\omega\tilde{B}_y = 0, \; (i\omega\mu\epsilon + \mu\sigma)\tilde{E}_z + ik\tilde{B}_y = 0. \tag{24}$$

From eqs. (22) we see that the only way to get nonzero $x$ components for **E** or **B** is to have $k = 0$. Such solutions are spacially uniform and not very interesting (see exercise 2).

The matrix form of eqs. (23) is

$$\begin{bmatrix} -ik & i\omega \\ i\omega\mu\epsilon + \mu\sigma & -ik \end{bmatrix} \begin{bmatrix} \tilde{E}_y \\ \tilde{B}_z \end{bmatrix} = \begin{bmatrix} 0 \\ 0 \end{bmatrix},$$

and the zero-determinant condition enabling nontrivial solutions gives the dispersion law

$$\omega^2\mu\epsilon - i\omega\mu\sigma = k^2 \tag{25}$$

For boundary value problems where we seek the response to a stimulus of frequency $\omega$ applied at one location, we solve (25) for $k$ and find the complex-valued square roots $k = k_{Re} + ik_{Im}$, where $k_{Re}$ and $k_{Im}$ have opposite signs when $\omega > 0$ (exercise 3). If we take $k = \beta - i\alpha$ with $\alpha, \beta > 0$ the solutions have the form $e^{-\alpha x}e^{i(\omega t - \beta x)}$, which meets the finiteness condition and the Sommerfeld condition for $x > 0$ and positive $\omega$. (See exercise 3 for the other cases.)

For initial value problems $k$ is real and we solve (24) for $\omega$:

$$\omega = \frac{i\mu\sigma \pm \sqrt{-\mu^2\sigma^2 + 4\mu\epsilon k^2}}{2\mu\epsilon}$$

The factor $e^{i(\omega t - kx)}$ then becomes

$$e^{-(\sigma/2\epsilon)t}e^{i(\pm\omega' t - kx)}, \; \omega' = \frac{\sqrt{4\epsilon\mu k^2 - \mu^2\sigma^2}}{2\mu\epsilon} \tag{26}$$

For short wavelengths (large $k$) (26) describes traveling waves which die out exponentially with the "relaxation time constant" $\tau = 2\epsilon/\sigma$. For long wavelengths $\omega'$ is imaginary and the disturbances die out even faster, without oscillating.

Exercise 4 shows that the dispersion law permitting nontrivial solutions for (23) is identical with (24), so the same considerations apply for the field components $\tilde{E}_z$, $\tilde{B}_y$. ∎

**Example 3.** Let us reconsider the waves on a string from this point of view. In Sec. 4.3 we derived the equation of motion (eq. (1))

$$\frac{\partial^2\Psi}{\partial t^2} = \frac{T}{\rho}\frac{\partial^2\Psi}{\partial x^2}$$

where $T$ is the tension and $\rho$ is the mass density. Substitution of $\Psi(x,t) = Ce^{i(\omega t - kx)}$ into this equation yields

$$-\omega^2 C = -\frac{T}{\rho}k^2 C,$$

so the dispersion law is simply

$$\omega^2 = \frac{T}{\rho}k^2 \tag{27}$$

For initial value problems we attach the time factor $e^{i\omega t} = e^{\pm ik\sqrt{T/\rho}\,t}$ to the eigenfunction $e^{-ikx}$; for boundary value problems we attach the factor $e^{-ikx} = e^{\pm i\omega\sqrt{\frac{\rho}{T}}\,x}$ to the sinusoid $e^{i\omega t}$. In the case $T/\rho = 1$ our solutions derived earlier,

$$\sum_{n=1}^{\infty} c_n \cos\frac{n\pi t}{a} \sin\frac{n\pi x}{a} \quad \text{(eq. (6), Sec. 7.3)}$$

and

$$\int_{-\infty}^{\infty} \frac{\sin\omega x}{\sin\omega} H(\omega)e^{i\omega t}\,d\omega \quad \text{(eq. (16), Sec. 8.2)} \tag{28}$$

conform to this prescription when the functions are expressed in exponential form (exercise 5). ∎

Now when $\omega$ and $k$ are real the function

$$e^{i(\omega t - kx)} = e^{-ik(x - [\omega/k]t)}$$

describes a sinusoid moving to the right at the (positive or negative) velocity $v_0 = \omega/k$. When one or the other is complex,

$$e^{i(\omega t - kx)} = e^{-Im(\omega)t}e^{Im(k)x}e^{i[Re(\omega)t - Re(k)x]},$$

and the sinusoid is damped and attenuated, but it still migrates at the velocity

$$v_0 = Re(\omega)/Re(k). \tag{29}$$

The phase of the sinusoid, $[Re(\omega)t - Re(k)x]$, is invariant on paths obeying $x - v_0 t =$ constant; consequently $v_0$ is known as the *phase velocity* of the waveform. Obviously the phase velocity is readily obtainable from the dispersion law:

$$v_0 = \pm\sqrt{v_s^2 + \frac{q^2\rho_0}{\epsilon_0 k^2}} \quad \text{for Bohm-Gross plasma waves (eq. (18)),}$$

$$v_0 = \pm\sqrt{\frac{1}{\epsilon\mu} - \frac{\sigma^2}{4\epsilon^2 k^2}} \quad \text{for electromagnetic waves (eq. (25)),}$$

$$v_0 = \pm\sqrt{\frac{T}{\rho}} \quad \text{for waves on a string (eq. (26)).}$$

Dispersion laws typically relate $\omega^2$ to $k^2$, indicating a choice of sign when we solve for one in terms of the other. This is an artifact of our using complex exponentials to model real functions; recall that we need both $e^{i\omega t}$ and $e^{-i\omega t}$ to construct, say, $\cos\omega t$.

If we attempt to keep track of all possibilities ($\pm\omega$ and $\pm k$) in the subsequent exposition, the notation will become excessive and obscure the main ideas. So instead we shall presume $+$ signs in all the computations that follow and accept the consequences - i.e., occasional unexpected complex answers.

There are two other interesting "wave velocities" which are governed by the dispersion law. To see this, recall that the *complete* solution to a boundary value problem involves decomposing the boundary disturbance into sinusoids $e^{i\omega t}$, finding the response to each sinusoid, and then reassembling these responses to match the given disturbance. Therefore the complete, assembled, solution to a boundary value problem takes a form resembling

$$\Psi(x,t) = \int_{-\infty}^{\infty} \Phi(\omega)e^{i(\omega t - k(\omega)x)}\, d\omega, \tag{30}$$

where $\Phi(\omega)$ is the Fourier transform of the boundary disturbance $\Psi(0,t)$. Obviously $\Phi(\omega)e^{-ik(\omega)x}$ is the Fourier transform of $\Psi(x,t)$.

Suppose that the physical system is quiescent prior to $t = 0$ - i.e., $\Psi(x,t) = 0$ for $t < 0$. Then, as was shown in Sec. 3.8, the *Laplace* transform of $\Psi(x,t)$ is obtained from its Fourier transform by replacing $\omega$ by $-is$ and multiplying by $2\pi$:

$$\mathcal{L}[s; \Psi(x,t)] = 2\pi\Phi(-is)e^{-ikx},\ k = k(-is).$$

$\Psi$ itself is recovered from its Laplace transform by the Bromwich integral

$$\Psi(x,t) = \frac{2\pi}{2\pi i}\int_{\alpha-i\infty}^{\alpha+i\infty} e^{st}\Phi(-is)e^{-ikx}\, ds, \tag{31}$$

where $\alpha$ is any real positive number satisfying $\Psi(x,t)e^{-\alpha t} \to 0$ as $t \to \infty$ (Sec. 3.8).

To continue our derivation we have to use the tools of analytic function theory (see Saff and Snider). First we rewrite (31) as

$$\Psi(x,t) = -i\lim_{R\to\infty}\int_L e^{s[t-xk/(-is)]}\Phi(-is)\, ds$$

where $L$ is the segment in the complex $s$-plane depicted in Fig. 1. It can be shown that the Laplace transform $\Phi(-is)$ is an analytic function of $s$ for $Re(s) \ge \alpha$, approaching zero as $s \to$

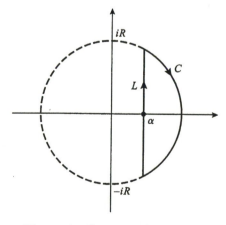

***Figure 1***   *Contours of integration*

$\infty$. Thus if $k = k(-is)$ can also be taken to be analytic for $Re(s) \geq \alpha$, the closed contour integral around $L$ and the circular arc $C$ vanishes by Cauchy's theorem:

$$\int_L e^{s[t-xk/(-is)]}\Phi(-is)\,ds$$
$$+ \int_C e^{s[t-xk/(-is)]}\Phi(-is)\,ds = 0. \quad (32)$$

If $k/(-is)$ has a real, positive limit (which we denote as $1/v_1$) at $\infty$, then the contribution of the circular portion of the contour can be estimated by

$$\int_C e^{s[t-x/v_1]}\Phi(-is)\,ds$$
$$= \int_C e^{Re(s)[t-x/v_1]}e^{iIm(s)[t-x/v_1]}\Phi(-is)\,ds \quad (33)$$

Now if $\Phi$ goes to zero at infinity sufficiently rapidly, then for $t < x/v_1$ the integral (33) vanishes as $R \to \infty$ since both the exponentials appearing therein are bounded (recall $Re(s) = \alpha > 0$). From (32) and (31) we conclude, then, that $\Psi(x,t)$ *remains zero for all times $t$ less than $x/v_1$.*

In other words the leading edge of a disturbance originating at $x = 0$ cannot reach the point $x$ until $t$ exceeds $x/v_1$, where the "signal-front" velocity $v_1$ is given by

$$v_1 = \lim_{s \to \infty} \frac{-is}{k(-is)} = \lim_{\omega \to \infty} \frac{\omega}{k(\omega)} \quad (34)$$

when this limit exists. From the dispersion laws, we see that for Bohm-Gross plasma waves (18) the signal-front speed is the sound speed $v_s$; for electromagnetic waves (24) it is $1/\sqrt{\mu\epsilon}$ (the speed of light); and for waves on a string (26) it is simply the phase velocity $v_0$.[3]

The other velocity that can be gleaned from the dispersion laws is the "group velocity." We have seen that a sinusoidal wave of frequency $\omega$ propagates at the phase velocity $v_0 = \omega/k$ (eq. (29)) if there is no damping or attenuation. Therefore if a disturbance is initially described as $\Psi(x,0)$, and we decompose it into sinusoids $e^{-ikx}$ and let each one migrate accordingly, we obtain $\Psi(x,t)$ by reassembling the shifted sinusoids at time $t$. Mathematically, we are saying that if

$$\Psi(x,0) = \int_{-\infty}^{\infty} \Phi(k)e^{-ikx}\,dk \quad (35)$$

---

[3]Einstein's theory of relativity postulates that the signal-front velocity can never exceed the speed of light. This is not true of the phase velocity.

then

$$\Psi(x,t) = \int_{-\infty}^{\infty} \Phi(k)e^{i[\omega(k)t-kx]}\,dk \tag{36}$$

Now if the dispersion relation is simply $\omega = v_0 k$, the result is a shifted replica of $\Psi(x,0)$:

$$
\begin{aligned}
\Psi(x,0) &= \int_{-\infty}^{\infty} \Phi(k)e^{i[v_0 kt-kx]}\,dk \\
&= \int_{-\infty}^{\infty} \Phi(k)e^{-ik[x-v_0 t]}\,dk \\
&= \Psi(x-v_0 t,0).
\end{aligned}
$$

We observed this behavior for the stretched string in Sec. 4.3. However if the phase velocity $\omega/k = v_0$ *varies* with $k$ - as for the Bohm-Gross plasma waves (18) or the electromagnetic waves in a conductor (25) - then each sinusoid travels a different distance in the time $t$ and the reassembled $\Psi(x,t)$ may bear little resemblance to the original $\Psi(x,0)$. The different Fourier components have "dispersed" (hence the terminology).

Nonetheless there would still be some coherence to the evolving waveforms if the dispersion relation took the form $\omega = a + bk$. For then we would have from (36)

$$
\begin{aligned}
\Psi(x,t) &= \int_{-\infty}^{\infty} \Phi(k)e^{i[\omega(k)t-kx]}\,dk \\
&= \int_{-\infty}^{\infty} \Phi(k)e^{i[at+bkt-kx]}\,dk \\
&= e^{iat}\int_{-\infty}^{\infty} \Phi(k)e^{-ik[x-bt]}\,dk \\
&= e^{iat}\Psi(x-bt,0).
\end{aligned}
\tag{37}
$$

If we rewrite this as

$$e^{iat}\Psi([t-\frac{x}{b}][-b],0),$$

the interpretation is the following: at the (remote) point $x_0$ there is an oscillation $e^{iat}$, which is *modulated in time* by the waveform of the initial disturbance, delayed by $(x_0/b)$ seconds. The speed at which the modulating ewnvelope propagates, $v_2 = b$, is called the *group velocity*. See Figure 2, observing that the independent variable is $x$ in Fig. 2a and $t$ in Fig. 2b.

Now the simplified dispersion law $\omega = a + bk$ is seldom met in practice. But suppose the Fourier transform of the initial waveform, $\Psi(k)$, happens to be concentrated around some wave number $k_0$, so that (36) can be ap-

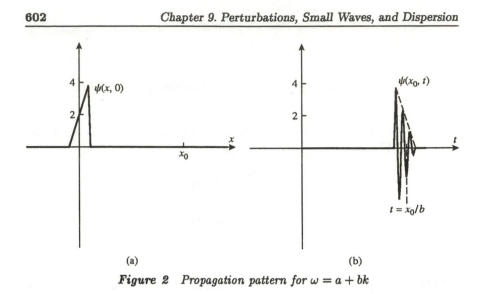

(a)                                                    (b)

**Figure 2**   *Propagation pattern for $\omega = a + bk$*

proximated

$$\Psi(x,t) = \int_{-\infty}^{\infty} \Phi(k)\ldots$$

$$\approx \int_{k_0-\Delta k}^{k_0+\Delta k} \Phi(k)\, e^{i[w(k)t-ikx]}dx$$

- and suppose also the first-order Taylor approximation $\omega(k) \approx \omega(k_0) + \omega'(k_0) \times (k - k_0)$ is accurate in the range $k_0 - \Delta k < k < k_0 + \Delta k$. Then the evaluation of the Fourier integral proceeds as in (37) and (within the accuracy of the approximations) we again observe a "carrier" oscillation modulated by the delayed envelope $\Psi(x,0)$:

$$\Psi(x,t) \approx \int_{k_0-\Delta k}^{k_0+\Delta k} \Phi(k)e^{i[\omega(k_0)t+\omega'(k_0)\cdot(k-k_0)t-kx]}\,dk$$

$$\approx e^{i[\omega(k_0)-\omega'(k_0)\cdot k_0]t}\int_{-\infty}^{\infty}\Phi(k)e^{-ik[x-\omega'(k_0)]t}\,dk$$

$$\approx e^{i[\omega(k_0)-\omega'(k_0)\cdot k_0]t}\Psi(x-\omega'(k_0)t,0].$$

For such so-called "wave packets," then, the modulating waveform propagates with the group velocity

$$v_2 = \omega'(k_0) = \left.\frac{d\omega}{dk}\right|_{k_0} \tag{38}$$

and the carrier wave has frequency $\omega(k_0) - k_0\omega'(k_0)$.

**Example 4.** Consider an initial waveform (Figure 3)

$$\Psi(x,0) = e^{-x^2+i\pi x}$$

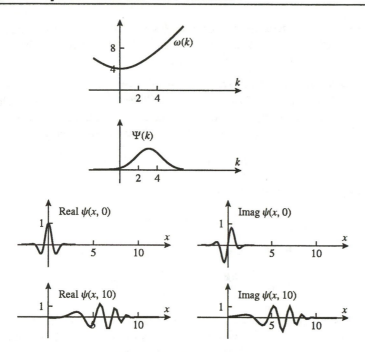

**Figure 3   Group velocity**

whose Fourier transform (see Table 3.1)

$$\Phi(k) = e^{-(k-\pi)^2/4}/2\sqrt{\pi}$$

is concentrated around $k_0 = \pi$. If $\Psi$ evolves according to eq. (36) with the dispersion law

$$\omega = \sqrt{16 + k^2}, \tag{39}$$

where we consider only the positive square root for simplicity, then $\Psi(x, 10)$ looks as depicted in the figure.

Its group velocity is

$$v_2 = \frac{d\omega}{dk} = \frac{k}{\sqrt{16 + k^2}} \approx .617 \quad \text{for } k_0 = \pi$$

and the carrier frequency is

$$\omega(k_0) - k_0\omega'(k_0) \approx 3.148.$$

The dispersion relation, $\Phi(k)$, $\Psi(x, o)$, and $\Psi(x, 10)$ are all displayed in Fig. 3. Sure enough, the initial wave "group" has drifted about $v_2 t \approx 6$ units to the right. The carrier-frequency factor is $e^{i3.148 \times 10} \approx 1$, so the transported wave shapes are roughly similar to the originals; distortion, however, is clearly evident. ∎

The computation of the group velocities for the plasma, electromagnetic, and stretched-string waves is left to exercise 6. The wave velocities for the Schrödinger equation are worth noting:

**Example 5.** As we saw in Sec. 4.7, the solutions of the Schrödinger equation for the free particle have the form $e^{i2\pi(px-Et)/h}$, where $p$ is the momentum, $E$ is the energy, and $h$ is Planck's constant. If we interpret this as $e^{i(kx-\omega t)}$, the frequency and wave number are respectively

$$\omega = 2\pi E/h, \ k = 2\pi p/h.$$

The identity $E = p^2/2m$ for a free particle then provides the dispersion law

$$\omega = hk^2/4\pi m$$

The phase, signal-front, and group velocities for the wave are thus

$$v_0 = \frac{hk}{4\pi m}, \ v_1 = \infty, \ v_2 = \frac{hk}{2\pi m}.$$

Recalling deBroglie's relation $p = mv = hk/2\pi$, we see that the group velocity equals the particle velocity. This is consistent with the probabilistic interpretation of the wave function. The phase velocity is half the particle velocity (recall eq. (7), Sec. 4.7). The fact that the signal-front velocity is infinite is a sign that the Schrödinger formulation is nonrelativistic. ■

In higher dimensions the interpretation of the dispersion law is more subtle. The wave number $k$ is associated with a particular direction in space, which is considered as the direction of propagation. For radiation in a waveguide (Sec. 7.3, Examples 4, 6 and 7), the waves are considered to propagate in the longitudinal ($z$) direction,[4] and the dispersion laws are mathematically similar to those for the Bohm-Gross waves:

$$\omega^2 = \left(\frac{m\pi}{a}\right)^2 + \left(\frac{n\pi}{b}\right)^2 + k^2 \quad \text{for the rectangular waveguide}$$
$$\omega^2 = j_{n,p}^2/b^2 + k^2 \quad \text{for the cylindrical waveguide}$$
$$\omega^2 = -\mu_{n,p} + k^2 \quad \text{for the coaxial waveguide}$$

These equations only to waves exhibiting a common transversal dependence ("waveguide mode") through the factors

$$\sin\frac{m\pi}{a}x \sin\frac{n\pi}{b}y,$$
$$[a_n \cos n\theta + b_n \sin n\theta] J_n(j_{n,p}\rho/b), \text{ or}$$
$$[a_n \cos n\theta + b_n \sin n\theta]$$
$$\times \{Y_n(\sqrt{-\mu_{n,p}}a)J_n(\sqrt{-\mu_{n,p}}\rho) - J_n(\sqrt{-\mu_{n,p}}a)Y_n(\sqrt{-\mu_{n,p}}\rho)\}$$

---

[4]There is merit, however, to regarding waveguide radiation as propagating in a "skew" direction and bouncing off the walls. See Kraus's textbook.

respectively.

The sinusoidal factor for three dimensional plane-wave solutions to the wave equation has the form (Example 5, Sec. 7.3) $e^{i(\mathbf{K}\cdot\mathbf{r}\pm\omega t)}$. If we restrict our attention to waves propagating in the direction $\mathbf{n}$, so that $\mathbf{K} = k\mathbf{n}$, we can write the dispersion law

$$\omega^2 = k^2.$$

Thus the signal-front, phase, and group velocities are all unity (in these units) and there is no dispersion of such waves.

For radial waves (Example 2, Sec. 8.5) the solutions look like

$$Y_{\ell m}(\theta, \phi) h_\ell^{(2)}(\omega r) e^{i\omega t} \approx i^{\ell+1} Y_{\ell m}(\theta, \phi) e^{i(\omega t - \omega r)} / \omega r. \qquad (40)$$

Identifying the factor $e^{i(\omega t - \omega r)}$ with the form $e^{i(\omega t - kr)}$, we conclude that the dispersion law is again

$$\omega^2 = k^2$$

Thus all initial disturbances in a given spherical harmonic mode $(\ell, m)$ propagate without dispersion as long as $r$ is large enough for the approximation (40) to be valid. There is attenuation, however, due to the factor $1/r$.

> Clearly the approach of inserting a solution form $e^{i(\omega t - kx)}$ into a system of partial differential equations is very powerful and convenient. However, it may not reveal all the possible solutions, since we are thereby restricting the functional forms to be considered. For example, insertion of $e^{i(\omega t - kr)}$ into the spherical wave equation will fail, since the $1/r$ factor is necessary even for an approximate solution! And insertion of $e^{i(\omega t - kx)}$ into the Schrödinger equation for the constant-force field (Sec. 7.4 Example 2) won't work, because Airy functions simply are not sinusoids. For coupled partial differential equations with constant coefficients, however, the method is quite effective.

# Exercises 9.3

**1.** Re-solve the plasma example by first Fourier-analyzing the unknowns in space - $\rho_1(x, t) = \int_{-\infty}^{\infty} P_1(t, k) e^{-ikx} \, dk$, etc. - and then seeking solutions of the form $P_1(t, k) = A e^{i\omega t} \ldots$. Show that the necessary condition for nontrivial solutions is eq. (17).

**2.** Derive the spacially uniform solutions of Maxwell's equations in a conductor.

3. Show that the roots of eq. (25) lie in the second and fourth quadrants of the complex plane when $\omega > 0$, and in the first and third quadrants when $\omega < 0$. Interpret the directions of migration and attenuation of the waveforms $e^{i(\omega t - kx)}$ for each quadrant.

4. Show that the dispersion law for eqs. (23) is identical with that for (24).

5. Write out the solutions (27) in terms of exponentials and confirm the forms described in Example 3.

6. Compute the group velocities for the Bohm-Gross plasma, electromagnetic, and stretched-string waves.

7. At low temperatures the electronic sound speed $v_s$ in eq. (1) goes to zero. The dispersion law (18) then becomes $\omega = \sqrt{q^2 \rho_0 / \epsilon_0}$, the "plasma frequency." What can you say about the phase, group, and signal-front velocities when $\omega$ does not depend on $k$? Why are these called "static waves"? Starting from eqs. (4-6), show that in this case each unknown $\rho_1, v_1$, and $E_1$ satisfies an *ordinary* differential equation in $t$. (Hint: what is the consequence of assigning, to each eigenfunction $e^{-ikx}$ comprising the initial state, the *same* time factor $e^{i\omega t}$ in (36))?

8. Consider a uniform plasma at low temperature ($v_s = 0$) with a constant magnetic field $\mathbf{B}_0$ in the $z$ direction. The quiescent conditions are as in Example 1. Newton's law now contains the Lorentz force:

$$\rho \frac{d\mathbf{v}}{dt} = -\nabla p - q\rho(\mathbf{E} + \mathbf{v} \times \mathbf{B}_0).$$

The equation of continuity and Maxwell's first equation hold in their vector form:

$$\frac{d\rho}{dt} = -\nabla \cdot \rho \mathbf{v}, \nabla \cdot \epsilon_0 \mathbf{E} = n - q\rho.$$

We are seeking "longitudinal wave perturbations perpendicular to $\mathbf{B}_0$." This means the perturbed electric field has only an $x$ component, all perturbed quantities vary as $e^{i(\omega t - kx)}$, and the perturbed velocity is perpendicular to $\mathbf{B}_0$. Derive the dispersion law

$$\omega^2 = q^2 \left[ \frac{\rho_0}{\epsilon_0} + B_0^2 \right].$$

This (constant) frequency is known as the "upper hybrid frequency."

9. Derive the dispersion law for longitudinal wave perturbations *parallel* to $\mathbf{B}_0$ in a quiescent low temperature plasma (see the previous exercise). This means the perturbed electric field has only a $z$ component, all perturbed quantities vary as $e^{i(\omega t - kz)}$, and the perturbed velocity is parallel to $\mathbf{B}_0$.

10. "Transverse electromagnetic waves in a low-temperature plasma" are governed by Maxwell's equations with charge and current sources

$$\nabla \cdot \epsilon_0 \mathbf{E} = n - q\rho, \nabla \cdot \mathbf{B} = 0,$$

$$\nabla \times \mathbf{E} = -\frac{\partial \mathbf{B}}{\partial t}, \nabla \times \mathbf{B} = \mu_0 \epsilon_0 \frac{\partial \mathbf{E}}{\partial t} + \mu_0 q\rho\mathbf{v};$$

Newton's law with the Lorentz force

$$\rho\frac{d\mathbf{v}}{dt} = -\nabla p - q\rho(\mathbf{E} + \mathbf{v} \times \mathbf{B}_0);$$

and the equation of continuity

$$\frac{d\rho}{dt} = -\nabla \cdot \rho\mathbf{v}.$$

Here the magnetic field $\mathbf{B}$, as well as the electric field $\mathbf{E}$, is zero in the quiescent state. Look for solutions of the form $e^{i(\mathbf{K}\cdot\mathbf{r}-\omega t)}$ where $\mathbf{K} = k\mathbf{n}$ and $\mathbf{n}$ is perpendicular or "transversal" to $\mathbf{E}$. Derive the dispersion law

$$\omega^2 = \frac{q^2\rho_0}{\epsilon_0} + \frac{k^2}{\epsilon_0\mu_0}.$$

What are the phase, signal-front, and group velocities?

11. The *index of refraction* $n$ of a material is the ratio of "$c$", the (fixed) velocity of light waves in a vacuum, to the phase velocity of light waves in the material: $n = c/v_0$. $n$ is therefore a function of frequency $\omega$. Show that the group velocity can be obtained from $n$ via the formula

$$v_2(\omega) = \frac{c}{n(\omega) + \omega n'(\omega)}.$$

12. The relativistic energy of a particle of rest mass $m$ and velocity $v$ is given by $E = mc^2/\sqrt{1 - v^2/c^2}$, and its relativistic momentum is $p = mv/\sqrt{1 - v^2/c^2}$. Use the Planck-Einstein relation $E = h\omega/2\pi$ and the deBroglie relation $p = hk/2\pi$ (Sec. 4.7) and eliminate $v$ to derive the dispersion relation for relativistic quantum waves

$$\omega^2 = \frac{2\pi mc^2}{h} + c^2 k^2.$$

Show that the phase velocity is $c^2/v$, the signal-front velocity is $c$, and the group velocity is $v$.

# Appendix A

# Some Numerical Techniques

**Summary** The *limit* concept and its underlying notion of convergence are fundamental to the whole calculus process; but to a practical analyst the *issue* of convergence is often less crucial than the *rate* of convergence. Also, although the elegant substitutions and transformations which enable the evaluation of many indefinite integrals are enlightening and edifying, they are wholly inadequate for the majority of integrals which arise in the solution of differential equations for engineering systems.

In this appendix we conduct a review of the mechanics of calculus from the perspective of their computational implementation. Our basic tool is the order-of-magnitude concept; coupled with the standard calculus machinery it lends insight into numerical approximation, numerical differentiation, numerical integration, and the numerical solution of equations.

## A.1 Order of Magnitude

The notion of the order of magnitude of an expression is a common theme in the older calculus books. Modern texts downplay its role, because its indiscriminate use can lead one to nonrigorous - if not misleading - arguments. However, the insight that it offers far outweighs the danger of its abuse. With an occasional cautious glance over our shoulder, then, we proceed with the exposition.

The basic concept of calculus is the derivative, $f'(x) = \frac{df}{dx}$, of a function $f$ of one variable $x$. The formal definition of the derivative at the point $x_0$

is

$$f'(x_0) = \frac{df(x_0)}{dx} = \lim_{h \to 0} \frac{f(x_0 + h) - f(x_0)}{h} \qquad (1)$$

It is instructive to review a derivative computation from first principles. Our example should look familiar; it appears in almost every calculus book just before the rule for differentiating $x^n$ is derived. If $f(x) = x^3$, and we want $f'(x_0)$, then the computation suggested by (1) goes as follows:

$$
\begin{aligned}
f'(x_0) &= \lim_{h \to 0} \frac{(x_0 + h)^3 - x_0^3}{h} = \lim \frac{(x_0^3 + 3x_0^2 h + 3x_0 h^2 + h^3) - x_0^3}{h} \\
&= \lim \frac{3x_0^2 h + [3x_0 h^2 + h^3]}{h} \qquad (2) \\
&= \lim 3x_0^2 + [3x_0 h + h^2] \qquad (3) \\
&= 3x_0^2 \qquad (4)
\end{aligned}
$$

The steps in this derivation are typical of all evaluations of a derivative from first principles. In step 2 we see in the numerator the term $3x_0^2 h$, which will eventually metamorphose into the final answer $f' = 3x_0^2$. Also there are "higher order" terms, $3x_0 h^2 + h^3$, which have no influence on $f'$ since they *go to zero even after being divided by $h$*. We are going to introduce some nomenclature for these terms.

**Definition.** The notation $o(h)$ is used to denote any expression which, after being divided by $h$, approaches zero as $h \to 0$:

$$\lim_{h \to 0} \frac{o(h)}{h} = 0. \qquad (5)$$

We read this symbol as "(small) oh of $h$" or "(small) order-of-magnitude $h$." (Capital $\mathcal{O}(h)$ will be introduced in the next section.) One can say that any expression which is $o(h)$ goes to zero *faster* than $h$, since by (5) $o(h)$ equals $h$ *times* a function which approaches zero. Indeed, condition (5) will be taken as the quantitative interpretation of the phrase "$o(h)$ goes to zero faster than $h$."

**Example 1.** (a) Any power $h^p$ of $h$, of degree 2,3, ... , is $o(h)$:

$$\lim_{h \to} \frac{h^p}{h} = \lim h^{p-1} = 0. \qquad (6)$$

In fact, $h^p$ is $o(h)$ for any $p > 1$.

(b) Constant multiples of $h^p$, for $p > 1$, are $o(h)$:

$$\lim_{h \to 0} \frac{ch^p}{h} = \lim ch^{p-1} = 0.$$

(c) If $f_1(h)$ is $o(h)$ and $f_2(h)$ is *bounded* as $h \to 0$, then $f_1(h)f_2(h)$ is $o(h)$:

$$\frac{f_1(h)f_2(h)}{h} = f_2(h)\frac{f_1(h)}{h} \to (bounded) \times (zero) = 0$$

(d) Constants and linear functions are *not* $o(h)$:

$$\lim_{h \to 0} \frac{c}{h} = \infty; \quad \lim_{h \to 0} \frac{ch}{h} = c \qquad \blacksquare$$

Now obviously $h^3$ goes to zero after being divided by $h$, so $h^3 = o(h)$. But it *still* goes to zero after being divided by $h^2$ (or, indeed, $h^p$ for any $p < 3$). To express this faster rate of convergence we extend the definition:

**Definition.** The notation $o(h^p)$ $(p \geq 0)$ is used to denote any expression which, after being divided by $h^p$, approaches zero as $h \downarrow 0$:

$$\lim_{h \downarrow 0} \frac{o(h^p)}{h^p} = 0 \qquad (7)$$

(The technical restriction that $h$ approach zero through *positive* values ensures that $h^p$ can be taken to be real even when $p$ is not an integer.) Thus any expression which is $o(h^p)$ goes to zero faster than $h^p$. The following display may give you a feel for these rates of convergence:

| $h$ | $h^2$ | $h^3$ | $h^4$ |
|------|----------|--------------|----------------|
| 1 | 1 | 1 | 1 |
| 0.1 | 0.01 | 0.001 | 0.0001 |
| 0.01 | 0.0001 | 0.000001 | 0.00000001 |
| 0.001 | 0.000001 | 0.000000001 | 0.000000000001 |

**Example 2.** (a) Notice that the statement $g(h) = o(1) = o(h^0)$ is just another way of saying that $g(h)$ simply approaches zero:

$$\lim_{h \to 0} \frac{g(h)}{1} = 0.$$

(b) Any expression which is $o(h^p)$ is automatically $o(h^q)$ for any $q < p$. This opens the door to some dangerous situations if we interpret the notation too literally. For example, from the true statements

$$h^3 = o(h), \quad h^3 = o(h^2),$$

we must not leap mindlessly to the conclusions

$$o(h) \quad = o(h^2) \quad (false) \qquad (8)$$
$$o(h^2) \quad = o(h) \quad (true) \qquad (9)$$

*Statements using order-of-magnitude notation do not have the logical status of equations*; they are abbreviations for assertions about limiting processes. As such, they are properly read only from left to right. Statement (9) is true - "a function which goes to zero faster than $h^2$ goes to zero faster than $h$" - but (8) is false.

(c) In the spirit of the above, the following "equations" are valid; be sure you understand them: for $p, q > 0$

$$o(h^p)o(h^q) = o(h^{p+q}) \tag{10}$$

$$h^p o(h^q) = o(h^{p+q}) \tag{11}$$

and if $p \geq q$ then also

$$o(h^p) + o(h^q) = o(h^q) \tag{12}$$

$$o(h^p)/h^q = o(h^{p-q}) \tag{13}$$

∎

In closing we mention that the order of magnitude notion can be extended in two ways. First of all a "comparison" function other than $h^p$ may be used; we say $o(\phi(h))$ designates a function which goes to zero faster than the function $\phi(h)$:

$$\lim_{h \to 0} \frac{o(\phi(h))}{\phi(h)} = 0 \tag{14}$$

For example $o[\ln(1 + h)]$ must go to zero faster than $\ln(1 + h)$, as $h$ goes to zero. Exercise 7 of Sec. 2.2 shows that any function which is $o(e^{-1/h^2})$ goes to zero very fast indeed.

Second, we may use a limit point other than $h = 0$; in particular the statement

$$f(x) = o(x^p) \ as \ x \to \infty$$

means simply

$$\lim_{x \to \infty} f(x)/x^p = 0.$$

## Exercises A.1

1. If the right hand side of eq. (12) is replaced by $o(h^p)$, is the resulting statement true?

2. If $f(x)$ is continuous and $f(x) = o(x^2)$ near $x = 0$, is the function $f(x)/x^2$ continuous? How about $f(x)/x$ and $f(x)/x^3$? Discuss the possibilities.

## A.2 The Accuracy of Taylor Polynomials

We have seen how the order of magnitude notation arises rather naturally in the concept of a derivative. Note that the defining equation (1) of the previous section can be rewritten, for the point $x = x_0$, as

$$\lim_{h \to 0} \left[ f'(x_0) - \frac{f(x_0 + h) - f(x_0)}{h} \right] = 0 \tag{1}$$

or

$$\lim_{h \to 0} \frac{f(x_0 + h) - f(x_0) - f'(x_0)h}{h} = 0 \tag{2}$$

and (2) is precisely the statement that

$$f(x_0 + h) = f(x_0) + f'(x_0)h + o(h). \tag{3}$$

In fact, (2) can be used to *characterize* the derivative (if there is one); $f'(x_0)$ *is the unique number $\alpha$ which endows the approximation*

$$f(x_0 + h) \approx f(x_0) + \alpha h \tag{4}$$

*with an error that goes to zero faster than $h$!*

(*Proof*: If

$$\frac{f(x_0 + h) - f(x_0) - \alpha h}{h} \to 0$$

then

$$\alpha = \lim_{h \to 0} \frac{f(x_0 + h) - f(x_0)}{h} = f'(x_0).$$

This characterization is sometimes used in abstract settings to generalize the concept of a derivative.)

The approximation in (4) has a simple graphical interpretation. As a function of $x = x_0 + h$, the right hand side is a straight line with slope $\alpha = f'(x_0)$. Thus (4) expresses the nearness of the curve $y = f(x) = f(x_0 + h)$ to its tangent line at $x = x_0$. For $x = x_0 + h$ the difference between the values on the curve and on the tangent line goes to zero faster than $h$, and this is true of no other straight line. For this reason (4) is called the "linear approximation" and we say we have "linearized $f(x)$ in the neighborhood of $x = x_0$."

**Example 1.** (*Linear approximation*)
a. For $x$ near $2$ ($x_0 = 2, x = 2 + h$), (3) and (4) state

$$x^3 = x_0^3 + 3x_0^2 h + o(h) \approx 2^3 + 3 \times 2^2 (x - 2) = 8 + 12(x - 2)$$

b. For $x$ near $\pi/4$,

$$\sin x = \sin x_0 + h \cos x_0 + o(h) \approx \frac{1}{\sqrt{2}} + \frac{x - \pi/4}{\sqrt{2}}$$

c. For $x$ near $1$,

$$e^x = e^{x_0} + h e^{x_0} + o(h) \approx e + (x - 1)e = xe$$

d. For $x$ near zero,

$$e^x = e^{x_0} + he^{x_0} + o(h) \approx 1 + x/1 = 1 + x$$

$$\sin x = \sin x_0 + h \cos x_0 + o(h) \approx 0 + (x - 0) = x$$

$$\cos x = \cos x_0 - h \sin x_0 + o(h) \approx 1$$

$$x^3 = x_0^3 + 3x_0^2 h + o(h) \approx 0.$$

Now that we have seen how the first derivative is used to construct a linear approximation to $f(x)$ near $x = x_0$ with an error which goes to zero faster than $h = x - x_0$, it is natural to seek better approximations; for example, let's try to get a *quadratic* approximation whose error goes to zero faster than $h^2$. In other words, can we find numbers $\alpha$ and $\beta$ so that

$$f(x) = f(x_0) + \alpha(x - x_0) + \beta(x - x_0)^2 + o([x - x_0]^2) ?$$

The answer turns out to be *yes*, with the parameters chosen as follows:

$$f(x) = f(x_0) + f'(x_0)(x - x_0) + f''(x_0)(x - x_0)^2/2 + o([x - x_0]^2) \quad (5)$$

(assuming the second derivative exists).

*Derivation of (5)*: The estimate (3) is valid for any differentiable function, so we can apply it to $f'$ to state

$$f'(x) = f'(x_0) + f''(x_0)(x - x_0) + o(x - x_0) \quad (6)$$

To get (5) we employ an integral representation:

$$f(x) = f(x_0 + h) = f(x_0) + \int_{x_0}^{x_0+h} f'(x)\, dx$$

$$= f(x_0) + \int_{x_0}^{x_0+h} [f'(x_0) + f''(x_0)(x - x_0) + o(x - x_0)]\, dx$$

$$= f(x_0) + f'(x_0)h + f''(x_0)h^2/2 + \int_{x_0}^{x_0} o(x - x_0)\, dx. \quad (7)$$

Since $x$ runs from $x_0$ to $x_0+h$, the integrand $o(x-x_0)$ is certainly $o(h)$:

$$\left| \frac{o(x - x_0)}{h} \right| = \left| \frac{o(x - x_0)}{x - x_0} \right| \left| \frac{x - x_0}{h} \right| \leq \left| \frac{o(x - x_0)}{x - x_0} \right| \to 0. \quad (8)$$

Therefore the last integral is estimated by $h \times o(h)$, which is $o(h^2)$.

This argument can easily be extended to derive the generalization of (5) to higher-order approximations:

$$f(x) = f(x_0) + f'(x_0)(x - x_0) + f''(x_0)(x - x_0)^2/2! + f'''(x_0)(x - x_0)^3/3!$$
$$+ \ldots + f^n(x_0)(x - x_0)^n/n! + o([x - x_0]^n) \quad (9)$$

The approximant in (9) is the *Taylor polynomial* of order $n$; see Sec. 2.2. Thus the order of magnitude notation enables us to quantify the degree of approximation of the Taylor polynomials.

Now we close this section by demonstrating how the latter can, in turn, provide a more specific characterization of the order of magnitude concept, itself.

If the function $f(x)$ has a derivative at $x = x_0$, its linear approximation by Taylor polynomials is expressed

$$f(x) = f(x_0 + h) = f(x_0) + f'(x_0)h + \varepsilon_1, \quad \varepsilon_1 = o(h) \qquad (10)$$

If $f$ has a second derivative, it has a quadratic approximation given by

$$f(x) = f(x_0 + h) = f(x_0) + f'(x_0)h + f''(x_0)h^2/2 + \varepsilon_2, \quad \varepsilon_2 = o(h^2).$$

The latter display gives us additional information about the error $\varepsilon_1$ in the *linear* approximation: clearly

$$\varepsilon_1 = f''(x_0)h^2/2 + o(h^2).$$

Thus we see that $\varepsilon_1$ is *not* merely $o(h)$; it is $o(h^p)$ for any $p < 2$:

$$
\begin{aligned}
\lim_{h \to 0} \frac{\varepsilon_1}{h^p} &= \lim_{h \to 0} \left[ \frac{f''(x_0)h^2/2}{h^p} + \frac{o(h^2)}{h^2} \frac{h^2}{h^p} \right] \\
&= \lim \left[ \frac{f''(x_0)h^{2-p}}{2} + \frac{o(h^2)}{h^2} h^{2-p} \right] \\
&= 0.
\end{aligned}
$$

For convenience we extend the order-of-magnitude notation and use the symbol $\mathcal{O}(h^q)$ (read "capital Oh of $h$ to the $q$") to designate an expression that, when divided by $h^q$, approaches a *finite limit* (but not necessarily zero) as $h$ goes to zero:

$$\lim_{h \to 0} \frac{\mathcal{O}(h^q)}{h^q} = L, \quad |L| < \infty$$

(A more precise definition is given below.) Then it is clear that $\varepsilon_1$ in (10) is $\mathcal{O}(h^2)$ when $f''$ exists:

$$\lim_{h \to 0} \frac{\varepsilon_1}{h^2} = \lim \frac{f''(x_0)h^2/2 + o(h^2)}{h^2} = f''(x_0)/2$$

The use of the capital $\mathcal{O}$ notation will often spare us the necessity of writing the awkward statement "$g(h)$ is $o(h^q)$ for any $q$ less than $p$," when all we need to say is "$g(h)$ is $\mathcal{O}(h^q)$." *Then* we can say that the error in the linear approximation (10) - which in general is only $o(h)$ - is $\mathcal{O}(h^2)$, *when we know that the function f has a second derivative.* Similar statements hold for the higher order Taylor polynomials:

$$
\begin{aligned}
f(x) &= f(x_0 + h) = f(x_0) + f'(x_0)h + f''(x_0))h^2/2! + f'''(x_0)h^3/3! \\
&\quad + \ldots + f^n(x_0)h^n/n! + \mathcal{O}(h^{h+1})
\end{aligned}
\tag{11}
$$

if $f^{n+1}(x_0)$ exists. This convergence is demonstrated for $f(x) = e^x$, expanded around $x = 0$, in Table 2.2, Sec. 2.2.

There is a slight generalization of the definition of $\mathcal{O}(h^q)$ which occurs in standard usage. Rather than insisting that $\mathcal{O}(h^q)/h^q$ approach a *limit* as $h \to 0$, one requires only that this ratio remain *bounded* for small $h$.

DEFINITION We say that a function $g(h)$ is $\mathcal{O}(h^q)$ if, for some number $M$,

$$
\left| \frac{g(h)}{h^q} \right| < M
\tag{12}
$$

for $h$ sufficiently small. $\mathcal{O}(\phi(h))$ is defined by the analogous condition for any function $\phi$:

$$
|\mathcal{O}(\phi(h))| < M, \ h \text{ sufficiently small.}
\tag{13}
$$

Thus if $g(h)/h^q$ oscillates between two "limiting" values without approaching either one, we shall still say that $g(h) = \mathcal{O}(h^q)$. The rules for manipulating the $\mathcal{O}$ symbol are similar to those for $o$, as listed in the previous section. Note also the following:

$$
o(h^p)\mathcal{O}(h^q) = o(h^{p+q}).
\tag{14}
$$

# A.3   Numerical Differentiation

In this section we will see how, by combining the Taylor developments and the order of magnitude estimates, one can derive numerical approximations for derivatives of all orders, and assess their accuracy as well.

Equation (1) in the previous section provides us with a difference-quotient or *finite difference* approximation for the first derivative:

$$
\begin{aligned}
f'(x_0) = \lim_{h \to 0} \frac{f(x_0 + h) - f(x_0)}{h} \ &= \ \frac{f(x_0 + h) - f(x_0)}{h} + o(1) \\
&\approx \ \frac{f(x_0 + h) - f(x_0)}{h}
\end{aligned}
\tag{1}
$$

It is fruitful to explore the finite difference approximations for higher derivatives. The second derivative is defined by applying (1) to $f'(x)$:

$$
f''(x_0) = \lim_{h \to 0} \frac{f'(x_0 + h) - f'(x_0)}{h}
\tag{2}
$$

Thus we are led to insert (1) into (2) and surmise

$$f''(x_0) \approx \frac{1}{h}\left\{\frac{f(x_0+2h)-f(x_0+h)}{h} - \frac{f(x_0+h)-f(x_0)}{h}\right\}$$

$$f''(x_0) \approx \frac{f(x_0+2h)-2f(x_0+h)+f(x_0)}{h^2} \tag{3}$$

To check the accuracy of (3), we use the Taylor developments of each of the terms and assemble them:

$$f(x_0+2h) = f(x_0) + f'(x_0)(2h) + f''(x_0)\frac{(2h)^2}{2} + o(h^2)$$
$$f(x_0+h) = f(x_0) + f'(x_0)h + f''(x_0)\frac{h^2}{2} + o(h^2)$$
$$f(x_0) = f(x_0)$$

Combining the first equation, minus twice the second, plus the third, yields

$$f(x_0+2h) - 2f(x_0+h) + f(x_0) = f''(x_0)h^2 + o(h^2) \tag{4}$$

From (4) we see that the estimate in (3) is valid because

$$f''(x_0) = \frac{f(x_0+2h)-2f(x_0+h)+f(x_0)}{h^2} + \frac{o(h^2)}{h^2}$$

$$= \frac{f(x_0+2h)-2f(x_0+h)+f(x_0)}{h^2} + o(1) \tag{5}$$

Higher-derivative approximates can be constructed and analyzed the same way. (See exercises.)

Before we look at some examples of these estimates, it is instructive to examine a trick used frequently in numerical analysis. For $h > 0$ in (1) the "sampled" values of $f(x)$ run to the *right* of $x = x_0$; the expression is called a *forward difference* formula. If we interpret $h$ as being negative we get the *backward difference* formula

$$f'(x_0) \approx \frac{f(x_0+h)-f(x_0)}{h} = \frac{f(x_0)-f(x_0-|h|)}{|h|} \tag{6}$$

The *centered difference* approximation uses symmetrically sampled values:

$$f'(x_0) \approx \frac{f(x_0+h)-f(x_0-h)}{2h}, \tag{7}$$

and if $f$ has two derivatives, the centered difference possesses higher accuracy:

$$f(x_0+h) = f(x_0) + f'(x_0)h + \frac{f''(x_0)h^2}{2} + o(h^2) \tag{8}$$
$$f(x_0-h) = f(x_0) + f'(x_0)(-h) + \frac{f''(x_0)(-h)^2}{2} + o(h^2) \tag{9}$$

The first minus the second yields

$$f(x_0 + h) - f(x_0 - h) = f'(x_0)(2h) + o(h^2),$$

or

$$f'(x_0) = \frac{f(x_0 + h) - f(x_0 - h)}{2h} + o(h) \qquad (10)$$

(Compare with the $o(1)$ approximation of Eq. (1)).

Note that it was necessary to carry out the Taylor developments in (8, 9) to $o(h^2)$ in order to discover the true accuracy of (10). If we had only used $o(h)$ developments -

$$f(x_0 + h) = f(x_0) + f'(x_0)h \qquad + \qquad o(h)$$
$$f(x_0 - h) = f(x_0) + f'(x_0)(-h) \qquad + \qquad o(h)$$
$$f(x_0 + h) - f(x_0 - h) = f'(x_0)(2h) + o(h),$$

we would have concluded that the accuracy of (10) was merely $o(1)$.

In this vein, one really should carry out the developments even further; perhaps (10) is even more accurate than $o(h)$! Assuming $f$ has three derivatives at $x_0$ we have

$$f(x_0 + h) = f(x_0) + f'(x_0)(h) + \frac{f''(x_0)h^2}{2} + \frac{f'''(x_0)h^3}{6} + o(h^3)$$
$$f(x_0 - h) = f(x_0) - f'(x_0)(h) + \frac{f''(x_0)h^2}{2} - \frac{f'''(x_0)h^3}{6} + o(h^3)$$
$$f(x_0 + h) - f(x_0 - h) = f'(x_0)(2h) + \frac{f'''(x_0)h^3}{3} + o(h^3),$$

Since the third derivative terms *don't* cancel, we find only a modest refinement of (10):

$$f'(x_0) = \frac{f(x_0 + h) - f(x_0 - h)}{2h} + \mathcal{O}(h^2)$$

The centered-difference approximation for $f''(x_0)$ is (see exercise 1)

$$f''(x_0) = \frac{f(x_0 + h) - 2f(x_0) + f(x_0 - h)}{h^2} + o(h), \qquad (11)$$

if $f'''(x_0)$ exists.

Table A.1 illustrates the accuracy of these numerical differentiation schemes for the derivatives of certain functions.

A few remarks are in order.

i. The cosine and exponential functions demonstrate the predicted behavior. The centered differences are quite superior to the forward differences, and all differences become more accurate as $h$ decreases.

Table A.1: FINITE DIFFERENCE APPROXIMATIONS FOR DERIVA-
TIVES

| $f(x)$ | $x_0$ | $h$ | Eq. (1) (forward diff.) | $f'(x_0)$ (exact) | Eq. (7) (centered diff.) | Eq. (3) (forward diff.) | $f''(x_0)$ (exact) | Eq. (9) (centered diff.) |
|---|---|---|---|---|---|---|---|---|
| $\cos x$ | $\pi/6$ | .1 | $-.542$ | | $-.4992$ | $-.811$ | | $-.8653$ |
| | | | | $-.5$ | | | $-.866025$ | |
| | | .01 | $-.504$ | | $-.499992$ | $-.861$ | | $-.86601$ |
| $e^x$ | 0 | .1 | 1.05 | | 1.002 | 1.106 | | 1.0008 |
| | | | | 1 | | | 1 | |
| | | .01 | 1.005 | | 1.00002 | 1.01 | | 1.000008 |
| $e^x x^{5/3}$ | 0 | .1 | .238 | | .212 | (3.59) | | (.431) |
| | | | | 0 | | | $(\pm\infty)$ | |
| | | .01 | .0469 | | .0464 | (5.66) | | (.0928) |
| $e^x x^{8/3}$ $+x^2$ | 0 | .1 | .124 | | .002 | 3.2 | | 2.4 |
| | | | | 0 | | | 2 | |
| | | .01 | .010 | | .000005 | 2.2 | | 2.09 |

ii. The function $e^x x^{5/3}$ does not possess a second derivative at $x_0 = 0$, so the rationale for expecting higher accuracy from the centered difference for $f'(0)$ is undermined; and, sure enough, the centered differences show no marked increase in accuracy. (The extra factor $e^x$ was included to highlight the effect.)

iii. For the function $e^x x^{8/3} + x^2$, the *third* derivative is faulty. Thus we expect centering to improve the accuracy of the first derivative, but not the second. The data confirm this.

   Always keep in mind that the derivative is, at bottom, a limit of difference quotients. Sometimes students emerge from the basic calculus course so inundated with rules and formulas for differentiating that this basic interpretation gets blurred. Calculus books are full of ingenious rules and identities which enable the evaluation of the derivative for virtually any function one can write down, but the reason $f'(x)$ is important in applied mathematics is because it has physical significance - it measures rate of change - and it can be computed (or at least estimated) from data.

# Exercises A.3

1. Derive eq. (11) for the centered difference approximation of the second derivative. (Hint: there are equivalent ways to write (11), with "$h$" rescaled as, for instance, "$2h$".)

2. Construct finite difference approximations for the third derivatives of a function. One way to proceed would be to solve

$$f(x) = f(x_0) + f'(x_0)h + f''(x_0)\frac{h^2}{2} + f'''(x_0)\frac{h^3}{6} + \mathcal{O}(h^4)$$

for $f'''(x_0)$ and use finite difference approximations for the lower derivatives.

   (a) Assess the accuracy if forward differences are used for $f'$ and $f''$.

   (b) Assess the accuracy if centered differences are used for $f'$ and $f''$

   (c) Numerically test the accuracy of the above approximations for the case $f(x) = e^x, x_0 = 0$.

3. Construct a finite difference approximation for $f^{(4)}(x)$ as in exercise 2, using centered differences for the lower derivatives. Test numerically with $f(x) = e^x, x_0 = 0$

## A.4   Difference Quotients in Two Dimensions

The key to constructing and analyzing the Taylor developments for functions of two or more variables is to treat the variables one at a time. Again the order of magnitude notation is enormously helpful. (Its indiscriminate use for multivariable functions can lead to incorrect estimates in certain patheological cases; however if the higher derivatives of $f$ are continuous no such problems arise. For this exposition we shall blithely presume such smoothness conditions are met, occasionally inserting observations pertaining to the delimiting cases.)

If we are given a function $f(x, y)$ and wish to approximate its values at the point $x = x_0 + h, y = y_0 + k$ in terms of the "base" values $x = x_0, y = y_0$, we begin by holding $y$ *fixed* and expanding in $x$ alone:

$$f(x_0 + h, y_0 + k) = f(x_0, y_0 + k)$$
$$+ f_x(x_0, y_0 + k)h + f_{xx}(x_0, y_0 + k)h2/2! + o(h^2).$$

Here we have used the notation $f_x$ and $f_{xx}$ to denote the partial derivatives $\partial f/\partial x$ and $\partial^2 f/\partial x^2$. Next we expand the *coefficients* in this expression, as functions of $y$, around the point $y = y_0$ (retaining only the second order terms):

$$f(x_0 + h, y_0 + k) =$$

$$f(x_0, y_0 + k) + f_x(x_0, y_0 + k)h + f_{xx}(x_0, y_0 + k)\frac{h^2}{2!} + o(h^2)$$

$$\parallel \qquad\qquad \parallel \qquad\qquad \parallel \qquad\qquad \parallel$$

$$f(x_0, y_0) \qquad f_x(x_0, y_0)h \qquad f_{xx}(x_0, y_0)\frac{h^2}{2} \qquad o(h^2)$$
$$+ \qquad\qquad + \qquad\qquad +$$
$$f_y(x_0, y_0)k \qquad f_{xy}(x_0, y_0)hk \qquad h^2 o(1)$$
$$+ \qquad\qquad +$$
$$f_{yy}(x_0, y_0)\frac{k2}{2} \qquad ho(k)$$
$$+$$
$$o(k^2) \tag{1}$$

$$f(x_0 + h, y_0 + k) = f(x_0, y_0) + f_x(x_0, y_0)h + f_y(x_0, y_0)k$$
$$+ f_{xx}(x_0, y_0)\frac{h^2}{2} + f_{xy}(x_0, y_0)hk + f_{yy}(x_0, y_0)\frac{k^2}{2}$$
$$+ o(h^2 + k^2) \tag{2}$$

Here $f_{xy}$ stands for the $y$ derivative of $f_x$:

$$f_{xy} = \frac{\partial}{\partial y}\frac{\partial f}{\partial x} = \frac{\partial^2 f}{\partial y \partial x}; \tag{3}$$

and we have taken advantage of the fact that all four error terms go to zero faster than $(h^2 + k^2)$.

Obviously $o(h^2)$, $h^2 o(1)$ and $o(k^2)$ approach zero faster than $(h^2 + k^2)$. To show rigorously why $ho(k)$ does so also, first expand the inequality $0 \le (h - k)^2$:

$$0 \quad \le \quad h^2 + k^2 - 2hk;$$
$$2hk \quad \le \quad h^2 + k^2 \tag{4}$$

(regardless of the sign of $hk$). Now substitute;

$$\frac{ho(k)}{h^2 + k^2} \le \frac{ho(k)}{2hk} = \frac{o(k)}{2k}$$

Since $o(k)/2k \to 0$, $ho(k)$ goes to zero faster than $(h^2 + k^2)$.[1]

Of course the expansion in (2) can be extended to more variables; see exercise 5. It can also be extended to higher derivatives, but usually the form (2) suffices for applications. The notation in the mixed partial derivative (3) may seem treacherous because the order of $x$ and $y$ switches when we go from subscript form to "fraction" form. Actually this seldom causes

---

[1] In a rigorous analysis one should account for the possibility that the rate at which $o(h^2)$ goes to zero in (1) may depend on the location of the point $(x_0, y_0 + k)$ (and thus on the value of $k$). In fact by exploiting this "loophole," violations of (2) can be concocted with functions possessing discontinuous second partial derivatives. We will alert the reader to situations where functional smoothness is a crucial issue in applications.

trouble because the reader may recall that the mixed partial derivatives are the same in any order:

$$f_{xy} = f_{yx} \tag{5}$$

under very general conditions.[2] A good way of remembering this is to consider a finite difference approximation for $f_{xy}$. Refer to Fig. 1.

The $y$ derivative of $f_x$ is approximated by the finite difference

$$f_{xy}(x_0, y_0) \approx \frac{f_x(x_0, y_0 + k) - f_x(x_0, y_0)}{k}$$

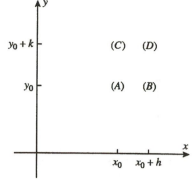

**Figure 1** *Finite difference geometry for $f_{xy}$*

If the $x$ partials therein are similarly approximated,

$$f_{xy}(x_0, y_0) =$$
$$\frac{1}{k} \left\{ \frac{f(x_0 + h, y_0 + k) - f(x_0, y_0 + k)}{h} - \frac{f(x_0 + h, y_0) - f(x_0, y_0)}{h} \right\}$$

we have, in the nomenclature of the figure,

$$f_{xy}(x_0, y_0) \approx \frac{f(D) - f(C) - f(B) + f(A)}{hk}. \tag{6}$$

By exchanging the order of $x$ and $y$ we arrive at the estimate for $f_{xy}$:

$$f_{yx} \approx \frac{f(D) - f(B) - f(C) + f(A)}{kh}. \tag{7}$$

But the difference quotients in (6) and (7) are identical - hence we conclude (5).

To estimate the accuracy of (6), combine the following:

$$
\begin{aligned}
f(D) &= f + f_x h + f_y k + f_{xx} h^2/2 + f_{xy} hk + f_{yy} k^2/2 + o(h^2 + k^2) \\
F(C) &= f \phantom{+ f_x h} + f_y k \phantom{+ f_{xx} h^2/2} + f_{xy} hk + f_{yy} k^2/2 + \phantom{+} o(k^2) \\
F(B) &= f + f_x h \phantom{+ f_y k} + f_{xx} h^2/2 \phantom{+ f_{xy} hk + f_{yy} k^2/2} + o(h^2) \\
F(A) &= f \phantom{+ f_x h + f_y k + f_{xx} h^2/2 + f_{xy} hk + f_{yy} k^2/2} \text{(all $f$'s evaluated at $x_0, y_0$)}
\end{aligned}
$$
and

$$f(D) - f(C) - f(B) + f(A) = f_{xy}(x_0, y_0) hk + o(h^2 + k^2)$$

thus

$$f_{xy}(x_0, y_0) = \frac{f(D) - f(C) - f(B) + f(A)}{hk} + o(1). \tag{8}$$

In exercise 1 you will be invited to construct and analyze a centered difference formula for $f_{xy}$.

---

[2] A painstaking proof of (5) reveals the necessity of imposing the condition that $f$'s first derivatives and at least one of its mixed partials exist and be continuous. The classic counterexample is the function

$$f(x, y) = xy \frac{x^2 - y^2}{x^2 + y^2}$$

with $f(0,0)$ taken to be zero. The mixed partials of $f$ are discontinuous and, in fact, $f_{xy}(0,0) = -1$ but $f_{yx}(0,0) = 1$. (This is discussed in most advanced calculus books.)

## Exercises A.4

1. Construct a centered difference approximation for $f_{xy}$ and estimate its accuracy.

2. Using centered differences construct and assess the accuracy of the five-point approximation (or five-point "molecule") for the Laplacian in two dimensions,

$$\nabla^2 f = \frac{\partial^2 f}{\partial x^2} + \frac{\partial^2 f}{\partial y^2},$$

which employs data at the five points illustrated in Fig. 2. Observe that for convenience we take $k = h$. In particular, show that coefficients can be chosen so that

$$\nabla^2 f = a_1 f(A) + a_2 f(B) + a_3 f(C) + a_4 f(D) + a_5 f(E) + \mathcal{O}(h^2)$$

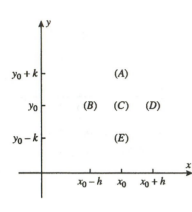

**Figure 2**   *Laplacian molecule*

3. Show that the Taylor development of degree two in eq. (2) can be expressed in matrix notation as

$$
f(x_0 + h, y_0 + k) = f(x_0, y_0) + \begin{bmatrix} h & k \end{bmatrix} \begin{bmatrix} f_x \\ f_y \end{bmatrix}
$$
$$
+ \frac{1}{2} \begin{bmatrix} h & k \end{bmatrix} \begin{bmatrix} f_{xx} & f_{xy} \\ f_{xy} & f_{yy} \end{bmatrix} \begin{bmatrix} h \\ k \end{bmatrix} + o(h^2 + k^2)
$$

4. Show that by setting $s = 1$ in the (*one* dimensional) Taylor development of the function

$$\mathcal{F}(s) = f(x_0 + sh, y_0 + sk),$$

as a function of $s$, we obtain eq. (2).

5. (a) Conjecture the Taylor development for a function $f(x, y, z)$ of three variables from the display in exercise 3.

   (b) Confirm your answer to part (a) by extending the computational scheme of eq. (1).

   (c) Confirm your answer to part (a) by extending the computational scheme of exercise 4.

## A.5   Solution of Algebraic Equations

There are known formulas for the solution of polynomial equations of low degree. The linear equation $x + b = 0$ has an obvious solution. The quadratic equation $ax^2 + bx + c = 0$ has the well-known quadratic formula for its solutions. The cubic and quadratic equations

$$ax^3 + bx^2 + cx + d = 0, \quad ax^4 + bx^3 + cx^2 + dx + e = 0$$

also have solution formulas, or rather solution procedures - see the *CRC Mathematical Handbook*, for example. However the quintic equation

$$ax^5 + bx^4 + cx^3 + dx^2 + ex + f = 0$$

admits no general solution formula or procedure; this was proved by Galois and Abel two centuries ago.

Most other equations, such as

$$x^{\frac{5}{6}} + 13x^{\frac{2}{5}} - 4x^{\frac{1}{2}} - 9 = 0$$

$$x - \sin x - .1 = 0 \tag{1}$$

for example, have no solution formulas. Thus we have to resort to numerical methods for their solution. These take the form of *iterative* methods whose iterates converge to solutions, and therefore provide *approximate* solutions after finite computation time. The easiest iterative method to implement is called, simply, "iteration". It requires that the equation to be solved,

$$f(x) = 0, \tag{2}$$

first be recast into the format

$$x = g(x) \tag{3}$$

Next, starting from some initial guess $x_{old}$, a new estimate for the solution is constructed via

$$x_{new} = g(x_{old}). \tag{4}$$

Then the new value of x overwrites the old, and (4) is repeated. *If* the sequence of iterates $x_1, x_2, x_3, ...$ converges to, say, $x_\infty$, we have a solution:

$$x_\infty = \lim x_{n+1} = \lim g(x_n) = x_\infty$$

(as long as $g$ is continuous).

**Example 1.** If we rewrite equation (1) as

$$x = g(x) = \sin x + .1 \tag{5}$$

and then iterate $x_{new} = \sin x_{old} + .1$, the iterates, starting from $x_1 = .5$, are listed in the *left* hand column below:

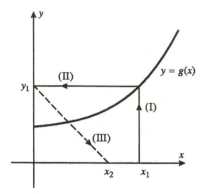

**Figure 3**  *Iteration path*

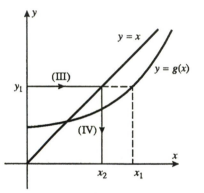

**Figure 4**

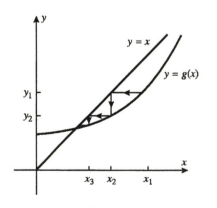

**Figure 5**

| $x_{new} = \sin x_{old} + .1$ | $x_{new} = \arcsin(x_{old} - .1)$ |
|---|---|
| .5000000000 | .8537501565 |
| .5794255386 | .8537501564 |
| .6475433309 | .8537501563 |
| .7032288670 | .8537501561 |
| .7466838985 | .8537051499 |
| .7792086621 | .8537501017 |
| .8027166262 | .8537494745 |
| .8493004114 | .8537416873 |
| .8532012068 | .8536810615 |
| .8537059049 | .8538168701 |
| .8537465922 | .8491849740 |
| .8537497198 | .8183956299 |
| .8537501031 | .6437100866 |
| .8537501501 | |
| .8537501561 | |
| .8537501566 | |
| .8537501566 | |

and we have a solution.                                                          ∎

**Example 2.** If equation (1) is recast in the form $x = \arcsin(x - .1)$ and the iterations are started again from $x_1 = .5$, they diverge! In fact, they diverge even when they are initiated from the near- exact solution .8537501565, as shown in the *right* hand column above.                                        ∎

As you can see, sometimes straighforward iteration works, sometimes not. When it works, it's usually fairly slow. But it has the advantage of simplicity - very little time is wasted by trying it out and seeing if it is suitable for the application at hand.

Actually, there's a charming argument as to why this method works or fails. The sequence of iterates is illustrated graphically in Figures 3 through 5. We start with an initial guess $x_1$ and locate it on the $x$-axis (Fig. 3). Then we move up until we hit the graph of $g(x)$ (route I) and slide to the left to read off $y_1 = g(x_1)$ (route II). Next we set $x_2 = y_1$ and locate $x_2$ on the $x$-axis (route III), and we are ready to begin again.

In Fig. 4 the process of setting $x_2 = y_1$ can be depicted, somewhat clumsily perhaps, by sliding to the right from the $y_1$-level on the $y$ axis to the line $y = x$ (the new route III), and then dropping down to the $x$-axis (route IV).

But this suggests a simpler way of graphing the steps of the algorithm. As Fig. 5 shows, there is no need to go all the way over to the axes each time; we can accomplish the same thing by going from the $g(x)$ graph *horizontally* to the $y = x$ line, and then *vertically* to the $g(x)$ graph, and so on.

Now in Fig. 5 we use this procedure to depict the iterates for a function $g(x)$ with a positive slope, less than $+1$. Clearly the iterates converge.

But if the slope is greater than $+1$, Fig. 6 shows that the iterates *diverge*.

Exercise 1 invites the reader to reconstruct these pictures for functions $g(x)$ with negative slope, first with $g'(x) > -1$, and then with $g'(x) < -1$.

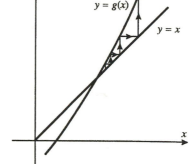

**Figure 6**

<u>*Conclusion.*</u> *Straightforward iteration converges if* $|g'(x)| < 1$ *within the interval of iteration.*

*This reasoning is used in higher mathematics for more abstract situations. A solution to the equation (3) is called a* **fixed point** *of $g(x)$. A function $g$ which has the property*

$$|g(a) - g(b)| < |a - b| \qquad (6)$$

*for all $a$ and $b$ under consideration is called a* **contracting map**. *Note that, because of the mean value theorem, any function satisfying $|g'| < 1$ is a contracting map. The contracting map principle states that every contracting map has a fixed point, under very general circumstances.*

As these examples show, straightforward iteration suffers from two deficiencies: its convergence is not guaranteed, and when it does converge it does so fairly slowly. On the other hand its ease of implementation makes it worthwhile in many situations. Next we are going to study *Newton's method*, which converges much more rapidly under more general circumstances, but costs more in computational complexity.

*Newton's method* approaches the problem in the format of eq. (2), rather than eq. (3). It's a minor point, but the method seeks zeros of $f(x)$, not fixed points of $g(x)$. Refer to Fig. 7, which depicts $f(x)$ and its zero crossing.

As indicated in the figure, the strategy of this method is to generate an improvement on an initial estimate $x_{old}$ by constructing the tangent line to the curve $y = f(x)$ at $x_{old}$, and letting $x_{new}$ be the point where the line crosses the $x$-axis. If the curve is well approximated by its tangent lines, then the sequence of crossings should converge rapidly to the solution point $x_{\infty}$.

To get the formula for $x_{new}$, note that slant of the tangent line is given by $f'(x_{old})$. Thus the geometry of the triangle dictates that

$$\frac{f(x_{old})}{x_{old} - x_{new}} = f'(x_{old})$$

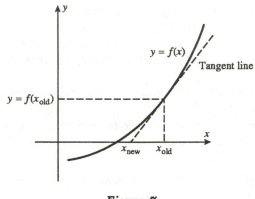

**Figure 7**

or

$$x_{new} = x_{old} - f(x_{old})/f'(x_{old}) \tag{7}$$

**Example 3.** The Newtonian iterates for eq. (1) are given by

$$x_{new} = x_{old} - \frac{x_{old} - \sin x_{old} - .1}{1 - \cos x_{old}} \tag{8}$$

Starting from $x_1 = .5$, they converge much faster than the previous methods:

$$
\begin{aligned}
&.5 \\
&1.148809025 \\
&.9175653406 \\
&.8578040551 \\
&.8537681045 \\
&.8537501570 \\
&.8537501566
\end{aligned}
\tag{9}
$$

∎

An analysis of this convergence can be conducted using the Taylor expansions. We expand $f(x_{old})$ and $f'(x_{old})$ around the actual solution $x_\infty$ :

$$
\begin{aligned}
f(x_{old}) = {}& f(x_\infty) + f'(x_\infty)[x_{old} - x_\infty] \\
& + f''(x_\infty)[x_{old} - x_\infty]^2/2! \\
& + \mathcal{O}([x_{old} - x_\infty]^3)
\end{aligned}
\tag{10}
$$

$$
\begin{aligned}
f'(x_{old}) = {}& f'(x_\infty) + f''(x_\infty)[x_{old} - x_\infty] \\
& + \mathcal{O}([x_{old} - x_\infty]^2
\end{aligned}
\tag{11}
$$

Next we subtract $x_\infty$ from both sides of eq. (8), insert (11) and (12), and make the simplifications

$$f(x_{old}) = 0, \quad \Delta x_{old} = x_{old} - x_\infty, \quad \Delta x_{new} = x_{new} - x_\infty,$$
$$f' = f'(x_\infty), \quad f'' = f''(x_\infty)$$

to find

$$\Delta x_{new} = \Delta x_{old} - \Delta x_{old} \frac{f' + f'' \frac{\Delta x_{old}}{2} + \mathcal{O}(\Delta x_{old}^2)}{f' + f'' \Delta x_{old} + \mathcal{O}(\Delta x_{old}^2)} \qquad (12)$$

To perform the division we employ the geometric series for $1/(1+h)$ (Sec. 2.2):

$$
\begin{aligned}
&\frac{1}{f' + f'' \Delta x_{old} + \mathcal{O}(\Delta x_{old}^2)} \\
&= \frac{1}{f'} \times \frac{1}{1 + f'' \Delta x_{old}/f' + \mathcal{O}(\Delta x_{old}^2)} \\
&= \frac{1}{f'} \times \frac{1}{1 + \cdots} \\
&= \frac{1}{f'} \times \{1 - \cdots + \mathcal{O}([\ldots]^2)\} \\
&= \frac{1}{f'} \times \{1 - f'' \Delta x_{old}/f' + \mathcal{O}(\Delta x_{old}^2)\}
\end{aligned}
\qquad (13)
$$

(merging all the $\mathcal{O}(\Delta x_{old}^2)$ terms) and insert into (12) to derive

$$
\begin{aligned}
\Delta x_{new} &= \Delta x_{old} - \Delta x_{old} \{1 - f'' \Delta x_{old}/2f' + \mathcal{O}(\Delta x_{old}^2)\} \\
&= \frac{f''}{2f'} \Delta x_{old}^2 + \mathcal{O}(\Delta x_{old}^3)
\end{aligned}
\qquad (14)
$$

The fact that $\Delta x_{new}$ is proportional to the *square* of $\Delta x_{old}$ (and, thus, $\Delta x_{n+1}$ would be proportional to the square of $\Delta x_n$) means that convergence will proceed very rapidly, once the iterates get close to $x_\infty$. For example if $f''/2f'$ is around unity and $\Delta x_i$ is .1, $\Delta x_{i+1}$ will be about .01, $\Delta x_{i+2}$ about .0001, $\Delta x_{i+3}$ about .00000001 - in other words, *the number of correct decimal digits will double every iteration.* Look back; this phenomenon occurred in Example 3.

From this analysis we conclude that Newton's method will converge rapidly as long as $f'$ (in the denominator of (14)) is not near zero, and as long as the iterates stay in a neighborhood of $x_\infty$ in which the graph of $f$ is approximately parabolic (so that neglecting the higher order terms in (11) and (12) is justified). Figure 8 demonstrates the danger of selecting an iterate near a zero of $f'$. Figure 9 illustrates "cycling", one of the hazards of

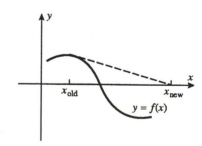

**Figure 8**

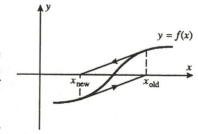

**Figure 9**

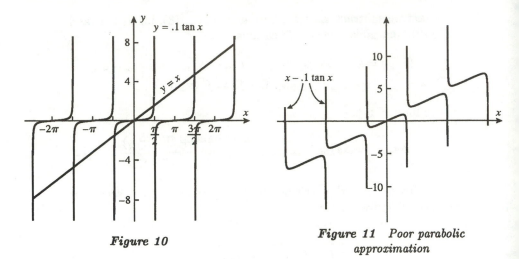

*Figure 10*

*Figure 11   Poor parabolic approximation*

straying into a region where $f$ is nonparabolic. In practice one has to rely on judgement, experience, graphical information, and blind luck to select a first guess which is in the range of convergence for the algorithm.

**Example 4.** We shall attempt to find the solutions to

$$x = .1 \ \tan x \tag{15}$$

This equation arises in Sec. 6.3, Example 1.

For this problem it is almost imperative that we begin by sketching the graph. Figure 10 demonstrates some interesting qualitative features:

(1) There are an infinite number of solutions.

(2) Zero is a solution, and if $x$ is a solution so is $-x$.

(3) "Most" of the positive solutions are very near, but just under, odd multiples of $\pi/2$.

It should be immediately clear that iteration of the equation $x = .1 \tan x$ is futile, since the graph demonstrates that the slope of $.1 \tan x$ far exceeds unity at all the solution points. The "inverse function" scheme

$$x = \arctan 10x \tag{16}$$

has possibilities if good initial guesses are used, but the quadranting ambiguity introduces complications (see exercise 2). Let's go to Newton's method.

From the graph of $f(x) = x - .1 \tan x = 0$ displayed in Fig. 11 it is quite obvious that the starting point is very crucial in determining to which root the algorithm will converge. The iteration equation (8) takes the form

$$x_{new} = x_{old} - \frac{x_{old} - 0.1 \ \tan x_{old}}{1 - .1 \ \sec^2 x_{old}} \tag{17}$$

As starting points we take $(\pi/2) - .01, (3\pi/2) - .01, (5\pi/2) - .01, (7\pi/2) - .01, \ldots$ The results are

| Iteration | Root 1 | Root 2 | Root 3 | Root 4 |
|-----------|----------|----------|----------|-----------|
| 0 | 1.560796 | 4.702389 | 7.843982 | 10.985574 |
| 1 | 1.552349 | 4.697087 | 7.841824 | 10.986561 |
| 2 | 1.539143 | 4.692774 | 7.841257 | 10.986473 |
| 3 | 1.522762 | 4.691211 | 7.841229 | 10.986472 |
| 4 | 1.509606 | 4.691076 | 7.841229 | 10.986472 |
| 5 | 1.504843 | 4.691075 | | |
| 6 | 1.504426 | 4.691075 | | |
| 7 | 1.504423 | | | |

The sluggish convergence is a symptom of the extreme irregularity of the graph; at the zero crossings in Fig. 11 the curve is nearly vertical, and this undermines the parabolic approximation inherent in Newton's method. ■

Newton's method can be extended to *systems* of equations; the extension is known as the *Newton-Raphson* algorithm. The elegant geometric picture that motivated the single-variable form (Fig. 7) is difficult to visualize in higher dimensions, so we'll use a Taylor series approach to derive the iteration equations. The problem is expressed in the format

$$f(x, y) = 0$$
$$g(x, y) = 0 \tag{18}$$

Let $(x_{old}, y_{old})$ denote the current value of the iterates. Expanding around $(x_{old}, y_{old})$ to express the values of $f$ and $g$ at $(x_{new}, y_{new})$, we have (recall Sec. A.4)

$$f(x_{new}, y_{new}) = f(x_{new}, y_{new}) + f_x(x_{old}, y_{old})[x_{new} - x_{old}]$$
$$+ f_y(x_{old}, y_{old})[y_{new} - y_{old}] + \mathcal{O}([x_{new} - x_{old}]^2 + [y_{new} - y_{old}]^2)$$

$$g(x_{new}, y_{new}) = g(x_{old}, y_{old}) + g(x_{old}, y_{old})[x - x_{old}] + g(x_{old}, y_{old})[y_{new} - y_{old}]$$
$$+ \mathcal{O}([x_{new} - x_{old}]^2 + [y_{new} - y_{old}]^2).$$

Now the objective is to pick $(x_{new}, y_{new})$ so that

$$f(x_{new}, y_{new}) = g(x_{new}, y_{new}) = 0$$

Thus we ignore the second order terms and choose $x_{new}, y_{new}$ to satisfy the equations

$$f_x(x_{old}, y_{old})[x_{new} - x_{old}] + f_y(x_{old}, y_{old})[y_{new} - y_{old}] = -f(x_{old}, y_{old})$$
$$g_x(x_{old}, y_{old})[x_{new} - x_{old}] + g_y(x_{old}, y_{old})[y_{new} - y_{old}] = -g(x_{old}, y_{old})$$
$$\tag{19}$$

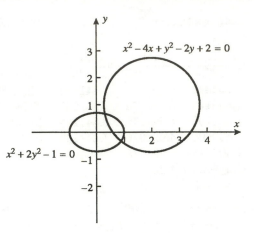

**Figure 12** *Graphs for Example 5*

Equations (19) are linear and can be solved by standard methods. *Cramer's rule*, for example, displays the solutions in terms of determinants as

$$
x_{new} = x_{old} + \frac{\begin{Vmatrix} -f & \partial f/\partial y \\ -g & \partial g/\partial y \end{Vmatrix}}{\begin{Vmatrix} \partial f/\partial x & \partial f/\partial y \\ \partial g/\partial x & \partial g/\partial y \end{Vmatrix}}
$$

$$
y_{new} = y_{old} + \frac{\begin{Vmatrix} \partial f/\partial x & -f \\ \partial g/\partial x & -g \end{Vmatrix}}{\begin{Vmatrix} \partial f/\partial x & \partial f/\partial y \\ \partial g/\partial x & \partial g/\partial y \end{Vmatrix}}
$$

(20)

where all functions are to be evaluated at $(x_{old}, y_{old})$.

**Example 5.** The graphs of the two equations

$$
\begin{aligned}
f(x,y) &= x^2 + 2y^2 - 1 & &= 0 \\
g(x,y) &= x^2 - 4x + y^2 - 2y + 2 & &= 0
\end{aligned}
$$

are shown in Fig. 12. There are a *pair* of simultaneous solutions.

To obtain convergence to the first-quadrant solution an initial guess of $(x_{old}, y_{old})=(0, .5)$ was tried; $(x_{old}, y_{old})= (0, -.5)$ was used for the fourth-quadrant intersection. Equations (20) were implemented with

$$
\begin{aligned}
\partial f/\partial x &= 2x, & \partial f/\partial y &= 4y; \\
\partial g/\partial x &= 2x - 4, & \partial g/\partial y &= 2y - 2.
\end{aligned}
$$

The results of the iterations were as follows:

| Iteration | x | y | x | y |
|-----------|--------|--------|---------|----------|
| 0 | .5 | 0 | -.5 | 0 |
| 1 | .25 | .75 | 1 | -.75 |
| 2 | .295732 | .679878 | .942308 | -.413462 |
| 3 | .298734 | .674840 | .890537 | -.333480 |
| 4 | .298750 | .674814 | .886755 | -.326930 |
| 5 | .298749 | .674814 | .886730 | -.326887 |
| 6 | | | .886730 | -.326887 |

The complications which can confound the convergence of Newton's method in one dimension (Figs. 8,9) are, as one might expect, even worse in higher dimensions. Consider the innocent-looking equation $z^2 = (x + iy)^2 = -1$; although its roots are complex ($z = \pm i$), it can be broken down into a pair of real equations $x^2 - y^2 = -1$, $2xy = 0$ for the real and imaginary parts. But if the initial guess $z_{old}$ for the Newton-Raphson iterations is real ($y_{old} = 0$), then all the subsequent iterates are real and convergence is denied (exercise 9). A complex initial guess, however, will produce convergence.

# Exercises A.5

1. Experiment with the techniques discussed in this section to find the lowest solution to the equation $e^{-x} = \sin x$.

   (a) Determine by numerical experimentation which of the two obvious iterative schemes converges. Explain on the basis of the contracting map principle.

   (b) Experiment with Newton's method.

2. Find the *fifth smallest* positive solution of (16) by direct iteration.

3. Find all solutions of the simultaneous system: $x^2 + 2x + y^2 - 3 = 0$, $x^2 + y^2 - 2y = 0$.

4. Find all solutions to the system

$$x^2 + y^2 + z^2 - 4 = 0$$
$$x^2 + y + z - 2 = 0$$
$$x + 2y - 3z + 6 = 0.$$

5. Find, to 6 decimals, the solution near $x = 1$ to $x^4 + x^3 = 2.1$.

6. Find, to 6 decimals, the solution to $x = 10 \sin x$ for $\pi/2 < x < \pi$.

7. For the system $x^3 - 3xy^2 + 1 = 0$, $3x^2y - y^3 = 0$, perform one iteration of the Newton-Raphson algorithm, starting from the initial guess $x = 2/3, y = 5/6$.

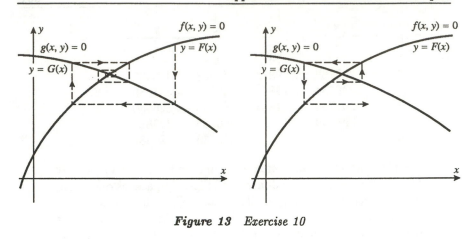

**Figure 13**   *Exercise 10*

8. Find a solution in the first quadrant, to 6 decimals: $x^2 + y^2 = 1$, $xy = x - y$ .

9. Confirm the statement that real initial guesses to the solution of $z^2 = x^2 - y^2 + 2xyi = -1$ will only lead to real iterates. Experiment with iterations starting from complex initial guesses.

10. One can adapt the "straightforward iteration" method so that it can be applied to simultaneous equations. Given the task of solving $f(x, y) = 0$ and $g(x, y) = 0$, rewrite the first equation as $x = F(y)$ and the second as $y = G(x)$. Then iterate in accordance with $x_{new} = F(y_{old})$, $y_{new} = G(x_{new})$.

   (a) Try out this procedure with the equations $x + y = 1$, $xy = x - y$, recast as

   $$x + y = 1 \leftrightarrow x = 1 - y, \quad xy = x - y \leftrightarrow y = \frac{x}{x + 1}.$$

   Start from $x = .6$, $y = .4$. Does it converge?

   (b) Repeat part (a) with the equations recast in the oppposite sense: $x + y = 1 \leftrightarrow y = 1 - x$, $xy = x - y \leftrightarrow x = \frac{y}{1 - y}$.

   (c) Interpret the diagrams in Figure 13 and explain the convergence behavior you observed in parts (a) and (b).

11. Experiment with straightforward iteration on the equations of exercises 3 and 4.

## A.6    Numerical Integration

The Fundamental Theorem of Calculus implies that one can evaluate a definite integral by using an antiderivative:

$$\int_a^b f(x)\,dx \ = F(b) - F(a) \quad \text{where } F'(x) = f(x). \tag{1}$$

However it is important in applied mathematics to keep in mind that the integral is *defined*, not by (1), but by a limit of sums. In theory one partitions the interval $[a, b]$ into $n$ subintervals

$$a = x_0 < x_1 < x_2 < \cdots < x_{n-1} < x_n = b, \tag{2}$$

then samples the function $f(x)$ at a point inside each subinterval

$$\{f(\xi_i); \ x_{i-1} < \xi_i < x_i\}, \tag{3}$$

and forms the sum of these sampled values, each weighted by the length of the corresponding subinterval:

$$\Sigma_{i=1}^n \ f(\xi_i \,(x_i - x_{i-1}) = \Sigma_{i=1}^n \ f(\xi_i)\,\Delta x_i \tag{4}$$

If the function is continuous (and sometimes even if it is not[3]) these sums approach a limit as the maximum interval length $\Delta x_i$ goes to zero (and, thus, the number of intervals $n$ goes to infinity). This limit is *defined* to be the integral:

$$\int_a^b f(x)\,dx = \lim_{\substack{n \to \infty \\ max\,\Delta x_i \to 0}} \Sigma_{i=1}^n f(\xi_i)\,\Delta x_i. \tag{5}$$

Figure 14 illustrates the *area* interpretation of the definite integral.

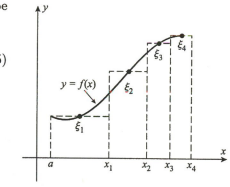

*Figure 14     Area interpretation*

Keep in mind that the integral is defined through this limiting process; the antiderivative substitution (1), "fundamental" though it may be, is not a definition. Every continuous function has an integral, even if - like $e^{-x^2}$ - it appears to have no antiderivative. In fact, one antiderivative for $e^{-x^2}$ is the function $F(x)$ *defined by the definite integral* :

$$F(x) \ = \int_0^x e^{-\xi^2}d\xi \tag{6}$$

The Fundamental Theorem says that the derivative of $F$ with respect to the upper limit $x$ is the integrand:

$$F'(x) \ = e^{-x^2} \tag{7}$$

---

[3] See the reference by Royden for details of the theory of integration.

Of course, this is unsatisfactory. How can we compute $F$ if we don't have a formula for it? The answer is, we approximate the integral with sums - just as the theory says! To compute $F(1)$, say, we partition the unit interval [0,1] into 4 parts:

$$[0,.25], \quad [.25,.5], \quad [.5,.75], \quad [.75,1] \ ;$$

sample the integrand inside each subinterval:

$$e^{-.2^2} \approx .96079, \quad e^{-.4^2} \approx .85214, \quad e^{-.7^2} \approx .61263, \quad e^{-.9^2} \approx .44486;$$

and form the sum:

$$F(1) \approx .96079 \times .25 + .85214 \times .25 + .61263 \times .25 + .44486 \times .25$$
$$\approx .69927.$$

To get a more accurate value we could take a finer partition; for 10 equal subintervals a possible sum would be

$$F(1) \approx \{e^{-.1^2} + e^{-.2^2} + e^{-.3^2} + e^{-.4^2} + e^{-.5^2} + e^{-.6^2} + e^{-.7^2} + e^{-.8^2}$$
$$+ e^{-.9^2} + e^{-1^2}\} \times \{.1\} \tag{8}$$
$$\approx .71460.$$

In (8) we have taken the sampled values at the right end point of each interval. Left-end samples would yield the approximation

$$F(1) \approx \{e^{0^2} + e^{-.1^2} + e^{-.2^2} + e^{-.3^2} + e^{-.4^2} + e^{-.5^2} + e^{-.6^2} + e^{-.7^2}$$
$$+ e^{-.8^2} + e^{-.9^2} + e^{-1^2}\} \times \{.1\} \tag{9}$$
$$\approx .77782.$$

Figure 15 illustrates how these sums give, in effect, the results of approximating the area under the $e^{-x^2}$ graph by rectangles.

The figure suggests a way of accelerating this laborious sequence of computations. Suppose, instead of using rectangles to approximate the area, we used *trapezoids* as in Figure 16. Since the area under a trapezoid equals the base times the mean height, the previous formulas would be replaced by

$$F(1) \approx \{ \frac{e^{-0^2} + e^{-.1^2}}{2} + \frac{e^{-.1^2} + e^{-.2^2}}{2} + \frac{e^{-.2^2} + e^{-.3^2}}{2} + \frac{e^{-.3^2} + e^{-.4^2}}{2}$$
$$+ \frac{e^{-.4^2} + e^{-.5^2}}{2} + \frac{e^{-.5^2} + e^{-.6^2}}{2} + \frac{e^{-.6^2} + e^{-.7^2}}{2} + \frac{e^{-.7^2} + e^{-.8^2}}{2}$$
$$+ \frac{e^{-.8^2} + e^{-.9^2}}{2} + \frac{e^{-.9^2} + e^{-1^2}}{2}\} \times \{.1\},$$

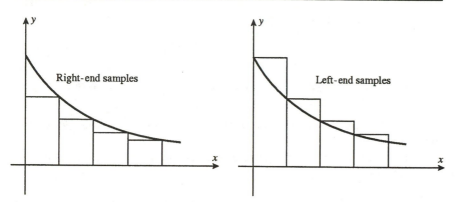

**Figure 15**  *Different sums approximating integrals*

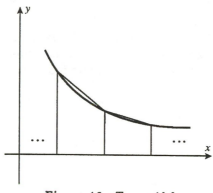

**Figure 16**  *Trapezoidal approximation*

which is no more laborious than (8) if it is reorganized as

$$F(1) \approx \frac{1}{2}\{e^{-0^2} + 2e^{-.1^2} + 2e^{-.2^2} + 2e^{-.3^2} + 2e^{-.4^2} + 2e^{-.5^2} + 2e^{-.6^2}$$

$$+ 2e^{-.7^2} + 2e^{-.8^2} + 2e^{-.9^2} + e^{-1^2}\} \times \{.1\}$$

$$\approx .74621.$$

$$(10)$$

To generalize: the *Trapezoid Rule* for numerical integration is given by the formula

$$\int_a^b f(x)\,dx \approx \frac{\Delta x}{2}\{f(a) + 2f(a + \Delta x) + 2f(a + 2\Delta x)$$

$$+ \cdots + 2f(b - \Delta x) + f(b)\}. \quad (11)$$

The accurate value for $F(1)$ obtained by using very fine subdivisions on high speed computers is

$$F(1) = .7468241328\ldots \qquad (12)$$

$F(x)$ is related to the "Error function",

$$Erf(x) = \frac{2}{\sqrt{\pi}} \int_0^x e^{-\xi^2} d\xi , \qquad (13)$$

which arises frequently in statistics. (See also Sec. 7.2, Example 4.)

The trapezoid rule can be motivated in another way. We ask the following question:

*What combination of the end-point values f(a) and f(a+$\Delta$x) will make the approximation*

$$\int_a^{a+\Delta x} f(\xi)\,d\xi \approx \{c_1 f(a) + c_2 f(a + \Delta x)\} \qquad (14)$$

**exact** *for linear functions?*

The answer, of course, is $c_1 = c_2 = 1/2$, since the graph of a linear function produces trapezoidal-shaped areas. But more importantly, this question suggests a philosophy for designing more accurate integration algorithms:

*What combination of the end-point* **and** **midpoint** *values f(a), f(a+$\Delta$x), f(a+2$\Delta$x) will make the approximation*

$$\int_a^{a+2\Delta x} f(\xi)\,d\xi \approx \{c_1 f(a) + c_2 f(a + \Delta x) + c_3 f(a + 2\Delta x)\}\Delta x \qquad (15)$$

*exact for* **quadratic** *functions?*

A little algebra (exercise 2) reveals that the correct choice is $c_1 = 1/3, c_2 = 4/3, c_3 = 1/3$, producing the formula

$$\int_a^{a+2\Delta x} f(\xi)\,d\xi \approx \frac{\Delta x}{3}\{f(a) + 4f(a + \Delta x) + f(a + 2\Delta x)\} \qquad (16)$$

Stringing together the approximation (16) along the entire interval from $a$ to $b$ we arrive at *Simpson's rule*:

$$\int_a^b f(\xi)\,d\xi \approx \frac{\Delta x}{3}\{f(a) + 4f(a + \Delta x) + 2f(a + 2\Delta x) + 4f(a + 3\Delta x)$$
$$+ 2f(a + 4\Delta x) + \cdots + 2f(b - 2\Delta x) + 4f(b - \Delta x) + f(b)\}$$
$$(17)$$

(Note from the construction that the number of subintervals $n = \frac{b-a}{\Delta x}$ must be even for this formula to apply.)

For the function considered above, Simpson's rule yields the value

$$F(1) \approx \frac{-.1}{3}\{e^{-0^2} + 4e^{-.1^2} + 2e^{-.2^2} + 4e^{-.3^2} + \cdots + 2e^{-.8^2}$$
$$+ 4e^{-.9^2} + e^{-1^2}\}$$
$$\approx .7468249. \qquad (18)$$

The astonishing improvement in accuracy ((18), as compared to (10)) in Simpson's rule at such a modest cost in computational complexity ((17), as compared to (11)) merits further attention. To assess the accuracy of (16) we expand $f(x)$ around $x = a$ in a Taylor development and evaluate both sides. We have

$$f(a + h) = f(a) + f'(a)h + f''(a)\frac{h^2}{2} + f^{(3)}(a)\frac{h^3}{6} + f^4(a)\frac{h^4}{24} + o(h^4)$$

or, for notational convenience,

$$f(a + h) = f_0 + f_1 h + f_2 \frac{h^2}{2} + f_3 \frac{h^3}{6} + f_4 \frac{h^4}{24} + o(h^4) \tag{19}$$

Using (19) in the integral on the left side of (16) we obtain, after some simplification,

$$\int_{\xi=a}^{a+2\Delta x} f(\xi)d\xi = \int_{h=0}^{2\Delta x} f(a + h)\,dh$$

$$= 2 f_0 \Delta x + 2 f_1 \Delta x^2 + 4\frac{4}{3} f_2 \Delta x^3 + \frac{2}{3} f_3 \Delta x^4 \tag{20}$$

$$+ \frac{4}{15} f_4 \Delta x^5 + o(\Delta x^6)$$

On the other hand assemblage of the right hand side of (16) yields

$$f(a) = f_0 \qquad\qquad\qquad \text{times } \frac{\Delta x}{3}$$

$$f(a + \Delta x) = f_0 + f_1 \Delta x + f_2 \frac{\Delta x^2}{2} + f_3 \frac{\Delta x^3}{6} + f_4 \frac{\Delta x^4}{24}$$

$$+ o(\Delta x^4) \qquad\qquad \text{times } \frac{4\Delta x}{3}$$

$$f(a + 2\Delta x) = f_0 + f_1(2\Delta x) + f_2 \frac{(2\Delta x)^2}{2} + f_3 \frac{(2\Delta x)^3}{6} + f_4 \frac{(2\Delta x)^4}{24}$$

$$+ o(\Delta x^4) \qquad\qquad \text{times } \frac{\Delta x}{3}$$

----

$$= 2 f_0 \Delta x + 2 f_1 \Delta x^2 + \frac{4}{3} f_2 \Delta x)^3 + \frac{2}{3} f_3 \Delta x^4$$

$$+ \frac{5}{18} f_4 \Delta x^5 + o(\Delta x^5) \tag{21}$$

Observing the mismatch in the coefficient of $f_4 = f^{(4)}(a)$, we conclude that for functions having a fourth derivative the accuracy of (16) is $\mathcal{O}(\Delta x^5)$:

$$\int_a^{a+2\Delta x} f(\xi)\,d\xi = \frac{\Delta x}{3}\{f(a) + 4f(a + \Delta x) + f(a + 3\Delta x)\} + \mathcal{O}(\Delta x^5) \tag{22}$$

And since formula (16) is pieced together $\frac{n}{2} = \frac{b-a}{2\Delta x}$ times in Simpson's rule, *the error in Simpson's rule is of order $\mathcal{O}(\Delta x^4)$ for functions with four derivatives:*

$$\int_a^b f(\xi)\, d\xi \approx \frac{\Delta x}{3} \{ f(a) + 4f(a + \Delta x) + 2f(a + 2\Delta x) + 4f(a + 3\Delta x)$$
$$+ 2f(a + 4\Delta x) + \ldots + 2f(b - 2\Delta x) + 4f(b - \Delta x) + f(b) \}$$
$$+ \mathcal{O}(\Delta x^4)$$

$$(23)$$

By comparison, the error in the trapezoid rule is $\mathcal{O}(\Delta x^2)$ (but the function only needs two derivatives to achieve this; Exercise 3).

A surprising benefit of Simpson's rule is that, although it was designed only to integrate *quadratics* exactly, it also integrates *cubics* exactly - because their fourth (and higher) derivatives $f^4(a)$ are zero (so the $f_4$ terms in (20, 21) disappear).

Clearly one can continue to devise higher order schemes by choosing coefficients $c_i$ so that, say, the approximation

$$\int_a^{a+4\Delta x} f(\xi)\, d\xi \approx \{ c_1 f(a) + c_2 f(a + \Delta x) + c_3 f(a + 2\Delta x)$$
$$+ c_4 f(a + 3\Delta x) + c_5 f(a + 4\Delta x) \quad (24)$$

integrates *quartics* exactly. Simpson's rule, however, seems to be the method of choice in engineering analysis. With functions having less than four derivatives, however, its performance is not so spectacular.

**Example 1.** The function $f(x) = x^{1/3}$ has no derivative at $x = 0$. Its integrals can be performed exactly with the antiderivative $.75x^{4/3}$, and

$$\int_0^1 x^{1/3}\, dx = .75.$$

$$(25)$$

The results of using the Trapezoid rule and Simpson's rule, for various sized $\Delta x$, are as follows:

| $\Delta x$ | Trapezoid rule | Simpson's rule |
|---|---|---|
| .1 | .7374 | .7436 |
| .05 | .74496 | .74748 |
| .01 | .74941 | .74971 |
| | (exact: .75) | |

These numbers are very inferior to the performance for the smooth function $e^{-x^2}$, where Simpson's rule gave 6-digit accuracy for $\Delta x = .1$ (see (12), (18)). ■

The great mathematician Gauss recognized that the strategy described so far - that of choosing coefficients in a formula such as (15) to make it exact for, say, quadratics - could be exploited further. If we don't insist that the evaluation points $(a,\ a + \Delta x,\ a + 2\Delta x)$ be *equally spaced*, he noted that we could choose them so as to achieve additional accuracy for the approximation. For example we can find values for the *six* parameters $c_1$, $c_2$, $c_3$, *and* $\xi_1$, $\xi_2$, and $\xi_3$ so that

$$\int_a^b f(\xi)\, d\xi \ \approx c_1 f(\xi_1) + c_2 f(\xi_2) + c_3 f(\xi_3) \qquad (26)$$

is exact for quintics (which have six terms $a_5 x^5 + a_4 x^4 + a_3 x^3 + a_2 x^2 + a_1 x + a_0$). This procedure is known as *Gaussian quadrature*, and tables of values for the parameters, as well as error estimates, are discussed in the references by Davis and Rabinowitz, Isaacson and Keller, and Abramowitz and Stegun (see also exercises 4-6). The ten-point Gauss quadrature approximation for the integral $\int_0^1 e^{-x^2}\, dx$ considered in (6) is .7460241327, is off by 1 in the tenth digit (see (12)).

Gaussian quadrature can be made even more flexible by considering weighted integral formulas - i.e. formulas of the form (24) which integrate

$$\int_a^b w(x) f(x)\, dx \qquad (27)$$

exactly when $f$ is a polynomial of the appropriate degree and $w$ is a fixed "weighting" function. Thus we have

Gauss-Legendre

Gauss-Chebyshev

Gauss-Hermite, and

Gauss-Laguerre

quadrature, with $w$ defined through the associated Sturm-Liouville problem (Chapter 6). Again we direct the reader to the references for details on this specialized topic.

The numerical evaluation of *improper* integrals requires more care. Remember that if the interval of integration is infinite then the integral itself is defined as a limit of finite integrals:

$$\int_a^\infty f(x)\, dx = \lim_{b\to\infty} \int_a^b f(x)\,dx; \int_{-\infty}^b f(x)\,dx = \lim_{a\to-\infty} \int_a^b f(x)\, dx \quad (28)$$

The numerical implementation of (28), taken literally, entails using (say) the trapezoidal rule over increasingly longer intervals and extrapolating the result. Note that longer intervals require more functional evaluations, if $\Delta x$ remains the same.

A better procedure is to *change variables* so as to make the interval finite. For example the interval $0 \leq x < \infty$ becomes, through the transformation $x = \tan t$, the *finite* interval $0 \leq t < \pi/2$ :

$$\int_0^\infty f(x)\,dx = \int_0^{\pi/2} f(\tan t)\,\frac{dx}{dt}\,dt = \int_0^{\pi/2} f(\tan t)\sec^2 t\,dt \qquad (29)$$

If the resulting integrand does not blow up at $\pi/2$, the techniques discused earlier can be used to approximate (29). Details and other transformations are discussed in the references.

**Example 2.** In Sec. 2.1 exercise 5 we revealed a "trick" which enables the exact evaluation of the definite integral

$$\int_0^\infty e^{-x^2}\,dx = \frac{\sqrt{\pi}}{2} = .8862269255\ldots. \qquad (30)$$

The change of variable (29) produces

$$\int_0^\infty e^{-x^2}dx = \int_0^{\pi/2} e^{-\tan^2 t}\sec^2 t\,dt \qquad (31)$$

L'Hopital's rule (exercise 5, Section 2.2) - or simple numerical experimentation - reveals that this integrand is finite as $t \to \pi/2$, and a ten-point trapezoidal rule approximation of (31) gives the value .88629. ∎

If $f(x)$ becomes infinite at some point - say, $\gamma$ - inside the interval of integration then the improper integral is again defined as a limit of integrals -

$$\int_a^b f(x)\,dx = \lim_{\xi_1 \uparrow \gamma} \int_a^{\xi_1} f(x)\,dx + \lim_{\xi_2 \downarrow \gamma} \int_{\xi_2}^b f(x)\,dx. \qquad (32)$$

The literal interpretation of (32) once more entails a sequence of applications of the trapezoidal rule (say) with an extrapolation to the limit. Better results can be achieved if the "singular part" of $f$ can be isolated:

**Example 3.** The integrand in $\int_0^1 x^{-1/2}\cos x\,dx$ is infinite at $x = 0$. However, the decomposition

$$\int_0^1 x^{-1/2}\cos x\,dx = \int_0^1 x^{-1/2}\,dx + \int_0^1 x^{-1/2}[\cos x - 1]\,dx \qquad (33)$$

is helpful. From the Taylor development

$$\cos x = 1 - \frac{x^2}{2} + \frac{x^4}{4!} + \cdots$$

it is clear that

$$x^{-1/2}[\cos x - 1] = \mathcal{O}(x^{3/2}),$$

and thus the second integrand remains finite in (33). Therefore we can accurately estimate (33) by integrating the first term $x^{-1/2}$ *exactly* and approximating $\int_0^1 x^{-1/2}[\cos x - 1]\,dx$ by a trapezoidal rule:

$$\int_0^1 x^{-1/2}\,dx = 2,$$

$$\int_0^1 x^{-1/2}[\cos x - 1]\,dx \approx -.191 \quad \text{(ten-point trapezoidal rule)};$$

$$\int_0^1 x^{-1/2}\cos x\,dx \approx 2 - .191 = 1.809. \qquad \blacksquare$$

If the singular part of the integrand $f$ can be *factored* out - $f(x) = g(x)\,h(x)$ with $h(x)$ finite, then a weighted Gauss quadrature scheme could possibly be devised with $g(x)$ serving as the weight. Again, we defer to the specialized literature.

Finally let us consider multiple integrals. Although the double integral $\int_a^b \int_c^d f(x,y)\,dy\,dx$ is *defined* as the limit of sums of sampled values of $f$ times incremental *areas* (analogous to (5)), the theory shows that for almost all functions encountered in practice the integral can be evaluated iteratively:

$$\int_a^b \int_c^d f(x,y)\,dy\,dx = \int_a^b \left[ \int_c^d f(x,y)\,dy \right] dx = \int_a^b g(x)\,dx \qquad (34)$$

where $g(x) = \int_c^d f(x,y)\,dy$. Equation (34) directs us to use, say, Simpson's rule to approximate the integral of $g$, with the requisite functional values of $g$ computed in a "subroutine" using, again, Simpson's rule. An example will make this clear.

**Example 4.** The numerical evaluation of

$$\int_0^1 \int_1^3 (x+y)^2\,dy\,dx = \int_0^1 \left[ \int_1^3 (x+y)^2\,dy \right] dx$$

via a 5 point Simpson rule for the $x$ integration calls for the computation of

$$\int_0^1 \int_1^3 (x+y)^2\,dy\,dx = \int_0^1 g(x)\,dx$$
$$\approx \frac{.25}{3}\{g(0) + 4g(.25) + 2g(.5) + 4g(.75) + g(1)\}$$

where $g(x) = \int_1^3 (x+y)^2 dy$. For comparable accuracy ($\Delta x = \Delta y$), a 9 point Simpson rule for the $y$ integration produces

$$g(0) = \frac{.25}{3}\{(0+1)^2 + 4(0+1.25)^2 + 2(0+1.5)^2$$
$$+ 4(0+1.75)^2 + 2(0+2)^2 + 4(0+2.25)^2$$
$$+ 2(0+2.5)^2 + 4(0+2.75)^2 + (0+3)^2\}$$
$$\approx 8.666,$$

$$g(.25) \approx \frac{.25}{3}\{(.25+1)^2 + 4(.25+1.25)^2 + 2(.25+1.5)^2 + 4(.25+1.75)^2$$
$$+ 2(.25+2)^2 + 4(.25+2.25)^2 + 2(.25+2.5)^2 + 4(.25+2.75)^2$$
$$+ (.25+3)^2\}$$
$$\approx 10.791,$$

and so on. The net result is $\int_0^1 \int_1^3 (x+y)^2 \, dy \, dx \approx 13.333\ldots$. ∎

Integrals with variable endpoints can be treated the same way.

**Example 5.** For integration over the domain depicted in Fig. 17, a 5 point Simpson's rule for $x$ gives

$$\int_{x=0}^1 \int_{y=x^2}^{1+x} (x+y)^2 \, dy \, dx \approx \{g(0) + 4g(.25) + 2g(.5) + 4g(.75) + g(1)\}.$$

But note that if we continue to use a 5 point rule for $y$ the size of $\Delta y$ changes:

$$g(x) = \int_{x^2}^{1+x} (x+y)^2 dy, \quad \Delta y = \frac{(1+x) - x^2}{4}$$

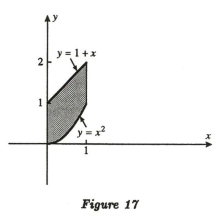

**Figure 17**

For $x = 0, y$ runs from 0 to 1 in increments of $\Delta y = .25$ and we get

$$g(0) \approx \frac{.25}{3}\{(0+0)^2 + 4(0+.25)^2 + 2(0+.5)^2 + 4(0+.75)^2 + (0+1)^2\}$$
$$= .333.$$

For $x = .25, y$ runs from $.25^2$ to 1.25 in increments of $\Delta y = .296875$ and

$$g(.25) \approx \frac{.296875}{3}\{(.25+.25^2)^2 + 4(.25+.25^2+.296875)^2 + 2(.25+.25^2$$
$$+ 2\times.296875)^2 + 4(.25+.25^2+3\times.296875)^2 + (.25+1.25)^2\}$$
$$\approx 1.115$$

For $x = .5$, $y$ runs from $.5^2$ to 1.5 in increments of $\Delta y = .3125$ and

$$g(.5) \approx \frac{.3125}{3}\{(.5 + .5^2)^2 + 4(.5 + .5^2 + .3125)^2 + 2(.5 + .5^2 + 2 \times .3125)^2$$
$$+ 4(.5 + .5^2 + 3 \times .3125)^2 + (.5 + 1.5)^2\}$$
$$\approx 2.526$$

and so on. The net result for the integral is 2.839 . . . .

Obviously Gaussian-type schemes can be developed for double integrals, and the extensions to higher dimensions is clear. ∎

> The art of numerical integration is a well established discipline; it has been so refined, in fact, that in this author's opinion the pursuit of exact antiderivative formulas is not a cost-effective activity in the practice of engineering analysis except in the simplest cases. Therefore throughout the main part of this book we have invoked the doctrine of discontinuing the analysis of any problem once its solution has been reduced to definite integrals.

# Exercises A.6

1. Using the methods in this section, experiment with the numerical evaluation of the following integrals:

   a. $\int_0^{\pi/2} \sin^2 x\, dx$    b. $\int_1^2 \ln x\, dx$    c. $\int_0^1 x^{3/2}\, dx$

   d. $\int_0^1 \frac{e^x}{\sqrt{x}}\, dx$    e. $\int_0^\infty xe^{-x^2}\, dx$    f. $\int_0^\infty \frac{1}{1+x^2}\, dx$

   g. $\int_0^1 \int_0^1 x^2 y^3\, dy\, dx$    h. $\int_0^1 \int_0^x x^2 y^3\, dy\, dx$

2. Verify that Simpson's rule (16) integrates quadratics exactly.

3. Show that the error in the trapezoid rule (11) is $\mathcal{O}(\Delta x^2)$ for functions with bounded second derivatives.

4. Verify that the two-point Gauss-Legendre formula

   $$\int_{-1}^1 f(x)\, dx \approx f(-1/\sqrt{3}) + f(1/\sqrt{3})$$

   integrates cubics exactly. (Hint: can you see a way to reduce this exercise to a verification for $f(x) = x^2$ only? Use symmetry.)

5. Derive the *one*-point Gauss-Legendre formula which integrates

   $$\int_{-1}^1 f(x)\, dx$$

   exactly for linear functions. (Answer: $2f(0)$.)

6. Derive the three-point Gauss-Legendre formula for integration over the interval [-1,1].

# Appendix B

# Evaluation of Bromwich Integrals

**Summary**  An evaluation of two Bromwich integrals arising in radiation examples is demonstrated through contour integration.

## B.1    The Radiating Sphere

In Example 2 of Sec. 8.5 we derived the following expression for the Laplace transform of the radiation emitted by a pulsating sphere:

$$\Phi(r,\phi,\theta,s) = \frac{\Omega}{\Omega^2 + s^2}\,\frac{h_p^{(1)}(isr)}{h_p^{(1)}(isb)}Y_{pq}(\phi,\theta).$$

The radiation itself has the (prescribed) form $\Psi(r,\phi,\theta,t) = Y_{pq}(\phi,\theta)\sin\Omega t$ at $r = b$. Since we would not expect the radiation intensity to increase as it spreads out, the statement $\Psi(r,\phi,\theta,t)e^{\alpha t} \to 0$ (as $t \to \infty$) should be valid for any $a > 0$. Thus the inverse transform, as given by the Bromwich integral (Sec. 3.8), is expressed $\Psi(r,\phi,\theta,t) = f(r,t)\,Y_{pq}(\phi,\theta)$, where

$$f(r,t) = \frac{1}{2\pi i}\int_{\alpha-i\infty}^{\alpha+i\infty}\frac{\Omega}{\Omega^2 + s^2}\,\frac{h_p^{(1)}(isr)}{h_p^{(1)}(isb)}e^{st}\,ds \tag{1}$$

and $\alpha > 0$. The path of integration is indicated in Fig. 1.

Now the spherical Hankel functions are analytic everyhere except the origin; but the approximations listed in Table 2.2 can be manipulated to show that for small $s$, $h_p^{(1)}(isr)/h_p^{(1)}(isb) \approx (b/r)^{p+1}$; thus the integrand is analytic at $s = 0$. Moreover, since the Hankel function is a multiple of $J_{p+\frac{1}{2}} + iY_{p+\frac{1}{2}}$, and since $J$ has only real zeros - none of which it shares with $Y$ (otherwise their Wronskian would vanish) - the spherical Hankel function

644

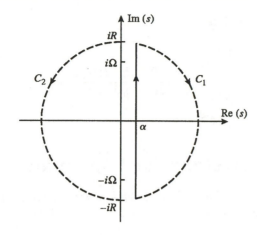

**Figure 1   Integration contours**

has no zeros. Therefore the integrand is, in fact, analytic everywhere except for its poles at $s = \pm i\Omega$. Finally, the approximations listed in eq. (34), Sec. 8.5, show that for large $s$, $h_p^{(1)}(isr)/h_p^{(1)}(isb) \approx e^{-s(r-b)}$.

It follows that on the circular arc $C_1$, where $Re(s) > 0$, we can estimate

$$\frac{1}{2\pi i} \int_{C_1} \frac{\Omega}{\Omega^2 + s^2} \frac{h_p^{(1)}(isr)}{h_p^{(1)}(isb)} e^{st}\, ds$$
$$\approx \frac{1}{2\pi i} \int_{C_1} \frac{\Omega}{\Omega^2 + s^2} e^{-s(r-b-t)} ds \to 0 \text{ as } R \to \infty \text{ if } t < r-b.$$

Thus if we close the contour in Fig. 1 with $C_1$, the integral (1) vanishes for $t < r - b$ since no singularities are enclosed.

On the other hand for $t > r - b$ the integral along $C_2$ can be estimated by

$$\frac{1}{2\pi i} \int_{C_2} \frac{\Omega}{\Omega^2 + s^2} \frac{h_p^{(1)}(isr)}{h_p^{(1)}(isb)} e^{st}\, ds$$
$$\approx \frac{1}{2\pi i} \int_{C_2} \frac{\Omega}{\Omega^2 + s^2} e^{s[t-(r-b)]}\, ds \to 0 \text{ as } R \to \infty \text{ if } t > r-b$$

since $Re(s) \leq \alpha$. Closing the contour with $C_2$, then, equates the integral (1) with the contribution of the residues at $s = \pm i\Omega$:

$$f(r,t) = \frac{2\pi i}{2\pi i} \Big\{ \frac{\Omega}{2i\Omega} \frac{h_p^{(1)}(-\Omega r)}{h_p^{(1)}(-\Omega b)} e^{i\Omega t} + \frac{\Omega}{-2i\Omega} \frac{h_p^{(1)}(\Omega r)}{h_p^{(1)}(\Omega b)} e^{-i\Omega t} \Big\} \qquad (2)$$

From the symmetry properties listed in Table 2.2 we derive $h_p^{(1)}(-a) =$

$(-1)^p h_p^{(2)}(a)$. Therefore

$$f(r,t) = \begin{cases} 0 & , t < r - b \\ \dfrac{h_p^{(2)}(\Omega r)}{h_p^{(2)}(\Omega b)} \dfrac{e^{i\Omega t}}{2i} - \dfrac{h_p^{(1)}(\Omega r)}{h_p^{(1)}(\Omega b)} \dfrac{e^{-i\Omega t}}{2i} & , t > r - b \end{cases} \tag{3}$$

as stated in eq. (33) of Sec. 8.5.

## B.2   The Rectangular Waveguide

In Example 1 of Sec. 8.5 we derived the following expression for the Laplace transform of the radiation transported for $z > 0$ along a waveguide:

$$\Phi(x,y,z,s) = \sin\frac{p\pi}{a}x \; \sin\frac{q\pi}{b}y \; \frac{\Omega}{\Omega^2 + s^2} \exp\{-z\sqrt{s^2 + \omega_{pq}^2}\}.$$

The radiation itself has the (prescribed) form

$$\Psi(x,y,0,t) = \sin\frac{p\pi}{a}x \; \sin\frac{q\pi}{b}y \; \sin\Omega t$$

at $z = 0$. As in the previous section, we would not expect the radiation intensity to increase as it propagates, so the statement $\Psi(x,y,z,t)e^{-\alpha t} \to 0$ (as $t \to \infty$) should be valid for any $\alpha > 0$. Suppressing the modal factors, then, we need to estimate the Bromwich integral

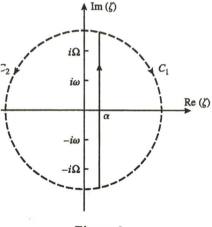

*Figure 2*

$$f(z,t) = \int_{\alpha-i\infty}^{\alpha+i\infty} \frac{\Omega}{\Omega^2 + s^2} e^{-\sqrt{s^2+\omega^2}z} e^{st} \, ds \tag{1}$$

where $\alpha > 0$ and we abbreviate $\omega_{pq}$ as simply $\omega$.

We analyze the integral (1) for two separate cases.

**Case 1:** $\Omega > \omega$. The path of integration is depicted in Fig. 2. We need to construct a branch of $\sqrt{s^2 + \omega^2}$ which is analytic and has a positive real part for $Re(s) > 0$. This is accomplished by taking

$$\sqrt{s - i\omega} = \sqrt{|s - i\omega|}\,e^{i\theta_1/2}, \quad \sqrt{s + i\omega} = \sqrt{|s + i\omega|}\,e^{i\theta_2/2},$$
$$\sqrt{s^2 + \omega^2} = \sqrt{|s^2 + \omega^2|}\,e^{i(\theta_1+\theta_2)/2} \tag{2}$$

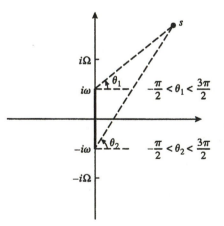

*Figure 3*

where the angles $\theta_1$ and $\theta_2$ are defined in Fig. 3. As $s$ crosses the imaginary axis below $-i\omega$, the jumps in the angles add up to $2\pi$, so that $\sqrt{s^2 + \omega^2}$ is analytic everywhere except on the segment from $-i\omega$ to $i\omega$. In fact for large $|s|$, $\theta_1 \approx \theta_2 \approx \arg(s)$ if $-\pi/2 < \arg(s) < 3\pi/2$; therefore $\sqrt{s^2 + \omega^2} \approx s$.

For $t < z$ we close the contour to the right with the circular arc $C_1$. On $C_2$, $e^{-\sqrt{s^2+\omega^2}z}\,e^{st} \approx e^{-s(z-t)}$ is bounded since $Re(s) > 0$, and the integral over $C_1$ goes to zero. There being no singularities inside the closed contour, the original integral (1) is thus also zero.

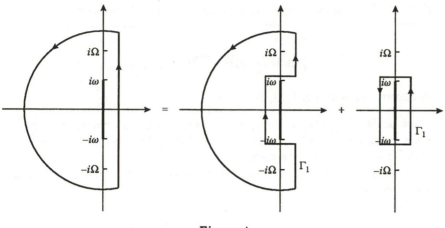

**Figure 4**

For $t > z$ we close the contour to the left with $C_2$. Here $e^{-\sqrt{s^2+\omega^2}z}e^{st} \approx e^{s(t-z)}$ is bounded since $Re(s) \leq \infty$, and the integral over $C_2$ goes to zero. We replace the closed contour by the two contours in Fig. 4. The integral over $\Gamma_1$ is given by the residue contributions

$$\frac{2\pi i}{2\pi i}\{\frac{\Omega}{2i\Omega}e^{i\Omega t}e^{-\sqrt{\Omega^2-\omega^2}\,z}e^{i(\pi/2+\pi/2)/2}$$
$$+ \frac{\Omega}{-2i\Omega}e^{-i\Omega t}e^{-\sqrt{\Omega^2-\omega^2}\,z}e^{i(3\pi/2+3\pi/2)/2}\}$$
$$= \frac{1}{2i}\{e^{i[\Omega t-\sqrt{\Omega^2-\omega^2}z]} - e^{-i[\Omega t-\sqrt{\Omega^2-\omega^2}z]}\}$$
$$= \sin(\Omega t - \sqrt{\Omega^2 - \omega^2}z). \tag{3}$$

The integral over $\Gamma_1$ just involves the jump in the factor $e^{-\sqrt{s^2+\omega^2}z}$ across the cut, and contributes

$$\frac{1}{2\pi i}\int_{-\omega}^{\omega}e^{i\eta t}\frac{\Omega}{\Omega^2-\eta^2}[e^{-\sqrt{\omega^2-\eta^2}\,z}e^{i(-\pi/2+\pi/2)/2}$$
$$- e^{-\sqrt{\omega^2-\eta^2}\,z}e^{i(3\pi/2+\pi/2)/2}]i\,d\eta$$
$$= \frac{1}{2\pi}\int_{-\omega}^{\omega}e^{i\eta t}\frac{\Omega}{\Omega^2-\eta^2}[e^{-\sqrt{\omega^2-\eta^2}\,z} - e^{\sqrt{\omega^2-\eta^2}z}]\,d\eta$$
$$= -\frac{1}{\pi}\int_{-\omega}^{\omega}e^{i\eta t}\frac{\Omega}{\Omega^2-\eta^2}\sinh\sqrt{\omega^2-\eta^2}z\,d\eta\,.$$

As in exercise 4, Sec. 3.4, we integrate this by parts. The integrated term is zero and the remainder is

$$\frac{-i}{\pi t}\lim_{\beta\uparrow\omega}\int_{-\beta}^{\beta}e^{i\eta t}\{\frac{2\Omega\eta\,\sinh\sqrt{\omega^2-\eta^2}z}{(\Omega^2-\eta^2)^2} - \frac{\Omega z\eta\,\cosh\sqrt{\omega^2-\eta^2}z}{(\Omega^2-\eta^2)^2\sqrt{\omega^2-\eta^2}}\}\,d\eta \tag{4}$$
$$= \mathcal{O}(1/t)$$

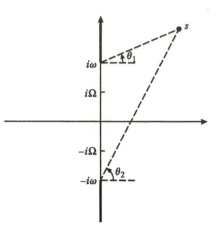

since the integral converges. Assembling these results we conclude

$$f(z,t) = \begin{cases} 0 & , t < z \\ \sin[\Omega t - \sqrt{\Omega^2 - \omega^2}\, z] + \mathcal{O}(1/t) & , t > z \end{cases}$$

in accordance with eq. (15), Sec. 8.5.

**Case 2**: $\Omega < \omega$. The geometry is depicted in Fig. 5. We define $\sqrt{s - i\omega}$, $\sqrt{s + i\omega}$, and $\sqrt{s^2 + \omega^2}$ by the same equations (2) as before, but the branch cuts for $\theta_1$ and $\theta_2$ are now different.

Then $\sqrt{s^2 + \omega^2}$ is analytic everywhere except on the indicated rays, and for large $|s|$

$$\sqrt{s^2 + \omega^2} \approx \begin{cases} s & \text{if } Re(s) > 0 \\ -s & \text{if } Re(s) < 0 \, . \end{cases}$$

**Figure 5**

If we close the contour to the right with the arc $C_1$ in Fig. 6 the exponential factor is roughly $e^{-s(z-t)}$, so for $t < z$ it is bounded and the integral is zero as before.

On the arc $C_2$ to the left the exponential factor is roughly $e^{+s(z-t)}$, which again is bounded since $Re(s) \leq \infty$. Thus the integral over $C_2$ goes to zero. We close the contour for $t > z$ with $C_2$, $\Gamma_1$, and $\Gamma_2$ in Fig. 7.

The closed contour integral equals the residue contributions

$$\frac{2\pi i}{2\pi i}\Big\{ \frac{\Omega}{2i\Omega} e^{i\Omega t} e^{-\sqrt{\omega^2 - \Omega^2}\, z}\, e^{i(-\pi/2 + \pi/2)/2}$$

$$+ \frac{\Omega}{-2i\Omega} e^{-i\Omega t} e^{-\sqrt{\omega^2 - \Omega^2}\, z}\, e^{i(-\pi/2 + \pi/2)/2} \Big\} \tag{5}$$

$$= \sin \Omega t \, e^{-\sqrt{\omega^2 - \Omega^2}\, z} \, .$$

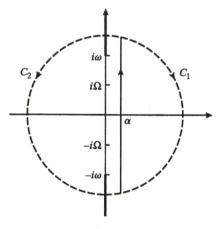

**Figure 6**

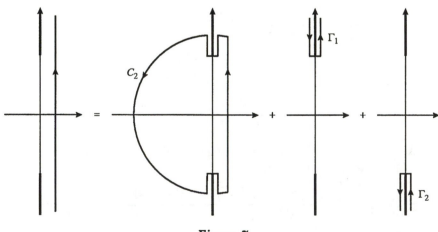

**Figure 7**

The integral over $\Gamma_1$ involves the jump in the square root, and contributes

$$\frac{1}{2\pi i} \int_{-\omega}^{\infty} e^{i\eta t} \frac{\Omega}{\Omega^2 - \eta^2} [e^{-\sqrt{\eta^2 - \omega^2} \, z \, e^{i(\pi/2 + \pi/2)/2}}$$
$$- e^{-\sqrt{\eta^2 - \omega^2} \, z \, e^{i(-3\pi/2 + \pi/2)/2}}] i \, d\eta$$
$$= -\frac{i}{\pi} \int_{\omega}^{\infty} e^{i\eta t} \frac{\Omega}{\Omega^2 - \eta^2} \sin \sqrt{\eta^2 - \omega^2} z \, d\eta \ .$$

As in the above, this integral is $\mathcal{O}(1/t)$. The contribution of $\Gamma_2$ is similarly $\mathcal{O}(1/t)$. Assembling these results, we have

$$f(z,t) = \begin{cases} 0 & , t < z \\ \sin \Omega t \, \exp\{-z\sqrt{\omega^2 - \Omega^2}\} + \mathcal{O}(1/t) & , t > z \end{cases}$$

as quoted in eq. (17), Sec. 8.5.

# References

1. Abramowitz, M., and Stegun, I. A., *Handbook of Mathematical Functions*, Dover Publications, New York, 1965.

2. Arfken, G., *Mathematical Methods for Physicists*, 2nd ed., Academic Press, New York, 1970.

3. Bellman, R., *Stability Theory of Differential Equations*, McGraw-Hill, New York, 1953.

4. Birkhoff, G., and Rota, G.-C., *Ordinary Differential Equations*, 2nd ed., Xerox College Publishing, Lexington, MA, 1969.

5. Bohm, D., *Quantum Theory*, Prentice Hall, Upper Saddle River, NJ, 1961.

6. Brown, E. W., and Shook, C. A., *Planetary Theory*, Dover, New York, 1964.

7. Braun, M., Coleman, C. S., and Drew, D. A., *Differential Equation Models*, Springer-Verlag, New York, 1983.

8. Chen, F. F., *Introduction to Plasma Physics*, Plenum Press, NY, 1974.

9. Cinelli, G., "An Extension of the Finite Hankel Transform and Applications," *Int. J. Eng. Sci.*, v. 3, 1965, p. 539.

10. Coddington, E. A., and Levinson, N., *Theory of Ordinary Differential Equations*, McGraw-Hill, New York, 1955.

11. Corben, H. C., and Stehle, P., *Classical Mechanics*, 2nd ed., Robert Krieger, Huntington, NY, 1974.

12. Courant, R. C. and Hilbert, D., *Methods of Mathematical Physics* (two vols.), Wiley-Interscience, New York, 1953.

13. Dahlquist, G., Bjorck, A., and Anderson, J., *Numerical Methods*, Prentice Hall, Upper Saddle River, NJ, 1974.

14. Davis, H. F., *Fourier Series and Orthogonal Functions*, Allyn and Bacon, Boston, 1963.

15. Davis, H. F., and Snider, A. D., *Introduction to Vector Analysis*, 7th ed., Quantum Pub., Charleston, SC, 1995.

16. Davis, P. J., *Interpolation and Approximation*, Dover, New York, 1975.

17. Davis, P. J. and Rabinowitz, P., *Methods of Numerical Integration*, Academic Press, New York, 1984.

18. Dorf, R. C., *Modern Control Systems*, 2nd ed., Addison-Wesley, Reading, MA, 1974.

19. Dwight, H. B., "Tables of Roots for Natural Frequencies in Coaxial Cavities," *J. Math. Phys.*, v. 27, no. 1, 1948, pp. 84-89.

20. Edwards, R. E., *Fourier Series: A Modern Introduction* (two vols.), Holt, Rinehart, and Winston, New York, 1967.

21. Erdelyi, A., Magnus, W., Oberhettinger., F., and Tricomi, F. G., *Tables of Integral Transforms* (two vols.), McGraw-Hill, New York, 1954.

22. Ewins, D. T., *Modal Testing: Theory and Practice*, Wiley, 1984.

23. Franklin, P., *A Treatise on Advanced Calculus*, Wiley, New York, 1940.

24. Friedman, B., *Principles and Techniques of Applied Mathematics*, Wiley, New York, 1956.

25. Galeev, A. A., "Neoclassical Theory of Transport Processes," in *Advances in Plasma Physics* v. 5, Wiley, 1974.

26. Garabedian, P., *Partial Differential Equations*, Wiley, New York, 1964.

27. Gear, C. W., *Numerical Initial Value Problems in Ordinary Differential Equations*, Prentice Hall, Upper Saddle River, NJ, 1971.

28. Goldstein, H., *Classical Mechanics*, Addison-Wesley, Reading, MA, 1951.

29. Greenberg, M. D., *Foundations of Applied Mathematics*, Prentice Hall, Upper Saddle River, NJ, 1978.

30. Hellwig, G., *Partial Differential Equations*, Blaisdell, New York, 1964.

31. Hildebrand, F. B., *Advanced Calculus for Applications*, 2nd ed., Prentice Hall, Upper Saddle River, NJ, 1976.

32. Ince, E. L., *Ordinary Differential Equations*, Dover, New York, 1926.

33. Isaacson, E., and Keller, H. B., *Analysis of Numerical Methods*, Wiley, New York, 1966.

34. Jackson, J. D., *Classical Electrodynamics*, Wiley, New York, 1962.

35. Jones, D. S., *Acoustic and Electromagnetic Waves*, Oxford University Press, New York, 1986.

36. Kammler, David, *A First Course in Fourier Analysis*, Prentice Hall, Upper Saddle River, NJ, 2000.

37. Keiser, G., *Optical Fiber Communications*, McGraw-Hill, New York, 1983.

38. Kraus, J. D., *Electromagnetics*, 3rd ed., McGraw-Hill, New York, 1984.

39. Levitan, B. M., and Sargsjan, I. S., *Introduction to Spectral Theory*, American Mathematical Society, Providence, RI, 1975.

40. Lighthill, M. J., *Introduction to Fourier Analysis and Generalized Functions*, Cambridge University Press, NY, 1958.

41. Luke, Y. L., *Integrals of Bessel Functions*, McGraw-Hill, New York, 1962.

42. Marcuvitz, N., *Waveguide Handbook*, Dover, New York, 1965.

43. Mehta, B. N., and Aris, R., "A Note on the Form of the Emden-Fowler Equation," *J. Math. Analysis Applications*, no. 36, 1971.

44. Messiah, A., *Quantum Mechanics* (two vols.), North-Holland, Amsterdam, 1961.

45. Morse, P. M., and Feshbach, H., *Methods of Theoretical Physics* (two vols.), McGraw-Hill, New York, 1978.

46. Nagle, R. K., Saff, E. B., and Snider, A. D., *Fundamentals of Differential Equations and Boundary Value Problems* (4th ed.), Addison-Wesley, Reading, MA, 2004.

47. Naylor, D., "On a Finite Lebedev Transform," *J. Math. Mech.*, v. 12, no. 3, 1963.

48. Naylor, D., "On a Finite Lebedev Transform. Part 2," *J. Math. Mech.*, v. 15, no. 3, 1966.

49. Panofsky, W. K. H., and Phillips, M., *Classical Electricity and Magnetism*, 2nd ed., Addison-Wesley, Reading, MA, 1962.

50. Papoulis, A., *The Fourier Integral and Its Applications*, McGraw-Hill, New York, 1962.

51. Papoulis, A., *Signal Analysis*, McGraw-Hill, New York, 1977.

52. Protter, M. H., and Weinberger, H. F., *Maximum Principles in Differential Equations*, Prentice Hall, Upper Saddle River, 1967.

53. Ramo, S., Whinnery, J. R., and van Duzer, T., *Fields and Waves in Communication Electronics*, Wiley, New York, 1965.

54. Royden, H. L., *Real Analysis*, Macmillan, New York, 1963.

55. Saff, E. B., and Snider, A. D., *Fundamentals of Complex Analysis for Mathematics, Science, and Engineering*, 3rd ed., Prentice Hall, Upper Saddle River, NJ, 2003.

56. Simmons, G. F., *Differential Equations with Applications and Historical Notes*, McGraw-Hill, New York, 1972.

57. Smith, D. R., *Variational Methods in Optimization*, Prentice Hall, Upper Saddle River, NJ, 1974.

58. Smith, D. R., *Singular Perturbation Theory*, Cambridge University Press, New Tork, 1985.

59. Sneddon, I. H., *The Use of Integral Transforms*, McGraw-Hill, New York, 1972.

60. Stewart, G. W., *Introduction to Matrix Computations*, Academic Press, NY, 1973.

61. Strang, G., *Introduction to Applied Mathematics*, Wellesley-Cambridge Press, Wellesley, MA, 1986.

62. Strang, G., and Fix, G. J., *An Analysis of the Finite Element Method*, Prentice Hall, Upper Saddle River, NJ, 1973.

63. Streetman, B. G., *Solid State Electronic Devices*, Prentice Hall, Upper Saddle River, NJ, 1980.

64. Watson, G. N., *A Treatise on the Theory of Bessel Functions*, Cambridge University Press, New York, 1922.

65. Weinberger, H. F., *A First Course in Partial Differential Equations*, Wiley, NY, 1965.

# Index